Joseph Meites

HORMONES IN DEVELOPMENT AND AGING

HORMONES IN DEVELOPMENT AND AGING

Edited by

Antonia Vernadakis, Ph.D.
Departments of Psychiatry and Pharmacology
University of Colorado
Denver, Colorado

Paola S. Timiras, M.D., Ph.D.
Department of Physiology-Anatomy
University of California
Berkeley, California

New York

Copyright © 1982 Spectrum Publications, Inc.

All rights reserved. No part of this book may be reproduced in any form, by photostat, microform, retrieval system, or any other means without prior written permission of the copyright holder or his licensee.

SPECTRUM PUBLICATIONS, INC.
175-20 Wexford Terrace
Jamaica, N.Y. 11432

Library of Congress Cataloging in Publication Data
Main entry under title:

Hormones in development & aging.

 Includes index.
 1. Endocrinology, Developmental. 2. Aging.
I. Vernadakis, Antonia, 1930- [DNLM: 1. Aging.
2. Growth. 3. Hormones—Metabolism. 4. Nervous
system—Drug effects. WK102 H8123]
QP187.6.H67   599.03   80-24180
ISBN 0-89335-140-7

Contributors

HOWARD A. BERN
Department of Zoology
and Cancer Research Laboratory
University of California
Berkeley, California

DOUGLAS B. CARTER
Department of Psychiatry
University of Colorado School of
  Medicine
Denver, Colorado

GREGORY M. COLE
Department of Physiology-Anatomy
University of California
Berkeley, California

JEAN M. CONS
Mathematical and Science Division
College of San Mateo
San Mateo, California

EVA B. CRAMER
Department of Anatomy and Cell
  Biology
Downstate Medical Center
State University of New York
Brooklyn, New York

BRUCE CULVER
School of Pharmacy
University of Wyoming
Laramie, Wyoming

EUGENE EISENBERG
Health and Medical Sciences Program
University of California
Berkeley, California

ROBERT FREEDMAN
Department of Psychiatry
University of Colorado School of
  Medicine
Denver, Colorado

DONALD H. FORD
The Council for Tobacco Research
New York, New York

JOHN GRIFFIN
Department of Psychiatry
Emory University School of
  Medicine
Atlanta, Georgia

FRANZ HALBERG
University of Minnesota
Minneapolis, Minnesota

SELNA L. KAPLAN
Department of Pediatrics
University of California
San Francisco, California

ARTHUR F. KOHRMAN
La Rabida Childrens' Hospital
  and Research Institute
  and Department of Pediatrics
University of Chicago
Chicago, Illinois

RICHARD P. MICHAEL
Department of Psychiatry
Emory University School of
  Medicine
Atlanta, Georgia

EVANGELOS PETROPOULOS
Department of Physiology
Godfrey Huggins School of Medicine
University of Zimbabwe
Mount Pleasant, Salisbury, Zimbabwe

JUDITH A. RAMALEY
Department of Physiology and
  Biophysics
University of Nebraska
College of Medicine
Omaha, Nebraska

EVELYN M. RIVERA
Department of Zoology
Michigan State University
East Lansing, Michigan

PAUL E. SEGALL
Department of Physiology-Anatomy
University of California
Berkeley, California

DOROTHY STRAUSS
Center for Human Sexuality and
  Department of Psychiatry
Downstate Medical Center
State University of New York
New York, New York

PAOLA S. TIMIRAS
Department of Physiology-
  Anatomy
University of California
Berkeley, California

ANTONIA VERNADAKIS
Departments of Psychiatry and
  Pharmacology
University of Colorado
Denver, Colorado

RICHARD F. WALKER
Department of Physiology-
  Anatomy
University of California
Berkeley, California

RICHARD E. WHALEN
The Long Island Reseach
  Institute
State University of New York
  at Stony Brook
Stony Brook, New York

Contents

1. Fertilization — 1
   Paola S. Timiras

2. Hormones During Prenatal and Neonatal Development — 25
   Paola S. Timiras and Jean M. Cons

3. Placental Hormones in Development — 81
   Evangelos Petropoulos

4. Somatic Growth — 125
   Selna L. Kaplan

5. The Anatomical Distribution of Hormones in the Central Nervous System — 151
   Donald H. Ford and Eva B. Cramer

6. Morphological Response of the Nervous System to Hormones — 181
   Eva B. Cramer and Donald H. Ford

7. Endocrine Regulation of Neural Development: Physiological and Biochemical Aspects — 207
   Antonia Vernadakis

8. Current Issues in the Neurobiology of Sexual Differentiation — 273
   Richard E. Whalen

9. The Neuroendocrinology of Puberty — 305
   Judith A. Ramaley

10.	Clinical Aspects of Puberty and Adolescence Arthur F. Kohrman	331
11.	Psychology and Psychopathology of Adolescence John Griffin and Richard P. Michael	349
12.	Neuroendocrine Control Systems in the Adult Antonia Vernadakis and Bruce Culver	371
13.	Biological Rhythms, Hormones, and Aging Franz Halberg	451
14.	Hormones During Aging Gregory M. Cole, Paul E. Segall and Paola S. Timiras	477
15.	Neuroendocrine Theories of Aging: Homeostasis and Stress Paola S. Timiras	551
16.	Physiological Aspects of Menopause: Clinical and Experimental Studies Eugene Eisenberg and Richard F. Walker	587
17.	Psychological Aspects of Menopause Dorothy Strauss	609
18.	Neuroendocrine Strategies in Psychiatric Research Robert Freedman and Douglas B. Carter	619
19.	Affective Disorders in Endocrine Disease Douglas B. Carter	637
20.	Hormones and Cancer Evelyn M. Rivera and Howard A. Bern	645

Preface

In this comprehensive treatise on endocrinology, we have covered such aspects of endocrine function as differentiation, development, maturation and aging of the endocrine glands, as well as the role of hormones in organizing cellular expression and regulating metabolic functions. Our underlying thesis throughout is that the endocrine system is integrally related to the nervous system and thus cannot be properly understood except in relation to it. On this basis, we continually integrate information on endocrine-nervous system relationships that are crucial to growth and adaptation, puberty, biological rhythms, the onset and aging of reproductive functions, general aging and neoplasia.

A primary objective is to identify the continuum of neuroendocrine events which occur at progressive ages, inasmuch as alterations in neuroendocrine function during development often underlie disturbances manifested in old age.

A secondary objective is to follow the effects of hormones on normal and abnormal brain development and function, as well as to study the role of hormones in regulating behavior throughout the span of life. Taking this longitudinal approach, we describe physiological (basic) aspects of development and aging of endocrine function, as well as pathological (clinical) alterations which affect health in later life and, ultimately, longevity. Whenever possible, interventive approaches are proposed, for it is our hope and expectation that by improving and strengthening physiologic function, we will be able to go beyond the disease-oriented therapy that characterizes current medicine, and formulate physiologic guidelines for optimal growth, development and aging.

Antonia Vernadakis
Paola S. Timiras

HORMONES IN DEVELOPMENT
AND AGING

Copyright©1982, Spectrum Publications, Inc.
Hormones in Development and Aging

Chapter 1
Fertilization
Paola S. Timiras

INTRODUCTION

The essential task of reproduction is to pass to each individual instructions for the life of the species. The first step in this passage occurs at fertilization. At the end of gestation, with the production of viable offspring, the life of the species in all its biological and social complexity, is ensured. Yet, the slow and continual change in hereditary instructions transmitted through reproduction could not operate in any species if already existing individuals remained alive indefinitely. Despite the perpetual struggle to maintain life and our increasing ability to prolong life, we know that, in the end, we all will die. Indeed, we cannot avoid the fundamental biological platitude that "we cannot consider birth without properly thinking of death." Thus, in a book that is concerned with the entire lifespan, an appropriate premise is that growth, development, and aging are stages along a continuum of events starting at fertilization and terminating at death, each having a positive value and serving some specific purpose. To some people, the view that aging and death are a positive contribution to life may seem paradoxical and absurd. The absurdity is perhaps justified when one contemplates the lifespan of the individual; nonetheless, the pattern of life is continued not in the single individual, but in the group or species of which the individual is a part, and birth and death are equally vital to the continuation and preservation of the species.

Within this framework, we shall trace in this chapter the process of fertilization, primarily in humans, seeking to describe how the genetic copy is achieved. In the next chapters, we will consider how each single member of the species comes to its adult state, how the adult individual can make the adjustments necessary to maintain life for a long time, and, finally, how progressive and irreversible changes of the control systems that regulate these adjustments lead to aging and death. In the context of the book, fertilization can be viewed as a first step in the life process where birth, growth, and development, adulthood and aging, and, finally, death are governed by specific rhythms—the so-called biological clocks— that regulate the vital functions of our lives (see Chapters 13 to 16).

Cell Structure and Division

The genetic information characteristic of the species is concentrated within specialized *germ cells* or *sex cells (gametes)* found in the gonads.

Within the nucleus of these cells are chromosomes where the hereditary information is recorded in the genes. The genes not only transmit the genetic code of the species but also control cell and tissue organization and ensure maintenance of structures and functions in the individual organism. Besides containing the maternal and paternal chromosomes, the gametes possess specialized structures with well-defined functions that make fertilization possible. For a detailed description of the structure and ultrastructure of the germ cells, and for a better understanding of their unique characteristics within the general context of cellular and molecular biology, the interested reader is referred to the numerous specialized textbooks on this subject. In this chapter we will consider the development of the gametes, primarily, the organs where they are formed and through which they pass, and some of the major events that occur at fertilization.

Cell division is both a prerequisite and a consequence of any increase in size, particularly during development. In most tissues of the body, cells undergo a continuous process of *death* or *regeneration* by division from parent cells. The epithelial layers of the skin that become cornified and are shed will be replaced from below by the cells of the germinative layer; similarly, the epithelial cells of the intestinal villus move up the apex, are cast off, and are replaced by division of epithelial cells in the crypt, and so on. In the young, growing organism, the enlargement of tissues and organs, and the increase in stature result primarily from multiplication of cells by a similar process of division. *Mitosis* is the process of division for all *somatic cells*, and is characterized by the production of two daughter cells, each with the same number of chromosomes as the parent cell (i.e., *diploid complement of chromosomes*). *Meiosis*, on the other hand, represents a specific type of cell division limited to the *germ cells* and results in the production of two daughter cells with *half* the number of chromosomes characteristic of a given species (i.e., *haploid complement of chromosomes*). Any chromosomal changes in somatic cells will be reflected only in the alteration of the cells of a particular individual, whereas chromosal changes in germ cells will be reflected in successive generations. This distinction between somatic cells, which will give rise only to subsequent cell generations in one tissue, and germ cells, capable of forming an entirely new individual, is of basic importance.

Chromosomal Number and Functions

The cells of every animal species contain a fixed number of chromosomes, a total of 46 in humans. This number is identical for all somatic cells of an animal and for its immature germ cells as well. With one exception, each member of a chromosomal pair is morphologically and functionally similar to its mate: in the female, there are 23 pairs of different kinds of chromosomes, in the male, where one the pairs contains 2 different chromosomes, the number of different kinds is 24.

Chromosomes are formed from the condensation of chromatin at the time of cell division and have two important functions: to regulate the synthetic reactions occurring in both the nucleus and cytoplasm of the cell and to transfer the characteristics of one cell to its daughter cells and thus ensure the continuity of specific cell types through countless generations. This controlling information is contained in the genes strung along the length of the chromosome, their general distribution being the same throughout each species (about 30,000 pairs or more in humans). The important component of the gene is DNA which acts as an organic catalyst and is said to produce a single immediate effect—such as the induction of a particular enzyme in the cytoplasm, the so-called one-gene-one-enzyme hypothesis, or the more recent one-operant-one-messenger theory. The DNA in the nucleus serves as a template for the synthesis of RNA found both in the nucleus (i.e., in the nucleolus) and the cytoplasm (i.e., in the ribosomes). The RNA moves from the nucleus into the cytoplasm where it regulates the synthesis of proteins by the cell. The synthesis of protein is the key to the control of cellular development.

In general, mitotic reproduction of a cell consists of the doubling of all components of the cell, followed by a division that distributes the components to the daughter cells. The mitotic events are grouped into four phases (prophase, metaphase, anaphase, and telephase), each charactertized by well-defined changes in the cytoplasm and nucleus. Again, the reader is referred to appropriate textbooks of cellular and molecular biology for a description of these phases. For present purposes, it is sufficient to recall that the most fundamental part of the process, responsible for preserving the character and the potentialities of each kind of cell, is the replication of the DNA molecule that carries the genetic code. The entire nuclear content of DNA is doubled prior to the initiation of cell division. The DNA is also responsible for synthesizing the protein that is combined with it. This synthesis is followed by a series of steps designed to permit accurate distribution of the chromosomes to the daughter cells.

In contrast with mitosis, division by meiosis results in the reduction from the diploid to the haploid number of chromosomes. This reduction is important in maintaining the chromosomal number characteristic of the species at fertilization when maternal and paternal nuclei fuse. Although each gamete must receive one of each pair of chromosomes, whether it receives the maternal or paternal chromosome is a matter of chance. Because this is true for each of the 23 pairs of chromosomes in humans, the number of chromosomal recombinations possible is enormous. Meiosis, then, accomplishes the biological necessity of evolution through controlled variability.

GAMETOGENESIS

Gametes—the ovum and the spermatozoon or sperm—are formed by the process of *gametogenesis*, which consists of three major phases: a period of proliferation during which the primitive germ cells divide repeatedly by mitosis (thereby producing cells of similar size and potentialities), a period of growth marked by rapid enlargement of the cells so produced, and a final period of maturation that involves fundamental nuclear changes and is limited to the final two (meiotic) divisions. Although these phases are common to both oogenesis (the process that results in the formation of the ovum) and spermatogenesis (the process that results in the formation of the sperm) the qualitative and quantitative differences between the two processes are sufficient to warrant separate presentations.

Oogenesis

During the fetal period, primordial cells or *oogonia* proliferate within the cortex of the ovary; by the third month of development, some of the oogonia differentiate into larger *primary oocytes* which, subsequently, become surrounded by epithelial cells to form the *primary follicles*. At birth, the primary oocytes have initiated the first meiotic division but do not proceed further until puberty (Biggers and Schuetz, 1972). By or before birth in most species, including humans, the oogonia have disappeared from the ovary by transformation to primary oocytes. From then on, the population of germ cells in the ovary is reduced over time; such reduction occurs, before puberty, primarily by *atresia* (i.e., involution and degeneration), and after puberty, by ovulation and atresia (see Chapter 2 also). Agents that damage or destroy the oocytes (e.g., radiation) will thus have a permanent and irrevocable effect on depleting the ovary of germ cells. This process is in strong contrast to the situation in the testis where mitotically active spermatogonia persist in the spermatogenic tubules and spermatocytes pass through the phases of meiosis continuously (see below). In each ovary, the total number of follicles (with corresponding oocytes) decreases from about one-half million at birth, to less than one-third of this number at puberty, and to about 30,000 by 30 years of age. During the 30 years or more that constitute the reproductive period in humans, follicles can always be found in various stages of growth. After menopause, when reproductive activity has ceased in the female, follicles are no longer seen in the "senile" ovary (see Chapter 16).

There is little advance beyond the stage of the primary follicle until puberty; at this age, under the influence of the cyclic release of hypothalamic and pituitary hormones, primarily follicle-stimulating hormone (FSH), a number of follicles (sometimes as many as 15 per ovary) start to grow during each cycle of ovarian activity. Most of these follicles fail to achieve maturity, however, and succumb to atresia. A few (usually one) follicles continue to grow. The epithelial cells surrounding the oocyte become cuboidal and begin to secrete an amorphous material, rich in mucopolysaccharides that forms the *zona pellucida* membrane. During this stage, fluid-filled spaces appear among the granulosa cells of the follicle and then come together to form a cavity of *antrum* (follicular antrum) while the cells surrounding the oocyte remain intact and form the *cumulus oophorus* (or egg-bearing hillock). At maturity, the follicle is known as the *Graafian follicle* and is surrounded by two layers of connective tissue, the *theca interna*, rich in blood vessels and considered to be, along with the granulosa cells, the source of *estrogens*, and the *theca externa*, which merges with the ovarian stroma (Figs. 1 (a) and (b)).

As the follicle matures and continues to grow under the combined influence of FSH and luteinizing hormone (LH), the other pituitary gonadotropin, the oocyte situated more and more off center in the follicle, due to the antrum formation, resumes the meiotic division that had been arrested during early development. This division leads to exchange of genetic material (e.g., during synapsis and crossing-over) and the production of two daughter cells of unequal size, but each with the haploid number of chromosomes. Of these two cells, one the *secondary oocyte* receives practically all of the cytoplasm; the other, the *first polar body*, situated between the zona pellucida and the cell membrane of the secondary oocyte, receives practically none. At the completion of the first maturational division, and before the nucleus of the secondary oocyte has returned to its resting stage, the cell enters, but does not complete, the second maturational division. The onset of the preovulatory maturation is marked by a sudden and dramatic rise in the release of gonadotropins from the pituitary, especially LH (the LH surge) triggered by the increasing levels of estrogens. (A classic example of positive feedback.) The interval between

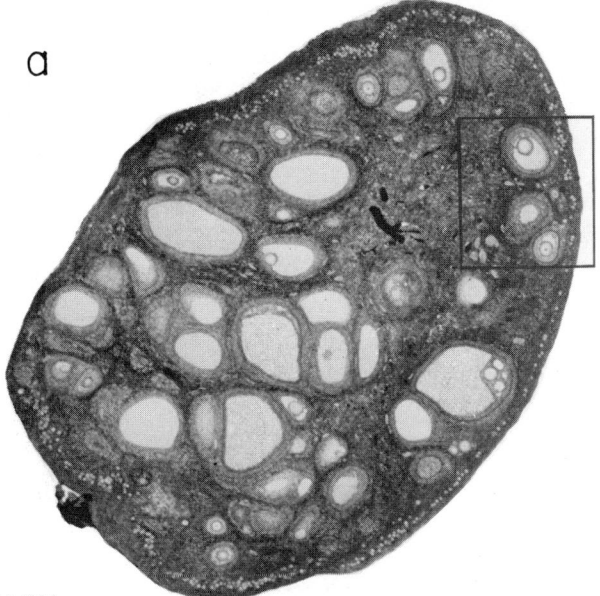

Fig. 1. OVARY
(a) Section of cat ovary with developing follicles containing ova. The rest of the organ consists of interstitial cells and connective tissue surrounding the follicles and supporting blood vessels. Mallory-azan stain (150 X). Courtesy of Dr. Herbert A. Srebnik.

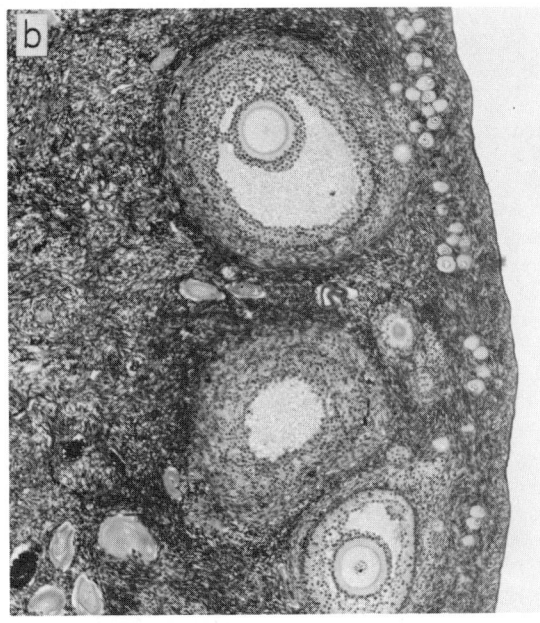

Fig. 1
(b) Enlarged section of ovary enclosed in the rectangle in (a). Primordial follicles occupy the area beneath the germinal epithelium. The growing follicles have cavities, antra, lined by zona granulosa which forms a stalk for the ovum, the cumulus oophorus. Separating the ovum from grandulosa is the zona pellucida (1,000 X).

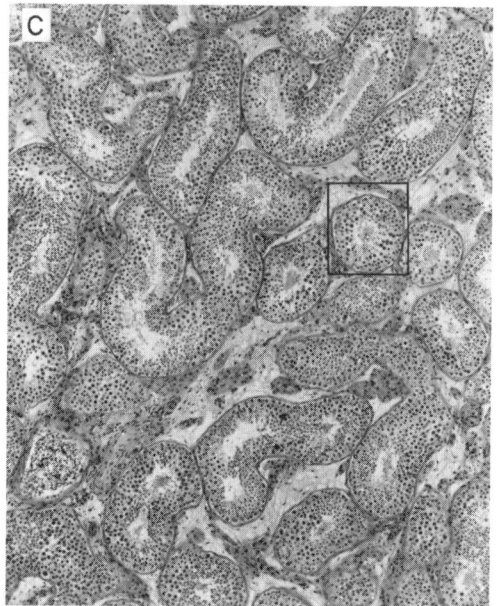

Fig. 1 TESTIS
(c) Section of human testis, composed of seminiferous tubules cut in various planes and interstitial tissue among tubules. Adjacent tubules are at different stages of spermatogenesis. Iron-hematoxylin stain (150 X).

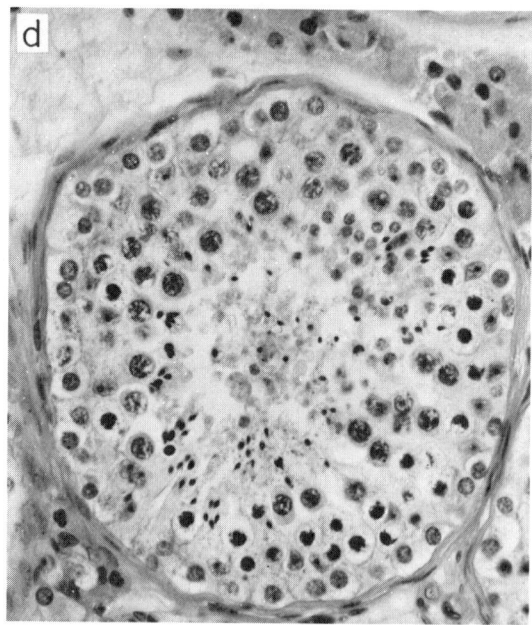

Fig. 1
(d) Cross section of seminiferous tubule enclosed in the rectangle in (c). All stages of the spermatogenic cycle are represented, from spermatogonia resting upon the basement membrane to maturing spermatozoa bordering the lumen and associated with Sertoli cells. Spermatocytes, some in meiotic division, occupy intermediate levels of seminiferous epithelium (1,000 X). *Courtesy of Dr. Herbert H. Srebnik.*

the LH peak and ovulation lasts approximately 36 hours in the human female. During this time, the follicle continues to enlarge greatly and the division of the oocyte progresses through the first meiosis and the beginning of the second; at this moment, ovulation occurs: the oocyte with its cumulus cells is shed from the ovary and is directed towards the Fallopian tube (oviduct) by ciliary currents and peristalsis.

The second maturational division (associated with the formation of a second polar body) is completed only if the oocyte is fertilized and, in this case, it occurs in the oviduct; if fertilization does not occur, the oocyte degenerates approximately 24 hours after ovulation. Whether or not the first polar body always undergoes a second division is uncertain, but fertilized ova accompanied by three polar bodies have been observed. As far as is known, the polar bodies are never fertilized or, if they are, the amount of cytoplasm available is insufficient to sustain a subsequent division.

Ovulation

Ovulation, i.e., the discharge of the oocyte and its surrounding cells from the ovary, represents a stage of the ovarian cycle. Ovarian cyclicity is regulated by neuroendocrine control and begins at puberty with the maturation of the hypothalamo-pituitary-gonadal axis (see Chapter 9). Releasing hormones (RH) from the hypothalamus (probably FSH-LH-RH) act on the cells of the anterior pituitary gland which secrete the gonadotropins, FSH and LH; these, in turn, stimulate and control the cyclic changes in the ovary. The ovarian cycle, and its corresponding menstrual cycle, lasts about 28 days, where day one is the first day of the menses. As indicated above, at the beginning of each cycle, several follicles start to grow under the influence of FSH, but only one reaches maturity. As the follicle matures, progressively higher levels of estrogens, by a positive feedback mechanism, stimulate the anterior pituitary gland to release LH, the hormone responsible for ovulation.

Under the influence of FSH and LH, the follicle continues to expand and forms a bulge on the surface of the ovary with a central avascular spot, the so-called stigma. As a result of the weakening of the ovarian surface above the expanding follicle, the stigma opens and the oocyte, while beginning its second meiotic division, is extruded from the ovary together with its surrounding cumulus oophorus cells and some of the follicular fluid. Ovulation takes place at midpoint in the ovarian cycle, approximately 14 days before the beginning of the following menstrual flow. Although the time between ovulation and the succeeding menstruation is constant, the time between ovulation and the preceeding menstruation is highly variable, since it depends on the length of time the follicle needs to mature.

After ovulation has occurred, the follicular cells remaining in the wall of the ruptured follicle become highly vascularized by surrounding vessels and, under the influence of LH, accumulate a yellowish pigment—hence the name of *corpus luteum* (yellow body) given to this structure. The cells of the corpus luteum secrete *progesterone* and this hormone together with the continuing secretion of estrogens cause the uterine mucosa to enter the *progestational or secretory* state in preparation for the implantation of the embryo (see Chapter 3).

Twinning

As already stated, in most cases, only one oocyte reaches maturation and is expelled from the ovary to be fertilized in the oviduct. In some cases, however, two oocytes become mature and are shed simultaneously from the ovary; they may then be fertilized by two different spermatozoa and result in *dizygotic* or *fraternal twins*. These twins are genetically different and have independent fetal membranes including the placenta. (In cattle, but not in humans, the two chorionic circulations may fuse and, if this fusion occurs between twins of different sex, male hormones from the male twin may disturb the sexual development of the female twin resulting in a female with masculinized genitalia, the so-called freemartin.) A second kind of twins, the *monozygotic* or *identical twins*, may result from the splitting of the fertilized egg (zygote) at various stages of development. In this case, the twins are genetically identical and may or may not share all or some of the fetal membranes. Consideration of the characteristics, incidence and distribution of twinning is well beyond the scope of this text. Nevertheless, it may be of interest to mention here in the context of hormonal regulation of fertilization, that administration of gonadotropins—either in the form of human chorionic gonadotropin, HCG, or of human pituitary FSH or of synthetic chemicals such as clomiphene—to sterile women, is capable of stimulating ovulation (in the case of anovulatory cycles) and restoring menstrual cyclicity (in the case of amenorrhea). Successful pregnancies have followed such treatments although the dangers of overstimulation and the possibility of producing multiple pregnancies (up to six fetuses) is a risk to be carefully considered. In comparison with gonadotropins, clomiphene is more easily administered, its response is more easily controlled, the risk of multiple birth is reduced, although chances of achieving pregnancy are not as great (Rosemberg, 1973).

The Mature Ovum

In precise terms, "ovum" refers to the end stage of cell maturation and "mature ovum" thus represents redundancy, however, it has become common usage to employ the term mature ovum to differentiate the ovum (or egg) during various stages of maturation from the ripe cell. In the process of fertilization, the role of the ovum is: (1) to contribute the maternal complement of genes to the nucleus of the fertilized egg (2) to reject all sperm but one and (3) to provide nutritional reserves until the embryo begins to feed upon exogenous materials.

The distinctive features of the egg include its relatively large size (in humans, 0.14 mm in diameter), its spheroidal shape, its lack of motility, and its protective membranes. The cytoplasmic body (or vitellus) of the egg is limited by a (plasma) *vitelline membrane* and the surface often exhibits microvilli. Lipoprotein granules situated just beneath the surface of the unpenetrated egg, are called "cortical granules" and participate to the "zona reaction" at the time of sperm penetration. Cytoplasmic organelles include mitochondria, multivesicular bodies, endoplasmic reticula, and Golgi bodies (Austin, 1961). Because the mammalian embryo soon becomes sheltered and nourished within the uterine wall of the mother, a large yolk reserve is not required for its development.

Outside of the ovum proper lies a thick, tough, and highly refractive capsule, the *zona pellucida* which increases the total diameter of the human egg to 0.15 mm. Beyond the zona pellucida are the cells of the *corona radiata;* these cells are of follicular origin and surround the ovum during its passage in the oviduct (Fig. 2).

Eggs possess two poles: the *animal pole,* where the polar bodies are pinched off, and the *vegetal pole,* which is involved in the development of nutritive organs (see Chapter 3). This polarity is particularly evident in those eggs containing large amounts of yolk, but is barely noticeable in human eggs.

Aging of the Mammalian Ova

The oocytes, once they have been expelled from the ovary at ovulation, enter the oviduct, and, in the absence of fertilization, undergo rapid degenerative changes which have been ascribed to cellular aging. Aging of the oocytes has been associated not only with their prolonged sojourn in the oviduct and with prolonged follicular maturation, but also with increasing maternal age; in all cases, whether follicular, tubal or chronological, the aging process does impair the physiological competence of the ovum (and perhaps of sperm also) and reduces reproductive capacity, resulting in spontaneous abortions of the fetus or congenital defects in the newborn.

Studies in women between 30 and 35 years of age have shown that if an egg is ovulated after day 14 of the menstrual cycle, the egg has only a 50% chance of developing normally (Hertig, 1967). Inasmuch as irregular menstrual cycles are characteristic of women in menopause (see Chapter 16), these irregularities could result in delayed ovulation and the subsequent release of defective oocytes which are less likely to be fertilized. In the mature rat, injections of pentobarbital have been used to block ovulation for 48 hours and arrest oocyte development in meiosis (Freeman et al., 1970), thus providing a convenient method for studying the effects of prolonged follicular aging on reproductive competence. Delayed ovulation had been associated with a variety of abnormalities: e.g., increased polyspermy, perhaps caused by interference with the synthesis of cortical granules; decreased fertilization and implanatation rates, probably related to impaired metabolism of the ovum; and increased incidence of gross developmental defects, possibly associated with the increased incidence of chromosomal aberrations (Peluso, 1976). Although these observations generally suggest that the oocyte is altered during the period of ovulatory delay, little is known of the specific factors responsible for these alterations. Exogenous estrogen appears to promote the formation of "gap" junctions (nexuses) among the granulosa cells which surround the ovum (Merk et al., 1972); it is possible that during the period of ovulatory delay the prolonged exposure to high levels of endogenous estrogen would alter intercellular communication not only among granulose cells but also between the granulosa cells and the oocyte (Peluso, 1976). Granulosa cells are thought to be involved in oocyte nutrition and metabolism and, consequently, any change in the oocyte-granulosa cell relationship could be detrimental to the oocyte. Additionally, estrogens are known to act on specific cytoplasmic and nuclear receptors and directly influence the cell genome (Gorski and Gannon, 1976), an action which could be responsible for some of the genetic abnormalities observed.

Studies of postovulatory deterioration and aging of the mammalian ovum reviewed by Austin (1970) show that when the time interval between ovulation and fertilization is prolonged beyond the optimal length characteristic for each animal species, the number of pregnancies and the size of the litter are reduced and, conversely, the number of preimplantation losses, abortions, and fetal reabsorptions is increased. In addition, the greater the time interval between ovulation and insemination, the greater the number of ovular abnormalities (e.g., polyspermy, failure to extrude the second polar body, absence or fragmentation of the pronucleus) and, if fertilization occurs despite these alterations, the greater the number of chromosomally aberrant embryos (e.g., triploidy, aneuploidy). Both cytogenic and ultramicroscopic observations suggest that the deleterious effects of tubular aging, at least in some mammalian species (e.g., rabbit, hamster, mouse) are attributable to cytoplasmic alterations [e.g., vacuolation and reduced volume, shrinkage of the cytoplasm, and discharge of cortical granules (Yanagimachi and Chang, 1961)] and nuclear alterations [e.g., deterioration of metaphase spindle and chromatid disjunction (Longo, 1974a,b,)].

Intrinsic changes occurring in the ova of the aged individual are very difficult to determine since the environment surrounding the maturing ovum is also deteriorating to a certain extent in aging women and female animals (Timiras and Meisami, 1972; Hafez, 1976; Talbert, 1977). For example, in rats and mice, a decline in fertility precedes the exhaustion of the oocyte "stores" and the ovulation rate does not appear to be decreased; rather, the cessation or reduction of fertility may involve alterations in the neuroendocrine regulation of pituitary-ovarian relationships or to age-related changes in secondary sex organs, particularly the oviduct and the uterus. Irregularities in menstrual cycles characteristic of premenopausal and menopausal women also may result from impairment of hypothalamo-pituitary-ovarian relationships that delay and produce "overripe" ova, which, if fertilized, would lead to developmental defects. Indeed, electrical (Clemens and Bennett, 1977) and pharmacological (Quadri et al., 1973) stimulation of the hypothalamus or dietary manipulation of those neuroendocrine functions which control ovulation in the young rat restore the capacity of old animals to ovulate, and promote continuation of reproductive function at least for a certain period of time (Segall and Timiras, 1976) (see Chapters 14 to 16.) It is also possible that the ova released during the immediate premenopausal and menopausal years have diminished viability and perhaps chromosomal abnormalities, both of which may contribute to their inability to be fertilized by the sperm or to develop into normal embryos (Peluso, 1976). The condition of mongolism (Down's syndrome), for example, is more common in children born of older women. While it was once believed that this condition was associated with the aging of the uterine environment, it has now been demonstrated to the primarily due to chromosomal abnormalities (trisomy 21 and 22) (Yamamoto et al., 1973a,b). Experiments designed to determine whether reproductive and developmental defects are caused by the aged ova, or the aged uterus were conducted in hamsters, rabbits, and mice by transfering embryos from young animals to aged recipients and *vice-versa*. The results obtained suggest that aging of both the ova and the uterus account for the impaired reproductive capability (Maurer and Foote, 1971; Stockton et al., 1973). From the observations in humans and the experiments in laboratory animals briefly discussed here, it may be inferred that aging of the ova involves a series of functional decrements from failure of neuroendocrine regulation of gonadal function to impairment of gonads and secondary sex organs.

Spermatogenesis

Spermatogenesis, the formation of the spermatozoa, differs markedly from oogenesis in the number of mature gametes produced and the length of its duration during the lifespan: whereas only a few hundred fertilizable ova are liberated from puberty to menopause, millions of motile spermatozoa are formed in the spermatogenic tubules of the male, starting at puberty and continuing, reportedly, until the eighth decade of life.

The primitive germ cells, or *spermatogonia*, are situated next to the basement membrane of the seminiferous tubules of the testis. These are the only elements to be seen until the time of puberty when a renewal of proliferative activity and the onset of spermatogenesis signal testicular maturation. Two kinds of spermatogonia have been distinguished: the so-called dormant cells that generate other spermatogonia and those that will mature themselves into sperm. The latter mature by an orderly and well-defined process: they differentiate into *primary spermatocytes*, which undergo the first maturational division, passing through several stages of a quite prolonged meiotic prophase (lasting from the initial leptotene to the final diplotene stage approximately 22 days); the resulting *secondary spermatocytes* contain the haploid numbers of chromosomes and divide to form *spermatids* (also with the haploid number of chromosomes) (Beatty and Gluekshon-Waelsch, 1972). Spermatids do not divide further, but undergo a complicated process of transformation or spermiogenesis that gives rise to the mature germ cell, the *spermatozoon* or *sperm*. In the rat, the different germ cell types mature in successive stages, with the least mature closest to the basement membrane and the mature spermatid near the tubular lumen. This maturational sequence characteristic of several mammalian species has been called the "cycle of seminiferous epithelium."

In man, spermatogenesis involves four such cycles, each lasting approximately 16 days for a total duration of about 64 days. Contrarily to what happens in rodents, in which the germinal elements follow a very orderly sequence throughout the seminiferous tubule, in man these cycles appear very irregular because the stages of spermatogenesis do not occupy the whole circumference of the tubule; rather, each cell stage occupies a small wedge-shaped area giving to the cross-section of the tubule the appearance of a mosaic made up of several stages of the cycle (Clermott, 1963) (Figs. 1 (c) and (d)).

Associated with the spermatogonia are the *Sertoli cells,* whose ramifications extend from the basement membrane of the tubule to the lumen; presumably these cells provide structural and metabolic stability to the germinal elements and are often referred to as "nurse cells." Spermatids are released from the Sertoli cells and become free in the lumen of the tubules. The Sertoli cells may secrete estrogens, and their development may be stimulated by FSH. The Sertoli cells seem also to be involved in the so-called *blood-testis barrier*, that is the protection of the testis from many substances (e.g., dyes) introduced in the bloodstream. This barrier would depend on the presence of special junctional complexes between two adjacent Sertoli cells near the base of the seminiferous tubules. These occluding (barrier-like) junctions are so situated that they divide the epithelium into a "basal compartment" containing the spermatogonia and an "adluminal compartment" containing the more advanced stages of germ cell differentiation. Substances from the blood could penetrate the basal but not the adluminal compartment.

The spaces between the tubules contain the *interstitial cells of Leydig*, responsible for the secretion of the male sex hormone, *testosterone*. As in oogenesis, spermatogenesis is under the control of pituitary gonadotropic hormones, FSH and LH (in the male, the latter is also referred to as interstitial cell stimulating hormone, ICSH); and these, in turn, are under the control of the respective hypothalamic hormones. The gonadotropins initiate and maintain spermatogenesis, and stimulate testosterone secretion from the Leydig cells; testosterone, itself, modulates several steps in the division and maturation of spermatogonia. FSH and, perhaps, also LH further support spermatogenesis by their tropic action on the Sertoli cells.

The view that gonadotropins and testosterone act "synergistically" to regulate spermatogenesis has been re-evaluated recently; it now appears that these hormones work "consecutively," each having its primary effect on a specific step of cell maturation (e.g., testosterone is essential for completion of meiotic division of primary spermatocytes and FSH, for the maturation of spermatids). Given this hormone-cell-type specificity, spermatogenesis would depend not only on adequate levels of the regulatory hormones but also on their presence at the appropriate time during the maturational cycle (e.g., in the absence of testosterone, meiotic division would not proceed to completion, and the spermatids would not be formed: thus the target cells for FSH would be absent and the tropic effect of this hormone would not occur).

Spermiogenesis, the last phase in the process of spermatogenesis, begins with the spermatid and terminates with the maturation of the spermatozoon. This phase is characterized by continuous maturation of both nucleus and cytoplasm, including: (1) formation of the *acrosome*, probably originating from the Golgi apparatus and extending over two thirds of the sperm head (2) condensation of the nucleus (3) formation of *neck, middle piece*, and *tail* characteristic of the mature spermatozoon, and (4) shedding of most of the cytoplasm (Phillips, 1974). These structural changes take place while the spermatid is engulfed in the cytoplasm of the Sertoli cells.

When fully formed, the spermatozoa enter the lumen of the seminiferous tubules. From here, they are driven toward the epididymis, possibly under the influence of contractile elements in the wall of the seminiferous tubules. Although only slightly motile initially, the spermatozoa attain full motility in the epididymis.

The Mature Spermatozoon

The role of the sperm in the process of fertilization is (1) to reach and to penetrate the egg (2) to activate the nuclear and cytoplasmic division of the egg necessary to embryonic development, and (3) to contribute the paternal complement of genes to the nucleus of the fertilized egg.

Whereas the egg is the largest cell of the female organism, the sperm is among the smallest cells in mass—in humans 50 µ in length—nearly half the diameter of the ovum. The head portion of the sperm contains the chromosomes and is constituted almost entirely of the nucleus; its anterior part is covered by the acrosome, rich in mucopolysaccharides and mucolytic and proteolytic enzymes; the middle piece, comprised of the neck and the body, has an axial filament as its central core, originating in the neck, continuing the entire length of the tail, and surrounded by double spiral strands of mitochondira; the flagellum-like tail, the longest portion of the sperm, enables it to accomplish the "swimming" motions, that transport it to the ovum in the oviduct (Fig. 2).

Human spermatozoa are extremely diverse in form, size, and structure in comparison with those of even closely related species (Pedersen, 1974; Afzelius, 1975; Baccetti and Afzelius, 1976). In addition to the diversity of normal sperm, large numbers of abnormal and immature sperm are present among the 300 to 500 million spermatozoa ejaculated at one time (Duckett and Racey, 1975).

STRUCTURES ASSOCIATED WITH GAMETOGENESIS AND FERTILIZATION

The function of the male and female reproductive systems is discussed in several chapters of this book: during fetal development (Chapter 2), at puberty (Chapter 9) and with aging (Chapter 16). Here, we shall restrict ourselves only to the information necessary to better understanding the processes of gametogenesis and fertilization in general.

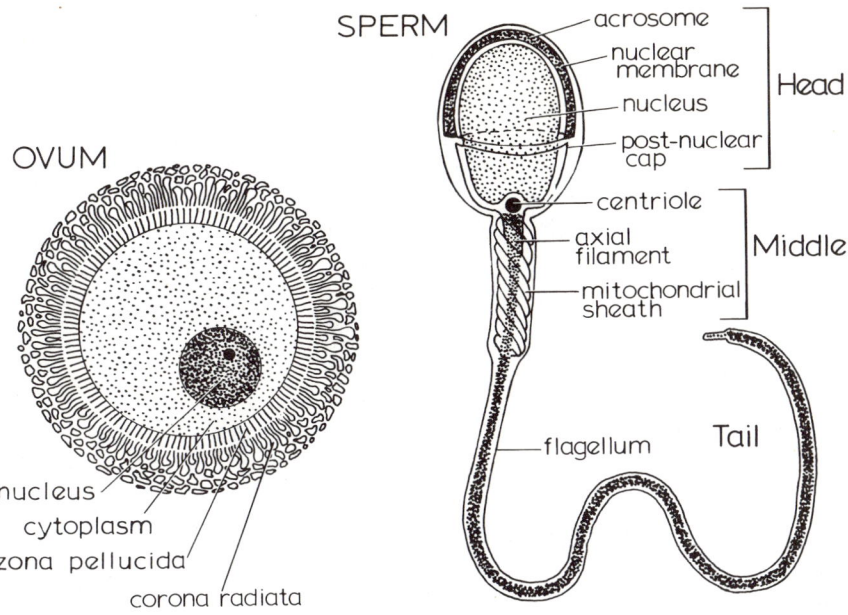

Fig. 2. Diagram of human ovum and sperm.

Female Reproductive System

The *ovaries*, the primary reproductive organs of the female, serve both gametogenetic and endocrine functions. In their reproductive role these functions include: (1) the development of the reproductive tract in preparation for fertilization, primarily through the production of estrogens; (2) the production, maturation and release of the ova, primarily through the influence of the gonadotropic hormones from the pituitary gland; and (3) the establishment of a favorable environment for the transport and implantation of the fertilized ovum in the uterus, as well as the development of the blastocyst, primarily through the continuing production of estrogens and the added production of progesterone. Estrogens and progesterone, the main steroid hormones of the ovary, play an important regulatory role in whole-body growth and in the development and maintenance of secondary sexual characteristics, such as the mammary glands, hair, fat distribution, and voice (see also Chapters 9 and 10). During pregnancy, the ovary also produces a polypeptide hormone called *relaxin*, responsible in mammals for widening the birth canal at parturition (e.g., loosening the ligaments of the pubic symphysis and dilating and softening the uterine cervix). This hormone has been clearly identified in a number of animal species (e.g., rats, rabbits, mice, guinea pigs, pigs) and its presence has been reported also in the serum of pregnant women at term (Weiss et al., 1976). Relaxin has been shown to chemically resemble insulin, (Schwabe and McDonald, 1977), to be produced by the corpus luteum of pregnancy, and to be correlated in its production with the secretion of luteal progesterone.

The secondary reproductive organs in the female—oviduct, uterus and vagina—not only transport the ovum into the oviduct, receive the sperm in the vagina and transport it to the oviduct as well, but also are involved in the fertilization of the ovum in the oviduct and with the housing, development, nutrition, and protection of the embryo and the fetus while in the uterus, (Fig. 3). Each ovary is in close connection with the fringed end of the *oviduct*, the Fallopian tube, that leads to the uterus. Rhythmic contractions of the ciliated fimbriae, the fringed opening of the oviduct,

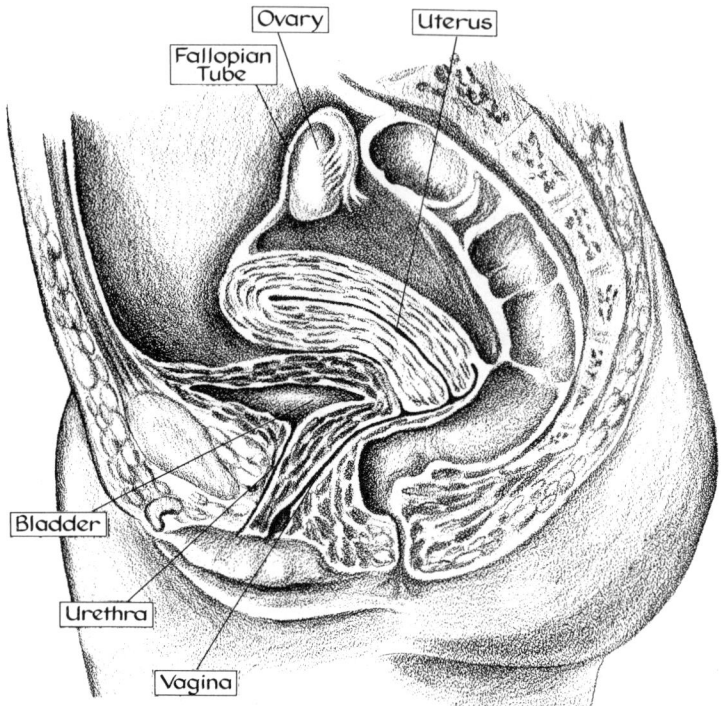

Fig. 3. Diagram of the female sex organs and connected structures.

are believed to sweep the ovum into the oviduct where the ova, for the first time, are exposed to conditions outside of the ovary. The ovum is transported downward by the contraction of the muscles of the tube walls combined with the motion of the cilia on the epithelial lining. Because there are no direct connections between the ovary and the opening of the oviduct, it is possible for the ovum, once freed from the ovary, to be released into the abdominal cavity—a rare occurence, however, because the sweeping movements of the fimbriae normally prevent it. The passage of sperm up the female genital tract to the oviduct depends not only on the motile powers of the sperm, but, at least in some species (e.g., humans), on uterine contractions as well. Within the oviduct, fertilization takes place, syngamy (i.e., union of the two gametes) occurs and the initial stages of cleavage and development occur. While it has been shown that many of these functions may occur under specialized conditions *in vitro*, the oviduct combines the essential components—environment, hormonal balances, and biochemical constituents—which make is possible for all of these functions to take place normally and in orderly fashion, thus making the process of reproduction possible (Johnson and Foley, 1974). The mucosal cells of the oviduct produce a fluid that increases the fertilizing capacity of the sperm. This process, called *capacitation,* takes place in the female reproductive tract and continues until the sperm is capable of penetrating the surface of th egg. *In vitro* studies have shown that, in humans, capacitation lasts about 7 hours. It has been suggested that capacitation may represent a change in the sperm membrane evoking, under appropriate conditions, the acrosome reaction and release of acrosomal enzymes; it also may destabilize the sperm membrane, remove blocking agents (e.g., a decapacitation factor), activate receptor sites and achieve competence to respond to other agents (Austin, 1974).

The *uterus* provides the site for the implantation of the fertilized ovum and, by developing the maternal side of the placenta, regulates the exchange

processes that ensure the growth and development of the embryo and
fetus. To accomplish this end, the uterus grows and adapts its shape,
size, and function to the rapidly growing fetus. At the termination of
gestation, it contracts in order to expel the mature fetus through the
birth canal. The uterus is held in place by the bony pelvis, ligaments,
and the strong muscles of the perineum. It has three openings; two
lateral ones that communicate with the oviducts and a single narrow opening situated at the extremity of the uterine neck or cervix at the upper
portion of the vagina. The *vagina* is a slightly curved muscular canal,
about 7-8 cm long in which the sperm are introduced at copulation; it is
lined by a thick layer of stratified squamous epithelium. The uterus has
a strong muscular layer, the *myometrium*, which is lined internally by a
mucosa, the *endometrium*. Both these layers are extremely responsive to
the hormonal secretions of the ovary and the endometrium undergoes cyclic
changes in parallel with ovarian and vaginal cycles (see Chapter 9).

Substances Affecting Uterine Motility

The myometrium is very sensitive to a number of substances capable of
influencing the contraction of smooth muscle; among these of particular
importance during pregnancy is the hormone *oxytocin*, which is synthesized
in the neuroendocrine cells of the supraoptic and paraventricular nuclei
of the hypothalamus and stored and released from the posterior lobe of
the pituitary. Oxytocin stimulates the contraction of the smooth muscle
of the uterus, particularly during late pregnancy and thereby initiates
and facilitates parturition; it may also act on the nonpregnant uterus to
facilitate sperm transport. The genital stimulation involved in coitus
releases oxytocin, but it has not yet been proved that oxytocin initiates
the rather specialized uterine contractions that assist in transporting the
sperm.

A series of closely related unsaturated fatty acids, the *prostaglandins*,
represent another group of substances capable of stimulating uterine
contractions. First detected in the semen and found in high levels in the
prostate (hence their name), they now have been detected in most tissues
and organs of the body. Prostaglandins have attracted considerable
attention because of their numerous actions—often synergistic or antagonistic
depending on the prostaglandins considered—on a multiplicity of systems
and functions (e.g., the immune system, inflammatory responses, fever,
platelet aggregation, blood pressure, renal excretion of sodium, brain
catecholamines, steroidogenesis). Because of their several actions in the
reproductive tract of both males and females, they are employed in various
clinical situations (Karim, 1975). For example, because of their stimulatory
action on the uterine muscle, they are being used both for inducing labor
at term and for inducing abortion in midtrimester pregnancy. Additionally,
because they are capable of directly relaxing the smooth muscle of the
female genital tract, and/or causing its contraction, they can be used to
enhance sperm migration to the uterus and oviduct and facilitate conception.
In the female, prostaglandins may also regulate reproductive cyclicity
either directly by causing luteolysis or indirectly by modifying pituitary
gonadotropic responses to hypothalamic hormones. The exact physiologic
role of these ubiquitous substances is still unknown as is their mechanism
of action, although it has been suggested that in response to various
stimuli they may modulate the generation of cyclic adenosine-3'-5'-monophosphate (cAMP), a key intracellular mediator for the effects of hormones
and other substances that modify cellular function.

Male Reproductive System

The primary reproductive organs of the male are the two *testes* which, like the ovaries, have both gametogenetic and hormonal functions. They are the site of production not only for spermatozoa but also for the male sex hormones, *androgens*, among which tesosterone is the most important. In addition, the testes secrete small amounts of estrogens. (Androgens and small amounts of estrogens are also secreted by the adrenal cortex in both sexes.)

The testes are composed predominantly of masses of seminiferous tubules and it is along the walls of these tubules that the primitive germ cells are converted into spermatozoa. The seminiferous tubules ultimately drain into the *epididymis*, an elongated cylindrical structure on the surface of the testes formed of a network of ducts that store the sperm. Contractions of the smooth muscle cells of the ducts, together with the ciliary-like action of the cells lining the lumen, advance the spermatozoa into the *vas deferens*. The vas deferens in turn, leads by way of the *ejaculatory ductus* into the *urethra* in the body of the *prostate gland;* prior to this, it dilates to form an enlargement or ampulla, at the distal end of which it forms a large blind glandular evagination, the *seminal vesicle*. Most of these structures serve for the progression, activation, nutrition, emission and ejaculation of the sperm. The two *seminal vesicles* secrete a fluid that mixes with the sperm during ejaculation; chemically, this fluid is high in potassium, low in sodium, with citrate as the major anion and fructose as the major metabolic substrate. The *prostate* also secretes a fluid which contributes to the formation of the semen and is rich in sodium and zinc. The *bulbourethral* (Cowper's) and *urethral glands,* as well as other smaller glands that are widely dispersed along the urethra, produce secretions rich in mucoproteins which, during ejaculation, are expelled before the spermatozoa. (Fig. 4).

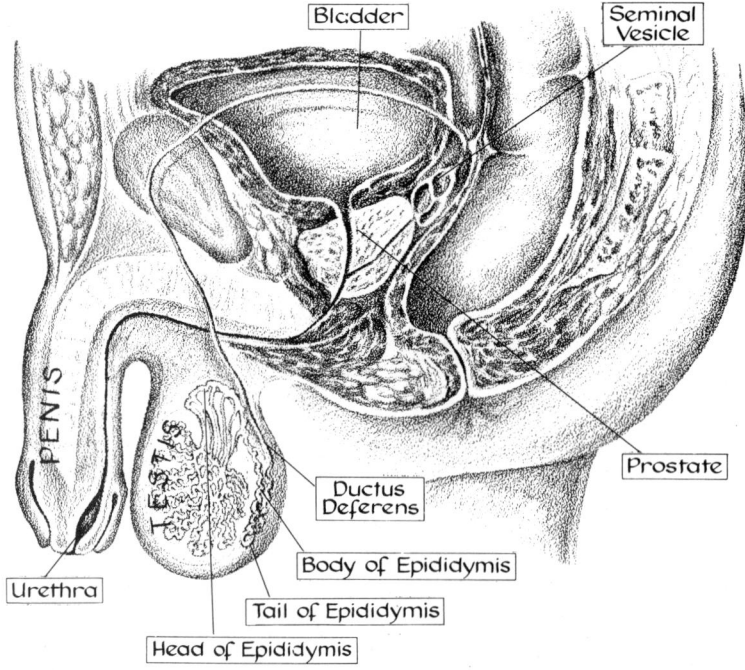

Fig. 4. Diagram of the male sex organs and connected structures.

Semen

Semen is the fluid ejaculated at the time of orgasm; it contains spermatozoa and the secretions of the seminal vesicles, prostate, Cowper's glands, and probably the urethral glands. The average volume ejaculated is 2-6 ml of which 60% is contributed by secretions of the seminal vesicles, 20% by the prostate, and the remainder by the other glands. To be optimally fertile, the semen must have a count of 100 million spermatozoa or more and contain fewer than 20% abnormal forms. Fifty percent of men whose spermatozoa counts are in the range of 20-40 million and essentially all men with less than 20 million spermatozoa are sterile. Even in normally fertile men, the volume of semen and sperm count decreases rapidly with frequent ejaculations. Besides sperm, semen contains inorganic components (primarily NA^+, K^+, Ca^{++} and zinc), organic acids (primarily, citric acid) and bases (primarily spermine and phosphorylcholine) as well as numerous enzymes, sugar (primarily fructose), and lipids (including prostaglandins). The semen is discharged from the male genital tract by a spinal reflex characterized by two stages: *emission*, the movement of the semen in the urethra, and *ejaculation*, the propulsion of semen out of the urethra at the time of orgasm; both reflexes are regulated by autonomic (sympathetic) impulses. The transfer of spermatozoa to the female reproductive tract occurs during *coitus*, or copulation, mainly by the introduction of the erected penis into the vagina and the subsequent ejaculation of the semen. *Erection* is induced by engorgement of the penis resulting from vasodilation in response to autonomic (para sympathetic) impulses.

Immunology and Reproduction

The concept that immune phenomena participate in human reproduction, although not new, has recently generated revived interest because of its potential use in fertility control. Clear-cut demonstration of the antigenic power of spermatozoa or of whole semen was reported at the end of the last century. Since then, sperm antigens and antibodies have been related to male and female infertility. The testicular tissue, spermatozoa, seminal plasma, and the male accessory glands possess enough antigenic potency to induce autoantibody formation capable of rendering the male organism temporarily or permanently sterile (Mancini, 1976). Similarly, in the female, unexplained infertility, early abortions, and some pathologic conditions, such as hydated mole and chorioncarcinoma, have been related to immunologic impairment. Controling fertility by immunization with placenta proteins (e.g., subunits of human chorionic gonadotropin) has been effective in the baboon (Stevens, 1975) and even though the applicability of these observations to humans is still remote, the possibility of developing an "immunity" to pregnancy is on the horizon (Shivers and Dunbar, 1977).

FERTILIZATION

The formation, maturation and meeting of a male and female sex cell are all prerequisite to their actual union into a combined cell, the zygote, that marks the beginning of a new individual. This penetration of an ovum by a spermatozoon and the fusion of their respective nuclei constitutes the essential process of fertilization. In practically all animals, fertilization also supplies the stimulus to start the ovum dividing and thus can be said to initiate development.

Parthenogenesis and Errors of Fertilization

Parthenogenesis, the normal mode of development of certain insects, such as the worker ants and lizards, represents the process by which the oocyte is activated to divide and a new organism to develop *without* sperm penetration (Cuellar, 1977). In most vertebrates, it can be induced by mechanical, physical, or chemical stimuli (Beatty, 1967). While these stimuli suffice to inititate nuclear and cytoplasmic divisions in most cases, the new individual will possess only the haploid (maternal) complement of chromosomes; in some cases the diploid number is re-established by retention of the second polar body. In mammals, parthenogenesis is known to proceed only to the stage of implantation; production of living young requires transfer of the dividing ovum into the uterus of a recipient animal of the same species (Graham, 1974).

Variations and anomalies in fertilization have been ascribed to chromosomal aberrations during gametogenesis, to aging of the ova and/or spermatozoa and to immaturity of the gametes. The most common "errors" include polyspermy (i.e., penetration of the ovum by more than one spermatozoon), polygyny (i.e., failure of emission of polar bodies) and immediate cleavage (i.e., cleavage of the unfertilized egg followed by fertilization of both daughter cells). The first two errors lead to triploidy and the third to mosaic development. Other variations of fertilization include superfetation, a condition in which ovulation and fertilization occur during an already established pregnancy, and superfecundation, a condition which may follow poly-ovulation, and in which one oocyte may be fertilized by the spermatozoon from a male and another oocyte is fertilized by a different male; these conditions have been described in various mammals but they have never been reported in humans.

Fertilization occurs in the ampulla, the section of the oviduct proximal to the ovary. As stated previously, the oocyte, in its cumulus mass, is released from the follicle at ovulation, picked up by the fimbriae, transported into the oviduct by fluid currents and reaches the ampulla within a few minutes. At this stage, the oocyte already has completed its first meiotic division and has initiated its second division which will continue to completion only upon fertilization. Spermatozoa may already be present in the ampulla or may reach it within 24 hours, the optimal period for fertilization.

The spermatozoa deposited at the time of coitus, reach the ampulla after passing rapidly through the vagina and the uterus and subsequently enter into the oviduct. This ascent—often within 30 minutes after coitus—is thought to be accelerated by the contractions of the musculature of the uterus and oviduct and the propulsion of the tail of the sperm. Once in the female genital tract, the sperm undergoes capacitation before it can fertilize the egg.

Passage to the site of fertilization is not an easy task for the spermatozoa, which face both mechanical and chemical barriers. While hundreds of millions of spermatozoa are deposited in the vagina, the numbers reaching the site of fertilization rarely exceed a few thousand; the majority are evacuated from the vagina or phagocytized. The magnitude of sperm numbers at the site of fertilization may still appear in excess of needs, but, in fact, their population is really quite sparse considering the small size of the spermatozoa and the comparatively large space within the oviduct.

The process of fertilization, itself, includes a series of well-coordinated steps which can be grouped arbitrarily into three phases: (1) the penetration of the egg by sperm; (2) the activation of the egg, and; (3) the union of the egg and sperm nuclei.

Penetration

Contact between sperm and ovum is generally the result of random collision. Although in some mammals the oocyte and spermatozoa appear to be attracted to each other by chemical influences (i.e., by such substances as fertilizins produced by the gametes to increase the probability of collision and promote adherence of sperm and egg), solid proof of this concept is lacking. *In vitro* studies have shown that human spermatozoa may swim quite close to the oocyte but pass by without any apparent attraction.

As indicated earlier, capacitation occurs when the spermatozoa enter the female genital tract where the subsequent changes in sperm membrane lead to the *acrosomal reaction;* both processes are required for fertilization. In the acrosomal reaction, the outer acrosomal and plasma membranes of the spermatozoa fuse to produce vesicles which release the acrosomal contents (hyaluronidase and other enzymes) and yet retain an intact limiting membrane, the inner acrosomal membrane, around the sperm head (Austin and Short, 1972a). The acrosomal reaction permits the acrosomal contents to escape from the acrosome and thereby facilitate the passage of the sperm through the membranes surrounding the oocyte (Austin, 1968).

As noted, in the great majority of mammals, only one sperm finds its way into the oocyte. Several processes block polyspermy, i.e., the occurrence of the entrance of more than one sperm into the egg. The *zona reaction* for example, is one of such processes wherein the cortical granules from beneath the cell membrane of the egg are transported into the perivitelline space (Barros and Yanagimachi, 1972). This rearrangement of cortical granules has been related to changes occurring in intracellular calcium that affect cell membrane permeability (Steinhardt, et al., 1977).

After passing through the cumulus oophorus and dispersing the cells of the corona radiata with the help of the enzymes liberated by the acrosome reaction, the spermatozoon penetrates the zona pellucida; penetration has been found to require 10 minutes *in vitro* and probably less time *in vivo*. The spermatozoon penetrates in a curved oblique path, leaving a slit in the zona pellucida. A lysin adhering to the inner acrosome membrane is thought to facilitate penetration. On entering the perivitelline space, the sperm head immediately lies flat to the cell membrane of the egg and its plasma membrane fuses with that of the egg. The sperm head then sinks into the cytoplasm of the egg. In humans, both the head and the tail of the spermatozoon enter the cytoplasm of the oocyte, but the plasma membrane is left behind on the oocyte surface (Fig. 5).

Activation

At the time of sperm penetration, the oocyte that had only started its maturational division becomes "activated," that is, undergoes a series of nuclear and cytoplasmic changes that inititate embryonic development. The current view is that the egg passes into an inhibited state after the beginning (i.e., the stage of spindle formation or metaphase) of the second meiotic division and that its metabolic processes continue at a subdued level, gradually declining after ovulation when the egg ages. Sperm entry, in some way, is believed to abolish this inhibition (Austin and Short, 1972a). The major changes, nuclear and cytoplasmic, include: (1) rearrangement of the chromosomes in a vesicular pronucleus, the *female pronucleus;* (2) *block to polyspermy* with the opening and evacua-

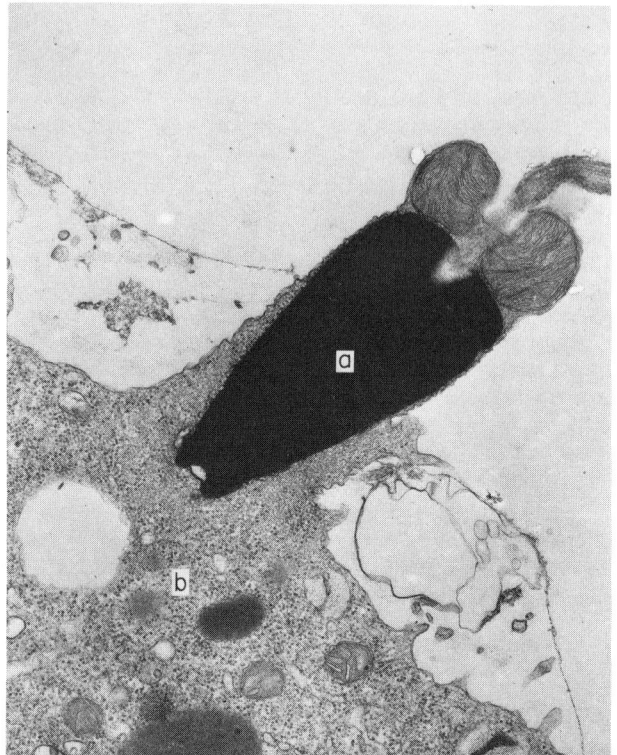

Fig. 5. A section through the fertilization cone of the sea urchin *Arbacia punctulata* demonstrating the continuity of the plasma membrane of the sperm (a) with that of the egg (b) (42,000 X). *Courtesy of Dr. Everett Anderson.*

tion of cortical granules; (3) *reduction in ooplasm* and *appearance of a perivitelline space* between the oocyte and the zona pellucida; and (4) *resumption of the second meiotic division* with emission of the second polar body. Meanwhile, the nucleus of the spermatozoon becomes swollen and forms the *male pronucleus*. Male and female nuclei, morphologically indistinguishable, are then surrounded by a new nuclear envelope. DNA is now replicated and transcription of maternal and paternal genes begins and becomes maximal during the subsequent cleavage divisions.

Fusion of Nuclei

The climax of fertilization is reached when the two pronuclei unite. Little RNA synthesis occurs during fertilization in asmuch as pronuclear nucleoli contain small amounts of RNA. As syngamy (union of the two gametes) approaches, the nuclear envelopes of the pronuclei break up, the nucleoli disappear and the chromosomes condense and promptly aggregate and mingle to form the first cleavage spindle. The egg then divides into the two first daughter cells or blastomeres. (In the rabbit, this is usually 12 minutes from sperm penetration.) In this manner, the full number of chromosomes, which had been temporarily halved in each gamete, is restored in the zygote. Fertilization is now complete and the subsequent divisions occur mitotically.

The fundamental results of fertilization are: (1) male and female sets of chromosomes are reassociated, restoring the full diploid number; bringing together an equivalent chromatin contribution from two parents furnishes the physical basis for inheritance as well as variation (2) the sex of the new individual is determined (see Chapter 2), and (3) the ovum is activated to divide and to produce the new organism. As a result of the first cleavage and the subsequent divisions, every cell of the developing body receives a sample of each kind of chromosome pooled at fertilization.

Artificial Reproduction

In vitro fertilization of a mammalian egg has been attempted over a long time (see Marx, 1973). After recognition of sperm capacitation, cytological evidence for *in vitro* fertilization of rabbit eggs with uterine spermatozoa, and the birth of young following the transfer to the uterus of *in vitro* fertilized ova has been achieved in several laboratory and domesticated animals (Austin and Short, 1972b; Austin, 1973; Edwards, 1973). Reimplantation of human embryos fertilized *in vitro* has now been accomplished (Steptoe and Edwards, 1978) as an answer to the problems of infertility, whether caused by oviduct occlusion or oligospermia. The method used involves collecting a preovulatory oocyte from the ovary under a simple laparoscopy lasting but a few minutes, fertilizing it *in vitro*, and reimplanting it in the woman via the cervix. While the procedure has been used successfully (success being defined as the birth of a viable infant) much work remains to be done before reimplantation can be considered a broadly applicable alternative for infertile couples. If it proves to be successful in a significant number of cases, this procedure could surpass the various surgical techniques now employed to repair damaged tissue in the oviduct, or to remove a blockage by shortening the oviduct. In addition, reimplantation has potential for women without an oviduct and for women who have been sterilized. Assurance that sterilization can, in effect, be reversed could have profound social implications that can be regarded as positive in the world-wide controversy surrounding birth control, abortions, and sterilization. (see Kolata, 1978.)

Another kind of artifical fertilization is the phenomenon of *cloning;* that is, the removal of the nucleus from an egg and its replacement with the nucleus of a somatic cell. This procedure initiates the formation of a new organism, the exact genetic copy of the cell donor. Ever since frogs were cloned in the early 1960s, attempts have been made to clone mammals but, despite some unsubstantiated claims to the contrary, this has not yet been successful. Nevertheless, current advances in test-tube fertilization of human eggs, genetic screening, and recombinant DNA technology suggest that cloning of mammals and eventually of humans may be possible "if not now, soon" (Cullinton, 1978). Finally, experiments in a number of mammals—especially dogs and sheep—have shown that it is possible to sustain growth and development in exteriorized, isolated fetuses, perfused with specialized blood-pumping circuitry and oxygenators. These studies are important not only for their future applicability to the artifical development of humans but also for their immediate contribution to the improvements of devices and procedures capable of supporting life of premature human infants and newborns at risk (Austin, 1973). However, progress in fetal research and experiments directed to improving the management of premature infants have created social, ethical and legal complications which must be resolved before the potential benefits of these studies can be fully exploited.

Fertility and Antifertility Agents

It would seem logical that a discussion of fertilization and the critical role of hormones in this complex process should include a full discussion of contraception and of our relatively recent capability of regulating and blocking conception by the use of exogenous hormones. In fact, this area has now become so extensive that a number of textbooks have been devoted to the subject. The reader is referred to some of the most comprehensive (Greep et al., 1976; Diamond and Korenbrot, 1978).

REFERENCES

Afzelius, B.A. *The Functional Anatomy of the Spermatozoa.* Permagon Press, New York (1975).
Austin, C.R. *The Mammalian Egg.* Charles C. Thomas Co., Springfield, Ill. (1961).
Austin, C.R. *Ultrastructure of Fertilization.* Holt, Rinehart and Winston, New York (1968).
Austin, C.R. Aging and reproduction: post-ovulatory deterioration of the egg. *J. Reprod. Fertil.* 12, 39-53 (1970).
Austin, C.R. (ed.) *The Mammalian Fetus in Vitro.* Chapman and Hall, Ltd., London (1973).
Austin, C.R. Recent progress in the study of eggs and spermatozoa: insemination and ovulation to implantation, in *Reproductive Physiology,* R.O. Greep, ed. Butterworths, London (1974), pp. 95-131.
Austin, C.R. and Short, R.V. (eds.) *Reproduction in Mammals, Book 1: Germ Cells and Fertilization.* Cambridge University Press, Cambridge (1972a).
Austin, C.R. and Short, R.V. (eds.) *Reproduction in Mammals, Book 5: Artificial Control of Reproduction.* Cambridge University Press, Cambridge (1972b).
Baccetti, B. and Afzelius, B.A. *The Biology of the Sperm Cell.* S. Karger, Basel (1976).
Barros, C. and Yanagimachi, R. Polyspermy-preventing mechanisms in the hamster egg. *J. Exp. Zool.* 180, 251-266 (1972).
Beatty, R.A. Parthenogenesis in vertebrates, in *Fertilization,* C.B. Metz and A. Monroy, eds. Academic Press, New York (1967), pp. 413-440.
Beatty, R.A. and Glueksohn-Waelsch, S. (eds.) *The Genetics of the Spermatozoon: Proceedings of an International Symposium.* Univ. of Edinburgh, Edinburgh (1972).
Biggers, J.D. and Schuetz, A.W. (eds.) *Oogenesis.* University Park Press, Baltimore (1972).
Clemens, J.A. and Bennett, D.R. Do aging changes in the preoptic area contribute to loss of cyclic endocrine function? *J. Gerontol.* 32, 19-24 (1977).
Clermont, Y. The cycle of the seminiferous epithelium in man. *Am. J. Anat.* 112, 35-51 (1963).
Cuellar, O. Animal parthenogenesis: a new evolutionary-ecological model is needed. *Science* 197, 837-843 (1977).
Culliton, B.J. Scientists dispute book's claim that human clone has been born. *Science* 199, 1314-1316 (1978).
Diamond, M.C. and Korenbrot, C.C. (eds.) *Hormonal Contraceptives, Estrogens and Human Welfare.* Academic Press, New York (1978).
Duckett, J.G. and Racey, P.A. *The Biology of the Male Gamete.* Academic Press, New York (1975).
Edwards, R.G. Physiological aspects of human ovulation, fertilization and cleavage. *J. Reprod. Fertil.* Suppl. 18, 87-101 (1973).
Freeman, M.E., Butcher, R.L., and Fugo, N.W. Alteration of oocytes and follicles by delayed ovulation. *Biol. Reprod.* 2, 209-215 (1970).

Gorski, J. and Gannon, F. Current models of steroid hormone action: a critique. *Ann. Rev. Physiol.* 38, 425-450 (1976).
Graham, C.F. The production of parthenogenetic mammalian embryos and their use in biological research. *Biol. Rev.* 49, 399-422 (1974).
Greep, R.O., Kobinsky, M.A., and Jaffe, F.S. *Reproduction and Human Welfare: A Challenge to Research.* The MIT Press, Cambridge, Mass. (1976).
Hafez, E.S.E. (ed.) *Aging and Reproductive Physiology.* Ann Arbor Science Publishers, Inc., Ann Arbor, Michigan (1976).
Hertig, A.T. The overall problem in man, in *Comparative Aspects of Reproduction Failure,* K. Benirschke, ed. Springer-Verlag, New York (1967), pp. 11-35.
Johnson, A.D. and Foley, C.W. (eds.) *The Oviduct and its Functions.* Academic Press, New York (1974).
Karim, S.M.M. *Prostaglandins and Reproduction.* University Park Press, Baltimore (1975).
Kolata, G.B. In vitro fertilization: is it safe and repeatable? *Science* 201, 698-699 (1978).
Longo, F.J. An ultrastructural analysis of spontaneous activation of hamster eggs aged in vivo. *Anat. Rec.* 179, 27-56 (1974a).
Longo, F.J. Ultrastructural changes in rabbit eggs aged in vivo. *Biol. Reprod.* 11, 22-39 (1974b).
Mancini, R.E. *Immunologic Aspects of Testicular Function.* Springer-Verlag, Berlin, New York (1976).
Marx, J.L. Embryology: out of the womb—into the test tube. *Science* 182, 811-814 (1973).
Maurer, R.R. and Foote, R.H. Maternal aging and embryonic mortality in the rabbit. I. Repeated superovulation, embryo culture and transfer. *J. Reprod. Fertil.* 25, 329-341 (1971).
Merk, F.B., Botticelli, C.R., and Albright, J.T. An intercellular response to estrogen by granulosa cells in the rat ovary: an electron microscope study. *Endocrinology* 90, 992-1007 (1972).
Pedersen, H. *The Human Spermatozoon.* Costers Bogtrykkeri, Copenhagen (1974).
Peluso, J.J. Aging of mammalian ova, in *Aging and Reproductive Physiology,* E.S.E. Hafez, ed. Ann Arbor Science Publishers, Ann Arbor, Michigan (1976), pp. 35-50.
Phillips, D.M. *Spermiogenesis.* Academic Press, New York (1974).
Quadri, S.K., Kledzik, G., and Meites, J. Reinitiation of estrous cycles in old constant-estrous rats by central acting drugs. *Neuroendocrinology* 11, 248-255 (1973).
Rosemberg, E. (ed.) *Gonadotropin Therapy in Female Infertility.* Proceedings of the Conference held at Worchester, Massachusetts, May 12 and 13, 1972. Excerpta Medica, Amsterdam (1973).
Schwabe, C. and McDonald, J.K. Relaxin: a disulfide homolog of insulin. *Science* 197, 914-915 (1977).
Segall, P.E. and Timiras, P.S. Pathophysiologic findings after chronic tryptophan deficiency in rats: a model for delayed growth and aging. *Mech. Ageing Dev.* 5, 109-124 (1976).
Shivers, C.A. and Dunbar, B.S. Autoantibodies to Zona Pellucida: a possible cause for infertility in women. *Science* 197, 1082-1084 (1977).
Steinhardt, R., Zucker, R., and Schatten, G. Intracellular calcium release at fertilization in the sea urchin egg. *Dev. Biol.* 58, 185-196 (1977).
Steptoe, P.C. and Edwards, R.G. Birth of a baby following the reimplantation of a cleaving embryo. *Lancet* ii, 366-000 (1978).
Stevens, V.C. Fertility control through active immunization using placenta

proteins, in *Immunological Approaches to Fertility Control,* 7th Karolinska Symposium, E. Diczfalusy, ed. *Acta Endocrinol.* Suppl. 194 Vol. 78, 357-375 (1975).

Stockton, B.A., Parkening, T.A., and Soderwall, A.L. Blastocyte transfer study in the senescent golden hamster. *J. Reprod. Fertil.* 32, 145-147 (1973).

Talbert, G.B. Aging of the reproductive system, in *Handbook of the Biology of Aging,* C.E. Finch and L. Hayflick, eds. Van Nostrand Reinhold Co., New York (1977), pp. 318-356.

Timiras, P.S. and Meisami, E. Changes in gonadal function, in *Developmental Physiology and Aging,* P.S. Timiras, Macmillan, New York (1972), pp. 527-541.

Weiss, G., O'Byrne, E.M., and Steinetz, B.G. Relaxin: a product of the human corpus luteum of pregnancy. *Science* 194, 948-949 (1976).

Yamamoto, M., Endo, A., and Watanabe, G. Maternal age of dependence of chromosome anomalies. *Nature, New Biology* 241, 141-142 (1973a).

Yamamoto, M., Shimada, T., Endo, A., and Watanabe, G. Effects of low-dose x irradiation on the chromosomal non-disjunction in aged mice. *Nature, New Biology* 244, 206-208 (1973b).

Yanagimachi, R. and Chang, M.C. Fertilization life of golden hamster ova and their morphological changes at the time of losing fertilizability. *J. Exp. Biol.* 148, 185-203 (1961).

Copyright©1982, Spectrum Publications, Inc.
Hormones in Development and Aging

Chapter 2
Hormones During Prenatal and Neonatal Development
Paola S. Timiras
Jean M. Cons

INTRODUCTION

Not only are hormones responsible for the regulation of gametogenesis and fertilization (see Chapter 1), of implantation of the fertilized egg and of the early developmental stages (see Chapter 3), but they also play a crucial role in differentiation and growth throughout embryonal and fetal development. Prenatally their function is manifested by actions similar to those exerted in the adult (e.g. regulation of metabolism) as well as by actions which are unique to the physiologic requirements of the developing organism (e.g., "organizing" or differentiation-promoting actions). Yet, despite its distinct characteristics, fetal endocrinology represents but the first stage in a continuum of hormonal adjustments accompanying and perhaps regulating the unfolding of the lifespan. This spectrum of endocrine activities highlighted, before birth by regulation of differentiation and maturation, is continued after birth, by such actions as regulation of growth in the infant and child (Chapter 4), maturation of the reproductive function in the adolescent (Chapter 9), cessation or decline of this function in old age (Chapter 16), and establishment of optimal adaptive capacity in the adult and its impairment in old age, (Chapter 15); all manifestations that can be interpreted as part of a biological program starting at fertilization and terminating at death (Timiras, 1978). In this chapter, we will summarize the embryogenesis and fetal development of the endocrine glands, identify the major differences between fetal and adult hormones in terms of levels, metabolism and actions, emphasize the importance of the maturation of neuro-interactions for the functional maturation of most endocrines, and, finally, relate fetal to placental and maternal hormones as part of a single, homogeneous unit integrating the major endocrine functions necessary for survival, and growth and differentiation of the new organism.

General Considerations

Fetal hormones differ from adult hormones both quantitatively and qualitatively in all species that have been observed. In quantitative terms, prenatal levels of the hormones in the respective endocrine glands, in blood, and in tissues are, for the most part, lower than the corresponding postnatal levels; they are, however, not stationary but undergo significant changes during the fetal period; for example, hormonal levels either increase progressively until birth (e.g., cortisol), or reach a peak sometime during midgestation and then decline until birth (e.g., growth hormone) or increase rapidly at the time of parturition to then decline during perinatal and early infancy periods (e.g., thyroid hormones). In qualitative terms, the secretory pattern of endocrines varies considerably with the fetal age (e.g., in the adrenal cortex, sex hormones are secreted early during development, whereas cortisol and aldosterone are secreted only towards the end of gestation) but fails to show the cyclicity which is inherent to some of these secretions in adulthood (e.g., circadian cyclicity of cortisol; estral and menstrual cyclicity of estrogen, progesterone and gonadotropins). Fetal hormones may possess the same actions as adult hormones (e.g., effects of insulin in regulating blood glucose) or may have actions which are specific to fetal life (e.g., the organizing actions of androgens in directing the differentiation of secondary sex organs) or may lack the action(s) they have postnatally (e.g., growth hormone, even though present in fetal blood and pituitary, does not appear to be necessary for fetal growth). In some cases, the action of the hormone differs depending on the degree of maturation of the target tissue, that is, the hormone is effective at a so-called critical age, generally characterized by accelerated growth and differentiation as well as high susceptibility to internal and external factors, but lacks this effectiveness prior or subsequent to the critical age (e.g., effectiveness of thyroid hormones on brain maturation at a specific developmental age).

Differences between fetal and adult endocrine function may depend on a number of factors, for the most part still not completely known. Among these factors, the most obvious are those related to the stage of maturation of the individual endocrine gland, the onset of appearance and function of the enzymes necessary for the biosynthesis of the hormones (e.g., in the adrenal cortex, cortisol and aldosterone are secreted at a later age than dehydoepiandrosterone because the enzymes necessary for the synthesis of the former hormones appear ontogenetically later than those for the synthesis of the latter). However, in some cases, an endocrine gland that appears capable of secreting does not because it has not received the appropriate stimulus from the controlling feedback system in the brain, primarily operating through the hypothalamo-pituitary axis. In the brain, as in other parts of the nervous system, the development of synaptic structure and function appears to be as essential component of neuronal and neurotransmission maturation (Chapter 7). Enzymes responsible for neurotransmitter synthesis develop during the fetal and perinatal periods; however, ingrowth of axons to the proper site for neurotransmitter action may occur later on the ontogenetic scale and thereby activation of a neural system will be delayed. Maturation of an endocrine function depends not only on the maturation of the endocrine gland itself but also on the maturation of the hypothalamic and other higher brain centers which regulate endocrine secretions in the adult (e.g., whereas the ovary differentiates relatively early in the embryo and is potentially capable of secreting estrogens and progesterone at birth, it will remain quiescent until puberty when the appropriate signals from the hypothalamic gonadotropin-releasing hormones and pituitary gonadotropins will trigger the rhythmic secretion of these hormones) (see also Chapter 9). In this respect a parallelism may be drawn between the events that unfold during early development and those that accompany old age; at both age extremes, certain endocrine glands are capable of secreting hormones in a pattern and in amounts comparable to those of the adult, but often do not. The absence of optimal function may be ascribed to early immaturity and to late functional decline of the necessary central neuroendocrine controls and feedbacks as well as to age-related changes in central and peripheral hormonal metabolism and hormone-receptor interactions (see also Chapters 12, 14 and 15).

Another factor to be considered in the development of endocrine function is represented by the sensitivity of the target tissues to hormonal secretions. Receptors for most hormones have been identified but the study of their development has just begun. Among the hormone receptors so far investigated, some progressively increase in density and/or binding affinity with age and reach stable levels in adulthood; some are "down-regulated" (i.e., decrease in density and affinity) with age to adult levels which are lower than during development. Identification of hormone receptors and their changes with age and/or physiological requirements is a rapidly advancing area of research; the findings reported so far show again some parallelism between developmental and aging changes. Endocrine function either ceases (ovary) or increases (pituitary secretion of gonadotropins) or shows alterations not in the secretory ability of the endocrine gland but in the capacity of the tissues to respond to the hormonal stimulus (thyroid hormones and decreased cell metabolism with aging) (see also Chapters 12 and 14). Such a decline in responsiveness could reflect a decrease in density and/or binding affinity or some other alteration (e.g., conformation, location, transport) of the receptors; alternatively, alterations in intracellular metabolism could lead to increasing production of hormone metabolites with the same or greater binding affinity for the receptor as the hormone itself but with a reduced biologic activity. Either or both of these alterations could explain the observed changes in hormone action during development and in old age under conditions in which hormonal levels in the gland and in plasma are apparently unchanged (see also Chapters 14 and 15).

As the development of the endocrine system cannot be separated from that of the nervous centers which regulate neuroendocrine interrelations, it is impossible to dissociate the development of endocrines from that of other organs such as liver and kidney which intimately participate in hormonal metabolism and excretion. The liver, for example, is important in the metabolic degradation of most hormones; it is also the site for the synthesis of those proteins which act either as hormone carriers in the blood (thereby separating the free, biologically active hormone from the protein-bound, biologically inactive hormone) or as hormone receptors at the membrane, cytosol or nucleus level in target cells. Thus, the development of the endocrine system, as of all other systems, must be viewed within a framework comprising the total organism where the appropriate signals designed for optimal differentiation and growth are reciprocal among all the organs and systems of the body.

Another outstanding feature of the fetal endocrines is that they are not only closely dependent for their own development on the growth and maturation of the entire embryo and fetus but also they are connected intimately with placental function and maternal endocrines. Indeed, it has often been suggested that the fetus and the placenta should be considered as an integrated endocrine unit that exhibits a rather high degree of autonomy. For example, neither the placenta nor the fetus, separately, are capable of carrying out steroid biosynthesis, but, together, they accomplish this function efficiently. Furthermore, both the fetus and the placenta interact with the endocrine system of the mother to promote, in the mother, those changes necessary for the maintenance of gestation, for the preparation for termination of pregnancy and for parturition, and after birth, for the promotion of lactation to ensure the growth of the new organism. Thus, from conception through old age an organism is dependent for its development and survival on the environment which influences and shapes the individual at all stages of the lifespan.

Difficulties and Limitations of Studies of Fetal Endocrinology

In attempting to identify some of the characteristics of fetal endocrinology, certain limitations inherent to the technical difficulties of prenatal measurements must be kept in mind in interpreting the data obtained, particularly in human fetuses: difficulty in estimating fetal age; status of the fetus at the time of measurement (most of the fetuses studied in the past and now are obtained after therapeutic abortion); the small number of samples available and, therefore, inability to determine the magnitude of variation of hormone levels in the same fetus; lack of knowledge of fetal hormone metabolism, distribution and dynamics of secretion; possible presence of heterogeneous forms of the hormones (and pro-hormones) at different developmental ages; and different timetables of maturation of the enzymes involved in hormone synthesis and degradation. Despite these limitations, the study of fetal endocrinology is advancing rapidly and the information so far accumulated represents a relatively large body of knowledge from which comparative considerations may be drawn.

THE ADRENAL GLAND

This compound gland is formed of two histologically and functionally distinct tissues; the cortex and the medulla. The hormones produced by these tissues exercise unique roles in the fetus before they assume their adult functions in postnatal life.

The Adrenal Cortex

The fetal adrenal cortex starts to secrete early in embryogenesis in response both to internal regulating signals and to environmental cues. Its function, essential for survival after birth, is not indispensable for the development of the fetus.

Embryogenesis. The tissue that will form the cortex initially develops from the celomic mesoderm of the posterior wall of the abdomen. At about four weeks of fetal life, a ridge of tissue rises from this retroperitoneal region and differentiates to form the urogenital primordia. In the 37-day old human embryo, cells from the superior pole of this ridge proliferate, break free and migrate caudally, then meet the ascending kidneys, coalesce and form the first nucleus of the future gland. A few days later, (40-day old embryo) a second, later wave of migrating cells, smaller and more acidophilic than the cells of the primary wave will add to the developing organ; both groups of cells will eventually be arranged in zones or layers to form a thick capsule around the medulla.

In the fetal adrenal, the three zonations found in the "adult" or "definitive" gland are present but are not well developed until after birth. The outermost layer, the zona glomerulosa, is quiescent throughout the entire gestation although modest amounts of aldosterone, the most important sodium-regulating hormone, are secreted at late gestation. The cellular inactivity of this zone has been confirmed by electron microscopic studies which reveal cells with low secretory activity (e.g., little, smooth endoplasmic reticulum, sparse mitochondria with poor cristae development, low cytoplasmic volume, and undifferentiated plasma membrane) (Ross, 1962). The next layer, the zona fasciculata and the innermost layer, the zona reticularis, secrete sex hormones (primarily dehydroxyepiandrosterone) and glucocorticoids (cortisol), respectively. Nonsecretory during the first trimester of pregnancy, these zones begin to secrete steroids much earlier than the glomerulosa and their secretory activity is associated with all morphological signs of active synthesis (Idelman, 1978).

Despite the relative structural immaturity of its cortex, the fetal adrenal is much larger than the adult gland: at four weeks, the fetal adrenal is of the same size or larger than the kidney, a relation which is approximately 1:28 in the adult (Lanman, 1968). The large size of the fetal gland is due to a unique layer of cells, the so-called fetal or transient zone, located between the medulla and the zona reticularis. The embryological origin, function and regression of this fetal zone have been the subject of considerable research, particularly to identify differences with the other zones.

Located between the "definitive" cortex and the medulla, the cells of this fetal zone are large and develop an acidophilic staining in early gestation; they are rich in smooth endoplasmic reticulum, free ribosomes and contain numerous mitochondria with tubular cristae, all signs of intense secretory activity. This is in contrast with the relative hypoactivity of the future adult zones. Comparative studies (in the armadillo, hampster, leopard, lion, rat, tiger, and others, see Lanman, 1968) revealed the presence of cells similar to those of the human fetal zone in a variety of species. However, the cells of this "X-zone" resemble only superficially those of the fetal zone, and in some species (e.g., rat), they do not regress at birth.

In humans, regression of the fetal zone after birth is rapid and its involution is frequently associated with hemorrhage and necrosis (Lanman, 1968). As a consequence of this regression at two weeks of postnatal age, the adrenal gland has reduced its weight by 50%. The adrenal gland regains its birth weight at puberty.

What is the function of the fetal zone?

On the basis of the replacement of the X-zone by connective tissue—a most dramatic transformation in the armadillo in which the X-zone is replaced by a fibrous sheet separating the cortex from the medulla—the hypothesis was formulated that the cells of the fetal or X-zone represent undifferentiated fibroblasts. Another hypothesis suggested that these cells represent an early form of the more permanent cells from the other zones. A more current hypothesis is based on the observations that the placenta, the fetal adrenal and the maternal organism form a functional unit to produce the estrogens of pregnancy. Accordingly, the cells of the fetal zone would be engaged in the collaborative production of these steroids (Bloch and Benirschke, 1962). In the second and third trimester of gestation, coincident with progressively increasing estrogen levels, these cells represent approximately 80% of the fetal gland weight and appear to be particularly active in steroidogenesis (Villee, 1969; Milkovic et al., 1976; Vinson and Kenyon, 1978).

Functional Development

Adrenocortical steroids are not secreted simultaneously during development, rather, their gradual appearance reflects the maturation of key enzymes in the synthetic pathway and provides a characteristic profile of the biochemical maturation of the fetal adrenal cortex. Since all the steroids share the same core structure, only slight modifications of chemical groups will give rise to the different adrenocortical hormones (Bloch, 1968; Bloch and Benirschke, 1962). Approximately 50 different steroids have been identified in the biosynthetic pathway and can all be classified as belonging to one of three classes of steroids. The sex hormones are the first class of steroids to be secreted by the human fetus at about the 10th gestational week. The chief fetal androgen, dehydroxyepiandrosterone, is a weak androgen (20% of the potency of testicular androgens) which assists testicular hormones in the virilization of the male fetus. More important, this androgen along with the androgens produced by the fetal cortex, will serve as the "coin of exchange" between the fetus and the placenta and will be released to the placenta to be converted into estrogen which will support pregnancy (see Chapter 3). Around the 15th week of gestation, the glucocorticoids, important to carbohydrate metabolism, begin to appear. In humans, the first measurable glucocorticoid is corticosterone and, as the fetus matures, cortisol is produced in greater quantities to approach the adult ratio of cortisol to corticosterone (10:1). The mineralocorticoids, the steroids responsible for regulating salt and water balance, are the last class of adrenal cortical hormones to be synthesized. The major representatives of this class in the human, aldosterone and deoxycorticosterone, have a minimal role as long as the placenta functions; for example, fetuses who suffer dysgenesis of the adrenal cortex have a normal ionic and water balance until the time of birth. The possibility that the placenta fulfills the water and electrolyte functions of the mineralocorticoids is further supported by the histologically hypotrophic appearance of the zona glomerulosa (e.g., absence of lipid inclusions and of mitochondria with tubular cristae), indicating that the cells are physiologically inactive (Lanman, 1968).

Hypothalamo-Pituitary-Adrenocortical Regulation

In the adult, adrenocortical secretion depends on multiple controls: secretion of glucocorticoids and sex steroids is regulated by the levels of corticotropin or adrenocorticotropic hormone (ACTH) released from the anterior pituitary and this release is regulated, in turn, by the secretion from the hypothalamus of a corticitropin releasing factor (CRF); mineralocorticoid secretion is regulated jointly by ACTH, by renin from the kidney indirectly, by angiotensin II and by a rise in plasma potassium (or a drop in plasma sodium). As indicated above, the secretion of mineralocorticoids is considerably less in the fetus, even near term, than in newborns and infants, the placenta supplying the necessary regulation for water and electrolyte balance. Little is known presently of the development of mineralocorticoid secretion except for what concerns the development of ACTH secretion and the development of the juxtaglomerular apparatus (the site of renin production) associated with renal maturation (Mott, 1975).

ACTH, synthesized and stored in the pituitary early in gestation, is detectable in the gland by immunofluorescence at 7 weeks and its content rises steadily throughout gestation to values near term comparable to those of the newborn (Kaplan et al., 1976). The increasing pituitary levels are associated with increasing plasma ACTH levels, these levels remaining higher in late fetal life and until the first week after birth than in childhood and adult life. Despite the existence of a human placental ACTH (Genazzani et al., 1975; Rees et al., 1975), a functional hypothalamo-pituitary system is essential, as demonstrated by clinical and animal studies, for maintenance, growth and secretory activity of the fetal adrenal gland. In the anencephalic fetus of less than 20 weeks, for example, adrenal weight is comparable to that of the normal fetus; at later fetal ages, however, adrenal growth ceases in the absence of ACTH but is re-established after ACTH administration (Honnebier et al., 1974). Normal growth of the fetal cortex seems to depend not only on ACTH but also on other hormones such as human chorionic gonadotropin and possibly growth hormone (GH) and prolactin (PRL) (Honnebier et al., 1974). The cause(s) for the perinatal elevation of ACTH levels are not clearly understood; these high levels have been ascribed to immaturity of the hypothalamo-pituitary feedback system or to relatively low fetal levels of free corticosteroids in plasma or, conversely, to activation of the hypothalamo-pituitary-adrenocortical system with high neonatal ACTH and corticosteroids levels accompanying the stress of birth (see below) or to other unknown factors.

Studies on the origin of pituitary and brain polypeptides indicate that ACTH originates, together with other peptides such as β-endorphin and β-lipotropin from a large pituitary glycoprotein (Mains et al., 1977). The concept of a common precursor which is cleaved into smaller molecules underlines the importance of those enzymes—proteases—necessary for the optimal sculpturing of pituitary hormones and brain peptides (see also Chapter 5). It also raises the possibility that the biologically active forms of ACTH and β-endorphin might be secreted concomittantly, particularly in response to CRF and stress (Guillemin et al., 1977). The interrelations between the secretion of pituitary hormones (such as ACTH and GH) involved in somatotropic (metabolic) adaptive reactions and a variety of peptides (such as enkephalins, endorphins, melanocyte stimulating hormone or lipotropins) involved in behavioral adaptive responses have important homeostatic implications but remain for the moment more suggestive than proven. While the development and physiological mechanisms of ACTH secretion have been elucidated over the last 30 years, brain and pituitary peptides have been discovered only over the last five years and the study of their ontogenetic development still awaits the availability of highly sensitive methods capable of measuring the low fetal levels. Nevertheless, in view of the numerous advances already achieved, it is likely that studies in this area may disclose significant functional relations between the fetal development of pituitary hormones and neurotropic peptides (Bayon et al., 1979).

Little information is available on the development of CRF secretion and its role in regulating fetal pituitary-adrenocortical function. In rats, CRF content in the hypothalamus has been measured postnatally and the regulatory role of the hormone in response to stress has been demonstrated during the first postnatal week and in the establishment of the circadian rhythmicity of corticosteroids during the first three weeks after birth (Hiroshige and Sato, 1971). If we correlate (with due reservations on the validity of extrapolating data from one to another animal species) neonatal brain development in rats with brain development in humans during the last fetal trimester, then we can assume that CRF secretion is initiated prenatally in humans. That this is the case is supported by the very fact that the newborn organism must and does adapt to the new environmental conditions after birth, an adaptation that requires efficient control of ACTH secretion. It appears, therefore, that ACTH secretion from the fetal pituitary depends upon the presence of adequate levels of CRF, at least towards the end of gestation. On the other hand, observations in rat show that the fetal pituitary has also some autonomous, nonhypothalamic-dependent corticostimulatory activity which ceases by day 19 (of gestation) with the maturation of the negative feedback of corticosteroids (Jost et al. 1974). The increased efficiency of this negative feedback on and after day 19 may depend on a change in the sensitivity of the pituitary to free plasma corticosteroids as well as on the increasing levels of corticosteroids or to changes in the number and binding affinity of corticosteroid receptors in target cells including neuronal (hypothalamic) and endocrine (hypophyseal) cells (see below).

Role of the Adrenal Cortex in Fetal Homeostasis

One of the unique characteristics of the adrenal cortex is its capacity not only to regulate the homeostatic adjustments of the organism to the environment but also to adapt its own function to environmental demands. Inasmuch as environmental demands differ greatly before and after birth, it is to be expected that adrenocortical responses in the fetus differ considerably—quantitatively, if not qualitatively—from those in the adult. It is generally agreed that the fetus develops in a well-protected environment, shielded from external, physical, and chemical stimuli by the placenta, a barrier the impenetrability of which is currently being questioned (see Petropoulos and Timiras, 1974; also Chapter 3). Even if we accept the view that the fetus is affected little by the external environment, we must concede that the continuing differentiation and growth of the various organs and tissues, their dynamic interrelations and their changing metabolic demands may, themselves, represent sufficient stimuli to challenge the adaptive function of the fetal adrenal. Thus, the relatively mild environmental demands which prevail *in utero,* by activating the developing hypothalamo-pituitary-adrenocortical system might serve as practice stimuli so-to-speak, preparatory for the maximal responses elicited by the extremely severe demands the new organism will face during its passage from intrauterine to atmospheric life. Indeed, it is during this period of fetal development, when the capacity for adaptation becomes established, that optimal maturation of homeostatic mechanisms and their neuroendocrine regulation may be a determining factor not only with respect to survival of the fetus and later of the newborn, but also the competence of the adult individual to respond to stress and, finally, the decline of such competence in old age (see Chapter 15).

The above considerations are still, for the most part, of a more theoretical than practical nature. There are, however, a number of observations which point to the important role of the fetal adrenal cortex in fetal homeostasis with respect to growth, differentiation and metabolism of specific tissues and organs. Some of the most dramatic among the effects of the fetal adrenocortical hormones are those related to lung maturation. Numerous studies have shown that a population of specific, secretory cells in the lung (the great septal cells, type II) is particularly sensitive to glucocorticoid levels. These cells produce a phospholipid lecithin surfactant that coats the developing alveolar cells and decreases the surface tension at the fluid-gaseous interface. Such a decreased surface tension prevents both collapse and overinflation of the alveoli and facilitates the entrance of air in the pulmonary tree with the initiation of pulmonary respiration. Surfactant begins to be produced by the lung around midgestation and its levels increase progressively until term, this increase being dependent on adequate glucocorticoid levels. A deficiency in surfactant leads to respiratory complications in the newborn, one of the most serious of these complications being the so-called respiratory distress syndrome most frequent in premature infants. (Avery and Fletcher, 1974.)

Additional effects of the adrenocortical hormones on the development of cellular systems and organs include: regulation of glycogen deposition in the liver; acceleration of development of retinal proteins, particularly marked in birds; promotion of induction of specific enzymes such as alkaline phosphatase in the cells of the intestinal mucosa; hastening of the age-related regression of lymphoid tissue, particularly the thymus and regulation of the hematopoietic tissue; interaction with other endocrines and the placenta (for a review, see Nathanielsz, 1976). A complex effect of glucocorticoids in the fetus and infant leads to morphologic and biochemical alterations in the development of the central nervous system. This effect varies depending on the age of the animal at the time of adrenocortical hyperfunction or glucocorticoid administration and on the region of the nervous system considered; it is often manifested by changes in the number of neurons and glial cells, amino acid and protein metabolism, myelination, and the establishment of optimal feedback controls in the hypothalamo-pituitary axis. (see Chapter 12.)

The Adrenal Medulla

The adrenal medulla derives from the nervous tissue and can be considered as a specialized sympathetic ganglion. The catecholamine secretions from this tissue, epinephrine and norepinephrine, have numerous and potent actions which not only mimic the effects of adrenergic neuron discharge but also exert metabolic effects that include glycogenolysis (in liver and skeletal muscle), mobilization of free fatty acids, and stimulation of the metabolic rate.

Embryogenesis. The presumptive medullary tissue arises in the 7-week old embryo from the primitive ganglia of the celiac plexus of the autonomic nervous system. The differentiated neuroectodermal cells migrate caudally until they meet the developing adrenal cortical cells, invade the medial side of the cortical anlage where they take up residence. By the midfetal period, the medullary cells are arranged in cords and masses in the central portion of the adrenal gland. The medullary cells produce not only norepinephrine as do the postganglionic cells of the sympathetic nervous system, but also epinephrine derived from norepinephrine by the activity of the enzyme phenylethanolamine-N-methyltransferase (PNMT). Besides these specific medullary cells, the adrenal medulla also contains the sympathetic neurons that innervate the adrenal vasculature.

The catecholaminergic cells stain with chromium salts and, because of this affinity, are designed as chromaffin cells. Chromaffin cells, the chief cellular component of the medulla, are also present in several other (extramedullary) locations (e.g., intestinal mucosa, prevertebral plexuses, around such organs as ovaries). Indeed, this extramedullary chromaffin tissue is quantitatively more abundant and matures earlier in the fetal period than the medullary tissue itself (Greenberg, 1975). A positive chromaffin reaction (indicative of the presence of secretory granules containing catecholamines) is first demonstrable by midgestation and catecholamines can be detected by fluorescent techniques as early as 10-12 weeks in human fetuses and at a corresponding gestational period in several other animal species. It is to be noted, however, that the presence of secretory granules by themselves does not conclusively establish that the prenatal medulla is capable of actively responding to an appropriate stimulus with the secretion of the catecholamines present in the gland. Rather, the efficiency of a secretory system, must be considered in terms not only of the amounts of secretory products available in the cells but also of their turnover (i.e., synthesis, release, catabolism or re-uptake) as well as the sensitivity of the specific receptors (i.e., number of binding sites and binding affinity). Within this more global context, the onset of functional development of the medullary activity remains still uncertain in humans and other animal species studies so far.

Functional Development

In contrast to the high epinephrine content in the adult medulla, the medullary hormone in embryonic and fetal life is largely norepinephrine.

Paralleling the progressive increase in the activity of the methylating enzyme, PNMT, epinephrine content increases towards the end of the fetal period and reaches values similar to those of the adult at the end of the first year of postnatal life (Villee, 1969). Development of medullary function has been followed not only in terms of the content and ratio of the two major catecholamines but also of maturation of their biosynthetic and catabolizing enzymes. It has been shown in experimental animals (usually the rat) that the age-related progressive increase in medullary hormones is associated with parallel changes in the activity of their synthesizing (e.g., tyrosine hydroxylase, DOPA-decarboxylase) and catabolizing (e.g., monoamine exidase) enzymes (Mirkin, 1972; Vaccari et al., 1978; Garvey et al., 1980). In this respect, pre- and postnatal developmental profiles are quite distinct for each enzyme. Some enzymes (e.g., DOPA-decarboxylase) show a slow but steady rise from day 19 of gestation (length of gestation in rat is 21-22 days) until the end of the first postnatal week; others (e.g., tyrosine hydroxylase) remain low during most of fetal development and then rapidly increase just around birth and rise more slowly in the first week after birth; still others (e.g., monoamine oxidase) remain low during the entire fetal period and then increase postnatally until reaching adult age (Vaccari et al., 1978; Garvey et al., 1980).

Even though catecholamines and catecholaminergic enzymes are present in increasing amounts and activity in the fetal medulla, their release into the circulation under conditions of steady-state or enhanced functional demand seems to occur only near term or neonatally. Slotkin (1975) suggests that the release of these hormones cannot take place until the chromaffin tissue has been functionally innervated by the splachnic nerve, an event which, in the rat, occurs during the first 10 postnatal days (Daikoku et al., 1977). Thus, in the medulla as in the sympathetic ganglia, maturation of the chromaffin cells is facilitated by the action of local growth factors (e.g., the nerve growth factor) capable of promoting nerve growth and synaptic development (Otten et al., 1978).

Hormones of the adrenal cortex are another factor that influences the development of the adrenal medulla. The cortico-medullary relationship characteristic of the adult animal is already manifest at early ages. Experimental evidence suggests that a high concentration of corticosteroids may increase the number of neuroblasts that differentiate into chromaffin cells (Eränkö et al., 1966; Costa et al., 1974; Hervonen, 1971). Should the fetus become stressed, the ensuing cortical hypersecretion would stimulate a larger population of neuroblasts to differentiate into chromaffin cells which, in turn, would initiate catecholamine secretion. On the other hand, the reverse may also be true: impaired or delayed development of the cortex, may result in a deficient development of the medulla as well. The resulting adrenal (both cortical and medullary) insufficiency, while not endangering survival in the fetus, may lead to failure of adaptation and death, postnatally.

The close functional cortico-medullary relationship in the adrenal is made possible by the presence of a specialized capillary net, similar to the portal system of the pituitary gland (see below) through which the blood from the cortical veins is drained into the medullary sinuses (Coupland, 1975). This portal system literally inundates the medulla with levels of corticosteroids that are one-hundred fold higher than those found in the systemic circulation. Several investigators have demonstrated that the enzymes for catecholamine metabolism are sensitive to corticoids (Cohen, 1973, 1976; Garvey et al., 1979; Wurtman and Axelrod, 1966). Corticosteroids are also thought to stimulate the PNMT activity and thereby increase the synthesis of epinephrine from norepinephrine. In young rabbits, for example, the portion of ganglion cells that has penetrated the adrenal cortex contains epinephrine whereas the portion remaining outside the cortex does not. Similarly, in rats, hypophysectomy produces adrenal atrophy and concomitant decrease in epinephrine (but not norepinephrine) whereas, in developing rats, adrenocortical stimulation induces an increased section of corticosteroids and epinephrine (Jost, 1966). In all of these examples, the adrenal cortex and epinephrine seem to be interrelated.

While increasing levels of the catecholamines (especially a higher proportion of epinephrine as compared to norepinephrine) and their metabolic enzymes are indicative of progressive maturation of the adrenal medulla, little information is as yet available on the onset of hormonal secretion, release and turnover, and on the degree of sensitivity of target cells to the hormones. In some animals (e.g., lambs), asphyxia *in utero* excites the medulla to discharge its hormones, especially norepinephrine. The responsiveness of the fetal medulla, however, varies from species to species and is probably related to the severity of the stress and the degree to which innervation has developed. This may possibly account for the conflicting reports in the literature (Comline and Silver, 1966; Girard and Zeghal, 1975; Nathanielsz, 1976). In the developing rat, hypoxia increases catecholamine enzyme activity beyond the age-related increase which normally accompanies development, this hypoxia-induced increase in activity is already manifest at birth and continues through adulthood (Vaccari et al., 1978). In the fetus, the development of PNMT activity and that of catecholamine enzymes appears to be only selectively (i.e., depending on fetal age and enzyme considered) affected by environmental hypoxic conditions despite adrenocortical hypofunction (Garvey et al., 1980). These and other observations suggest that cortex and medulla can be affected differentially and independently by environmental demands (Philpott et al., 1969; Garvey et al., 1979, 1980; Akana and Timiras, 1981). Factors other than those already considered here (nerve growth factor, corticosteroids, stress) might be operative in the developing animal to regulate medullary maturation. The identification of such factors and the clarification of the timetable of cortico-medullary relationships represent a concern to be pursued actively, not only because of theoretical interest in this aspect of fetal endocrinology, but also because of the practical importance in determining optimal conditions for survival of the newborn and infant.

THE GONADS

In mammals, the sex of an individual is determined at the time of fertilization and depends upon whether or not the Y chromosome is present in the fertilizing sperm. During normal embryonic development, the genetic sex of an individual determines whether the cellular components of the sexually indifferent gonads will differentiate into either ovaries or testes. With the establishment of the gonads (gonadal sex), the stage is set for the future production of either ova or sperm and a source of appropriate sex steroids is assured for the control of all subsequent sexual differentiation. Under the influence of gonadal hormones, the duct system of the fetal reproductive tract and the external genitalia differentiate in conformity with the genetic sex along either male or female lines (genital sex). During subsequent growth and development, the gonadal sex hormones will continue the process of sex differentiation of somatic tissues and will have profound lifelong effects on metabolism, body structure, behavior and reproductive capacity.

Sex Determination

Of the 46 chromosomes in human cells, 2 are called sex chromosomes (designated as XX in normal females and XY in normal males) because they exert genetic control over sex determination. The remaining 44 chromosomes are referred to as somatic chromosomes (autosomes). Meiotic division of germ cells in ovaries and testes results in one X chromosome and 22 autosomes in each normal ovum and 22 autosomes and either an X or Y chromosome in each normal sperm. When an X-bearing sperm fertilizes an ovum, a genetic female (XX) will result. When a Y-bearing sperm fertilizes an ovum, the XY zygote develops into a genetic male. The genetic sex of an individual, therefore, depends on the presence or absence of the Y chromosome in the fertilizing sperm and is determined at the time of fertilization.

Y and X Chromosomes. The Y chromosome has potent male determiners but is otherwise considered to be genetically inert. There is equivocal evidence that the only nonsexual gene known to be on the Y chromosome of human males causes hypertrichosis (hairiness) of the pinna of the ear. In contrast to the Y chromosome, X chromosomes are concerned not only with sex determination, but also are known to carry many other genes for various somatic characteristics. It is now known that the only cells in the body of the female where *both* X chromosomes are apparently functional are the female primordial germ cells. The importance of the genetic information from two X chromosomes in these cells is shown in XO females (Turner's syndrome) in which the primordial germ cells begin to develop normally but undergo atresia by the time of birth. Since primordial germ cells induce granulosa cell development, necessary for follicle formation and future production of estrogen and progesterone, XO females are sterile, and also fail to develop steroid-dependent secondary sexual characteristics.

Female mammals seem to have developed a mechanism to compensate for the wealth of genetic information carried on the X chromosomes. This "dosage compensation" is accomplished by the partial or complete inactivation of one of the two X chromosomes which does not uncoil after mitosis but remains in a compact mass in close apposition to the nuclear membrane of most somatic cells. The presence of this nuclear mass (called sex chromatin) was first reported in 1949 (Barr and Bertram, 1949) and is often referred to as Barr bodies. In 1961, Mary Lyon proposed the widely accepted notion of X chromosome inactivation. According to the Lyon hypothesis, dosage compensation is a normal process that begins early in embryonic development and is passed on to all progeny of somatic cells. The inactivation takes place in all somatic cells that have more than one X chromosome so that XY, XO, XYY individuals have no sex chromatin bodies; XX, XXY, XXYY, individuals possess one sex chromatin body; XXX, XXXY and XXXYY individuals possess two sex chromatin bodies, and so on. It should be noted that while two X chromosomes represent too much genetic information in cells, the absence of *any* X chromosome is incompatible with human embryonic survival, hence, OY is a lethal condition. An unresolved central issue relating to X-inactivation is concerned with the mechanisms and consequences of two X chromosomes that are present in the very young embryo before dosage compensation takes place (Lyon, 1971).

Development of the Gonads

The Indifferent Gonads. During the first six weeks of human development, the primitive gonads are histologically identical and it is not possible to differentiate between the male and the female gonad until about the seventh week of development. In both sexes the primordial sex cells (or germ cells) possess pseudopodia which allow cells to migrate from their origin in the wall of the yolk sac to the wall of the hindgut and then along the mesentery to each side of the embryo into the region of the developing kidneys. A primitive gonad arises when nearby epithelial cells are stimulated to multiply and arrange themselves in cords which surround the sex cells, a process that results in the condensation of tissue called the genital ridge. Once the primordial germ cells have completed the colonization of the presumptive gonad, their numbers greatly increase by mitosis. The gonads morphologically develop a cortex and a medulla and in the course of subsequent development undergo important modifications based on the genetic sex of the embryo. The differentiation of the sex glands, which established gonadal sex, is controlled by unknown factors and is determined by the somatic cells of the gonads and not by the gonocytes (Hamilton et al., 1972).

The Ovary

Embryogenesis. In genetic females, the medulla of the developing ovary is nearly devoid of germ cells and undergoes regression. The sex cords develop and then break into isolated masses in the gonadal cortex which is the progenitor of the ovary. The cortical masses are organized so that primordial germ cells are encapsulated by epithelial cells (the pregranulosa cells) thereby forming primordial ovarian follicles. The transition from indifferent gonad to definitive ovary occurs in the human fetus at about the 12th week of gestation (Grumbach and Van Wyk, 1974). At about this time, the descent of the ovaries from their origin in the abdomen to their position in the pelvis is accomplished (Snell, 1973). The caudal end of each ovary is connected to the genital swelling by a gubernaculum. As development proceeds, the gubernaculum also attaches to the uterus at the junction between the uterus and the uterine tube. This part of the gubernaculum will persist as the round ligament of the ovary and the part that joins the uterus to the labium persists as the round ligament of the uterus. During the growth of the trunk of the fetus, the ovaries enter the true pelvis and come to lie posterior to the right and left uterine tubes.

The first stages of meiotic division of the oogonia occur between 3 and 7 months postconception. The development of oocytes occurs along a gradient with the most advanced oocytes located near the corticomedullary boundary and the less advanced deeper in the ovarian cortex. During this period of fetal development, the oocytes begin a prolonged suspension of the meiotic process, possibly due to the influence of surrounding granulosa cells (Ohno and Smith, 1964). Oocytes remain in an arrested stage of prophase until just before ovulation, i.e., for about 12 to 50 years (from menarche to menopause) in the human. Studies in mammals indicate that oocytes are not formed in adult life and it is generally accepted that the lifetime source of oocytes is restricted to the period of fetal development (Kennelly and Foote, 1966; Peters, 1970) (also see Chapters 1 and 16).

Of the millions of ollicles that form in the fetal ovary, only a few will survive to full maturity and ovulation. The majority are destined to be destroyed by a degenerative process known as *atresia* which begins in the fetus and continues in adult life. The rate of atresia is controlled by gonadotropins, possibly by their effects on steroid production. The mechanism of atretic destruction is unknown but it always begins with the

ovum and is histologically apparent by the intense staining reaction of
the zona pellucida and the accumulation of fat droplets and granules within
the ovum (Nalbandov, 1976). Following hypophysectomy, atresia is
greatly reduced and replacement therapy with gonadotropins resumes the
atretic depletion of follicles (Baker, 1963). The large number of follicles
that perish by atresia in fetal life is probably caused by the direct action
of gonadotropins or to estrogen from fetal and/or placental origin.

Functional Development

During fetal life, the ovaries are small and show little evidence of
endocrine or exocrine functional activity (Kaplan et al., 1976; Short, 1972;
Resko, 1970). The female gonads have a passive role in promoting sex
differentiation. Even the feminization of the internal duct system seems
to proceed without endocrine support by the ovaries and ovariectomy of
the fetus brings about little modification of sexual development in genetic
females (Short, 1972). The lack of a hormonal role of the ovaries in the
normal differentiation of the internal genitalia has been demonstrated by
the autonomous development of Mullerian ducts in *in vitro* cultures
(Short, 1972) and by the development of female somatic sex structures
by agonadal fetuses (Grumbach and Van Wyk, 1974). During fetal life,
however, large amounts of estrogenic hormones are synthesized by the
maternal-feto-placental unit, and are available to fetal tissues and the
possibility that this extragonadal source of estrogen has physiological
importance cannot be discounted.

The role of gonadotropins in ovarian morphogenesis is still subject to
investigation and there is very little information concerning human
fetuses. Numerous investigations of other species, particularly rodents,
have provided important information about the ontogenesis of gonadotropin
secretion, ovarian receptivity and response to these hormones (See
Chapter 12). Although it is very difficult to extrapolate information
gathered from one species to another species much can be learned from
animal studies since it is probable that some of the developmental sequelae
seen postnatally in rodents are prenatal events in humans. Grumbach
and Kaplan (1973) report a correlation between the patterns of luteinizing
hormone (LH) and follicle-stimulating hormone (FSH) in fetal serum and
the formation of primordial follicles, with gonadotropin levels and follicle
development peaking at about 22 weeks of gestation. By the end of the
first week in neonatal rats, ovarian receptors for gonadotropins can be
demonstrated and ovaries show early follicular maturation (Peluso et al.,
1976; Siebers et al., 1977). FSH levels (Cons and Timiras, 1975) and
estradiol levels (Dohler and Wuttke 1975) in circulation surge in the first
few neonatal days and again at day 9-17 indicating an estrogenic effect
of FSH. These levels of estrogen and FSH exceed the quantities seen
later in life during puberty and in the sexually mature adult. LH levels
in the circulation during postnatal rat development do not exceed the
preovulatory surge seen in adulthood, show large individual differences,
and do not seem to correspond to changes in estradiol levels (Dohler and
Wuttke, 1975). Odell and Swerdloff (1976) suggest that the developmental
period of changing ovarian (and testicular) response to tropic hormone
is characterized by the induction of LH receptors by FSH. This may
possibly explain the developmental patterns of high FSH release in advance
of and concurrent with the LH secretion in immature rats (Cons and
Timiras, 1975) and in young humans (Faiman and Winter, 1974).

In addition to estradiol, production of testosterone by females has been
demonstrated in neonatal rats (Dohler and Wuttke, 1975), in the fetal
Rhesus monkey (Resko, 1970) and in humans (Mizuno et al., 1968; Rivarola

et al., 1968). Surprisingly, circulating levels of this androgen do not differ markedly between young females and young male rats, although males have slightly higher levels (Dohler and Wuttke, 1975). Female Rhesus fetuses produce minute amounts of testosterone and androstenedione in the first half of gestation but the remaining gestational period is marked by amounts of androstenedione nearly as high as male levels. The similar amounts of androgen produced by Rhesus and human fetuses of both sexes is measurable as testosterone in maternal blood (Resko, 1970; Mizuno et al., 1968; Rivarola et al., 1968). The physiological significance during development of circulating testosterone in females (and of estrogen in males) is uncertain. Research in this area must deal with the complication of the body's ability to convert steroids from one form to another, changes with age and different cellular actions (Reddy et al., 1974). Furthermore, the source of fetal and neonatal steroids is unresolved. It has been shown that some estrogen in the male (Baird et al., 1969) and androstenedione in females (Block and Benirschke, 1959) are of adrenal origin. It is interesting to note that estrogen receptors have been demonstrated in the adrenal cortex of adult female rats (Cutler et al., 1978) which suggests a direct effect of estrogen on adrenal function. Heretofore, the numerous stimulatory effects of estrogen on adrenal function have been ascribed to the known action of estrogen to increase hepatic clearance of corticosterone thereby causing acceleration of ACTH release from the anterior pituitary (Colby and Kitay, 1974). Clarification of gonadal-adrenal interactions during fetal development awaits further investigation.

Alpha Fetoprotein. To further complicate the interpretation of the developmental impact of circulating estrogens is the presence of alpha fetoprotein (AFP), a feto-neonatal estradiol-binding plasma protein which circulates in high amounts during the first 25 days of postnatal life in the rat (Andrews and Ojeda, 1977). In the presence of AFP, the availability of estradiol is compromised and the physiological action of this important steroid is hindered, even though fetal or infatile tissue may have appropriate receptors and are potentially receptive to estrogenic hormones. This reduction in the biological activity of circulating estradiol may be of critical importance during the organization of the female hypothalamus and development of feedback mechanisms which include estrogen as an integral component.

Unlike estrogens, androgens do not bind to AFP and therefore have freer access to somatic tissues during critical periods of development. In animals such as the rat, which is born with a relatively under-developed nervous system, the neonatal presence of testosterone is available to act on the androgen-sensitive hypothalamus. This is an important aspect of normal male development, and the permanent results of interference with this process is apparent in numerous studies of this critical period in sex differentiation (See Chapter 8 and also below).

The Testis

Embryogenesis. During the seventh and eighth weeks of human development in genetic males, the testes emerge as the cells of the gonadal medulla organize into distinct cords, the cortex regresses, and the gonads increase in size. The mesenchyme in the developing gonad forms the connective tissue, fibrous septa, and interstitial cells of Leydig. The seminiferous tubules are formed from the sex cords which contain primordial sex cells (which form the spermatogonia) and epithelial cells (which form the Sertoli cells). During the fourth month of fetal life, the system of tubules of the testes is fully organized but they remain as relatively solid cords until puberty when the hormonal influence of anterior pituitary gonadotropins stimulates spermatogonia proliferation, the increased size of Sertoli cells, and the formation of lumena in the seminiferous tubules.

Like the ovaries, the gonads in males originate within the abdomen but, later in fetal life, the testes come to occupy different anatomical positions. Owing to the attachment of a column of mesenchyme (gubernaculum) from the caudal end of the testis to the genital swelling during the differential growth of the trunk of the body, the testes descend from the abdominal cavity through the abdominal wall into the scrotum. The descent into the scrotum is accomplished by the eighth gestational month in humans and is under hormonal control, probably by androgen from the fetal testes. Administration of gonadotropins and/or androgens can usually induce the completion of this process in cases of undescended testes, provided the normal pathway of descent is not obstructed, and thereby prevent the deleterious effects of body temperature on testicular tissue (Hamilton et al., 1972; Snell, 1973).

Functional Development

The two main functions of the testes, virility, which is achieved primarily by endocrine messengers, and fertility, which involves the production and transport of spermatozoa, begin during fetal development. In prenatal life the testes are highly active endocrine glands whose steroid and polypeptide hormones provide the appropriate milieu for normal male differentiation. The capability of fetal testes to produce steroid hormones during critical periods of sexual differentiation has been demonstrated in the rat (Feldman and Block, 1978; Resko et al., 1968), Rhesus monkey (Resko, 1970), and human (Serra et al., 1970). The source of testicular androgens is the Leydig cells and the steroidogenic capacity of the testes parallels the increased number of cells in fetal life (Roosen-Runge and Anderson, 1959).

Prior to their appearance in the human at about 60 days of gestation, there is no production of testosterone. The increased number of these steroidogenic cells between 10 and 18 weeks is paralleled by the appearance and increasing concentration of testosterone in the testes and circulation of the male fetus (Süteri and Wilson, 1974). During this time the masculinization of the external genitalia and the preferential development of the Wolffian anlage occurs in response to androgen, probably testosterone (Payne and Jaffee, 1975). Following this early period of rapid proliferation and increasing function of the Leydig cells, there is a sharp decline in testosterone formation that persists through the rest of gestation. Following a transient elevation of plasma testosterone in the neonatal period, levels decline to prepubertal values (Kaplan et al., 1976). The gestational pattern of testosterone and androstenedione production in the Rhesus monkey coincides with the developmental timetable of male differentiation for this species (Resko, 1970). It has been posutlated by Kaplan and her collegues (1976) that the presence of testosterone during human gestation marks the onset of a complicated negative feedback mechanism of androgens to the central nervous system which advances through several more stages in growing children and reaches final maturity at puberty. It should be pointed out that there is some debate about which hormone is the active androgenic substance which promotes masculinization of the fetus. There is evidence that there is peripheral and/or cellular conversion of testosterone and other steroids to different steroidal forms (Reddy et al., 1974). This compounds the interpretation of data showing serum or organ content of sex steroids.

Studies of rat testes (Feldman and Block, 1978) and rabbit testes (George et al., 1978) have shown that the *initiation* of Leydig cell differentiation and of androgen synthesis can occur independently of extragonadal hormone stimulation. However, responsiveness and sensitivity to LH or human chorionic gonadotropin (hCG) increase with advancing age and development. There is some evidence that in the neonatal rat, growth hormone synergizes with LH thereby facilitating testosterone production and release (Odell and Swerdloff, 1976). The response of the testes to gonadotropins is apparently linked to the development of receptors, cAMP responsiveness and the appearance of enzymes capable of synthesizing testosterone from C_{21} precursors and from cholesterol (George et al., 1978; Warren et al., 1973). The increased steroidogenesis and testicular response to LH is seen in fetal life during masculine differentiation and then again at puberty (Odell et al., 1974). Although it has been postulated that FSH increases sensitivity of the prepubertal rat testes to LH (Odell et al., 1973) and in the human (Odell and Swerdloff, 1976), the cellular mechanisms for the changing responsiveness of the testes during critical periods of maturation are still obscure.

There is indirect evidence that the functional changes in human fetal Leydig cells may arise from the combined action of placental chorionic gonadotropin and fetal pituitary LH and changes in their appearance and concentrations in fetal blood. Human CG is a glycoprotein produced by syncytiotrophoblast cells of the placenta. This hormone reaches both maternal and fetal circulation, although the fetal circulation contains only about 1% of maternal levels. HCG is demonstrable in maternal circulation by 14 days after conception with peak levels between 54 and 104 days, and declines thereafter. The transient elevation of placental production results in increased gonadotropin levels in the fetus that coincide with genital organogenesis. Serum LH in the human male fetus reaches its peak at 22-24 weeks of gestation (Kaplan et al., 1976). The apparent relationship in humans between choroinic gonadotropin and fetal androgen production is not seen in the Rhesus monkey since neither pituitary gonadotropins nor chorionic gonadotropin are present in maternal circulation 35 days after mating (Tullner and Hertz, 1966).

The decline in testicular steroidogenesis during late human gestation may be influenced by the elevation of circulating prolactin. Testosterone and prolactin are known to be inversely related in the male circulation and this pattern of prolactin depression of testosterone synthesis is seen in the human fetus and neonate (Aubert et al., 1975). Further actions of prolactin are revealed in studies of rats, indicating that prolactin elicits growth of sex accessory organs in this species. Hypophysectomized immature animals given prolactin will have increased growth of testes, prostate and seminal vesicles (Negro-Vilar et al., 1973). There is evidence that prolactin may also be a regulator of gonadotropin binding capacity by synergizing with LH in both the ovaries and testes of the rat (Richards and Williams, 1976; Aragona et al., 1977).

The exocrine function of the testes is accomplished by the production of gametes in the seminiferous tubules and their transport by a circuitous system of ducts that leads, finally, to the exterior of the body. Although the fetal testes are quiescent with respect to the proliferation of germ cells in the seminiferous tubules, the tubular epithelium produces an important polypeptide hormone that causes regression of the Mullerian ducts. This hormone, which is probably produced by Sertoli cells (Josso et al., 1975), eliminates the further development of female internal genitalia. The increasing length of the seminiferous tubules and the proliferation of Sertoli cells during testicular development correspond to the circulating levels of FSH in fetal blood (Kaplan et al., 1976; Odell and

Swerdloff, 1976). It is possible that the Sertoli function is brought about by both FSH and testosterone. Means et al., (1976) have shown that in immature postnatal rat, the Sertoli cells have receptors for both these hormones.

Inhibin. The seminiferous tubules are probably the source of another hormone, or factor(s), called "inhibin" that selectively lowers the FSH levels in circulating blood. Preliminary evidence by Baker et al., (1976) suggests that "inhibin" has a protein component, and may be produced by the Sertoli cells. The point of entry from testicular ducts into the vascular system as well as the site and mechanism of action of "inhibin" is known. "Inhibin" has not been demonstrated in the female, but indirect evidence infers that it may be secreted by ovarian follicles and contributes to the control of FSH secretion in adult humans (Sherman and Koreman, 1975). It is not known whether "inhibin" is produced by fetal gonads. The resolution of the questions surrounding this interesting compound awaits further research.

As was mentioned previously, the establishment of the gonads in mammals is followed by a period of relative inactivity. With advancing age and somatic development, the pituitary support of the gonads is activated. This culminates in the final maturation of hormonal and gametogenic potential. In the male, this represents the beginning of adult levels of testosterone secretion (and the accompanying somatic effects) and the lifelong production of large numbers of spermatozoa.

Differentiation And Development Of Secondary Sex Organs: Genital Organogenesis

Experiments by Jost (1976) show that regardless of genetic sex, sexual differentiation of the internal and external genitalia of the rabbit fetus develops as a phenotypic female unless a functional fetal testes is present. Moreover, the ovary is not necessary for female morphogenesis. The bipotential nature of somatic sex structures also exists in the human fetus (Grumbach and Van Wyk, 1974). Accordingly, by the seventh week of gestation, the human embryo has the primordia of both the male and female reproductive tracts. These consist of two paired duct systems known as the Wolffian (or mesonephric) and Mullerian (or paramesonephric) ducts. At this time, the embryonic ducts (i.e., the internal genitalia) are in a neutral, indifferent or bipotential state. In response to the systemic secretion of androgen from the fetal testes, the Wolffian ducts develop into the definitive male internal genitalia consisting of efferent ductules of the testes, the ducts of the epididymides, ductus deferens, and the seminal vesicles. Concurrent with the development of these ducts, a protein hormone synthesized by the seminiferous tubules (probably by Sertoli cells), which is known as Mullerian regression factor, acts locally to retard growth of potentially female structures (Jost, 1970; Josso et al., 1975). In the absence of these potent hormones, the development of a female reproductive tract is favored. In the normal female, therefore, the paired Mullerian ducts persist and form the Fallopian tubes (uterine tubes or oviducts), uterus, cervix, and the proximal vagina while the heterologous anlagen, the Wolffian ducts, retrogress.

The external genitalia are derived from the same embryonic structures in the two sexes and, as seen with the gonads and the internal duct system, the external genitalia are initially bipotential structures. At the seventh week of human gestation, the genitalia are identical and consist of a single genital tubercle, a pair of lateral genital folds, and a pair of genital swellings that appear lateral to the genital folds. At about the eighth week of gestation, testosterone from the fetal testes induces the virilization of the genital tubercle (phallus) and the genital folds to form the structures of the penis, and the genital swellings form the scrotum. In the absence of testosterone, the external genitalia of the normal female fetus specialize: the genital tubercle is bent caudally and forms the clitoris, the genital folds form the labia minora and the genital swellings enlarge to form the labia majora.

Urogenital Structures. The embryonic development of the urinary and genital systems are closely linked, and in the early stages of embryogenesis their ducts open into a common dilated terminal part of the hindgut called the cloaca. In the fully developed human male, the urinary and genital systems continue to use a common channel (the lower part of the prostatic urethra, the membranous urethra and the penile urethra) to discharge their products, urine and semen. The prostrate gland forms from a series of endodermal buds growing from the prostatic urethra. In the female, the primitive excretory ducts undergo marked retrogression and take no part in the formation of the functional reproductive system.

Gonadal Hormones In Sexual And Somatic Differentiation

Genetic and Hormonal Factors in Sex Differentiation. As was mentioned earlier, the genetic sex of an individual is established at conception and depends upon whether or not the fertilizing sperm contains a Y chromosome. The Y chromosome is a potent masculinizer and plays a critical role in the differentiation of the bipotential gonads into testes. In the absence of a Y, the embryo will develop as a female.

The exact role of the genetic information contained in the X and Y chromosomes is not fully understood, but is gradually being elucidated. In a review of genetic data collected from studies of experimental animals, domestic animals and humans, Federman (1973) suggests that the genetic information for testicular morphogenesis is borne on the X chromosome and that the Y carries a regulatory locus which controls the expression of these genes for maleness. More research is necessary to clarify the genetic impact of the X chromosome, since a male is inadequately virilized if he has more than one X. Two Xs (XXY, or Klinefelter's syndrome) result in sterility and some feminization. Larger numbers of Xs increase the abnormalities, particularly, mental retardation (Winchester, 1977). Conversely, at least one X is essential for life—and an embryo with the normal number of somatic chromosomes, but a single Y—will not develop.

The presence of two X chromosomes is necessary for the differentiation of normal, functional ovaries and plays a significant role in fetal viability and somatic differentiation of human females. The importance of two X chromosomes to the viability of female fetuses is shown in the statistical finding that only one of every 40 XO conceptuses survive to term and that the XO karyotype (Turner's syndrome) is one of the most common findings in spontaneously aborted human fetuses (Larson and Titus, 1970). The characteristic reproductive anomalies of the surviving XO individuals are rudimentary or absent ovaries (gonadal dysgenesis), amenorrhea, failure of breast development, and juvenile external genitalia. A description of somatic aberrations are discussed later in this chapter.

Steroidogenesis and Steroid Receptors. The steroidogenic pathway of sex steroids is analagous in the ovaries and testes, and both types of gonads are capable of producing androgens and estrogens during development and in adulthood (Dohler and Wuttke, 1975). The adult human testes produce daily, but small, amounts of estrone and estradiol (Lipsett, 1970) and human ovaries at all ages produce androgens as an obligatory step in the biosynthesis of estrogens (Federman, 1973). The regulation of gonadal enzymes which determines the fate of androgens, i.e., release into the circulation or conversion to estrogens, is decided at the time of gonadal differentiation by the genetic action of the Y or a second X chromosome on the steroid-synthesizing somatic cells of the genital ridge. With the establishment of gonadal sex, the continuation of sexual differentiation is determined by the secretion of sufficient quantities of the appropriate steroid hormones which have specific action on responsive somatic tissues.

In a discussion of steroids, or any hormone, it is necessary to emphasize the importance of tissue responsiveness as an essential correlate to hormonal action. Contemporary endocrinology has shown that the presence of appropriate amounts of receptors is requisite to the specific response of cells to hormones. Several examples are presented in the sections of this Chapter concerning gonadal function in the fetus. Studies of somatic tissue in rats indicate that perinatal testosterone may be necessary to the induction of its own receptor sites during development of the external genitalia. The reduction of androgen synthesis at days 1, 3, and 5 by the injection of antisera to gonadotropins demonstrate the importance of this early effect of testosterone. Interference with neonatal receptor induction hinders normal development of the penis causing a reduction of its size that cannot be restored even by large doses of androgen given in adulthood (Goldman et al., 1972).

The relationship of receptors to cell response has been well-demonstrated in an inherited human disorder called testicular feminization. This syndrome is characterized by an XY genotype, bilateral testes, absence of internal genital tract development, development of female external genitalia, feminization of body contour and breast development at puberty. The clinical complaint is lack of menarche. In these genetic males, the somatic tissues are resistant to the action of testosterone, although plasma levels are in the normal or high male range (Judd et al., 1972). Studies of intracellular metabolism have shown a diminished binding of testosterone, or its major metabolite, dihydrotestosterone, to nuclear chromatin (Northcutt et al., 1969; Strickland and French, 1969). Lacking this critical step in steroid hormone action, the tissues are resistant to virilizing hormone and this results in the development of a phenotypic female. The genetic mechanism for this disorder in humans is uncertain, but experiments in mice suggest that testicular feminization is an X-linked defect (Lyon and Hawkes, 1970).

Somatic Correlates of Sex Differentiation

The establishment of sexual differentiation of body structures involves a complicated series of developmental events during specific periods of critical tissue sensitivity. As maturation continues, an increasing variety of structures responds to the presence of gonadal steroids. During these critical periods of structural and functional maturation, important hormonal relationships are established among the hypothalamus, the anterior pituitary and the gonads (see Chapters 8 and 9).

Somatic sexual differentiation depends upon the availability of specific receptors which allow cellular recognition of sex hormones that are produced in accordance with the genetic and gonadal sex of the fetus. In the sexually mature male, testosterone and appropriate tissue receptors are necessary for the maturation and availability of spermatozoa, for metabolic activity and contractility of the epididymis and for maintenance of the prostate, seminal vesicles, penis, and scrotum. In the sexually mature female, tissue receptors respond to estrogens and progesterone which control the muscular and secretory functions of the uterus, uterine tubes, and vagina, thereby determining tubal fluid quantity and quality (glucose, amino acid, electrolyte content and pH) as well as ovum transport and uterine support of a fetus (Greep, 1975). The menstrual cycle is regulated by the feedback mechanisms involving the gonads, hypothalamus and anterior pituitary.

Gonadal sex hormones initiate cellular responses that affect almost all somatic tissues resulting in sexual differences in metabolism, in bone and muscle mass, hair distribution, cartilage growth (particularly in the larynx and epiphyses), red blood cell production and the size of visceral organs such as the heart, kidneys and lungs. Recent studies now indicate that fetal sex steroids may play a critical role in the relative rates of maturation of the body leading to lateral asymmetries of the brain and of the rest of the body (Levy and Levy, 1978).

Aberrant Sex Differentiation

Anomalies of Sex Chromosomes. Anomalies of sex chromosomes in humans can result in marked deviations from normal sexual differentiation and development. The sex chromosome anomalies probably result from errors of cell division during gamete formation or in the early stages of zygote development. Aberrations may involve the loss of a sex chromosome (XO or Turner's syndrome), the addition of either the X or Y chromosomes (XXY, XYY, XXYY, XXX, XXXY, XXXX, XXXXY, and XXXXX), mosiacs (an individual with two cell lines, e.g., XX/XO, XY/XX, etc.) or the translocation of a portion of a sex chromosome to a somatic chromosome. The most common of these aberrations, as seen in individuals who survive to adulthood, involves the presence of too many chromosomes.

Studies of sex chromosome aberrations allow us to draw a few general conclusions about the actions of the X and Y chromosomes. Although the Y chromosome is not necessary for survival, it is a potent masculinizer and the genes located on the Y probably regulate genes located on other chromosomes. The presence of an extra Y (XYY) in a phenotypic male results in increased linear growth, and may be associated with lower than normal mental development and aggressive behavior in some, but not all, XYY males. The presence of the single X in the normal male (XY) is necessary for normal male development (YO and YY are lethal conditions) and more than one X will cause some feminization (XXY or Klinefelter's syndrome) and additional Xs (XXXY, XXXXY) will cause feminization and other abnormalities. Each extra X chromosome increases the severity of mental retardation in either females (XXX, XXXX, XXXXX) or males (XXY, XXXY, XXXXY). Two X chromosomes are necessary for normal sexual and somatic development of females. The XO condition is usually, but not always, lethal and those embryos that survive to term usually live into adulthood. The typical sexual characteristic of surviving XO females is sterility. Somatic abberations may include all or some of the following: short stature, webbing of the neck, a wide chest with broad-spaced nipples, multiple pigmented nevi, short 4th metacarpals, hypoplasia of the nails, and coarctation of the aorta. Frequently, renal anomalies and hemangiomas occur, and occasionally, telangiectasis occurs in the gastrointestinal tract. Mental deficiency is rare, but many XO females suffer from spatial disorientation and thus do poorly in performance sections of I.Q. tests (Holvey, 1976; Ferguson-Smith, 1973). The multiple abnormalities found in XO females and the lethal nature of YY and YO inheritance gives an indication of the amount of important genetic information located in the X chromosomes. The normal complement of sex chromosomes and the balance of sex chromosomes and autosomes are essential for normal sexual, somatic and mental development.

Sex-linked Inheritance. In addition to being a sex chromosome, the X carries genetic determinants for traits having no connections with sex differentiation. The sex-linked conditions that are known are all X-linked and are expressed almost exclusively in males and are transmitted on the mother's X chromosome. There are more than fifty known X-linked conditions including the pathological mutations found in hemophilia A and B, G-6-PD deficiency, ectodermal dysplasia–anhydrotic type, Duchenne muscular dystrophy, Lesch-Nyhan syndrome, nephrogenic diabetes insipidus, ocular albinism, X-linked ichtyosis, and Hunter's syndrome (Fraser, 1973). Fortunately, advances in medical technology allow the detection of many sex-linked and autosomal genetic diseases. With expert genetic counseling and prenatal testing it is now possible for individuals to prevent the devastating manifestations of some of the serious genetic disorders for which therapy is not yet available.

THE PANCREAS

The pancreas is an organ of both exocrine and endocrine functions, essential for the maintenance of life. The exocrine function leads to the production of digestive enzymes by the acinar tissue which comprises most of the weight of the gland. Lying embedded in the acinar tissue are the islets of Langerhans formed of endocrine cells that secrete predominantly two polypeptide hormones, insulin (secreted by the B cells) and glucagon (secreted by the A cells), both of which have important metabolic functions; a third type of cells (D cells) secrete two other hormones, gastrin and somatostatin; the latter may suppress the secretion of insulin and glucagon. Insulin is anabolic, increasing the storage of glucose in liver and muscle (and thereby inducing hypoglycemia) and of fatty acids and amino acids. Glucagon is catabolic, mobilizing glucose (and thereby inducing hyperglycemia), fatty acids, and amino acids from stores into the blood. The two hormones are thus reciprocal in their overall action and are reciprocally secreted in most circumstances. Somatostatin, originally isolated from the hypothalamus as a growth hormone inhibitor, has been found to inhibit not only the endocrine but also the exocrine secretions of the pancreas as well as the secretion of other gastrointestinal substances and gastric motility.

Embryogenesis. The fetal pancreas originates as two outpocketings from the endodermal lining of the gut at the level of the duodenum (Falkmer and Patent, 1972). At about seven weeks in the human embryo, the two pancreatic buds fuse to form the pancreatic ducts and the parenchyma, which differentiates into exocrine and endocrine cells. The islets of Langerhans are developed by the third month and both insulin (and its precursor, pro-insulin) and glucagon can be detected at this time in the fetal pancreas. At what exact time A and B cells begin to specialize is not known, although A cells seem to be more developed and in greater number than B cells during early developmental stages–a relationship that is reversed in the adult. Early studies had proposed that, similarly to all other polypeptide-hormone producing cells of the digestive tract, pancreatic cells as well would originate from the neural crest (Pearse, 1969); this neural origin, however, was refuted and more recent studies suggest that all cells of the pancreas derive from a pluripotential endodermal cell from which they proliferate and differentiate under the influence of a "mesenchymal factor" which would preferentially stimulate the proliferation of acinar (exocrine) cells *in vitro* (Rutter et al.,

1978; Wessells, 1977); other factors that influence the differentiation and development of the pancreas include the glucocoticoids and the amino acids of the culture medium. With respect to glucocorticoids, their action would be primarily one of modulation rather than one essential for differentiation; thus, these hormones would promote the expression of certain acinar genes while inhibiting B cells differentiation (Rutter et al., 1978).

Functional Development

Insulin is secreted by the human fetal pancreas (as measured by radio immunoassay and fluorescent antibody techniques) as early as day 80 of gestation, approximately one week after the first histologic demonstration of islet formation. Like the exocrine enzymes, the development of insulin in the pancreas follows a diphasic pattern: levels of the hormone are relatively low until about ten weeks of age and then progressively rise until 24 weeks, at which age they become temporarily and relatively stationary; from 34 to 40 weeks of gestation, pancreatic insulin content further increases and, not only is the relative percentage of islet tissue per total pancreas higher in the fetus at this age and in the newborn than in the adult, but the insulin concentration is also higher than in the adult (Kaplan et al., 1972). This diphasic developmental profile is not limited to humans, but is also found telescoped in several animals such as the rat in which pancreatic insulin is relatively low until the 12th fetal day, remains constant until the 14th day, and then rises sharply until birth, at 22 days (Pictet and Rutter, 1972).

Even though pancreatic insulin content increases progressively with development, the concentration of insulin in blood does not appear to rise with advancing fetal age nor to be significantly affected by administration of glucose *in vivo* and *in vitro* (see Kaplan et al., 1972). What, then, are the factors that influence insulin secretion in the fetus? It is well established that there is little or no transfer of insulin across the placenta in most animal species studied and, therefore, the insulin in fetal blood primarily reflects the secretion of the hormone by the fetal gland (Goodner and Freinkel, 1961; Clark and Soeldner, 1967). In the adult, a slight increase in blood glucose is a powerful stimulant for insulin secretion. In the fetus, the levels of blood glucose are lower than those in the maternal blood; in some species (man, sheep, and monkey) they remain half the maternal level until term, whereas in other species (rat, guinea-pig, and rabbit), glycemia rises during the latter part of gestation to levels comparable to those in the adult. At birth, probably as a consequence of hypoxia and other stress factors, the glycemia of the newborn rises to adult range (probably under the influence of sympatho-adreno-medullary stimulation), but it soon falls again to lower levels within a few hours after birth and remains low for several days (7-10) in full-term, normal infants and as long as two to three weeks in prematures. In early studies, the low blood glucose levels in humans before birth were attributed either to a high rate of insulin secretion or to hypersensitivity of tissues to insulin action, two suggestions that are being refuted today. For example, administration of glucose to pregnant women increases significantly insulin blood levels in the mother but not in the fetus, thus reinforcing the view that, prenatally, glucose has little stimulatory effect on insulin secretion and refuting the possibility that fetal hypoglycemia may be due to fetal hyperinsulinism. Another example is the failure of fetal insulin to increase uptake of glucose in several tissues normally insulin-responsive in the adult; this failure has been ascribed to the immaturity of insulin-dependent transport enzymes, an explanation that argues against the suggestion that fetal tissues are more sensitive to insulin than adult tissues. Although the exact mechanisms underlying fetal hypoglycemia remain elusive, they appear to depend more on the specific patterns of pre- and perinatal metabolism (e.g., increased energy requirements for growth and differentiation, carbohydrate as the main metabolic fuel, relative inefficiency of gluconeogenesis and fatty acid degradation as source of energy) than on alteration in the responsiveness of glucose metabolism to hormonal influences (see Dawes, 1968; Timiras, 1972).

If glucose is not the primary stimulator of insulin secretion, then insulin release must be regulated in the fetus by different mechanisms than those operative in the adult. This seems to be the case and current evidence points to a series of successive controls unfolding as the animal develops. In several mammals, including humans, the first control to become effective would be mediated through the stimulatory action of glucagon, the other hormone secreted by the pancreas. We know that, in the adult, besides glucose, glucagon, as well as such amino acids as leucine and arginine, are potent stimulators of insulin release. In the case of glucagon, the mechanism of this action would depend on the activation in the hepatic cells of cyclic AMP, a classic example of mediation of a hormonal effect via cyclic AMP. A developmental study of glucagon in the rat shows that this hormone follows a pattern quite different from that of insulin: levels are already high during early development, reach a peak during the second half of gestation and remain high until and after birth (Pictet and Rutter, 1972). Accordingly, it has been suggested that during early development, glucagon represents the primary stimulus for insulin secretion (Espinosa et all, 1970). Later, insulin release would be responsive to stimulation by the amino acids, leucine first, followed by arginine. Finally, the responsiveness of insulin to glucose would develop towards the end of gestation or shortly after delivery. Injections of glucagon in relatively high (pharmacological) doses in the fetal rat have been reported not only to induce insulin release but also to exert several metabolic effects on fetal liver comparable to those produced by the hormone in the adult. Because of the peculiarities of fetal metabolism for which no endogenous glucose is required (e.g., glycogenesis from glucose of maternal origin, no glycogenolysis nor gluconeogenesis), glucagon is relegated to a role of minor importance from a metabolic point of view; it cannot be excluded, however, that it may play a role in differentiation of fetal tissues, perhaps in the early development of the fetal pancreas (Girard et al., 1974).

The regulatory role of blood glucose on the hormonal secretions (both insulin and glucagon) of the pancreas has been related also to the presence of specific glucose receptors and/or to an intrinsic metabolic system (Matschinsky et al., 1971). The existence of "glucose receptors" has long been bypothesized on the basis that certain compounds such as galactose and galactosamine are capable of stimulating insulin release even though they are not metabolized; these compounds would exert their action by binding with receptors specific for glucose and closely related substances, independently of their metabolic effects. In parallel with the presence of these glucose receptors, insulin synthesis and release would also be regulated by the flux of glucose metabolized through the glycolytic and, perhaps, pentose-phosphate pathways, the enzymes involved in such pathways being extremely sensitive to relatively small fluctuations in extracellular glucose levels. Studies on the development of the glucose receptors and on the enzymes of glucose metabolism, not available at present, would be highly desirable for a better understanding of prenatal glucose metabolism and its relationships to hormonal controls.

The mechanism(s) of action of insulin are still unclear; some views hold that insulin increases cellular glucose uptake by acting on some glucose-carrier system (Morgan et al., 1964); if this were the case, then, the effectiveness of insulin during development would depend not only on the functional maturation of the endocrine cells of the pancreas but also on the efficiency of the carrier system, the maturation of which might coincide or not with that of the synthesis and release of the hormone. Other views hold that insulin acts by increasing the permeability of the cell (muscle) membrane to glucose, glucose uptake being an essentially passive phenomenon (Naftalin, 1970; Zierler, 1972). Insulin receptors have been identified on the cell surface and partially characterized biochemically and functionally (Narahara, 1972). Whether the number and/or binding affinity of these receptors vary with age remains to be determined. Still other investigators hold that insulin has no or little effect on the membrane transport of glucose in the liver but, rather would act primarily by stimulating intracellular enzymes (Gorden et al., 1978); in this case, the maturation of the hepatocytes and their enzymes would be necessary for the effect of insulin to take place. Indeed, in the case of insulin, as well as many of the other hormones discussed here, maturational patterns involve not only the development of the synthesis and secretion of the hormone and of the control systems that regulate its release, but also the maturation of those factors that mediate the hormone action at the target cell, that is, the binding of insulin to its specific receptor, its transport into the cell and ultimately its activation of intracellular metabolism.

The Diabetic Mother and Her Infant. Diabetes mellitus represents one of the commonest medical causes of problems to the fetus and the newborn. If a diabetic, whether overt (i.e., with manifest diabetes before initiation of pregnancy), latent (i.e., with diabetes becoming manifest only under the stress of pregnancy), or potential (i.e., with normal glucose tolerance but with a strong family history of diabetes) becomes pregnant, she will undergo periods of hyperglycemia even under the best therapeutic regimen; this hyperglycemia will be transmitted to the fetus in whom it may have some untoward metabolic effects *per se* and also stimulate the fetal pancreas to increase its secretion of insulin. Thus, the infant of the diabetic mother must sustain the effects of a possible diabetic genotype as well as of the unfavorable environment engendered by the maternal diabetes.

Complications in the diabetic mother with insufficient control of the disease include hydramnios (i.e., accumulation of excessive amounts of amniotic fludi), preeclampsia (i.e., rise in blood pressure with edema and proteinuria), ketoacidosis (i.e., accumulation of ketone bodies consequent to increased fatty acid levels in the liver), and dystocia (i.e., difficult and painful delivery caused by the large size of the baby).

Complications in the infant of the diabetic mother include: (a) high incidence of fetal (stillbirths) and perinatal mortality generally associated with immaturity of the fetus, the degree of immaturity being somewhat proportional to the severity of the maternal disease; (b) the characteristic large size (over 4,000 gm in weight) and the moon-faced and overall puffy appearance of these babies apparently caused by accumulation of body fat and increased size of the viscera, particularly heart and liver; (c) hypertrophy and hyperplasia of the islets of Langherans with high insulin levels; (d) a rapidly developing (but generally transient) hypoglycemia in the newborn, presumably caused by the preexisting fetal hyperinsulinism consequent to maternal and fetal hyperglycemia (Mølsted-Pedersen, 1974). Inasmuch as the pancreas seems to be relatively unresponsive to glucose until term, in normal fetuses, one must conclude that during intrauterine exposure to hyperglycemia, the pancreas of the fetus becomes more responsive and at an earlier age than does that of a fetus of a nondiabetic mother; (e) a high incidence of congenital malformation (e.g., malformations of lower limbs and ventricular septal defect of the heart) and of the complications of immaturity (e.g., respiratory distress syndrome).

From the above list of complications it is clear that babies from diabetic mothers belong in the high-risk category. The best treatment is preventive, that is control of maternal diabetes before and throughout pregnancy. After birth, appropriate therapeutic regimens are available to combat each of the eventual complications. Generally, no long-term or permanent impairment in growth and development has been reported in the surviving infants but the probability that these children will develop diabetes, although small is real.

Role of Insulin in Fetal Growth. The large body size of infants born to diabetic women has been related to an increase in the release of growth hormone from the fetal pituitary as a result of impaired glucose metabolism. The relationship between a fall in blood glucose levels and a rise in growth hormone secretion is well documented in both developing and adult individuals. In the normal fetus (and eventually also in the fetus from a diabetic mother in whom hyperinsulinism may lead to episodes of hypoglycemia, the so called "insulin reactions") the persisting hypoglycemia would represent a constant stimulus to pituitary growth hormone secretion. Fetal growth hormone, however, does not seem to be necessary to fetal body growth (see below). Therefore, whether or not fetal hypoglycemia stimulates growth hormone secretion does not seem to be important to body size. Rather, one should consider the important role of insulin itself as a promotor of growth. Not only does insulin stimulate protein formation (anabolic action) but, in its absence, protein catabolism is accelerated and protein synthesis is depressed. The anabolic action of insulin is explained in part by the protein-sparing effect of adequate intracellular glucose supplies and in part by an increased incorporation of amino acids into proteins. The latter action, independent from the effects of the hormone on glucose metabolism, has been related to facilitated amino acid transport and activation of ribosomes. Failure to grow is a symptom of diabetes in children and, on the other hand, acromegaly is often associated with diabetes, probably as an expression of the anti-insulin action of growth hormone. In view of its well-known anabolic action and its stimulatory effect on growth postnatally, it has been suggested that insulin exerts a similar growth-promoting action prenatally. If that were the case—and it is not proven at present that indeed it is so—hormonal control of growth would involve a number of endocrine interrelationships with different endocrines playing a predominant role at specific periods during development: insulin and prolactin, primarily during fetal life; growth and thyroid hormones, primarily during childhood, and sex hormones during adolescence.

Somatostatin

This polypeptide first isolated from the hypothalamus and subsequently found also in specific cells of the pancreatic islets (D cells) and of the gastrointestinal mucosa, inhibits the secretion of both insulin and glucagon as well as that of growth hormone. Radioimmunoassay measurements of somatostatin in the human fetus reveal that the hormone is present in the hypothalamus and, in lesser amounts, in other brain areas as early as ten weeks of gestational age; from very low levels at this age, hormonal levels progressively increase until birth, but remain always significantly lower than in the adult (Aubert et al., 1977). In the fetal pancreas of the rat, somatostatin levels are also very low until about ten days of age when they start to increase, the rate of accumulation following an essentially biphasic pattern similar to that of insulin but differing from it with respect to the age and duration of the peak of accelerated increase (McIntosh et al., 1977). At present it is not known whether hypothalamic and pancreatic somatostatin participates in the regulation of the secretion of growth hormone, insulin and glucagon in the fetus as it does in the adult or whether it has some other distinct function specific to fetal metabolism.

THE PARATHYROID

The parathyroids are essential to life–after parathyroidectomy, plasma calcium levels steadily decline, leading to a condition of hypocalcemia, and, eventually to death.

Embryogenesis. Parathyroids (usually four in number but sometimes varying between two and six) are derived from the third and fourth pharyngeal pouches. The cells that will form the parathyroid tissue start to differentiate by the fifth and sixth week into light, water-clear and dark cells, depending on the degree of their secretory activity; these cells will give origin to the "chief" cells, the secretory cells of the gland (the active "dark" cells containing numerous secretory granules that are thought to be the intracellular form of the parathyroid hormone). The other cell type of the parathyroid, the large acidophilic cells which will give rise to the "oxyphilic" cells usually do not appear until the time of puberty. These latter cells have not been associated with any function; it is believed that they are derived from chief cells possibly as the result of aging (Roth, 1971). Studies of the relative size of the parathyroids at various fetal ages, shows that their rate of growth exceeds that of whole body between 14 and 20 weeks (a period of active growth for many endocrines) but slows down thereafter. At birth, the size of the gland is less than half that of the adult and will not reach its final length and weight until the third and fourth decades.

Functional Development

Little is known about the stage of fetal development at which the parathyroid gland becomes functional, i.e., begins its secretory activity. The fact that the cell types in fetal parathyroids are similar to those found in the postnatal parathyroids (except for the absence of the oxyphilic cells) is considered histologic evidence in support of the prenatal function of these glands (Anast, 1975a). The appearance of the water-clear chief cells (which usually predominate in the hyperfunctioning gland in the adult) during the first half of gestation is especially considered to be associated with increasing parathyroid function. Thus, on the basis of these histologic criteria, the fetal parathyroids would become functional before midgestation. Several studies suggest that secretion may be initiated as early as the 12th or 13th week and that calcium-parathyroid hormone feedback mechanisms are functional during fetal life (at least in the sheep) (Smith et al., 1972). Clinical evidence also shows that the parathyroid glands are capable of producing and releasing parathyroid hormone before birth in humans and that their secretory activity is regulated in the fetus as in the adult by blood calcium levels. Thus, high blood calcium levels in the mother would be capable of inducing hypocalcemia in the fetus and lead to parathyroid insufficiency in the neonate. Such infants display marked tetany (characterized by convulsions, hyperirritability and vomiting) one of the symptoms of hypoparathyroidism, which, however, generally subsides as glandular activity normalizes. Tetany of the newborn may also stem from a number of other causes such as prematurity (manifested in early neonate hypocalcemia), or a high-phosphorus diet, hypomagnesemia, refractoriness to parathyroid hormones (manifested in late neonatal hypocalcemia) and other causes (see Anast, 1975b).

The Parathyroid Hormone in humans and other animals has been identified and purified as a single-chain polypeptide containing 84 amino acids. It derives from a precursor pro-hormone containing 109 amino acids and having one-third the activity of the hormone. Both synthesis and catabolism of parathyroid hormone appear to depend on a series of specific cleavages from the point of initial cellular biosynthesis to ultimate disappearance from circulation (Habener and Kronenberg, 1978). The first cleavage occurs in cells of the parathyroid gland where pro-parathyroid hormone is converted to parathyroid hormone which is then secreted into the blood. After secretion, the hormone rapidly undergoes further cleavage to smaller peptide fragments which represent the predominant species of the hormone found in the peripheral circulation.

Little is known of the way in which hormonal biosynthesis and cleavage are initiated during development except for some studies on the levels of parathyroid hormone, calcium, magnesium and phosphorus in the cord blood and in blood of neonates (Anast, 1975b). Parathyroid hormone levels seem to be undetectable to low (by radioimmunoassay techniques) in cord blood and in neonatal blood but to increase slowly within 48 hours and in the first weeks after birth. The prevalence of such low levels appears to be inconsistent with the evidence of fetal parathyroid gland activity demonstrated by the morphologic observations (see above). However, it is important to point out that undetectable levels do not mean that the hormone may not be present but rather that the level of sensitivity of our detecting techniques is insufflcient. Indeed, important changes in calcium homeostasis and its regulation occur in both mother and fetus during gestation. In human studies, plasma calcium (both total and iodized) decreases while parathyroid levels increase during the last trimester of pregnancy. The fact that at birth, low calcium levels are associated with low parathyroid hormone levels might not suggest hypofunction of the parathyroid gland but rather implicate extraparathyroid factors responsible for reducing calcium levels (e.g., complications of labor and pregnancy Anast 1975b). For example, it has been postulated that increased adrenocorticoids during stress could account for the high incidence of hypocalcemia in babies born after traumatic or difficult labor. As yet, however, no direct evidence has been presented to support this hypothesis.

Calcitonin (thyrocalcitonin).

In contrast to the parathyroid hormone which elevates blood calcium levels, the hormone calcitonin, secreted by distinct C-cell imbedded in the thyroid gland (hence the alternate name of thyrocalcitonin) and the thymus, lowers circulating calcium levels. In nonmammalian vertebrates, calcitonin is secreted by the ultimobranchial bodies, a pair of glands derived embryologically from the fifth branchial arches; however, studies in both birds and mice suggest that the calcitonin-secreting cells migrate from the neural crest and are neuroectodermal in origin. In humans, C-cells can be identified histologically in the thyroid by the 12th week of gestation. A polypeptide, calcitonin is released in response to increased concentration of calcium in the blood and exerts its hypocalcemic action by suppressing the resorption of bone; hypocalcemia, in turn, limits further secretion of calcitonin.

The hormone is most effective in lowering calcium levels in young, rapidly growing animals. It is not known, however, what role calcitonin may have in the human fetus nor when its secretion is initiated. Although the fetal skeleton takes form by the 8th to 10th week of intrauterine life, it is only in the last trimester that appreciable amounts of calcium are required for osteogenesis. If calcitonin suppresses the resorption of bone in the fetus, the reportedly high values for serum calcitonin in human fetus may have functional significance. In considering the control of mineral metabolism, it must be kept in mind, however, that besides parathyroid hormone and calcitonin, vitamin D also plays an important role; in addition, whereas some of these hormones act not only on the bone, but also on renal reabsorption and intestinal absorption of calcium, other factors, such as adequate maturation of the kidney and intestinal mucosa, may be necessary to insure optimal calcium regulation (Anast, 1975b, Arnaud, 1978).

THE PITUITARY OR HYPOPHYSIS

The pituitary gland or hypophysis is formed of three lobes, anterior, intermediate, and posterior, each of which secretes specific hormones. In humans and large mammals, however, the intermediate lobe is rudimentary and is generally described with the anterior lobe. The present discussion, therefore, is divided into two sections concerned with the anterior (including the intermediate) and the posterior lobe, respectively.

The Anterior Pituitary or Adenohypophysis

This lobe is responsible for the secretion of six hormones, adrenocorticotropic hormone (ACTH), thyrotropic hormone (TSH) the gonadotropic hormones, follicle-stimulating hormone (FSH), luteinizing hormone (LH), prolactin (PRL), and growth hormone (GH); all of these hormones, except growth hormone and prolactin, regulate the function of other, so-called target, endocrine glands. The activity of the anterior pituitary gland is in turn regulated by the hormonal secretions of the target endocrines and of the hypothalamus as well as by environmental stimuli acting on the brain. It has been claimed from time to time that the anterior pituitary secretes additional hormones besides those indicated above. No definite proof of such hormones, however, exists and, therefore, our discussion will be confined only to those hormones which are well recognized.

Embryogenesis. The adenohypophysis consisting of the anterior and intermediate lobes is derived from the primitive oropharynx as an evagination of the roof of the mouth called Rathke's pouch, already distinct in the three-week old embryo. By the end of the second month, the connection of the Rathke's pouch with the oral cavity has disappeared and the gland comes into close contact with the infundibulum, a downward extension of the floor of the diencephalon which will give rise to the posterior lobe of the pituitary. At this same time, the front of the pouch thickens into an important secretory mass that represents the anterior lobe while the intermediate lobe remains thin and closely adherent to the anterior lobe.

Cytologic differentiation of the anterior pituitary begins at seven to eight weeks of gestation, and by the tenth week, all specialized cells are distinguishable in terms of affinity (chromophiles) or lack of affinity (chromophobes) for histological dyes (i.e., acidophil and basophil cells) and structural and ultrastructural characteristics (e.g., cell shape, size, arrangement and types of intracellular secretory granules and structures) (Hartemann, et al., 1972). Together with morphological differentiation, each group of cells begins its specific secretion. Early studies have indicated that the time of the onset of endocrine secretion varied with the hormone considered, and, in the case of gonadotropins, with the sex of the fetus. In humans, measurable levels of GH and ACTH could be detected in the pituitary and serum at 10 weeks of age, whereas TSH could be detected at 12 to 13 weeks; of the gonadotropins, FSH could be found in the pituitary of the fetus at 13-14 weeks in females, but at 20 weeks in males and LH secretion would extend over a period of 12-20 weeks, irrespective of sex. Prolactin levels are measurable by 10-14 weeks in the pituitary and by 12-15 weeks in the serum, but they are very low at these ages compared to those of more advanced fetuses and of newborns. The differences in the onset of secretion among pituitary hormones may be more apparent than real and may be ascribed to technical difficulties inherent in the measurement of each individual hormone; they may also reflect a true differential timetable of development similar to that reported for the adrenocortical hormones and dependent on the sequential maturation of the enzymes for the synthesis and degradation of the hormones and hormone precursors.

Development of the Hypothalamic-Portal-Pituitary Complex in the Fetus. It is well known that the anterior pituitary depends for its function on the stimulation or inhibition it receives from the hormones secreted in the hypothalamus (Reichlin et al., 1976). These hypothalamic hormones are carried to the anterior pituitary primarily through a characteristic network of capillaries which form a portal system (by analogy with the portal system of the liver), that is, capillaries carry the blood directly from the median eminence in the hypothalamus to the anterior pituitary without passing through the systemic circulation. Inasmuch as an intact and mature hypothalamic-hypophyseal-portal complex is essential for optimal anterior pituitary function, the development of the anterior pituitary cannot be studied adequately without also considering the development of the hypothalamus and the portal system (Jost et al., 1974).

The development of the hypothalamus, an integral part of the nervous system, is considered in many speciality textbooks and also briefly in Chapter 6. With respect to its regulatory role on pituitary secretion, the hypothalamus has been shown, by radioimmunoassay methods, to contain, in measurable amounts, several of the releasing hormones (thyrotropin-releasing and gonadotropin-releasing hormones and somatostatin) early during gestation—by 10 weeks in humans—and these amounts increase progressively until 22 weeks, the last age at which these hormones were studied (Kaplan et al., 1976). Similarly the presence of such neurotransmitters as norepinephrine and dopamine, implicated in the release of these hypothalamic hormones, has been detected (by fluorescence methods) in the hypothalamus at corresponding ages (10-24 weeks) (Hyyppa, 1972; Nobin and Bjorklund, 1973); however, little is known of the metabolism and behavior of monoamines in human fetal hypothalamus. In the infant rat, studies of the development of monoaminergic and cholinergic systems show some interesting relations between maturation of these systems and the maturation of endocrine functions, particularly those of the thyroid (Geel and Timiras, 1967; Valcana, 1974; Vaccari, et al., 1977b) and the gonads (Vaccari, et al., 1977a).

The development of the portal vessels begins at an age approximating the onset of neurosecretory function or, according to some investigators, somewhat later, and its timetable of maturation appears rather slow. Portal vessels have been described in 12-week old human fetuses, but the characteristic capillary loops cannot be observed earlier than 16 weeks and the capillary network becomes well-established only by 22 weeks. Incompleteness of the portal system and, therefore, inefficiency of hypothalamic control, may explain, at least in part, the relative lack of respon-

siveness of the anterior pituitary to environmental stimuli during this
early period of development. On the other hand, hypothalamo-hypophyseal
communications seem to be established before maturation of the portal
system is achieved, probably through other routes. This seems to be
the case in the rat, in which the portal system matures postnatally
(Florsheim and Rudko, 1968) (within the first week after birth), but in
which stimulation of hypophyseal secretions occurs pre- and neonatally
by diffusion of hypothalamic hormones (Glydon, 1957) or through an
early neural component (Fink and Smith, 1971; Daikoku, et al., 1971).
Such also may be the case in humans, in whom recent anatomical evidence
has identified at least seven routes of communication among the pituitary,
hypothalamus, brain and systemic circulation (Bergland and Page, 1978).
It is possible to assume from the above considerations and by analogy
with other species that have a less developed portal system than mammals,
that the hypothalamic control of the pituitary is initiated at an early age
and then greatly improved with further development of both neurosecretory
function and portal capillary system.

Functional Development

Several of the hormones of the anterior pituitary have been discussed
together with their target endocrines. We will consider here growth
hormone, prolactin and melanocyte-stimulating hormone, only.

Growth Hormone (GH). Available evidence suggests that neither fetal
nor maternal growth hormone is essential for normal fetal growth. The
length at birth of the apituitaric, anencephalic newborn and of the
hypopituitaric infant is usually within the normal range. Similarly, infants
born to women with GH deficiency or to mothers hypophysectomized during
gestation do not show evidence of growth retardation. In experimental
animals, hypophysectomy (by decapitation) of fetuses does not prevent
continuation of fetal life nor does it reduce body (i.e., trunk) weight and
length at birth (Jost, 1953, 1966, 1975). Yet the capacity of the fetal
pituitary to synthesize and secrete GH is established early during
gestation. In the gland the hormone is measurable at around 70 days of
age, its content increases progressively until birth at which time it reaches
levels which remain constant during the first postnatal year; its
concentration, on the other hand reaches a peak at 25-29 weeks of age,
declines moderately towards birth and then rises again within the first
year (Kaplan et al., 1972). Coincident with the presence of GH in the
pituitary, GH levels can be measured in serum at 70 days of age, increase
to a peak by 22-25 weeks and then decrease slowly by 30-34 weeks and
persist at this relatively low level until after birth. Fetal GH resembles
the adult hormone in terms of its immunologic and biologic properties as
well as the heterogeneity of its major forms (Frohman et al., 1972). The
pattern of GH secretion in the fetus has been correlated by Kaplan and
associates (1972, 1976) with the differentiation and development of the
pituitary gland, the hypothalamic monoaminergic network, the hypothalamic
neurosecretory neurons and median eminence, the hypophyseal portal
system and the maturation of the brain. According to these authors, the
synthesis and secretion of GH by the fetal pituitary is regulated early
during development by a growth hormone-releasing hormone (GRH) from
the hypothalamus even before full differentiation of the portal system,
probably by diffusion or by a primitive vascular connection. By midgestation maturation of the hypothalamus and portal system leads to increased
secretion of GRH and GH. During late gestation, GH decreases following
a decrease in GRH or an increase in somatostatin or both.

If GH is not necessary for fetal growth and yet is present in the fetal pituitary and plasma in levels comparable to those of the infant, then what role does it play in fetal life? It has been suggested that injections of GH to pregnant rats may specifically promote growth of fetal brain (Zamenoff, et al., 1966). Whether such an effect does indeed occur under normal conditions or is limited to unfavorable conditions (e.g., malnutrition) and whether it results from a direct action of GH on the fetal brain or as an indirect consequence of hormonal imbalance in the pregnant rat receiving GH, remains a subject of controversy (Croskerry and Smith, 1975). It is possible that appropriate levels of GH might be necessary to insure the maturation of those hypothalamic-hypophyseal interrelations which will regulate GH secretion and release at later ages (Kaplan et al., 1972, 1976).

Prolactin (PRL) can be measured in the pituitary at about 70 days of age and follows the general pattern of development of GH hormone, but its levels rise more slowly especially during late gestation. In serum, PRL shows a progressive increase from early (12-15 weeks) fetal age to birth, a pattern quite different from that of GH. Regulation of fetal PRL remains to be clarified but the suggestion has been made that increasing PRL levels during late gestation can be correlated directly with concomitantly increasing levels of estrogens of placental origin (Aubert, et al., 1975). The effect of estrogens on plasma PRL is well documented in the adult (Robyn et al., 1973). PRL is the only pituitary hormone that shows the same serum concentration in the normal and anencephalic infant, indicating that hypothalamic control is not essential for its secretion in the human fetus (Kaplan et al., 1976). Indeed, explants of human chorion-decidual tissue are capable of synthesizing and secreting a PRL undistinguishable (by chromatographic, electrophoretic, immunologic and receptor assay techniques) from the *in vivo* secreted PRL (Golander et al., 1978). This additional source of PRL might explain the presence of this hormone in anencephalic infants and its high levels in the amniotic fluid (see Chapter 3).

Melanocyte-Stimulating Hormone (MSH). Of the two forms of MSH, the α form is the same in many mammalian species but has not been found in humans; the β form varies from species to species and has been found in the cells of both the intermediate and anterior pituitary lobe in humans In amphibians, fish, and reptiles, MSH has a well-established regulatory function on skin coloration, but its role in mammals remains unsettled. It has been suggested that it has become vestigial. Alternatively, it has been suggested by recent studies that β-MSH is a large molecule which splits into ACTH and a β-lipotropin, a peptide related to other brain peptides such as enkephalins and endorphins (Bradbury et al., 1976; Mains et al., 1977). β-MSH is demonstrable in the human fetal pituitary by 10-11 weeks of gestation as well as in plasma and the amniotic fluid and its level increases progressively throughout gestation (Levina, 1968; Ances and Pomerantz, 1974). According to some investigators, β-MSH would influence body growth in the fetus. These investigators feel that the current evidence is not sufficient to exclude completely the influence of some hypothalamic-hypophyseal factor on fetal growth; they argue that GH, even though not indispensible to fetal growth, is present in high levels at a period of development also characterized by an accelerated rate of growth; such a spurt of growth would not occur in anencephalic humans nor in rat fetuses in which the hypothalamus and pituitary had been destroyed (despite a normal body size at birth). In these latter animals, α-MHS was the only one, among many pituitary hormones and hypothalamic extracts capable of stimulating intrauterine growth (Swaab and Honnebier, 1974). In the rat, α-MSH is present in the pituitary by the 18th fetal day and its levels increase gradually until birth when they level off; administration of antibodies to MSH (anti α-MSH) on fetal day 18, would induce a drop in fetal weight (Swaab et al., 1976).

The Posterior Pituitary Or Neurohypophysis

This endocrine secretes in the blood two hormones, arginine vasopressin or antidiuretic hormone (ADH) and oxytocin. The former is important for maintenance of water balance and the latter for smooth-muscle contraction. The two hormones are produced in nuclei in the hypothalamus and are stored in the neurohypophysis bound to polypeptides called neurophysins which are released in the blood together with the hormones upon stimulation of the gland.

Embryogenesis. The posterior pituitary derives from the diencephalon. By the 3rd or 4th week, a projection of the floor (infundibulum) of the diencephalon flattens against the Rathke's pouch to form, with the median eminence of the hypothalamus, the neurohypophysis.

Functional Development

Both ADH and oxytocin are secreted in the fetus although the onset and characteristics of the secretory activity are still uncertain. The findings vary according to the technique used for determination of the neurosecretory ganules. With respect to ADH, neurosecretion can be detected in the human pituitary with histochemical techniques by 19 weeks and by 12 weeks with radioimmunoassay techniques (Rinne et al., 1962; Chard et al., 1971). Vasotocin, an analogue of vasopressin (or ADH) with physiological activity intermediate between ADH and oxytoxin has also been determined at 12 weeks of age by radioimmunoassay in the human fetus. In newborns and infants, earlier studies suggested that such aspects of renal function as difficulty in eliminating water load and concentrating urine were caused by immaturity of ADH secretion or lower sensitivity of the renal tubule to ADH in the young as compared with the adult. Current observations, however, indicate that immaturity of the nephron rather than blunted responsiveness to vasopressin or decreased secretion of ADH in response to a water load may be responsible for the decreased concentrating ability of the newborn infant (Valcana, 1972).

Pituitary Development And A Masterplan For Development And Aging. Despite the limitations of the techniques available for human fetuses (see Introduction) and the difficulty of extrapolating from one species to another because of the many interspecies differences, some general conclusions emerge from the studies of early differentiation and prenatal function of the hypothalamo-pituitary complex.
First, there is evidence that the synthesis and secretion of hypothalamic and hypophyseal hormones begins early in gestation, the developmental patterns of the hormones being markedly diverse. Most of the hypophyseal hormones appear to develop independently from one another but their secretion is closely related to the maturation of the hypothalamic neurosecretions even before the portal system achieves definitive maturation.

Second, after their early onset, hormonal secretions continue gradually to mature towards a level of activity, optimal for fetal growth and capable of supporting the increased demands for survival and adaptation that challenge the newborn and infant. This progression of competence depends not only on the maturation of the neurosecretory and endocrine cells themselves, but also on the maturation of those brain centers which regulate hypothalamic function, on the development of neurotransmitters systems that trigger the release of hypothalamic hormones, on the growth of the portal vasculature and on the establishment of efficient feedback controls between peripheral endocrines and hypothalamo-pituitary axis. Of the fetal pituitary hormones, some (e.g., GH), although in elevated levels, have not been linked to any definitive physiological action; others (e.g., ACTH, TSH, gonadotropins) are important for the differentiation and maturation of the target endocrines as well as other target organs.

A clear knowledge of the development of the hypothalamo-pituitary is important not only in itself, but also because of the continuing crucial role of this system in regulating the homeostatic adjustments of the organism throughout the lifespan. If we accept the working hypothesis that the various biologic stages of the lifespan—fertilization, birth, adolescence, adulthood, aging and death—are part of a programmed continuum of events, a so-called masterplan, regulated by neuroendocrine signals and controls, then an understanding of the development of these neuroendocrine controls is essential if we are to advance our understanding, not only of development, but also of aging (see also Chapter 15).

THE THYROID GLAND

The thyroid gland concentrates iodide from the blood to form thyroid hormones, thyroxine (tetraiodothyronine) and triiodothyronine (as well as reverse triiodothyronine), through the control of the thyroid-stimulating hormone (TSH) from the anterior pituitary. Thyroid hormones, even though not essential for life, have important functions such as stimulation of cellular oxygen consumption, regulation of lipid and carbohydrate metabolism, and promotion of growth and maturation. The thyroid gland also secretes calcitonin, a plasma calcium lowering hormone, discussed in relation with the parathyroid gland and the hormonal control of calcium metabolism.

Embryogenesis. The first sign of differentiation of the thyroid gland is the appearance (in humans, at approximately four weeks) of a thyroglossal duct as an outpouching from the floor of the developing buccal cavity. With further development, continuity with the buccal cavity is lost and, as the neck elongates, the gland descends along the hyoid bone and laryngeal cartilages to reach its final position in front of the trachea by the seventh week. By then, it has acquired a small isthmus and two lateral lobes and histological differentiation has begun. Initially, in the so-called precolloid phase, the gland is composed of a solid mass of cells with no follicular structure and no colloid. Subsequently, with the ingrowth of connective tissue and blood vessels, the cell mass is broken up into isolated cords that proliferate until, by the 10th to 12th week, they become arranged as follicles around a central stage which becomes filled with colloid material secreted by the follicular cells. The colloid is first observed in the three-month old embryo and the gland seems to become functional at approximately this age.

Functional Development

Because of the many "maturational" actions of thyroid hormones, studies in the fetus have been concerned with several aspects of thyroid development important for optimal differentiation and growth. Among these, some of particular interest are: the onset of thyroid hormone secretion, the degree of responsiveness of the thyroid gland to stimulation by hypothalamic thyrotrophic-releasing hormone (TRH) and hypophyseal thyroid stimulating hormone (TSH), and by specific stimuli (e.g., cold), the permeability of the placenta to thyroid hormones, the consequences of thyroid dysgenesis on fetal growth and maturation, and finally, the changes in thyroid function occurring at birth and during the neonatal period.

The human thyroid has been reported first capable of concentrating iodide between the third and fourth fetal month. Studies in animals, have shown that the ability of the gland to trap iodide precedes the onset of synthesis of iodinated amino acids (mono- and diiodothyrosine, the precursors of the hormones) and then, the synthesis of thyroxine (Shepard, 1968). This chronology is similar to that currently accepted for the biosynthesis of thyroid hormones in the adult. In humans, however, the capacity to manufacture thyroxine has been detected at the same time that iodide-concentrating capacity is observed. Likewise, the ratio of thyroxine to triiodothyronine content in the thyroid is similar in fetal and adult humans (Fisher et al., 1973).

Because the thyroid gland is capable of concentrating iodide and making thyroxine even in anencephalic fetuses (Yamazaki et al., 1959), it has been presumed that the biosynthetic mechanisms of the thyroid gland can develop in the absence of TSH from the pituitary. Several animal experiments, however, demonstrate that the pituitary-thyroid system is necessary in the fetus, as in the adult, for optimal concentration of oidide and efficient hormone synthesis and that, in fact, this system becomes active before birth. For example, in the rat, thyroid growth and ^{131}I uptake are decreased after hypophysectomy (by decapitation) in utero and may be restored to normal by TSH administration. In this species, fetal pituitary TSH undergoes a 10-fold increase between days 18 and 22 (birth occurring on day 22) and TRH from the fetal hypothalamus is also present and released although in apparently smaller amounts than in the adult (Conklin et al., 1973). In the sheep, the activity of the fetal hypothalamo-pituitary-thyroid axis appears to be considerably greater than in the adult and plasma TSH levels are three to five times higher in the fetus than in the neonate (Nathanielsz, 1975). However, the rise in fetal TSH and thyroid hormones elicited by administration of TRH seems to be less than in the newborn even though TRH has been detected at an early fetal age suggesting that the thyroid may respond somewhat differently to TSH and TRH pre- and postnatally or, alternatively, that these tropic hormones differ somewhat in their biologic (and possibly immunologic) activity at different stages in the lifespan (see also Chapter 15).

Extensive studies of the maturation of the hypothalamo-pituitary control of the fetal thyroid in humans have shown that the concentration of TSH (as measured by radioimmunoassay) in fetal serum and pituitary is very low until 18 weeks, that it increases abruptly between 18 and 22 weeks and that the level of activity reached at 22 weeks persists until term. The marked increase in TSH between 18 and 22 weeks is accompanied by a progressive elevation in total and free thyroxine in the serum, the changes in TSH and thyroxine being independent of maternal levels of these hormones (Fisher et al., 1970; Fisher and Dussault, 1974). In fact, neither TSH nor thyroxine show any correlation between maternal and fetal serum at any time during gestation, indicating that the fetal pituitary-thyroid axis functions autonomously. The minimal function of this axis before 18 weeks and its subsequent rapid increase in activity suggest that in human the fetal hypothalmo-pituitary-thyroid system begins to mature at this critical period. It seems likely that this maturation involves either an increase in hypothalamic secretion and release of TRH and/or an increase in pituitary responsiveness to TRH. This maturation has been compared to amphibian metamorphosis which is characterized by histological and functional maturation of the hypothalamus and median eminence and a progressive increase in thyroxine secretion; the hypothalamus, being particularly sensitive, during this period, and dependent for its maturation on the rising levels of thyroxine. The intrauterine period of development of the hypothalamo-pituitary-thyroid axis described in humans has also been reported for sheep and pig, however, in the rat and rabbit, the timetable of maturation develops exclusively postnatally, within the first week after birth (Cons et al., 1975). The increase in TSH (and thyroxine) levels near mid-gestation in the human fetus coincides with the corresponding increase in content and release of other hormones from the anterior pituitary (see below). Thus, a general maturation of the neural and neuroendocrine systems controlling the secretion of adenohypophyseal hormones seems to occur at approximately 20 weeks (Levina, 1968; Pavolva et al., 1968) and is manifested by hypothalamic activation followed by stimulation of pituitary TSH synthesis and activity and the consequent initiation and progressive increase of thyroid activity.

Studies of the fetal secretion and metabolism of thyroid hormones in humans and sheep have found that the low levels of serum thyroxine (associated with the low levels of TSH) before 18 weeks may be due to the low binding capacity of thyroxine-binding globulin. It should be noted here that, postnatally, thyroid hormones are almost completely bound to plasma proteins—albumin, primarily prealbumin, and, in larger amounts, globulin. The free thyroid hormones in plasma are in equilibrium with the protein-bound thyroid hormones in the tissues; the free hormones are added to the circulating pool by secretion from the thyroid gland and represent the physiologically active hormones, capable also of inhibiting pituitary secretion of TSH. After the period of hypothalamo-pituitary maturation, described above, between the 18th and 22nd fetal weeks in humans, the concentrations of thyroxine, both total and free, increase progressively and fetal free thyroxine values at term may even exceed maternal levels (Fisher et al., 1969).

The continuing rise in fetal serum thyroxine may be partially ascribed to a progressive increase in serum thyroxine-binding globulin concentration which in turn is due to the progressive increase in placental estrogen secretion during the latter half of pregnancy (see Chapter 3); it has been demonstrated that high levels of estrogen, either endogenous (as in pregnancy) or exogenous (after administration of diethylstilbestrol) significantly elevate thyroxine-binding protein and thereby may influence the kinetics of iodine metabolism (Dowling et al., 1956). In the fetus, however, the concomitant increase in free thyroxine concentration indicates that thyroxine secretion rises more rapidly than the level of the binding protein, so that a progressive saturation of protein-binding sites occurs. Inasmuch as it has been possible to exclude the intervention of such factors as maternal thyroxine and TSH, and fetal thyroid stimulation by human chorionic gonadotropin, the conclusion can be reached that the progressive increase in fetal thyroxine during the latter half of gestation occurs in response to fetal TSH stimulation.

Not only are thyroid hormones secreted in the fetus at the same or higher rate than in the mother, but also their metabolism, related to body weight, exceeds that in the mother as demonstrated by the higher turnover of thyroxine (but not of triiodothyronine) in the fetal sheep. The increased thyroid hormone turnover has been interpreted as reflecting the greater secretory activity of the fetal gland as compared to that of the mother (Dussault et al., 1971, 1972). However, the thyroid picture of high plasma TSH and thyroxine after mid-gestation does not seem to be accompanied by a similar elevation in triiodothyronine, but rather by low levels of this hormone. We know that, in the adult, plasma triiodothyronine derives both from the thyroid gland, from which it is secreted in about half the amounts as thyroxine, and from deiodination of thyroxine by tissues. The relatively low levels of triiodothyronine in the fetus have been explained in several ways—decreased thyroidal secretion, decreased monodeiodination in tissues, increased clearance from plasma—all of which might occur (Nathanielsz, 1976). That the metabolism of thyroid hormones might be quantitatively (if not qualitatively) different in the fetus than the adult is supported not only by the unusually high levels of thyroxine and low levels of triiodothyronine, but also by the presence of high levels of reverse triiodothyronine found in fetal sheep and umbilical cord blood in humans (Chopra et al., 1975). Reverse triiodothyronine represents another hormone of the thyroid gland; in the adult it is present in very small amounts, but in the fetus, its concentration is markedly elevated. The reasons for this fetal elevation of reverse triiodothyronine are not clear, but it has been suggested that it results from the differential maturation of the enzymes (deiodinases) deputed to the deiodination of thyroxine into triiodothyronine and reverse triiodothyronine. The rapid increase in serum triiodothyronine concentration in the early hours of postnatal life and the progressive decrease in the levels of reverse triiodothyronine during the first three days in the newborn suggest that the fetal pattern of thyroxine metabolism is rapidly altered after parturition and approaches the adult pattern within a few days after birth.

Thyroid Hormone Receptors: Distribution, Function and Development.
A better understanding of the mechanisms of action of thyroid hormones, has been provided by the demonstration of the presence of limited capacity, high affinity binding sites for T_4 and especially T_3 in several tissues (Surks and Oppenheimer, 1976). Thus, using the rat as the experimental animal of choice, receptors for these hormones have been recognized in the nucleus, cytoplasm and mitochondria of all so-called hormone-sensitive tissues (e.g., liver, kidney), i.e., which respond to administration of thyroid hormones with increased O_2 consumption, as well as the brain, despite the apparent unresponsiveness of the adult and perhaps also developing nervous tissue to thyroid hormones in terms of this response (Schwartz and Oppenheimer, 1978).

Nuclear receptors are considered mediators of the effects of thyroid hormone on growth, development and cell maintenance in most tissues including the brain where they also seem to play an important role in regulating the transmission of the nervous impulse by influencing myelinogenesis and synaptic function. In the developing brain, nuclear receptors are relatively low in number before birth, increase in number in the first two weeks after birth—a critical period for the effects of thyroid hormones on brain maturation — and then decline to adult values, the latter remaining higher in the cerebral hemispheres than in liver (Eberhardt et al., 1978; Valcana, 1979; Valcana and Timiras 1978; Schwartz and Oppenheimer, 1978). The developmental pattern in brain differs from that of liver where the number of receptors, low prenatally, increases uniformily to adult values (without showing any peak during the first two postnatal weeks). (Valcana and Timiras, 1979.)

Thyroid hormone receptors have also been identified in the cytosol where they are more abundant but have lower T_3 affinity than the nuclear receptors. Cytosol binding proteins may contain more than one binding site per molecule with a cooperative effect among the different sites. Unlike steroid hormones, cytosol binding is apparently not a prerequisite for the interaction of thyroid hormones with nuclear receptors; rather cytosol proteins would play a regulatory role in the retention and supply of the hormone to the various cell components, primarily mitochondria and nucleus. During early development, the affinity of rat brain (cerebellum) cytosol for T_3 is high when cytodifferentiation is most intense and declines with age with not apparent change in the number of binding sites. In sharp contrast, the liver shows a marked increased in the number of binding sites with age, however, with little change in affinity. (Geel, 1977.)

Receptors for thyroid hormones have also been localized on the mitochondrial membrane where they have been held responsible for the effects of thyroid hormones on energy metabolism. These mitochondrial receptors bind T_3 in many tissues including the developing but not the adult brain (Sterling et al., 1978).

The number of receptors in cerebral hemispheres and liver seems to be regulated by the thyroid state and is significantly higher in nuclei isolated from hypothyroid than from euthyroid animals. In the neonatally thyroidectomized rats, the increase in receptors is observed as early as 13 days after thyroidectomy, whereas in the adult, no significant difference is found until 4 weeks after thyroidectomy. (Valcana, 1979). These data are not in agreement with previous studies in which no alterations in the binding capacity or in the affinity of nuclear T_3 receptors after thyroidectomy were reported. Such a descrepancy in results may derive from differences in the severity and duration of the hypothyroidism. Hyperthyroidism, on the other hand, does not seem to alter the number or affinity of receptors. The increase in the number of nuclear receptors after thyroidectomy has been interpreted as being indicative of a regulatory action of thyoid hormones on their nuclear receptors analagous to the well-known effects of estrogens and insulin in regulating their receptors in their own target tissues. A similar time-and-dose-relation between thyroid hormones and their receptors has been demonstrated in cultured tumor cells from rat pituitary (Samuels et al., 1976).

In any tissue, the biologic response to the hormone may depend not only on the density of the receptors and their affinity for the hormone but also on the degree and duration of their occupancy. In this respect some thyroid hormone metabolites are of interest inasmuch as they may have similar affinity for the receptors as the hormone but a different biologic potency. For example, triiodothyroacetic acid (Triac) has an affinity for nuclear receptors similar to that of T_3 although its biological potency is about one-sixth of that of $T3$. When such a compound is generated within the target cell, it could interfere with T_3 binding to nuclear receptors provided it were not rapidly degraded further or removed from the cell. Current studies of the metabolism of T_3 and T_4 in brain in vitro and in vivo show significant differences in the metabolism of these hormones during development (Naidoo et al., 1978; Naidoo and Timiras, 1979; Valcana et al., 1979).

Neonatal Thyroid Hyperactivity. Immediately after birth and lasting for several days, the serum levels not only of triiodothyronine but also of thyroxine (both total and free) increase in the neonate and this increase is associated with other signs (e.g., thyroid radioactive iodine clearance) of thyroid hyperactivity. This hyperthyroid state results from a sharp increase in serum TSH due to increased pituitary TSH release during the late stages of labor and early hours of life, followed by a more chronic hypersecretion of TSH which results in elevated TSH serum levels throughout the first 2-3 postnatal days. On the basis of our knowledge that thyroid activity in the adult is stimulated by changes in environmental temperature, it is assumed that the sudden exposure to ambient temperature at birth "cools" the infant and, thus, stimulates the pituitary-thyroid axis. However, the TSH surge occurs even when the newborn is maintained in an incubator at warm temperatures, but the increment of increase rises sharply when these infants are exposed to cooler temperatures (Fisher and Odell, 1969). The thyroid hyperactivity of the neonate is not limited to humans but has been found to occur in a number of mammals (e.g., pigs, calves, lambs, etc.) characterized by a relatively well-developed hypothalamic-pituitary integration at birth; in contrast, early postnatal thyroid stimulation is less marked in those species (e.g., rats, rabbits) in which the development of the hypothalamo-pituitary system occurs comparatively late.

Clinical symptoms of hyperthyroidism do not occur in the newborn, probably because of the transient nature of the hyperthyroxinemia. On the other hand, lack of increased thyroid stimulation at this age may be indicative of a prenatal, hypothyroid condition; more importantly, lack of such stimulation may contribute to disturbances in neonatal cardio-respiratory adjustments. Thyroid hormones have been shown to stimulate brain excitability and, because initiation of respiration depends, in large part, on the degree of excitability of the medullary respiratory centers, it is possible that the neonatal surge of thyroid activity may lower the threshold of these medullary centers to the environmental stimuli that impinge on the newborn and thereby facilitate the onset and maintenance of breathing; in this case, neonatal hyperthyroidism would represent a highly significant factor in promoting perinatal survival.

The Placenta And Thyroid Hormone: Transfer And Metabolism. The placenta is known to transport iodide actively from maternal to fetal circulation in several mammalian species. Thyroid hormones, in contrast, do not traverse the placental barrier early in pregnancy; near term, they are transported from the mother to the fetus in most species studied, including man, but the extent of such transfer has not been systematically investigated and from the available evidence, is generally considered quite limited (Fisher et al., 1964). Studies in small mammals indicate that the placenta is impermeable to TSH (Knobil and Josimovich, 1958) and that, in general, the fetal pituitary-thyroid axis may be considered independent from the maternal thyroidal status (Varma et al., 1978). Direct data are not available in man but the difference between TSH concentrations in maternal and fetal serum throughout pregnancy (i.e., levels higher in maternal than fetal serum in the first half of gestation and vice-versa in the second half) would support this view.

The placenta does not appear to synthesize iodotyrosines or iodothyronines nor does it contain enzymes for their deiodination. Whether it can secrete TSH or a TSH-like substance is discussed in a subsequent chapter; at present, the significance of this eventual placental thyrotropin, its distribution among mammalian species and its role in maternal and fetal thyroid metabolism in pregnancy remain to be clarified (see Chapter 3).

The Role Of The Fetal Thyroid In Development. Even though body size of anencephalic human fetuses and neonates and of some laboratory animals hypophysectomized (by decapitation) in utero does not seem to differ from that of normal individuals of comparable age (see below), clinical and experimental evidence supports the view that normal thyroid function is necessary pre- as well as postnatally for optimal growth and development. Thus, administration of the goitrogen, propylthiouracil, to gestating rats induces thyroid hypertrophy and reduction of whole-body and brain weight, already evident during late fetal development and becoming more marked in the early postnatal period (Oklund and Timiras, 1977). The severity of growth retardation appears to be directly related to the doses of the goitrogen and the time and duration of its administration (i.e., the greater the dose, the earlier and longer the administration, the more severe the effects). Similar developmental problems have been reported in human fetuses with congenital absence of the thyroid or with thyroid insufficiency. The latter condition occurs most frequently in individuals living in isolated regions in the Andes, Himalayas, and New Guinea, to the less remote areas of western Europe in which there is a chronic deficiency of iodine in the diet (Stanbury and Kroc, 1972; Dunn and Medeiros-Neto, 1974). In these iodine-deficient populations and in experimental animals with induced hypothyroidism, the age of onset and the degree and type of developmental deficiencies vary considerably from individual to individual. These defects include, singly or in association stunted growth, a variety of neurologic abnormalities and postnatal intellectual impairments (cretinism). The crucial role of thyroid hormones in brain maturation is discussed in detail in Chapter 7. It remains to be emphasized here that varying degrees of thyroid insufficiency may occur in utero not only as a consequence of iodine deficiency as indicated above, but also of a variety of other factors such as infections, abnormal (autoimmune) immunologic reactions, irradiation, drugs, etc. Thus, mild degrees of hypothyroidism, often undetected at birth, may have extremely profound immediate and long-term effects on growth of the entire organism (see Chapter 4) and on the development of the central nervous system (see Chapter 7).

GASTROINTESTINAL HORMONES

Little information is currently available on the embryogenesis and the time of onset of function of the gastrointestinal hormones known to be secreted by entero-endocrine cells. This lack of information is the consequence in part of the complexity of the polypeptide structure of these hormones and in part of the difficulty of localizing their source of origin. Nevertheless, because of increasing interest in the regulation of growth processes in the gut, and because some (e.g., gastrin) of the gastrointestinal hormones seem to influence such growth, a number of recent investigations have initiated the study of the development of these hormones in man and other mammalian species as well as in organ and tissue culture (Enochs and Johnson, 1977).

Cells secreting gastrin, a 17-amino-acid polypeptide from the pyloric mucosa, have been found in human fetuses as well as in fetuses of rats, guinea pigs, opossum and dogs; the number of these cells and the amount of the secretory granules they contain seem to increase with development—in the rat, at least until weaning. However, at what exact stages of their development, gastric cells are capable of releasing the hormone has not been clarified.

The hypothesis that gastrointestinal hormones have a direct and important trophic effect on gastrointestinal tissues is generally well accepted; accordingly, it is important to identify at what age, that is, at what embryonic, fetal or postnatal stage of development, the growth-promoting action of these hormones becomes operative. This question, still unanswered is further complicated by the increasing evidence of a close interrelation between gastrointestinal and other hormones. Thus, it has been shown, at least, for gastrin, that the secretion of this hormone is depressed in hypophysectomized adult rats and stimulated after injection of growth hormone, corticosterone and thyroxine, singly or in association. The nature of these interrelations during fetal development remains currently unknown even though it would be extremely interesting to ascertain whether growth hormone—which, as discussed in a previous section, has no effect on whole-body growth in the fetus—would have a growth-promoting action on gastrin-secreting cells as well as stimulate the secretion of this hormone. Such a regulatory action, if proven, would reinforce the concept that fetal growth hormone does promote growth and maturation of specific tissues such as the hypothalamus, perhaps the cerebral cortex and some enteroendocrine cells. On the other hand, almost every tissue that is trophically stimulated by gastrin, secretes a hormone or peptide that has been shown to inhibit gastrin release. For example, the oxyntic gland secretes acid and contains a growth-hormone-release-inhibitory hormone which appears to be a nonspecific inhibitor of the release of gastrin and several other hormones, in addition to growth hormone; the small intestine releases secretin (another gastrointestinal hormone) and the pancreas secretes glucagon which, together with other local peptides inhibit gastrin secretion. The physiologic significance of these inhibitory effects remains to be determined, but their presence in association with the presence of the trophic effects, points to a complex and hitherto neglected interaction of the gastrointestinal hormones with the remainder of the endocrine system. At present, this interaction has been well established for gastrin while little is known of the development of function of the other hormones of this group. However, on the basis of the current observations on gastrin, it may be postulated that gastrointestinal hormones can no longer be isolated as a group of digestive hormones, but rather must be considered to play an important, integrated role in the endocrine regulation of growth and metabolism (Enochs and Johnson, 1977; Zimmerman, 1979).

MATERNAL HORMONES

During pregnancy the maternal organism undergoes important adaptations necessary for triggering and supporting physiologic and psychologic adjustments required for optimal growth of the fetus and continuing well-being of the mother. Such adaptations include changes in circulatory and respiratory functions and profound shifts in metabolism that are in large part under hormonal control. Thus, it is not surprising that alterations in the maternal endocrine system are observed at all stages of pregnancy from fertilization (Chapter 1) to implantation, placentation, maintenance and termination of pregnancy (Chapter 3). Major maternal changes in hormonal secretions involve the pituitary, adrenal cortex and thyroid, and lesser changes, the pancreas, parathyroid and adrenal medulla.

The pituitary increases in size (as represented mainly by an increase in the anterior lobe) and its enlargement is accompanied by cytologic arrangement and the appearance of a specific, transitory type of cells, the so-called pregnancy cells, and by increased secretory activity. With respect to the secretion of gonadotropins, significantly increased during pregnancy, the high levels of these hormones represent an anomaly in view of the known negative feedback between pituitary and ovarian hormones: high blood levels of estrogens and progesterone, as occur during pregnancy, would be expected to inhibit gonadotropin secretion, but do not. It would appear that, for these hormones, as well as for ACTH, TSH, and prolactin, the secretory function of the pituitary gland in the pregnant woman is reset at a different (higher) level than that of the nonpregnant one. One exception would be the secretion of GH which seems to be markedly decreased in late pregnancy. At the same time, the concentrations of prolactin (from the pituitary) and the somatomammotropin (from the placenta) in maternal blood are high (Chapter 3). The suggestion of an inverse reciprocal relationship among somatotropic hormones, however, does not offer a satisfactory explanation for the low GH levels, inasmuch as these levels remain low during the first weeks postpartum (despite the disappearance of placental somatomammotropin) and return to normal after about two months (despite the continuing high levels of prolactin during lactation). As discussed in detail in the next chapter (Chapter 3), the maternal pituitary and ovaries are not necessary for maintenance of pregnancy which continues until parturition, even in the absence of these two endocrine glands.

In parallel with the increased secretory activity of the anterior pituitary, the posterior and intermediate lobes increase their secretion as well. The levels of MSH increase fivefold during pregnancy, but the functional significance of this increase is uncertain. In view of the relation of MSH to other brain proteins (e.g., enkephalins) and hormones (e.g., ACTH), its presence in high levels suggest a resetting of its controls in the hypothalamus and higher brain centers in response to the increased requirements of pregnancy. The levels of oxytocin and ADH are also elevated in the maternal blood. It is difficult to understand why oxytocin shows an increase during the early months of pregnancy when its main role appears to surface much later at parturition. On the other hand, the increasingly high levels of ADH may have a direct supporting role in the regulation of water and electrolyte balance in response to the changing requirements of the pregnant organism.

Although the blood levels of adrenocortical hormones and their metabolites are markedly increased in the pregnant woman, the functional consequences of such an increase remain controversial. Of the two main forms of adrenocortical hormones of the cortisol type present in the blood—one free, biologically active and one protein-bound, biologically inactive—the bound form seems to increase proportionally more than the free form. Such a differential increase may explain the cases of hypercorticoidism observed in a small number of pregnant women. However, increased levels of free cortisol have also been reported and have been associated with the presence of purple striae, diabetic glucose tolerance curves, easily bruised skin, hypertension, and fluid retention: all signs present in some pregnant women and compatible with hypercorticoiolism. Similarly, the levels of aldosterone in blood are significantly and consistently increased, especially in the last months of pregnancy. High aldosterone levels have been ascribed in part to an increase in plasma renin activity, induced by the stimulatory action of high blood levels of estrogens on this enzyme, and in part to increased maternal requirements for electrolyte and water retention.

The thyroid gland is moderately enlarged in pregnancy owing to hyperplasia of the gland and increased vascularity. However, as for cortisol, thyroid hormones (both thyroxine and to a much lesser extent, triiodothyronine) are increased in terms of the total hormone but not of the free form which remains practically unchanged; hence, symptoms of hyperthyroidism are usually absent. This increase in circulating thyroxine in apparently euthyroid individuals has been related to alterations in the protein (alpha-globulin) which binds thyroxine (TBP). Such TBP is increased in total amount and in ability to bind the hormone during pregnancy.

The serum concentration of insulin in pregnancy is also elevated, despite its rapid degradation, unlike other hormones, by a placental enzyme (insulinase). A maternal glucose sparing effect is noted in pregnancy that may be related to the demands of the fetus. Although there is an increased secretion of insulin from the maternal pancreas, there is also a decreased utilization (uptake) of glucose by the tissues. This apparent increased resistance to insulin has been attributed to the presence of anti-insulin factors, primarily, the diabetogenic effect of the chorionic somatomammotropin from the placenta (Chapter 3); this hormone would increase lipolysis, elevate free fatty acids in blood and thereby provide noncarbohydrate calories for the mother while conserving carbohydrates and amino acids for the fetus. As for insulin, parathyroid hormone levels in serum show a definite rise during the latter part of pregnancy, a period when fetal demands for maternal calcium are maximal. Although more parathyroid hormone is secreted at this time, many pregnant women show a relative deficiency or a predisposition to parathyroid deficiency in late gestation.

Adrenal medullary activity appears to be only slightly increased or to remain normal depending on the experiments considered; catecholamines, and particularly epinephrine and norepinephrine have been found to be increased in the urine but are normal in the blood. While eventual increases in epinephrine secretion during the stress of labor might be expected, that does not seem to happen, and catecholamine levels remain for the most part unchanged during late pregnancy, labor and the immediate post-partum period.

Role of Fetal Hormones on Growth and Development of the Fetus

The analysis of the immediate effects of fetal hormones on growth and development of the fetus and of their long-term effects on growth and development of the child and adolescent represents an area of endocrinology still relatively new but nevertheless of crucial importance for the expression of the physiologic potential with which the individual is born. The absence of any clearly demonstrable effect of fetal GH on total body growth (weight and size) discussed in this chapter, has often been invoked to deny a significant role of fetal hormones in the regulation of fetal growth. That fetal hormones deficiency does not seem to affect growth prenatally or that it does affect it only slightly, does not preclude a possibility of its causing long-term impairment of growth and development. Thus, the importance of fetal androgens in bringing about differentiation and growth of male sex organs and that of thyroid hormones in insuring brain maturation is amply documented. Less spectacular, or perhaps less well known is the role of other hormones.

Regardless of whether they regulate fetal whole-body growth or not and to what extent, fetal hormones influence fetal physiology through a number of regulatory actions which can be grouped as follows: (1) *Regulatory action on endocrine glands* (endocrine-endocrine relationship), as demonstrated for example, by the regulatory role of the tropic hormones of the anterior pituitary on the target endocrine glands (e.g., adrenal cortex, thyroid). The hormones from the target endocrines in turn control the release of the pituitary tropic hormones through negative and positive feedbacks at the level of the pituitary or hypothalamus or both. (2) *Regulatory action on target organs and systems* (endocrine-organ relationship) as illustrated by the influence of fetal hormones on liver glycogen, thymus size and brain development. Often the action of the hormone is most effective at so-called "critical periods" characterized by accelerated growth and/or differentiation and by great vulnerability to internal (e.g., hormonal) and external (e.g., environmental) factors. The timetable of critical periods varies with the hormone and the target tissue as well as the animal species. (3) *Regulatory role on the placenta* (fetal-placental relationship) as discussed in detail next chapter (Chapter 3). (4) *Regulatory action of fetal hormones on the mother* (fetal-maternal relationship) still little understood. It is well-known that such hormones as chorionic somatomammotropin act on the mother as well as the fetus, but more uncertain is the direct influence of fetal hormones on the mother.

As indicated above, fetal hormones influence not only the fetus and, directly or indirectly, the mother, but may also have long-term effects postnatally. Differentiation of target organs (e.g., of secondary sex organs by androgens) or of specific functions (e.g., maturation of the brain by thyroid hormones) occurs prenatally and, by affecting the development of cells, tissues or organs, has profound consequences on the individual throughout the lifespan (Chapters 8, 9 and 15).

In addition to hormonal factors, genetic and/or environmental conditions affect fetal growth. One genetic influence on growth is seen in the sex differences in birth weight. The male infant weighs approximately 100 g more than the female at birth, perhaps because of the genetically-determined development of the fetal testis and the anabolic effects of testosterone on placental weight and function. Other genetic factors that may affect fetal growth are stature and birth weight of the mother, race (although socioeconomic factors complicate the genetic versus environmental role of race), and parity (the multipara generally having larger babies than the primipara). The role of each genetic factor on fetal growth may be considered insignificant in itself, but in combination and together with environmental factors, their additive influence becomes quite substantial. Of the environmental factors, the single most important is perhaps fetal nutrition, primarily dependent on placental transfer function, but other important environmental factors are the metabolic activity of the fetus (in part regulated by fetal hormones) the competence of the fetal circulation (which regulates tissue oxygenation) and maternal health (particularly with respect to nutrients, drugs, and "lifestyle" during pregnancy).

REFERENCES

Akana, S.F. and Timiras, P.S. Neuroendocrine strategies for adaptation to high altitude, in *Environmental Physiology: Aging, Heat and*

Altitude, S.M. Horvath and M.K. Yousef, eds. Elsevier/North-Holland, Amsterdam (1981) pp. 351-362.

Anast, C. Development of the normal embryonic fetal and neonatal parathyroid, in *Endocrine and Genetic Diseases of Childhood and Adolescence*, second ed. L.I. Gardner, ed. W.B. Saunders Co., Philadelphia (1975a), pp. 355-377.

Anast, C. Tetany of the newborn, in *Endocrine and Genetic Diseases of Childhood and Adolescence*, second ed. L.I. Gardner, ed. W.B. Saunders Co., Philadelphia (1975b), pp. 377-399.

Ances, I.G. and Pomerantz, S.H. Serum concentrations of β-melanocyte stimulating hormone in human pregnancy. *Am. J. Obstet. Gynec.* 119, 1062-1068 (1974).

Andrews, W.W. and Ojeda, S.R. On the feedback actions of estrogen on gonadotropin and prolactin release in infantile female rats. *Endocrinology* 101, 1517-1523 (1977).

Aragona, C., Bohnet, H.G., and Friesen, H.C. Localization of prolactin binding in prostate and testes: the role of serum prolactin concentration on the testicular LH receptor. *Acta Endocrinol.* 84, 402-409 (1977).

Arnaud, C.D. Calcium homeostasis: Regulatory elements and their integration. *Fed. Proc.* 37, 2557-2560 (1978).

Aubert, M.L., Grumbach, M.M., and Kaplan, S.L. The ontogenesis of human fetal hormones. *J. Clin. Invest.* 56, 155-164 (1975).

Aubert, M.L., Grumbach, M.M., and Kaplan, S.L. The ontogenesis of human fetal hormones. IV. Somatostatin, luteinizing hormone releasing factor, and thyrotropin releasing factor in hypothalamus and cerebral cortex of human fetuses 10-22 weeks of age. *J. Clin. Endocr. Metab.* 44, 1130-1141 (1977).

Avery, M.E. and Fletcher, B.D. *The Lung and its Disorders in the Newborn Infant*. W.B. Saunders Co., Philadelphia (1974).

Baird, D.T., Horton, R., Longcope, C., and Tait, J.F. Steroid dynamics under steady state conditions. *Recent Prog. Horm. Res.* 25, 611-656 (1969).

Baker, H.W.G., Bremner, W.J., Burger, H.G., de Krester, S.M., Dulmanis, A., Eddie, L.W., Hudson, B., Keogh, E.J., Lee, V.W.K., and Rennie, G.C. Testicular control of follicle-stimulating hormone secretion. *Recent Prog. Horm. Res.* 32, 429-476 (1976).

Baker, T.G. A quantitative and cytological study of germ cells in human ovaries. *Proc. R. Soc. Lond. (Biol.)* 158, 417-433 (1963).

Barr, M.L. and Bertram, E.G. A morphological distinction between neurones of the male and female, and the behavior of the nucleolar satellite during accelerated nucleoprotein synthesis. *Nature* 163, 676-677 (1949).

Bayon, A., Shoemaker, W.J., Bloom, F.E., Mauss, A., and Guillemin, R. Perinatal development of the endorphin- and enkephalin-containing systems in the rat brain. *Brain Res.* 179, 93-101 (1979).

Bergland, R.M. and Page, R.B. Can the pituitary secrete directly to the brain? (Affirmative Anatomical Evidence.) *Endocrinology* 102, 1325-1338 (1978).

Bloch, E. Fetal adrenal cortex: function and steroidogenesis, in *Functions of the Adrenal Cortex*, Vol. 2 K.W. McKerns, ed. Appleton-Century-Crofts, New York (1968), pp. 721-772.

Bloch, E. and Benirschke, K. Synthesis *in vitro* of steroids by human fetal adrenal gland slices. *J. Biol. Chem.* 234, 1085-1089 (1959).

Bloch, E. and Benirschke, K. Steroidogenic capacity of foetal adrenals *in vivo*, in *The Human Adrenal Cortex*. A.R. Currie, T. Symington, and J.K. Grant, eds. E. and S. Livingston LTD., London (1962), pp. 589-595.

Bradbury, A.F., Smyth, D.G., and Snell, C.R. Lipotropin: precursor to two biologically active peptides. *Biochem. Biophy. Res. Commun.* 69, 950-1956 (1976).

Chard, T., Hudson, C.N., Edwards, C.R.W., and Boyd, N.R.H. Release of oxytocin and vasopressin by the human foetus during labour. *Nature* 234, 352-354 (1971).

Chopra, I.F., Scak, J., and Fisher, D.A. 3,3',5'-Triiodothyronine (reverse T_3) and 3, 3', 5-Triiodothyronine (T_3) in fetal and adult sheep: studies of metabolic clearance rate, production rate, serum binding and thyroidal content relative to thyroxine. *Endocrinology* 97, 1080-1088 (1975).

Clark, D.M. and Soeldner, J.S. Fetal-maternal insulin and glucose interrelationships. *Diabetes* 16, 516-517 (1967).

Cohen, A. Plasma corticosterone concentration in the fetal rat. *Horm. Met. Res.* 5, 66 (1973).

Cohen, A. Adrenal and plasma corticosterone levels in the pregnant, foetal and neonatal rat, in the perinatal period. *Horm. Met. Res.* 8, 474-478 (1976).

Colby, H.D. and Kitay, J.I. Interaction of estradiol and ACTH in the regulation of adrenal corticosterone production in the rat. *Steroids* 24, 527-531 (1974).

Comline, R.S. and Silver, M. Development of activity in the adrenal medulla of the foetus and new-born animal. *Br. Med. Bull.* 22, 16-20 (1966).

Conklin, P.M., Schinkler, W.J., and Hull, S.F. Hypothalamic thyrotrophin releasing factor: activity and pituitary responsiveness during development in the rat. *Neuroendocrinology* 11, 197-211 (1973).

Cons, J.M. and Timiras, P.S. Developmental patterns of FSH and LH in female rats deprived of light before puberty. *Environ. Physiol. Biochem.* 5, 355-360 (1975).

Cons, J.M., Umezu, M., and Timiras, P.S. Developmental patterns of pituitary and plasma TSH in the normal and hypothyroid female rat. *Endocrinology* 97, 237-240 (1975).

Costa, M., Eränkö, O., and Eränkö, L. Hydrocortisone-induced increase in the histochemically demonstrable catecholamine content of sympathetic neurons of the newborn rat. *Brain Res.* 67, 457-466 (1974).

Coupland, R.E. Blood supply of the adrenal gland, in *Handbook of Physiology, Section 7: Endocrinology, Vol. VI. Adrenal Gland.* H. Blaschko, G. Sayers, and D.A. Smith, eds. Amer. Physiol. Soc., Washington, D.C. (1975), pp. 283-294.

Croskerry, P., and Smith, G.K. Prolongation of gestation by growth hormone: a confounding factor in the assessment of its prenatal action. *Science* 189, 648-650 (1975).

Cutler, G.B., Barnes, K.M., Sauer, M.A., and Loriaux, D.L. Estrogen receptor in rat adrenal gland. *Endocringology* 102, 252-257 (1978).

Daikoku, S., Kinutani, M., and Sako, M. Development of the adrenal medullary cells in rats in reference to synaptogenesis. *Cell Tissue Res.* 179, 77-86 (1977).

Daikoku, S., Kotsu, T., and Hasimoto, M. Electron microscopic observations on the development of the medial eminence in perinatal rats. *Z. Anat. EntwGesch.* 134, 311-327 (1971).

Dawes, G.S. *Foetal and Neonatal Physiology.* Year Book Medical Publishers, Chicago (1968).

Dohler, K.D. and Wuttke, W. Changes with age in levels of serum gonadotropins, prolactin, and gonadal steroids in prepubertal male and female rats. *Endocrinology* 97, 898-907 (1975).

Dowling, J.T., Freinkel, N., and Ingbar, S.H. Effect of diethylstibestrol on the binding of thyroxine in serum. *J. Clin. Endocr. Metab.* 16, 1491-1506 (1956).

Dunn, J.T., and Medeiros-Neto, G.A., eds. *Endemic Goitre and Cretinism: Continuing Threats to World Health.* WHO Scientific Publication No. 292. WHO, Washington, D.C. (1974).

Dussault, J.H., Hobel, C.J., and Fisher, D.A. Maternal and fetal thyroxine secretion during pregnancy in the sheep. *Endocrinology* 88, 47-51 (1971).

Dussault, J.H., Hobel, C.J., DiStefano, J.J., III, Erenberg, A., and Fisher, D.A. Triiodothyronine turnover in maternal and fetal sheep. *Endocrinology* 90, 1301-1308 (1972).

Eberhardt, N.L., Valcana, T., and Timiras, P.S. Triiodothyronine nuclear receptors: an in vitro comparison of the binding of triiodothyronine to nuclei of adult rat liver, cerebral hemisphere and anterior pituitary. *Endocrinology* 102, 556-561, 1978.

Enochs, M.R., and Johnson, L.R. Trophic effects of gastrointestinal hormones: physiological implications. *Fed. Proc.* 36, 1942-1947 (1977).

Eränkö, O., Lempinen, M., and Raisanen, L. Adrenaline and noradrenaline in the organ of Zuckerkandl and adrenals of newborn rats treated with hydrocortisone, *Acta Physiol. Scand.* 66, 253-254 (1966).

Espinosa, A., Driscoll, S.G., and Steinke, J. Insulin release from isolated human fetal pancreatic islets. *Science* 168, 1111-1112 (1970).

Faiman, C. and Winter, J.S.D. Gonadotropins and sex hormone patterns in puberty: clinical data, in *The Control of the Onset of Puberty.* M.M. Grumbach, G.D. Grave, and F.E. Mayer, eds. Wiley, New York (1974), pp. 32-61.

Falkmer, S., and Patent, G.J. Comparative and embryological aspects of the pancreatic islets, in *Handbook of Physiology, Section 7: Endocrinology, Vol. I. Endocrine Pancreas.* D.S. Steiner, and N. Frankel, eds. Am. Physiol. Soc. Washington, D.C. (1972), pp. 1-23.

Federman, D.D. Genetic control of sexual difference, in *Progress in Medical Genetics.* A.G. Steinberg, and A.G. Bearn, eds. Grune and Stratton, New York (1973), pp. 215-235.

Feldman, S.C. and Bloch, E. Developmental pattern of testosterone synthesis by fetal rat testes in response to luteinizing hormone. *Endocrinology* 102, 999-1007 (1978).

Ferguson-Smith, M.A. Chromosomal abnormalities II: sex chromosome defects, in *Medical Genetics.* V. McKusick, and R. Claiborne, eds. HP Publishing Co., New York (1973), pp. 16-26.

Fink, G. and Smith, G.C. Ultrastructural features of the developing hypothalamo-hypophysial axis in the rat, a correlative study. *Z. Zellforsch.* 119, 208-226 (1971).

Fisher, D.A., and Dussault, J.H. Development of the mammalian thyroid gland, in *Handbook of Physiology, Section 7: Endocrinology, Vol. III. Thyroid.* M.A. Greer and D.H. Solomon, eds. Am. Physiol. Soc., Washington, D.C. (1974), pp. 21-38.

Fisher, D.A., and Odell, W.D. Acute release of thyrotropin in the newborn. *J. Clin. Invest.* 48, 1670-1677 (1969).

Fisher, D.A., Lehman, H., and Lackey, C. Placental transport of thyroxine. *J. Clin. Endocr. Metab.* 24, 393-400 (1964).

Fisher, D.A., Dussault, J.H., Hobel, C.J., and Lam. R. Serum and thyroid gland triiodothyronine in the human fetus. *J. Clin. Endocr. Metab.* 36, 397-400 (1973).

Fisher, D.A., Hobel, C.J., Garza, R., and Pierce, C. Thyroid function in the pre-term fetus. *Pediatrics* 46, 208-216 (1970).

Fisher, D.A., Odell, W.D., Hobel, C.J., and Garza, R. Thyroid function in the term fetus. *Pediatrics* 44, 526-535 (1969).

Florsheim, W.J., and Rudko, P. The development of portal system in the rat, *Neuroendocrinology* 3, 89-98 (1968).

Fraser, F.C. Genetic counseling, in *Medical Genetics*. V.A. McKusick, and R. Claiborne, eds. HP Publishing Co., New York (1973), pp. 221-228.

Frohman, L.A., Burek, L., and Stachura, M.E. Characterization of growth hormone of different molecular weights in rat, dog and human pituitaries. *Endocrinology* 91, 262-269 (1972).

Garvey, D.J., Akana, S. Weisman, A., and Timiras, P.S. Alterations in adrenal growth and corticosteroid content in foetal and neonatal rats developing at high altitude. *J. Endocrinol.* 80, 333-342 (1979).

Garvey, D.J., Vaccari, A., and Timiras, P.S. Developmental profiles of catecholaminergic enzymes in adrenals of prenatal rats: Effect of a hypoxic environment. *Horm. Met. Res.* 12, 318-322 (1980).

Geel, S.E. Development-related changes of triiodothyronine binding to brain cytosol receptors. *Nature* 269, 428-430 (1977).

Geel, S.E. and Timiras, P.S. Influence of neonatal hypothyroidism and of thyroxine on the acetycholinesterase and cholinesterase activities in the developing central nervous system of the rat. *Endocrinology* 80, 1069-1074 (1967).

Genazzani, A.R., Fraioli, F., Hurlimann, J., Fioretti, P., and Felber, J.P. Immunoreactive ACTH and cortisol plasma levels during pregnancy. Detection the partial purification of corticotrophin-like placental hormone: the human chorionic corticotrophin (HCC). *Clin. Endocrinol.* (Oxford) 4, 1-14 (1975).

George, F.W., Catt, K.J., Neaves, W.B., and Wilson, J.D. Studies on the regulation of testosterone synthesis in the fetal rabbit testis. *Endocrinology* 102, 665-673 (1978).

Girard, J.R. and Zeghal, N. Adrenal catecholamines content in fetal and newborn rats. *Biol. Neonate* 26, 205-213 (1975).

Girard, J.R., Assan, R., and Marliss, E.B. Glucogon and perinatal metabolism in the rat in *Hormones and Embryonic Development (Advances in the Bio-Sciences 13)*. Permagon Press, New York (1974), pp. 5-16.

Glydon, R. St. J. The development of the blood supply of the pituitary in the albino rat, with special reference to the portal vessels. *J. Anat.* 91, 237-244 (1957).

Golander, A., Hurley, T., Barrett, J., Hizi, A., and Handwerger, S. Prolactin synthesis by human chorion-decidual tissue: A possible source of prolactin in the amniotic fluid. *Science* 202, 311-313 (1978).

Goldman, B.D., Quadagno, D.M., Shryne, J., and Gorski, R.A. Modification of phallus development and sexual behavior in rats treated with gonadotropin antiserum neonatally. *Endocrinology* 90, 1025-1031 (1972).

Goodner, C.J. and Freinkel, N. Carbohydrate metabolism in pregnancy. IV. Studies in permeability of the rat placenta to insulin. *Diabetes* 10, 383-392 (1961).

Gorden, P., Carpentier, J.L., Freychet, P., LeCam, A., and Orci, L. Intracellular translocation of iodine-125-labeled insulin: direct demonstration in isolated hepatocytes. *Science* 200, 782-785 (1978).

Greenberg, R.E. The physiology and metabolism of catecholamines, in *Endocrine and Genetic Diseases of Childhood and Adolescence*. L.I. Gardner, ed., second ed. W.B. Saunders Co., Philadelphia (1975) pp. 886-898.

Greep, R.O. *Reproductive Physiology*. MTP International Review of

Science, Series 1, Vol. 8. Butterworth and Co., London (1975).
Grumbach, M.M. and Kaplan, S.L. Ontogenesis of growth hormone, insulin, prolactin and gonadotropin secretion in the human foetus, in *Textbook of Endocrinology*. K.W. Cross and P.W. Nathanielsz, eds. Cambridge University Press, New York (1973), pp. 462-487.
Grumbach, M.M. and Van Wyk, J.J. Disorders of sex differentiation, in *Textbook of Endocrinology*. R.H. Williams, ed. W.B. Saunders Co., Philadlphia (1974), pp. 423-501.
Guillemin, R., Vargo, T., Rossier, J., Minick, S., Ling, N., Rivier, C., Vale, W., and Bloom, F. β-Endorphin and adrenocorticotropin are secreted concomitantly by the pituitary gland. *Science* 197, 1367-1369 (1977).
Habener, J.F. and Kronenberg, H.M. Parathyroid hormone biosynthesis: Structure and function of biosynthetic precursors. *Fed. Proc.* 37, 2561-2566 (1978).
Hamilton, W.J., Boyd, J.D., and Mossman, H.W. *Human Embryology: Prenatal Development of Form and Function*. Williams and Wilkins Co., Baltimore (1972).
Hartemann, P., Malaprade, D., Lemoine, D., Grignon, G., Nabel, P., and Pierson, M. Mise en evidence des elements morphologiques et des activities secretoires STH et LH dans ℓ-hypophyse de foetus humain au cours du development. *C.R. Soc. Biol.* 167, 105-110 (1972).
Hervonen, A. On the innervation and differentiation of human fetal chromaffin tissue, in *Histochemistry of Nervous Transmission, Progress in Brain Research*, Vol. 34. O. Eränkö, ed. Elsevier, Pub., New York (1971), pp. 445-454.
Hiroschige, T. and Sato, T. Changes in hypothalamic content of corticotropin-releasing activity following stress during neonatal maturation in the rat. *Neuroendocrinology* 7, 257-270 (1971).
Holvey, D.N., ed. *The Merck Manual of Diagnosis and Therapy*, Merck Sharp and Dohme Research Laboratories, Rahway, New Jersey (1976).
Honnebier, W.J., Jobsis, A.C., and Swaab, D.F. The effects of hypopseal hormones and human chorionic gonadotropin (HCG) on the anencephalic fetal adrenal cortex and on parturition in the human. *J. Obstet. Gynaecol. Br. Commonw.* 81, 423-438 (1974).
Hyyppa, M. Hypothalamic monoamines in human fetuses. *Neuroendocrinology* 9, 257-266 (1972).
Idelman, S. The structure of the mammalian adrenal cortex, in *General Comparative and Clinical Endocrinology of the Adrenal Cortex*. I. Chester Jones and I.W. Henderson, eds. Vol. 2. Academic Press, New York (1978), pp. 1-200.
Josso, N., Forest, M.G. and Picard, J. Muellerian-inhibiting activity of calf fetal testes: relationship to testosterone and protein synthesis. *Biol. Reprod.* 13, 163-167 (1975).
Jost, A. Problems of fetal endocrinology. The gonadal and hypophyseal hormones. *Recent Prog. Horm. Res.* 8, 379-418 (1953).
Jost, A. Anterior pituitary function in foetal life, in *The Pituitary Gland*, Vol. 2. G.W. Harris and B.T. Donovan, eds. University of California Press, Berkeley (1966), pp. 299-323.
Jost, A. Hormonal factors in the sex differentiation of the mammalian fetus. *Phil. Trans. R. Soc. Lond. (Biol.)* 259, 119-130 (1970).
Jost, A. The fetal adrenal cortex, in *Handbook of Physiology, Section 7: Endocrinology, Vol. VI. Adrenal Gland*. H. Blaschko, G. Sayers, and D.A. Smith, eds. Am. Physiol. Soc., Washington, D.C. (1975) pp. 107-115.
Jost, A. Sexual differentiation, in *Fetal Physiology and Medicine: The*

Basis of Perinatology. R.W. Beard, and P.W. Nathanielsz, eds. W.B. Saunders Co., Philadelphia, (1976), pp. 1-16.

Jost, A., Dupouy, J.P., and Rieutort, M. The ontogenetic development of hypothalamo-hypophyseal relations, in *Integrative Hypothalamic Activity*. D.F. Swaab, and J.P. Schade, eds. Progress in Brain Research, Vol. 41. Elsevier, Amsterdam (1974), pp. 209-219.

Judd, H.L., Hamilton, C.R., Barlow, J.J., Yen, S.S.C., and Kliman, B. Androgen and gonadotropin dynamics in testicular feminization syndrome. *J. Clin. Endocr.* 34, 229-234 (1972).

Kaplan, S.L., Grumbach, M.M., and Aubert, M.L. The ontogenesis of pituitary hormones and hypothalamic factors in the human fetus: maturation of central nervous system regulation of anterior pituitary function. *Recent Prog. Horm. Res.* 32, 161-233 (1976).

Kaplan, S.L., Grumbach, M.M., and Shepard, T.H. The ontogenesis of human fetal hormones, I. Growth Hormone and insulin. *J. Clin. Invest.* 51, 3080-3093 (1972).

Kennelly, J.J., and Foote, R.H. Oocytogenesis in rabbits. The role of neogenesis in the formation of the definitive ova and the stability of oocyte DNA measured with tritiated thymidine. *Am. J. Anat.* 118, 573-589 (1966).

Knobil, E. and Josimovich, J.B. Placental transfer of thyrotropic hormone, thyroxine, triiodothyronine and insulin in the rat. *Ann. N. Y. Acad. Sci.* 75, 895-904 (1958).

Lanman, J.T. An interpretation on foetal adrenal structure and function, in *The Human Adrenal Cortex*, Vol. 2. K.W. McKerns, ed. Appleton-Century-Crofts, New York (1968), pp. 547-558.

Larson, S. and Titus, J. Chromosomes and abortions. *Mayo Clin. Proc.* 45, 60-72 (1970).

Levina, S.E. Endocrine features in development of human hypothalamus, hypophysis and placenta. *Gen. Comp. Endocrinol.* 11, 151-159 (1968).

Levy, J. and Levy, J.M. Human lateralization from head to foot: sex-related factors. *Science* 200, 1291-1292 (1978).

Lipsett, M.B. Steroid secretion by the human testis, in *The Human Testis*. E. Rosenberg and C.A. Paulsen, eds. Plenum Press, New York (1970) pp. 407-421.

Lyon, M.F. Gene action in the X-chromosome of the mouse *(Mus musculus L.)*. *Nature* 190, 372-373 (1961).

Lyon, M.F. Possible mechanisms of X-chromosome inactivation. *Nature, New Biol.* 232, 229-232 (1971).

Lyon, M. and Hawkes, S. X-linked gene for testicular feminization in the mouse. *Nature* 227, 1217-1219 (1970).

Mains, R.E., Eipper, B.A., and Ling, N. Common precursor to corticotropins and endorphins. *Proc. Natl. Acad. Sci. USA* 74, 3014-3018 (1977).

Matschinsky, F.M., Ellerman, J.E., Krzanowski, J., Kotler-Brajtburg, J., Landgraf, R., and Fertel, R. The dual function of glucose in islets of Langerhans. *J. Biol. Chem.* 246, 1007-1011 (1971).

McIntosh, N., Pictet, R.L., Kaplan, S.L., and Grumbach, M.M. The developmental pattern of somatostatin in the embryonic and fetal rat pancreas. *Endocrinology* 101, 825-829 (1977).

Means, A.R., Fakunding, J.L., Huckins, C., Tindall, D.J., and Vitale, R. Follicle-stimulating hormone, the Sertoli cell and spermatogenesis. *Recent Prog. Horm. Res.* 32, 477-527 (1976).

Milkovic, K., Joffe, J., and Levine, S. The effect of maternal and fetal corticosteroids on the development and function of the pituitary-adrenal cortical system. *Endokrinologie* 68, 60-65 (1976).

Mirkin, B.L. Ontogenesis of the adrenergic nervous system: functional and pharmacologic implications. *Fed. Proc.* 31, 65-73 (1972).

Mizuno, M., Lobotsky, J., Lloyd, C.W., Kobayashi, T., and Murasawa, Y. Plasma androstenedione and testosterone during pregnancy and in the newborn. *J. Clin. Endocr.* 28, 1133-1142 (1968).

Mølsted-Pedersen, L. *Studies on Carbohydrate Metabolism in Newborn Infants of Diabetic Mothers.* Bogtrykkeriet Forum, Copenhagen (1974).

Morgan, H.E., Regen, D.M., and Park, C.R. Identification of a mobile carrier-mediated sugar transport system in muscle. *J. Biol. Chem.* 239, 369-374 (1964).

Mott, J.C. The place of the renin-angiotensin system before and after birth. *Br. Med. Bull.* 31, 44-49 (1975).

Naftalin, R.J. A model for sugar transport across red cell membrane without carriers. *Biochim. Biophys. Acta.* 211, 65-78 (1970).

Naidoo, S. and Timiras, P.S. Effects of age on the metabolism of thyroid hormones by rat brain tissue *in vitro*. *Dev. Neurosci.*, 2, 213-224 (1979).

Naidoo, S., Valcana, T., and Timiras, P.S. Thyroid hormone receptors in the developing rat brain. *Am. Zool.* 18, 545-552 (1978).

Nalbandov, A.V. *Reproductive Physiology of Mammals and Birds.* W.H. Freeman and Co., San Francisco (1976).

Narahara, H.T. Binding of insulin to tissues in relation to biological action of the hormone, in *Handbook of Physiology, Section 7: Endocrinology, Vol. I. Endocrine Pancreas.* D.S. Steiner and N. Frankel, eds. Am. Physiol. Soc., Washington, D.C. (1972), pp. 33-324.

Nathanielsz, P.W. Thyroid function in the fetus and newborn mammal. *Br. Med. Bull.* 31, 51-56 (1975).

Nathanielsz, P.W. *Fetal Endocrinology: An Experimental Approach.* Elsevier/North-Holland, Amsterdam (1976).

Negro-Vilar, A., Krulich, L., and McCann, S.N. Changes in serum prolactin and gonadotropins during sexual development of the male rat. *Endocrinology* 93, 660-664 (1973).

Nobin, A. and Bjorklund, A. Topography of the human brain as revealed in fetuses. *Acta Physiol. Scand.*, Suppl. 338, (1973), pp. 1-40.

Northcutt, R.C., Island, D.P., and Liddle, G.W. An explanation for the target organ unresponsiveness to testosterone in the testicular feminization syndrome. *J. Clin. Endocr.* 29, 442-425 (1969).

Odell, W.D. and Swerdloff, R.S. Etiologies of sexual maturation: a model system based on the sexually maturing rat. *Recent Prog. Horm. Res.* 32, 245-288 (1976).

Odell, W.D., Swerdloff, R.S., Jacobs, H.S., and Hescox, M.A. FSH induction of sensitivity of LH: one cause of sexual maturation in the male rat. *Endocrinology* 92; 160-165 (1973).

Odell, W.D., Swerdloff, R.S., Bain, J., Wollesen, F., and Grover, P.K. The effect of sexual maturation on testicular response to LH stimulation of testosterone secretion in the intact rat. *Endocrinology* 95, 1380-1384 (1974).

Ohno, S., and Smith, J.B. The role of fetal follicular cells in meiosis of mammalian oocytes. *Cytogenetics* 3, 324-333 (1964).

Oklund, S. and Timiras, P.S. Influences of thyroid levels in brain ontogenesis *in vivo* and *in vitro*, in *Thyroxine and Brain Development.* G.D. Grave, ed. Raven Press, New York (1977), pp. 33-47.

Otten, U., Hatanaka, H., and Thoenen, H. Role of cyclic nucleotides in NGF-mediated induction of tyrosine-hydroxylase in rat sympathetic ganglia and adrenal medulla. *Brain Res.* 140, 385-389 (1978).

Pavlova, E.B., Pronina, T.S., and Skebelskaya, Y.B. Histo-structure of adenohypophysis of human fetuses and contents of somatotropic and adrenocorticotropic hormones. *Gen. Comp. Endocrinol.* 10, 269-276 (1968).

Payne, A.H. and Jaffe, R.B. Androgen formation from pregnenolone sulfate in fetal, neonatal, prepubertal and adult human testes. *J. Clin. Endocr. Metab.* 40, 102-107 (1975).

Pearse, A.G.E. The cytochemistry and ultrastructure of polypeptide hormone-producing cells of the apud series and the embryologic, physiologic and pathologic implication of the concept. *J. Histochem. Cytochem.* 17, 303-313 (1969).

Peluso, J.J., Steger, R.W., and Hafez, E.S.E. Development of gonadotropin-binding sites in the immature rat ovary. *J. Reprod. Fertil.* 47, 55-58 (1976).

Peters, H. Migration of gonocytes into the mammalian gonad and their differentiation. *Phil. Trans. R. Soc. London (Biol.)* 259, 91-101 (1970).

Petropoulos, E.A. and Timiras, P.S. Effects of hypoxic environment on prenatal brain development: Recent evidence versus earlier dogma, in *Drugs and the Developing Brain.* A. Vernadakis and N. Weiner, eds. Plenum Press, New York (1974), pp. 429-449.

Philpott, J.E., Zarrow, M.X., and Denenberg, V.H. The presence of the adult-type pattern of adrenal steroids in the one day old rat. *Steroids* 14, 21-31 (1969).

Pictet, R. and Rutter, W.J. Development of embryonic endocrine pancreas, in *Handbook of Physiology, Section 7: Endocrinology, Vol. I. Endocrine Pancreas.* D.S. Steiner and N. Frankel, eds. Am. Physiol. Soc., Washington, D.C. (1972), pp. 25-66.

Reddy, V.V.R., Naftolin, F., and Ryan, K.J. Conversion of androstenedione to estrone by neural tissues from fetal and neonatal rats. *Endocrinology* 94, 117-121 (1974).

Rees, L.H., Burke, C.W., Chard, T., Evans, S.W., and Letchwarth, A.T. Possible placental origin of ACTH in normal human pregnancy. *Nature* 254, 620-622 (1975).

Reichlin, S., Saperstein, R., Jackson, I.M.D., Boyd, A.E. III, and Patel, Y. Hypothalamic hormones, in *Annual Review of Physiology.* E. Knobil ed., R.R. Sonnenschein and I.S., Edleman, eds. Annual Reviews, Inc., Palo Alto, California (1976), pp. 389-424.

Resko, J.A. Androgen secretion by the fetal and neonatal Rhesus monkey. *Endocrinology* 87, 680-687 (1970).

Resko, J.A., Feder, H.H., and Goy, R.W. Androgen concentrations in plasma and testis of developing rats. *J. Endocrinol.* 40, 485-491 (1968).

Richards, J. and Williams, J.J. Luteal cell receptor content for prolactin and luteinizing hormone: Regulation by LH and PRL. *Endocrinology* 99, 1571-1581 (1976).

Rinne, U.K., Kivalo, E., and Talanti, S. Maturation of Human Hypothalamic Neurosecretion. *Biol. Neonate* 4, 351-364 (1962).

Rivarola, M.A., Forest, M.G., and Migeon, C.J. Testosterone, androstenedione and dehydroepiandrosterone in plasma during pregnancy and at delivery: concentration and protein binding. *J. Clin. Endocrinol.* 28, 34-40 (1968).

Robyn, C., Delvoye, P., Nokin, J., Vekemans, M., Badawi, M., Perez-Lopez, F.R., and L'Hermite, M. Prolactin and human reproduction, in *Human Prolactin.* J.L. Pasteels and C. Robyn, eds., F.J.G. Ebling, co-ed. Excerpta Medical Amsterdam, American Elsevier, Publ., Co., New York (1973), pp. 167-187.

Roosen-Runge, E.C. and Anderson, D. The development of the interstitial cells in the testis of the albino rat. *Acta Anat.* 37, 125-129 (1959).

Ross, M.H. Electron microscopy of the human foetal adrenal cortex, in *The Human Adrenal Cortex.* A.R. Currie, T. Symington, and J.K. Grant, eds. E. and S. Livingston, LTD., London (1962), pp. 558-568.

Roth, S.I. Recent advances in parathyroid gland pathology. *Am. J. Med.* 50, 612-622 (1971).

Rutter, W.J., Pictet, R.L., Harding, J.D., Chirgwin, J.M., MacDonald, R.J. and Przybyla, A.E. An analysis of pancreatic development; role of mesenchymal factor and other extracellular factors, in *Molecular Control of Proliferation and Differentiation.* J. Papaconstantinou and W.J., Rutter, eds. Academic Press, New York (1978), pp. 205-227.

Samuels, H.H., Stanley, F., and Shapiro, L.E. Dose dependent depletion of nuclear receptors by L-triiodothyronine: evidence for a role in induction of growth hormone synthesis in cultured GH cells. *Proc. Natl. Acad. Sci. USA* 73, 3877-3881 (1976).

Schwartz, H.L. and Oppenheimer, J.H. Ontogenesis of 3,4,3'-triiodothyronine receptors in neonatal rat brain: dissociation between receptor concentration and stimulation of oxygen consumption by 3,4,3'-triiodothyronine, *Endocrinology* 103, 943-948 (1978).

Serra, G.B., Perez-Palacios, G., and Jaffe, R.B. De Novo testosterone biosynthesis in the human fetal testis. *J. Clin. Endocr. Metab.* 30, 128-130 (1970).

Shepard, T.H. Development of the human fetal thyroid. *Gen. Comp. Endocr.* 10, 174-181 (1968).

Sherman, B.M. and Korenman, S.G. Hormonal characteristics of the human menstrual cycle throughout reproductive life. *J. Clin. Invest.* 55, 699-706 (1975).

Short, R.V. Sex determination and differentiation, in *Reproduction in Mammals: Embryonic and Fetal Development.* C.R. Austin and R.V. Short eds. University Press, Cambridge (1972), pp. 43-71.

Siebers, J.W., Peters, F., Zenes, M.Y., Schmidtke, J., and Engel, W. Binding of human chorionic gonadotrophin to rat ovary during development. *J. Endocrinol.* 73, 491-496 (1977).

Siiteri, P.K. and Wilson, J.D. Testosterone formation and metabolism during male sexual differentiation in the human embryo. *J. Clin. Endocr. Metab.* 38, 113-125 (1974).

Slotkin, T.A. Maturation of the adrenal medulla-III. Practical and theoretical considerations of age-dependent alterations in kinetics of incorporation of catecholamines and non-catecholamines. *Biochem. Pharmac.* 24, 89-97 (1975).

Smith, F.G. Jr., Alexander, D.P., Buckle, R.M., Britton, H.G., and Nixon, D.A. Parathyroid hormone in foetal and adult sheep: The effect of hypocalcaemia. *J. Endocrinol.* 53, 339-348 (1972).

Snell, R.S. *Clinical Anatomy for Medical Students,* Little Brown and Co., Boston (1973).

Stanbury, J.B. and Kroc, R.L., eds. *Human Development in the Thyroid Gland: Relation to Endemic Cretinism.* Plenus Press, New York (1972).

Sterling, K., Lazarus, J.H., Milch, P.O., Sakurada, T., and Brenner, M.A. Mitochondrial thyroid hormone receptor localization and physiological significance. *Science* 201, 1126-1128 (1978).

Strickland, A.L., and French, F.S. Absence of response to dihydrotestosterone in the syndrome of testicular feminization. *J. Clin. Endocr.* 29, 1284-1286 (1969).

Surks, M.I. and Oppenheimer, J.H. Isolation and characterization of thyroid hormone receptors, in *Hormone-Receptor Interaction.* G.S. Levy, ed. Marcel Dekker, Inc., New York (1976), pp. 373-384.

Swaab, D.F. and Honnebier, W.J. The role of the fetal hypothalamus in development of the feto-placental unit and in parturition, in

Integrative Hypothalamic Activity. D.F. Swaab and J.P., Schade, eds. Progress in Brain Research, Vol. 41. Elsevier, Amsterdam (1974), pp. 255-280.

Swaab, D.F., Visser, M., and Tilders, F.J.H. Stimulation of intra-uterine growth in rat by α-melanocyte-stimulating hormone. *J. Endocrinol.* 70, 445-455 (1976).

Timiras, P.S. *Developmental Physiology and Aging.* Macmillan Company. New York (1972).

Timiras, P.S. Biological Perspectives on aging: In search of a masterplan. *Am. Sci.* 66, 605-613 (1978).

Tullner, W.W., and Hertz, R. Chorionic gonadotropin levels in the Rhesus monkey during early pregnancy. *Endocrinology* 78, 204-207, (1966).

Vaccari, A., Brotman, S., Cimino, J., and Timiras, P.S. Sex differences of neurotransmitter enzymes in central and peripheral nervous system. *Brain Res.* 132, 176-780 (1977a).

Vaccari, A., Valcana, T., and Timiras, P.S. Effects of hypothyroidism on the enzymes for biogenic amines in the developing rat brain. *Pharmacol. Res. Comm.* 9, 763-780 (1977b).

Vaccari, A., Cimino, J., Brotman, S., and Timiras, P.S. High altitude hypoxia and adrenal development in the rat: enzymes for biogenic amines, in *Environmental Endocrinology*. I. Assenmacher and D.S., Farner, eds. Springer-Verlag, Berlin (1978), pp. 283-289.

Valcana, T. Development of kidney function, in *Developmental Physiology and Aging.* Timiras, P.S., Macmillan, New York (1972), pp. 246-272.

Valcana, T. Developmental changes in ionic composition of the brain in hypo- and hyperthyroidism, in *Drugs and the Developing Brain*. A. Vernadakis and N. Weiner, eds. Plenum Press, New York (1974), pp. 289-304.

Valcana, T. The role of triiodothyronine (T_3) receptors in brain development, in *Neural Growth and Development*. E. Meisami and M., Brazier, eds. Raven Press, New York (1979) pp. 39-57.

Valcana, T., Geel, S.E., Miller, C., and Timiras, P.S. T_4 and T_3 conversion and relative binding to nuclear receptors in developing brain. *Fed. Proc.* 38 (ii), 1029 (1979).

Valcana, T. and Timiras, P.S. Nuclear triiodothyronine receptors in the developing rat brain. *Mol. Cell. Endocrinol.* 11, 31-41 (1978).

Valcana, T. and Timiras, P.S. Changes in rat liver nuclear triiodothyronine receptors with age and thyroid activity, in *Hormones and Development*. L. Macho and V. Strback, eds. Solvak Academy of Sciences, Bratislava, CSSR (1979) pp. 47-76.

Varma, S.K., Murray, R., and Stanburn, J.B. Effect of maternal hypothroidism and triiodothyronine on the fetus and newborn in rats. *Endocrinology* 102, 24-30 (1978).

Villee, D.B. Development of endocrine function in the human placenta and fetus. *New Engl. J. Med.* 281, 473-484 (1969).

Vinson, G.P. and Kenyon, C.J. Steroidogenesis in the zones of the mammaliam adrenal cortex, in *General, Comparative and Clinical Endocrinology of the Adrenal Cortex*. I.C. Jones and I.W., Henderson, eds. Vol. 2 Academic Press, New York (1978), pp. 201-264.

Warren, D.W., Haltmeyer, G.C., and Eik-Nes, K.B. Testosterone in the fetal rat testis. *Biol. Reprod.* 8, 560-564 (1973).

Wessells, N.K. *Tissue Interactions and Development.* W.A. Benjamin, Inc., Menlo Park, California (1977), pp. 104-121.

Winchester, A.M. *Genetics: A Survey of the Principles of Heredity.* Houghton Mifflin Co., Boston (1977).

Wurtman, R.J. and Axelrod, J. Control of enzymatic synthesis of adrenaline in the adrenal medulla by adrenal cortical steroids. *J. Biol. Chem.* 241, 2301-2305 (1966).

Yamazaki, E., Noguchi, A., and Slingerland, D.W. The development of hormonal biosynthesis in human fetal thyroids. *J. Clin. Endocr. Metab.* 19, 1437-1439 (1959).

Zamenhof, S., Mosley, J., and Schuller, E. Stimulation of the proliferation of cortical neurons by prenatal treatment with growth hormone. Science 152, 1396-1397 (1966).

Zierler, K.L. Insulin, ions, and membrane potentials, in *Handbook of Physiology, Section 7: Endocrinology, Vol. I. Endocrine Pancreas.* D.S. Steiner and N. Frankel, eds. Am. Physiol. Soc., Washington, D.C. (1972), pp. 347-368.

Zimmerman, E. (ed.) Peptides of the brain and gut, *Fed. Proc.* 38, 2286-2354 (1979).

Copyright©1982, Spectrum Publications, Inc.
Hormones in Development and Aging

Chapter 3
Placental Hormones in Development
Evangelos A. Petropoulos

INTRODUCTION

In the course of evolution various forms of placentation and viviparity have appeared in lower animals. As the eons went by, structural and functional changes were gradually wrought upon these early prototype placentas, culminating eventually in the intricate placenta of higher mammals. The mammalian placenta, in providing the requisite interphase between fetus and mother, must of necessity be a highly developed but constantly changing, pluripotent and yet ephemeral, organ. The placental functions are many and diverse including the secretion of various hormones necessary for proper fetal development (Thomsen and Hiersche, 1969). This latter aspect of placental physiology will be discussed here; the other placental functions have been reviewed elsewhere (Ginsburg and Jeacock, 1964; Laga et al., 1974; Mossman, 1965; Szabo and Grimaldi, 1970; Timiras, 1972; Villee, 1967; Winick, 1968; Wynn, 1968).

The placenta was not always thought of as an endocrine gland. Bouchacourt (1902) was the first to suggest that a "placental secretion" (the term "hormone" had yet to be coined by Starling in 1905) may be responsible for the "witch's milk," a mammary discharge seen in human newborns of either sex. Subsequent investigations by such pioneers as Ascheim, Aschner, Fellner, Halban, Phillip, Starling, Vos, and Zondek provided experimental evidence that the placenta was indeed an endocrine gland. An excellent historic account of these investigations may be found in the article by Simmer (1968).

During these early experiments investigators soon realized that the classical endocrinological experiment could not be applied to the investigation of the endocrine function of the placenta. The placenta is unique, in that it is located within one of its target organs, the uterus, and is inseparably associated with its other functional target, and partner, the fetus. In the classical endocrinological experiment the "gland" under investigation is removed and then transplanted back or, instead, its hormones are injected into the animal, and any relevant changes manifested at any of these stages are recorded. It is evident that such a procedure is not feasible in the case of the placenta. Removal of the placenta inevitably leads to fetal death and hence elimination of one target organ on which post-endocrinectomy (placentectomy) responses would have been recorded. Further, grafting back the placenta into the uterus does not re-establish fully the original functional arrangements, in that the fetus, supplier of precursors for estrogen synthesis by the placenta, would not be present as well.

In spite of these difficulties, an impressive number of elegant experiments accumulated over the years to unequivocally demonstrate that the placenta is indeed a multipotentially differentiated endocrine gland. Placental hormones are secreted in huge quantities, as compared to those of their counterparts in the nonpregnant female; they constitute a wide variety of chemical species, such as oligopeptides, polypeptides, proteins, glycoproteins and steroids; finally, some of them are novel molecules to the maternal organism. Such functional versatility and prolificacy are not encountered in any other endocrine gland in the body. Thus, after the seemingly innocuous act of fertilization and implantation, the maternal organism is gradually flooded with huge quantities of old and new hormonal species. Such a hormonal flood would lead to a state of disease in a nonpregnant female; that this is not the case in the pregnant animal shows that certain safeguard mechanisms must be operating during pregnancy. In this context it should be noted that the secretion of some maternal hormones is almost completely shut off, the secretory rate of others is radically changed, and the secretion of novel maternal hormones during pregnancy has been proposed (Lajos et al., 1959; Lajos et al., 1976).

These data clearly indicate that new negative/positive feedback interrelations are established, and new regulators are brought into action during pregnancy to orchestrate hormone secretion in the maternal, placental and fetal compartments. As a result of this orchestration the prolific and novel hormonal secretion of the placenta is contained in amount and modulated in time, according to the developmental needs of the fetus and the metabolic and hormonal state of the mother. Thus, although pregnancy is a stress and a challenge to the maternal organism, these regulatory mechanisms ensure that the health of the mother is maintained, that the normal development of the fetus is secured, and that parturition is triggered timely and accomplished successfully. Although direct experimental evidence for these hypothetically inferred feedback mechanisms is scarce, such evidence is steadily accumulating and is reviewed later in this chapter (see the section entitled "Interactions and Integration of the Endocrine Placenta with the Fetomaternal Endocrine Continuum").

PLACENTAL HORMONES

The human placenta is capable of synthesizing a variety of hormones ranging from oligopeptides (hCTRH, hCGnRH), polypeptides (β-endorphin, reportedly: International Symposium, 1979) and proteins/glycoproteins (hCS, hCG, hCT), to steroids (estrogens and progesterone). Claims have been presented for placental secretion of other hormones, such as androgens, corticosteroids, relaxin, oxytocin, insulin, prolactin, ACTH, MSH and renin. These claims were based mainly on the isolation of these hormones from placental extracts, or on their localization in the trophoblast by means of immunomicroscopy; however, presence of these hormones in the placenta does not imply placental synthesis and the evidence for the latter remains inconclusive (see Chatterjee and Munro, 1977; Frame et al., 1979; Healy et al., 1979; Rees et al., 1975; Saxena, 1971; Simmer, 1968).

Human Chorionic Gonadotropin (hCG)

Secretion of this glycoprotein hormone is initiated very early in pregnancy by the implanting blastocyst. If a specific radioimmunoassay (RIA) is used, hCG can be detected in peripheral blood samples as early as nine days after fertilization. This specific RIA was introduced by Vaitukaitis et al. (1972) and employs an anti-hCGβ serum with ^{125}I-hCG as a label. With regard to the placental site of hCG production, existing evidence from electronmicroscope and immunofluorescence studies points, though inconclusively, to the syncytiotrophoblast (Simmer, 1968). The physicochemical and biological properties of hCG are summarized in Table 1. The molecule of hCG can be dissociated into two nonidentical, biologically inactive subunits, a smaller alpha and a longer beta, by exposure to 8M urea (Swaminathan and Bahl, 1970). The primary structure of hCGα would have been almost identical to those of the hypophysial glycoprotein hormones (hLH, hFSH and hTSH), except for its extra three amino acids (Ala-Pro-Asx) at the N-terminus, the inversion of amino acid sequences at positions 84 and 85, the different sites of carbohydrate chain attachment on the peptide chain, and the different carbohydrate sequences in the carbohydrate units. In spite of these differences, α subunits of hypophysial glycoprotein hormones can fully substitute for hCGα in recombination experiments with hCGβ, and the resultant molecule is fully active biologically. Such recombination experiments helped establish that full biological activity is manifest when the alpha and beta subunits are combined and that the β-subunit confers specificity of hormonal action. On the other hand, removal of the sialic acid moiety from hCG by neuraminidase causes almost total loss of biologic activity, without affecting immunologic activity (Hall et al., 1971). hCGβ shows considerably less homology when compared to hLHβ (and the β-subunits of the other glycoprotein hormones): hCGβ is longer by 32 amino acids, and is different in 18 other amino acid residues, in the number of carbohydrate units attached and, possibly, in the carbohydrate sequence of these units. The full amino-acid sequences of hCGα and hCGβ can be found in Frieden's book (1976).

The immunologic properties of hCG are similar to the other human glycoprotein hormones. Antibodies to the hCGα subunit crossreact fully with the α subunits, as well as with the complete molecules, of the other human glycoprotein hormones (Vaitukaitis and Ross, 1972). Immunologic specificity rests with the β subunit, and by raising antibodies against hCGβ it became possible to develop a specific RIA, which measures primarily hCG in biological samples containing this and other glycoprotein hormones, especially hLH. Recently, an ultraspecific RIA for hCG has been developed by raising and using an antibody against the unique C-terminal peptide (residues 123-145) of hCGβ (Chen et al., 1976). A substance displaying hCG antigenic, but not biologic, activity has been demonstrated in liver and colon extracts from three male patients, who did not suffer from neoplastic disease (Yoshimoto et al., 1977); the assays used in this study were hCGβ RIA and hCG radioreceptor assay. An hCG-like substance was also found in pituitary and urinary extracts from nonpregnant women (Chen et al., 1976) and in the serum of women using IUDs for contraception, although in the latter the finding is thought to indicate the existence of occult pregnancies (Seppala et al., 1978). Immunologic similarities have been observed among CGs from various primate species. Chen and Hodgen (1976), using hCGβ and hCGβ-C-terminal RIAs, showed that hCG is antigenicly very similar to chimpanzee-CG and gorilla-CG, but dissimilar to baboon-CG, macaque-CG, and marmoset-CG.

The biologic actions of hCG are generally regarded as similar to those of hLH and, in this context, it is interesting to note that ovarian and testicular cellular receptors for hLH show binding affinity to hCG as well (Catt et al., 1971). Yet, this is a rather simplistic assertion in view of the complex and diverse roles of hLH now recognized, and the fact that the pharmacologic actions of hCG have been investigated mostly in rodents, men and nonpregnant women, whereas very little is known about its precise physiologic role in pregnancy (Waitukaitis, 1977). Existing evidence is

TABLE 1. SOME PHYSICOCHEMICAL AND BIOLOGICAL PROPERTIES OF PLACENTAL PEPTIDE HORMONES

Property	Human Chorionic Gonadotropin	References	Human Chorionic Somatomammotropin	References	Human Chorionic Thyrotropin	References
Molecular weight, daltons	about 35,000	Frieden, 1976; Vaitukaitis, 1977	about 22,000	Josimovich, 1977	about 28,000	Hennen & Freychet, 1974
Number of peptide chains	two: alpha & beta	Frieden, 1976; Vaitukaitis, 1977	one	Josimovich, 1977	?	
Number of aminoacid residues	α 92 (MW 15,000) β 147 (MW 25,000)	Frieden, 1976; Vaitukaitis, 1977	190	Josimovich, 1977	?	
Carbohydrate content	25%	Frieden, 1976; Vaitukaitis, 1977	none	Josimovich, 1977	most likely a glycoprotein hormone	
Chemical similarity with	hLH, hFSH, hTSH	Frieden, 1976; Vaitukaitis, 1977	hGH, hPRL	Handwerger and Sherwood, 1974; Shome & Parlow, 1977	hTSH, bTSH	Hennen & Freychet, 1974
Immunologic similarity with	mostly hLH	Vaitukaitis, 1977	hGH	Spellacy, 1977; Josimovich, 1974	bTSH	Hennen & Freychet, 1974
Crossreactivity with antisera for other hormones	mostly anti-hLH also anti-hFSHα anti-hTSHα	Vaitukaitis, 1977	anti-hGH	Spellacy, 1977; Josimovich, 1974	bTSH, pTSH	Hennen & Freychet, 1974
Site of production	Syncytiotrophoblast	Simmer, 1968	Syncytiotrophoblast	Simmer, 1968	?	

Property	Human Chorionic Gonadotropin	References	Human Chorionic Somatomammotropin	References	Human Chorionic Thyrotropin	References
Earliest appearance in plasma	9th day after presumed ovulation & fertilization	Vaitukaitis, 1977	3-4 weeks after presumed ovulation & fertilization	Josimovich and Venning, 1975	?	
Biological activity of highly purified preparations	12,000-20,000 IU/mg	Frieden, 1976	—		350 mIU/mg	Hennen & Freychet, 1974
Bioassays used	in vivo or in vitro	Mitchell and Bagshawe, 1976	in vivo or in vitro	Josimovich, 1977; Josimovich, 1974	in vivo	Hennen & Freychet, 1974
Specific Radio-immunoassays	with anti-hCGβ serum anti-hCGβ C-terminal peptide serum	Vaitukaitis et al., 1972 Chen et al., 1976	highly specific anti-hCS sera	Spellacy, 1977		

discussed below in the section on "Effects of Placental Hormones on the Mother"; however, it should be stated here that highly purified hCG preparations display intrinsic FSH (Albert, 1969) and TSH (Nisula et al., 1974) activities. It has been shown further that hCG binds to the TSH receptor of the thyroid follicular cell membrane (Azukizawa et al., 1977), that it can displace TSH from that receptor (Silverberg et al., 1978) and that its thyrotropic activity is 1/4000 of that of hypophysial TSH (Kenimer et al., 1975).

The actions of hCG at the cellular level (ovarian tissue) are brought about by activation of adenyl cyclase and formation of cAMP followed by activation of cellular protein kinases and phosphorylation of chromosomal proteins (Williams et al., 1979; see also Chatterjee and Munro, 1977). As will be discussed later under "Intraplacental Interrelations," hCG stimulates cAMP formation also in the placenta.

Human Chorionic Somatomammotropin (hCS)

This placental hormone is a simple protein synthesized, most likely, by the syncytiotrophoblast. It is a single chain of 190 amino-acid residues and its primary strucutre (Handwerger and Sherwood, 1974) shows only 13% sequence identity with hPRL (Shome and Parlow, 1977), but a high degree of homology with hGH; 163 of the 190 residues are identical between hCS and hGH (Handwerger and Sherwood, 1974) and many of the others are "highly acceptable" or "acceptable" substitutes in terms of modern genetic coding theory (Bewley and Li, 1974). hCS may assume monomeric or dimeric forms depending on the concentration and the pH of the solution (Josimovich and Venning, 1975). In dilute concentrations and alkaline pH hCS appears to exist as a dimer, whereas in acidic solutions it assumes the monomeric form. Further, the somatotropic effects of hCS are manifest only when acidic solutions are injected into bioassay animals (Li, 1970). It is thought that circulating hCS is in the monomeric form, although existing evidence indicates that initially hCS is synthesized, and perhaps released into the circulation, as a larger molecule, known as big hCS (Schneider et al., 1975). The physicochemical and biological properties of hCS are summarized in Table 1.

The immunological properties of hCS are very similar to those of hGH and as a result crossreaction between the two hormones is manifest during radio- and immunoassays. However, in practice this crossreaction does not present a problem in hCS measurements during pregnancy; on the one hand, the use of highly purified materials allows for a high degree of specificity in hCS RIAs (Spellacy, 1977), and on the other, hCS circulating levels are very high from early pregnancy, and, therefore, present little problem in terms of sensitivity of the RIAs used. Vinik et al., (1973) reported that monkey placentae secrete chorionic somatomammotropins and that the macaque-CS has partial crossreactivity with hCS.

Human Chorionic Thyrotropin (hCT)

Twenty-five years ago, Akasu et al., (1955) reported that the human placenta contained a thyrotropin-like material. Additional evidence was reported later by Hennen (1965) and Hershman and Starnes (1969). In accordance with nomenclature adopted for other placental peptide hormones this material was named hCT. Extraction, purification and preliminary characterization of hCT were further pursued in subsequent years by both Hennen's and Hershman's groups. The physicochemical and biological properties of hCT are summarized in Table 1. Although direct proof of hCT secretion by the placenta is still missing, inferential evidence supports the chorionic origin of this hormone; not only its placental levels are much higher than those expected for hypophysial TSH trapped in the placenta, but also its cross-reaction with antisera against hypophysial hTSH is minimal. In contrast, hCT cross-reacts strongly with antibovine-TSH sera and its biological activity is neutralized if incubated with such antisera, but not with anti-hTSH sera. The cross-reaction between hCT and anti-bTSH sera is not inhibited by hCG. Recent studies dissociated hCT from the high thyrotropic activity observed in blood of women with hydatidiform mole; it is now known that this activity resides with hCG (Kenimer et al., 1975; Nisula et al., 1974). hCT produces responses similar to those of hTSH in the thyroid ^{131}I-uptake and the chick thyroid ^{32}P-uptake bioassay systems. Yet, its physiologic role on the maternal and/or fetal organism remains entirely conjectural. In a recent review by Hennen and Freychet (1974) more details and pertinent references can be found; on the other hand, a more recent report from Hershman's group (Harada and Hershman, 1978) casts doubt on their earlier observations on this hormone.

Human Chorionic Thyrotropin Releasing Hormone (hCTRH) and
Gonadotropin Releasing Hormone (hCGnRH)

Gibbons et al. (1975) reported that human placental extracts contain substances with the same elution volume as hypothalamic TRH and GnRH in carboxylmethylcellulose ion exchange columns; that placental homogenates display similar activity to that of hypothalamic TRH and GnRH when bioassayed in rats; and that placental tissue is capable of synthesizing *in vitro* biologically active TRH and GnRH, if incubated with suitable radiolabeled aminoacid precursors. Similar findings were reported by Khodr and Siler-Khodr (1980) with regard to hCGnRH. Regarding hCTRH, Shambaugh et al. (1979) have confirmed that peptides with TRH bioactivity may be synthesized de novo by the placenta, and that such peptides have immunologic and chromatographic properties identical to, but lower bioactivity than, the synthetic TRH. Although the physiologic function of these peptides during pregnancy is unknown, it has been suggested that hCTRH may be implicated in fetal thyroid regulation (Shambaugh et al., 1979), and hCGnRH in hCG secretion by the placenta (pp. 106-108); maternal prolactin secretion may also be influenced by hCTRH (pp. 95-98).

Steroid Hormones

That the human placenta elaborates and secretes progestagens (progesterone) and estrogens (estradiol, estrone and estriol) is fully established (see Josimovich and Venning, 1975; Simmer, 1968). With regard to progesterone, the placenta is able to synthesize this hormone de novo from cholesterol, in yields as high as 31% (Morrison et al., 1965). Moreover, the placenta can synthesize cholesterol from acetate, but in very low yields (Zelewski and Villee, 1966), insufficient to secure the large amounts required for placental progesterone production (Telegdy et al., 1970b). Existing evidence indicates that the placenta synthesizes progesterone predominantly from cholesterol, and to a certain extent from pregnenolone, supplied primarily by the mother, but also by the fetus; in turn, placental progesterone is secreted into both, maternal and fetal compartments (see Klopper and Fuchs, 1977). With regard to estrogens, the placenta alone cannot convert cholesterol (or pregnenolone) to estradiol, estrone or estriol, because it lacks enzymes necessary at certain steps of the biosynthetic sequence from cholesterol to estrogens. Thus, the placenta is able to synthesize estrogens only when supplied by the mother and the fetus with suitable precursors (androgens) at the penultimate step of the estrogen biosynthetic sequence. This functional arrangement is detailed below in the discussion of the fetoplacental unit. Histochemical evidence indicates that the site of placental steroid hormone production is the syncytiotrophoblast.

The Concept of Fetoplacental Unit

Diczfalusi (1962) first introduced the term "fetoplacental unit" to indicate that the fetus and the placenta complement each other functionally in the production of steroid hormones during pregnancy. The term is now largely accepted, but this fetoplacental collaboration applies only to steroid synthesis and, moreover, the mother is also a significant contributor to it. The functional interrelations among mother, placenta and fetus, with regard to steroid hormone biosynthesis are shown in Fig. 1. The data used in constructing this chart are reviewed by Beling (1977), Diczfalusi (1969), Klopper and Fuchs (1977) and Tulchinsky (1976). These data were derived mainly from perfusion studies during legal abortions, but also from *in vitro* experiments involving incubation of placental or fetal explants. However, results from these two experimental approaches are sometimes at variance with each other.

Evidently, from Fig. 1, the placenta and to a certain extent the fetus depend on the mother for their supply of basic precursors for steroid biosynthesis. The placenta is then able to provide independently the large amounts of progesterone circulating during pregnancy and thus to supplant the maternal ovaries in this function; yet the placenta cannot proceed beyond progesterone without the help of the fetus and the mother. On the fetal site, initial evidence had indicated that in the early fetal adrenal cortex all biosynthetic steps from acetate to progesterone are operating at low yields (Diczfalusi, 1969). Consequently, the fetal adrenals would depend on placental progesterone and pregnenolone for their synthesis of gluco-/mineralo-corticoids and androgens, respectively. In return, the fetal adrenals would supply the placenta with androgen precursors suitable for placental estrogen production. These considerations constituted the basis on which the concept of fetoplacental unit was formulated. However, other reports have indicated that the efficiency of the fetal biosynthetic steps from acetate to progesterone is higher than previously thought (Telegdy et al., 1970a), that pregnenolone levels are higher in the umbilical arteries than in the unbilical veins, that the placenta appears to extract considerable amounts of pregnenolone sulfate (Scommegna et al., 1972) and 17αOH-pregnenolone (Belisle et al., 1978) from the umbilical arteries, and that human adrenals in organ culture synthesize appreciable amounts of pregnenolone de novo from cholesterol (Simpson et al., 1979).

It is evident from Fig. 1 why measurements of estriol, and more recently estetrol (see Tulchinski, 1976), in maternal plasma and/or urine, have gained such importance in the evaluation of fetal development and wellbeing. Ninety percent of placental estriol is derived from fetal precursors synthesized in the fetal adrenals and liver, whereas estetrol itself is synthesized in toto by the fetal liver. Accordingly, their levels in the maternal circulation reflect the functional state of fetal adrenals and liver, as well as that of the placenta, although the latter can be assessed independently with measurements of placental peptide hormones.

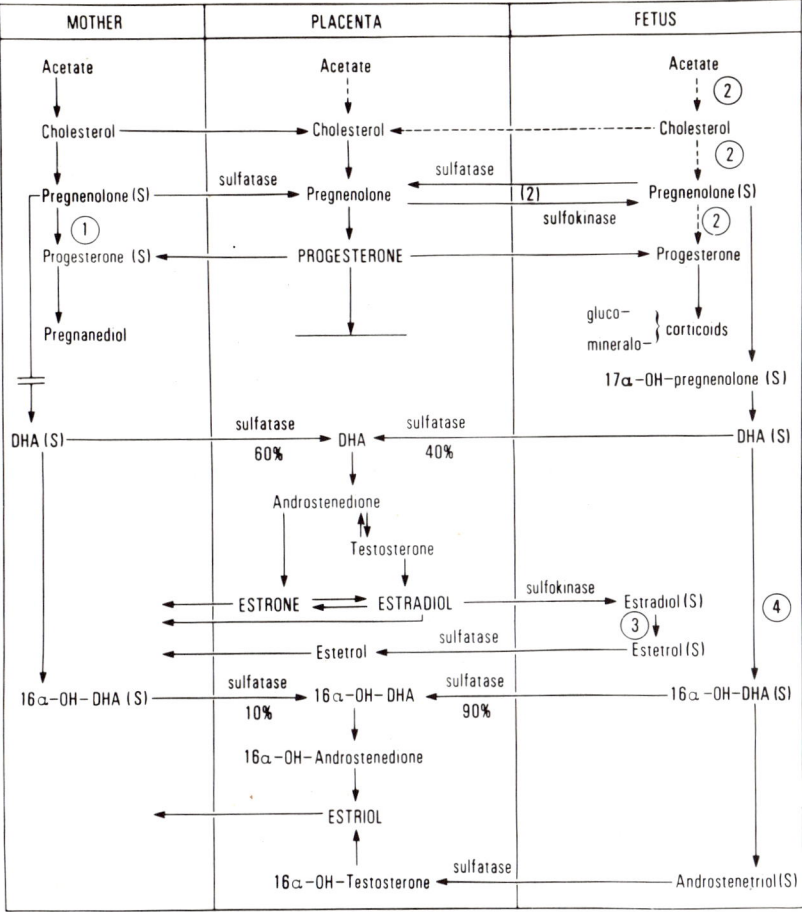

Fig. 1. Steroid-Hormone Interrelations in the Materno-Fetoplacental Unit.

Key: ⟶ High yield pathway or transport; --> Low yield pathway or transport; ↓ Enzymatic block; 1. Although this is normally a high yield pathway, the maternal contribution to total circulating progesterone during pregnancy is insignificant; 2. Although these are shown as low yield pathways, there is some evidence to the contrary (see text: "The Concept of Fetoplacental Unit"). 3. This hydroxylation takes place in the fetal liver; 4. This hydroxylation occurs in both the fetal liver and adrenals; DHA: Dehydroepiandrosterone; (S): Sulfate. Percentage figures indicate the percent amount of estrogen (estradiol, estrone, estriol) derived from either maternal or fetal precursors. Similarly to estradiol, placental estriol and estrone are secreted also into the fetal compartment; however, for reasons of simplicity, this is not shown in this figure.

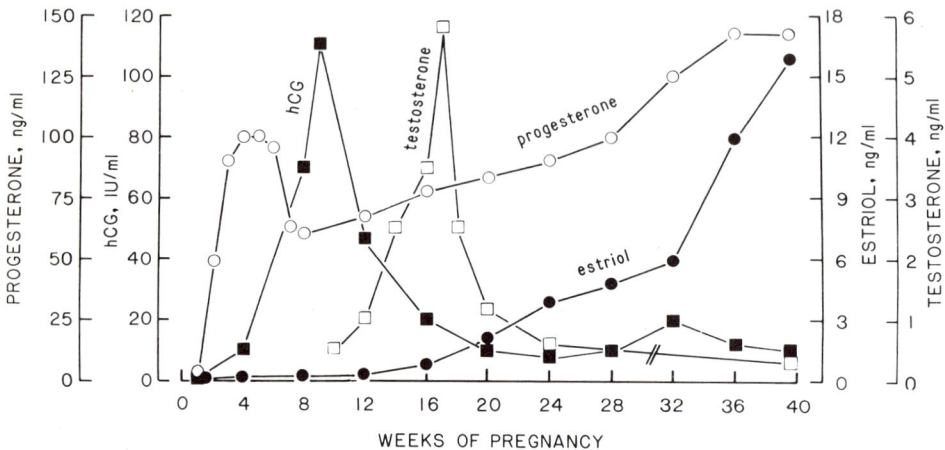

Fig. 2. Developmental profile of hCG, estriol and progesterone levels in maternal plasma, and of testosterone in male fetal plasma during pregnancy. This is a composite picture assembled from data reported by various authors (hCG: Brody, 1969; Vaitukaitis, 1977; Varma et al., 1971. Estriol: Beling, 1977; De Hertog, 1975; Simmer, 1968. Progesterone: Cooke, 1976; MacNaughton, 1976; Yoshimi et al., 1969. Testosterone: Abramovich and Rowe, 1973; Nagamani et al., 1979; Reyes et al., 1974). Emphasis was placed on the developmental pattern and the hormonal interrelations rather than the fiducial limits of individual values. For the latter, the reader is referred to the original papers. Around the fourth week, progesterone secretion by the ovary begins to decline and it shifts gradually from the corpus luteum to the placenta (see text, p. 95). Interrelations between hCG and testosterone, as well as between hCG, estriol and progesterone, are discussed in the text: pp. 101-102 and 106-109, respectively.

DEVELOPMENT OF PLACENTAL ENDOCRINE FUNCTION

Secretion Rates And Circulating Levels Of Placental Hormones At Various Stages Of Gestation

Levels of placental hormones circulating in the mother and the fetus reflect the continuous developmental changes the placenta and the fetus undergo with the advancement of pregnancy. Consequently, levels of placental hormones are useful in signaling the normal succession of various developmental stages, or the appearance of developmental aberrations, during ontogenesis. Thus, hormonal levels may be directly related to, among other factors, the size of the placenta (Josimovich, 1977), the smooth functioning of the placental enzymatic systems (pp. 105-106), the presence or absence of the fetal pituitary (Beling, 1977), the proper functional development of the fetal zone of the adrenal cortex, and even the sex of the fetus (Boroditsky et al., 1975; Brody and Carlstrom, 1965). Normal values of placental hormones in various compartments and at various stages of gestation are given in Table 2. The developmental pattern through pregnancy of some of these hormones is shown in Fig. 2.

All placental hormones are secreted into both the maternal and the fetal compartment, although the amounts for each hormone differ. hCG levels in maternal plasma are roughly 700 times higher than those in the plasma of newborns. Similarly, hCS levels in the fetal circulation are roughly 1.5% of those found in the maternal circulation. Levels in the maternal compartment are influenced by the activity of the liver, the rate of hepato-intestinal circulation of the hormone and the activity of the placental sulfatase system (in the case of steroids), the renal clearance of the hormone, and the rate of its uptake by the fetus. For example, hCG levels are higher in the umbilical vein than the umbilical artery (Effer et al., 1973), suggesting consumption by the fetus; very little hCS, as contrasted to the large amounts of hCG, is excreted in the urine; and, finally, placental sulfatase deficiency, resulting in decreased placental steroid synthesis, has been reported (pp. 105-106).

Maternal plasma levels of hCS and placental total content show a direct correlation with the weight of functioning trophoblast mass at various stages of gestation (see Josimovich, 1977; Saxena, 1971). Similarly, total placental content of hCG increases fourfold during the second and third trimester, (see Simmer, 1968), although hCG placental concentration does not change (Table 2) during this period. Yet, hCG levels in maternal plasma do not rise, as one would expect in view of this increase in the absolute hCG amount in the placenta. It is possible, that this discrepancy might be accounted for by an increased extraction of hCG by the fetus during the last two trimesters of gestation. However, to substantiate this hypothesis serial hCG determinations in the umbilical vein and artery during this period are needed.

Placental hormones circulate in various forms in the maternal organism. hCS is believed to circulate as a monomer (see Josimovich, 1977), whereas hCG, and perhaps hCT, circulate in the form of a dimer of two noncovalently bonded subunits. Their α and β free subunits are also found in plasma. Steroids circulate in free, conjugated (as sulfates and glucuronates), hydroxylated, or bound-to-protein, forms. Only the unbound and unconjugated forms display biologic activity.

Secretory Rhythms Of Placental Hormones

Circadian or episodic secretory rhythms have been reported for many hormones in the normal subject. If placental hormones are subject to the regulatory feedback mechanisms surmized in the introduction, one might expect that some of the placental hormones will perhaps display rhythmic secretory patterns, as well. Alternatively, if present, such rhythms would in themselves constitute an argument in favor of a feedback control on placental hormone secretion.

Pujol-Amat et al. (1973) have reported a circadian variation of hCG levels in maternal serum during the last trimester of pregnancy. The

TABLE 2. SECRETORY RATES, HALFLIVES AND CIRCULATING LEVELS OF PLACENTAL HORMONES[a]

Tri-Mester	Parameter	Human Chorionic Gonadotropin	Human Chorionic Somatomammotropin	Human Chorionic Thyrotropin	Progesterone	Estradiol	Estrone	Estriol
1st	Placental concentration	600 IU/g wet weight	100-200 μg/g wet weight	—	5-6 μg/g wet weight	—	—	—
	Daily placental secretion rate	26 mg/day	—	—	? 92 mg/day	—	—	—
	Halflife	12-36 hr	15-30 min, 1 hour	—	—	—	—	—
	Maternal plasma levels[b]	0.1-100 IU/ml	0.3 μg/ml	20-1300 μIU/ml	13-44 ng/ml	1-2 ng/ml	1-2 ng/ml	1-2 ng/ml
2nd	Placental concentration	20 IU/g wet weight	100-200 μg/g wet weight	—	2.5 μg/g wet weight	32 ng/g wet weight	64 ng/g wet weight	119 ng/g wet weight
	Daily placental secretion rate	1.4 mg/day	—	—	75 mg/day	—	—	—
	Halflife	12-36 hr	15-30 min, 1 hour	—	—	—	—	—
	Maternal plasma levels[b]	30-60 IU/ml	2-5 μg/ml	? 7.5 μU/ml	50-80 ng/ml	7 ng/ml[c]	3 ng/ml[c]	2.7 ng/ml

Tri-Mester	Parameter	Human Chorionic Gonadotropin	Human Chorionic Somatomammotropin	Human Chorionic Thyrotropin	Progesterone	Estradiol	Estrone	Estriol
	Placental concentration	20 IU/g wet weight	100–200 μg/g wet weight	0.2–4 mU/g wet weight	2 μg/g wet weight	105 ng/g wet weight	44 ng/g wet weight	274 ng/g wet weight
	Daily placental secretion rate	1.4 mg/day	0.3–1 g/day	—	190–280 mg/day	15–20 mg/day	15–20 mg/day	30–40 mg/day
3rd	Halflife	12–36 hr	15–30 min, 1 hour	—	6 and 95 min	—	—	—
	Maternal plasma levels[b]	50–70 IU/ml	4–6 μg/ml	40–80 μIU/ml	125–190 ng/ml	23 ng/ml[c]	5–10 ng/ml[c]	6–15 ng/ml[c]
	Umbilical artery levels	165–250 mIU/ml	10–100 ng/ml	—	536 ng/ml	4 ng/ml	19 ng/ml	30 ± 9 ng/ml[c]
	Umbilical vein levels	250–375 mIU/ml	10–100 ng/ml	—	966 ng/ml	3–55 ng/ml[c]	5–38 ng/ml[c]	95 ± 19 ng/ml[c]

[a] This table was composed from data discussed by various reviewers (Beling, 1977; Effer et al., 1973; Hennen and Freychet, 1974; Josimovich, 1977; Josimovich and Venning, 1975; Klopper and Fuchs, 1977; Mitchell and Bagshaw, 1976; Saxena, 1971; Tulchinsky et al., 1977; Vaitukaitis, 1977). It should be noted that normal hormonal parameters are constantly changing throughout pregnancy; recording one value for such long periods as trimesters may be misleading. Accordingly, ranges of values, or mean ± SE, are given whenever possible. It should also be pointed out that the values recorded above were secured through a variety of methods prevailing in each author's laboratory, and, therefore, are not directly comparable with each other. ?, Value not certain, but indicative.

[b] The urinary levels of these hormones, recorded at various stages of gestation, show the same developmental patterns to those found in plasma.

[c] Unconjugated form.

lowest value was observed at 0800 hours and the highest at 1000 hours; thereafter, hCG levels decreased progressively throughtout the day and night, and no significant differences were recorded between 1900 and 0800 hours. The mechanism responsible for this diurnal variation is obscure. With regard to hCS no significant variation in circulating levels during a normal 24-hour period was found by most investigators. Pavlou et al. (1972), however, using relatively large numbers of pregnant subjects and more frequent sampling, and subjecting their data to full statistical analysis in relation to the precision of the assay itself, reported a significant diurnal variation in hCS plasma levels; this variation, however, did not follow any recognizable pattern. The implication of these findings is that serial hCS determinations rather than a single hCS sample are necessary for the assessment of fetal well-being in complicated pregnancies.

Reports on progesterone rhythmicity during pregnancy have appeared. Peak levels were reported to occur at 0400 and 1600 hours (Craft et al., 1969; Johanson, 1969), whereas troughs were variably found at 0800 and 2000 hours (Craft et al., 1969; Vadora et al., 1974). However, results were conflicting from day to day in the same patient, the population size and sampling frequency were usually inadequate, the range of the subjects' pregnancy stage in each group was too wide, and specific statistical analysis was not undertaken. Accordingly, nothing definite can be presently said about progesterone circadian secretory rhythms during pregnancy.

Estrogen diurnal secretory rhythms during gestation, have been investigated more extensively. Yet, the same drawbacks as with progesterone are encountered here. The pregnancy stage investigated is sometimes not stated (Duhring et al., 1973), statistical pooling of values collected from pregnancy stages more than five weeks apart is practiced (Dickey et al., 1966), rhythms are investigated in pathological pregnancies (Dickey et al., 1966), blood sampling frequency is sometimes only 2-3 times per 24 hours, the number of subjects assessed is often very small, and serious discrepancies between papers by the same authors (Selinger and Levitz, 1969; Levitz et al., 1974), or between plasma and urine values are encountered. Levitz et al. (1974) and Masson and Wilson (1972) reported no diurnal rhythms for plasma estriol in the last few weeks of pregnancy. According to Duhring et al. (1973) a diurnal variation exists, but it is complex and unpredictable. Sometimes the rhythm reverses itself in the same subject from day to day (Levitz et al., 1974). Others have reported that estrogens, and more specifically estriol, reach a high level in the afternoon or early evening and a low level in the morning (Goebel and Kuss, 1974; Tulchinsky et al., 1971). Yet, others found exactly the opposite; peaks in the morning and troughs in the evening (Munson et al., 1972; Selinger and Levitz, 1969; Townsley et al., 1973). It is hoped that this controversy will be resolved, when investigators use: frequent daily sampling, normal subjects in the same gestational week, modern radioimmunoassay techniques, and currently adopted techniques for statistical processing of biorhythms.

INTERACTIONS AND INTEGRATION OF THE ENDOCRINE PLACENTA WITH THE FETOMATERNAL ENDOCRINE CONTINUUM

Placenta And Mother

Effects Of Placental Hormones On The Mother

Human chorionic gonadotropin directly inhibits LH release from the maternal pituitary (Hirono et al., 1972; Miyake et al., 1976) and, in turn, assumes a luteotropic effect on the maternal ovary resulting in a prolongation of the luteal function and progesterone secretion for up to 6-8 weeks after conception (see Vaitukaitis, 1977). However, about the fourth week of pregnancy, steroid synthesis in the corpus luteum starts declining, in spite of the still rising levels of hCG during this period (Yoshimi et al., 1969). At that time a gradual shift of progesterone production from the corpus luteum to the placenta occurs and thus after the sixth week, the ovaries become expendable without any harm to the continuation of pregnancy. The findings of Yoshimi et al. (1969) were recently corroborated by Williams et al., (1979) who found that hCG stimulated progesterone synthesis *in vitro* by luteal cells obtained from nonpregnant women, whereas it had no such effect on luteal cells taken from pregnant women beyond the sixth week of pregnancy; furthermore, addition of estrogens in the incubation medium inhibited this hCG effect on the luteal cells from nonpregnant women. The importance of hCG's FSH-like activity in the pregnant woman remains obscure; however, from Albert's studies in the rat (1969), it may be inferred that hCG has a stimulatory action on the human ovarian follicle. Recently, it has been suggested that hCG suppresses the immune response of the mother and thus not only secures the nonrejection of a semiallograft, the blastocyst, but also permits its malignant-like invasion and implantation. Existing evidence in this respect is still circumstantial (see Mitchell and Bagshawe, 1976) and further studies are needed to establish such a role for hCG. But rather than hCG, placentotropin, a new hormone purportedly secreted by the maternal pituitary during pregnancy, is also suggested to be responsible for the lowering of the maternal immune resistance to the conceptus (Lajos et al., 1959; Lajos et al., 1976). hCG may interact also with the development of the fetal immune response; the hCG uptake by the fetal thymus is four times higher than that of other fetal tissues (gonads and kidneys excluded) (Huhtaniemi et al., 1978). hCG does not cross the human blood-brain barrier readily (Assies et al., 1978; McCormick, 1954). Yet, Uemura (1968) reported that hCG injected into rabbits caused significant changes in the excitability of the central nervous system, although proper controls were not included and interference from other hormones was not excluded. Hall and Heintz (1969) reported that hCG injections into rhesus monkeys stimulated production of estrogens by the adrenals, although this effect was manifest only in the presence of ACTH; on the other hand, Scurry and Bruton (1968) reported no stimulation of glucocorticoid and androgen synthesis by the adrenal cortex after administration of hCG to rhesus monkeys. The importance of hCG's thyrotropic activity during pregnancy has not yet been fully assessed. In the mouse bioassy 1 IU of hCG has intrinsic thyrotropic activity equivalent to 0.5 μU of TSH. Accordingly, hCG plasma levels during the first trimester of pregnancy would be equivalent to 20 μU of TSH approximately; this thyrotropic activity is substantially greater than serum TSH activity in normal subjects. The thyrotropic activity of hCG was investigated by Hershman and Burrow (1976) and by

Sowers et al., (1978) in normal pregnant women and euthyroid men respectively. There was a delayed stimulation of thyroidal ^{125}I release, but no significant changes in T3, T4, or TSH serum levels. It is concluded that under normal circumstances hCG is a weak thyroid stimulator; in cases of molar pregnancies, however, this stimulatory effect may cause severe hyperthyroidism (Higgins and Hershman, 1978). The extent of thyroid stimulation by hCT during pregnancy remains uncertain (Harada and Hershman, 1978; Harada et al., 1979). However, irrespective of the type of thyrotropic substances, circulating during pregnancy, thyroid stimulation does occur, as evidenced by the elevated serum free thyroxin and triodothyronin levels recorded in pregnant women by Harada et al. (1979).

Human chorionic somatomammotropin displays a variety of physiologic actions. Most of these actions were investigated in experimental animals and only a few are documented in humans. hCS has been called the "metabolic hormone of pregnancy", because it assures a normal supply of nutrients to the fetus and an appropriate metabolic state in the mother. hCS displays growth promoting activity in specific somatotropic-hormone bioassays, if injected at a suitable pH (see p. 86). hCS stimulates insulin release but also inhibits peripheral glucose uptake in the maternal organism. Thus, maternal glucose is spared and channelled to the fetus, which primarily uses glucose for its energy needs. Conversely, maternal energy needs are increasingly met by burning fat; hCS mobilizes free fatty acids from body fat depots and elevates ketone bodies. Maternal starvation during pregnancy leads to a drastic increase of ketone bodies and fetal harm may ensue. Protein synthesis is enhanced and a positive nitrogen, potassium and phosphorus balance occurs. These effects of hCS are modulated by estrogens and progesterone and by feasting or fasting. Growth hormone levels in maternal plasma are low, presumably because of a negative feedback on the hypothalamic-hypophysial axis by hCS. The changes in the somatomedin levels observed in maternal (Bala et al., 1978; Furlanetto et al., 1978), and umbilical venous (Gluckman and Brinsmead, 1976) serum during human pregnancy may be related to hCS activity (Furlanetto et al., 1978). In suitably primed rats, hCS has mammotropic (increases mammary gland weight), lactogenic (increases milk production), luteotropic and erythropoietic activity. In the human male and nonpregnant female, hCS may stimulate aldosterone secretion, which, interestingly enough, is elevated during pregnancy (Smeaton et al., 1977). The physiological and pharmacological actions of hCS have been extensively reviewed by Josimovich (1977), Kaplan and Grumbach (1974), Saxena (1971), and Simmer (1968).

The effects of estrogens and progesterone on the maternal organism are diverse. One maternal endocrine gland that may be influenced by estrogens is the pituitary. It is known since 1909 that the volume and weight of the maternal pituitary increases twofold during pregnancy, and that the anterior lobe is swarmed with the so-called pregnancy cells, whereas the relative number of somatotrophs declines. It has been suggested that the pregnancy cells are prolactin secreting cells (Goluboff and Ezrin, 1969) and that the somatotrophs decline in response to a negative feedback by hCS. Alternatively, it has been postulated that the high levels of estrogens circulating during pregnancy stimulate the maternal pituitary to produce a new hormone, placentotropin, which, in turn, stimulates hCG synthesis and release by the placenta (Lajos, 1960; Lajos et al., 1959; Lajos et al., 1976). Whether placentotropin is indeed secreted by the pregnancy cells remains to be substantiated. This hormone is further discussed in the next subsection. Although the existence of a placentotropin remains at present an attractive hypothesis, the changes in hypophysial prolactin secretion during pregnancy are increasingly being documented. Data accumulated in recent years indicate that prolactin levels in the maternal plasma increase progressively throughout pregnancy (20-fold at term as compared with normal nonpregnancy levels), that this increase parallels the increase in estrogen and hCS levels, and that estrogen administration induces transformation of pituitary somatotrophs to mammotrophs in experimental animals and increases prolactin secretion in nonpregnant women (see Pozo del et al., 1977; Ylikorkala et al., 1979b); furthermore, intravenous infusion of dehydroepiandrosterone in pregnant women brings about an increase in serum prolactin levels, presumably as a response to an increase in the levels of circulating endogenous estrogens (Ylikorkala et al., 1979a) (pp. 105-106). Although estrogens stimulate hypophysial prolactin secretion, they nevertheless antagonize prolactin's actions in the periphery, and more specifically at the mammary gland level. (It is of interest that estrogens have a similar action on the secretion and the peripheral actions of growth hormone in normal subjects). Thus, in spite of the tremendous increase in the levels of circulating prolactin, lactation does not occur during pregnancy, but after the source of estrogens is eliminated by the expulsion of the conceptus at the time of parturition. In view of the above, the physiological role of hCS during pregnancy should be closely re-examined. It seems that more emphasis should be placed on the metabolic actions of hCS and its inhibitory effect on growth hormone secretion, than on its mammotropic actions. The latter were demonstrated *in vitro*, or *in vivo* in nonpregnant animals suitably primed with exogenous hormones (e.g., estrogens), which might have had stimulated prolactin secretion as well. No evidence exists as yet for a mammotropic action of hCS in the human (see Josimovich et al., 1974). Thus, the predominant mammotropic role during pregnancy should be vested with prolactin rather than with hCS; a more appropriate name for the latter hormone might then be necessary. Yet another change in the maternal pituitary, effected by the sex steroids circulating in high levels during pregnancy, is the inhibition of gonadotropin (FSH, LH) secretion; in pregnant women, LH content in the pituitary is very low or undetectable (De la Lastra and Llados, 1977), plasma FSH and LH basal levels are very low, and the gonadotropin secretory response to GnRH is very sluggish or abolished (see Kulin et al., 1979; Reyes et al., 1976).

Estrogens are also responsible for the increased levels of various hormone binding proteins (e.g., thyroxin-binding globulin and transcortin) in the maternal circulation. Estrogens may increase the excitability of the central nervous system and of the uterine muscle. In contrast, progesterone opposes these estrogen actions. Uterine metabolism, growth and blood supply are all stimulated by estrogens. The contribution of the individual estrogens to these effects is not clear; nor is it clear why estriol is produced in such large quantities during pregnancy. Progesterone's roles in preparing the endometrium for implantation, in promoting breast growth and in antagonizing aldosterone's action on the renal tubule are well known. This antagonism might be one of the causes of aldosterone's threefold elevation in the maternal plasma during pregnancy (Smeaton et al., 1977). The importance of these steroid effects on the pregnant uterus is illustrated by the fact that bilateral ovariectomy in many pregnant animals and in early human pregnancy results in abortion.

Human chorionic thyrotropin releasing hormone may synergize with estrogens in stimulating prolactin release from the maternal and fetal pituitary; a significant rise of prolactin, in both maternal and fetal plasma, follows an intravenous administration of synthetic TRH into the maternal or fetal circulation in Rhesus monkeys (Azukizawa et al., 1976).

Maternal Endocrine Influences On Placenta

One such influence has been touched upon previously in this chapter; that the maternal pituitary may control the hormonal secretion of the placenta, which is an attractive hypothesis. If true, it will provide the first evidence for the long sought-after control of placental endocrine function and will mark the beginning of a new era in the field of reproductive endocrinology. Lajos et al. in some 13 papers from 1951 through 1976 provided evidence that a placentotropin circulates in women during pregnancy. However, their evidence that this placentotropin emanates from the maternal pituitary is indirect and inconclusive. It appears that placentotropin is a glycoprotein with a molecular weight of about 36,000 (Butt, 1967), its secretion is stimulated by estrogens, and, in turn, it stimulates hCG production by the placenta. Eventually, hCG stimulates estrogen synthesis in the placenta (see the subsection on "Intraplacental Interrelations"). Such an arrangement reflects a positive feedback loop between placenta and maternal pituitary. However, if such a feedback indeed exists, it is violated after the first trimester of pregnancy, since although circulating estrogen levels rise gradually, the levels of placentotropin decrease in tandem with those of hCG (Lajos et al., 1976). These results have not yet been confirmed by other laboratories and, further, hCG plasma levels do not decrease substantially after maternal hypophysectomy early in pregnancy (see Saxena, 1971). However, these latter studies were conducted in the 1950s, when available methods for hCG determinations in plasma were neither sensitive nor specific. Clearly, confirmation of Lajos's observations by other laboratories is needed to substantiate the placentrotropin hypothesis. It should be mentioned, however, that when placentrotropin is injected during pregnancy in the rat, it stimulates placental protein synthesis (Petropoulos and Cons, 1974).

Recent studies from various laboratories have also focused on possible maternal hormonal controls of hCG and hCS secretion. Administration of thyrotropin-releasing hormone *in vivo* (Hershman and Burrow, 1976) and of somatostatin *in vitro* (Macaron et al., 1978b) failed to stimulate or inhibit hCG and hCS secretion. Dopamine, however, inhibited hCS secretion by normal trophoblastic cells *in vitro* (Macaron et al., 1978a). On the other hand, epidermal growth factor (Beneviste et al., 1978), dibutyrylcyclic AMP (Hussa et al., 1978) and gonadotropin-releasing hormone (Macaron et al., 1976) all stimulate hCG production from choriocarcinoma cells *in vitro*. Similarly hCG, but not hCS, production by normal term placentas incubated *in vitro* is significantly stimulated by dibutyrylcAMP and theophylline, alone or in combination (Handwerger et al., 1973). It should be noted that urinary secretion of cAMP and cGMP is significantly increased during pregnancy (Kopp et al., 1977). Finally, whereas maternal blood glucose levels may provide an appropriate negative feedback effector controlling the rate of hCS secretion (Kim and Felig, 1971; Gaspard et al., 1977), circulating free fatty acids (Gaspard et al., 1977) and amino acids (Mochizuki et al., 1976) seem to have no such effect.

The maternal ovaries have been implicated in the control of placenta growth and metabolism in the rat. That pregnancy is terminated in the rat after bilateral ovariectomy is well documented. Fetuses are squeezed to death by the nonexpanding myometrium, whereas placentas become atrophic and eventually are resorbed or expelled (Petropoulos, 1973). Similarly, fetectomy, performed in such a way as to leave the placenta in situ undisturbed, leads eventually to placental atrophy and resorption (Petropoulos, 1973). However, when ovariectomy is coupled with simultaneous fetectomy, again leaving the placenta in situ, placental protein synthesis is increased, followed by hypertrophy and increased placental weight (Petropoulos, 1973). These studies suggest that in the rat, placental growth and development is under a dual inhibitory control by the fetus and the maternal ovaries; however, neither of these controllers can effect this inhibition alone. Similar studies in the rabbit showed that placental weight increased in ovariectomized animals, but remained normal in ovariectomized animals injected with estradiol (Abdul-Karim et al., 1971). In the pregnant woman, ovariectomy before the sixth week of gestation leads to abortion, whereas it has no such effect if performed afterwards (see p. 95, and Simmer, 1968). Human fetectomy for therapeutic reasons leads eventually to placental degeneration (Friedman et al., 1969) similar to that observed in the rat. Ovariectomy and fetectomy combined, leaving the placenta in situ, have not been performed in the human.

Placental Function In States Of Maternal Endocrine Dysfunction

Human pregnancy may sometimes coincide with various maternal endocrine dysfunctions. Such natural "experimental" setups, that would have otherwise been impossible in the human, provide unique opportunities for studying hormonal interrelations during pregnancy. Almost all possible maternal endocrinopathies have been associated with pregnancy in the human. Such cases involved, for example, hypophysectomized and acromegalic women, and women with diabetes insipidus; hyper-, hypo-, and pseudopseudohypoparathyroidism; hypo- and hyperthyroidism; adrenal gland hyperfunction (Cushing's, or Conn's syndrome, adrenogenital syndrome, adrenogenital syndrome, pheochromocytoma) or hypofunction (Addison's disease). The various cases have been reviewed extensively by Gabrilove (1965), Gerbie (1969) and Hytten and Leitch (1971). In most of these endocrine dysfunctions pregnancy advancement is not affected and parturition occurs on time at term. Experimental evidence indicates, however, that excess corticosteroids may inhibit normal placental growth in the rat (Blackburn et al., 1965), and the human (Brown et al., 1968) and may cause a decrease in the synthesis of placental estrogens in women (Brown et al., 1968). Surprisingly, low placental secretion of estrogens and progesterone was observed also in an adrenalectomized pregnant woman during three successive pregnancies (Charles et al., 1970). Maternal diabetes mellitus may also affect carbohydrate metabolism and glycogen deposition in the placenta (Gabbe et al., 1972). On the other hand, maternal starvation brings about a sustained increase (25-48%) in hCS secretion by the placenta (Kim and Felig, 1971) (p. 109).

Placenta And Fetus

Effects Of Placental Hormones On Fetal Endocrine Development

In view of the fact that hCG is primarily secreted into the maternal compartment (Table 2), it was suggested that few, if any, physiological actions of hCG were directed towards the fetus. Yet, hCG levels are higher in the umbilical vein than in the umbilical arteries (Table 2), indicating that the fetus is actively metabolizing, and perhaps using, this hormone, even during the latter part of pregnancy. Evidence has now accumulated indicating that hCG may play an important role in the normal development of the fetal adrenals and gonads.

The fetal adrenal cortex comprises two sections, the fetal zone and the definitive cortex. The former atrophies rapidly after birth (Benirschke, 1956), whereas the latter remains functional throughout life. The fetal zone secretes primarily dehydroepiandrosterone and very little cortisol; on the other hand, the definitive cortex secretes primarily gluco- and mineralocorticoids and very little dehydroepiandrosterone during fetal life. Tropic hormones controling the endocrine function of the fetal adrenal could originate in the maternal pituitary, the placenta, the fetal pituitary itself, or any combination of these. To further complicate matters, the two cortical sections of the fetal adrenal may have a differential responsiveness to any or all of the tropic hormones above. Indeed, in experiments of fetal adrenal stimulation by ACTH, the definitive cortex produces gluco- and mineralocorticoids, whereas the fetal zone secretes only androgens; further, the response of the fetal zone is less consistent than that of the definitive cortex, indicating that ACTH may not be the primary regulator of the fetal zone (Serón-Ferré et al., 1978b). This differential responsiveness is further evident in the rapid involution of the fetal zone after birth (Benirschke, 1956), in spite of the presence of normal levels of circulating ACTH and presumably because of the withdrawal of a specific placental (?) stimulator.

With regard to maternal tropic hormones, maternal hypophysial ACTH is an unlikely stimulator of the fetal adrenals, because it does not cross the placenta readily (Miyakawa et al., 1974).

With regard to fetal tropic hormones, Benirschke reported in 1956 that in anencephalic and apituitary fetuses the development of the fetal adrenal gland proceeds normally up to 20 weeks of gestation, but that the fetal zone undergoes rapid involution thereafter. In subsequent years evidence about the onset of fetal hypophysial ACTH secretion was presented. Secretory granules are detectable in the human fetal pituitary by the end of the first trimester (Villee, 1969), and ACTH is detectable in fetal circulation as early as the 12th week of gestation (Winters et al., 1974); however, the origin of the hormone, whether fetal hypophysial, or perhaps placental (Rees et al., 1975), remains to be ascertained. On the other hand, the ACTH content of the fetal pituitary is relatively low throughout gestation, in comparison to α-melanocyte stimulating hormone (α-MSH) and corticotropin-like intermediate-lobe peptide (CLIP) content (Silman et al., 1976); only near parturition, however, there is a sharp increase in hypophysial ACTH content, triggering perhaps the full functional maturation of the fetal definitive adrenal cortex and the cortisol surge necessary for labor initiation (p. 111).

With regard to placental tropic hormones, chorionic ACTH (Rees et al., 1975) may have an active role in fetal adrenal regulation; however, confirmation of this hypothesis must await definite proof of such placental secretion, as well as direct experimental demonstration of its stimulatory effect on the fetal adrenal. Irrespective of what future research will unravel about placental ACTH, presently existing evidence points strongly to hCG as a regulator of the fetal adrenals (especially the fetal zone), at least during the first half of pregnancy. Johannisson (1968) found that intra-amniotic injections of hCG at midgestation produced ultrastructural changes in the fetal zone, similar to those induced by ACTH injections.

In the context of Benirschke's findings, this observation was interpreted as indicating that the function of the fetal zone is regulated by hCG in the early, and by fetal ACTH in the later, stages of gestation. Further evidence supports this hypothesis: peak circulating hCG levels occur around the 9th-10th week of gestation (Table 2 and Figure 2); administration of hCG to newborn infants brings about a significant increase in urinary dehydroepiandrosterone excretion (Lauritzen and Lehmann, 1967); Serón-Ferré et al. (1978a) reported that dehydroepiandrosterone sulfate production by isolated adrenal fetal zones (from early fetuses of 12-17 weeks of gestation) superfused in vitro is significantly stimulated in the presence of hCG; similar fetal zone stimulation occurs with ACTH (Serón-Ferré et al., 1978a), although inconsistently (Serón-Ferré et al., 1978b). On the other hand, Huhtaniemi et al. (1978) have reported that hCG content in the adrenals of 12-20 week human fetuses is not significantly higher than that found in other body organs. On the basis of the preceding and particularly Benirschke's observation that the fetal zone declines after the 20th week of gestation in anencephalic fetuses and the fact that the same zone involutes after birth in normal fetuses (ACTH present), the following are proposed: (1) the adrenal fetal zone is under the control of hCG up to 20 weeks of gestation; (2) thereafter, this zone is under the dual control of both hCG and fetal ACTH, either of which, if withdrawn, brings about atrophy of the zone; (3) the main stimulators of the fetal definitive adrenal cortex are corticotropic peptides secreted by the fetal pituitary; the role, if any, of hCG on this cortex has not been investigated, as yet.

The possibility that hCG may regulate androgen production by the fetal zone of the fetal adrenal ties together with hCG's proposed control on placental steroid hormone synthesis (Fig. 3, and pp. 106-109). If fully substantiated, such a combined effect will make hCG the main tropic hormone in control of placental estrogen production, at least during the first half of gestation. Under such a scheme, hCG will be regulating both the supply rate of the placenta with fetal precursors for estrogen biosynthesis, and the placental estrogen biosynthetic process per se, a very efficient arrangement.

The fetal gonads, testes or ovaries, are also influenced by hCG. It has been generally believed that in the human fetus, the ovaries differentiate somewhat later than the testes (Villee, 1969). However, recent evidence indicates that the capacity for estrogen and testosterone synthesis in the fetal ovary and the fetal testis respectively, appears at about the same time (8-10 weeks of gestation), (George and Wilson, 1978). This observation raises the possibility that the same factor(s) triggers the onset of sex steroid synthesis in both gonads. LH/FSH from the fetal pituitary could have plausibly been such gonad-stimulating factors, if they were secreted in sufficient quantities at the time of the functional initiation of the gonads. Such timely LH/FSH secretion was, however ruled out recently by Clements et al. (1976), who reported that LH and FSH concentrations, measured by specific beta-chain RIAs in the fetal pituitary, fetal serum (obtained by cardiac puncture) and amniotic fluid, were unmeasurable or very low, prior to 12 weeks fetal age. After the 12th week, LH levels began to rise, but remained, nevertheless, below hCG levels up to 20 weeks of gestation in male fetuses. The possibility that LH/FSH emanating from the maternal pituitary could have been the gonad-initiating tropic hormones must be discounted; their secretion is effectively shut off during pregnancy, their crossing the placental barrier is unlikely, and maternal hypophysectomy in Rhesus monkeys has no discernible effects on fetal testicular (Gulyas et al., 1977a) or ovarian (Gulyas et al., 1977b) development.

On the other hand, accumulated evidence indicates that the functional initiation of the fetal gonads is triggered by hCG. Morphological differentiation and functional maturation of the Leydig cells in the testis begin at 8 weeks of gestation (Niemi et al., 1967), coincident with the rising levels of hCG in the maternal and fetal (Clements et al., 1976) circulation. Eventually, maximal testosterone secretion by the fetal testis occurs at 11-17 weeks fetal age (Reyes et al., 1974), subsequent to peak hCG secretion at 9-11 weeks (Fig. 2). Intra-arterial infusion of hCG to male Rhesus monkey fetuses in utero during late pregnancy brings about a significant increase in fetal serum testosterone (and 5α-dihydrotestosterone) levels measured by RIA (Huhtaniemi et al., 1977b). Human fetal testis in vitro can specifically bind hCG at physiologic concentrations (Huhtaniemi et al., 1977a); further, hCG at such concentrations can stimulate testosterone synthesis and secretion by fetal testis (gestational age 14-19 weeks) incubated (Huhtaniemi et al., 1977a) or cultured (Abramovich et al., 1974) in vitro. Similar stimulation was observed in fetal Rhesus monkey testes (gestational age 20-23 weeks) incubated in vitro in the presence of hCG (Huhtaniemi et al., 1977b). Moreover, other studies have shown that endogenous hCG net uptake by the human fetal testes in vivo is higher than that found in other tissues, such as lung, liver, spleen, and muscle (Huhtaniemi et al., 1978). Interestingly, fetal ovaries show by far the highest net hCG uptake than any other organ tested (Huhtaniemi et al., 1978). Unfortunately, the significance of this finding cannot be fully assessed, because no other experimental evidence exists at present regarding the functional interrelations, if any, between fetal ovaries and hCG. Although the studies discussed above show clearly that the placenta may control early testicular, and perhaps ovarian, development, other studies, and "experiments" of nature, such as fetal anencephaly or apituitarism, have demonstrated that fetal hypophysial LH/FSH are important in the later stages of fetal gonadal development. Sexual differentiation proceeds initially unimpeded in both male and female anencephalic and apituitary fetuses (see: Clements et al., 1976; Huhtaniemi et al., 1977a) indicating that the gonads are independent from the fetal pituitary at this stage; eventually, however, in such fetuses the Leydig cell number is decreased (Bearn, 1959); the external male genitalia become hypotrophic (see: Clements et al., 1976) and the ovaries show reduced weights, a paucity of interstitial cells and arrested follicular development (Ross, 1974). Further, hypophysectomy late in gestation of Rhesus monkey fetuses in utero, without terminating the pregnancy, leads to marked testicular (Gulyas et al., 1977a) or ovarian atrophy (Gulyas et al., 1977b). Evidently, the gonads, like the adrenals, depend on hCG for differentiation and functional initiation early in pregnancy, and on fetal hypophysial LH/FSH for further development after midgestation. When assessing hCG physiologic actions on the fetal adrenals and gonads, it should be remembered that, on a weight basis, hCG is 2-6 times more potent than LH (see Clements et al., 1976) and it has intrinsic FSH activity, as well (p. 86).

In view of these tropic actions of hCG on fetal adrenal and gonadal hormone secretion, it may be eventually pertinent to investigate whether a negative feedback connection operates between these fetal glands and the placenta. This hypothesis is further discussed below (pp. 106-109).

If hCTRH is indeed produced by the placenta and secreted into the fetal circulation, it may stimulate the release of fetal hypophysial prolactin and TSH; in turn, the latter will regulate thyroid hormone secretion. Intravenous injection of synthetic TRH into the Rhesus monkey fetus, brings about a substantial increase in prolactin, TSH, thyroxine and triodothyronine levels in the fetal plasma (Azukizawa et al., 1976).

Estrogens circulating during pregnancy may regulate placental growth and metabolism as discussed earlier. Changes in placental growth can, in turn, influence fetal size, as shown by Abdul-Karim et al. (1971) in ovariectomized pregnant rabbits treated with progesterone, or progesterone and estradiol. Placental and fetal weight was higher in the former than in the latter and the control groups. Although many other factors, such as litter size, parity, maternal nutritional state and uterine blood supply, can also affect placental growth and, in turn, fetal size, an estrogen inhibitory effect on rodent placenta growth cannot be excluded (Croskerry and Dobbing, 1978). The mechanism(s) by which experimental placental hyperplasia/hypertrophy leads to an increased fetal weight in rodents remains to be elucidated.

Fetal Endocrine Influences On Placenta

The fetoplacental partnership in the production of estrogens and progesterone (Fig. 1) has been discussed earlier in the chapter. From that discussion it is evident that any changes in the flow of fetal steroid precursors to the placenta would influence the rate of synthesis of placental steroids. This assertion is documented both by experiments of man and by "experiments" of nature. Thus, placental estrogen secretion decreases drastically after fetectomy (Friedman et al., 1969; Kim et al., 1971), and, in anencephalic fetuses, placental estriol secretion is decreased by as much as 90% (Frandsen and Stakeman, 1961; Breborowicz and Biniszkiewicz, 1967), presumably because of insufficient supply of precursors (Fig. 1) from the fetal adrenals, which are underdeveloped (Frandsen and Stakeman, 1961). Primary maldevelopment of the fetal adrenals, also leads to low estriol secretion by the placenta (Roberts and Cawdery, 1970). Berge (1965) argues that low estrogen levels in pregnancies with anencephalic fetuses may reflect also placental degenerative lesions; yet, dynamic functional studies have shown that the estrogen metabolic capacity of the placenta is unimpaired in fetal anencephaly (see Beling, 1977).

The response of the other placental hormones to fetectomy may be different from that of the steroids. Urinary excretion of hCG decreases gradually after removal of the fetus; however, this downhill course can be interrupted by secondary significant rises of hCG, followed by similar increases in the excretion of pregnanediol (Friedman et al., 1969). Although relatively insensitive methods for hCG determination were used in this study, this observation correlates with the proposition that an intraplacental feedback mechanism may exist between progesterone and hCG (pp. 106-109).

That hCG levels in the maternal circulation may be influenced by the fetus is shown by accumulated inferential or direct evidence. Not only are the hCG maternal levels higher when the fetus is absent (molar pregnancy, choriocarcinoma) or pathological (diabetes, Rh isoimmunization, toxemia of pregnancy) (see the next subsection), they also vary according to the sex of the fetus; Brody and Carlstrom (1965) have shown that hCG maternal levels in the last trimester of pregnancy are significantly higher in pregnancies associated with female than with male fetuses. Although this observation was based on hCG determinations by the complement fixation method, the finding was subsequently confirmed by other investigators using radioimmunoassay methods (Boroditsky et al., 1975; Crosignani et al., 1972; Penny et al., 1974). However, Spellacy et al. (1975) found no such hCG differences at term. These findings may reflect a direct functional (feedback) control of hCG production by the fetus, or sex differences in the fetal metabolism of hCG, or, still, a sex-determined difference in the synthesis of hCG by the placenta. The first possibility is discussed elsewhere in the chapter (pp. 99-102, 106-109). The second cannot be overlooked in view of the fact that the fetus consumes hCG (see the subsection on secretion rates and hormone levels and Table 1); however, higher hCG consumption by male fetuses has not been reported as yet. Finally, the last possibility is supported by the finding that term placentas of male fetuses contain half the hCG found in placentas of female fetuses (Hobson and Wide, 1974).

As for hCG, evidence that the fetal genotype influences the rate of progesterone synthesis and/or metabolism by the placenta, and the fate of progesterone in the fetus, has been reported. Progesterone levels, in both the umbilical vein and artery of female rhesus fetuses, are significantly higher than those in male fetuses; further the difference between progesterone levels in the umbilical vein and the umbilical artery is significantly higher in female than in male fetuses (Hagemenas and Kittinger, 1972). Yet Spellacy et al., (1975) and Boroditsky et al., (1975) found no such differences in pregnant women at term or during the third trimester respectively.

Differences related to fetal sex have been reported also for estradiol. Robinson et al. (1977) reported that estradiol levels in the amniotic fluid at midgestation are higher with female than with male fetuses. On the other hand, Warne et al. (1978) did not find such differences.

That the fetus may influence placental growth and metabolism has been shown in rats (Petropoulos, 1973) and other animals (see Petropoulos, 1973). Although such experimental evidence cannot be obtained in humans, various indirect observations indicate that some fetal control on placental growth is present also in our species.

Placental Endocrine Indices Of Fetal Welfare Or Dysfunction

Placental hormonal levels, both in the mother and the conceptus, can be useful indices in monitoring normal fetal development. All placental hormones have been intensively investigated in relation to the various fetal distress states. As a result, each of these hormones has been proposed, at one time or another, as a specific diagnostic or prognostic means for any one of the various fetal or placental dysfunctions. Although such hormonal assays are undoubtedly valuable in assessing these dysfunctions, their usefulness can be maximized only when both their physiologic basis and their limitations are fully understood by the working obstetrician (see Klopper, 1970).

A variety of endogenous, exogenous or methodological conditions, as well as the commonly used clinical terminology, contribute to the limitations of these assays as useful indices of fetal well-being. In the endogenous group one may include the intricate and multifactorial interactions between maternal, placental and fetal hormones, the biorhythmicity of their secretion (Buster et al., 1978), the volume of their distribution, their renal clearance rates (Alexander et al., 1979), their metabolism by the maternal liver and their affinity to certain body tissues (e.g., steroids to adipose tissue); it should also be borne in mind that placental hormone assays reflect the biosynthetic function of the placenta rather than its, equally important, two-way transfer and barrier functions. Exogenous conditions may be those related to individual habits such as smoking (Spellacy et al., 1977), alcohol consumption, aversion to certain foods, and so forth. Methodological factors include: specificity and sensitivity of the assay method, establishment of a normal range of values for a specific pregnancy period in a truly normal population of pregnant women, stress of sampling, and the cost of serial hormonal determinations, when these are warranted. Finally, clinical terms such as, for example, "placental insufficiency" contribute to the difficulty in the interpretation of the assay results; many diseases cause placental insufficiency and each of them affects placental hormone secretion in a different way and to a different extent. These overall problems in the assessment of fetal welfare by placental hormone assays have been extensively discussed in excellent reviews by Chard (1976) and by Klopper (1976).

Placental hormone determinations are performed in various pathological pregnancy conditions, such as diabetes, Rh isoimmunization, suspected anencephaly, toxemia of pregnancy, hydatidiform mole, and choriocarcinoma. Only a sketchy outline of placental hormone changes in these conditions will be given here, because a detailed discussion of the subject is beyond the scope of this chapter.

Human chorionic gonadotropin levels in maternal plasma and/or urine are low in early, but not late, threatened abortion, normal to high in multiple pregnancy, and high in diabetic and toxemic pregnancy, in Rh isoimmunication, in hydatidiform mole and choriocarcinoma (see Van Leysden 1976). However, hCG values in these conditions show such a large scatter and overlap with normal pregnancy values, that they should be interpreted with caution.

Human chorionic somatomammotropin maternal levels are low in trophoblastic disease (molar pregnancy, choriocarcinoma), in threatened abortion (after the 10th week of pregnancy), in toxemia of pregnancy and in intrauterine growth retardation, but variably high in multiple pregnancy, in Rh isoimmunization, in diabetic pregnancy (see Letchworth, 1976) and in heavy smokers (Spellacy et al., 1977). As with hCG, the same degree of caution must be exercised when interpreting hCS measurements.

Estrogen measurements are valuable in pathological pregnancies. In view of the fetal involvement in the production of estriol and estetrol, these are the estrogens of choice in the assessment of fetal viability or distress. Estriol and estetrol maternal levels are normal in uncomplicated diabetic pregnancies, but low in toxic pregnancies and pregnancies with dead fetuses and fetuses suffering from anencephaly, primary adrenal hypoplasia (Roberts and Cawdery, 1970), Rh isoimmunization, and intrauterine growth retardation (see Baird, 1976). Interestingly, estriol excretion in women with hydatidiform mole is almost as high as in women with normal pregnancies of comparable duration (Macdonald and Siiteri, 1964). However, low estriol levels do not always signal fetal distress or death; this is the rare case of pregnancies with placental sulfatase deficiency, in which the fetus develops normally, although it may have to be delivered by cesarean section, because of labor difficulties (Taylor and Shackleton, 1979). Placental estriol production is drastically reduced in such cases, because the placenta lacks the enzyme necessary to hydrolyze the sulfate from the 16α-OH-dehydroepiandrosterone-sulfate and androstenetriol-sulfate supplied by the fetus (Fig. 1); hence, the placenta is unable to utilize the main precursors of estriol biosynthesis. Maternal plasma unconjugated estradiol levels have been found significantly higher in diabetic than normal women during the third trimester of pregnancy (De Hertogh et al., 1976).

Progesterone measurements in pathological pregnancies are less informative than other placental hormones. Maternal levels are high in molar pregnancy, high to normal in diabetic pregnancy, normal in toxemic pregnancy, in threatened abortion and in Rh isoimmunization, but low in fetal death (see Cooke, 1976).

From the foregoing, it is evident that although the levels of all placental hormones show a wide fluctuation in the various abnormalities of pregnancy, individual hormones may be more specific in the screening of one pathological condition than of another. Because of this wide fluctuation, and in order to attain greater diagnostic and/or prognostic accuracy with these hormone assays, serial rather than single, and dynamic rather than static, hormonal tests are increasingly being adopted.

Precursor-loading, stimulation or inhibition dynamic tests are used to elicit a response from the placenta, or the conceptus as a whole, and thus to obtain useful and accurate information about their functional state. Injection of 50-200 mg dehydroepiandrosterone sulfate (DHEAS) into a normal pregnant woman causes a significant rise of free estradiol (30-90 min after injection) and of free estrone (2-3 hr after injection) in the maternal plasma (Lauritzen, 1967; Lauritizen et al., 1976; Fraser et al., 1976); such a rise is not observed in pregnant women with placental sulfatase deficiency (Fraser et al., 1976). When DHEAS is injected into the amniotic fluid of normal fetuses, the levels of estradiol, estrone and also estriol, increase significantly in the maternal plasma (Lauritzen et al., 1976). With the exception of estriol, similar increases of the other two estrogens are observed after intramniotic injection also of ACTH or hCG to normal fetuses (Lauritzen et al., 1976). Estriol failed to respond also after intramuscular injection of ACTH to the fetus (Burd et al., 1970). On the other hand, dexamethasone administration to the mother brings about a drastic decrease in maternal urinary estriol levels (Burd et al., 1970); that dexamethasone can cross the placenta and effect fetal adrenal suppression has been established (see Baird, 1976; Beling, 1977).

These dynamic tests are more sensitive than the simple static determinations of estrogens, and provide a more reliable diagnostic/prognostic tool for pregnancies at risk.

Intraplacental Interrelations

In secreting both tropic (peptide) and steroid hormones, the placenta would have the rather unique opportunity to autoregulate its own hormone production. This hypothesis was proposed as early as 1934, and it has been investigated since by many researchers, who obtained evidence in support of at least a partial autoregulation. Early perfusion studies of human placentas provided evidence that hCG stimulates conversion of estradiol to estriol and accelerates the aromatization of androgens in the placenta (see: Saxena, 1971). However, it was later proven that the methods employed in these earlier studies did not measure estriol as such, but rather 6α-hydroxy-estradiol-17β (see: Saxena, 1971). Thus, doubt was cast on the existence of placental autoregulation. Yet, more recently, Villee and Gabbe (1972), in a review of their own work and that of others, concluded that hCG: enchances conversion of cholesterol to pregnenolone in the placenta (in vitro, Villee et al., 1966), increases aromatization of androgen precursors to estrogens, and stimulates glycogen phosphorylase activity resulting in glycogenolysis and increased content of free glucose in the placenta. When term placentas were perfused in vitro with testosterone and cAMP, the synthesis of estrone and estradiol from testosterone was significantly stimulated by either hCG or cAMP (Cedard et al., 1970). Further, the addition of theophylline had similar results, presumably by inhibiting the hydrolysis of endogenous cAMP by phosphodiesterase. In parallel with the estrogens, placental free glucose content was increased in this experiment. Placental estriol synthesis from 16α-OH-androstenedione is also enhanced by hCG (Varangot et al., 1965). In subsequent studies Auguy et al., (1976) confirmed that in term placentas perfused in vitro, hCG, dibutyryl cAMP, or prostaglandin F_{2a}, all stimulate phosphorylase b activity, and thus increase the rate of glycogenolysis and the release of free glucose in the placenta. Some of this glucose escapes into the perfusate and some is metabolized through the pentose phosphate pathway generating NADPH, a cofactor necessary in the hydroxylating steps of estrogen synthesis. That prostaglandins may be the second messenger, and cAMP the third, mediating the action of various hormones in various body tissues is well known. Consequently, the findings of Auguy et al. (1976) above may be interpreted as indicating that hCG's mode of action at the cellular level is through prostaglandins, as a second messenger, and cAMP, as a third. Indeed, hCG increases the formation of cAMP, and so do prostaglandins, in term placentas perfused in vitro; further, prostaglandins also stimulate aromatiation of androgens (Alsat et al., 1974). Similar hCG actions at the cellular level have been recorded with ovarian tissue, as well (see pp. 82-86). Finally, hCG is reported to significantly stimulate incorporation of labeled amino acids into proteins by homogenates of human placentas at term (Orlandi et al., 1966).

In the context of evidence presented earlier in the chapter and of the present discussion, hCG emerges as the tropic hormone having overall control on sex steroid synthesis in the fetal testis (and perhaps ovary), in the fetal adrenal (fetal zone), and in the placenta, at least during the first half of gestation. In Fig. 2 a correlation of hCG, progesterone, estriol, and testosterone levels during pregnancy is attempted. From the shape of the curves it is obvious that, when ovarian progesterone begins to decline at about five weeks of pregnancy, hCG continues to rise to reach eventaully its peak value at about nine weeks. Moreover, fetal testosterone rises constantly from the ninth week onwards, but after the 17th week declines gradually to reach low levels at birth (see pp. 99-103). On the basis of the above, it is proposed that the, unexplainable thus far, physiologic function of hCG's sharp peak around the 9th week of pregnancy is to provide the prime boost for the functional initiation of the placenta, the fetal adrenals, and testes into the synthesis of progesterone (and estrogen), dehydroepiandrosterone and testosterone respectively. Once these endocrine glands are stimulated to full scale production of their hormones, hCG levels decline to maintenance tonic levels through the rest of gestation. From the findings of Goodman and Hogden (1979), who interrelated the secretory patterns of macaque-CG and progesterone in ovariectomized pregnant rhesus monkeys, a role similar to hCG may be envisaged for mCG, especially in the functional initiation of the placenta into progesterone secretion. The mechanism for the post-peak decline of hCG might be a negative feedback between the tropic hormone and its target hormones: a short feedback loop in the case of placental progesterone and estrogens, and a long feedback loop in the case of fetal androgens. This hypothesis is summarized in Fig. 3:

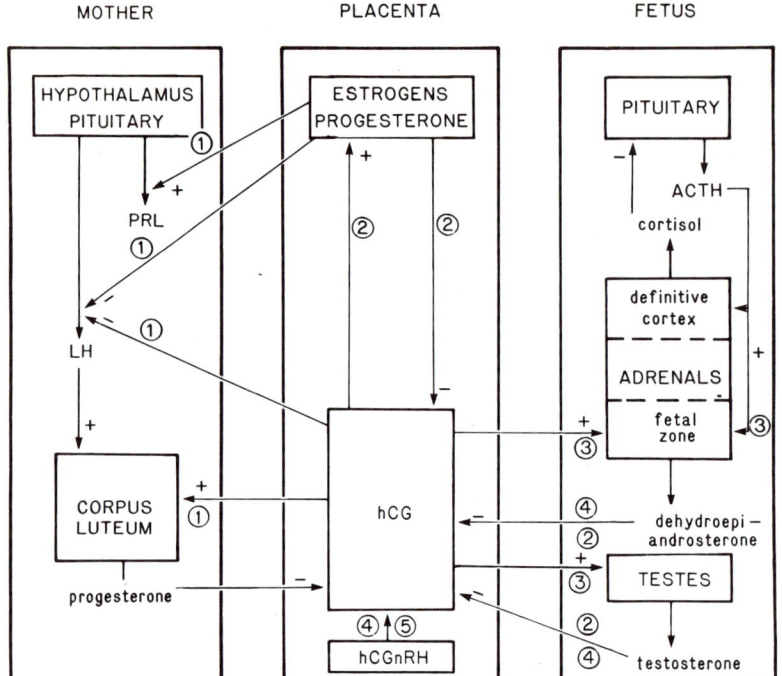

Fig. 3. Proposed feedback interrelations among maternal, placental and fetal hormones.

+, stimulation. -, negative feedback.

For discussion see text:

1, pp. 95-98

2, pp. 106-109

3, pp. 99-103

4, pp. 103-104

5, p. 87

from the foregoing discussion it is evident that the stimulatory effects of hCG are more substantiated than the negative feedback actions of its target hormones shown in this figure. Future research may prove or disprove the existence of such a negative feedback. Presently, only inferential evidence may be cited in support of this concept: (1) hCG levels are lower and testosterone levels higher in pregnancies with male than with female fetuses (p. 103); (2) the rate of increase of estriol levels in plasma/urine accelerates abruptly at about 33 weeks of pregnancy; as shown in Fig. 2, the slope of the rising curve becomes much steeper at that time (see: Beling, 1977; De Hertogh et al., 1975; Simmer, 1968) coincident with, or slightly after, the appearance of the secondary hCG peak at about 32 weeks (Brody, 1969; Vaitukaitis, 1977; Varma et al., 1971); (3) when measurements of urinary hCG and pregnanediol were made in a case of fectomized placenta left in situ, rises observed in hCG levels were always followed by rises in pregnanediol excretion (Friedman et al., 1969). Conversely, there seem to be no feedback inter-relations between corpus luteum and hCG; after corpus luteum removal early in pregnancy in rhesus monkeys, the secretory pattern of mCG was unaffected (Goodman and Hogden, 1979).

A three-level feedback interaction, patterned after the relationship among the hypothalamus, the pituitary and their target glands, might also be present in the placenta (Fig. 3); this possibility should be further investigated in the light of evidence for hCGnRH synthesis by the placenta (p. 87) and of reports that release of hCGα and hCGβ by the human placenta in vitro (Khodr and Siler-Khodr, 1980), and of monkey-CG by the monkey placenta in vivo, is stimulated after administration of synthetic GnRH (Siler-Khodr et al., 1979).

DEVELOPMENT OF PLACENTAL ENDOCRINE FUNCTION IN ADVERSE ENVIRONMENTS

Exposure to adverse environmental conditions, such as high altitude (hypoxia), maternal undernutrition or malnutrition, and toxic substances, during pregnancy causes pathological changes in the placenta and endangers the normal development of the fetus. Although many aspects of environmental fetoplacental pathology have been extensively investigated, surprisingly little research has gone into the endocrine function of the placenta in adverse environments.

High Altitude

At high altitude placental weight is increased, the number of placental cotyledons is decreased, abnormal placental shapes are more frequent, placental thickness is decreased (presumably to provide more surface area for gas exchange), small, or massive, vascular placental infarcts and degenerative changes are routinely observed (see Petropoulos and Timiras, 1974). With regard to placental hormones, there is one report indicating that, at 38 weeks of pregnancy, estrogen, and in particular estriol, excretion is significantly lower at high altitude (Peruvian Andes) than at sea level (Sobrevilla et al., 1968). It should be noted, however, that in this study both placental and newborn weights were found significantly decreased at high altitude.

Maternal Malnutrition

The relationship between hCS plasma levels and the state of maternal nutrition was touched upon briefly earlier in the chapter. Acute maternal starvation in midpregnancy brings about a significant and sustained increase in plasma hCS levels (Kim and Felig, 1971; Tyson et al., 1971); levels of circulating glucose were lowered and there was rapid mobilization of free fatty acids and accumulation of ketone bodies in starved pregnant women, as compared to starved nonpregnant controls. Inasmuch as, acidosis has a deleterious effect on the fetus, the physiologic metabolic activities of hCS during pregnancy must be reconsidered under a new light. Cahill (1967) has suggested that this hCS reaction represents a primitive type of adaptation to starvation to secure the survival of the mother and the placenta itself, even at the expense of the fetus. There are as yet no systematic studies of hCS secretion in undernutrition or malnutrition; yet, it is in these conditions rather than in starvation, that the beneficial effect of hCS on the fetus is likely to be manifested. Undernutrition or malnutrition is usually associated with protein deprivation, whereas carbohydrate intake is sufficient. Under these circumstances, the protein sparing effect of hCS is beneficial and the mobilization of fat is moderate resulting in an acceptable degree of keto-acidosis.

Exposure To Toxic Substances

Ampicillin, a penicillin widely used as an antibiotic for the treatment of infections during pregnancy, depresses estriol levels both in plasma and in urine (William and Pulkkinen, 1971). The decrease is due to a reduced placental secretion, but also to a decreased maternal estriol renal clearance (Alexander et al., 1979), and to alterations in the maternal enterohepatic estriol cycle, because of the destruction of the intestinal microflora by the antibiotic. Similar findings are recorded after treatment with neomycin (Pulkkinen and Willman, 1973).

PLACENTAL HORMONES AND PARTURITION

The mechanisms triggering the onset of labor and leading to parturition are not only complex but also species specific. They have been extensively investigated in the sheep, but are not completely understood in the human. These mechanisms may include hormonal control of excitability of the myometrium (at any one level, neuronal or physicochemical, between relaxation and contraction), mechanical factors (e.g., stretch of the myometrial cells) or neurogenic factors, or, still, specific fetal signals. An extensive discussion of the subject of labor is beyond the scope of this review; however, the possible involvement of human placental hormones in the onset of labor and parturition is briefly outlined below, and then a sketchy general model of hormonal interrelations, that purportedly are at work during labor, is given.

Estrogens

All three estrogens show a progressive increase during pregnancy, but a wide range of values at term. When a highly uniform group of healthy primigravidae with normal term deliveries was followed by weekly RIA determinations of estradiol, values showed a sustained increase, and actually doubled during the last six weeks of gestation (Turnbull et al., 1974). That pregnancies with anencephalic fetuses go usually beyond term, might not indicate a deficient production of placental estrogens caused by scarcity of fetal precursors, because an equal number of such pregnancies end in premature labor. On the other hand, it has been already mentioned that pregnant women with placental sulfatase deficiency (and thus inability to synthesize estrogens from fetal precursors) have difficulties in labor (pp. 105-106).

Progesterone

Many years ago, Csapo (1961) advanced the hypothesis that withdrawal of progesterone triggers the onset of labor in the human. Since then, many conflicting studies have appeared in support of or against this hypothesis. Turnbull et al. (1974), in the same group of primigravidae described above, found a continuous and significant decrease in progesterone levels during the last six weeks of pregnancy before spontaneous labor. Yet, not all subjects showed the same pattern and the findings have not yet been confirmed by other laboratories. Obviously, more studies employing sensitive and specific steroid assay methods and studying highly homogeneous groups of normal pregnant women, are needed to clarify the role of both estrogens and progesterone in the initiation of labor.

Other Placental Hormones

The peptide placental hormones have not yet been implicated in the mechanism of labor initiation.

A Model Of Possible Endocrine Interrelations During Labor

Hypotheses and some facts about the mechanism of labor initiation in the human are listed below in sequential order, as they probably occur in the course of parturition:

(a) At the appropriate time the fetus emits a signal, which will start the avalanche of events that lead irreversibly to parturition. This signal is initiated in the fetal brain and then, through the hypothalamic releasing factors and, in turn, the tropic hypophysial hormones, it reaches the peripheral endocrine glands of the fetus.

(b) In particular, the fetal adrenals are stimulated to produce more cortisol and perhaps more DHEA, the precursor of placental estrogens.

(c) Placental estrogen synthesis is increased; estradiol and estriol rise in the maternal circulation.

(d) Placental progesterone secretion decreases; also progesterone is gradually displaced from its cellular receptors in the myometrium by the competing fetal cortisol, which reaches the uterus now in increased amounts.

(e) The progesterone/estrogen (P/E) ratio is now effectively lowered at the myometrial level and, as a result, the uterine excitability is increased.

(f) The fetal posterior pituitary releases now oxytocin and vasopressin, which upon reaching the myometrium through various ways, cause segmented uterine contractions, which, in turn, cause stretching of other myometrial fibers.

(g) The decreased P/E ratio and increased cortisol levels in the myometrium, as well as the stretching of some myometrial fibers, may facilitate uterine prostaglandin synthesis. Prostaglandins, in turn, will potentiate the effect of oxytocin on the myometrium.

(h) The maternal hypothalamus and pituitary now enter the scene. Stretch of the uterine cervix initiates a neuroendocrine reflex that triggers oxytocin release from the maternal hypothalamus. Further, the low P/E ratio may activate the hypothalamus/hypophysis to secrete ACTH, which, in turn, may be responsible for the rise of the maternal cortisol levels during labor.

(i) Whether the human uterine cervix and the symphysis pubis are influenced at this stage by relaxin, as is the case in other mammals, remains entirely conjectural at present. O'Byrne et al. (1978), using a heterologous RIA, measured immunoreactive relaxin in human maternal plasma during pregnancy. Relaxin was present as early as the fourth gestational week, was detectable throughout pregnancy, showed no peak before parturition (as in many other species) and its developmental pattern through gestation resembled closely that of hCG.

(j) Maternal oxytocin and uterine intracellular prostaglandins maintain rhythmic contractions of the myometrium, until the uterus is evacuated.

Existing evidence related to the above model has been discussed in an excellent review by Fuchs (1977).

EPILOGUE

The intricate interrelations between placental and fetomaternal hormones have been reviewed. Much more awaits discovery in this field. What has been revealed so far tells a fascinating story of efficient mechanisms designed to protect a helpless zygote at the time of inception, through its development to a highly complex organism at birth; an organism ready and capable of eventually realising its full potential in adulthood. From the past achievements new vistas open to future research in this field. Through more research, an intimate knowledge of the endocrine developmental patterns of pregnancy will be acquired. This knowledge will enable us to secure, through rational pharmacotherapy when necessary, a "milieu interieur" conducive to a normal fetal development and a timely delivery. Intrauterine fetal damage and the high perinatal morbidity and mortality from the complications of childbirth would then be avoided.

ACKNOWLEDGMENTS

I wish to thank Mrs. Denise P. Woods for her expert typing of this manuscript. Bibliography searches (Nos. S2044103, P0034045) were undertaken for this review by the Medlar Search Station at the Institute for Medical Literature of the South African Medical Research Council, Tygeberg, S.A., and are gratefully acknowledged.

REFERENCES

Abdul-Karim, R.W., Nesbitt, R.E.L., Jr., Drucker, M.H., and Rizk, P.T. The regulatory effect of estrogens on fetal growth. *Am. J. Obstet. Gynecol.* 109, 656-661 (1971).

Abramovich, D.R. and Rowe, P. Foetal plasma testosterone levels at mid-pregnancy and at term: relationship to foetal sex. *J. Endocrinol.* 56, 621-622 (1973).

Abramovich, D.R., Baker, T.G., and Neal, P. Effect of human chorionic gonadotropin on testosterone secretion by the foetal human testis in organ culture. *J. Endocrinol.* 60, 179-185 (1974).

Akasu, F., Kawahara, S., Ohki, H., Harano, M., and Tejima, Y. Thyroid stimulating hormone extracted from human placenta. *Endocrinol. Jap.* 2, 297-307 (1955).

Albert, A. Follicle-stimulating activity of human chorionic gonadotropin. *J. Clin. Endocrinol. Metab.* 29, 1504-1509 (1969).

Alexander, S., Stavric, V., Smuk, M., Sugar, J., and Schwers, J. Renal clearance of estriol and its conjugates in normal and abnormal pregnancies. *J. Clin. Endocrinol. Metab.* 49, 588-593 (1979).

Alsat, E., Levilliers, J., Auguy, A., and Cedard, L. Kinetic study of the action of gonadotropins and prostaglandins on oestrogen biosynthesis, glycogen phosphorylase and cyclic AMP levels in human full term placentas perfused "in vitro." *J. Steroid Biochem.* 5, 401-402 (1974).

Assies, J., Schellekens, A.P.M., and Touber, J.L. Protein hormones in cerebrospinal fluid: evidence for retrograde transport of prolactin from pituitary to the brain in man. *Clin. Endocrinol.* 8, 487-491 (1978).

Auguy, A., Alsat, E., and Cedard, L. Activité phosphorylasique du placenta humain perfusé in vitro: action des hormones gonadotropes, du dibutyryl AMPc et des prostaglandines. *C.R. Acad. Sci. (D) (Paris)* 282, 897-900 (1976).

Azukizawa, M., Kurtzman, G., Pekary, A.E., and Hershman, J.M. Comparison of the binding characteristics of bovine thyrotropin and human chorionic gonadotropin to thyroid plasma membranes. *Endocrinology* 101, 1880-1889 (1977).

Azukizawa, M., Murata, Y., Ikenoue, T., Martin, C.B., Jr., and Hershman, J.M. Effect of thyrotropin-releasing hormone on secretion of thyrotropin, prolactin, thyroxine, and triiodothyronine in pregnant and fetal rhesus monkeys. *J. Clin. Endocrinol. Metab.* 43, 1020-1028 (1976).

Baird, D.T. Oestrogens in clinical practice, in *Hormone Assays and their Clinical Application*, J. Loraine and E.T. Bell, eds. Churchill Livingstone, Edinburgh (1976), pp. 408-446.

Bala, R.M., Wright, C., Bardai, A., and Smith, G.R. Somatomedin bioactivity in serum and amniotic fluid during pregnancy. *J. Clin. Endocrinol. Metab.* 46, 649-652 (1978).

Bearn, J.G. The male anenceplalic foetus. *Lancet* 2, 464-465 (1959).

Beling, C. Estrogens, in *Endocrinology of Pregnancy*, 2nd Edition, F. Fuchs and A. Klopper, eds. Harper & Row, New York (1977), pp. 76-98.

Belisle, S., Fencl, M. de M., Osathanondh, R., Tulchinsky, D. Sources of 17α-hydroxypregnenolone and its sulfate in human pregnancy. *J. Clin. Endocrinol. Metab.* 46, 721-728 (1978).

Benirschke, K. Adrenals in anencephaly and hydrocephaly. *Obstet. Gynecol.* 8, 412-425 (1956).

Benveniste, R., Speeg, K.V. Jr., Carpenter, G., Cohen, S., Lindner, J., and Rabinowitz, D. Epidermal growth factor stimulates secretion of human chorionic gonadotropin by cultured human choriocarcinoma cells. *J. Clin. Endocrinol. Metab.* 46, 169-172 (1978).

Berge, B.S. Ten. The placenta in anercephaly *Gynaecologia (Basel)* 159, 359-364 (1965).

Bewley, T.A. and Li, C.H. Structural similarities between human pituitary growth hormone, human chorionic somatomammotropin, and ovine pituitary growth and lactogenic hormones, in *Problems of Human Reproduction: Vol. 2, Lactogenic Hormones, Fetal Nutrition and Lactation*, J. B. Josimovich, M. Reynolds, and E. Cobo, eds. J. Wiley & Sons, New York (1974) pp. 19-32.

Blackburn, W.R., Kaplan, H.S., and McKay, D.G. Morphologic changes in developing rat placenta following prednisolone administration. *Am. J. Obstet. Gynecol.* 92, 234-246 (1965).

Boroditsky, R.S., Reyes, F.I., Winter, J.S., and Faiman, C.F. Serum human chorionic gonadotropin and progesterone patterns in the last trimester of pregnancy: relationship to fetal sex. *Am. J. Obstet. Gynecol.* 121, 238-241 (1975).

Bouchacourt, M.L. Nouvelles recherches sur l'opotherapie placentaire. *C.R. Soc. Biol.* 54, 133-135 (1902).

Breborowicz, H. and Biniszkiewicz, W. The relation between the urinary oestriol and 17Ks levels and the state of maternal and foetal adrenals, in *Intra-uterine Dangers to the Foetus*, J. Horsky and Z. K. Stempera, eds. Excerpta Medica Foundation, Amsterdam (1967), pp. 283-286.

Brody, S. Protein hormones and hormonal peptides from the placenta, in *Foetus and Placenta*, A. Klopper and E. Diczfalusy, eds. Blackwell Scientific Publications, Oxford (1969), pp. 299-411.

Brody, S. and Carlström, G. Human chorionic gonadotropin pattern in serum and its relation to the sex of the fetus. *J. Clin. Endocrinol. Metab.* 25, 792-797 (1965).

Brown, J.B., Beischer, N.A., and Smith, M.A. Excretion of urinary oestrogens in pregnant patients treated with cortisone and its analogues. *J. Obstet. Gynaecol. Br. Commonw.* 75, 819-828 (1968).

Burd, L.I., Bieniarz, J., Nedoss, B.R., Charles, A.G., and Scommegna, A. Maternal urinary estriol excretion after administration of ACTH to the fetus. *Obstet. Gynecol.* 36, 574-581 (1970).

Buster, J.E., Meis, P.J., Hobel, C.J., and Marshall, J.R. Subhourly variability of circulating third trimester maternal steroid concentrations as a source of sampling error. *J. Clin. Endocrinol. Metab.* 46, 907-910 (1978).

Butt, W.R. *The Chemistry of the Gonadotrophins.* Charles C. Thomas, Co. Springfield Ill. (1967), pp. 47-48.

Catt, K.J., Dufau, M.L., and Tsuruhara, J.J. Studies on a radioligand-receptor assay system for luteinizing hormone and chorionic gonadotropin. *J. Clin. Endocrinol. Metab.* 32, 860-863 (1971).

Cahill, G.F., Jr. Diabetes mellitus in man and experimental animals. *Ann. Int. Med.* 66, 227-230 (1967).

Cedard, L., Alsat, E., Urtasun, M.-J., and Varangot, J. Studies on the mode of action of luteinizing hormone and chorionic gonadotropin on estrogenic biosynthesis and glycogenolysis by human placenta perfused in vitro. *Steroids* 16, 361-376 (1970).

Chard, T. Normality and abnormality, in *Plasma Hormone Assays in Evaluation of Fetal Wellbeing*, A. Klopper, ed. Churchill Livingstone, Edinburgh (1976), pp. 1-19.

Charles D., Harkness, R.A., Kenny, F.M., Menini, E., Ismail, A.A.A., Durkin, J.W., and Loraine, J.A. Steroid excretion patterns in an adrenalectomized woman during three successive pregnancies. *Am. J. Obstet. Gynecol.* 106, 66-74 (1970).

Chatterjee, M. and Munro, H.N. Structure and biosynthesis of human placental peptide hormones. *Vitam. Horm.* 35, 149-208 (1977).

Chen. H.-C. and Hodgen, G.D. Primate chorionic gonadotropins: antigenic similarities to the unique carboxyl-terminal peptide of hCGβ subunit. *J. Clin. Endocrinol. Metab.* 43, 1414-1417 (1976).

Chen, H.-C., Hodgen, G.D., Matsuura, S., Lin, L.J., Gross, E., Reichert, L.E., Jr., Birkin, S., Canfield, R.E., and Ross, G.T. Evidence for a gonadotropin from non-pregnant subjects that has physical, immunological and biological similarities to human chorionic gonadotropin. *Proc. Natl. Acad. Sci.* 73, 2885-2889 (1976).

Clements, J.A., Reyes, F.I., Winter, J.S.D., and Faiman, C. Studies on human sexual development. III. Fetal pituitary and serum, and amniotic fluid concentrations of LH, CG and FSH. *J. Clin. Endocrinol. Metab.* 42, 9-19 (1976).

Cooke, I.D. Progesterone and its metabolites, in *Hormone Assays and their Clinical Applications.*, J. A. Loraine and E. T. Bell, eds. Churchill Livingstone, Edinburgh (1976), pp. 447-507.

Craft, I., Wyman, H., and Somerville, I.F. Serial analysis of plasma progesterone and pregnanediol in human pregnancy. *J. Obst. Gynaecol. Br. Commonw.* 76, 1080-1089 (1969).

Crosignani, P.G., Noncioni, T., and Brambati, B. Concentration of chorionic gonadotropin and chorionic somatomammotropin in maternal serum, amniotic fluid and cord blood serum at term. *J. Obstet. Gynaecol. Br. Commonw.* 79, 122-126 (1972).

Croskerry, P.G. and Dobbing, J. Placental inhibition of foetal growth enhancement in the rat. *Nature* 273, 147-149 (1978).

Csapo, A.I. The onset of labor. *Lancet* 2, 277-280 (1961).
De Hertogh, R., Thomas, K., Bietlot, Y., Vanderheyden, I., and Ferin, J. Plasma levels of unconjugated estrone, estradiol, and estriol and of hCS throughout pregnancy in normal women. *J. Clin. Endocrinol. Metab.* 40, 93-101 (1975).
De Hertog, R., Thomas, K., and Vanderheyden, I. Quantitative determination of Sex Hormone-Binding Globulin Capacity in the plasma of normal and diabetic pregnancies. *J. Clin. Endocrinol. Metab.* 42, 773-777 (1976).
De la Lastra, M. and Llados, C. Luteinizing hormone content of the pituitary gland in pregnant and non-pregnant women. *J. Clin. Endocrinol. Metab.* 44, 921-923 (1977).
Dickey, R.P., Besch, P.K., Vorys, N., and Ullery, J.C. Diurnal excretion of estrogen and creatinine during pregnancy. *Am. J. Obstet. Gynecol.* 94, 591-594 (1966).
Diczfaluzi, E. Endocrinology of the foetus. *Acta Obstet. Gynecol. Scand.* 41, Suppl. 1, 45-55 (1962).
Diczfalusi, E. Steroid metabolism in the foeto-placental unit, in *The Foeto-placental Unit*, A. Pecile and C. Finzi, eds. Excerpta Medica Foundation, Amsterdam (1969), pp. 65-109.
Duhring, J.L., McKean, H.E., and Green, J.W. Diurnal variation of estriol excretion in human pregnancy. *Am. J. Obstet. Gynecol.* 115, 875-880 (1973).
Effer, S.N., Gupta, K., and Younglai, E.V. Concentrations of human chorionic gonadotropin, progesterone, and unconjugated and total estriol in umbilical artery and vein plasma at term. *Am. J. Obstet. Gynecol.* 116, 643-647 (1973).
Frame, L.T., Wiley, L., and Regol, A.D. Indirect immunofluorescent localization of prolactin to the cytoplasm of decidua and trophoblast cells in human placental membranes at term. *J. Clin. Endocrinol. Metab.* 49, 435-437 (1979).
Frandsen, V.A. and Stakeman, G. The site of production of oestrogenic hormones in human pregnancy. Hormone excretion in pregnancy with anencephalic foetus. *Acta Endocrinol.(Kbh)* 38, 383-391 (1961).
Fraser, I.S., Leask, R., Drife, J., Balcon, L., and Michie, E.A. Plasma estrogen response to dehydroepiandrosterone sulphate injection in normal and complicated late pregnancy. *Obstet. Gynecol.* 47, 152-158 (1976).
Frieden, E.H. *Chemical Endocrinology*. Academic Press, New York (1976).
Friedman, S., Gans, B., Eckerling, B., Goldman, J., Kaufman, H., and Rumny, M. Placental hormone activity after removal of the fetus in a case of advanced abdominal pregnancy. *J. Obstet. Gynaecol. Br. Commonw.* 76, 554-558 (1969).
Fuchs, F. Endocrinology of labor, in *Endocrinology of Pregnancy*, 2nd Edition, F. Fuchs and A. Klopper, eds. Harper & Row, New York (1977), pp. 327-349.
Furlanetto, R.W., Underwood, L.E., Van Wyk, J.J., and Handwerger, S. Serum immunoreactive somatomedin-C is elevated late in pregnancy. *J. Clin. Endocrinol. Metab.* 47, 695-698 (1978).
Gabbe, S.G., Demers, L.M., Greep, R.O., and Villee, C.A. Placental glycogen metabolism in diabetes mellitus. *Diabetes* 21, 1185-1191 (1972).
Gabrilove, J.L. The adrenal cortex, in *Medical, Surgical and Gynecologic Complications of Pregnancy*. 2nd Edition, J. J. Rovisky and A. F. Guttmacher, eds. Williams and Wilkins, Baltimore (1965), pp. 571-580.
Gaspard, U.J., Luyckx, A.S., George, A.N., and Lefebvre, P.J. Relationship between plasma free fatty acid levels and human placental

lactogen secretion in late pregnancy. *J. Clin. Endocrinol. Metab.* 45, 246-254 (1977).

George, F.W. and Wilson, J.D. Conversion of androgen to estrogen by the human fetal ovary. *J. Clin. Endocrinol. Metab.* 47, 550-555 (1978).

Gerbie, A.B. Endocrine diseases complicated by pregnancy, in *Medical Complications during Pregnancy*, D M. Haynes, ed. McGraw-Hill, New York (1969), pp. 335-377.

Gibbons, J.M., Mitnick, M., and Chieffo, V. In vitro biosynthesis of TSH- and LH- releasing factors by the human placenta. *Am. J. Obstet. Gynecol.* 121, 127-131 (1975).

Ginsburg, J. and Jeacock, M.K. The placental barrier, in *Absorption and Distribution of Drugs*, T. B. Binns, ed. Livingstone, Edinburgh (1964), pp. 86-102.

Gluckman, P.D. and Brinsmead, M.W. Somatomedin in cord blood: relationship to gestational age and birth size. *J. Clin. Endocrinol. Metab.* 43, 1378-1381 (1976).

Goebel, R. and Kuss, E. Circadian rhythm of serum unconjugated estriol in late pregnancy. *J. Clin. Endocrinol. Metab.* 39, 969-972 (1974).

Goluboff, L. G. and Ezrin, C. Effect of pregnancy on the somatotroph and the prolactin cell of the human adenohypophysis. *J. Clin. Endocrinol. Metab.* 29, 1533-1538 (1969).

Goodman, A.L. and Hogden, G.D. Corpus luteum - conceptus - follicle relationships during the fertile cycle in rhesus monkeys: pregnancy maintenance despite early luteal removal. *J. Clin. Endocrinol. Metab.* 49, 469-471 (1979).

Gulyas, B.J., Tullner, W.W., and Hogden, G.D. Fetal or maternal hypophysectomy in Rhesus monkeys (*Macaca mullata*): effects on the development of the testes and other endocrine organs. *Biol. Reprod.* 17, 650-660 (1977a).

Gulyas, B.J., Hogden, G.D., Tullner, W.W., and Ross, G.T. Effects of fetal or maternal hypophysectomy on endocrine organs and body weight in infant monkeys (*Macaca mullata*): with particular emphasis on oogenesis. *Biol. Reprod.* 16, 216-227 (1977b).

Hagemenas, F.C. and Kittinger, G.W. The influence of fetal sex on plasma progesterone levels. *Endocrinology* 91, 253-256 (1972).

Hall, E.V. van and Heintz, A.P.M. Influence of hCG on the production of estrogens by the adrenal gland. *J. Endocrinol.* 43, xxxiv (1969).

Hall, E.V. van., Vaitukaitis, J.L., Ross, G.T., Hickman, J.W., and Aschwell, G. Immunological and biological activity of hCG following progressive desialylation. *Endocrinology* 88, 456-464 (1971).

Handwerger, S. and Sherwood, L.M. Comparison of the structure and lactogenic activity of human placental lactogen and human growth hormone, in *Problems of Human Reproduction: Vol. 2, Lactogenic Hormones, Fetal Nutrition and Lactation*, J. B. Josimovich, M. Reynolds and E. Cobo, eds. J. Wiley & Sons, New York (1974), pp. 33-47.

Handwerger, S., Barrett, J., Tyrey, L., and Schomberg, D. Differential effect of cyclic adenosine monophosphate on the secretion of human placental lactogen and human chorionic gonadotropin. *J. Clin. Endocrinol. Metab.* 36, 1268-1270 (1973).

Harada, A. and Hershman, J.M. Extraction of human chorionic thyrotropin (hCT) from term placentas: failure to recover thyroptropic activity. *J. Clin. Endocrinol. Metab.* 47, 681-685 (1978).

Harada, A., Hershman, J.M., Reed, A.W., Braunstein, G.D., Dignam, W.J., Derzko, C., Friedman, S., Jewelewicz, R., and Pekary, A.E. Comparison of thyroid stimulators and thyroid hormone concentrations in the sera of pregnant women. *J. Clin. Endocrinol. Metab.* 48, 793-797 (1979).

Healey, D.L., Kimpton, W.G., Muller, H.K., and Burger, H.G. The synthesis of immunoreactive prolactin by decidua-chorion. *Br. J. Obstet. Gynaecol.* 86, 307-313 (1979).

Hennen, G.P. Detection and study of a human chorionic thyroid-stimulating factor. *Arch.Int. Physiol. Biochim.* 73, 689-695 (1965).

Hennen, G.P. and Freychet, P. Human chorionic thyrotropin: its relationship to thyroid stimulators from chorionic neoplasms and nonendocrine cancers. *Isr. J. Med. Sci.* 10, 1332-1334 (1974).

Hershman, J.M. and Burrow, G.N. Lack of release of human chorionic gonadotropin by thyrotropin releasing hormone. *J. Clin. Endocrinol. Metab.* 42, 970-972 (1976).

Hershman, J.M. and Starnes, W.R. Extraction and characterization of a thyrotropic material from the human placenta. *J. Clin. Invest.* 48, 923-929 (1969).

Higgins, H.P. and Hershman, J.M. The hyperthyroidism due to trophoblastic hormone. *Clin. Endocrinol. Metab.* 7, 167-175 (1978).

Hirono, M., Igarashi, M., and Matsumoto, S. The direct effect of hCG upon pituitary gonadotropin secretion. *Endocrinology* 90, 1214-1219 (1972).

Hobson, B.M. and Wide, L. Chorionic gonadotropin in the human placenta in relation to the sex of the foetus at term. *J. Endocrinol.* 60, 75-80 (1974).

Huhtaniemi, I.T., Korenbrot, C.C., and Jaffe, R.B. hCG binding and stimulation of testosterone biosynthesis in the human fetal testis. *J. Clin. Endocrinol. Metab.* 44, 963-967 (1977a).

Huhtaniemi, I.T., Korenbrot, C.C., and Jaffe, R.B. Content of chorionic gonadotropin in human fetal tissues. *J. Clin. Endocrinol. Metab.* 46, 994-997 (1978).

Huhtaniemi, I.T., Korenbrot, C.G., Serón-Ferré, M., Foster, D.B., Parer, J.T., and Jaffe, R.B. Stimulation of testosterone production in vivo and in vitro in the male rhesus monkey fetus in late gestation. *Endocrinology* 100, 839-844 (1977b).

Hussa, R.O., Pattillo, R.A., Ruckert, A.C.F., and Scheuermann, K.W. Effects of butyrate and dibutyryl cyclic AMP on hCG-secreting trophoblastic and non-trophoblastic cells. *J. Clin. Endocrinol. Metab.* 46, 69-76 (1978).

Hytten, F.E. and Leitch, I. *The Physiology of Human Pregnancy*, 2nd Edition. Blackwell, Oxford (1971), pp. 179-233.

International Symposium on "Human Placenta - Proteins and Hormones," Siena (Italy), July 1979. Proceedings to be published by Academic Press.

Johannisson, E. The fetal adrenal cortex in the human. Its ultrastructure at different stages of development and in different functional states. *Acta Endocrinol. (Kbh)* 58, Suppl. 130, 1-107 (1968).

Johanson, E.D.B. Plasma levels of progesterone in pregnancy measured by a rapid competitive protein binding technique. *Acta Endocrinol. (Kbh)* 61, 607-617 (1969).

Josimovich, J.B. Human placental lactogen: Further evidence of placental mimicry of pituitary function. *Am. J. Obstet. Gynecol.* 120, 550-552 (1974).

Josimovich, J.H. Human placental lactogen, in *Endocrinology of Pregnancy*, 2nd Edition, F. Fuchs and A. Klopper, eds. Harper & Row, New York (1977), pp. 191-205.

Josimovich, J.B. and Venning, E.H. Hormonal physiology of the placental polypeptide and steroidal hormones, in *Gynecologic Endocrinology*, J. J. Gold, ed. Harper & Row, New York (1975), pp. 78-98.

Josimovich, J.B., Stock, R.J., and Tobon, H. Effects of primate placental lactogen upon lactation, in *Problems of Human Reproduction: Vol. 2, Lactogenic Hormones, Fetal Nutrition, and Lactation*, J. B. Josimovich, M. Reynolds, and E. Cobo, eds. J. Wiley & Sons, New York (1974), pp. 335-350.

Kaplan, S.L. and Grumbach, M.M. Effects of primate chorionic somatomammatropin on maternal and fetal metabolism, in *Problems of Human Reproduction: Vol. 2 Lactogenic Hormones, Fetal Nutrition and Lactation*, J.B. Josimovich, M. Reynolds, and E. Cobo, eds. J. Wiley & Sons, New York (1974), pp. 183-191.

Kenimer, J.G., Herschman, J.M., and Higgins H.P. The thyrotropin in hydatidiform moles is human chorionic gonadotropin. *J. Clin. Endocrinol. Metab.* 40, 480-489 (1975).

Khodr, G.S. and Siler-Kodr, T.M. Placental luteinizing hormone-releasing factor and its synthesis. *Science* 207, 315-317 (1980).

Kim, Y.J. and Felig, P. Plasma chorionic somatomammotropin levels during starvation in mid-pregnancy. *J. Clin. Endocrinol. Metab.* 32: 864-867 (1971).

Kim, M.H., Borth, R., McCleary, P.H., Woolever, C.A., and Young, P.C.M. Sex hormone secretion of the placenta left in situ after ovarian pregnancy. *Am. J. Obstet. Gynecol.* 110, 658-662 (1971).

Klopper, A. Assessment of fetoplacental function by hormone assay. *Am. J. Obstet. Gynecol.* 107, 807-827 (1970).

Klopper, A. Criteria for the selection of steroid assays in the assessment of fetoplacental function, in *Plasma Hormone Assays in Evaluation of Fetal Wellbeing*, A. Klopper, ed. Churchill Livingstone, Edinburgh (1976), pp. 20-35.

Klopper, A. and Fuchs, F. Progestagens, in *Endocrinology of Pregnancy*, 2nd Edition, F. Fuchs and A. Klopper, eds. Harper & Row, New York (1977), pp. 99-122.

Kopp, L., Paradiz, G., and Tucci, J.R. Urinary excretion of cyclic 3',5'-adenosine monophosphate and cyclic 3',5'-guanosine monophosphate during and after pregnancy. *J. Clin. Endocrinol. Metab.* 44, 590-594 (1977).

Kulin, H.E., Santner, S.J., and Mann, W.J. Urinary folicle-stimulating hormone during pregnancy: relationship to sex of fetus. *J. Clin. Endocrinol. Metab.* 48, 736-738 (1979).

Laga, E.M., Driscoll, S.G., and Munro, H.N. Human Placental Structure: *Problems of Human Reproduction: Vol. 2. Lactogenic Hormones, Fetal Nutrition and Lactation*, J. B. Josimovich, M. Reynolds, and E. Cobo, eds. J. Wiley and Sons, New York (1974), pp. 143-181.

Lajos, L. Unknown endocrinological properties of the adenohypophysis. *Gynaecologia* 150, 366-378 (1960).

Lajos, L., Szabo, T., Egyed, R., and Vereczey, G. Les relations entre la placentotrophine pituitaire, le tissu trophoblastique normal et les tumeurs du chorion. *Vie Med. Canad. Franc.* 4, 162-171 (1976).

Lajos, L., Csaba, I., Domany, S., Szekely, J., and Breila, I. Experimental examination of the conditions of choriongonadotropin production. *Gynaecologia* 147, 152-164 (1959).

Lauritzen, Ch. A clinical test for placental functional activity, using DHEA-Sulfate and ACTH-injections in the pregnant woman. *Acta Endocrinol. (Kbh)* 56, Suppl. 119, 188 (1967).

Lauritzen, Ch. and Lehmann, W.-D. Levels of chorionic gonadotropin in the newborn infant and their relationship to adrenal dehydroepiandrosterone. *J. Endocrinol.* 39, 173-182 (1967).

Lauritzen, Ch., Strecker, J., and Lehmann, W.-D. Dynamic tests of placental function: some findings on the conversion of DHAS to oestrogens, in *Plasma Hormone Assays in Evaluation of Fetal Wellbeing*, A. Klopper, ed. Churchill Livingstone, Edinburgh (1976), pp. 113-135.

Letchworth, A.T. Human placental lactogen assay as a guide to fetal well-being, in *Plasma Hormone Assays in Evaluation of Fetal Wellbeing*, A. Klopper ed. Churchill Livingstone, Edinburgh (1976), pp. 147-173.

Levitz, M., Slyper, A.J. and Selinger, M. On the lack of periodicity in plasma estriol in human pregnancy. *J. Clin. Endocrinol. Metab.* 38, 698-700 (1974).

Li, C.H. On the characterization of human chorionic somatomammotropin. *Ann. Sclavo.* 12, 651-662 (1970).

Macaron, C., Famuyiwa, O., and Singh, S.P. In vitro effect of dopamine and pimozide on human chorionic somatomammotropin (hCS) secretion. *J. Clin. Endocrinol. Metab.* 47, 168-170 (1978a).

Macaron, C., Freinkel, N., Brewer, J., Wilber, J.F., Kahn, B., and Halpern, B. Gonadotropin-releasing hormone (GnRH) stimulates human chorionic gonadotropin (hCG) release in trophoblastic disease. *Clin. Res.* 24, 274 (1976).

Macaron, C., Kyncl, M., Rutsky, L., Halpern, B., and Brewer, J. Failure of somatostatin to affect human chorionic somatomammotropin and human chorionic gonadotropin secretion in vitro. *J. Clin. Endocrinol. Metab.* 47, 1141-1143 (1978b).

Macdonald, P.C. and Siiteri, P.K. Study of estrogen production in women with hydatidiform mole. *J. Clin. Endocrinol. Metab.* 24, 685-690 (1964).

MacNaughton, M.C. Hormone assays in early pregnancy, in *Plasma Hormone Assays in Evaluation of Fetal Wellbeing*, A. Klopper, ed. Churchill Livingstone, Edinburgh (1976), pp. 36-47.

Masson, G.M. and Wilson, G.R. Variability of total plasma oestriol in late human pregnancy. *J. Endocrinol.* 54, 245-250 (1972).

McCormick, J.B. Gonadotropin in urine and spinal fluid: quantitative studies for chorionic moles and choriocarcinomas. *Obstet. Gynecol.* 3, 58-66 (1954).

Mitchell, H.D.C. and Bagshawe, K.D. Human chorionic gonadotropin, in *Hormone Assays and their Clinical Application*, J. A. Loraine and E. T. Bell, eds. Churchill Livingstone, Edinburgh (1976), pp. 141-174.

Miyakawa, I., Ikeda, I., and Maeyama, M. Transport of ACTH across human placenta. *J. Clin. Endocrinol. Metab.* 39, 440-442 (1974).

Miyake, A., Tanizawa, O., Aono, T., Yasuda, M., and Kurachi, K. Suppression of luteinizing horome in castrated women by the administration of human chorionic gonadotropin. *J. Clin. Endocrinol. Metab.* 43, 928-932 (1976).

Mochizuki, M., Morikawa, H., Kawaguchi, K., and Tojo, S. Growth hormone, prolactin and chorionic somatomammotropin in normal and molar pregnancy. *J. Clin. Endocrinol. Metab.* 43, 614-621 (1976).

Morrison, G., Meigs, R.A., and Ryan, K.J. Biosynthesis of progesterone by the human placenta. *Steroids* 6, Suppl. 2, S177-S188 (1965).

Mossman, H.W. The principal interchange vessels of the chorioallantoic placenta of mammals, in *Organogenesis*, R. L. De Haan, and H. Ursprung, eds. Holt-Tinehart-Winston. New York (1965), pp. 771-786.

Munson, A.K., Yannone, M.E., and Mueller, J.R. The diurnal pattern of 17β-estradiol in pregnancy. *Acta Endocrinol.(Kbh)* 69, 410-412 (1972).

Nagamani, M., McDonough, P.G., Ellegood, J.O., and Mahesh, V.B. Maternal and amniotic fluid steroids throughout human pregnancy. *Am. J. Obstet. Gynecol.* 134, 674-680 (1979).

Niemi, M., Ikonen, M., and Hervonen, A. Histochemistry and fine structure of interstitial tissue in human foetal testis, in Ciba Foundation, *Colloquia on Endocrinology*, Vol 16: *Endocrinology of the Testis*. G. E. W. Wolstenholme and M. O'Connor, eds. Churchill, London (1967), pp. 31-52.

Nisula, B.C., Morgan, F.J., and Canfield, R.E. Evidence that chorionic gonadotropin has intrinsic thyrotropic activity. *Biochem. Biophys. Res. Commun.* 59, 86-91 (1974).

O'Byrne, E.M., Carriere, B.T., Sorensen, L., Segaloff, A., Schwabe, C., and Steinetz, B.G. Plasma immunoreactive relaxin levels in pregnant and non-pregnant women. *J. Clin. Endocrinol. Metab.* 47, 1106-1110 (1978).

Orlandi, C., Segata, L., and Becca, B. Effeto dell' HMG e dell' HCG sull' incorporazione di aminoacidi radioattivi in omogenati di placenta umana a termine. *Riv. Ital. Ginecol.* 50, 233-239 (1966).

Pavlou, G., Chard, T., and Letchworth, A.T. Circulating levels of human chorionic somatomammotropin in late pregnancy: disappearance from the circulation after delivery, variation during labor and circadian variation. *J. Obstet. Gynaecol. Br. Commonw.* 79, 629-634 (1972).

Penny, R., Olambiwonnu, N.O., and Frasier, S.D. Follicle stimulating hormone (FSH) and luteinizing hormone-human chorionic gonadatropin (LH-hCG) concentrations in paired maternal and cord sera. *Pediatrics* 53, 41-47 (1974).

Petropoulos, E.A. Maternal and fetal factors affecting the growth and function of the rat placenta. *Acta Endocrinol. (Kbh)* 72, Suppl. 176, 1-69 (1973).

Petropoulos, E.A. and Cons. J.M. Regulation of placental growth and metabolism by ovarian and hypophyseal hormonal stimuli. *Eur. J. Obstet. Gynecol. Reprod. Biol.* 4, Suppl. 1, S115-S123 (1974).

Petropoulos, E.A. and Timiras, P.S. Biological effects of high altitude as related to increased solar radiation, temperature fluctuations and reduced partial pressure of oxygen, in *Progress in Human Biometeorology, Vol. 1, Part IA: Micro- and Macro-environments in the Atmosphere and their Effects on Basic Physiological Mechanisms of Man*, S. W. Tromp, ed. Swets and Zeitlinger, Amsterdam (1974), pp. 295-328 and 646-662.

Pozo del, E., Hiba, J., Lancranjan, I., and Künzig, H.J. Prolactin measurements throughout the life cycle: endocrine correlations, in *Prolactin and Human Reproduction*, P. G. Crosignani and C. Robyn, eds. Academic Press, New York, (1977), pp. 61-69.

Pujol-Amat, P., Perez-Lopez, F.R., Calaf, J., Camissans, O., and Robyn, C. Circadian periodicity of human chorionic gonadotropin (HCG) concentration in serum during the last trimester of pregnancy. *Acta Endocrinol. (Khb)* 73, Suppl. 177, 235 (1973).

Pulkkinen, M.O. and Willman, K. Reduction of maternal estrogen excretion by neomycin. *Am. J. Obstet. Gynecol.* 115, 1153-1154 (1973).

Rees, L.H., Burke, C.W., Chard, T., Evans, S.W., and Letchworth, A.T. Possible placental origin of ACTH in normal human pregnancy. *Nature* 254, 620-622 (1975).

Reyes, F.I., Winter, J.S.D., and Faiman, C. Pituitary gonadotropin function during human pregnancy: serum FSH and LH levels before and after LHRH administration. *J. Clin. Endocrinol. Metab.* 42, 590-592 (1976).

Reyes, F.I., Boroditsky, R.S., Winter, J.S.D., and Faiman, C. Studies on human sexual development. II. Fetal and maternal serum gonadotropin and sex steroid concentrations. *J. Clin. Endocrinol. Metab.* 38, 612-617 (1974).

Roberts, G. and Cawdery, J.E. Congenital adrenal hypoplasia. *J. Obstet. Gynaecol. Br. Commonw.* 77, 654-656 (1970).

Robinson, J.D., Judd, H.L., Young, P.E., Jones, O.W., and Yen, S.S.C. Amniotic fluid androgens and estrogens in midgestation. *J. Clin. Encocrinol. Metab.* 45, 755-761 (1977).

Ross, G.T. Gonadotropins and preantral follicular maturation in women. *Fertil. Steril.* 25, 522-543 (1974).

Saxena, B.N. Protein-polypeptide hormones of the human placenta. *Vitam. Horm.* 29, 95-151 (1971).

Schneider, A.B., Kowalski, K., and Sherwood, L.M. Identification of "big" human placental lactogen in placenta and serum. *Endocrinology* 97, 1364-1372 (1975).

Scommegna, A., Bard, L., and Bienarz, J. Progesterone and pregnenolone sulfate in pregnancy plasma. *Am. J. Obstet. Gynecol.* 113, 60-65 (1972).

Scurry, M.T. and Bruton, J. Steroid hormone production following adrenal stimulation in the rhesus monkey. *Acta Endocrinol. (Kbh)* 58, 637-642 (1968).

Selinger, M. and Levitz, M. Diurnal variation of total plasma estriol levels in late pregnancy. *J. Clin. Endocrinol. Metab.* 29, 995-997 (1969).

Seppala, M., Rutanen E.-M., Jalanko, H., Lehtovirta, P., Stenman, U.-H., and Engvall, E. Pregnancy-specific-β_1-glycoprotein and chorionic gonadotropin-like immunoreactivity during the latter half of the cycle in women using intrauterine contraception, *J. Clin. Endocrinol. Metab.* 47, 1216-1219 (1978).

Serón-Ferré, M., Lawrence, C.C., and Jaffe, R.B. Role of hCG in regulation of the fetal zone of the human fetal adrenal gland. *J. Clin. Endocrinol. Metab.* 46, 834-837 (1978a).

Serón-Ferré, M., Lawrence, C.C., Siiteri, P.K., and Jaffe, R.B. Steroid production by definitive and fetal zones of the human fetal adrenal gland. *J. Clin. Endocrinol. Metab.* 47, 603-609 (1978b).

Shambaugh, G. III, E., Kubek, M., and Wilber, J.F. Thyrotropin-releasing hormone activity in the human placenta. *J. Clin. Endocrinol. Metab.* 48, 483-486 (1979).

Shome, B. and Parlow, A.F. Human pituitary prolactin (hPRL): the entire linear amino acid sequence. *J. Clin. Endocrinol. Metab.* 45, 1112-1115 (1977).

Siler-Khodr, T.M., Khodr, G.S., Eddy, C.A., and Pauerstein, C.J. LRF stimulation of chorionic gonadotropin and estrogen release in the pregnant monkey. *Endocrinology* 104, 199A (1979).

Silman, R.E., Chard, T., Lowry, P.J., Smith, I., and Young, I.M. Human foetal pituitary peptides and parturition. *Nature* 260, 716-718 (1976).

Silverberg, J., O'Donnell, J., Sugenoya, A., Row, V.V., and Volpe, R. Effect of human chorionic gonadotropin on human thyroid tissue in vitro. *J. Clin. Endocrinol. Metab.* 46, 420-424 (1978).

Simmer, H.H. Placental hormones, in *Biology of Gestation: Vol. 1, The Maternal Organism,* N. S. Assali, ed. Academic Press, New York (1968), pp. 290-354.

Simpson, E.R., Carr, B.R., Parker, C.R. Jr., Milewich, L., Porter, J.C., and Macdonald, P.C. The role of serum lipoproteins in steroidogenesis by the human fetal adrenal cortex. *J. Clin. Endocrinol. Metab.* 49, 146-148 (1979).

Smeaton, T.C., Andersen, G.J., and Fulton, I.S. Study of aldosterone levels in plasma during pregnancy. *J. Clin. Endocrinol. Metab.* 44, 1-7 (1977).

Sobrevilla, L.A., Romero, I., Kruger, F., and Whittembury, J. Low estrogen excretion during pregnancy at high altitude. *Am. J. Obstet.*

Gynecol. 102, 828-833 (1968).

Sowers, J.R., Hershman, J.M., Carlson, H.E., and Pekary, E. Effect of human chorionic gonadotropin on thyroid function in euthyroid men. J. Clin. Endocrinol. Metab. 47, 898-901 (1978).

Spellacy, W.N. Insulin, glucagon and growth hormone in pregnancy, in Endocrinology of Pregnancy, 2nd Edition, F. Fuchs and A. Klopper, eds. Harper & Row, New York (1977), pp. 206-221.

Spellacy, W.N., Conly, P.W., Cleveland, W.W., and Buhi, W.C. Effects of fetal sex and weight and placental weight on maternal serum progesterone and chorionic gonadotropin concentrations. Am. J. Obstet. Gynecol. 122, 278-282 (1975).

Spellacy, W.N., Buhi, W.C., and Birk, S.A. The effect of smoking on serum placental lactogen levels. Am. J. Obstet. Gynecol. 127, 232-234 (1977).

Swaminathan, N. and Bahl, O.P. Dissociation and recombination of the subunits of human chorionic gonadotropin. Biochem. Biophys. Res. Commun. 40, 422-427 (1970).

Szabo, A.J., and Grimaldi, R.D. The metabolism of placenta. Adv. Metab. Disord. 4, 185-228 (1970).

Taylor, N.F. and Shackleton, C.H.L. Gas chromatographic steroid analysis for diagnosis of placental sulfatase deficiency: a study of nine patients. J. Clin. Endocrinol. Metab. 49, 78-86 (1979).

Telegdy, G., Weeks, J.W., Archer, D.F., Wiqvist, N., and Diczfalusi, E. Acetate and cholesterol metabolism in the human foeto-placental unit at midgestation. 3. Steroids synthesized and secreted by the foetus. Acta Endocrinol. (Kbh) 63, 119-133 (1970a).

Telegdy, G., Weeks, J.W., Lerner, U., Stakemann, G., and Diczfalusy, E. Acetate and cholesterol metabolism in the human foetoplacental unit at midgestation. 1. Synthesis of cholesterol. Acta Endocrinol. (Kbh) 63, 91-104 (1970b).

Thomsen, K. and Hiersche, H.D. The functional morphology of the placenta, the foetus, the membranes and the umbilical cord, in Foetus and Placenta. A. Klopper, and E. Diczfalusi, eds. Blackwell Scientific Publications, Oxford (1969) pp. 61-137.

Timiras, P.S. Developmental Physiology and Aging. The Macmillan Co., New York (1972), pp. 103-113.

Townsley, J.D., Dubin, N.H., Grannis, G.F., Gartman, L.J., and Crystle, C.D. Circadian rhythms of serum and urinary estrogens in pregnancy. J. Clin. Endocrinol. Metab. 36, 289-295 (1973).

Tulchinsky, D. The value of oestrogen assays in obstetric disease, in Plasma Hormone Assays in Evaluation of Fetal Wellbeing, A Klopper, ed. Churchill Livingstone, Edinburgh (1976), pp. 72-86.

Tulchinsky, D., Hobel, C.J., and Korenman, S.G. A radioligand assay for plasma unconjugated estriol in normal and abnormal pregnancies. Am. J. Obstet. Gynecol. 111, 311-318 (1971).

Tulchinsky, D., Osathanondh, R., Belisle, S., and Ryan, K.J. Plasma estrone, estradiol, estriol and their precursors in pregnancies with anencephalic fetuses. J. Clin. Endocrinol. Metab. 45, 1100-1103 (1977).

Turnbull, A.C., Patten, P.T., Flint, A.P.F., Keirse, M.J.N.C., Jeremy, J.Y., and Anderson, A.B.M. Significant fall in progesterone and rise in oestradiol levels in human peripheral plasma before onset of labour. Lancet 1, 101-104, (1974).

Tyson, J.E., Austin, K.L., and Farinholt, J.W. Prolonged nutritional deprivation in pregnancy: changes in human chorionic somatomammotropin and growth hormone secretion. Am. J. Obstet. Gynecol. 109, 1080-1082 (1971).

Uemura, T. The effect of human chorionic gonadotropin upon the central nervous system viewing from EEG, particularly in its relation with pregnancy. *J. Jap. Obstet. Gynecol. Soc.* 15, 28-39 (1968).

Vadora, E., Bacchi, A., Calestani, V., Salvarani, C., and Salvatori, B. Diurnal variations of plasma progesterone levels during the last trimester of pregnancy. *IRCS* (Endocrine System) 2, 1369 (1974).

Vaitukaitis, J.L. Human chorionic gonadotropin, in *Endocrinology of Pregnancy*, 2nd Edition, F. Fuchs and A. Klopper, eds. Harper & Row New York (1977), pp. 63-75.

Vaitukaitis, J.L. and Ross, G.T. Antigenic similarities among the human glycoprotein hormones and their subunits, in *Gonadotropins*, B.B. Saxena, C.G. Beling, and H.M. Gandy, eds. J. Wiley & Sons, New York (1972), pp. 435-443.

Vaitukaitis, J.L., Braunstein, G.D., and Ross, G.T. A radioimmunoassay which specifically measures human chorionic gonadotropin in the presence of human luteinizing hormone. *Am. J. Obstet. Gynecol.* 111, 751-758 (1972).

Van Lausden, H.A. Chorionic gonadotropin in pathological pregnancy, in *Plasma Hormone Assays in Evaluation of Fetal Wellbeing*, A. Klopper, ed. Churchill Livingstone, Edinburgh (1976), pp. 48-71.

Varangot, J., Cedard, L., and Yannotti, S. Perfusion of the human placenta in vitro: study of the biosynthesis of estrogens. *Am. J. Obstet. Gynecol.* 92, 534-547 (1965).

Varma, K., Larranga, L., and Selenkow, H.A. Radioimmunoassay of serum human chorionic gonadotropin during normal pregnancy. *Obstet. Gynecol.* 37, 10-18 (1971).

Villee, C.A. Biochemical aspects of mammalian placenta, in *The Biochemistry of Animal Development; Vol. 2, Biochemical Control Mechanisms and Adaptation in Development*, R. Weber, ed. Academic Press, New York (1967), pp. 383-412.

Villee, C.A. and Gabbe, S.G. Effects of gonadotropins on placental steroidogenesis, in *Gonadotropins*. B.B. Saxena, C.G. Beling, and H.M. Candy, eds. Wiley-Interscience, New York (1972), pp. 309-326.

Villee, C.A. Van Leusden, H.A., and Zelewski, L. The regulation of the biosynthesis of sterols and steroids in the placenta. *Adv. Enzyme Regul.* 4, 161-179 (1966).

Villee, D.B. Development of endocrine function in the human placenta and fetus. *New Engl. J. Med.* 281, 473-484; 533-541 (1969).

Vinik, A.I., Kaplan, S.L., and Grumbach, M.M. Purification, characterization and comparison of immunological properties of monkey chorionic somatomammotropin with human and monkey growth hormone, human chorionic somatomammotropin and ovine prolactin. *Endocrinology* 92, 1051-1064 (1973).

Warne, G.L., Reyes, F.I., Faiman, C., and Winter, J.S. Studies on human sexual development. VI. Concentrations of unconjugated dehydroepiandrosterone, estradiol and estriol in amniotic fluid throughout gestation. *J. Clin. Endocrinol. Metab.* 47, 1363-1367 (1978).

Williams, M.T., Roth, M.S., Marsh, J.M., and LeMaire, W.J. Inhibition of human chorionic gonadotropin-induced progesterone synthesis by estradiol in isolated human luteal cells. *J. Clin. Endocrinol. Metab.* 48, 437-440 (1979).

Willman, J. and Pulkkinen, M.O. Reduced maternal plasma and urinary estriol during ampicillin treatment. *Am. J. Obstet. Gynecol.* 109, 893-896 (1971).

Winick, M. Cellular growth of the placenta as an indicator of abnormal fetal growth, in *Diagnosis and Treatment of Fetal Disorders*, K. Adamsons, ed. Springer-Verlag, New York (1968), pp. 83-101.

Winters, A.J., Oliver, C., Colston, C., MacDonald, P.C., and Porter,

J.C. Plasma ACTH levels in the human fetus and neonate as related to age and parturition. *J. Clin. Endocrinol. Metab.* 39, 269-273 (1974).

Wynn, R.M. Morphology of the placenta, in *Biology of Gestation; Vol. 1, The Maternal Organism*, N.S. Assali, ed. Academic Press, New York (1968) pp. 93-184.

Ylikorkala, O., Kauppila, A., and Viinikka, L. Intraamniotic or intravenous injection of dehydroepiandrosterone sulfate in midgestation: effect on prolactin level in maternal serum and amniotic fluid. *J. Clin. Endocrinol. Metab.* 49, 452-455 (1979a).

Ylikorkala, O., Kivinen, S., and Reinila, M. Serial prolactin and thyrotropin responses to thyrotropin-releasing hormone throughout normal human pregnancy. *J. Clin. Endocrinol. Metab.* 48, 288-292 (1979b).

Yoshimi, T., Strott, C.A., Marshall, J.R., and Lipsett, M.B. Corpus luteum function in early pregnancy. *J. Clin. Endocrinol. Metab.* 29, 225-230 (1969).

Yoshimoto, Y., Wolfsen, A.R., and Odell, W.D. Human chorionic gonadotropin-like substance in nonendocrine tissues of normal subjects. *Science* 197, 575-577 (1977).

Zelewski, C. and Villee, C.A. The biosynthesis of squalene, lanosterol and cholesterol by minced human placenta. *Biochemistry* 5, 1805-1814 (1966).

Copyright©1982, Spectrum Publications, Inc.
Hormones in Development and Aging

Chapter 4
Somatic Growth
Selna L. Kaplan

INTRODUCTION

Growth is a multifaceted process which entails increments in trunk, limb, skull, and organ size and is integrated with developmental maturation of enzyme, secretory and receptor systems from fetal life through adulthood.

A specialized pattern of organ and tissue growth has been described; brain growth is most rapid during fetal and infantile (two years) period whereas heart, liver, and kidney parallel growth of body trunk. Lymphoid growth peaks in the peripubertal period and declines subsequently. Reproductive organs do not achieve maximum size until pubertal development occurs. Spatial differences in rate of growth occur, particularly during fetal life with more rapid growth in cephalocaudal and distoproximal directions, i.e., arm before leg, head before trunk, hand before upper arm. (Thompson, 1942).

The most active growth phase in man occurs during fetal life. A sixfold increase in body length is seen by four months gestation with a two-fold increment during the last four months of gestation. At 10 weeks of gestation, the crown-rump length is 40 mm, 112 mm at 16 weeks, 160 mm at 20 weeks, 2100 mm at 34 weeks, and 3600 mm at 40 weeks (Fig. 1). The birth length of a normal full-term infant is 50.4 ± 2 cm for males and 49.7 ± 1.9 cm for females which represents 30% of final height. (Nicolson and Hanley, 1953).

This accelerated growth rate continues during the first postnatal year with an average increment of 25-30 cm. By three years of age, the child has attained 46% of final height (Fig. 2). A relative deceleration in growth ensues with a yearly increment of 6-8 cm between 3 and 5 years and 4.5-6 cm between 6 and 10 years (females) and 6 and 12 years (males). (Kaplan and Reiter, 1977). In the immediate peripubertal period in males (11-13 years), a transient (six months) period of decreased cessation of growth may be observed (Fig. 3). (Tanner et al., 1966; Nicolson and Hanley, 1953).

The final accelerative growth phase is associated with pubertal development. In females, the onset of the pubertal growth spurt parallels the initiation of physical signs of sexual development at 10 years. In the normal female, 84% of final height has been attained at 10 years and 95% at 12.5 years (mean age of menarche in U.S.). Postmenarchal growth is limited to 2-3 inches (Fig. 2). The accumulative growth during puberty

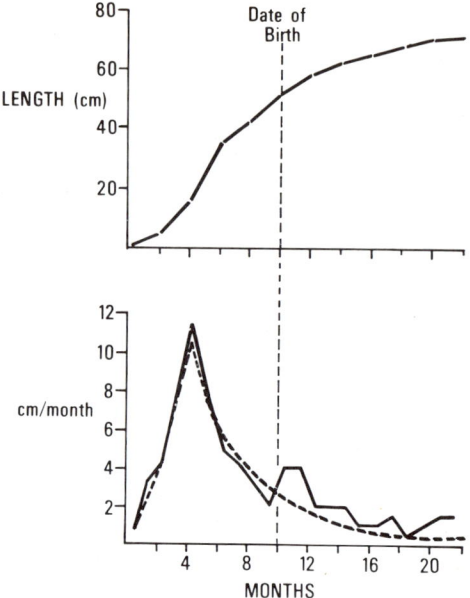

Fig. 1. The growth curve (distance and incremental) during fetal life at one year postnatally (from Thompson, 1942).

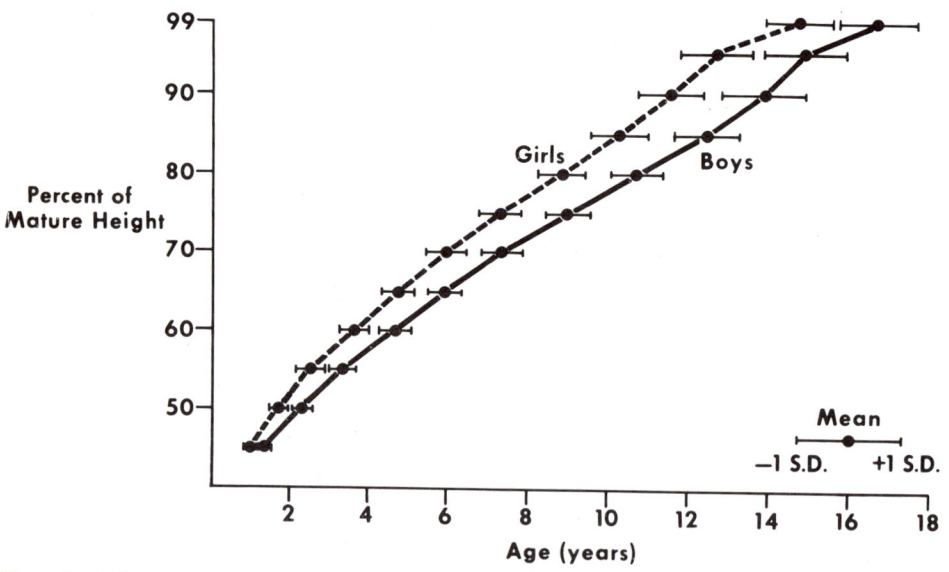

Fig. 2. The percent of mature height (mean ±1 SD) attained in males, females from one to eighteen years of age (from Nicholson and Hanley, 1953).

in the female is 23-28 cm (9-11 inches) (Tanner and Whitehouse, 1976). The onset of pubertal development is 2 years later in the male than in the female. At 12 years of age, only 84% of final height has been attained and 95% of final height is not achieved until 15 years of age.

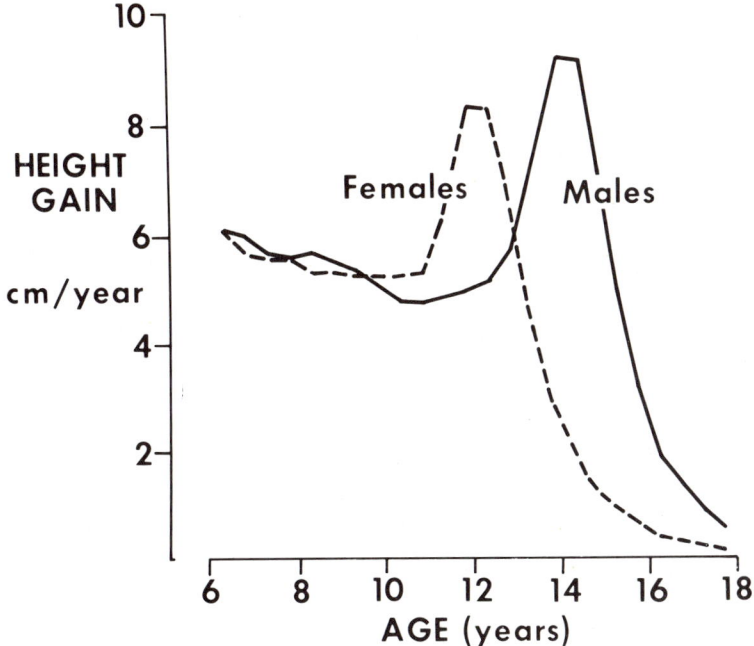

Fig. 3. The mean height gain (cm/year) in males and females from the peripubertal to pubertal period. Note the slight deceleration between 11 and 12 years in males (from Tanner, 1960).

The cumulative growth during puberty is usually only 2 to 5 cm greater in the male than in the female (Fig. 3). In the U.S., the mean height of adult women is 65.1 ± 1.7 inches and of adult men 69.8 ± 2.7 inches. (National Center for Health Statistics, 1970). This difference in final height is attributable primarily to the 2-3 year additional peripubertal growth period (Kaplan and Reiter, 1977).

LIMB GROWTH

The length of lower limbs is determined by measurement from symphysis pubis to heel (lower segment) or from sitting height; the upper segment of body denotes measurement from head to symphysis pubis. Upper limb length is determined by measurement of arm span. The sitting height and limb length of males is greater than that of females during adolescence and adulthood. Racial differences in bodily proportions have been reported; Black American and African adult males have longer limbs than Caucasian adult males, with upper to lower segment ration (U/L) of 0.85, whereas Oriental males generally have proportionally shorter limbs for total length. (Evelath and Tanner, 1976).

The ratio of trunk to limb length progresses from fetal life through adolescence. In the fetus, trunk length increases at a more rapid pace than limb length. At birth, the U/L is 1.7 but decreases to 1.0 by 10 years. During adolescence limb length exceeds trunk length growth with a further decrease in U/L to 0.95 (Fig. 4). Upper limb growth usually parallels lower limb growth; arm span is usually comparable to total body length. Determination of sitting height has been recommended by Tanner and associates as a more precise assessment of lower limb growth (Fig. 4).

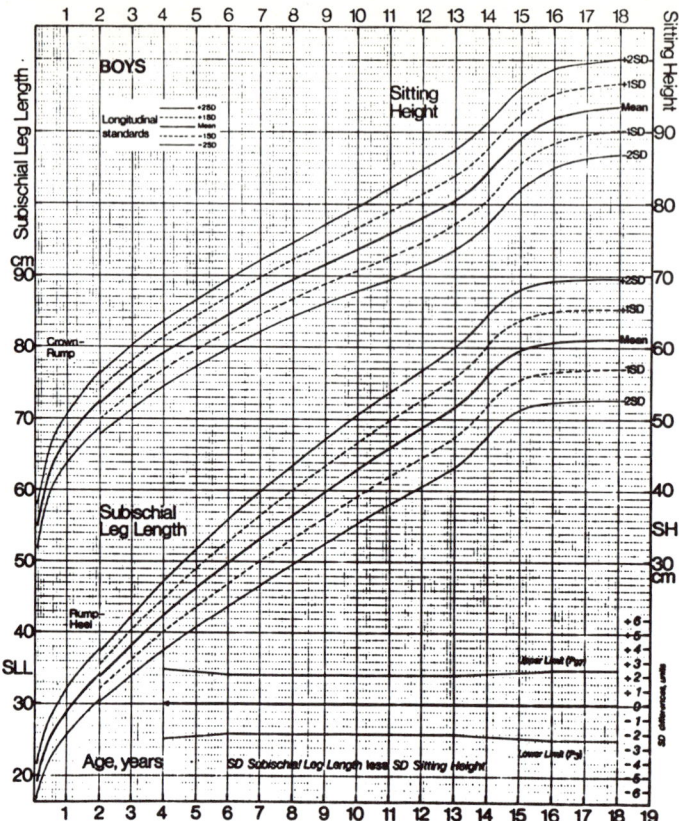

Fig. 4. The graphic representation of changes in sitting height and subischial leg length (cm) from birth to 19 years of age (courtesy of Dr. James M. Tanner).

BODY WEIGHT

During gestation the proportional increments in weight are greater than those of length. From the 10th to 17th weeks of gestation there is approximately a thirty-fold increase in weight from 5 g to 150 g. In mid-gestation (19 to 28 weeks) there is a six-fold increase in weight and from 28 weeks to term there is a three-fold increase in weight from 1000 g to 3400 g (Fig. 5). A slight deceleration in weight increment occurs during the last week of gestation. At the time of birth approximately 5% of final adult weight has been attained. A parallel increase in organ weight (brain, liver, heart, lung, kidneys) to that of body weight occurs during fetal life. At a fetal body weight of 1000-3500 g, the ratio of organ to brain weight is constant. (Sinclair, 1973).

The mean birth weight is 3.4 kg (range of 2.5-4.6 kg). In the first few days of life, a 5%-10% loss of weight occurs which is primarily attributable to a decrease in body water content. Within the second week of life the birth weight is generally regained. An increase in weight of one to two pounds per month occurs during the first six months of life so that by one year of age, tripling of birth weight is achieved. The weight gain is considerably less during the second year, 0.5 pound per month, and is sustained at this rate from three to five years of age. Between the third to fifth year of age the weight gain is approximately 4.5 lbs. or 2 kg per year. During the peripubertal period the weight gain increases to 3 to 3.5 kg per year (Fig. 6). During puberty the incremental weight for girls is 30 kg and for males 36 kg (Tanner and Whitehouse, 1976; Kaplan and Reiter, 1976). The proportional contribution of various organs

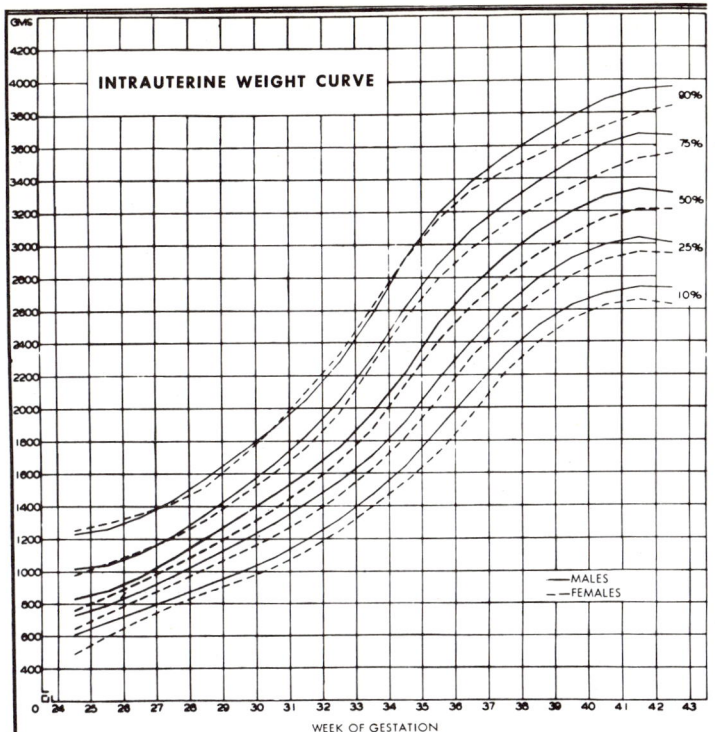

Fig. 5. Changes in weight (grams) of the male and female fetus from 25 weeks to birth.

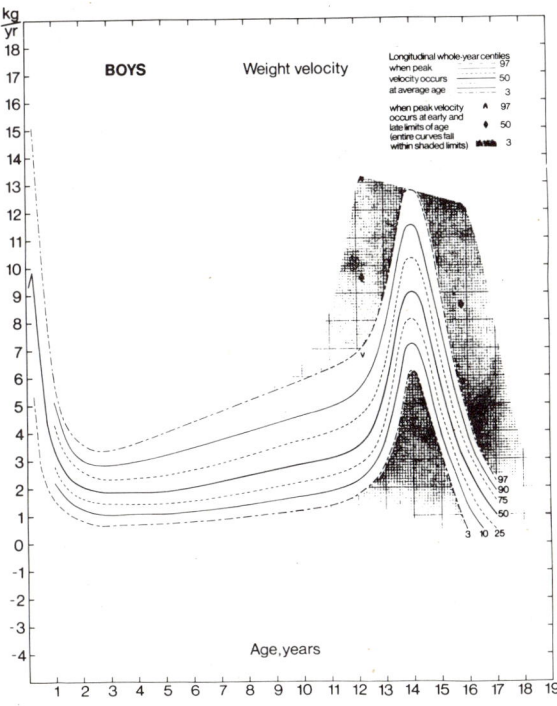

Fig. 6. The distance curve for incremental weight gain for males from birth to 18 years. The measurements are plotted as median with the range of 3rd to 97th percentile (from Tanner and Whitehouse, 1976).

and tissues to body weight changes from the fetal to adult state in human (Table 1) (Widdowson, 1974).

Table 1

CONTRIBUTION OF ORGANS AND TISSUES TO BODY WEIGHT AT VARIOUS AGES (% BODY WEIGHT)[a]

Organs and Tissues	Fetus 20-24 Weeks	Full term Newborn	Adult
Skeletal muscle	25.0	25.0	40.0
Skin	13.0	4.0	6.0
Skeleton	22.0	18.0	14.0
Heart	0.6	0.5	0.4
Liver	4.0	5.0	2.0
Kidneys	0.7	1.0	0.5
Brain	13.0	12.0	2.0

[a] Widdowson, 1974

BODY COMPOSITION

Lean Body Mass

Lean body mass may be a better index than body weight and its variable adipose content. Underwater weighing or measurements of total body water or total body distribution of the isotope 40 K+ can be used to determine lean body mass. According to the data of Forbes (1972), the ratio of lean body mass to height increases yearly in males and females from age 7.5 to 12.5. During the adolescent period the incremental rise is greater in males with a peak ratio attained at 19-20 years. In females the maximum ratio is attained at 16 years and is two-thirds that of the male (Fig. 7).

Skeletal Muscle

At birth, skeletal muscle constitutes 25% of body weight with an increase to 40% in adult man. A two-fold increase in muscle cell size occurs from midgestation to term. An additional three-fold increase is seen during childhood (Cheek et al., 1966; Widdowson, 1974). Maximal cell size and number are achieved in females at the onset of puberty at 10.5 years. In males, the muscle cell number continues to increase after ten years of age and at adolescence is approximately three times greater than that of adolescent female. The muscle cell size increases until adulthood in the male. Thus, the differences in lean body mass, muscle cell size and number may reflect the increased muscle mass of adolescent males compared with that of adolescent female (Fig. 8).

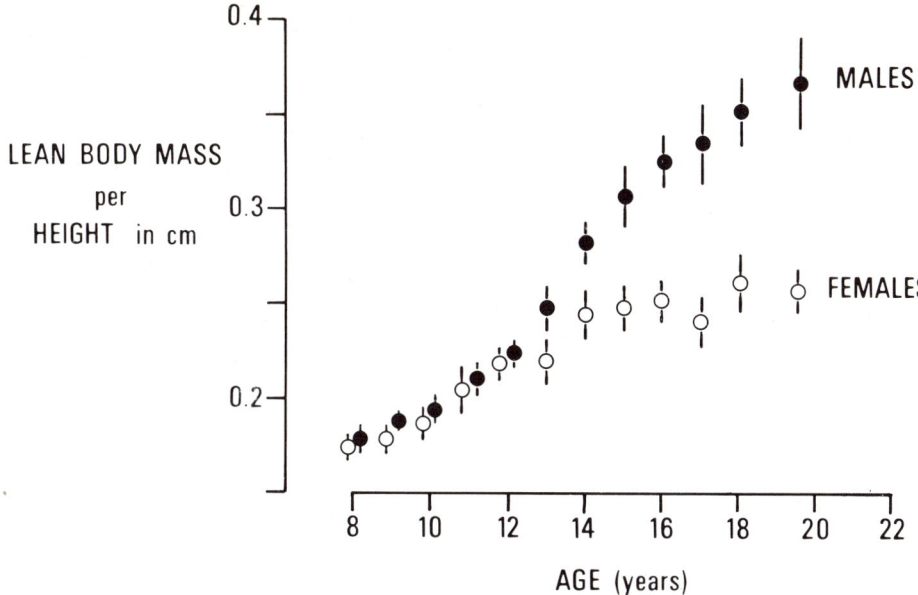

Fig. 7. Lean body mass/height ratio for age eight to twenty-two years in females and males (from Cheek, 1968).

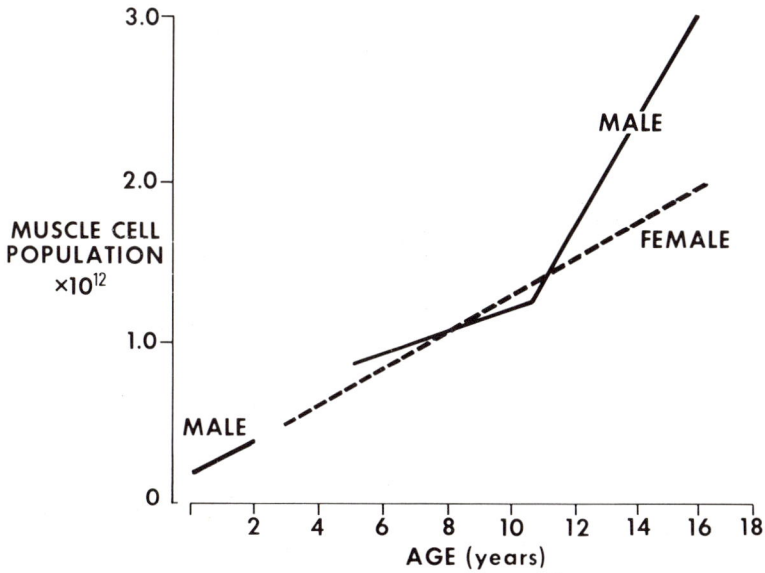

Fig. 8. Changes in muscle cell population ($\times 10^{12}$) with age in years. Note the deviation of the lines at ten years of age (from Cheek, 1968).

Adipose Tissue

Adipose content in the fetus is minimal early in gestation (0.1%/g body weight) and accumulates exponentially with fetal body weight. At 33 weeks of gestation, adipose content is 5%/g body weight and increases markedly at term to 16%/g of body weight. During the first year of life the fat content increases to 22% but by five years of age decreases to 12.5 to 15% (Widdowson, 1974). There is a higher percentage of body fat in females than males. In the peripubertal period, the proportion of body fat decreases in the male and increases in the female so that the adolescent female has twice the fat content 24.6% as compared with 12.3% in the adolescent male. The peak adipose cell number is observed at one year of age with a five-fold increase thereafter to adolescence (Brook, 1972). Adipose cell size doubles from midgestation to term with a slight increase during the first year postnatally. Adipose cell size remains unchanged from 4 to 12 years of age. An increase in adipose cell number is seen in children who are obese by one year of age whereas obesity during childhood is associated with an increase in cell size (Fig. 9).

Ossification

Bone formation is initiated at 8-10 weeks of gestational age in the clavicle and mandible. Ossification of the long bones occurs by 12 weeks of gestation. The primary centers appear in the shaft; the shaft ossifies and secondary centers appear and become epiphyseal centers. The epiphysis of the os calcis is the first to develop at 23 to 26 weeks of gestation. By late gestation, distal femoral, proximal tibial, and proximal humeral epiphyses appear. The amjor accretion of skeletal calcium during late gestation and parallels the increase in fetal weight.

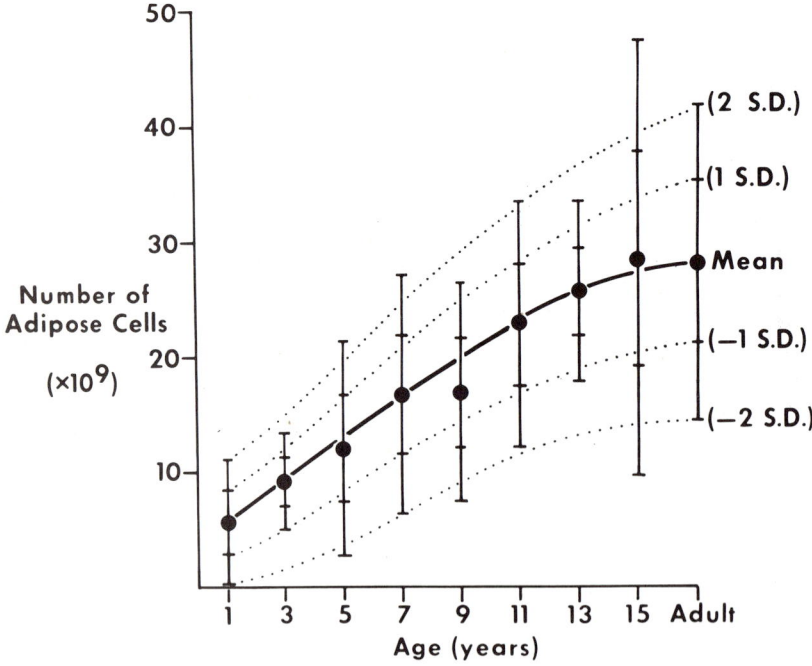

Fig. 9. Changes in adipose cell number ($\times 10^9$) from infancy to adulthood (from Brook, 1972).

Assessment of maturational state at birth has been described utilizing epiphyseal development but wide individual variations limit its usefulness. Absence of epiphyseal centers at birth is seen in prematurely born and in hypothyroid neonates.

During the postnatal period and throughout childhood, progression in length and maturation of the skeleton ensues. The age of appearance of the secondary ossification centers (epiphyses), alterations in epiphyseal size and molding of long bones and ultimate fusion of epiphyses with the diaphysis provide indices of skeletal age.

The described methods for determination of skeletal age include:

1) Greulich and Pyle (1959) Atlas Standards

Radiograph of left hand and wrist of male and female children from 1 to 18 years are used for comparison with that of index subject. The degree of variability in development varies in different age groups. The standard deviation in skeletal age is 2-4 months at 1-2 years, 6 months at 3-4 years, and over eight years of age is 12 months (Fig. 10).

2) Maturity Scoring System (TW2)

This method assigns eight stages of development to each epiphysis in the hand and wrist (Fig. 11). This maturity score provides a more quantitative assessment of skeletal age as described by Tanner and associates (1975).

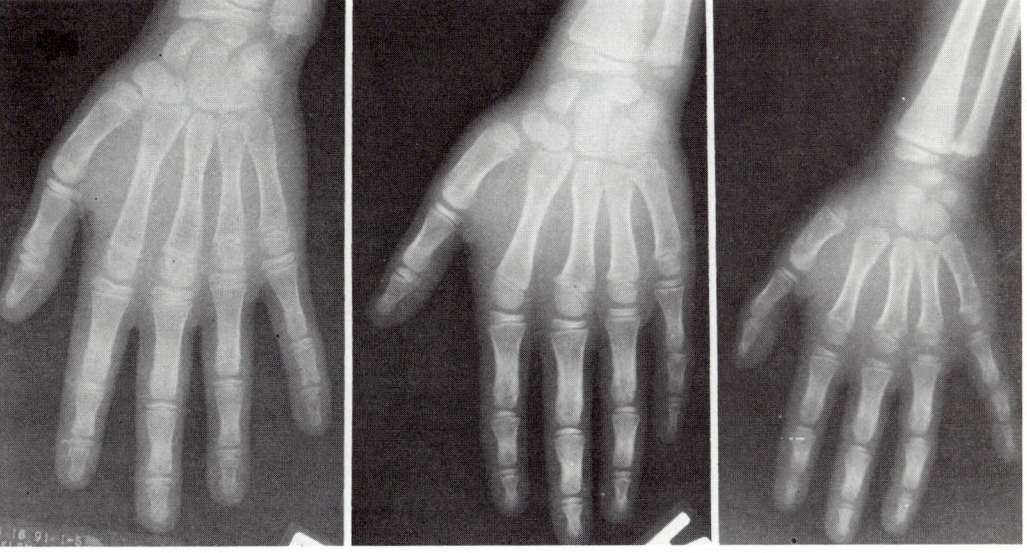

Fig. 10. A comparison of the skeletal age of three male children aged 4.5, 7, and 12 years, respectively, according to standards of Greulich and Pyle, (1959).

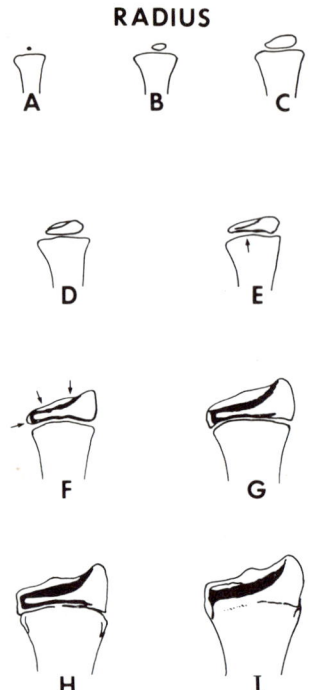

Fig. 11. The maturity scoring system for the epiphysis of the radius (from Tanner et al., 1975).

Epiphyseal maturation may be variable within an individual so that a radiograph of the hand may not be representative. Atlas standards are available for the elbow and knee but are used less frequently. A hemiskeleton epiphyseal score may provide a better assessment of skeletal age but is no longer used in view of increased exposure to irradiation.

Skeletal age correlates significantly with percentage of ultimate height achieved per year as described by Nicolson and Hanley in 1953. Bayley and Pinneau (1952) utilized this as the basis for their height prediction charts. Thus, a normal five-year-old male with a skeletal age of five years has attained 61.8% of his final height. Whereas a five-year-old male with a skeletal age of three years has attained only 53.8% of his final height.

Currently available height prediction methods include, midparental height, skeletal age by maturity scoring, and weight of child. In the average child, most of these methods provide a similar ultimate height prediction which does not deviate significantly from the ultimate achieved height.

Dental maturation as assessed by tooth eruption parallels skeletal maturation. Timing of tooth formation as judged by dental radiographs provides a more sensitive index of maturation. (Filipsson, 1975).

CRANIO-FACIAL GROWTH

The growth of the cranial vault parallels that of the brain. At birth, the skull represents one-fourth of total body length but only one-eighth by adulthood. At birth, the circumference of the head is 35.3 ± 1.2 cm with an 11-cm increase during the first year of life but only 2 cm a year thereafter until 12 years of age. The head circumference is greater than the chest circumference at birth but approaches unity by one year of age (Sinclair, 1973) (Fig. 12).

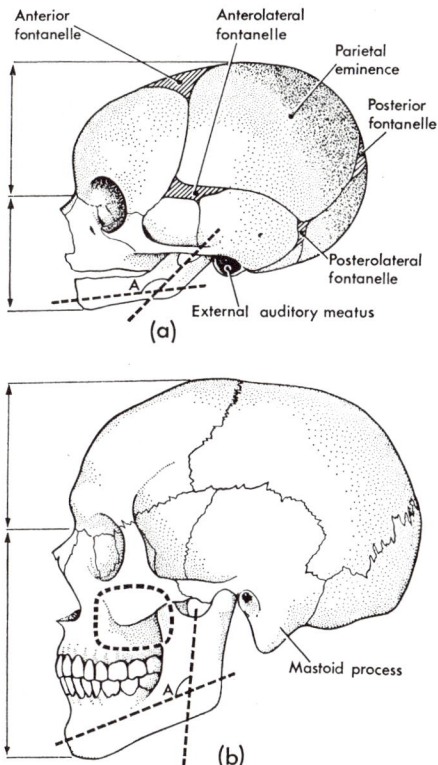

Fig. 12. Alterations in the skull and face from newborn to adulthood (from Sinclair, 1973).

Proportional changes occur in the cranio-facial relationship from a ratio of 8:1 at birth to 2:1 in the adult. The mandible is small at birth and the rami do not fuse until the first year of life. Forward thrust of the mandible and increased growth of the maxilla is associated with dental development during childhood. Further maturational changes in facial appearance are induced during pubertal development; mandibular growth is completed and an increase in A-P diameter of the nose with sharpening of facial features and a less rounded profile. (Tanner, 1960).

HORMONAL EFFECT ON GROWTH

Stimulation of growth can be induced by physiologic doses of some hormones (growth hormone and thyroid hormone) only when deficiencies exist but at elevated concentrations, growth acceleration can be induced, i.e., gigantism. Sex steroids (estradiol and testosterone) can accelerate growth and epiphyseal maturation when administered at any prepubertal age.

The mechanism of action of the major hormonal stimulants of growth are discussed below.

Growth Hormone

Growth hormone is a pituitary polypeptide hormone whose release is under alpha adrenergic control and regulated by a stimulatory (GRF) and an inhibitory (SRIF) hypothalamic factor. It enhances protein synthesis and synthesis of messenger and ribosomal RNA, stimulates cellular transport of amino acids, enhances fatty acid release by adipocytes, and affects metabolism of growth cartilage indirectly by tissue growth factors (somatomedins). Vascularization and mitotic activity of the germinal and proliferative cells of the epiphyseal cartilage are dependent on the presence of growth hormone (Trueta, 1974). Growth hormone is the major hormonal stimulant for growth during childhood.

Thyroid Hormones

Thyroxine and thiodothyronine accelerate DNA and RNA synthesis. Decreased muscle cell number and DNA content has been described in thyroid deficient animals and children (Cheek et al., 1965). Thyroxine and thiodothyronine primarily affect osteogenesis rather than chondrogenesis with resultant decreased columnar formation of cartilage and hypertrophy of chondrocytes. In hypothyroid animals the cartilaginous columns of the epiphysis are erratic and nonvertical. Vascularization is polytopic rather than central which results in several fragmented centers of ossification referred to as epiphyseal dysgenesis (Royer, 1974). Synergism of biologic action of thyroid hormones with that of growth hormone is essential for normal ossification and linear growth.

Somatomedins and Other Putative Growth Factors

These peptides may be the intracellular mediator of bioactivity attributed to growth hormone. They are synthesized primarily by the liver and degraded in the kidney.

These peptides with approximate MW in range of 7600 include designated somatomedin A and C and IGF-1 and have minor differences in aminoacid composition and significant differences in isoelectric point. All share similar biologic actions: a) incorporation of $^{35}SO_4$ into glucosoaminoglycans of epiphyseal cartilage; b) stimulate uptake of ^3H uridine into RNA and of ^3H thymidine into DNA of rat cartilage; c) stimulate glucose uptake by fat cells; d) increase amino acid and glucose uptake into muscle; e) administration of growth hormone in vivo leads to increased release of these peptides (Chochinov and Daughaday, 1976). A direct stimulatory effect of somatomedins or IGF-1 on linear growth of animals has not been demonstrated to date.

Sex Steroids

Testosterone

Testosterone has both anabolic and virilizing biologic activity; only the former will be reviewed (Kochakian, 1975; Ashman-Williams, 1975). Retention of nitrogen, phosphorous, and calcium, similar to the effect of growth hormone occurs following administration of testosterone. Androgens enhance DNA and protein synthesis and induce cell multiplication. The stimulatory effects on muscle cell size and mass during puberty were discussed earlier. High dose androgens have been utilized in athletes to increase muscle mass.

At low doses, it enhances development of growth cartilage cells and at high doses it decreases cellular proliferation and enhances the disappearance of epiphyseal cartilage (Trueta, 1974).

The growth acceleration induced by testosterone in the prepubertal child is associated with skeletal maturation and advancement in epiphyseal fusion (Royer, 1974). It is not recommended as a growth stimulant in the young prepubertal child but may be beneficial in adolescents with delayed pubertal development.

Estrogens

The stimulation of tissue growth particularly of the reproductive tract and mammary glands by estrogens has been described. The immediate effect of estrogen is stimulation of DNA synthesis with a subsequent inhibitory or refractory phase of DNA synthesis (Gorski et al., 1977). The protein anabolic effect of estrogen has been demonstrated by the moderate increases in weight and fat content induced in estrogen-treated chickens and cows.

The magnitude of growth acceleration induced by estrogen administration in prepubertal girls relative to epiphyseal maturation is thought to be less than that of testosterone. This is based on the assumption that the attained growth during puberty is significantly less in females than in males; the data of Marshall (1974) and Tanner (1960) indicate only a 2-5 cm greater increment in total growth of males compared with females. Although suppression of somatomedin release in response to growth hormone occurs in estrogen-treated individuals, the anabolic action of growth hormone is not inhibited (Wiedemann and Schwartz, 1972). The stimulatory effect of estrogens on linear growth must be reassessed in view of available clinical data. Low dose estradiol in prepubertal females with primary hypogonadism or delayed adolescence accelerates growth without a marked advancement in bone age (Alexander et al., 1978). In contrast, high dose estrogen administered to prepubertal females with tall stature accelerates both linear growth and bone maturation (Wettenhall et al., 1975).

The Assessment of Growth Retardation

The assessment of short stature in children requires the demonstration that the height is inappropriate for age and for familial pattern. Utilization of growth charts provide the best screening procedure for the assessment of abnormalities of growth (Kaplan and Reiter, 1977). Three types of growth charts are available.

1) The Distance Growth Chart

The absolute height is plotted and compared with cross-sectional and longitudinal data from a large group of normal children; variation from the mean is expressed either as standard deviation or as percentiles. Since these data compare children of the same age but of mixed parental stature, the range for normal falls within -2.5 to +2.5 standard deviations from the mean or from the 3rd to 97th percentile (Fig. 13) (Kaplan and Reiter, 1977; Tanner et al., 1966).

2) Incremental Height Chart

The absolute increase in height is plotted at yearly or six month intervals and compared with the normal incremental height achieved from infancy through adolescence (Fig. 14). A rapid increment of growth occurs during the first two years of life and a steady increase from 4 to 10 years

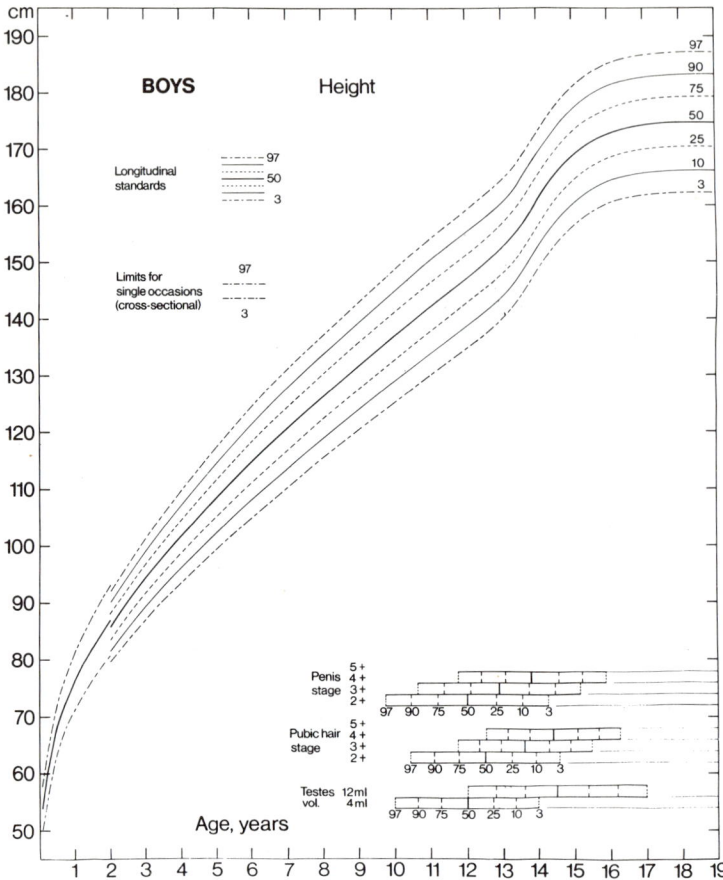

Fig. 13. Distance growth curves for height of males from birth to 17 years of age. The measurements are plotted as median (50th percentile) with a range of 3rd to 97th percentile (from Tanner and Whitehouse, 1976).

of age with an additional increment of height at puberty (Tanner and Whitehouse, 1976). An increment in height of less than 4 cm per year from two years of age until the peripubertal period is significantly less than normal.

3) *Midparental Height Chart*

This chart accommodates for the differences in genetic endowment. It allows the comparison of appropriateness of the child's height for parental heights (Tanner et al., 1970). The midparental height (average of both parents' height) correlates best with the normalcy of a child's height (Fig. 15).

On this basis, significant retardation in growth may be defined as 1) a height less than 2.5 SD for mean for age; 2) a growth rate per year less than 2.5 SD for age; and 3) a height less than 2.5 SD for midparental height. The evaluation should include laboratory studies listed on Table 2.

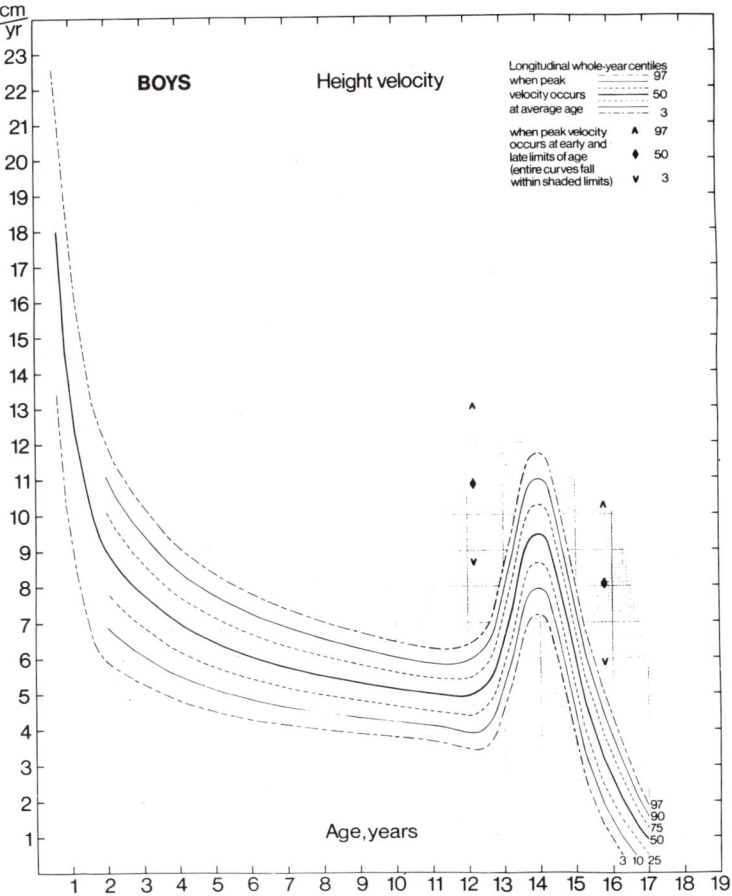

Fig. 14. The incremental growth charts for height (cm/year) of males from birth to 19 years of age plotted as median with a range of 3rd to 97th percentile (from Tanner and Whitehouse, 1976).

Hormonally Induced Growth Retardation

Primary Hypothyroidism

Maldevelopment or failure of development of the thyroid gland is the most common endocrine disorder during childhood. The incidence of this disorder is high at one in 5,000 births based on several neonatal screening programs (Klein et al., 1972; Dussault et al., 1978). Inborn errors of thyroxine metabolism, an autosomal recessive abnormality, is a less common cause of primary hypothyroidism in infancy. Low serum thyroxine and elevated serum TSH levels provide laboratory evidence compatible with primary hypothyroidism.

The newborn with primary hypothyroidism is of normal birthweight and length which contrasts with the reported effect of in utero hypothyroidism in animals. Maturation of the nervous system and of the skeleton is delayed significantly by in utero hypothyroidism. Retardation in height is a common manifestation of primary hypothyroidism in late infancy and during childhood. The body proportions remain immature for age with a more marked retardation in skeletal maturation than in height. These findings are in contrast to those observed in children with growth

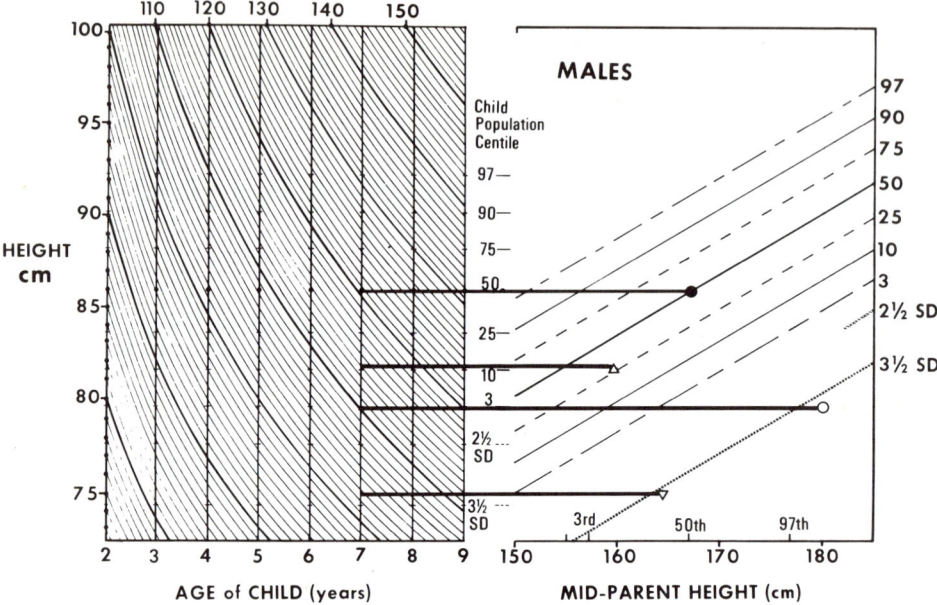

Fig. 15. The appropriateness of the height of the child with that of the parents (average or midparental height) as shown on the Tanner-Whitehouse chart. The point of intersection of the height of the child with the midparental height is shown for four seven-year-old male children (from Kaplan and Reiter, 1977).

hormone deficiency (see below). Other physiologic changes associated with hypothyroidism have been well-discussed (Smith et al., 1975).

Intellectual impairment is greater in infants not treated with thyroid hormone until after four months of age than in those treated during the neonatal period (Klein et al., 1972). Intellectual impairment is less commonly associated with arrested development of the thyroid gland due to the retention of some functional thyroid tissue.

Growth Hormone Deficiency

The estimated incidence of growth hormone deficiency in childhood is approximately 1:10,000 to 1:50,000. The birth weight and length of these children are within the normal range which has led to the conclusion that fetal growth hormone does not contribute to in utero growth (Kaplan et al., 1972). The incidence in males is 2.5-3 times greater than in females.

Confirmation of the diagnosis is dependent on demonstration of a deficient immunoreactive growth hormone response to more than two stimulatory agents (Youlton et al., 1969). The growth hormone response to stimuli of a normal child is more than 7 ng/ml. Growth hormone deficient children do not show a rise above 3 ng/ml whereas those with partial GH deficiency have a maximum GH response of >3 ng/ml and <7 ng/ml even after estradiol or propanolol priming.

During the first 2 years of life, 50% of these children have evidence of growth retardation (Goodman et al., 1968). A decreased rate of growth may not be documented in the remainder of the group until 4-10 years

TABLE 2
LABORATORY EVALUATION FOR GROWTH RETARDATION

1. CBC and sedimentation rate
2. Renal function tests—overnight specific gravity, BUN, creatinine, CO_2, K, Cl
3. Serum thyroxine and TSH
4. Plasma FSH and LH
5. Buccal smear and chromosomal analysis (females)
6. Radiologic examination: left hand for bone age, lateral view of skull for fossa size; long bone is appropriate
7. Growth hormone stimulation tests
8. Hypothalamic releasing factor stimulation tests—LRF, TRF if indicated

of age (Fig. 16). The latter group may include predominantly those with partial growth hormone deficiency. A variable degree of short stature has been observed from -2.5 SD to -10 SD from mean height for age.

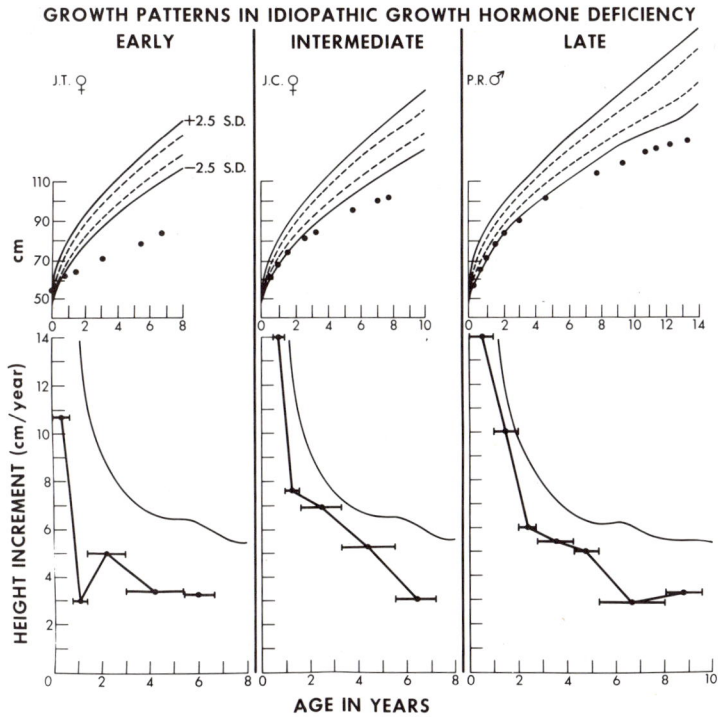

Fig. 16. The distance and incremental growth curves of three children with growth hormone deficiency. The growth rate is less than normal by one year of age (left panel), by four years of age (middle panel), and by eight years of age (right panel).

They tend to be pudgy with immature doll-like facies and normal body proportions for age (Fig. 17). Skeletal maturation is delayed but of a lesser degree in those with isolated GH deficiency than in those with GH and TSH deficiency.

The etiology of growth hormone deficiency is diverse and a classification based on data from our laboratory is shown on Table 3 (Grumbach and Kaplan, 1978). Purified growth hormone obtained from human pituitary glands at autopsy is the only effective treatment. An acceleration in growth rate and changes in physical appearance are induced following administration of human growth hormone (0.2-0.3 IU/kg/week) (Fig. 17). At present, there is adequate growth hormone available to treat most GH deficient children to induce stimulation of growth prior to epiphyseal fusion.

Glucocorticoid Excess

Growth retardation and delayed skeletal maturation is a prominent feature of hypersecretion of adrenal corticoids associated with Cushing's syndrome during childhood. Obesity, hypertension and cushinoid facies are concomitant physical findings. Exogenous administration of glucocorticoids for allergic disorders or skin diseases can lead to growth retardation and iatrogenic Cushing's syndrome. Removal of the tumor or discontinuation of exogenous glucocorticoids induces an acceleration in growth.

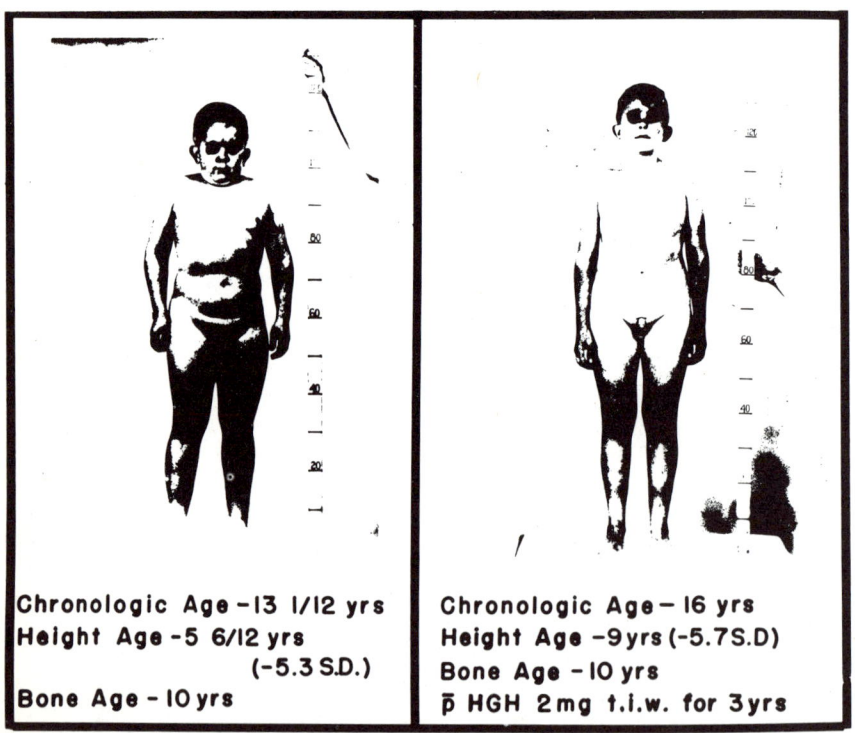

Fig. 17. The physical appearance of a boy with growth hormone deficiency is shown on left before treatment and on right after treatment with human growth hormone.

Table 3

CLASSIFICATION OF HYPOTHALAMIC-HYPOPITUITARISM

1. Isolated Growth Hormone Deficiency
   a. Sporadic
   b. Genetic deficiency
2. Multiple Pituitary Hormone Deficiencies
   a. Idiopathic
   b. Genetic
   c. Developmental defects

   Optic-septo dysplasia
   Cleft lip and/or palate
   Single giant incisor

   d. Expanding lesions

   Neoplasms, histiocystosis, granulomas

   e. Irradiation
   f. Trauma

In contrast, children with exogenous obesity as a consequence of intake of food are generally of normal or advanced height for age with slight advancement in skeletal age.

Excess Secretion of Sex Steroids

Elevated levels of testosterone or estradiol associated with precocious sexual development induce rapid growth during childhood with premature epiphyseal fusion and ultimate short stature. Precocious puberty due to premature CNS activation, sex steroid secreting tumor of gonad, or adrenal or virilizing adrenal hyperplasia are most common disease entities. Suppression of excess sex steroid secretion before advanced skeletal maturation may lead to attainment of normal height.

Altered Secretion of Insulin

Insulin may be critical for fetal growth as judged by decreased birth length and weight of neonates with insulin deficiency and increased birth weight and length of infants of diabetic mothers (Kaplan et al., 1972).
Short stature in older diabetic children is usually a consequence of poorly controlled insulin replacement therapy and/or dietary inadequacy.

Somatomedin Deficiency—Laron Dwarfism

A familial form of growth retardation associated with hypoglycemia and elevated levels of growth hormone was described by Laron and associates in 1966. Subsequent studies have demonstrated low basal levels of somatomedin with no response to administered growth hormone. This contrasts with growth hormone deficient patients who have low basal levels of somatomedin which rise briskly following administration of human growth hormone. No treatment is available at present so that final adult height is between 3.5 and 4.5 feet.

NONHORMONAL CAUSES OF GROWTH RETARDATION

Intrauterine Growth Retardation

This designation is applied to neonates whose birth weight is lower than appropriate for gestational age. Etiologic factors include 1) maternal malnutrition, alcoholism, drug addiction or cigarette smoking; 2) intrauterine infections such as rubella; 3) autosomal chromosome anomalies such as mongolism and other trisomies; 4) sex chromosome defects (see below); 5) environmental factors such as high altitude (Kaplan and Reiter, 1977).

These children are slender with craniofacial dysproportion and retardation in height of (-3.5 to 5.0 SD). The yearly growth rate is usually normal for age and their final height is in the range of 54-60 in.

Gonadal Dysgenesis

Primary hypogonadism in girls due to loss of X chromosome is associated with marked shortness of stature. Intrauterine growth retardation is common with sustained severe shortness of stature (-4 to -6 SD) throughout childhood. Other physical features such as webbed neck, lymphedema of hands and feet and cardiovascular and renal malformation occur less frequently. The final height is in range of 4'2"-4'10" (Grumbach and Van Wyk, 1974).

Malnutrition

Deficient protein-caloric intake is the most common cause of growth retardation from a worldwide perspective. It has been suggested that 60% of all preschool children in the world have some degree of malnutrition. The long-term effects of growth, physical activity, and development have been discussed in detail in numerous publications.

Malnutrition and diminished nutrient utilization is the most likely basis for the growth retardation associated with chronic diseases in childhood. These include renal disease, congenital heart disease particularly cyanotic form, pulmonary dysfunction, gastrointestinal malabsorption syndrome, and hematologic disorders (Kaplan and Reiter, 1977).

Bone Diseases

Developmental defects of cartilage or collagen result in decreased long bone formation. Children with chondrodystropic bone disease including achondroplasia and hypochondroplasia have normal trunk lengths but shortened arms and legs. Final height is generally between 3.5 and 4.5 feet (Rimoin and Horton, 1978).

Chronic rickets caused by vitamin D deficiency or metabolism of vitamin D associated with deformities of lower extremities frequently results in shortened adult stature.

Genetic Short Stature and Constitutional Delay in Growth

The height of children with genetic short stature is decreased for their peers (-2.5 to -3.5 SD) but appropriate for midparental height. They have a normal growth rate and a slight retardation in bone age. The final height is usually 5'-5'4" for males and 4'11"-5'2" for females (Fig. 18).

The degree of retardation in height in children with constitutional delay may be similar to that of the children with genetic short stature.

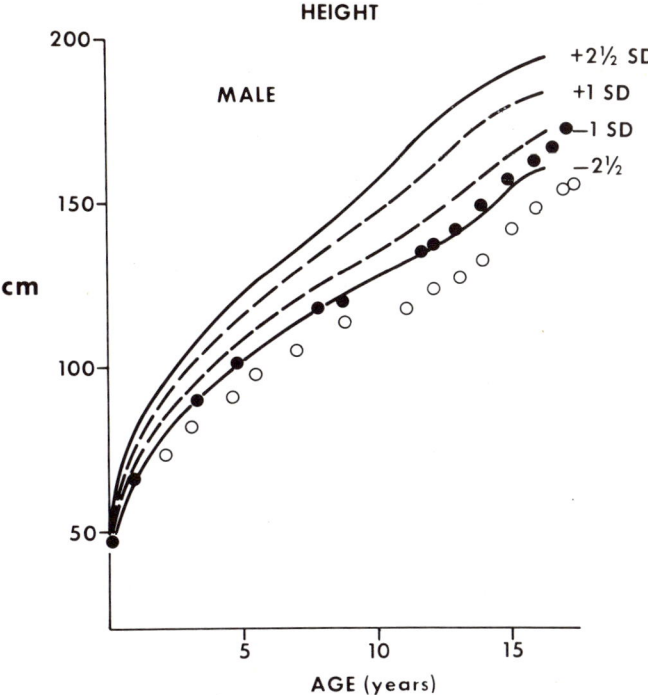

Fig. 18. The growth curve is shown for a male with genetic short stature (open circles): final height five feet, and for a male with constitutional short stature (closed circles): final height 5'7".

Generally, there is a more marked retardation in bone age (two to three years less than chronologic age). The height of these constitutionally smaller children is appropriate for midparental height. The height of the parent is within the average range for U.S. adults as is the final height attained by these children (70±2.7 inches for men; 65±1.7 inches for women) (Fig. 18).

Psychologic Disorders

The association of emotional disorders with delayed growth was first described by Talbot and associates in 1948. Powell and colleagues in 1967 described a syndrome in children with severe growth retardation and a bizarre behavioral pattern; pica, gorging of food with induced vomiting, stealing and foraging of food and other abnormal eating patterns. Laboratory evaluation may be misleading since at least one-third may have a response to stimulus. Placement of the child in a foster home or institution, or following prolonged hospitalization leads to a reversal of symptomatology and a rapid acceleration in growth rate (Fig. 19).

A summary of factors affecting growth is presented in Table 4. The precise cellular sequence of action of growth hormone remains to be elucidated. The role of somatomedins and other growth factors in normal and retarded growth is still unknown. The existence of other dysfunctional states of growth hormone metabolism can be postulated such as abnormal forms of growth hormone, excess somatostatin, or growth

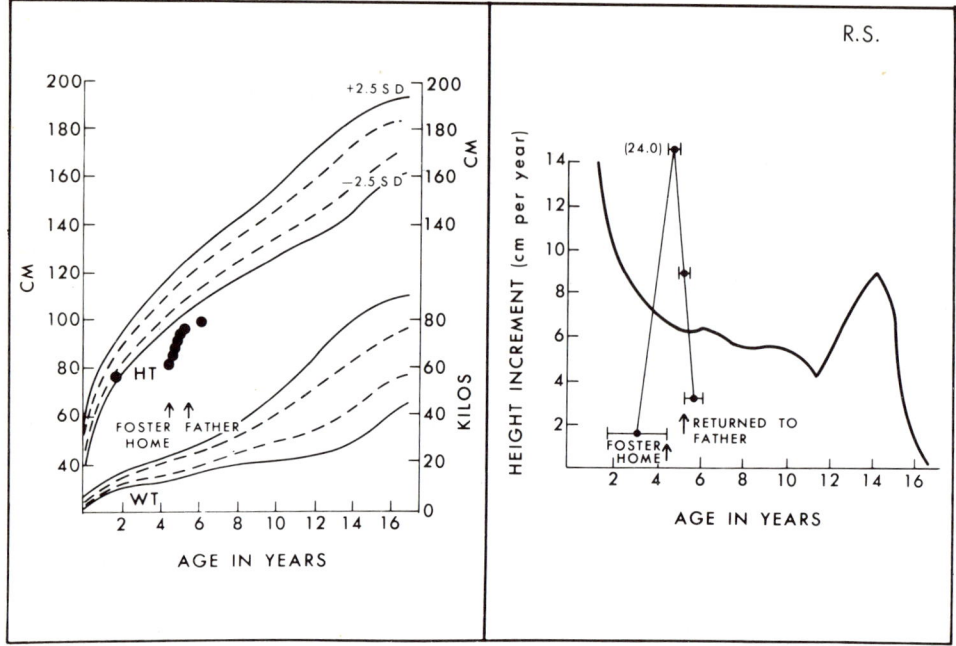

Fig. 19. The distance and incremental growth curve is shown for a child with psychosocial dwarfism. Note the marked acceleration in growth following placement in foster home with deceleration on return to natural home.

hormone and/or somatomedin receptor defects which could result in growth retardation.

Table 4

CAUSES OF GROWTH RETARDATION

1. Constitutional or genetic
2. Nutritional
3. Intrauterine factors (including chromosomal defects)
4. Cardiovascular abnormalities (cyanotic heart disease)
5. Gastrointestinal (malabsorption, regional enteritis)
6. Renal disease (renal tubular acidosis; renal insufficiency)
7. Bone disease (chondrodystrophy; rickets)
8. Endocrine and metabolic diseases
   a. Hypothyroidism
   b. GH deficiency
   c. Somatomedin deficiency
   d. Sex chromosomal abnormalities
   e. Hypercortisolism
   f. Psychologic disorders

ACKNOWLEDGMENTS

This work was supported in part by grants from the National Institutes of Child Health and Human Development and the National Institute of Arthritis, Metabolism, and Digestive Diseases, NIH, USPHS.

REFERENCES

Alexander, R.A., Conte, F.A., Kaplan, S.L., and Grumbach, M.M. The effect of estrogen treatment on height in patients with gonadal dysenesis. *Clin. Res.* 26, 174a (1978).

Bayley, N. and Pinneau, S.R. Tables for predicting adult height from skeletal age: revised for use with the Greulich-Pyle standards. *J. Ped.* 40, 423-000 (1952).

Brook, C.G.D. Evidence for a sensitive period in adipose cell replication in man. *Lancet* 2, 624-627 (1972).

Cheek, D.B., Mellits, D., and Elliott, D. Body water, height and weight during growth in normal children. *Amer. J. Dis. Child.* 112, 312-317 (1966).

Cheek, D.B., Powell, G.K., and Scott, R.E. Growth of muscle cells (size and number) and liver DNA in rats and Snell-Smith mice with insufficient pituitary, thyroid or testicular function. *Bull. Johns Hopkins Hosp.* 117, 306-321 (1965).

Chochinov, R.H., and Daughaday, W.H. Current concepts of somatomedin and other biologically related growth factors. *Diabetes* 25, 994-1004 (1976).

Dussault, J.H., Morissette, J., Letarte, J., Guyda, H., and Laberge, C. Modification of a screening program for neonatal hypothyrodism. *J. Ped.* 92, 274-277 (1978).

Evelath, P.B. and Tanner, J.M. *Worldwide Variation in Human Growth*, Cambridge University Press, Cambridge (1976).

Filipsson, R. A new method for assessment of dental maturity using the individual curve of number of erupted permanent teeth. *Ann. Human Biol.* 2, 13-24 (1975).

Forbes, G.B. Relation of lean body mass to height in children and adolescents. *Ped. Res.* 6, 32-37 (1972).

Goodman, H.G., Grumbach, M.M., and Kaplan, S.L. Growth and growth hormone. II. A comparison of isolated growth hormone deficiency and multiple pituitary hormone deficiencies in 35 patients with idiopathic hypopituitary dwarfism. *New Engl. J. Med.* 278, 57-68 (1968).

Gorski, J., Stormshak, R., Harris, J., and Wertz, N. Hormone regulation of growth: stimulatory and inhibitory influences of estrogen on DNA synthesis. *J. Toxic. and Environ. Health* 3, 271-279 (1977).

Greenberg, A.H., Najjar, S., and Blizzard, R.M. Effects of thyroid hormone on growth, differentiation and development, in *Handbook of Physiology, Vol. III, Thyroid.* M. Greer and D.H. Solomon, eds. Williams & Wilkins, Baltimore (1974), pp. 377-389.

Greulich, W.W. and Pyle S.I. *Radiographic Atlas of Skeletal Development of the Hand and Wrist,* 2nd Edition. Stanford University Press, Stanford (1959).

Grumbach, M.M. and Kaplan, S.L. Unpublished data, 1978.

Grumbach, M.J. and Van Wyk, J.J. Disorders of sex differentiation, in *Textbook of Endocrinology,* 5th ed. R.H. Williams, ed. W.B. Saunders Co., Philadelphia (1974), pp. 423-501.

Kaplan, S.L. and Reiter, E.R. Normal and abnormal growth, in *Pediatrics* 16th Ed. A. Rudolph, ed. Appleton-Century-Crofts, New York (1977), pp. 95-120.

Kaplan, S.L., Grumbach, M.M., and Shepard, T.H. The ontogenesis of human fetal hormones. I. Growth hormone and insulin. *J. Clin. Invest.* 51, 3080-3093 (1972).

Klein, A.H., Meltzer, S. and Kenny, F.M. Improved prognosis in congenital hypothyroidism treated before age 3 months. *J. Ped.* 81, 912-915 (1972).

Kochakian, D.C. Definition of androgens and protein anabolic steroids. *Pharm. Therap. B.* 1, 149-177 (1975).

Laron, Z., Pertzelan, A., and Mannheimer, S. Genetic pituitary dwarfism with high serum concentration of growth hormone, new inborn error of metabolism? *Israel J. Med. Sci.* 2, 152-155 (1966).

Marshall, W.A. Interrelationships of skeletal maturation, sexual development and somatic growth in man. *Ann. Human Biol.* 1, 29-40 (1974).

National Center for Health Statistics, Series 11, No. 124. *Height and Weight of Children in the United States.* U.S. Dept. of Health, Education and Welfare (1970).

Nicolson, A.B. and Hanley, C. Indices of physiological maturity: derivation and interrelationships. *Child Development* 24, 3-38 (1953).

Powell, G.F., Brasel, J.A. and Blizzard, R.M. Emotional deprivation and growth retardation simulating idiopathic hypopituitarism. *New Engl. J. Med.* 276, 1271-1283 (1967).

Rimoin, D.L. and Horton, W.A. Short stature. *J. Ped.* 92, 697-704 (1978).

Royer, P. Growth and Development of bony tissue, in *Scientific Foundations of Pediatrics.* J. Davis and J. Dobbing, eds. W.B. Saunders, Philadelphia (1974) pp. 376-398.

Sinclair, D. *Human Growth After Birth,* 2nd Edition. University Press, London, Oxford (1973).

Smith, D.W., Klein, A.M., Henderson, J.R., and Myrianthipoulous, N.L. Congenital hypothyroidism: signs and symptoms in the newborn period. *J. Ped.* 87, 958-000 (1975).

Talbot, N.B., Sobel, E.H., Burke, B.S., Lindemann, E., and Kaufman, S.B. Dwarfism in health children: its possible relationship to emotional, nutritional and endocrine disturbances. *New Engl. J. Med.* 236, 783-793 (1947).

Tanner, J.M. Genetics of human growth, in *Human Growth,* Vol. III. J.M. Tanner, ed. Symposium Publications, Division of Pergamon Press, New York (1960), p. 43.

Tanner, J.M. and Whitehouse, R.H. Clinical longitudinal standards for height, weight, height velocity, weight velocity and the stages of puberty. *Arch. Dis. Child.* 51, 170-179 (1976).

Tanner, J.M., Goldstein, H., and Whitehouse, R.H. Standards for children's height at ages 2 to 9 years, allowing for height of parents. *Arch. Dis. Child.* 45, 755-762 (1970).

Tanner, J.M., Whitehouse, R.H., and Takaishi, M. Standards from birth to maturity for height, weight, height velocity and weight velocity: British children, 1965. Parts I and II. *Arch. Dis. Child.* 41, 454-471 (1966).

Tanner, J.M., Whitehouse, R.H., Marshall, W.A., Healy, M.J.R., and Goldstein, H. *Assessment of Skeletal Maturity and Prediction of Adult Height: TW2 Method.* Academic Press, New York (1975).

Thompson, D'Arcy W. *On Growth and Form.* Cambridge University Press, Cambridge (1942).

Trueta, J. The growth and development of bones and joints: orthopedic aspects, in *Scientific Foundations of Pediatrics*. J. Davis and J. Dobbing, eds. W.B. Saunders Co., Philadelphia (1974), pp. 399-419.

Van der Werff ten Bosch, J.J. Testosterone as growth stimulant in man. *Pharm. Therap.* 2, 17-32 (1977).

Wettenhall, H.N.B., Cahill, C., and Roche, A.F. Tall girls: a survey of 15 years of management and treatment. *J. Ped.* 86, 602-610 (1975).

Widdowson, E.M. Changes in body proportions and composition during growth, in *Scientific Foundations of Pediatrics*. S.A. Davis and J. Dobbing, eds. W.B. Saunders Co., Philadelphia (1974), p. 153.

Wiedemann, E. and Schwartz, E. Suppression of growth hormone-dependent human serum sulfation factor by estrogen. *J. Clin. Endocrinol. Metab.* 34, 51-58 (1972).

Williams-Ashman, H.G. Metabolic effects of testicular androgens, in *Handbook of Physiology Endocrinology—Male Reproductive System*, Vol. 5. D.H. Hamilton and R.O. Greep, eds. Williams & Wilkins Co., Baltimore (1975), pp. 473-490.

Youlton, R., Kaplan, S.L., and Grumbach, M.M. Growth and growth hormone: IV. Limitations of the growth hormone response to insulin and arginine and of the immunoreactive insulin response to arginine in the assessment of growth hormone deficiency in children. *Pediatrics* 43, 989-1004 (1969).

Copyright©1982, Spectrum Publications, Inc.
Hormones in Development and Aging

Chapter 5
The Anatomical Distribution of Hormones in the Central Nervous System
Donald H. Ford
Eva B. Cramer

INTRODUCTION

Before one can consider whether or not hormones have any influence on the growth, structural organization or function of the cental nervous system (CNS), it is necessary to ascertain if hormones do indeed enter the tissues of the CNS and whether they subsequently become associated with neurons, glial cells or both upon leaving the confines of the vascular system. Extensive evidence has demonstrated that thyroid hormones enter brain tissue, become associated with neurons and undergo degradation. The principal products of thyroid hormone breakdown (iodide, mono- and diiodotyrosine) leave the brain rapidly so that accumulation of these products rarely occur. Steroid hormones (androgens, estrogen, progesterone, and corticosterone) have also been observed to accumulate in brain tissue; specific sites of uptake are the nuclei of neurons in specific brain nuclei associated mostly with the limbic system and hypothalamus. Accumulation of LH-RH has also been detected by immunocytochemistry in neuronal teminals in the basal hypothalamus and in neurons of the median eminence. With the rapid rate of development of this area of cytochemistry, one may anticipate reports in the near future of the localization of the other regulating hormones within the CNS.

Uptake of hormones by neurons in some instances, as is the case with the hormones of the reproductive organs, is associated to a considerable degree with those parts of the nervous system concerned with reproductive function or behavior. The hypothalamic peptide hormonal factors are to a considerable degree found in those parts of the hypothalamus which correspond with the physiologic locus for the so-called hypophysiotropic centers. Thyroid hormone appears to be localized in nerve cell rich areas without regional differences other than those related to the population density of the neurons, that is, the more neurons present in a nucleus, the higher the accumulation of hormone. Such a nonspecific neuronal uptake of a hormone would reasonably correspond with the type of hormone distribution which one might expect for a substance which influences neuronal excitability. The observations on corticosterone accumulation in nerve cells by biochemical procedures (Ford et al., 1971) suggests that probably all neurons accumulate this hormone, which again would correlate with its possible function in nerve cell excitability. On the other hand, the specific association of corticosterone with the hippocampus (McEwen et al., 1969), which appears to follow different binding parameters than with other neurons could reasonably be associated with the role played by this hormone in behavior.

Recent findings have shown that the neuroendocrine and neurotransmitter systems overlap anatomically as well as in their influencing each other (Grant and Stumpf, 1975). Some neurons appear to be both aminergic and hormonergic. Such maps are not yet available to demonstrate a similar relationship between hormonally receptive neurons and other neurotransmitter systems such as acetylcholine. However, a number of peptide putative transmitters are active in brain areas involved in the hypothalamic-hypophyseal regulation.

LOCALIZATION OF THYROID HORMONES IN THE CSN

In 1932, Schittenhelm and Eisler demonstrated that the highest concentrations of iodide were to be found in the hypothalamus and midbrain. These observations were later confirmed using radioactive iodide by Courrier et al. (1949) and Jensen and Clark (1951). More recently, autoradiographic studies with rabbit, guinea pig, and rat (Ford and Gross, 1958a, b; Ford et al., 1959) have demonstrated distribution of triiodothyronine (T_3) and thyroxine (T_4) in the adult CNS which was within brain nuclear and cortical areas (Fig. 1). The more densely populated a region was with neurons, the greater the degree of autoradiographic response (Ford and Gross, 1958a). Ascending paper chromatographic separation of ammoniacal ethanolic extracts of brain tissue from animals injected with ^{131}I-T_3 or ^{131}I-T_4 demonstrated that the primary labeled material present in the brain tissue corresponded with the labeled hormone injected (Fig. 2). Investigations of the relative amounts of T_3 and T_4 accumulated by the brain indicated that T_3 was 3-5 times more concentrated in brain gray matter than T_4 (Ford and Gross, 1958a). It seems likely that this difference in brain accumulation of T_3 and T_4 may be due in part to the higher levels of binding to plasma proteins for T_4 (Robbins and Rall, 1960). Metabolites of thyroid hormones (iodide, mono- and diiodotyrosine) were found in only minute amounts in adult brain. Iodide, the principal metabolite of thyroid hormone in the adult, is very rapidly removed from the brain and apparently leaves the brain tissue as fast as the thyroid hormone molecule is deiodinated. The iodotyrosines appear as metabolites of thyroid hormone primarily in neonatal brains (Cohan et al., 1969). Additional studies by Ford and Rhines (1967) on ^{131}I-T accumulation in the spinal cord of rat demonstrated that the hormone was concentrated primarily in neurons, with relatively little of the radioactivity being present in the surrounding neuropil (Fig. 3).

Dry-mount autoradiography (Stumpf and Roth, 1966), which demonstrates the total amount of a radioactive material present (free and bound), illustrated that the concentration of ^{125}I-T_3 and T_4 in brain after injection of labeled hormones was most evident in the neuron rich areas (Stumpf and Sar, 1975b), confirming the previous observations of Ford and Rhines. They further demonstrated particularly high concentrations of labeled material over brain areas adjacent to the third ventricle which lack a blood-brain barrier. Thus, label was observed over the choroid plexus, over the ependyma in the region of the optic and infundibular recesses, and over the region of the vascular organ of the lamina terminalis. In the tuberal region, the hormonal associated radioactivity may have been in ependymal cells whose processes extend to the brain surface (tanycytes). These areas of radioactivity are congruent with the so-called thyrotrophic area in the anterior hypothalamus (see review by Szentagothai et al., 1968) and are to a degree comparable with the site of high levels of radioactivity associated with T_3 reported in the tuberal region of the guinea pig (Ford and Gross, 1958b, see Fig. 1).

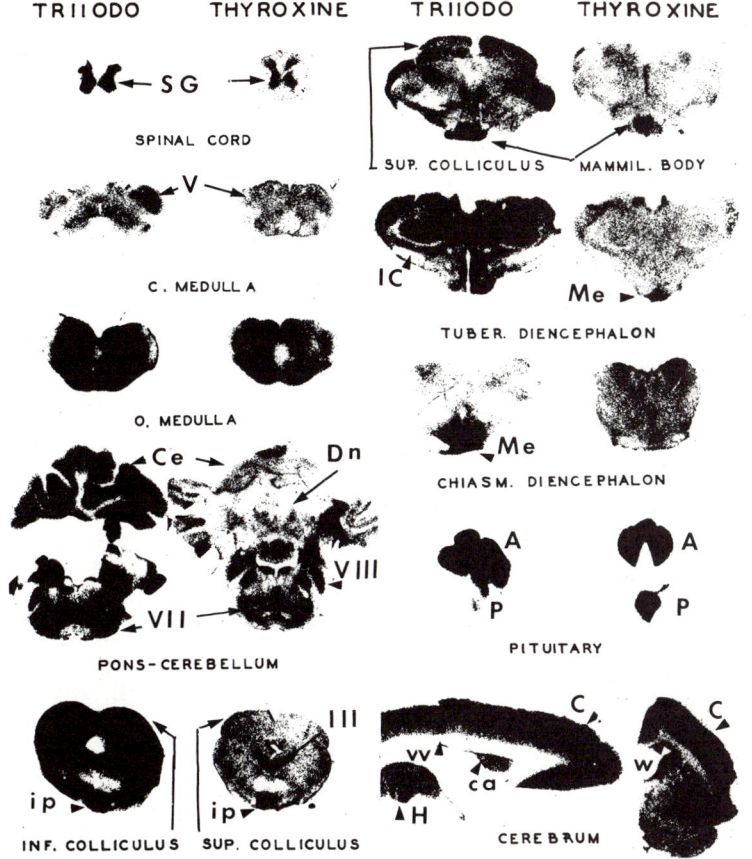

Fig. 1. X-ray autoradiographs of sections of guinea pig brain after intravenous injection of ^{131}I-triiodothyrone (T$_3$) or ^{131}I-thyroxine (T$_4$). The radioactive hormones concentrated primarily in the cerebral and cerebellar cortex and in nuclear aggregates of neurons. Areas of white matter showed virtually no radioactive response. Further, in the brain there was a more intense accumulation in animals injected with T$_3$ than with T$_4$. A = anterior pituitary lobe, C = cerebral cortex, Ca = caudate nucleus, Ce = cerebellar cortex, Dn = deep nuclei of cerebellum, H = hippocampus, IC = internal capsule, IP = interpeduncular nucleus, Me = median eminence, P = posterior pituitary lobe, III = oculomotor nucleus, V = descending spinal trigeminal nucleus, VII = facial nucleus, VIII = vestibular nuclei, SG = spinal gray matter, W = white matter (from Ford and Gross, 1958b.)

The localization of thyroid hormone (^{125}I-T$_4$) in the CNS has been further refined by electronmicroscopic studies of hormone uptake into cells in sections of cultured spinal cord and cerebellum (Manuelidis and Bornstein, 1970; Manuelidis, 1972, Manuelidis and Manuelidis, 1972). Positive reactions were observed in the emulsion over the nuclei of ventral horn neurons and cerebellar granule cells and to a lesser degree over astrocytes. Grains in the emulsion were also observed over the mitochondria and ribosomes of the rough endoplasmic reticulum. Scattered grains were noted over the ependymal cells of the fourth ventricle and over the myelin of cerebellar axons. In view of the much higher brain uptake of T$_3$ than T$_4$, it would

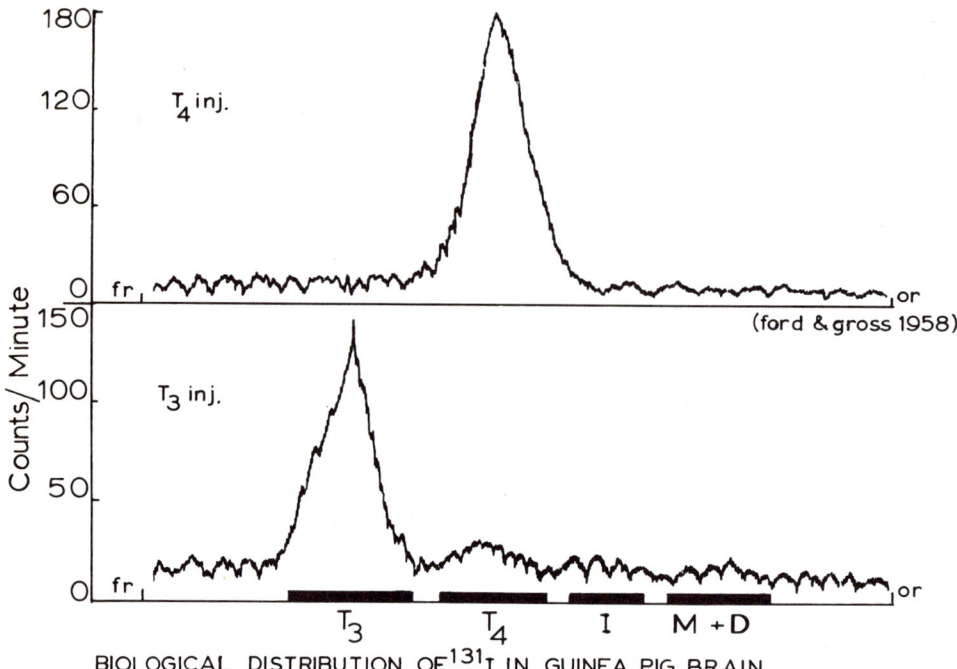

Fig. 2. The biological identity of ^{131}I-labeled compounds in extracts of brain tissue in guinea pig killed at various time intervals following intravenous injection of ^{131}I-T$_3$ or ^{131}I-T$_4$ as determined by ascending paper chromatography. OR = origin, fr = front, M + D = mono- and diiodotyrosine, I = iodide. (From Ford and Gross, 1958a.)

be interesting to determine what the distribution of T$_3$ would be like under the conditions of this experiment.

The amount of ^{131}I-T$_3$ accumulated in rat brain was shown to be age and sex related (Ford and Rhines, 1970). Brain tissues from male rats accumulated more hormone at 24 months than in three-month old animals, whereas with females there were no age-related differences in accumulation within this age range. However, female rats at the beginning of sexual maturity (seven weeks) accumulated more ^{131}I-T$_3$ in the cerebral cortex, white matter, hypothalamus, cerebellum and brain stem than did male rats of the same age. Prior to this age (2-4 weeks of life, before sexual maturation) accumulation of hormone was essentially the same in both sexes (Fig. 4, Bleecker et al., 1970).

LOCALIZATION OF ADRENOCORTICAL HORMONES IN THE CNS

While effects of cortical hormones were noted on the levels of electrolytes in the CNS by 1954 (Woodbury) the biochemical detection of corticosteroids in neural tissues was first described by Touchstone and co-workers (1963; 1966). They demonstrated that cortisol was present in human peripheral myelinated nerves as well as in brain. Henken et al. (1968) further showed that cortisol and corticosterone identified in cat brain was dependent on the presence of the adrenal glands and fluctuated with blood levels of the hormones. It was also apparent that accumulation of the hormone in brain was dependent on some sort of active concentrative process.

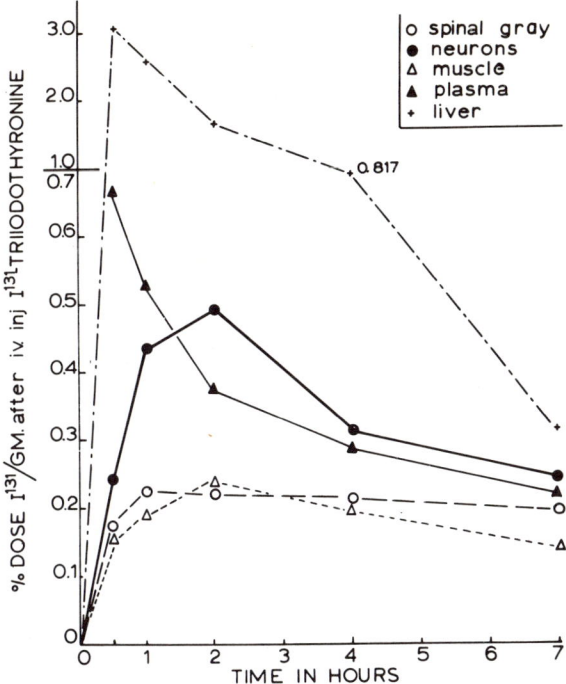

Fig. 3. The accumulation of ^{131}I-T_3 in ventral horn neurons, spinal cord gray, muscle, liver and plasma of the male rat at various time intervals after intravenous injection of labeled hormone. Note the low levels of hormone present in spinal cord gray matter and muscle as compared to that in neurons (from Ford and Rhines, 1967).

Cell fractionation studies of cat brain following intravenous injection of ^3H-cortisol (Fontana et al., 1970) demonstrated that the major localization of hormone was in cell nuclei. Autoradiographic studies (Grelich and McEwen, 1972) further confirmed this observation. In an earlier study, McEwen et al. (1969) demonstrated preferential accumulation of ^3H-corticosterone in the hippocampus and septal region of the rat. This was also suggested by Stevens et al. (1971; see also Fig. 5), although only the hippocampus had a significantly higher accumulation. Accumulation studies in neurons dissected from formalin-fixed rat brain after intravenous injection of ^3H-corticosterone also showed specific neuronal uptake (Fig. 6 and 7) with levels of accumulation varying with neuron type (Ford et al., 1971). Binding studies (McEwen and Wallach, 1973) demonstrated the high affinity of ^3H-corticosterone for neuronal nuclei. In descending order, nuclear binding by region proceeded as follows: hippocampus, amygdala, cerebral cortex, hypothalamic preoptic area, midbrain, and cerebellum. Autoradiographic studies of Stumpf and Sar (1975a), using the dry-mount technique demonstrated accumulations of radioactivity after treatment with ^3H-corticosterone in nuclei of certain nerves in the hippocampal complex (highest in CA_2), dentate gyrus, septum, induseum griseum, subiculum, amygdala, entorhinal, and piriform cortex. The diencephalon was unlabeled. The choroid plexus also concentrated label with the cells showing a high cytoplasmic content of hormone. Injection of adrenalectomized rats with a tritium labeled synthetic glucocorticoid, dexamethasone (Rees et al., 1975) demonstrated the highest levels of accumulation in the brain to be in the choroid plexus, ventricular lining

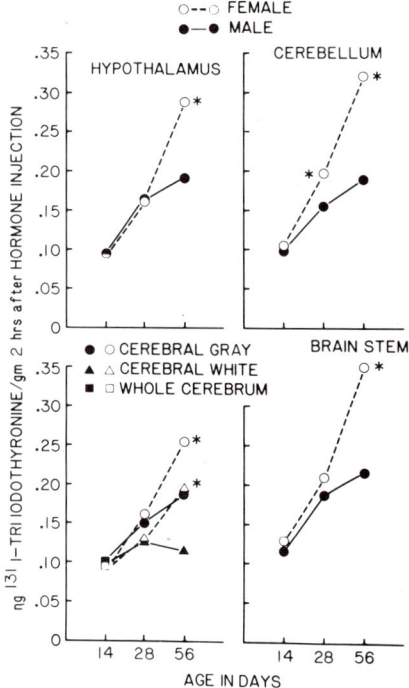

Fig. 4. The accumulation of ^{131}I-T_3 in various brain areas in male and female rats at two, four, and eight weeks of life (from Bleecker et al., 1971). Asterisk represents statistical significance of $P < 0.05$.

and in the periventricular tissue and along the cortical surface. While many of the cells of the brain showed nuclear labeling, the distribution of label appeared to be relatively nonspecific. Neurons, glia, endothelial cells of blood vessels, circumventricular organs and cells of the dura, pia, and arachnoid were all labeled to some degree 30 minutes after the injection. By three hours post-injection, the label was widely spread throughout the brain, and was still nonspecifically distributed.

LOCALIZATION OF SEX STEROID HORMONES IN THE CNS

The development of the dry-mount autoradiographic procedure by Stumpf and Roth (1966) provided a tool whereby lipid soluble compounds such as steroid hormones could be localized in the CNS and other tissues without or with minimal artifactual movement of the steroid from its site of accumulation. With this technic steroid hormones have been localized primarily within nuclei of neurons. However, ependymal and subependymal cytoplasmic localizations are also reported. (See Stumpf et al., 1975d.)

Numerous regional differences have been observed in intensity of labeling among different hormones and species. What is perhaps of particular interest is the high degree of consistency of labeling of neurons with sex steroid hormones in those areas of the hypothalamus generally associated with reproductive processes or behavior and in the limbic system. The number of investigations reporting hormone localization is large, as is the number of species surveyed. Although there are exceptions

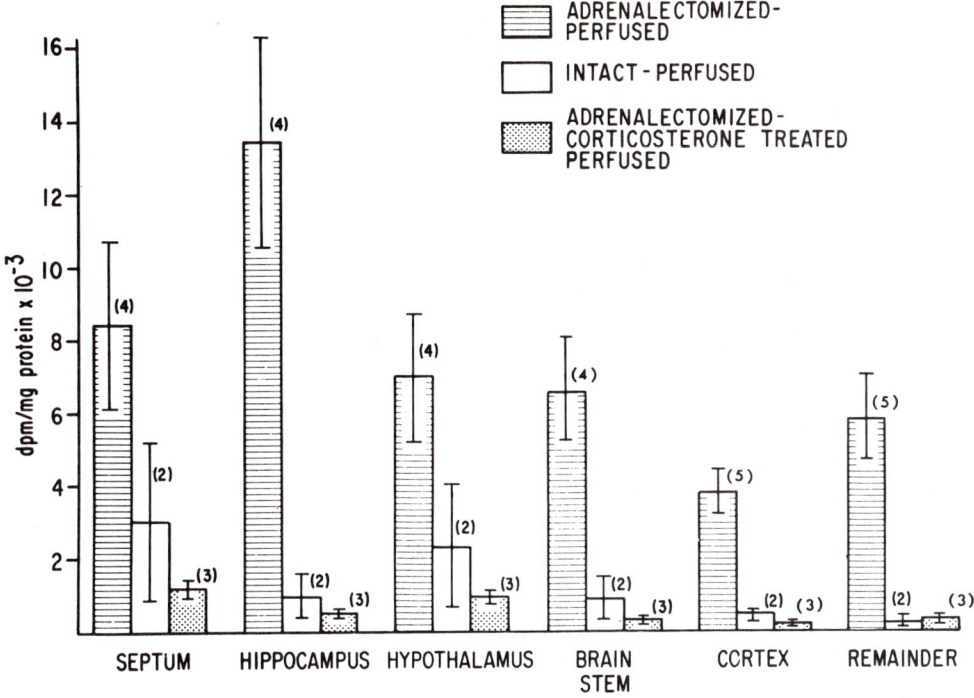

Fig. 5. The regional distribution of radioactivity following injection of ^{3}H-corticosterone in the rat brain. "Remainder" refers to those parts of the brain not assayed separately. Numbers in paraenthesis refer to the number of animals. DPM = disintegrations/min. (from Stevens et al., 1971).

to the concept that sex steroids localize in brain regions associated with reproductive processes (ventral horn and other motor neurons for dihydrotestosterone, see Sar and Stumpf, 1977), most reports emphasize the limbic-mesencephalic medial hypothalamic-medial preoptic locus (Krieger et al., 1976). That other neurons may respond in ways not associated with reproduction is not questioned; it is only that these other sites of hormone action or location appear minor in relation to those involving CNS components associated with reproductive function. Further, there are species differences in hormone requirements for induction of reproductive behavior in females; i.e., in the rat, estrogen alone will induce reproductive vaginal changes (Ford, 1954) whereas in the hamster (Ciaccio and Lisk, 1971) and the guinea pig (Ford and Young, 1951) progesterone is also required for vaginal and behavioral changes. As indicated above there are also neuroanatomical differences in the hormone sensitive sites associated with behavior within different species. Thus, one animal may show a higher affinity for a hormone at a particular locus than in another. A summary of reported sites of sex steroid hormone localization in a number of species without indicating concentration is presented in Tables 1-3. Autoradiographs illustrating sex hormone uptake are demonstrated in LOCALIZATION OF PEPTIDE HORMONE IN CNS, Figs. 8-10.

The introduction of immunocytochemical (immunoperoxidase) techniques has made it possible to identify and localize various peptides within the brain as depicted in Fig. 11 which illustrates labeling of LH-RH containing

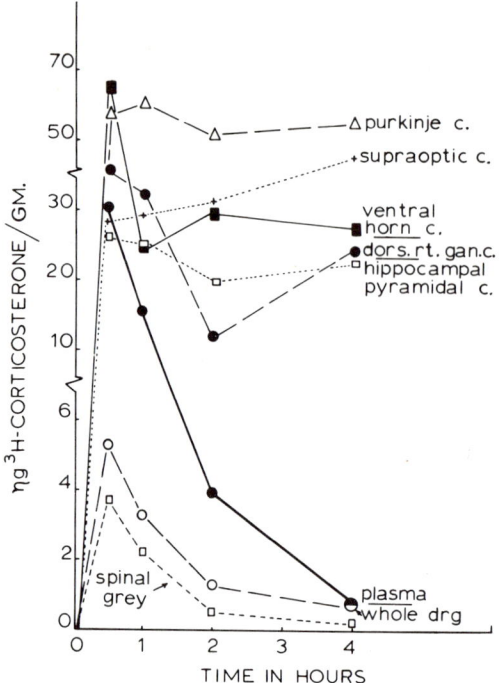

Fig. 6. The accumulation of chromatographically identified ^{3}H-corticosterone in ventral horn neurons, spinal gray matter, retina, dorsal root ganglia, muscle, liver, and plasma at various time intervals after intravenous injection of the labeled hormone. (From Ford et al., 1971.)

terminals in the tuberal region of the rat stalk and in the vascular organ of the lamina terminalis (Kordon and Ramirez, 1975). This immunocytochemical technique in conjunction with bioassay of extracts of specific hypothalamic regions have enable various investigators to demonstrate regional localization of hypothalamic releasing or inhibiting hormones of LH-RH, PIF, PRF, GRF, GIF, and TRF (McCann et al., 1975). The results of these localization studies are summarized in Table 4. Numerous observations have shown that the hypothalamic dopamine neurons are associated with the regulation of LH-RH release and may also represent the PIF factor. Positive and negative correlations between LH-RH and the enzymes involved in catecholamine synthesis are illustrated in Fig. 12.

As shown in Tables 1-4 the distribution of neurons capable of accumulating labeled androgens, estrogen, and progesterone as well as the location of neurons or nerve fibers containing LH-RH are often within the fields of termination of neurons of the amine system. In some instances the hormone may be concentrated within amine-containing neurons, as has been shown for estrogen (Heritage et al., 1977). This is of particular interest since dopamine, norepinephrine, and serotonin have all been implicated in the control of gonadal, adrenal, and thyroid function (Schaepdryver et al., 1969; Grimm and Reichlin, 1973; Scapagnini and Preziosi, 1973). The manner in which this control is exerted appears quite complicated, and is probably applied at many levels

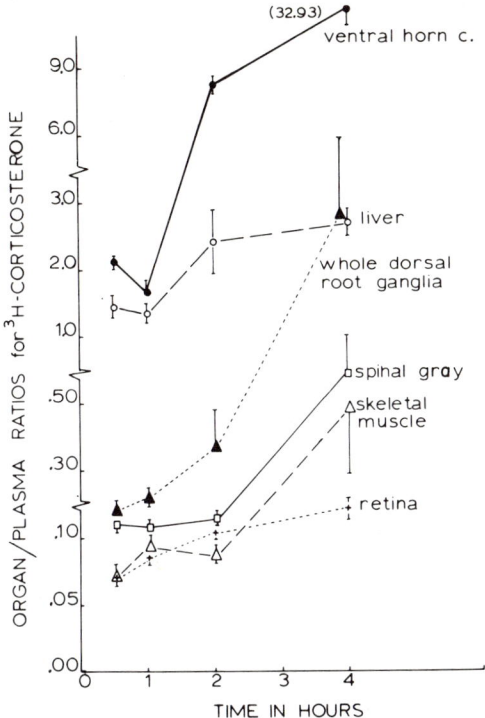

Fig. 7. A plot of the organ/plasma ratios of ^{3}H-corticosterone in neurons, dorsal root ganglia, spinal cord gray matter, liver, muscle, and retina to illustrate the significantly higher concentrations occurring in neurons. (From Ford et al., 1971).

within the brain-pituitary axis. In turn, monoaminergic neurons appear themselves to be influenced by various endocrine hormones (Ahern et al., 1971; Anton-Tay et al., 1970; Coppola, 1969; Greenglass and Tonge, 1972; Javoy et al., 1968; Prange et al., 1970). Further, hypothalamic peptides, in particular TRH may influence brain amine turnover or metabolism (Constantinidis et al., 1974; Keller et al., 1974). The details of the interrelationships between amine systems and the hormone-target neurons remain to be clarified. However, the juxtaposition of these neurons which appear to be both targets of hormones as well as capable of synthesizing amines, and the demonstration that amines regulate the function of terminal endocrine glands by acting on hypothalamic neurons, support the view that the neuroendocrine and monoamine systems are closely interrelated functionally (Figs. 13-15).

Table 1

ANDROGEN LOCALIZATION IN BRAIN

SPECIES	AGE	SEX	SITE	HORMONE	REFERENCE
Rat	Adult	Male and Female	Saturable binding in cytosol of hypothalamus, amygdala, and cerebral cortex	5-α-dihydrotestosterone	Barley et al., 1975
Rat	28 day	Male	Cytosol receptor proteins of hypothalamus; receptor-hormone complex translocated to nuclear compartment	5-α-dihydrotestosterone	Kato, 1975
Rat	3, 7 wks. 27 days & adult	Male	Cytosol macromolecule receptors in hypothalamus for 5-α-hydrotestosterone and testosterone with binding kinetics suggesting different receptors for each hormone	testosterone 5-α-dihydrotestosterone	Kato, 1976
Rat	Neonatal (7 days)	Female	Saturable receptors for testosterone in hypothalamus in 8S region of sucrose density gradients	testosterone	Kato, 1976
Rat	Adult	Male	Lateral septum, nuc. of stria terminalis, medial and lateral cortical amygdalar nuc., central gray (and lateral to central gray), amygdala, medial preoptic nuc.	testosterone	Morrell et al., 1975 Stumpf, 1972
Rat	26 day castrate adult	Male	Arcuate nuc., ventromedial hypothalamic nuc., medial preoptic nuc., interstitial nuc. of stria terminalis, lateral septal nuc., hippocampus and amygdala	testosterone	Sar and Stumpf, 1973a
Rat	Immature & adult	Male	Purkinje neurons, ventral horn neurons, and neurons of motor nuclei; widely scattered neurons in reticular formation of pons and medulla, also some hormone in ependymal and subependymal cells of IVth ventricle and spinal cord	5-α-dihydortestosterone	Sar and Stumpf, 1977

SPECIES	AGE	SEX	SITE	HORMONE	REFERENCE
Rat	2 day old	Female	Preoptic nuc., suprachiasmatic nuc., interstitial nuc. of stria terminalis, periventricular nuc., arcuate nuc., ventromedial and ventrolateral hypothalamic nuc., ventral premammillary nuc., medial amygdalar nuc.	testosterone	Sheridan et al., 1974
Rat	Adult castrate estrogen treated	Male	Was a decreased uptake of ^{3}H-testosterone in areas previously shown to take up the hormone	testosterone	Simmons, 1971
Chick	2-weeks old	Male	Medial preoptic area, other hypothalamic nuc., olfactory bulb, hippocampus, striatum, nuc. rotundus, optic tectum, medial mammillary nuc. and cerebellum	testosterone	Meyer, 1973 Morrell et al., 1975
Chicken	Adult	Male	Parolfactory lobe, lateral septum, medial preoptic area, unc. of stria terminalis, paraventricular nuc., anterior, posterior and medial hypothalamic nuc., tuberal nuc., nuc. of the taenia, 2/3 of the archistriatum, intercollicular nuc. and medial part of dorsolateral nuc. of thalamus	testosterone	Morrell et al., 1975
Chicken	Adult	Male	Dorsal hyperstriatum, nuc. intercalatus, preoptic and supraoptic nuc.	testosterone	Wood-Gush et al., 1977
King dove	Adult	Male	Cerebrum and hypothalamus	testosterone	Morrell et al., 1975
Zebra finch	Adult	Male	Lateral septal area, anterior preoptic area, paraventricular nuc., anterior and medial hypothalamic nuc., internal cellular stratum, tuberal nuc., nuc. of the taenia of the archistriatum, nuc. of the neostriatum, posterior nuc. of the ventral hyperstriatum, intercollicular nuc. of midbrain and nuc. intermedius of medulla	testosterone	Zigmond et al., 1973

SPECIES	AGE	SEX	SITE	HORMONE	REFERENCE
Chaffinch	Adult	Male	Lateral septum, medial preoptic area, paraventricular nuc., posterior and medial basal hypothalamus as well as the periventricular zone above the optic recess, central gray and reticular formation	testosterone	Zigmond et al., 1973
Frog	Adult	Male & Female	Anterior preoptic area, ventral part of tuber cinereum, torus semicircularis, tectum, ventral thalamus, brainstem ventral to cerebellum (less intense labeling of neurons than with 3H-estradiol)	testosterone	Morrell et al., 1975
Teleost	Adult	Male	Lateral tuberal nuc.	testosterone	Morrell et al., 1975

Table 2

ESTROGEN LOCALIZATION IN BRAIN

SPECIES	AGE	SEX	SITE	HORMONE	REFERENCE
Mouse	Weaning	Female	Hypothalamic and preoptic homogenates bound estradiol	Estradiol	Fox and Johnston, 1974
Mouse	Young	Female ovariect.	Medial preoptic nuc., supraoptic nuc., suprarhinal cortex, central gray, ventral hypothalamus, amygdala, organum vasculosum and ependyma of medulla	Estradiol	Stumpf and Sar, 1975
Rat	Adult	Male & Female	Lamina I and II ov dorsal horn and dorsal part of lamina X in lumbar spinal cord	Estradiol	Keefer et al., 1973
Rat	Adult	Female gonadect.	Invarious neurons of midbrain, pons, and medulla shown to also contain catecholamines	Estradiol	Heritage, et al., 1977
Rat	Adult	Female	Amygdala, hippocampus, nuc. accumbens, anterior hypothalamic area, arcuate nuc., central gray, diagonal band of Broca, dorsomedial nuc. of hypothalamus, lateral habenular nuc., medial preoptic area, mammillary body, medial preoptic nuc., nuc. of stria terminalis, parafascicular nuc. nuc., olfactory tubercle, ventromedial hypothalamic nuc.	Estradiol	Morrell et al., 1975
Rat	Adult	Male & Female	Preoptic area, anterior hypothalamus, medial basal hypothalamus, amygdala, dorsal hypothalamus, cerebral cortex (increased after gonadectomy)	Estradiol	Ogren and Woolley, 1976 Stumpf, 1972
Rat	Adult	Male & Female	Periventricular nuclei of brain	Estradiol	Stumpf, 1970

SPECIES	AGE	SEX	SITE	HORMONE	REFERENCE
Tree shrew	Adult	Female intact & ovariect.	Periventricular brain areas, (preoptic area, lateral spetal area, amygdala, bed nuc. of stria terminalis, infundibulum, median eminence	Estradiol	Keefer and Stumpf, 1975 Stumpf and Sar, 1971
Guinea pig	Adult	Female	Arcuate nuc., ventromedial nuc. of hypothalamus, amygdala, medial preoptic area, interstitial nuc. of stria terminalis, suprachiasmatic nuc., lateral septal nuc., paraventricular nuc., supraoptic nuc., mammillary body	Estradiol	Sar and Stumpf, 1975b
Guinea pig	Adult	Female ovariect.	3H activity found in midbrain, hypothalamus, hippocampus and cerebral cortex	Estradiol	Wade and Feder, 1972a
Rabbit	Immature	Female	3H activity associated with cell fractions (nuclei, mitochondria, microsomes and supernatant fractions) of hypothalamus & cerebellum	Estradiol	Chader and Villee, 1970
Hamster	Adult	Female	Medial preoptic area, medial anterior hypothalamus, arcuate nuc., ventromedial nuc., amygdala, bed nuc. of stria terminalis, ventrolateral septal area	Estradiol	Morrell et al., 1975
Cat	Adult	Female	Lateral septal area, preoptic area, hypothalamus	Hexoesterol	Morrell et al., 1975
Mink	Adult	Female	Bed nuc. of stria terminalis, lateral septal and preoptic areas	Estradiol	Morrell et al., 1975
Squirrel monkey	Young	Female intact & ovariect.	Medial preoptic nuc., interstitial nuc. of stria terminalis, infundibular nuc., ventromedial nuc. of hypothalamus, ventral premammillary nuc., amygdala and several scattered cells in various other nuc.	Estradiol	Keefer and Stumpf, 1975b

SPECIES	AGE	SEX	SITE	HORMONE	REFERENCE
Rhesus monkey	Adult	Female	Lateral septal area, bed nuc. of stria terminalis, medial preoptic area, medial anterior hypothalamic area, ventromedial and arcuate nuc., amygdala, subfornical organ, ependyma in the anterolateral walls of the tuber cinereum	Estradiol	Morrell et al., 1975
Frog	Adult	Female	Ventral striatum, ventrolateral septum, amygdala, preoptic area, ventral part of tuber cinereum	Estradiol	Morrell et al., 1975
Chicken	Adult	Female	Ventral hyperstriatum	Estradiol	Wood-Gush et al., 1977
Dove	Adult	Male & Female gonadect.	Arcuate nuc., posterior hypothalamus, taenial nuc., medial preoptic area	Estradiol	Martinez-Vargas et al., 1975
Ring Dove	Adult	Male & Female intact	Medial preoptic area, interstitial nuc. of stria terminalis, infundibular and ventromedial nuc. of hypothalamus, taenial nuc., intercollicular nuc.	Estradiol	Morrell et al., 1975

See also: Anderson and Greenwald, 1969; Attramadal and Aakvaag, 1970; Kato et al., 1971, 1974, 1975; Keefer and Stumpf, 1975c; Mauer and Woolley, 1971; Pfaff and Keiner, 1972, 1973; Sheridan et al., 1974; Stumpf, 1970, 1971, 1972; Stumpf and Grant, 1975c & d, (Anatomical Neuroendocrinology); Stumpf et al., 1975; Zigmond and McEwen, 1970.

Table 3

PROGESTERONE LOCALIZATION IN BRAIN

SPECIES	AGE	SEX	SITE	HORMONE	REFERENCE
Mouse	30 days	Female ovariect. adrenalect.	Cerebral cortex, cerebellar vermis, olfactory bulb, medial hippocampus, amygdala, septal area, preoptic-anterior hypothalamus, medial posterior hypothalamus, interpeduncular region, medulla	Progesterone	Luttge et al., 1973
Rat	Adult	Female	Anterior and posterior hypothalamus, cerebral and cerebellar cortex	Progesterone	Seiki et al., 1969
Rat	Adult	Male & Female, intact, ovariect. or adrenalectomized	Anterior and posterior hypothalamus, midbrain tegmentum, cerebral cortex	Progesterone	Whalen and Luttge, 1971
Guinea Pig	35 days	Female ovariect.	Arcuate nuc., ventromedial hypothalamus nuc., premammillary nuc., periventricular preoptic area, unc. of the diagonal band, suprachiasmatic nuc., (in the suprachiasmatic-preoptic area uptake was enhanced by pretreatment with estradiol)	Progesterone	Sar and Stumpf, 1973b
Guinea Pig	Adult	Female ovariect.	Cerebral cortex, hippocampus, hypothalamus, midbrain	Progesterone	Wade and Feder, 1972b
Guinea Pig	Adult	Female ovariect.	Cerebral cortex, hippocampus, hypothalamus, midbrain	20α-hydroxy-pregn-4-3n-3-one	Wade and Feder, 1972a

CENTRAL NERVOUS SYSTEM 167

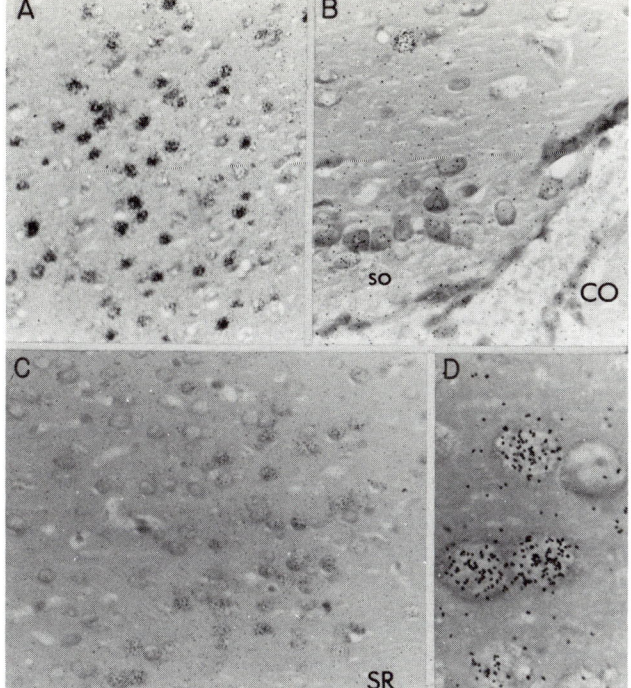

Fig. 8. Autoradiographs of (A) the medial preoptic nucleus, (B) supraoptic nucleus, and (C) and (D) suprarhinal cortex showing differences in nuclear concentration of radioactivity after injection of ^{3}H-estradiol. (Stempf and Sar, 1975c.)

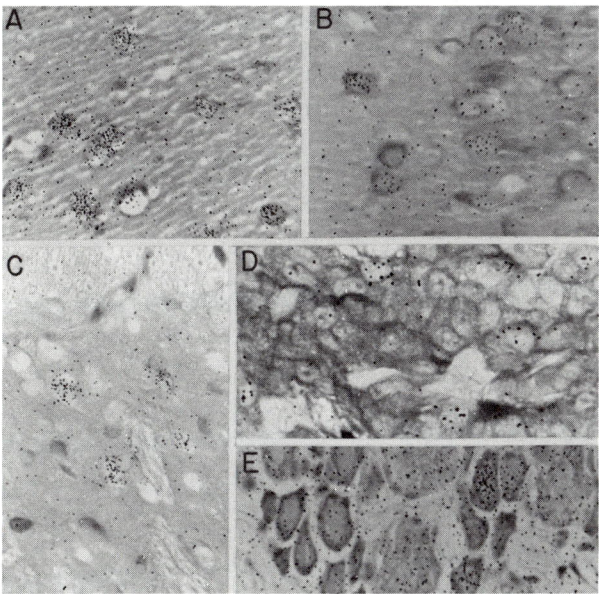

Fig. 9. Autoradiographs of ^{3}H-estradiol in (A) entorhinal cortex, (B) area CA$_3$ of the ventral hippocampus, (C) substantia nigra, (D) pineal, and (E) trigeminal ganglion of the rat. (Stumpf et al., 1975).

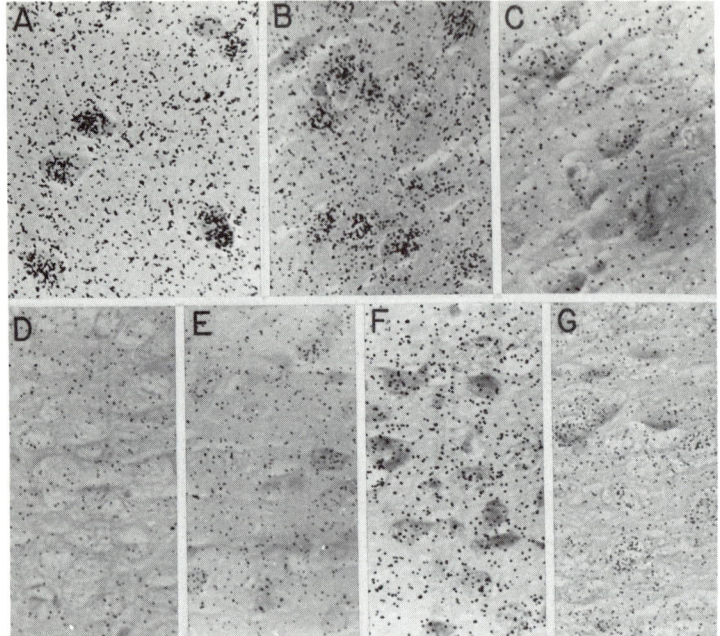

Fig. 10. Autoradiographs of ^{3}H-testosterone over cells in (A) the dentate gyrus, (B) hippocampus, (C) paraventricular nucleus, (D) ventromedial nucleus, (E) entorhinal cortex, and (F) prelateral mammilary and pontine nuclei in the castrated male rat. (Sar and Stumpf, 1975a).

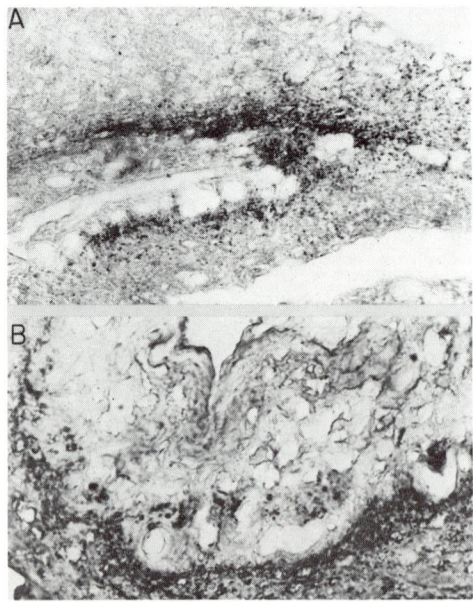

Fig. 11. Immunocytochemical illustration of the distribution of LH-RH containing terminals in (A) the tuberal region of the pituitary stalk and (B) in the vascular organ of the lamina terminalis. (Kordon and Ramirez, 1975; permission for publication granted).

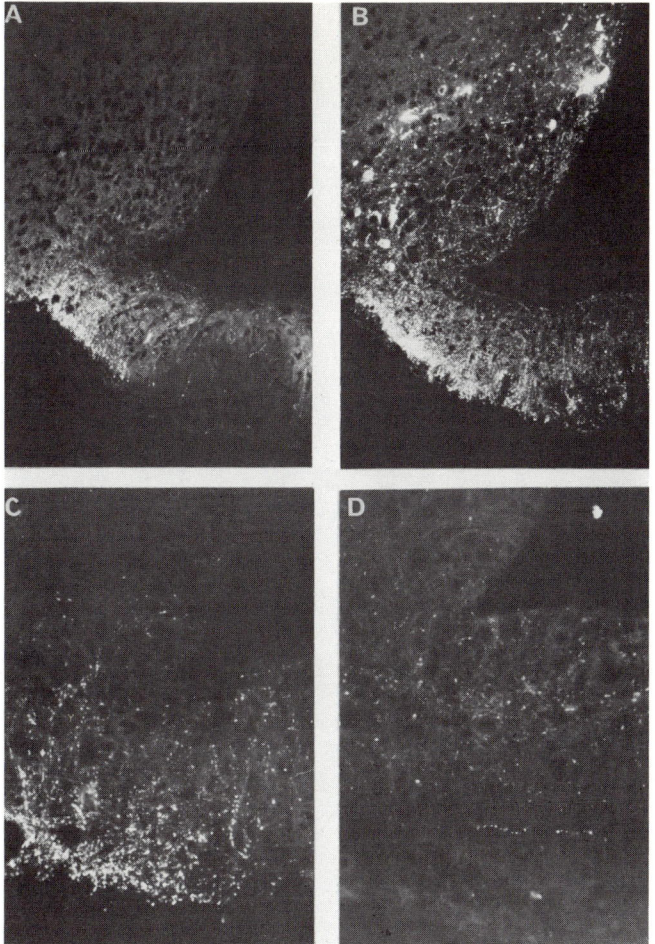

Fig. 12. Immunocytochemical illustration of (A) and (C) the distribution of LH-RH, (B) tyrosine hydroxylase and, (D) dopamine-β-hydroxylase. (A) and (B) show consecutive sections and demonstrate LH-RH terminals, mostly in the external layer of the lateral part of the median eminence. The positive reaction to tyrosine hydroxylase (B) indicates the distribution of the dopamine neuron system in relation to the LH-RH containing terminals. The very faint response in (D) illustrates the very low levels of norepinephrine and or epinephrine in this area. (Hökfelt et al., 1975.) Mag. (A) and (C) = 120X; (B) and (D) = 300X.

Table 4

HYPOTHALAMIC HORMONE LOCALIZATION

SPECIES	AGE	SEX	SITE	HORMONE	REFERENCE
Mouse	Adult	Male and Female	Arcuate nucleus, tanycytes of third ventricle, in axons	Gonadotrophic releasing hormone (LRF)	Kozlowski et al., 1975
			Supraoptic and paraventricular nic. in axons	Neurophysin	Kozlowski et al., 1975
Rat	Adult	Male	Throughout brain with highest concentrations in hypothalamus	Somatostatin, TRF	Brownstein, 1977
Rat	Adult	Male and Female	CSF of third ventricle, median eminence	LRF, TRF	Joseph et al., 1975
Rat	Adult	not known	Suprachiasmatic area (by bioassay)	LRF	McCann et al., 1975
			Arcuate median eminence area and pituitary stalk	FSH-RF	McCann et al., 1975
			Bed nuc. of stria terminalis extending caudally to the dorsomedial nuc. and ventrally into median arcuate area	TRF	McCann et al., 1975
			Lateral preoptic area	PIF	McCann et al., 1975
			Suprachiasmatic region	PRF	McCann et al., 1975
			Lateral part of ventromedial nuc.	GRF	McCann et al., 1975
			Median eminence	GIF	McCann et al., 1975
Rat	Adult	Male	Vascular region of the lamina terminalis, median eminence, around capillaries of hypothalamo-hypophyseal portal system in axons, pituitary stalk in axons	LH-RH (LRF)	Kordon and Ramirez, 1975 Kordon et al., 1974

SPECIES	AGE	SEX	SITE	HORMONE	REFERENCE
Rat	Young Adult	Female hypopy-sect.	Immunoreactive sites for HCG in arcuate nuc. third ventricle ependyma, and hypependymal cells, median eminence, ventromedial nuc. of hypothalamus, suprachiasmatic nuc., ventral premammillary nuc. choroid plexus, subcommissural organ	HCG	Petrusz, 1975
Sheep	Adult	not given	Infundibular neurons, supraoptic and paraventricular nuc.	Neurophysin	Kozlowski et al., 1975
Man	3 days to 88 years	Male and Female	Immunoreactive neurons in precommissural septum, retrochiasmatic area, lamina terminalis, pupraoptic region, infundibular nuc. posterior infundibular eminence. Immunoreactive axons run through infundibulum, along deep capillary loops of the portal plexus, among the capillaries of the vascular organ of the lamina terminalis. Reactive neurons were also present in the retromammillary area and rostral midbrain. These latter appear to give rise to extra-hypo-physeal tracts ending in telencephalon and brainstem.	LRF	Barr, 1977

See also: Barr et al., 1974; Barr and Carette, 1975; Barr and Dubois, 1974, 1975; Goldsmith and Ganong 1975; Krulich et al., 1977; Mulder et al., 1970; Choy and Watkins, 1977; King and Gerall, 1976; Kardon et al., 1977; Kastin et al., 1976; Pelletier et al., 1975; Pimstone et al., 1976; Silverman, 1976; Sokol et al., 1976; Stern et al., 1975.

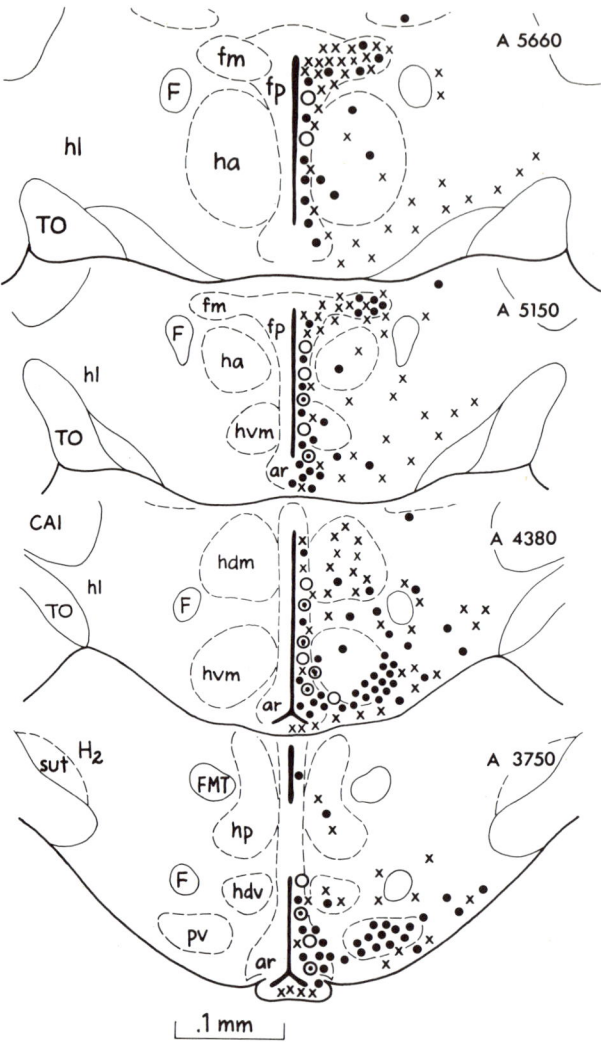

Fig. 13. An illustration depicting catecholamine-estrogen target neuron interrelationships in the rat hypothalamus. Solid circles indicate neurons with nuclear uptake of estradiol. The dopamine cells are indicated by the hollow circles. Hollow circles with a solid core represent estro-dopaminergic neurons. The xs refer to other catecholamine terminals, mostly norepinephrine. fmfp - paraventricular nucleus, ar = arcuate nucleus, ha = anterior hypothalamus, hvm = ventromedial hypothalamic nucleus, hdm = dorsomedial hypothalamic nucleus, hdv = nucleus dorsomedialis hypothalamis parsventralis, pv = ventral premammillary nucleus, hl = lateral hypothalamus, CA1 = tip of internal capsule, F = fornix, TO = optic tract, SUT = subthalamic nucleus, H_2 = lenticular fasciculus, FMT = mammilo thalamic tract, hp = posterior hypothalamus, fp = paraventricular nucleus, hdm = dorsomedial hypothalamus. (Grant and Stumpf, 1975.)

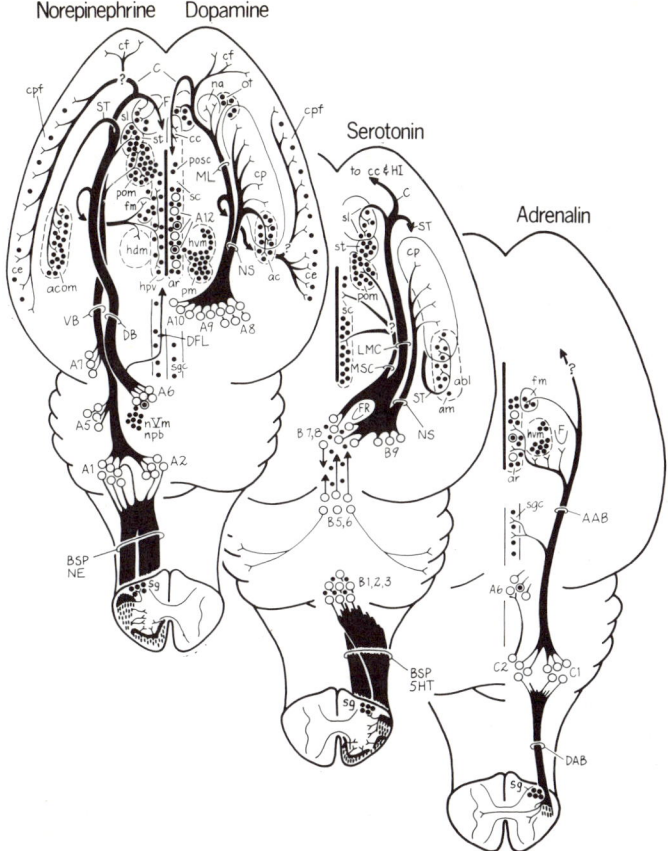

Fig. 14. The monoamine-estrogen target relationship in the central nervous system. The amine systems were identified by histochemical fluorescent methods and shown to contain norepinephrine, dopamine, serotonin and epinephrine. Hollow circles = monoamine neurons, solid circles = estrogen neurons, and hollow circles with solid cores = estromonoaminergic neurons. AAB = ascending epinephrine bundle, A1-7 = norepinephrine cell groups, A8-12 = dopamine cell groups, B1-9 = serotonin cell groups, BSP 5-HT = bulbospinal serotonin system, BSP NE = bulbospinal norepinephrine system, C = cingulum, CL 1-2 = epinephrine cell groups, DAB = descending epinephrine bundle, DB = dorsal ascending norepinephrine bundle, DFL = dorsal longitudinal fasciculus, F = fornix, FR = fasciculus retroflexus, HI = hippocampus, LMC = lateral mesencephalic-cortical serotonin tracts, ML = mesolimbic dopamine system, MSC = medial subcortical serotonin tract, NS = negrostriatal dopamine tract, ST = stria terminalis, VB = ventral ascending norepinephrine bundle; abl = lateral basal amygdaloid nuc., ad = central amygdaloid nuc., aco = cortical amygdaloid nuc., acom = amygdaloid nucleus, am = medial amygdaloid nuc., ar = arcuate nucleus, cc = cingulate cortex, ce = entorhinal cortex, cf = frontal cortex, cp = caudate-putamen, cpf = piriform cortex, fm = paraventricular nucleus, hdv = dorsomedial hypothalamic nuc., hdm = dorsomedial hypothalamus, hpv = periventricular hypothalamus, hvm = ventromedial hypothalamus, na = nucleus accumbens, npb = medial parabrachial nucleus, nVm = mesencephalic tract of trigeminal nerve, ot = olfactory tubercle, pm = ventral premammillary nucleus, pom = medial preoptic nucleus, posc = suprachiasmatic part of preoptic nucleus, sc = suprachiasmatic nucleus, sg = substantia gelatinosa, sgc = central gray substance, sl = lateral septal area, st = interstitial nucleus of stria terminalis. (Grant and Stumpf, 1975.)

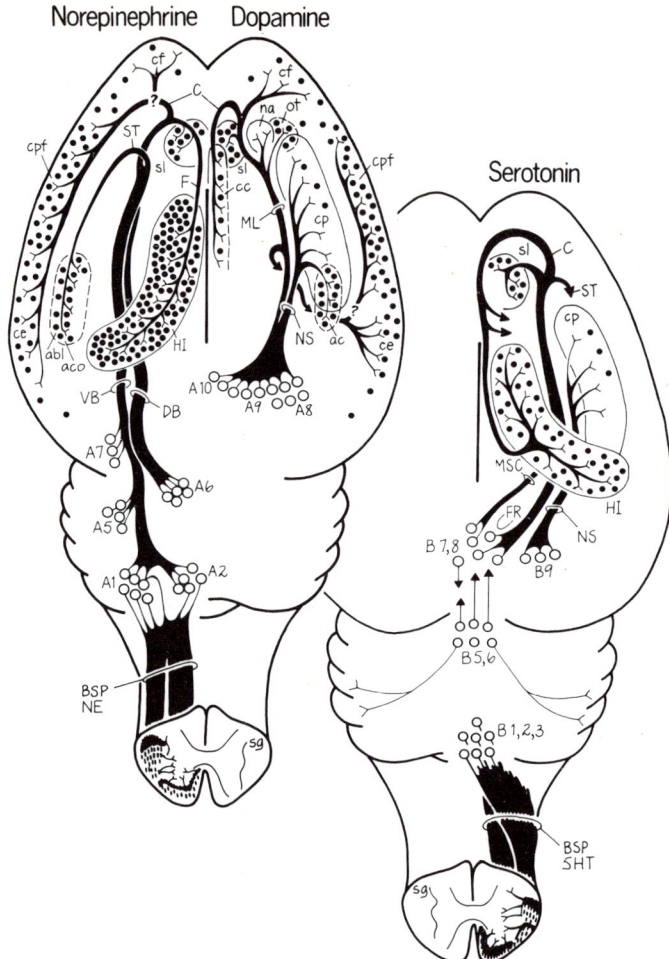

Fig. 15. An illustration of the monaminergic-corticosteroid relationship in the central nervous system. The solid circles represent corticosteroid neurons and the hollow circles the monoaminergic neurons. For abbreviations see Fig. 14. (Grant and Stumpf, 1975.)

REFERENCES

Ahren, K., Fuxe, K., Hamberger, L., and Hökfelt, T. Turnover changes in the tubero-infundibular dopamine neurons during the ovarian cycle of the rat. *Endocrinology* 88, 1415-1424 (1971).

Anderson, C.H. and G.S. Greenwald. Autoradiographic analysis of estradiol uptake in the brain and pituitary of the female rat. *Endocrinology* 85, 1160-1165 (1969).

Anton-Tay, F., Anton, M., and Wurtman, R.F. Mechanism of changes in brain norepinephrine metabolism after ovariectomy. *Neuroendocrinology* 6, 265-273.

Attramadal, A. and Aakvaag, A. The uptake of ^{3}H-oestradiol by the anterior hypophysis and hypothalamus of male and female rats. *Z. Zellforsch.* 104, 582-596 (1970).

Barley, J., Ginsburg, M., Greenstein, B.D., MacLusky, N.J., and Thomas, P.J. An androgen receptor in rat brain and pituitary. *Brain Res.* 100, 383-393 (1975).

Barr, J. Immunofluorescence study of LRF neurons in man. *Cell Tiss. Res.* 181, 1-14 (1977).

Barry, J. and Carette, B. Immunofluorescence study of LRF neurons in primates. *Cell Tiss. Res.* 164, 163-178 (1975).

Barry, J. and Dubois, M.P. Study of the preoptico-infundibular LH-RH neurosecretory pathway in female guinea pigs during gestation and the oestrous cycle, in *Neurosecretion–The Final Neuroendocrine Pathway.* VI. International Symposium on Neurosecretion, London, Springer-Verlag, Berlin (1974) pp. 148-153.

Barry, J. and Dubois, M.P. Immunofluorescence study of LRF-producing neurons in the cat and the dog. *Neuroendocrinol.* 18, 290-298 (1975).

Barry, J., Dubois, M.P. and Carette, B. Immunofluorescence study of the preoptico-infundibular LRF neurosecretory pathway in the normal, castrated or testosterone-treated male guinea pig. *Endocrinology* 95, 1416-1423 (1974).

Bleecker, M.L., Ford, D.H. and Rhines, R.K. Accumulation of ^{131}I-1-triiodothyronine in the rat brain: effect of age and sex, in *Influence of Hormones on the Nervous System,* D.H. Ford, ed. S. Karger,

Brownstein, M.J. Studies of the distribution of biologically active peptides in the brain. *Adv. Exp. Biol. Med* 87, 41-48 (1977).

Chader, G.J. and Villee, C.A. Uptake of oestradiol by the rabbit hypothalamus. Specificity of binding by nuclei *in vitro. Biochem. J.* 118, 92-97 (1970).

Choy, V.J. and Watkins, W.B. Immunohistochemical localization of thyrotropin-releasing factor in the rat median eminence. *Cell Tiss. Res.* 177, 371-374 (1977).

Ciaccio, L.A. and Lisk, R.D. The role of progesterone in regulating the period of sexual receptivity in the female hamster. *J. Endocrinol.* 50, 201-207 (1971).

Cohan, S.D., Ford, D.H., Rhines, R. and Thompson, D. The effect of neonatal x-irradiation on the accumulation and degradation of ^{131}I-1-triiodothyronine in the maturing rat central nervous system. *Acta Neurolog. Scandinav.* 45, 129-150 (1969).

Constantinidis, J., Geissühler, F., Gaillard, J.M., Hovaguimaian, T. and Tissot, R. Enhancement of cerebral norepinephrine turnover by thyrotropin-releasing hormone. Evidence by fluorescence histochemistry. *Experientia* 30, 1182-1183 (1974).

Coppola, J.A. Turnover of hypothalamic catecholamines during various stages of gonadotropin secretion. *Neuroendocrinology* 5, 75-80 (1969).

Courrier, R., Horeau, A., Marois, M. and Morel, F. Etude quantitative de la penetration de la radiothyroxine dans les cellules hypophysaires. *Compt. Rend. Soc. Biol.* 143, 935-937 (1949).

Fontana, J.A., Walker, M.D., Casper, A.G.T., Meret, S. and Henkin, R.I. Sequential subcellular localization of cortisol in cat brain. *Endocrinology* 86, 1469-1471 (1970).

Ford, D.H. The role of progesterone in the production of vaginal changes in ovariectomized female rats. *Endocrinology* 55, 230-231 (1954).

Ford, D.H. and Gross, J. The metabolism of ^{131}I-labeled thyroid hormones in the hypophysis and brain of rabbit. *Endocrinology* 62, 416-436 (1958a).

Ford, D.H. and Gross, J. The localization of ^{131}I-labeled triiodothyronine and thyroxine in the pituitary and brain of the male guinea pig. *Endocrinology* 63, 549-560 (1958b).

Ford, D.H. and Young, W.C. The role of progesterone in the production of cyclic vaginal changes in the female guinea pig. *Endocrinology* 49, 795-804 (1951).

Ford, D.H. and Rhines, R. Accumulation of (^{131}I) triiodothyronine in neurons and other tissues following intravenous injection of the labeled hormone. *Brain Res.* 6, 481-488 (1967).

Ford, D.H. and Rhines, R.K. Effect of age on the accumulation of ^{131}I-triiodothyronine in male and female rat brains and other tissues. *Brain Res.* 21, 265-274 (1970).

Ford, D.H., Kantounis, S., and Lawrence, R. The localization of ^{131}I-labeled triiodothyronine in the pituitary and brain of normal and thyroidectomized male rats. *Endocrinology* 64, 977-991 (1959).

Ford, D.H., Rhines, R.K. and Steig, C. Hormone localization in the nervous system, in *Influence of Hormones on the Nervous System* D.H. Ford, S. Karger, Basel (1971) pp. 2-16.

Fox, T.O. and Johnston, C. Estradiol receptors in mouse brain and uterus: binding to DNA. *Brain Res.* 77, 330-336 (1974).

Gerlach, J.L. and McEwen, B.S. Brain binds adrenal steroid hormones: radioautography of hippocampus with corticosterone. *Science* 175, 1133-1136 (1972).

Goldsmith, P.C. and Ganong, W.F. Ultrastructural localization of luteinizing hormone-releasing hormone in the median eminence of the rat. *Brain Res.* 97, 181-193 (1975).

Grant, L.D. and Stumpf, W.E. Hormone uptake sites in relation to CNS biogenic amine systems, in *Anatomical Neuroendocrinology*, W.E. Stumpf and L.D. Grant, eds. S. Karger, Basel (1975) pp. 445-463.

Greenglass, P.M. and Tonge, S.R. Effects of oestrogen and progesterone on brain monoamines: interactions with psychotropic drugs. 24: suppl. 149P (1972).

Grimm, Y. and Reichlin, S. Thyrotropin-releasing hormone (TRH): neurotransmitter regulation of secretion by mouse hypothalamic tissue *in vitro*. *Endocrinology* 93, 626-631 (1973).

Henkin, R.I., Casper, A.G.T., Brown, R., Harlan, A.B. and Bartter, F.C. Presence of corticosterone and cortisol in the central and peripheral nervous system of the cat. *Endocrinology* 82, 1058-1061 (1968).

Heritage, A.S., Grant, L.D., and Stumpf, W.E. ^{3}H-estradiol in catecholamine neurons of rat brain stem: formaldehyde-induced fluorescence. *J. Comp. Neurol.* 176, 607-630 (1977).

Hökfelt, T., Fuxe, K., Goldstein, M., Johansson, O., Park, D., Fraser, H., and Jeffcoate, S.L. Immunofluorescence mapping of central monoamine and releasing hormone (LRH) systems. In *Anatomical Neuroendo-*

crinology, W.E. Stumpf and L.D. Grant, eds. S. Karger, Basel (1975) pp. 381-392.

Javoy, F., Glowinski, J., and C. Kordon. Effects of adrenalectomy on the turnover of norepinephrine in the rat brain. *Eur. J. Pharmacol.* 4, 103-104 (1968).

Jensen, J.M. and Clark, D.E. Location of radioactive 1-thyroxine in the neurohypophysis. *J. Lab. and Clin. Med.* 38, 663-670 (1951).

Joseph, A.A., Sorrentino, S., Jr., and Sundberg, D.K. Releasing hormones, LRF and TRF in the cerebrospinal fluid of the third ventricle. In *Brain-Endocrine Interaction II*, K.M. Knigge, D.E. Scott, H. Kobayashi, Miura-shi, and S. Ishii, eds. S. Karger, Basel, (1975) pp. 306-312.

Kardon, F.C., Winokur, A., and Utiger, R.D. Thyrotropin-releasing hormone (THR) in rat spinal cord. *Brain Res.* 122, 578-581 (1977).

Kastin, A.J., Plotnikoff, N.P., Schally, A.V., and Sandman, C.A. Endocrine and CNS effects of hypothalamic peptides and MSH, in *Reviews of Neuroscience*, S. Ehrenpreis and I. Kopen, eds. Raven Press, New York, (1976) pp. 111-148.

Kato, J. Receptor proteins for androgen and estrogen in the hypothalamo-hypophyseal unit, in *Brain-Endocrine Interactions II*, K.M. Knigge, D.E. Scott, H. Kobayashi, Miura-Shi, and S. Ishi, eds. S. Karger, Basel (1975) pp. 217-229.

Kato, J. Cytosol and nuclear receptors for 5 α-dihydrotesterone and testosterone in the hypothalamus and hypophysis, testosterone receptors isolated from neonatal rat hypothalamus. *J. Steroid Biochem.* 7, 1179-1187 (1976).

Kato, J., Atsumi, Y., and Inaba, M. Estradiol receptors in female rat hypothalamus in the developmental stages and during pubescence. *Endocrinology* 94, 309-317 (1974).

Kato, J., Sugimara, N., and Kobayashi, T. Changing patterns of the uptake of estradiol by the anterior hypothalamus, the medial eminence, and the hypophysis in the developing female rat, in *Hormones in Development*, M. Hamburgh and E.J.W. Barrington, eds. Appleton-Century-Crofts, New York, (1971) pp. 689-704.

Keefer, D.A. and Stumpf, W.E. Estrogen-concentrating neuron systems in the brain of the tree shrew. *Gen. Comp. Endocrinol.* 26, 504-516 (1975a).

Keefer, D.A. and Stumpf, W.E. Atlas of estrogen-concentrating cells in the central nervous system of the squirrel monkey. *J. Comp. Neurol.* 160, 419-440 (1975b).

Keefer, D.A. and Stumpf, W.E. Estrogen localization in the primate brain, in *Anatomical Neuroendocrinology*, W.E. Stumpf and L.D. Grant, eds. S. Karger, Basel (1975c) pp. 153-165.

Keefer, D.A., Stumpf, W.E., and Sar, M. Estrogen-topographical localization of estrogen-concentrating cells in the rat spinal cord following ^{3}H-estradiol administration. *Proc. Soc. Exp. Biol. Med.* 143, 414-417 (1973).

Keller, H.H., Bartholini, G., and Pletscher, A. Enhancement of cerebral noradrenaline turnover by thyrotropin-releasing hormone. *Nature, London.* 248, 528-529 (1974).

Kordon, C. and Ramirex, V.D. Recent developments in neurotransmitter-hormone interactions, in *Anatomical Neuroendocrinology*, W.E. Stumpf and L.D. Grant, eds. S. Karger, Basel (1975) pp. 409-419.

Kordon, C., Kerdelhue, B., Pattou, E., and Justicz, M. Immunocytochemical localization of LHRH in axons and nerve terminals of the rat median eminence. *Proc. Soc. Exp. Biol. Med.* 147, 122-127 (1974).

Kozlowski, G.P., Nett, T.M., and Zimmerman, E.A. Immunocytochemical localization of gonadotropin-releasing hormone (Gn-RH) and neurophysin in the brain, in *Anatomical Neuroendocrinology*, W.E. Stumpf and L.G. Grant, eds. S. Karger, Basel (1975) pp. 185-191.

Krieger, M.S., Morrell, J.I., and Pfaff, D.W. Autoradiographic localization of estradiol-concentrating cells in the female hamster brain. *Neuroendocrinol.* 22, 193-205 (1976).

Krulich, L., Quijada, M., Wheaton, J.E., Illner, P., and McCann, S.M. Localization of hypophysiotropic neurohormones by assay of sections from various brain areas. *Fed. Proc.* 36, 1953-1959 (1977).

Luttge, W.G., Chronister, R.B., and Hall, N.R. Accumulation of ^{3}H-progestins in limbic, diencephalic, and mesencephalic regions of mouse brain. *Life Sci.* 12, 419-424 (1973).

Manuelidis, L. Studies with electron microscopic autoradiography of thyroxine ^{125}I. *Yale. J. Biol. Med.* 45, 501-518 (1972).

Manuelidis, L. and Bornstein, M. I-^{125}-labelled thyroid hormones in cultured mammalian nerve tissue. *Z. Zellforsch.* 106, 189-199 (1970).

Manuelidis, L. and Manuelidis, E.E. Studies with electron microscopic autoradiography of thyroxine ^{125}I in organotypic cultures of the CNS. 1. Fixation of thyroxine ^{125}I. *Yale J. Biol. Med.* 45, 487-500 (1972).

Martinez-Vargas, M.D., Stumpf, W.E., and Sar, M. Estrogen localization in the dove brain. Phylogenetic considerations and implications of nomenclature, in *Anatomical Neuroendocrinology*, W.E. Stumpf and L.D. Grant, eds. S. Karger, Basel (1975), pp. 166-175.

Maurer, R. and Woolley, D. Distribution of ^{3}H-estradiol in clomiphene-treated and neonatally androgenized rats. *Endocrinology* 88, 1281-1287 (1971).

McCann, S.M., Krulich, L., Quijada, M., Wheaton, J., and Moss, R.L. Gonadotropin-releasing factors. Sites of production secretion and action in the brain, in *Anatomical Neuroendocrinology*, W.E. Stumpf and L.D. Grant, eds. S. Karger, Basel (1975) pp. 192-199.

McEwen, B.S. and Wallach, G. Corticosterone binding to hippocampus: nuclear and cytosol binding *in vitro*. *Brain Res.* 57, 373-386 (1973).

McEwen, B.S., Weiss, J.M., and Schwartz, L.S. Uptake of corticosterone by rat brain and its concentration by certain limbic structures. *Brain Res.* 16, 227-241 (1969).

Meyer, C.C. Testosterone concentrations in the male chick brain: an autoradiographic study. *Science* 180, 1381-1383 (1973).

Morrell, J.I., Kelly, D.B., and Pfaff, D.W. Sex steroid binding in the brains of vertebrates, in *Brain-Endocrine Interactions II*, K.M. Kniggee, et al., eds. S. Karger, Basel (1975) pp. 230-256.

Mulder, A.H., Geuze, J.J., and de Weid, D. Studies on the subcellular localization of corticotrophin releasing factor (CRF) and vasopressin in the median eminence of the rat. *Endocrinology* 87, 61-79 (1970).

Ogren, L. and Woolley, D. Increase in ^{3}H-estradiol binding in brain and pituitary with time after gonadectomy in adult male and female rats. *Neuroendocrinol.* 22, 259-272 (1976).

Pelletier, G., Leclerc, R., and Puviani, R. Localization ultrastructurale d'hormones hypothalamiques. *Union. Med. Can.* 104, 355-362 (1975).

Petrusz, P. Localization of sites of action of gonadotropins. In *Anatomical Neuroendocrinology*, W.E. Stumpf and L.G. Grant, eds. S. Karger, Basel (1975) pp. 176-184.

Pfaff, D.W. and Keiner, M. Estradiol-concentrating cells in the rat amygdala as part of a limbic-hypothalamic hormone sensitive system, in *The Neurobiology of the Amygdala*, B.E. Eleftheriou, ed., Plenum Publishing Co., New York (1972) pp. 775-785.

Pfaff, D. and Keiner, M. Atlas of estradiol-concentrating cells in the central nervous system of the female rat. *J. Comp. Neurol.* 151, 121-158 (1973).

Pimstone, B.C., Berelowitz, M., and Kronheim, S. Somatostatin, *S. A. Med. J.* 50, 1471-1474 (1976).

Prange, A.J., Meek, J.L., and Lipton, M.A. Catecholamines: diminished rate of synthesis in rat brain and heart after thyroxine treatment. *Life Sci.* 9, 901-907 (1970).

Rees, H.D., Stumpf, W.E., and Sar, M. Autoradiographic studies with ^{3}H-dexamethasone in the rat brain and pituitary. In, *Anatomical Neuroendocrinology*, W.E. Stumpf and L.D. Grant, eds., S. Karger, Basel (1975) pp. 262-269.

Robbins, J. and Rall, J.E. Proteins associated with the thyroid hormone. *Physiol. Rev.* 40, 415-489 (1960).

Sar, M. and Stumpf, W.E. Autoradiographic localization of radioactivity in the rat brain after the injection of 1, 2-^{3}H-testosterone. *Endocrinology* 92, 241-256 (1973a).

Sar, M. and Stumpf, W.E. Neurons of the hypothalamus concentrate (^{3}H) progesterone or its metabolites. *Science* 183, 1266-1268 (1973b).

Sar, M. and Stumpf, W.E. Distribution of androgen-concentrating neurons in rat brain. In, *Anatomical Neuroendocrinology*, W.E. Stumpf and L.D. Grant, eds., S. Karger, Basel (1975a) pp. 120-133.

Sar, M. and Stumpf, W.E. Cellular localization of progestin and estrogen in the guinea pig hypothalamus by autoradiography, in *Anatomical Neuroendocrinology*, W.E. Stumpf and L.D. Grant, eds., S. Karger, Basel (1975b) pp. 142-152.

Sar, M. and Stumpf, W.E. Androgen concentration in motor neurons of cranial nerves and spinal cord. *Science* 197, 77-79 (1977).

Scapagnini, U., and Preziosi, P. Role of brain norepinephrine and serotonin in the tonic and phasic regulation of brain hypothalamic hypophyseal adrenal axis. *Arch. Int. Pharmacodyn. Ther.* 196, Suppl. 196, 205+ (1973).

Schaepdryver, A.F., de Preziosi, P., and Scapagnini, U. Brain monoamines and stimulation or inhibition of ACTH release. *Arch. Int. Pharmacodyn.* 180, 11-18 (1969).

Schittenhelm, A. and Eisler, B. Thyroxine und zentral nervensystem. *Klin. Wchnschr.* 11, 94 (1932).

Seiki, K., Miyamoto, M., Yamashita, A., and Kotani, M. Further studies on the uptake of labelled progesterone by the hypothalamus and pituitary of rats. *J. Endocr.* 43, 129-130 (1969).

Sheridan, P.J., Sar, M., and Stumpf, W.E. Autoradiographic localization of ^{3}H-estradiol or its metabolites in the central nervous system of the development rat. *Endocrinology* 94, 1386-1390 (1974).

Silverman, A.J. Ultrastructural studies on the localization of neurohypophyseal hormones and their carrier proteins. *J. Histochem. Cytochem.* 24, 816-827 (1976).

Simmons, J.E. Uptake of (1,2-^{3}H) testosterone in oestrogenized male rats. *Acta Endocrinol.* 67, 535-543 (1971).

Sokol, H.W., Zimmerman, E.A., Sawyer, W.H., and Robinson, A.G. The hypothalamic-neurohypophyseal system of the rat. Localization and quantitation of neurophysin by light microscopic immunocytochemistry in normal rats and in Brattleboro rats deficient in vasopressin and neurophysin. *Endocrinology*, 98, 1176-1188 (1976).

Stern, W.C., Miller, M., Resnick, O., and Morgaine, P.J. Distribution of ^{125}I-labeled rat growth hormone in regional brain areas and peripheral tissues of the rat. *Am. J. Anat.* 144, 503-507 (1975).

Stevens, W., Grosser, B.I., and Reed, D.J. Corticosterone binding molecules in rat brain. *Brain Res.* 35, 602-607 (1971).

Stumpf, W.E. Estrogen-neurons and estrogen-neuron systems in the periventricular brain. *Am. J. Anat.* 129, 207-217 (1970).

Stumpf, W.E. Autoradiographic techniques and the localization of estrogen, androgen and glucocorticoid in the pituitary and brain. *Am. Zool.* 11, 725-739 (1971a).

Stumpf, W.E. Estrogen, androgen, and glucocorticosteroid concentrating neurons in the amygdala, studied by dry autoradiography, in *The Neurobiology of the Amygdala*, B.E. Eleftheriou, ed. Plenum Publishing Co. (1972) pp. 763-774.

Stumpf, W.E., and Grant, L.E. *Anatomical Neuroendocrinology*, S. Karger, Basel (1975), pp. 2-472.

Stumpf, W.E. and Roth, L.J. High resolution autoradiography with dry-mounted freeze-dried frozen section. *J. Histochem. Cytochem.* 14, 274-287 (1966).

Stumpf, W.E. and Sar. M. Estradiol concentrating neurons in the amygdala. *Proc. Soc. Exp. Biol. Med.* 136, 102-106 (1971).

Stumpf, W.E. and Sar. M. Anatomical distribution of corticosterone-concentrating neurons in rat brain, in *Anatomical Neuroendocrinology*, W.E. Stumpf and L.D. Grant, eds. S. Karger, Basel (1975a) pp. 254-261.

Stumpf, W.E. and Sar, M. Localization of thyroid hormones in the mature rat brain and pituitary, in *Anatomical Neuroendocrinology*, W.E. Stumpf and L.D. Grant, eds. S. Karger, Basel (1975b) pp. 318-327.

Stumpf, W.E. and Sar, M. Hormone-architecture of the mouse brain with ^{3}H-estradiol, in *Anatomical Neuroendocrinology*, W.E. Stumpf and L.D. Grant, eds. S. Karger, Basel (1975c) pp. 82-103.

Stumpf, W.E., Sar, M., and Keefer, D.A. Atlas of estrogen target neurons in rat brain, in *Anatomical Neuroendocrinology*, W.E. Stumpf and L.D. Grant, eds., S. Karger, Basel (1975d) pp. 104-119.

Szentagothai, J., Flerko, B., Mess, B., and Halasz, B. *Hypothalamic Control of the Anterior Pituitary*, Akademiai Kiado, Budapest, pp. 15-399 (1968).

Touchstone, J.C., Griffin, J.E., and Kasparow, M. Cortisol from human nerve. *Science*, 141, 1275 (1963).

Touchstone, J.C., Kasparow, M., Hughes, P.A., and Horwitz, M.R. Corticosteroids in human brain. *Steroids* 7, 205-211 (1966).

Wade, G.N. and Feder, H.H. Uptake of (1,2-^{3}H) 20a-hydroxypregn-4-en-3-one, (1,2-^{3}H) corticosterone, and (6,7-^{3}H) estradiol-17B by guinea pig brain and uterus: comparison with uptake of (1,2-^{3}H) progesterone. *Brain Res.* 45, 545-554 (1972a).

Wade, G.N. and Feder, H.H. (1,2-^{3}H) Progesterone uptake by guinea pig brain and uterus: differential localization, time-course of uptake and metabolism, and effects of age, sex, estrogen-priming and competing steroids. *Brain Res.* 45, 525-543 (1972b).

Whalen, R.E. and Luttge, W.G. Differential localization of progesterone uptake in brain. Role of sex, estrogen pretreatment and adrenalectomy. *Brain Res.* 33, 147-155 (1971).

Woodbury, D.M. Effect of hormones on brain excitability and electrolytes. *Rec. Prog. Hormone Res.* 10, 65-107 (1954).

Wood-Gush, D.G.M., Langley, G.A.S., Leitch, A.F., Gentle, M.J., and Gilbert, A.B. An autoradiographic study of sex steroids in the chicken telencephalone. *Gen. Comp. Endocrinol.* 31, 161-168 (1977).

Zigmond, R.E., and McEwen, B.S. Selective retention of oestradiol by cell nuclei in specific brain regions of the ovariectomized rat. *J. Neurochem.* 17, 889-899 (1970).

Zigmond, R.E., Nottebohm, F., and Pfaff, D.W. Androgen-concentrating cells in the midbrain of a songbird. *Science* 179, 1005-1007 (1973).

Copyright©1982, Spectrum Publications, Inc.
Hormones in Development and Aging

Chapter 6
Morphological Response of the Nervous System to Hormones
Eva B. Cramer
Donald H. Ford

INTRODUCTION

An understanding of the effects of hormones on the developing and aging nervous system is complicated by the intricate interactions of the hormones themselves and the complex sequence of events in the developmental and aging process. Interpretation of experimental results can be difficult because of variation in ablative methods, choice of hormone or synthetic analog, dosage, age at treatment, age of observed effects, and finally the animal species. In spite of these problems it is clear that the developing brain appears particularly sensitive to a number of hormones as well as other environmental influences during certain critical periods of development. At these times, the transcription and/or translation of the genetic material of some neurons become irreversibly determined. The permanence of these effects imply that the consequences of these hormone-receptor interactions are very different from those which occur at other time intervals. The mechanisms which underlie the increased sensitivity of the nervous system to hormones at critical periods remain, for the most part, unknown, although the consequences of some of these interactions are anatomically apparent. However, at other time intervals, when hormonal influence appears to be primarily regulatory, structural alterations are less noticeable. As a result, most anatomical research has centered on the effects of hormones on the developing rather than the adult nervous system and relatively little information exists on the anatomical effects of hormones on the aging nervous system. Of the various hormones, thyroid hormone appears to have the most dramatic anatomical effect on the developing nervous system and has been the most extensively studied. In the following pages we have attempted to review the morphological response of the developing and aging nervous system to thyroid, adrenal, sex, and growth hormone as well as to insulin.

THYROID HORMONES

When thyroid deficiency occurs in humans early in life, it causes the mental retardation associated with cretinism. In humans, the anatomical findings associated with this developmental retardation include lowered brain weight, reduction in the size of cells in the cerebral cortex, malformed convolutions of the cortical layers, axonal degeneration, and delayed myelination (Mott, 1917; Marinesco, 1924; Benda, 1946; Beierwaltes et al., 1959; Meier and Bischoff, 1977). However, when thyroid imbalances occur later in life the effects on the brain are mainly metabolic and are much less severe.

Attempts to study the effects of thyroid hormone on the nervous system have led investigators to study animals such as the rat, which show a significant degree of brain development after birth (Balázs, et al., 1971a; Bass et al., 1977). Similar to the situation in humans, in the rat there appears to be a critical period of development during which thyroid hormone has been shown to influence proliferation, differentiation, and maturation (dendritic and axonal growth, synaptogenesis, and myelination) of brain cells (for reviews see Kollros, 1968; Hamburgh, 1969; Hamburgh and Barrington, 1971; Balázs, 1976; also see Chapter 7). As a result it has been suggested that thyroid hormones function as a biological timing mechanism for the complex sequence of normal brain maturation (Hamburgh et al., 1971).

Cerebral Cortex

The development of the rat cerebral cortex takes place to a great extent after birth. Normally, the weight of the cerebrum increases about eight times during postnatal development. This is due not only to the growth of cells existing at birth but also to an increase in cell numbers (estimated by DNA content) after birth which approach adult levels by 14 days (Balázs et al., 1971a).

The development of the cerebrum is influenced by the thyroid state of the animal. If rats are thyroidectomized at birth or shortly thereafter, there is a 40% impairment in their rate of body growth (Eayrs and Taylor, 1951; Bass and Young, 1973; Cramer and Ford, 1977), a reduction in the weight of the brain, a change in its shape which involves an increase in width and height relative to length (Eayrs and Taylor, 1951), a decrease in the microtubule initiating factor necessary for the formation of brain microtubules (Francon et al., 1977), a retarded behavioral development (Eayrs and Lishman, 1953), and an impaired ability to learn (Eayrs and Levine, 1963). The cerebral cortex of such animals exhibit normal DNA content (Balázs et al., 1968) but histologic changes include:

(1) an increase in size of blood vessels and a decrease in the number of capillaries (Eayrs, 1954),
(2) a decrease in size of pyramidal perikarya (Eayrs and Taylor, 1951),
(3) a decrease in the amount of neuropil (Eayrs, 1955; Eayrs and Horn, 1955) probably due to a reduction in the length and the amount of branching of dendrites as well as a reduced axon density, proportionally greatest in the internal granular layer (Eayrs, 1955),
(4) a significant decrease in the mean number of spines/50 μm segments of the apical shaft of pyramidal cells, as well as an abnormal pattern of spine distribution (Sanchez-Toscano et al., 1977),
(5) an impairment in the migration of glial cells from the subependymal zone (Bass and Young, 1973) and a reduced deposition of myelin (Barrnett, 1950; Bass and Young, 1973; Hamburgh et al., 1977).

The cerebral cortex appears to be sensitive during the same critical period of development (the first 10 days of postnatal life) to an excess of thyroid hormone as well as to its deficiency. Rats made hyperthyroid during the first 24 days of life have accelerated maturational changes such as the opening of the eyes and development of the pinnae, but, on the other hand, the animals grow more slowly, and this is reflected in a significant decrease in brain weight (Eayrs, 1964). This reduced brain weight is thought to be related in the cerebral cortex to a decrease in postnatal cell formation rather than a decrease in cell size (Balázs et al., 1971a).

Although innately organized responses (Eayrs, 1964), neurochemical maturation (Balázs et al., 1971b) and myelination (Hamburgh et al., 1977) are advanced, adaptive, cortically mediated behavior is impaired (Eayrs, 1964) and this may be due to the deficiency of postnatal cell formation (Balázs et al., 1971a).

Cerebellar Cortex

In the rat cerebellum 97% of the final cell number including the majority of the nerve cells are acquired during the first three weeks after birth (Patel et al., 1973). The germinal site of these cells is the external granular layer or EGL (Balázs, 1976).

In the hyperthyroid rat, the rate of cell acquisition in the cerebellum is accelerated during the first few days after birth (Nicholson and Altman, 1972a; Gourdon et al., 1973; Weichsel, 1974) and ends prematurely in the second week (Balázs et al., 1971a). This reaction appears to be a consequence of a shortening of the cell cycle as a result of a decrease in the length of the pre-DNA synthetic phase G_1 and the early onset of neuronal differentiation (Lauder, 1977a). In the hypothyroid state the number of cerebellar cells are initially reduced in the second week after birth due to a reduction of replicating cells in the EGL (Nicholson and Altman, 1972a) and the death of differentiated granule cells in the internal granular layer (Lewis et al., 1976). However, eventually the number of cells become similar to control levels due to the persistence of dividing cells in the EGL of the hypothyroid rat (Nicholson and Altman, 1972a).

Altered thyroid states also influence the time of onset of neuronal differentiation in the cerebellum (Lauder, 1977b). Purkinje cells, the only efferent cells in the cerebellar cortex, are differentiated prenatally. However, the inhibitory interneurons (Golgi cells: perinatal period; basket cells: first postnatal week; and stellate: second postnatal week) are formed by the end of the second postnatal week and approximately 50% of the excitatory granule cells are formed during the third week (Altman, 1969). Consequently, it is not surprising that thyroidal imbalance initiated at birth or during the first or second week would affect the time of onset of differentiation of basket, stellate, and granule cells and thereby influence the final composition of the neuron cell population (Nicholson and Altman, 1972a; Clos and Legrand, 1973).

Thyroid imbalance also affects the glial cell population (Clos et al., 1973). Hypothyroidism, which prolongs proliferation, causes a significant increase in the number of astroglia in the molecular layer and hyperthyroidism, which causes an early termination of cell proliferation, causes a slight decrease in the number of these cells (Nicholson and Altman, 1972a).

Thyroid hormone also affects neuropil development and synaptogenesis in the cerebrellar cortex. Hypothyroidism initiated at birth causes a severe retardation of the dendritic arborization of the Purkinje cells (Fig. 1) (Legrand, 1967). The hypoplasia of the dendritic tree results in a decreased availability of synaptic sites for the parallel fibers of the granule cells (Nicholson and Altman, 1972a; Clos and Legrand, 1973). In addition, there is a decrease in growth of the parallel fibers and this also contributes to the overall decrease in the number of parallel fiber-Purkinje cell dendritic spine synapses (Lauder, 1977b). On the other hand, hyperthyroidism has been shown to accelerate synaptogenesis in this circuit, because it causes an accelerated development of Purkinje cell dendritic spines (Rebière and Legrand, 1972) and an increase in length of parallel fibers (Lauder, 1977b).

As a consequence of hypothyroidism, delays in proliferation, differentiation and growth of neuronal processes occur and immature neuronal circuits

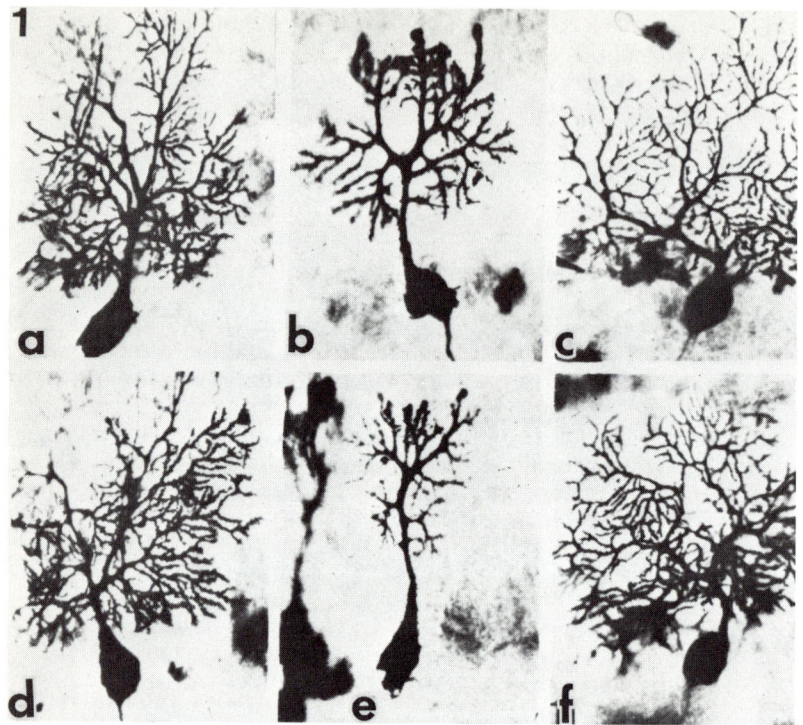

Fig. 1. A montage of illustrations of Golgi-Cox preparations demonstrating Purkinje cells in normal (a and b), hypothyroid rats (c and d), and in rats treated with propylthiouracil, (PTU) an antithyroid agent, but which had also received thyroxine during the first postnatal week (e and f). All animals were 14-days old when killed. The Purkinje cells of the PTU-treated animal receiving the thyroxine supplement are essential identical to the cells in the controls (From Legrand, 1967).

(i.e., climbing fiber-Purkinje cell) persist for a longer time, therefore the later developing neuronal circuits (mossy fiber-granule cell-Purkinje cell) are either delayed or formed in fewer numbers (Hajós et al., Balázs, 1976).

Finally, thyroid hormone is thought to have a direct action on the formation of myelin in the cerebellum. In vitro studies have shown that the onset of myelin formation in the cerebellum is accelerated by excess thyroxine (Hamburgh, 1966).

Subcortical Nuclei

Most investigations of thyroid effects on the nervous system have centered on cerebral and cerebellar cortical structure. However, subcortical nuclei such as the caudate nucleus have also been studied in developing rats of propylthiouracil-fed lactating dams. While at 14 days there is a decreased number of neurons, inhibition of dendritic arborization and spines and a reduced complexity of axonal plexuses, a compensatory spurt of neural growth and differentiation occurs and eventually a seemingly normal caudate cytoarchitecture is formed (Lu and Brown, 1977).

Hypothalamus

While the anatomical effect of thyroid hormones on the cerebral and cerebellar cortex have been extensively examined, the hypothalamus has received a limited amount of attention. A large portion of the work done in this area has centered on the effects of thyroid hormone on changes in neuronal organelles.

Thyroidectomy of adult rats causes a decrease in the nuclear size of neurons within sixteen hypothalamic nuclei and areas, while thyroxine treatment causes the opposite reaction (Talanti, 1965; Talanti, 1967b). A similar response to thyroid hormone has also been noted in the ependymal cells of the third ventricle and in the modified ependymal cells of the subcommissural organ (Talanti, 1967a). On the other hand, adult hypothroidism in the rat causes a significant increase in nucleolar size in various nuclear groups of the hypothalamus. This includes neurons in the preoptic area, periventricular-preoptic zone, supraoptic and paraventricular nuclei, posterior premammillary, and supramammillary area (Ifft, 1964).

Examination of the medial basal hypothalamus of 28-day-old rats injected with ^{131}I at birth revealed the accumulation of whorled formations of closely apposed concentric cisternae of smooth endoplasmic reticulum in the cell bodies of arcuate neurons. The whorled formations were continuous with the cisternae of rough endoplasmic reticulum (Fig. 2) (Cramer and Ford, 1977). It has been speculated that these neurons represent sites of synthesis of thyrotrophin releasing hormone (TRH) as labeled thyroid hormone concentrates in this area (Ford et al., 1959) and these whorls are morphological signs of enhanced synthetic activity (Christensen and Fawcett, 1966).

Peripheral Nervous System

Thyroid hormone appears necessary for normal development of the peripheral nervous system. Postnatal hypothyroidism impairs the maturation of the sympathetic nervous system. The superior cervical ganglion of 26-day-old rats injected with propylthiouracil at birth contain smaller neuronal cell bodies, thinner axons and reduced catecholamine content

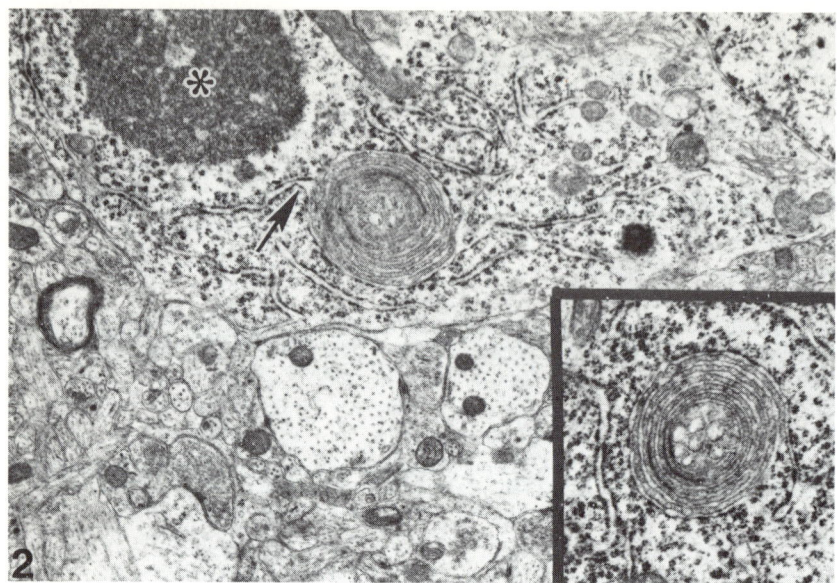

Fig. 2. Portion of cell body of an arcuate neuron containing the characteristic "chromatoid body" (*) from a hypothyroid 28-day-old female rat. The rough endoplasmic reticulum is continuous (arrow) with a whorled body. The outermost membrane of the whorl is studded with ribosomes. X13,600. The cytoplasmic core of the whorled body contains vacuoles (inset). X15,300. (From Cramer and Ford, 1977).

(Gresik, 1976). A similar decrease in fiber diameter was observed in nonmyelinated axons of the mouse sciatic nerve after neonatal thyroidectomy (Reier and Hughes, 1972). On the other hand, administration of thyroid hormone to rats (Cockett and Kiernan, 1973; McIsaac and Kiernan, 1975) or humans (McQuarrie, 1975) accelerates the rate of axonal regeneration and recovery from peripheral nerve injury.

Possible Complications of Undernutrition and Decreased Growth Hormone

A general problem of postnatal hypothyroid studies has been the possible contribution of neonatal undernutrition and a general endocrine imbalance initiated by lack of thyroid hormone.

Although, hypothyroid rats have a delayed development of dentition and eye-opening, and an impairment in food seeking behavior during the suckling period (Eayrs and Horn, 1955), the mother rats appear to compensate for these deficits with a prolongation in maternal care (Hamburgh et al., 1977). While postnatal starvation causes a significant amount of neuropathology, some of the findings are sufficiently different from that seen in the cretinoid brain to argue against the proposition that hypothyroidism is entirely mediated through the effects of undernutrition. In malnutrition there is a decrease in cell acquisition throughout the brain (Winick and Noble, 1966), whereas the effects of neonatal

thyroid deficiency seem confined to areas of significant neurogenesis
(Balázs, 1977). In the cerebellum of hypothyroid rats, the EGL increases
more slowly and is present longer reflecting a retarded but prolonged
period of cell proliferation and acquisition (Lauder, 1977a) whereas in
undernutrition similar changes occur but to a lesser degree and are less
prolonged (Lewis et al., 1975). In addition, cell death is at least twice
as prevalent in the EGL and subependymal layers in 12-day-old undernourished rats as in controls or hypothyroid animals (Lewis, 1975).
However, myelin formation is less affected by malnutrition than by
hypothyroidism (Rosman and Malone, 1977). Finally, the most prominent
difference between the two pathologic conditions is their effect on the
generation cycle of dividing cells. While in thyroid deficiency the
generation cycle of dividing cells is more or less normal, in starvation
there is a marked prolongation of the S-phase of the cell cycle and the
G_1 phase is drastically curtailed (Balázs, 1977).

It is not clear is thyroid imbalance has a direct effect on the nervous
system or whether its effects are secondary to its effect on the endocrine
system. Changes in thyroid function have been shown to be accompanied
by alterations in secretion and content of pituitary somatotrophin
(Earthy and Leblond, 1954; Contopoulos et al., 1958; Solomon and Greep,
1959; Schooley et al., 1966; Iwatsubo et al., 1967). Loss of thyroid
hormone at birth is associated with a decrease in mitotic activity in
somatotrophs and an increase in mitotic activity in the thyrotrophs of
developing anterior pituitary glands of both male and female rats (Cramer
and Ford, 1977). In addition to the shift in mitotic activity, there is
a degranulation of existing somatotrophs (Solomon and Greep, 1959;
Schooley et al., 1966; Cramer and Ford, 1977).

In view of the probable decrease in growth hormone secretion and the
effects of lack of growth hormone on the developing nervous system
(see section on growth hormone in this chapter) it is possible that some
of the neuronal effects of hypothyroidism may be accentuated by the
decrease in growth hormone.

ADRENAL CORTICAL HORMONES

In a number of mammalian species, including humans, levels of
corticosteroids in the fetus rise at term and there is some speculation
that this rise is associated with the maturational process of some fetal
organs (Liggins, 1976). Clinically, corticosteroid treatment of the human
fetus is used for the purpose of inducing rapid organ maturation (in the
prevention of respiratory distress syndrome) or in the treatment of a
variety of diseases during infancy (Blodgett et al., 1956; Klevit, 1970;
Baden et al., 1972; Liggins and Howie, 1972). While the benefit of such
treatment is obvious, the possibility of harmful effects on other organs
must be considered. In 1965, experimental neonatal corticosteroid
administration to mice was observed to result in an irreversible reduction
in brain weight (Howard, 1965). Subsequently a number of electrophysiological, biochemical and morphological alterations have been observed
in the developing brain following steroid administration.

Cerebral and Cerebellar Cortex

Although corticosteroids, like thyroid hormones, affect the behavior
of rats when administered at birth for a period of approximately five
days, the effects seem to be quite different (Eayrs, 1968; Howard and
Granoff, 1968; Schapiro et al., 1970). Thyroid hormone accelerates

behavioral development of the central nervous system (CNS) while cortisol delays CNS development (Schapiro et al., 1970).

In neonatal rats treated with cortisol for the first four days of life, there is a reduction of body weight up to 50% and of brain weight, depending on the region, of up to 30% by the 35th day of life. During the four days of cortisol treatment but not after, the normal increase in brain cell number is severely suppressed due to an inhibition of cell division rather than increased cell destruction. Hydrocortisone appears to inhibit DNA synthesis and causes a reduction in the formation of proteins associated with DNA (Burdman et al., 1975; Cotterrell et al., 1972). This results in a final deficit in cell number of approximately 20% in the cerebrum and 30% in the cerebellum (Cotterrell et al., 1972). Similar observations are observed in the mouse brain (Howard, 1965, 1968). Corticosteroid and thyroid hormone when administered at birth have their effect on brain formation at different times after birth. Cell multiplication is inhibited throughout the brain by corticosteroids, mainly during the time of treatment, whereas thyroid hormone affects cell multiplication only after a delay of approximately one week in the cerebrum or two weeks in the cerebellum (Cotterrell et al., 1972). As a result, different types of nerves which are formed at various times after birth will be affected.

Experimental neonatal corticosteroid administration in rats also results in a decrease in the number of dendrites, their length and number of branches in the cerebral cortex (Oda and Huttenlocher, 1974). Similarly in the cerebrum of almost half of the newborn monkeys born to mothers administered bethamethasone for two days prior to premature delivery, shrunken, densely stained nerve cell bodies and gliosis are observed (Epstein et al., 1977). In addition, there is a delay in myelination in the rat pyramidal tract with a net decrease in the number of myelin lamellae per axon (Gumbinas et al., 1973). A decrease in the number of microglia after corticosteroid administration has also been noted (Field, 1955).

The ability of animals to respond to hormonal signals is altered during aging and may be due to alterations in hormone binding to target tissue receptors (Roth and Adelman, 1975). Cytosol binding of corticosteroids in the rat hypothalamus is less than 50% of that observed in the cortex and hippocampus (Nelson et al., 1976). This regional pattern of cytosol binding differs from that of cell nuclear binding, which is highest in the hippocampus and markedly lower in both cerebral cortex and hypothalamus (McEwen et al., 1970; Gerlach and McEwen, 1972; McEwen and Pfaff, 1973; Rhees et al., 1975). In senescent male rats but not mice (Nelson et al., 1976), the cerebral cortex exhibit a 55-65% reduction in the concentration of glucocorticoid cytosol binding sites (Roth, 1976).

Hypothalamus

Although within the CNS, the hypothalamus does not have the highest affinity for glucocorticoids, it certainly has a significant number of binding sites and the morphological effects of lack of adrenal hormone on this region of the brain have recently been examined. Following adrenalectomy changes in the nucleus and endoplasmic reticulum of some hypothalamic nuclei have been observed. By 24 hours after adrenalectomy (and still present on the seventh day after surgery) an increase in nuclear volume is seen in the anterior medial cell group of the ventromedial nuclei and the most orally situated cell group of the arcuate nuclei. These findings are accompanied by nuclear volume shrinkage in anterior ventromedial and ventral premammillary nuclei 24 hours after adrenalectomy and in the supraoptic nucleus on the seventh postoperative day (Palkovits and Stark, 1972).

Specific cytological changes have also been observed within arcuate nuclei of the hypothalamus. By 14 days after unilateral or bilateral adrenalectomy the number of whorled bodies (rough endoplasmic reticulum becoming smooth and forming a concentric spiral) is significantly increased (Figs. 3-5) (Ford and Milks, 1978). It has been postulated that this represents increased neuronal activity as a similar result is also seen after castration and thyroidectomy. Another possible example of increased synthetic activity in the hypothalamus following adrenalectomy is the observed increase in neurophysin and vasopressin in the axons of both the internal and external zone of the rat median eminence (Vandesande et al., 1974; Watkins et al., 1974; Dube et al., 1976; Zimmerman, 1976).

Figs. 3-5. Structure and distribution of smooth endoplasmic reticular (SER) whorls in the arcuate nucleus of adrenalectomized male rats. Electron micrographs were stained with uranyl acetate and lead citrate.

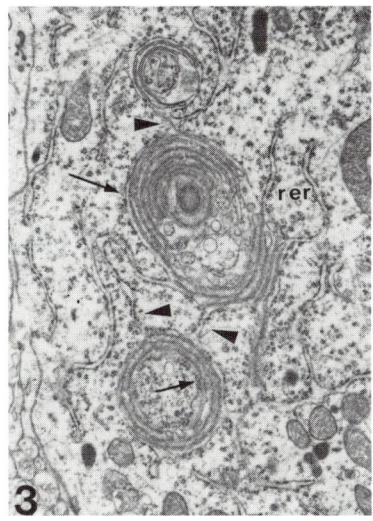

Fig. 3. Arcuate neuron containing three SER whorls interconnected (arrowheads) by rough endoplasmic reticulum (rer). Ribosomes are present on the inner and outer surfaces of the SER whorls (arrows). x 11,000.

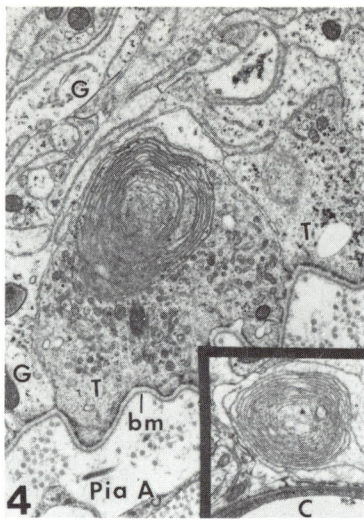

Fig. 4. SER whorls in tanycyte-like processes (T) located adjacent to the pial-arachnoid membrane (Pia A) and, in inset, abutting a non-fenestrated capillary (C) at the tubuloinfundibular region of the median eminence. bm = basement membrane, G = glial cell process. x 13,300; inset, x 7,400.

Peripheral Nervous System

The catecholamines stored in various neural tissues have been reported to be affected by glucocorticoids (Lempinen, 1964; Hellström and Koslow, 1976; Eränko and Eränko, 1972). Glucocorticoids cause hyperplasia (Eränko and Eränko, 1972; Lempinen, 1964), the appearance of adrenaline (Eränko et al., 1966) and the induction of phenylethanol-amine-N-methyl-transferase (Ciaranello et al., 1973) in extra-adrenal chromaffin tissue.

Hydrocortisone causes a significant increase in the number of intensely fluorescent nerve cell bodies and an increase in the degree of fluorescent intensity (Costa et al., 1974) as well as an increase in size of the nerve cell bodies in the superior cervical ganglia (Korochkin and Korochkina, 1970) of newborn rats. Dexamethasone treatment of newborn but not adult rats results in a 100-fold increase in epinephrine concentration of the superior cervical ganglion with only a minor increase in norepinephrine and dopamine concentrations (Koslow et al., 1975). In addition, the number and catecholamine content of small intensely fluorescent (SIF) cells in sympathetic ganglia of newborn rats, but not adult rats, increase after treatment of hydrocortisone (Eränko and Eränko, 1972). These findings are even more pronounced when hydrocortisone is added to cultures of sympathetic ganglia of newborn rats (Eränko et al., 1972a, 1972b).

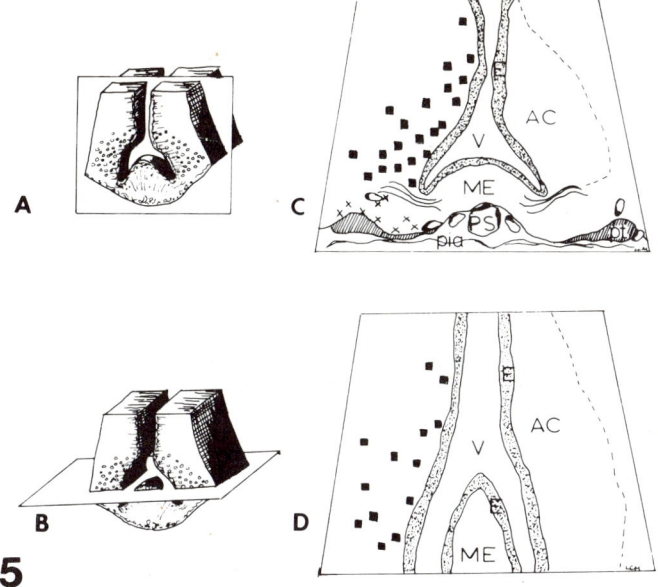

Fig. 5. Diagrammatic illustration indicating planes of section made through the median arcuate region of the hypothalamus. A and C = coronal sections, B and D = horizontal sections. The diagrams at C and D illustrate the location of whorl bodies in neurons (solid blocks), as seen in Fig. 3, and in tanycyte-like processes (Xs), as seen in Fig. 4. AC = arcuate nucleus, E = ependyma, ME = median eminence, pia = pia-arachnoid membrane, PS = portal system, pt = pars tuberalis of the hypophysis, V = third ventricle, (From Ford and Milks, 1978).

Glucocorticoid treatment also increases the catecholamine content of the carotid body in adult rats (Korkala, et al., 1973) and this corresponds to a 2.6- and 2.0-fold increase in the concentration of norepinephrine and dopamine, respectively (Hellström and Koslow, 1976). In contrast to the superior cervical ganglia, dexamethasone induced changes occur in carotid bodies of both adult and newborn rats (Hellström and Koslow, 1976).

SEX HORMONES

The gonadal steroids, secreted by the ovaries and testes during embryonic and early postnatal development directly effect the sexual differentiation of the brain, while in the adult they influence the functioning of genes in these neuronal circuits which are already permanently expressed (McEwen, 1976).

In spontaneously ovulating species, the female, but not the male, has a cyclic pattern of pituitary gonadotrophin secretion associated with periodic ovulations and behavioral estrus (Gorski, 1971). The determination of a male pattern of gonadotrophin secretion and male mating behavior does not depend on the genetic sex of the animal but upon whether the brain has been exposed to androgens (Pfeiffer, 1936; Harris and Jacobsohn, 1952; Harris, 1964; Harris and Levine, 1965; Barraclough, 1966) or estrogens (Wilson, 1943; Gorski, 1963; Whalen and Nadler, 1963; Levine and Mullins, 1964) during a critical period of development (a few days before birth to approximately 10 days after birth in rats, although there are species variability, see Gorski, 1971). Although the adult brain can convert testosterone to estradiol and dihydrotestosterone, in reality at least in the developing rat, the hormone which elicits the male brain pattern seems to be estradiol. Testosterone can be converted to estradiol by enzymes in the nerve cells of the newborn rat hypothalamus, whereas, the administration of dihydrotestosterone to the newborn rat has no effect on sexual differentiation (Naftolin et al., 1975; McEwen et al., 1977).

The highest density of estrogen concentrating cells in the newborn and in the adult rat brain is in the pituitary, hypothalamus, amygdala cerebral cortex and preoptic area. The only apparent difference between newborns and adults in sites of estrogen receptors is that estrogen receptors exist in the cerebral cortex of newborn rats for 2-3 weeks but are absent from the adult cortex. The function of this transitory receptor is unknown however, since the cortex lacks aromatizing activity, its function does not appear related to testosterone-derived estrogen (McEwen et al., 1977).

It is not surprising that estrogen receptors are present in the hypothalamus of new born and adult rats. A variety of different experiments have shown that the tuberal hypothalamus is essential for the maintenance of a basal level of secretion of anterior pituitary gonadtrophins and that the anterior hypothalamus is necessary for the cyclic surge of gonadotrophins preceding ovulation (Everett, 1964; Barraclough, 1966; Harris and Campbell, 1966). Although it is well-known that the hypothalamus of male and female rats becomes sexually differentiated during early postnatal development, the anatomical correlates of sexual dimorphism are just beginning to accumulate. These findings include larger neuronal nuclear size in the ventromedial, arcuate, suprachiasmatic, paraventricular, and supraoptic nuclei of young female vs. male rats (Hellman et al., 1976), ultrastructural differences in axon terminals in the arcuate nucleus (Rainer and Adamo, 1971), and an increase in the number of synapses of nonamygdalar origin on dendritic spines of the preoptic area of female as opposed to male rats (Raisman and Field, 1973).

Administration or loss of sex hormones during early development results in changes in nuclear size, myelinization, synaptic connections, and number of sympathetic neurons. In the hypothalamus changes in nuclear size are observed after neonatal castration (Pfaff, 1966; Borisova and Stefanov, 1976) and after postnatal androgen (Döcke and Smollich, 1968; Dörner and Staudt, 1968, 1969) or estrone administration (Arai and Kusama, 1968). Postnatal estradiol also accelerates myelinization in the developing rat hypothalamus (Curry and Heim, 1966). Moreover, gonadectomy of neonatal males results in typical female synaptology in the preoptic area of the adult while estradiol treatment of neonatal females results in male synaptology in this area (Raisman and Field, 1973). Postnatal treatment of female rats with estradiol doubled the number of axodendritic synapses in the arcuate nucleus (Matsumoto and Arai, 1976). Similarly, in in vitro studies, the addition of estradiol or testosterone to mouse hypothalamic explants resulted in accelerated and intense proliferation of neuronal processes (Toran-Allerand, 1976). Androgen levels during the critical period of brain differentiation cause permanent structural changes in the amygdalar region as well. The lower the androgen level, the larger the nuclear volumes in the adult neurons in the medial and central region of the amygdala (Staudt and Dörner, 1976). In addition, the number of preganglionic sympathetic neurons in the spinal cord are also influenced by sex hormones (Calaresu and Henry, 1971). In contrast to the previous findings, neonatal administration of progesterone showed no consistent or significant effects on brain development in rodents (Coyle et al., 1976).

In the adult, sex hormones modulate the functional activity of some neurons. In animals which show reproductive cyclicity, changes in the nuclear size of hypothalamic nuclei have been recorded. During the estrous cycle in the rat changes in nuclear size have been described in the neurons of the anterior hypothalamic area (Döcke and Koloczek, 1966) and in the ventromedial nucleus in relation to seasonal initiation of spermatogenesis in the adult squirrel monkey (Bubenik and Brown, 1973). Morphological changes in other organelles within neurons of the adult rat hypothalamus and pineal are observed after the loss of sex hormones. Following castration supraoptic and paraventricular nuclei hypertrophy and show signs of hyperactivity, characterized by dilated rough endoplasmic reticulum, and an increase in the number of ribosomes (Zambrano and DeRobertis, 1968b), oxytocin (Deis, 1959), and secretory granules in the neurohypophysis (Zambrano and DeRobertis, 1968a). After supplementary administration of sex hormones, all modifications produced by castration disappear. Following castration some cells of the arcuate nucleus show an enlarged nucleolus and an increase in dense core vesicles (Zambrano and DeRobertis, 1968c). In addition, whorled formations of closely apposed concentric cisternae of smooth endoplasmic reticulum which are continuous with the rough endoplasmic reticulum are also observed (Brawer, 1971). If castrated male rats are given testosterone the number of whorled formations decrease (Price et al., 1976). On the other hand, if male rats are treated with estradiol benzoate or cyproterone acetate, an antiandrogen thought to competitively inhibit testosterone binding, the incidence of whorl containing neurons in the arcuate nucleus is 2-4 times higher than controls (Price et al., 1977). It would appear that either the arcuate nucleus contains several different populations of neurons or that the same cells respond to decreases in testosterone (Brawer, 1971), thyroid (Cramer and Ford, 1977), and adrenal hormones (Ford and Milks, 1978), as well as increases in estradiol levels (Price et al., 1977).

Loss of sex hormones in the adult animals also affects the pineal. Following orchidectomy or after the administration of cyproterone acetate, pinealocytes exhibit morphological indications of increased protein synthesis as revealed by increased development of the rough endoplasmic reticulum, golgi apparatus, and an increase in number of lipid droplets and lysosomes (Gusek, 1976; Karasek et al., 1976). These changes were more marked if orchidectomized animals were then injected with LHRH (Karasek et al., 1976).

Reproductive capacity in mammalian species declines with age and the ability to produce live offspring usually ceases well before the death of the animal. The factors involved in this loss of reproductive capacity appear to vary to some degree from one species to another. For example, the primary cause for reproductive failure in old female rats appears to be due to functional changes in the hypothalamus and pituitary, whereas in women the primary cause appears to lie in the ovaries (Meites et al., 1975; McPherson et al., 1977; see also Chapter 15). In those species in which the brain appears to be primarily responsible for the declining reproductive capacity, some morphological changes have been reported (Azcoaga, 1963; Babichev, 1973; Frolkis et al., 1972; Machado-Salas et al., 1977). In the rodent, there appears to be a progressive deteriorative process which by golgi-staining techniques initially appears as localized irregularities of somatodendritic silhouettes and as a decrease in the spine-like processes of many neurons. With time there is a progressive loss of dendrites and increased deformities or swelling of the nerve cell bodies. These structural changes were unequally distributed throughout the aging hypothalamus but appeared more frequently in the rostral portion (Machado-Salas et al., 1977). Even though the hypothalamus is involved in the regulation of many endocrine systems, brain-gonadal functions normally decline well before other brain-endocrine relationships.

GROWTH HORMONE

It is generally agreed that growth hormone elicits a spectrum of effects on the major metabolic pathways of many cells, however, the role of growth hormone in the growth and development of the nervous system is still controversial. In cases of human growth hormone deficiency, some reports claim normal intelligence (Pollitt and Money, 1964; Rosenbloom et al., 1966; Drash et al., 1968) and school performance (Rosenbloom et al., 1966), while others find intellectual deficits (Laron et al., 1971). Of the few reports on the influence of growth hormone on the structure of the developing nervous system, most studies suuport an effect, although possibly indirect, on the central nervous system.

Subcutaneous injections of bovine pituitary growth hormone into pregnant rats result in a statistically significant increase in brain weight, number of brain cells, cortical cell density, and ratio of neurons to glia (Zamenhof et al., 1966; Sara et al., 1974). In addition there is a 23% increase in the mean number of dendrites associated with each neuron, a 22% increase in the mean length of dendrites and an enhancement in the performance of cortically mediated behavior (Clendinnen and Eayrs, 1961). Injections of growth hormone after birth to 38-day-old male rats hypophysectomized at 21 days and killed 11 days later, show an increase in body weight and an increase in DNA content of the cerebrum (Cheek and Graystone, 1969). In contrast, studies from another laboratory have shown administration of excess growth hormone after birth does not alter brain weight, brain protein content, thickness of the cerebral cortex or branching of basal dendrites of rat cortical pyramidal cells (Diamond et al., 1969). In normal young rats, injections of growth hormone do not accelerate the histologic differentiation of the developing cerebellum but gave rise to slight lengthening of the dendritic endings of Purkinje cells (Rebière and Legrand, 1970). On the other hand, chronic deficiency of growth hormone, produced in rats by injecting antibodies against rat somatotrophin during the first week of postnatal life, results in a significant decrease in the rate of body and brain growth. Within the cerebral cortex there is a 70-80% decrease in myelin lipids, a 65% reduction of DNA, a significant decline of RNA, and an abnormal accumulation of undifferentiated glial cells in the subependymal zone in association with decreased amounts of stainable myelin in the subcortical white matter (Pelton et al., 1977).

Many of the metabolic effects of growth may be mediated by secondary messengers termed somatomedins (Hall and Luft, 1974; Van Wyk et al., 1974; Sara et al., 1976; see also Chapter 4). The somatomedins, a group of relatively low molecular weight polypeptides produced primarily by the liver in the presence of growth hormone, have been shown to exert an anabolic action on somatic and neural cells (Fryklund et al., 1974; Westermark and Wasteson, 1975; Chochinov and Daughday, 1976). Although definitive proof that the effects of growth hormone on the developing nervous system are mediated by somatomedin is still lacking, the possibility that this may be the case should be considered.

INSULIN

The role of insulin in brain differentiation and growth has not been extensively investigated. In humans, the IQ of infants of diabetic mothers is reduced if ketosis is present during pregnancy (Churchill et al., 1969). Alloxan induced diabetes in pregnant rabbits results in congenital abnormalities in the fetal brains (Barashnev, 1964) and streptozotocin induced diabetes in pregnant rhesus monkeys results in a reduction in the cell population in the fetal cerebrum (Cheek et al., 1975).

Most anatomical studies on the influence of insulin on the nervous system are from studies of long term, adult human or animal diabetics. In humans, neurologic manifestations most frequently described in association with diabetes mellitus involve the peripheral nerves and nerve roots. However, both myelopathy and encephalopathy may be part of the diabetic process. In the former case this involves degeneration of the long tracts, demyelination, gliosis, and microinfarcts (Olsson et al., 1968; Slager and Webb, 1973; DeJong, 1977) while in encephalopathy this includes cell damage and loss, demyelination, gliosis, and severe angiopathy (Reske-Nielsen and Lundback, 1963; Reske-Nielsen et al., 1965; Olsson et al., 1968; DeJong, 1977).

Peripheral nerves of rats with alloxan diabetes for two years show evidence of Schwann cell and axonal injury, demyelination, and vascular abnormalities (Powell et al., 1977). However, animals with diabetes for less time (six months to one year) do not demonstrate these changes (Sharma and Thomas, 1974). In long-term studies, Schwann-cell alterations include disintegration and removal of myelin sheaths, the presence of supernumerary basal lamina and concentric arrays of attenuated Schwann-cell processes around an axon (Vracko, 1974; Bischoff, 1973; Powell et al., 1977). Abnormalities within the axons involve the accumulation of glycogen within mitochondria and inclusions resembling Lafora's bodies and the inclusions observed in the disease, glycogenosis type IV (Powell, et al., 1977). Ultrastructural studies of the retina of diabetic rats also reveal the accumulation of glycogen in neurons and glial cells (Sosula et al., 1974). It has been suggested that the neuropathy in alloxan diabetes is associated with a metabolic impairment which results in intra-cellular retention of insoluble sugar alcohols (Thomas and Eliasson, 1975; Ward, et al., 1972).

Pathologic alterations have been observed in the blood vessels associated with the central and peripheral nervous system of human and animal diabetics. Aggregations of periodic acid Schiff-positive material accumulate in the vessel wall as well as in the perivascular spaces in the vasa nervorum of adult diabetics (Fagerberg, 1959), in vessels associated with globus pallidus and dentate nucleus of juvenile diabetics (Reske-Nielsen et al., 1965), and in the vessels associated with peripheral nerves in diabetic rats (Powell, et al., 1977). These latter vessels also are occasionally surrounded by multiple layers of basal laminae (Powell, et al., 1977). Similar supernumerary basal laminae around vessels have been documented in human diabetes (Thomas and Eliasson, 1975; Vracko, 1974; Bischoff, 1973), and it has been suggested that this is a result of repeated endothelial lyses and regeneration with persistence of the old and the renewal of new endothelial basal lamina. Such changes may account for the increased vascular permeability demonstrated in alloxan diabetic rats (Seneviratne, 1972).

Adult rats with acute or chronic alloxan induced diabetes show vacuolation of the cytoplasm of supraoptic and paraventricular nuclei and a decrease in neurosecretory material in the conducting tracts and posterior lobe of the hypophysis. These findings are reversed following insulin administration (Rabkina, 1965). More recently, karyometry of the anterior, medial, and posterior hypothalamic nuclei of alloxan-diabetic male rats reveal that certain hypothalamic nuclei change the size of their mucleus in response to insulin deficiency. The size of the ventromedial and arcuate nuclei decrease while there is a concomitant increase in the size of the supraoptic nucleus (Akamyev and Rabkina, 1976a). Similar studies of the dorsal nucleus of the vagus nerve by the same laboratory demonstrate a decrease in nuclear volumes (Akmayev and Rabkina, 1976b).

While there are morphological studies indicating neural changes associated with diabetes in vivo and in vitro, studies on the uptake of insulin by CNS have been both positive and negative. Even though a number of laboratories have been unable to demonstrate the uptake of labeled insulin by the brain (Haugaard et al., 1954; Elgee et al., 1954; Goodner and Berrie, 1977) or cerebrospinal fluid (CSF) (Mahon, et al., 1962; Woods and Porte, 1976), others have demonstrated that insulin crosses the blood-CSF barrier (Margoles and Altszuler, 1967) and that the concentration of insulin in CSF parallels changes in plasma insulin concentration under steady-state conditions (Owen et al., 1974). Similar to the in vivo situation, studies of the insulin responsiveness of brain tissues in vitro have also been both negative (Beloff-Chain, et al., 1955; Goodner and Berrie, 1977) and positive (Rafaelsen, 1958; Field and Adams, 1964; Mellerup and Rafaelsen, 1969). Consequently, it is not entirely clear if the changes in the nervous system associated with diabetes are the direct result of lack of insulin or are related to other aspects of this complicated disease.

REFERENCES

Akmayev, G. and Rabkina, A. CNS-endocrine pancreas system. I. The hypothalamus response to insulin deficiency. *Endokrinologie* 68, 211-220 (1976a).

Akmayev, G. and Rabkina, A. CNS-endocrine pancreas system. II. Response of dorsal nucleus of the vagus nerves to insulin deficiency. *Endokrinologie* 68, 221-225 (1976b).

Altman, J. DNA metabolism and cell proliferation, in *Handbook of Neurochemistry*. A. Lajtha, ed. Plenum Press, New York (1969) pp. 137-182.

Arai, Y. and Kusama, T. Effect of neonatal treatment with estrone on hypothalamic neurons and regulation of gonadotrophin secretion. *Neuroendocrinology* 3, 107-114 (1968).

Azcoaga, J. Modificaciones gliales del hypotálamo senil. *Arch. Histol. Normal Patol.* 8, 278-289 (1963).

Babichev, V. Characteristics of hypothalamic neurons controlling the pituitary-gonadotropic function in old female and male rats. *Byull. Eksp. Biol. Med.* 75, 603-605 (1973).

Baden, M., Bauer, C., Colle, E., Klein, G., Taeusch, H., Jr., and L. Stern. A controlled trial of hydrocortisone therapy in infants with respiratory distress syndrome. *Pediatrics* 50, 526-534 (1972).

Balázs, R. Hormones and brain development, in *Perspectives in Brain Research*. Vol. 45, M. Corner and D. Swaab, eds. Elsevier Scientific Publishing Company, New York (1976) pp. 139-159.

Balázs, R. Effect of thyroid hormone and undernutrition on cell acqusition in the rat brain, in *Thyroid Hormones and Brain Development*. G. Grave, ed. Raven Press, New York (1977) pp. 287-302.

Balázs, R., Cocks, W., Eayrs, J., and Kovacs, S. Biochemical effects of thyroid hormones on the developing brain, in *Hormones in Development.* M. Hamburgh and E. Barrington, eds., Appleton-Century-Crofts, New York (1971a) pp. 357-379.

Balázs, R., Kovacs, S., Cocks, W., Johnson, A. and Eayrs, J. Effect of thyroid hormone on the biochemical maturation of the brain: postnatal cell formation. *Brain Res.* 25, 555-570 (1971b).

Balázs, R., Kovacs, S., Teichgraber, P., Cocks, W. and Eayrs, T. Biochemical effects of thyroid deficiency on the developing brain. *J. Neurochem.* 15, 1335-1349 (1968).

Barashnev, Y. Disturbances of the fetal brain development in maternal alloxan diabetes. *Ark. Patologii* 26(5), 63-71 (1964).

Barraclough, C. Sex modifications in the CNS regulation of reproduction after exposure of prepubertal rats to steroid hormones. *Recent Progr. Hormone Res.* 22, 503-539 (1966).

Barrnett, R. Experimental production of cretin-like rats. *Yale J. Biol. Med.* 22, 313-322 (1950).

Bass, N., Pelton, E., II, and Young, E. Defective maturation of cerebral cortex: An inevitable consequence of dysthyroid states during postnatal life, in *Thyroid Hormones and Brain Development.* G. Grave, ed. Raven Press, New York (1977) pp. 199-214.

Bass, N. and Young, E. Effects of hypothyroidism on the differentiation of neurons and glia in developing rat cerebrum. *J. Neurol. Sci.* 18, 155-173 (1973).

Beierwaltes, W., Carr, E., Raman, G., Spafford, W., Aster, R., and Lowrey, G. Institutionalized cretins in the state of Michigan. *J. Mich. State Med. Soc.* 58, 1077-1095 (1959).

Beloff-Chain, A., Cantanzaro, R., Chain, E., Masi, I., and Pocchiaria, F. Fate of uniformly labelled 14C glucose in brain slices. *Proc. R. Soc. B.* 144, 22-28 (1955).

Benda, C. *Mongolism and Cretism.* Greene and Stratton, New York (1946).

Bischoff, A. Ultrastructural pathology of peripheral nervous system in early diabetes. *Adv. Metab. Disord.* 2 (Suppl.), 441-449 (1973).

Blodgett, F., Burgin, L., Iezzoni, D., Gribetz, D., and Talbot, N. Effects of prolonged cortisone therapy on the structural growth, skeletal maturation, and metabolic status of children. *New Engl. J. Med.* 254, 636-641 (1956).

Borisova, N., and Stefanov, S. Karyometric investigation of the hypothalamus of neonatally castrated rats. *Soviet J. Develop. Biol.* 7, 314-317 (1976).

Brawer, J. The role of the arcuate nucleus in the pituitary-gonad axis. *J. Comp. Neurol.* 143, 411-446 (1971).

Bubenik, G., and Brown, G. Morphologic sex differences in primate brain areas involved in regulation of reproductive activity. *Experientia* 29, 619-621 (1973).

Burdman, J., Jahn, G., and Szijan, E. Early events in the effect of hydrocortisone acetate on DNA replication in the rat brain. *J. Neurochem.* 24, 663-666 (1975).

Calaresu, F., and Henry, J. Sex difference in the number of sympathetic neurons in the spinal cord of the cat. *Science* 173, 343-344 (1971).

Cheek, D., and Graystone, J. The action of insulin, growth hormone, and epinephrine on cell growth in liver, muscle, and brain of the hypophysectomized rat. *Pediat. Res.* 3, 77-88 (1969).

Cheek, D., Hill, D., Brayton, J., and Scott, R. Changes in growth in the fetal brain after ablation of the pancreatic beta cells, in *Fetal and Postnatal Cellular Growth,* D. Cheek, ed. John Wiley & Sons, New York (1975) pp. 155-165.

Chochinvo, R., and Daughaday, W. Current concepts of somatomedin and other biologically related growth factors. *Diabetes* 25, 994-1004 (1976).

Christensen, A., and Fawcett, D. The fine structure of testicular interstitial cells in mice. *Am. J. Anat.* 118, 551-572 (1966).

Churchill, J., Berendes, H., and Nemore, J. Neuropsychological deficits in children of diabetic mothers. *Am. J. Obstet. Gynecol.* 105, 257-268 (1969).

Ciaranello, R., Jacobowitz, D., and Axelrod, J. Effect of dexamethasone in phenylethanolamine-N-methyltransferase in chromaffin tissue of the neonatal rat. *J. Neurochem.* 20, 799-805 (1973).

Clendennen, B., and Eayrs, J. The anatomical and physiological effects of prenatally administered somatotrophin on cerebral development in rats. *J. Endocrin.* 22, 183-193 (1961).

Clos, J. and Legrand, L. Effects of thyroid deficiency on the different cell populations in the cerebellum in the young rat. *Brain Res.* 63, 450-455 (1973).

Clos, J., Rebière, A., and Legrand, J. Differential effects of hypothyroidism and undernutrition on the development of glia in the rat cerebellum. *Brain Res.* 63, 445-449 (1973).

Cockett, S. and Kiernan, J. Acceleration of peripheral nervous regeneration in the rat by exogenous triiodothyronine. *Exp. Neurol.* 39, 389-394 (1973).

Contopoulos, A., Simpson, M., and Koneff, A. Pituitary function in the thyroidectomized rat. *Endocrinology* 63, 642-653 (1958).

Costa, M., Eränkö, O., and Eränkö, L. Hydrocortisone-induced increase in the histochemically demonstrable catacholamine content of sympathetic neurons of the newborn rat. *Brain Res.* 67, 457-466 (1974).

Cotterrell, M., Balázs, R., and Johnson, A. Effects of corticosteroids on the biochemical maturation of rat brain: postnatal cell formation. *J. Neurochem.* 19, 2151-2167 (1972).

Coyle, I., Anker, R., and Cragg, B. Behavioral, biochemical and histological effects of prenatal administration of progesterone in the rat. *Pharmacol. Biochem. Behav.* 5, 587-590 (1976).

Cramer, E., and Ford, D. Ultrastructural changes in the hypothalamo-hypophyseal axis in rats thyroidectomized at birth, in *Thyroid Hormones and Brain Development*. G. Grave, ed. Raven Press, New York (1977) pp. 19-32.

Curry, J., III, and Heim, L. Brain myelination after neonatal administration of oestradiol. *Nature (London)* 209, 915-916 (1966).

Deis, R. Influencia de las hormonas sexuales en la concentracion de ocitocina en la neurohipófisis. *Rev. Soc. Argent. Biol.* 35, 315-320 (1959).

De Jong, R. CNS manifestations of diabetes mellitus. *Postgrad. Med.* 61, 101-107 (1977).

Diamond, M., Johnson, R., Ingham, C., and Stone, B. Lack of direct effect of hypophysectomy and growth hormone on postnatal rat brain morphology. *Exp. Neurol.* 23, 51-57 (1969).

Döcke, F., and Koloczek, G. Einfluss einer postnatalen androgenbehandlung auf den nucleus hypothalamicus anterior der weiblichen Ratte. *Endokinologie* 50, 225-230 (1966).

Döcke, R., and Smollich, A. Morphologic effect of a single postnatal administration of androgen on medial preoptic nucleus in prepubertal female rats. *Endocrinol. Experimentalis* 3, 107-112 (1968).

Dörner, G., and Staudt, J. Structural changes in the preoptic anterior hypothalamic area of the male rat, following neonatal castration and androgen substitution. *Neuroendocrinology* 3, 136-140 (1968).

Dörner, G., and Staudt, J. Structural changes in the hypothalamic ventromedial nucleus of the male rat following neonatal castration and androgen treatment. *Neuroendocrinology* 4, 278-281 (1969).

Drash, P., Greenberg, N., and Money, J. Intelligence and personality in four syndromes of dwarfism, in *Human Growth*. D. Cheek, ed. Lea & Febiger, Philadelphia (1968) pp. 568-581.

Dube, D., Leclerc, R., and Pelletier, G. Electron microscopic immunohistochemical localization of vasopressin and neurophysin in the median eminence of normal and adrenalectomized rats. *Am. J. Anat.* 147, 103-108 (1976).

Earthy, H., and Leblond, C. Identification of the effects of thyroxine mediated by the hypohysis. *Endocrinology* 54, 249-271 (1954).

Eayrs, J. The vascularity of the cerebral cortex in normal and cretinous rats. *J. Anat.* 88, 164-174 (1954).

Eayrs, J. The cerebral cortex of normal and hypothyroid rats. *Acta Anat.* 25, 160-183 (1955).

Eayrs, J. Effect of neonatal hyperthyroidism on maturation and learning in the rat. *Anim. Behav.* 12, 195-199 (1964).

Eayrs, J. Developmental relationships between brain and thyroid, in *Endocrinology and Human Behaviour*. R. Michael, ed. Oxford University Press, London (1968), pp. 239-255.

Eayrs, J. and Horn, G. The development of the cerebral cortex in hypothyroid and starved rats. *Anat. Rec.* 121, 53-62 (1955).

Eayrs, J. and Levine, S. Influence of thyroidectomy and subsequent replacement therapy upon conditioned-avoidance learning in the rat. *J. Endocr.* 25, 505-513 (1963).

Eayrs, J. and Lishman, W. The maturation of behavior in hypothyroidism and starvation. *Anim. Behav.* 3, 17-24 (1953).

Eayrs, J. and Taylor, S. The effect of thyroid deficiency induced by methylthiouracil on the maturation of the central nervous system. *J. Anat.* (Lond.) 85, 350-358 (1951).

Elgee, N., Williams, R., and Lee, N. Distribution and degradation studies with insulin-I^{131}. *J. Clin. Invest.* 33, 1252-1260 (1954).

Epstein, M., Farrell, P., Sparks, J., Pepe, G., Driscoll, S., and Chez, R. Maternal betamethasone and fetal growth and development in the monkey. *Am. J. Obstet. Gynecol.* 127, 261-263 (1977).

Eränkö, L. and Eränkö, O. Effect of hydrocortisone on histochemically demonstrable catecholamines in the sympathetic ganglia and extraadrenal chromaffin tissue of the rat. *Acta Physiol. Scand.* 84, 125-133 (1972).

Eränkö, O., Heath, J., and Eränkö, L. Effect of hydrocortisone on the ultrastructure of the small intensely fluorescent, granule-containing cells in cultures of sympathetic ganglia of newborn rats. *Z. Zellforsch.* 134, 297-310 (1972b).

Eränkö, O., Lempinen, M., and Raisanen, L. Adrenaline and noradrenaline in the organ of Zuckerkandl and adrenals of newborn rats with hydrocortisone. *Acta Physiol. Scand.* 66, 253-254 (1966).

Eränkö, O., Eränkö, L., Hill, C., and Burnstock, G. Hydrocortisone-induced increase in the number of small intensely fluorescent cells and their histochemically demonstrable catecholamine content in cultures of sympathetic ganglia of the newborn. *Histochem. J.* 4, 49-58 (1972a).

Everett, J. Central neural control of reproductive functions of the adenohypophysis. *Physiol. Rev.* 44, 373-431 (1964).

Fagerberg, S. Diabetic neuropathy: a clinical and histological study of the significance of vascular affections. *Acta Med. Scand.* 164 (Suppl. 345), 1-81 (1959).

Field, E. Observations on development of microglia together with a note on the influence of cortisone. *J. Anat.* (Lond.) 89, 201-208 (1955).

Field, R. and Adams, L. Insulin response of peripheral nerve. I. Effects on glucose metabolism and permeability. *Medicine* 43, 275-279 (1964).

Ford, D. and Milks, L. Smooth endoplasmic reticular whorls in neurons of the arcuate nucleus in male rats following adrenalectomy. *Psychoneuroendocrinology* 3, 65-83 (1978).

Ford, D., Kantounis, S. and Lawrence, R. The localization of I^{131} labeled triiodothyronine in the pituitary and brain of normal and thyroidectomized male rats. *Endocrinology* 64, 977-991 (1959).

Francon, J., Fellows, A., Lennon, A., and Nunez, J. Is thyroxine a regulatory signal for neurotubule assembly during brain development? *Nature* 266, 188-190 (1977).

Frolkis, V., Bezrukov, V., Duplenko, Y., and Genis, E. The hypothalamus in ageing. *Exp. Gerontol.* 7, 169-184 (1972).

Fryklund, L., Uthne, K., and Sieventsson, H. Isolation and characterization of polypeptides from human plasma enhancing the growth of human normal cells in culture. *Biochem. Biophys. Res. Commun.* 61, 950-956 (1974).

Gerlach, J. and McEwen, B. Rat brain binds adrenal steroid hormone: radioautography of hippocampus with corticosterone. *Science* 175, 1133-1136 (1972).

Goodner, C. and Berrie, M. The failure of rat hypothalamic tissues to take up labeled insulin *in vivo* or to respond to insulin *in vitro*. *Endocrinology* 101, 605-612 (1977).

Gorski, R. Modification of ovulatory mechanisms by postnatal administration of estrogen in the rat. *Amer. J. Physiol.* 205, 842-844 (1963).

Gorski, R. Gonadal hormones and the perinatal development of neuroendocrine function, in *Frontiers in Neuroendocrinology*. L. Martini and W. Ganong, eds. Oxford University Press, New York (1971) pp. 237-290.

Gourdon, J., Clos., J., Coste, C., Dainat, J., and Legrand, J. Comparative effects of hypothyroidism, hyperthyroidism and undernutrition on the protein and nucleic acid contents in the young rat. *J. Neurochem.* 21, 861-871 (1973).

Greski, E. Preliminary observations on the effects of chronic hypothyroidism on the development of the superior cervical ganglion of the rat. *Brain Res.* 110, 619-622 (1976).

Gumbinas, M., Oda, M., and Huttenlocher, P. The effects of corticosteroids on myelination of the developing rat brain. *Biol. Neonate* 22, 355-366 (1973).

Gusik, W. Die Feinstruktur der Rattenzirbel und ihr Verhalten unter Einfluss von Antiandrogen und nach Kastration. *Endokrinologie* 67, 129-151 (1976).

Hajós, R., Patel, A. and Balazs, R. Effect of thyroid deficiency on the synpatic organization of the rat cerebellar cortex. *Brain Res.* 50, 387-401 (1973).

Hall, K. and Luft, R. Growth hormone and somatomedin. *Adv. Metab. Dis.* 7, 1-36 (1974).

Hamburgh, M. Evidence for a direct effect of temperature and thyroid hormone on myelinogenesis *in vitro*. *Dev. Biol.* 18, 15-30 (1966).

Hamburgh, M. The role of thyroid and growth hormones in neurogenesis. in, *Current Topics in Developmental Biology*. A. Moscona and A. Monroy, eds., Vol. 4, Academic Press, New York (1969) pp. 109-148.

Hamburgh, M. and Barrington, E. (eds.) Hormones in Development Appleton-Century-Crofts, New York (1971).

Hamburgh, M., Mendoza, L., Burkart, J. and Weil, F. 1971 The thyroid as a time clock in the developing nervous system. in, *Cellular Aspects of Neural Growth and Differentiation.* D. Pease, ed., University of California Press, Berkeley (1971) pp. 321-328.

Hamburgh, M., Mendoza, L. Bennett, I., Krupa, P., So Kim, Y., Kahn, R., Hogreff, K. and Frankfort, H. Some unresolved questions on brain-thyroid relationships. in, *Thyroid Hormone and Brain Development.* G. Grave, ed. Raven Press, New York (1977) pp. 49-72.

Harris, G. Sex hormones, brain development and brain function. *Endocrinology* 75, 627-648 (1964).

Harris, G. and Campbell, H. The regulation of the secretion of luteinizing hormone and ovulation. in, *The Pituitary Gland.* G. Harris and D. Donovan, Vol. 2, Butterworths, London (1966) pp. 99-165.

Harris, G. and Jacobsohn, D. Functional grafts of the anterior pituitary gland. *Proc. Roy. Soc. (B)* 139, 263-276 (1952).

Harris, G. and Levine, S. Sexual differentiation of the brain and its experimental control. *J. Physiol. (London)* 181, 379-400 (1965).

Haugaard, N., Vaughan, M., Haugaard, E., and Stadie, W. Studies of radioactive injected labeled insulin. *J. Biol. Chem.* 208, 549-563 (1954).

Hellman, R., Ford, D. and Rhines, R. Growth in hypothalamic neurons as reflected by nuclear size and labelling with ^{3}H-uridine. *Psychoneuroendocrinology* 1, 389-397 (1976).

Hellström, S. and Koslow, S. Effects of glucocorticoid treatment on catecholamine content and ultrastructure of the adult rat carotid body. *Brain Res.* 102, 245-254 (1976).

Howard, E. Effects of corticosterone and food restriction on growth and DNA, RNA, and cholesterol contents of the brain and liver in infant mice. *J. Neurochem.* 12, 181-191 (1965).

Howard, E. Reduction in size and total DNA of cerebrum and cerebellum in adult mice after corticosterone treatment in infancy. *Exp. Neurol.* 22, 191-208 (1968).

Howard, E. and Granoff, D. Increased voluntary running and decreased motor coordination in mice after neonatal corticosterone implanatation. *Expl. Neurol.* 22, 661-673 (1968).

Ifft, J. 1964 The effect of endocrine gland extirpations on the size of nucleoli in rat hypothalamic neurons. *Anat. Rec.* 148, 599-604 (1964).

Iwatsubo, J., Miyai, K., Abe, H., Kumahara, Y., Omori, K., Okada, Y., and Fukuchi, M. Human growth hormone secretion in primary hypothyroidism before and after treatment. *J. Clin. Endocr.* 27, 1751-1754 (1967).

Karasek, M., Pawlikowski, M., Kappers, J. and Stepien, H. Influence of castration followed by administration of LH-RH on the ultrastructure of rat pinealocytes. *Cell Tiss. Res.* 167, 325-339 (1976).

Klevit, H. Corticosteroid therapy in the neonatal period. *Pediat. Clin. N. Amer.* 17, 1003-1013 (1970).

Kollros, J. Endocrine influences in neural development, in *Ciba Foundation Symposium on Growth of the Nervous System.* G. Wolstenholme and M. O'Connor, eds., J. and A. Churchill, London (1968) pp. 179-192.

Korkala, O., Eränkö, O., Partanen, S., Eränkö, L., and Hervonen, A. Histochemically demonstrable increase in the catecholamine content of the carotid body in adult rats treated with methylprednisolone or hydrocortisone. *Histochem. J.* 5, 479-485 (1973).

Korochkin, L. and Korochkina, L. Hormonal influence on the differentiation of nerve cells of sympathetic and parasympathetic nervous system. *Z. Mikr. Anat. Forsch.* 82, 293-321 (1970).

Koslow, S., Bjegovic, M. and Costa, E. Catecholamines in sympathetic ganglia of rat: effects of dexamethasone and reserpine. *J. Neurochem.*

24, 277-281 (1975).

Laron, Z., Pertzelan, A., and Frankel, J. Growth and Development in the syndromes of familila isolated absence of HGH or pituitary dwarfism with high serum concentration of an immunoreactive but biologically inactive HGH. In: *Hormones in Development*. Edited by M. Hamburgh and E. Barrington, Appleton-Century-Crofts, New York, (1971) pp. 573-585.

Lauder, J. The effects of early hypo- and hyperthyroidism on the development of rat cerebellar cortex. III. Kinetics of cell proliferation in the external granular layer. *Brain Res.* 126, 31-51 (1977a).

Lauder, J. Effects of thyroid state on development of rat cerebellar cortex. In: *Thyroid Hormones and Brain Development*. Edited by G. Grave, Raven Press, New York, (1976) pp. 235-254.

Legrand, J. Analyse de l'action morphogenetique des hormones thyroidiennes sur le cervelet du jeune rat. *Arch. Anat. Microsc. Morphol. Exp.* 56, 206-244 (1967).

Lempinen, M. Extra-adrenal chromaffin tissue of the rat and the effect of cortical hormones on it. *Acta Physiol. Scand.* 62 (Suppl. 231), 7-91 (1964).

Levine, S. and Mullins, R., Jr. Estrogen administered neonatally affects adult sexual behavior in male and female rats. *Science* 144, 185-187 (1964).

Lewis, P. Cell death in the germinal layers of the postnatal rat brain. *Neuropathol. Appl. Neurobiol.* 1, 21-29 (1975).

Lewis, P., Balázs, R., Patel, A., and Johnson, A. The effect of undernutrition in early life on cell generation in the rat brain. *Brain Res.* 83, 235-247 (1975).

Lewis, P., Patel, A., Johnson, A. and Balázs, R. Effect of thyroid deficiency on cell acquisition in the postnatal rat: a quantitative histological study. *Brain Res.* 104, 49-62 (1976).

Liggins, G. Adrenocortical-related maturational events in the fetus. *Am. J. Obstet. Gynecol.* 126, 931-941 (1976).

Liggins, G. and Howie, R. A controlled trial of antepartum glucocorticoid treatment for prevention of the respiratory distress syndrome in premature infants. *Pediatrics* 50, 515-525 (1972).

Lu, E. and Brown, W. The developing caudate nucleus in the euthyroid and hypothyroid rat. *J. Comp. Neurol.* 171, 261-284 (1977).

McEwen, B. Interactions between hormones and nerve tissue. *Sci. Amer.* 235, 48-58 (1976).

McEwen, B. and Pfaff, D. Chemical and physiological approaches to neuroendocrine mechanisms: attempts at integration, in *Frontiers in Neuroendocrinology*. W. Ganong and L. Martini, eds. Oxford University Press, New York (1973) pp. 267-335.

McEwen, B., Weiss, J., and Schwartz, L. Retention of corticosterone by cell nuclei from brain regions of adrenalectomized rats. *Brain Res.* 17, 471-485 (1970).

McEwen, B., Lieberburg, I., Maclusky, N., and Plapinger, L. Do estrogen receptors play a role in the sexual differentiation of the rat brain? *J. Steroid Biochem.* 8, 593-598 (1977).

McIsaac, G. and Kiernan, J. Accelerated recovery from peripheral nerve injury in experimental hyperthyroidism. *Exp. Neurol.* 48, 88-94 (1975).

McPherson, J., Costoff, A. and Mahesh, V. Effects of aging on the hypothalamic-hypophyseal-gonadal axis in female rats. *Fertil. Steril.* 28, 1365-1370 (1977).

McQuarrie, I. Nerve regeneration and thyroid hormone treatment. *J. Neurol. Sci.* 26, 499-502 (1975).

Machado-Salas, J., Scheibel, M., and Scheibel, A. Morphologic changes

in the hypothalamus of the old mouse. *Exp. Neurol.* 57, 102-111 (1977).
Mahon, W., Steinke, J., McKhann, G. and Mitchell, M. Measurement of I^{131}-insulin and of insulin-like activity in cerebrospinal fluid of man. *Metabolism* 11, 416-420 (1962).
Margolis, R. and Altszuler, N. Insulin in the cerebrospinal fluid. *Nature* 215, 1375-1376 (1967).
Marinesco, M. Contribution a l'etude des lesions du myxedema congenital (iodiotic myxodemateuse du Bourneville), *Encephale* 19, 265-273 (1924).
Matsumoto, A. and Arai, Y. Effect of estrogen on early postnatal development of synaptic formation in the hypothalamic arcuate nucleus of female rats. *Neurosci. Lett.* 2, 79-82 (1976).
Meier, C. and Bischoff, A. Polyneuropathy in hypothyroidism: clinical & nerve biopsy study of 4 cases. *J. Neurol.* 215, 103-114 (1977).
Meites, J., Huang, H. and Riegle, G. Relation of the hypothalamo-pituitary-gonadal system to decline of reproductive functions in aging female rats, in *Hypothalamus and Endocrine Functions*. F. Labrie, J. Meites, and G. Pelletier, eds., Plenum Press, New York and London (1975) pp. 3-20.
Mellerup, E. and Rafaelsen, O. Brain glycogen after intracisternal insulin injection. *J. Neurochem.* 16, 777-781 (1969).
Mott, F. The changes in the central nervous system in hypothyroidism. *Proc. Roy. Soc. Med.* 10, 51-55 (1917).
Naftolin, F., Ryan, K., Davies, I., Reddy, V., Fores, F., Petro, Z., Kuhn, M., White, R., Takaoka, Y., and Wolin, L. The formation of estrogens by central neuroendocrine tissues. *Recent Prog. Horm. Res.* 31, 295-319 (1975).
Nelson, J., Holinka, C., Latham, K., Allen, J., and Finch, C. Corticosterone binding in cytosols from brain regions of mature and senescent male C57BL/6J mice. *Brain Res.* 115, 345-351 (1976).
Nicholson, J. and Altman, J. The effects of early hypo- and hyperthyroidism on the development of rat cerebellar cortex. I. Cell proliferation and differentiation. *Brain Res.* 44, 13-23 (1972a).
Oda, M. and Huttenlocker, P. The effect of corticosteroids on dendritic development in the rat brain. *Yale J. Biol. Med.* 47, 155-165 (1974).
Olsson, Y., Säve-Söderbergh, J., Sourander, P and Angervall, L. A pathoanatomical study of the central and peripheral nervous system in diabetes of early onset and long duration. *Pathol. Eur.* 3, 62-79 (1968).
Owen, O., Reichard, G., Jr., Boden, G., and Shuman, C. Comparative measurements of glucose, beta-hydroxybutyrate, acetoacetate, and insulin in blood and cerebrospinal fluid during starvation. *Metabolism* 23, 7-14 (1974).
Palkovits, M. and Stark, E. Quantitative histological changes in the rat hypothalamus following bilateral adrenalectomy. *Neuroendocrinology* 10, 23-30 (1972).
Patel, A., Balázs, R., and Johnson, A. Effect of undernutrition on cell formation in the rat brain. *J. Neurochem.* 20, 1151-1165 (1973).
Pelton, E., Grindeland, R., Young, E., and Bass, N. Effects of immunologically induced growth hormone deficiency on myelinogenesis in developing rat cerebrum. *Neurology* 27, 282-288 (1977).
Pfaff, D. Morphological changes in the brains of adult male rats after nenonatal castration. *J. Endocr.* 36, 415-416 (1966).
Pfeiffer, C. Sexual differences of the hypophyses and their determination by the gonads. *Amer. J. Anat.* 58, 195-225 (1936).
Pollitt, E. and Money, J. Studies in the psychology of dwarfism. I. Intelligence quotient and school achievement. *J. Pediat.* 64, 415-421 (1964).
Powell, H., Knox, D., Lee, S., Charters, A., Orloff, M., Garrett, R., and Lampert, P. Alloxan diabetic neuropathy: electronmicroscopic

studies. *Neurology* 27, 60-66 (1977).

Price, M., Olney, J., and Cicero, T. Proliferation of lamellar whorls in arcuate neurons of the hypothalamus of castrated and morphine-treated male rats. *Cell Tiss. Res.* 171, 277-284 (1976).

Price, M., Olney, J., and Cicero, T. Proliferation of lamellar whorls in arcuate neurons of the hypothalamus of male rats treated with estradiol benzoate or cyproterone acetate. *Cell Tiss. Res.* 182, 537-540 (1977).

Rabkina, A. Effect of insulin on restoration of hypothalamic structure and neurosecretion in alloxan diabetes. *Fed. Proc. Transl.* 24 (Suppl), 379-381 (1965).

Rafaelsen, O. Action of insulin on isolated rat spinal cord. *Lancet* 2, 941-943 (1958).

Rainer, A. and Adamo, N. Arcuate nucleus region in androgen-sterilized female rats: ultrastructural observations. *Neuroendocrinology* 8, 26-35 (1971).

Raisman, G. and Field, P. Sexual dimorphism in the neurophil of the preoptic area of the rat and its dependence on neonatal androgen. *Brain Res.* 54, 1-29 (1973).

Rebière, A. and Legrand J. Absence d'effets marques de l'hormone hypophysaire de croissance sur la maturation histologique du cortex cerebelleux chez le jeune rat normal on hypothyroidien. *Brain Res.* 299-312 (1970).

Rebière, A. and Legrand J. Comparative effects of underfeeding, hypothyroidism and hyperthyroidism on the histological maturation of the molecular layer of the cerebellar cortex of the young rat. *Arch. Anat. Microsc. Morphol. Exp.* 61, 105-126 (1972).

Reier, R. and Hughes, A. An effect of neonatal radiothyroidectomy upon non-myelinated axons and associated Schwann cells during maturation of the mouse sciatic nerve. *Brain Res.* 41, 263-282 (1972).

Reske-Neilsen, E. and Lundback, K. Diabetic encephalopathy: Diffuse and focal lesions of the brain in long-term diabetes. *Acta Neurol Scand.* 39 (Suppl 4), 273-290 (1963).

Reske-Nielsen, E., Lundback, K. and Rafaelsen, O. Pathologic changes in the central and peripheral nervous system of young long term diabetics: 1. Diabetic encephalopathy. *Diabetologia* 1, 233-241 (1965).

Rhees, R., Grosser, B. and Stevens, W. Effect of steroid competetion and time on the uptake of (^{3}H) corticosterone in the rat brain; an autoradiographic study. *Brain Res.* 83, 293-300 (1975).

Rosenbloom, A., Smith, D. and Loeb, D. Scholastic performance of short-statured children with hypopituitarism. *J. Pediat.* 69, 1131-1133 (1966).

Rosman, N. and Malone, M. Brain myelination in experimental hypothyroidism: morphological and biochemical observations, in *Thyroid Hormones and Brain Development*. G. Grave, ed. Raven Press, New York (1977) pp. 169-198.

Roth, G. Reduced glucocorticoid binding site concentration in cortical neuronal perikaya from senescent rats. *Brain Res.* 107, 345-354 (1976).

Roth, G. and Adelman, R. Age related changes in hormone binding by target cells and tissues; possible role in altered adaptive responsiveness. *Exp. Gerontol.* 10, 1-11 (1975).

Sanchez-Toscano, F., Del Rey, F., De Escobar, G., and Ruiz-Marcos, A. Measurement of the effects of hypothyroidism on the number and distribution of spines along the apical shaft of pyramidal neurons of the rat cerebral cortex. *Brain Res.* 126, 547-550 (1977).

Sara, V., King, T., Stuart, M., and Lazarus, L. Hormonal regulation of fetal brain cell proliferation: presence in serum of a trophin responsive to pituitary growth hormone stimulation. *Endocrinology* 99, 1512-1518 (1976).

Sara, V., Lazarus, L., Stuart, M., and King, T. Fetal brain growth: selective action by growth hormone. *Science* 186, 446-447 (1974).

Schapiro, S., Salas, M., and Vukovich, K. Hormonal effects on ontogeny of swimming ability in the rat: assessment of central nervous system development. *Science* 168, 147-151 (1970).

Schooley, R., Friedkin, S., and Evans, E. Re-examination of the discrepancy between acidophil numbers and growth hormone concentration in the anterior pituitary following thyroidectomy. *Endocrinology* 79, 1053-1057 (1966).

Seneviratne, K. Permeability of blood nerve barriers in the diabetic rat. *J. Neurol. Neurosurg. Psychiatry* 35, 156-162 (1972).

Sharma, A. and Thomas, P. Peripheral nerve structure and function in experimental diabetes. *J. Neurol Soc.* 23, 1-15 (1974).

Slager, U. and Webb, A. Pathologic findings in the spinal cord. *Arch. Pathol.* 96, 388-394 (1973).

Solomon, J. and Greep, R. The effect of alterations in thyroid function in the pituitary growth hormone content and acidophil cytology. *Endocrinology* 65, 158-164 (1959).

Sosula, L., Beaumont, P., Hollows, F., Jonson, K. and Regtop, H. Glycogen accumulation in retinal neurons and glial cells of streptozotocin-diabetic rats. *Diabetes* 23, 221-231 (1974).

Staudt, J. and Dörner, G. Structural changes in the medial and central amydala of the male rat, following neonatal castration and androgen treatment. *Endokrinologie* 67, 296-300 (1976).

Talanti, S. Effect of thyroxine, thiouracil and thyroidectomy on the neurosecretory ganglion cells of the rat. *Life Sci.* 4, 2151-2156 (1965).

Talanti, S. Effect of thiouracil, excess thyroxine and thyroidectomy on the ependymal cells with special reference to the subcommissural organ. *Anat. Rec.* 159, 379-386 (1967a).

Talanti, S. The effect of thiouracil and excess thyroxine on the hypothalamus of the rat with special reference to neurosecretory phenomena. *Z. Zellforsch.* 79, 92-109 (1967b).

Thomas, P. and Eliasson, S. Diabetic neuropathy, in *Peripheral Neuropathy*. P. Dyck, P. Thomas, and E. Lambert eds., W.B. Saunders Co., Philadelphia (1975) pp. 956-981.

Toran-Allerand, C. Sex steroids and the development of the newborn mouse hypothalamus and preoptic area *in vitro*: implications for sexual differentiation. *Brain Res.* 106, 407-412 (1976).

Vandesande, F., DeMey, J., and Dierick, K. Identification of neurophysin producing cells. I. The origin of the neurophysin-like substance-containing nerve fibers of the external region of the median eminence of the rat. *Cell Tiss. Res.* 151, 187-200 (1974).

Van Wyk, J., Underwood, L., Hinz, R., Clemmons, D., Voina, S., and Weaver, R. The somatomedins: a family of insulin-like hormones under growth hormone control. *Rec. Prog. Horm. Res.* 30, 259-318 (1974).

Vracko, R. Basal lamina layering in diabetes mellitus: evidence for accelerated rate of cell death and cell regeneration. *Diabetes* 23, 94-104 (1974).

Ward, J., Baker, R. and Davis, B. Effect of blood sugar control on accumulation of sorbitol and fructose in nervous tissues. *Diabetes* 21, 1173-1178 (1972).

Watkins, W., Schwabedal, P. and Bock, R. Immunohistochemical demonstration of a CRF-associated neurophysin in the external zone of the rat median eminence. *Cell Tiss. Res.* 152, 411-421 (1974).

Weichsel, M., Jr. Effect of thyroxine on DNA synthesis and thymidine kinase activity during cerebellar development. *Brain Res.* 78, 455-465 (1974).

Westermark, B. and Wasteson, A. The response of cultural human normal glial cells to growth factors. *Adv. Metab. Dis.* 8, 85-100 (1975).

Whalen, R. and Nadler, R. Suppression of the development of female mating behavior of estrogen administered in infancy. *Science* 141, 273-274 (1963).

Wilson, J. Reproductive capacity of adult female rats treated prepuberally with estrogenic hormones. *Anat. Rec.* 86, 341-359 (1943).

Winick, M. and Noble, A. Cellular response in rats during malnutrition at various ages. *J. Nutr.* 89, 300-306 (1966).

Woods, S. and Porte, D., Jr. Insulin and the set-point regulation of body weight, in *Hunger: Basic Mechanisms and Clinical Implications.* D. Novins, W. Wyrwicka, and G. Bray, eds., Raven Press, New York (1976) pp. 273-280.

Zambrano, D. and de Robertis, E. Ultrastructural changes of the neurohypophysis of the rat after castration. *Z. Zellforsch.* 86, 14-25 (1968a).

Zambrano, D. and de Robertis, E. The effect of castration upon the ultrastructure of the rat hypothalamus. I. Supraoptic and paraventricular nuclei. *Z. Zellforsch.* 86, 487-498 (1968b).

Zambrano, D. and de Robertis, E. The effect of castration upon the ultrastructure of the rat hypothalamus. II. Arcuate nucleus and outer zone of the median eminence. *Z. Zellforsch.* 87, 409-421 (1968c).

Zamenhof, S., Mosley, J., and Schuller, E. Stimulation of the proliferation of cortical neurons by prenatal treatment with growth hormone. *Science* 152, 1396-1397 (1966).

Zimmerman, E. Localization of hypothalamic hormones by immunocytochemical techniques, in *Frontiers in Nueroendocrinology.* Vol. 4, L. Martin and W. Ganong, eds., Raven Press, New York (1976) pp. 25-62.

Chapter 7
Endocrine Regulation of Neural Development: Physiological and Biochemical Aspects
Antonia Vernadakis

INTRODUCTION

Critical periods of brain maturation have been defined according to the timetable of maturational events in the nervous system. One early critical period is during neurogenesis. During this period both genetic or programmatic events and epigenetic factors contribute to the final expression of neural differentiation. Although the genetic events which participate in neural differentiation are not understood some epigenetic factors are beginning to become evident. Hormones are appearing to be a group of epigenetic factors which may exert a regulatory influence in early neural expression.

As discussed in Chapters 2 and 8, the role of androgens on the differentiation of the hypothalamus provides a prime example of a critical neural period of hormone specificity. Although the exact molecular events leading to such hormone-induced genomic expression are not presently known for any hormone-sensitive cell, it seems extremely likely that the binding of the particular steroid hormone to a stereospecific binding protein in the cytoplasm and cell nucleus of the target cells plays a central role.

Another critical period of brain maturation is after differentiation when neurons and glial cells form their interrelationships: neurons form synaptic contacts and glial cells modulate the neuronal microenvironment by participating in neuronal growth, myelination process, ionic regulation, and synaptic expression. Evidence that hormones exert an influence at this level is accumulating. Of particular importance are the thyroid hormones and steroids.

It has been established that in humans a decrease or lack of thyroid hormones during early brain maturation, that is, in late fetal or early postnatal periods results in profound impairment of CNS development that is manifested clinically in cretinism, a condition characterized by arrested mental development and accompanied by other developmental abnormalities such as reduced body growth and skeletal maldevelopment. The arrest of mental development induced by hypothyroidism is underlined by anatomic, electrophysiologic and neurochemical alterations that are respectively reflected in the reduction of the number and size of neurons and their branching (Chapter 6), alterations in the development of spontaneous and evoked electrical activity and impaired protein and nucleic acid metabolism. It is also well-known that the mental retardation of the cretinoid human or animal can be prevented or effectively remedied if adequate doses of thyroid hormones are administered at an appropriate time; in humans within the first year of life, and in the rat, an animal often used for these studies, before 12 days of age. The timing is critical, for if hormone therapy is instituted beyond these critical periods, even high doses will be inaffective in restoring normal CNS development.

Because of the many maturational actions of the thyroid hormones, thyroxine (T_4) and triiodothyronine (T_3), the question of the onset of thyroid function is of particular interest. The human thyroid has been reported to be capable of concentrating iodide during early fetal months (Chapter 2). Studies of fetal thyroid in some animals have indicated that the ability to concentrate iodide is acquired first, followed by the ability to form monoiodothyronine (MIT), diiodothyronine (DIT), and finally, thyroxine. In addition, because the thyroid is capable of concentrating iodide and making thyroxine even in anencephalic fetuses it has been presumed that the biosynthetic mechanisms of the thyroid gland can develop in the absence of thyrotropin stimulating hormone (TSH) from the pituitary. In humans, evidence is accumulating that indicates that the pituitary-thyroid feedback system is already operating at 19 to 20 weeks and that by the 20th week the secretion of the hypothalamic TSH-releasing factor has begun. The presence, therefore, of thyroid hormones during early periods of maturation when the neurons are actively differentiating and glial cells are proliferating, supports the important role that these hormones play in neural growth.

The appearance of adrenocortical hormones during maturation of the human organism is not clearly delinated. It has been shown that at the end of fetal life, there are still quantitative but no qualitative differences between newborn and adult adrenals in the enzymes concerned with steroid biosynthesis (Chapter 2). Thus, during the first 10 to 15 week of embryonic activity, the pattern of adrenocortical secretion is characterized by high levels of androgens (androstenedione, dehydroepiandrosterone) and some mineralocorticoids. During the next five weeks, all the biosynthetic enzymes become active and cortisol and aldosterone begin to be secreted although the levels remain low throughout fetal life.

The role of glucocorticoids on neural growth are less well understood. The recent evidence, however, of steroid receptors in neural cells, both neuronal and glial, (see also Chapter 5, 12) in CNS areas other than target tissues, i.e., hypothalamus, pituitary, supports the view that steroid hormones are involved in cell growth processes.

From the foregoing discussion it is evident that intercommunication of neurons appears to be significantly regulated by hormones. This hormonal influence on neural substrates is reflected in both the electrical activity of the brain and in behavioral patterns of the organism. In this chapter an attempt has been made to discuss some of the biological mechanisms that may be involved in the influence of hormones on brain maturation. Thyroid hormones, glucocorticoids and gonadal hormones are the three types primarily discussed and the information is grouped under three main sections: Hormonal influence on cell growth and differentiation; hormones and biochemical aspects of brain maturation; and hormones and physiological aspects of brain maturation. Although reference to human studies will be made whenever available, the primary information on hormones and brain maturation has derived from animal studies and particularly the rat. Morphological, biochemical, and physiological investigations have provided detailed information of the maturation of the rat brain. Briefly, the brain of the rat is immature at birth. Neurogenesis is completed in most brain areas by birth but glial cells are actively proliferating during the first weeks after birth and slowly throughout the life span of the rat. Myelination, although is accomplished with the first 3-4 weeks in most brain areas, some myelinogenesis continues up to 3-6 months. Neurotransmission processes have matured by three weeks after birth and functional activity as assessed electrophysiologically has reached adult levels by this time. Sexual maturation in the rat is reached by 4 weeks in the female and five weeks in the male. In this chapter some normal maturational events will be recalled in some detail in order to more effectively describe the effects of hormones.

SECTION I: HORMONAL INFLUENCE ON CELL GROWTH AND DIFFERENTIATION

Several laboratories including ours have been focusing their research on the role of hormones on neural growth and differentiation. As discussed repeatedly in this book developing neuronal circuits in the CNS depend on the proliferation and differentiation of nerve cells being closely coordinated in time and space. In addition, normal development also depends on the formation and differentiation of both neurons and glial cells occuring in an interrelated fashion. Finally hormones and neurohumor substances appear to modulate this neuronal-glial interrelationship and the microenvironment of the neuron.

Cell Formation

Balazs and associates have extensively investigated the effects of thyroid hormones on the proliferation of cells in the rat brain and their work has been summarized in several reviews (Balazs 1971a,b; Balazs et al., 1975a,b; Balazs, 1977). In the brain, cell multiplication takes place mainly at certain sites, such as the subependymal layer lining the forebrain ventricles and the external granular layer covering the surface of the cerebellum although there are also stem cells, probably glial precursors, which are dispersed throughout the brain (see Balazs, et al., 1975a,b). Extensive cell formation in the brain is restricted to about the first three weeks after birth, although cell mutiplication does not cease completely after that age; the external granular layer disappears by about 24 days, but remnants of the subependymal layer are still present in the adult rat. In most of the species studied, neurogenesis during the early postnatal period leads to a selective increase in inteneurones in certain brain regions, such as the olfactory lobe, hippocampus and cerebellum. It seems that the formation of different nerve cell types occurs in a strictly chronological order, for example, in the rat cerebellum the origin of the basket cells is at 2-6 days, that of the stellate cells at the end of the second week, whereas approximately 50% of the granule cells are formed in the third week of life. There is also evidence suggesting the need to reconsider the belief that neurogenesis is completed in man before birth (see refs in Balazs, et al., 1975b). In the first 1-5 years after birth, the cell number increases by a factor of three in the cerebrum and six in the cerebellum. The great increase in cell number indicates that, at least in the cerebellum, neurogenesis must also occur in man after birth since in this part of the brain the most abundant cells are the granule cells. These cells originate from the external granular layer which is present in humans until about 1-5 years of age.

Thyroid hormones influence postnatal cell formation. For example, the rate of acquisition of cells in the rat cerebellum is significantly slower after neonatal thyroidectomy, although the final cell number is normal (Balazs et al., 1968). The attainment of the normal cell number results from prolongation of the period of extensive cell proflieration as indicated by the persistence of the external granular layer in the cerebellum of thyroid deficient rats for 1-2 weeks longer than in controls (Legrand,

1967a,b; Hamburg, 1968; Nicholson and Altman, 1972a,b).

Postnatal cell formation is also influenced in the rat brain by treatment with triiodothyronine (T_3) during infancy (Balazs et al., 1971; Nicholson and Altman, 1972a,b; Weichsel, 1974). Initially, the rate of cell acquisition is more or less normal, but cell proliferation is prematurely terminated resulting in a permanent deficit in cell number throughout the brain. The premature cessation of cell acquisition is the consequence of an earlier decline of cell proliferation in the hyperthyroid rats. In comparison with controls, the germinal site in the cerebellar cortex, the external granular layer (EGL), disappears sooner, and the activity of thymidine kinase, which seems to reflect closely the proliferative capacity of cells, also declines earlier. It has been observed that treatment with thyroid hormone leads to an acceleration in the increase in thymidine kinase activity in the first days after birth, and this is associated with an increase in the DNA content of the cerebellum during the same period (Weichsel, 1974). This, it would appear that thyroid hormones are involved in the regulation of cell proliferation in the brain; the effects in the first postnatal week are consistent with an increase in either the number of replicating cells or with the rate of cell replication.

Balazs and associates (Cotterrell et al., 1972) examine the effects of corticosteroids on postnatal cell formation in the rat brain. Brain weight was permanently reduced as a result of cortisol treatment early postnatally. The effect of cortisol in reducing the normal increase in cell number was observed mainly during the period of the administration of the hormone. Between days 2 and 5 the increase in cell number was inhibited by 86 percent in the cerebrum, by 70 percent in the cerebellum and by 60 percent in the olfactory lobes, but during the following eight days the rise in cell number was only little affected. The decrease in cell number by cortisol was a result of the effect of the hormone in inhibiting the formation of DNA. However, at 20 days of age, the mitotic activity was above the control level in both the cerebrum and cerebellum of the cortisol-treated animals, suggesting a tendency to compensate for the initial deficit.

More recently Bohn and Lauder (1978) have further studied the effects of corticosteroids on rat cerebellar development. As discussed earlier the EGL, a secondary germinal zone present on the surface of the developing cerebellum, gives rise to basket, stellate, and granule cells during the first three postnatal weeks. In rats receiving hydrocortisone neonatally, development of cerebellar foilation was affected and the growth of the EGL was inhibited during the hormone treatment and for a few days thereafter. The Purkinje cells, which are scattered in the cerebellar cortex of the newborn rat, form a distinct monolayer as the apical dendritic growth cone becomes directed toward the pial surface. This lining up of Purkinje cells was delayed in the hormone-treated animals. In agreement with Cotterrel et al. (1972), Bohn and Lauder also observed that cell proliferation in the EGL was inhibited by cortisol but a partial recovery occurred. The inhibition of proliferation of EGL cells during hormone treatment resulted in early birthdays for a greater proportion of basket and granule cells. Normally, the levels of corticosterone in rat plasma and brain, which are high immediately following birth, drop precipitously shortly thereafter and remain low until they rise to adult levels after the second postnatal week (Sze, 1976). Bohn and Lauder (1978) suggest that levels of glucocorticoids in the neonatal rat must be low in order to permit normal neurogenesis in the developing rat cerebellar cortex.

Neuronal Circuits in the Cerebellum

The functional plan of the cerebellar cortex is based on a population of large prenatally formed neurons, the Purkinje cells, which are the only afferent cells from the cortex and which are activated by two major afferent systems (climbing and mossy fibers, respectively). The functioning of both the input and the output systems is modulated by internal circuits. In the adult, the climbing fiber synaptic contact is on the Purkinje cell dendrites (climbing fiber-Purkinje cell circuit), whereas the mossy fiber input is transmitted first to the granule cells, which make synaptic contact with the spiny branchlets of the Purkinje cells dendrites through the parallel fibers (mossy fiber-granule cell-Purkinje cell circuit). In the modulation of both circuits, the inhibitory interneurons, Golgi, basket, and stellate cells, play an important role. In the rat, the Golgi cells are formed in the perinatal period, but all the other interneurons including the excitatory granule cells are predominantly acquired after birth.

Thyroid deficiency has a marked quantitative influence on the constituents of the neuronal circuits in the cerebellum (reviews, Balazs et al., 1975a,b). It would appear that with respect to cellular composition the major effect of thyroid deficiency is a reduction in the number of basket cells and an increase in glial cells. Although the number of the Purkinje and granule cells is not decreased, both the major postsynpatic area, the Purkinje cell dendrites, and the number of presynaptic structures in the molecular layer are substantially depressed in comparison with controls. Besides these quantitative alterations, thyroid deficiency in infancy also leads to qualitative changes in the "wiring pattern" in the cerebellar cortex. Hajos et al. (1973) reported that in thyroid deficiency the reorganization of the climbing fiber-Purkinje cell circuit significantly delayed: the Purkinje cell somatic spines persist longer than in controls, although they disappear ultimately. Furthermore, the development of the cerebellar glomeruli severely retarded. There are characteristic signs of the maturation of these structures. The huge mossy fiber rosette contains many mitochondria and synaptic vesicles, and each terminal digit of the electron translucent granule cell dendrite makes a single synaptic contact with the rosette. In thyroid deficiency the mossy fiber terminals were small, glial cells occupied a relatively large area of the glomerulus, and the granule cells were immature. Their dendrites were often not digitated, and the electron-dense dendritic trunks formed multiple synapses with the mossy fiber terminal. It seems therefore that in thyroid deficiency an immature wiring pattern persist for a longer time, and the balance between the two major circuits is shifted in favor the climbing fiber-Prukinje cell circuit.

Synaptic Development

It has already been established and discussed elsewhere that thyroid hormones influence the synthesis of protein in the CNS and particularly in the cerebellum. Legrand and associates (Rabie and Legrand, 1973), also examined the effects of thyroid hormones on the amount of synaptosomal fraction in the cerebellum of the young rat. In 10-day-old rats made hypothyroid by giving them propylthiouracil (PTU), the content of synaptosomal protein per cerebellum or per milligram of cerebellum was lower than in normal animals; this content was higher, however, in animals which have received both PTU and small doses of thyroxine and in animals receiving high doses of the hormone alone than in normal animals. Only in the case of animals made hyperthyroid, however, did the cerebellum have a higher total protein content. At 10 days of age, deficiency or excess of thyroid hormone has therefore, in the cerebellum, a more pronounced effect on the content of synaptosomal protein than on the total protein content. Later in life the effects of giving thyroxine were less marked. At 35 days of age, the synaptosomal protein concentration found in young receiving both PTU and thyroxine or high doses of hormone alone did not differ significantly from that in the normal, whereas it was still reduced in hypothyroid animals. These results show that thyroid hormones have a selective effect on the ontogenesis of nerve terminals and that this effect becomes apparent soon after birth.

More recent studies have more closely examined biochemical synaptic changes with neonatal hypothyroidism (Verity et al., 1976). Maturation profiles of total synaptosome fraction and specific activities of lactage dehydrogenase (LDH), Na^+-K^+ ATPase, cytochrome c oxidase, and protein were obtained from rats 6-32-days of age. The greatest changes were found in the total activities of enzymes isolated from the cerebellum.

To further examine synaptic changes with hypothyroidism changes in gangliosides and glycoproteins have been examined by Geel and Gonzales (1977). Gangliosides and glycoproteins are strucutral components of neurons and specifically associated with membranes of nerve terminals (DeKirmenjian and Brunngraber, 1969; Lapetina et al., 1967). The age interval during which cerebral gangliosides and glycoporteins are accumulating at a maximum rate in the rat (Holian et al., 1971; Suzuki, 1965; Vanier et al., 1971), coincides with the period of rapid axonal and dendritic proliferation (Eayrs and Goodhead, 1959), the establishment of synaptic contacts (Aghajanian and Bloom, 1967), and myelinogenesis (Norton and Pudoslo, 1973). In view of their structural and compositional heterogeneity, gangliosides and glycoproteins have a suspected function as surface recognition molecules (Hughes, 1973) and thus may be important in mediating cellular differentiation in the nervous system.

Neonatal hypothyroidism had no apparent effect on the concentration of cerebral ganglioside and glycoprotein N-acetylneuraminic acid (NANA) (Geel and Gonzales, 1977). In addition, the relative distribution of the 4 major gangliosides was essentially normal in hypothyroidism. The concentration of myelin-associated glycolipids (cerebrosides and sulfatides), however, was markedly depressed in thyroid hormone-deficient rats, indicating a deficiency in myelinogenesis. In normal rats the kinetics of incorporation of (^{14}C) glucosamine in vivo into brain gangliosides was considerably different from that in the lipid-free protein containing glycoprotein. Hypothyroidism produced a striking increase in the radioacitivty associated with both gangliosides and lipid-free protein and their constituent monosaccharides although the pattern of incorporation was similar to controls. Qualitative and quantitative changes in rat brain gangliosides and glycoporteins have been reported during development and the incorporation of radioactive precursors into their constituent monosaccharides is more prominent in immature rats than in adults (Holian et al., 1971; Margolis et al., 1975; Quarles and Brady, 1971). In the light of these observations, it is logical to consider that the increased incorporation of (^{14}C) glucosamine into gangliosides and glycoporteins of hypothyroid rats is a consequence of immaturity or retardation of cerebral development.

Studies of Hormonal Actions Using Neural Culture

In recent years neural tissue and cell culture has become a very useful tool with which to study cellular effects of hormones. We and collaborative laboratories have used various culture models to study the role of glial cells in hormonal actions in neural growth. Before describing the experimental findings a brief description will be presented on the various culture systems used by various investigators. A more detailed description will be found in a recent report by Vernadakis and Culver (1980).

Neural Culture Systems. Early neural culture systems have been the *organ culture* and *organotypic culture* (see review Vernadakis and Culver, 1979). In the *organ culture* the nervous tissue explant (cerebrum, cerebellum, spinal cord, hypothalamus) is oriented on a triangular stainless stell organ culture grid (Fig. 1). Platforms with explants are placed over the center well of organ culture dishes and the area surrounding the well is covered with an absorbent ring. The culture medium is added to the center well of the organ culture and does not reach the top of the platform. Humidity is maintained by saturating the absorbent with distilled water. Explants are incubated at 35°-36°C. In this culture system the explant maintains its cytoarchitecture for up to 48 hr and is usually used for studying acute 24 - 48-hr effects of hormones drugs or other experimental conditions.

For the *organotypic culture* system a Maximow double coverslip assembly is used (Fig. 2) described in detail by Murray (1965). Fragments of nervous tissue, 0.5 mm^3, are explanted on rectangular coverslips which are previously coated with a thin film of reconstituted rat tail collagen (Bornstein, 1958). A single drop (50 µℓ) of nutrient medium is added to the explants and then incorporated into a Maximow double coverslip assembly. In this culture system, the original organization of the tissue may be lost but the constituent cells emerge into the zone of outgrowth. As has been previously observed by Murray (1965) microglial cells are the first to emigrate from the explant. These cells are followed by neuroglial elements, notably oligodendroglia. As the culture advances in age and complexity of organization, oligodendroglia are overshadowed by a variety of astrocytes. This culture system is maintained for several weeks and has been used extensively for electrophysiological studies, myelination, neurotransmitter uptake.

Neuronal-glial interrelationships have been studied using the *dissociated brain cell culture* system. There are several methods for dissociation of nervous tissue (see review Vernadakis and Culver, 1980). We use that described by Sensenbrenner et al. (1971) and Booher and Sensenbrenner (1972). One of the advantages in using dissociated cell cultures is that cells can be maintained in culture for several weeks and thus can be used for chronic studies. Additionally, the cell cultures exhibit a specific growth pattern which is characterized first by neuronal elements, primarily during the first two weeks in culture followed by a marked proliferation of glial cells and the disappearance of neurons. Thus, this culture system can be useful when the responses of neurons and glial cells are to be studied individually.

Neoplastic clonal cell lines of either neurons or glial cells have been extensively used to study neuronal or glial properties and neuronal-glial interactions (see review Vernadakis and Culver, 1980).

Steroid Hormones: In early studies, we employed *organ* and *organotypic* culture systems to study the effects of steroid hormones and neurohumoral substances on neural growth in culture (Vernadakis and Timiras, 1967; Vernadakis, 1971, 1974; Vernadakis and Berni, 1973). As seen in Table 1, DNA content was higher in cerebellar explants following 24 hours in organ culture, compared to noncultured (in situ) tissue removed from 16-day-old chick embryos. This increase demonstrates the continued proliferation of neural cells throughout. Addition of cortisol or estradiol in the culture medium further increased DNA content. The increase in DNA induced by steroid hormones may reflect increased proliferation of glial cells, since these cells are actively dividing in the chick cerebellum

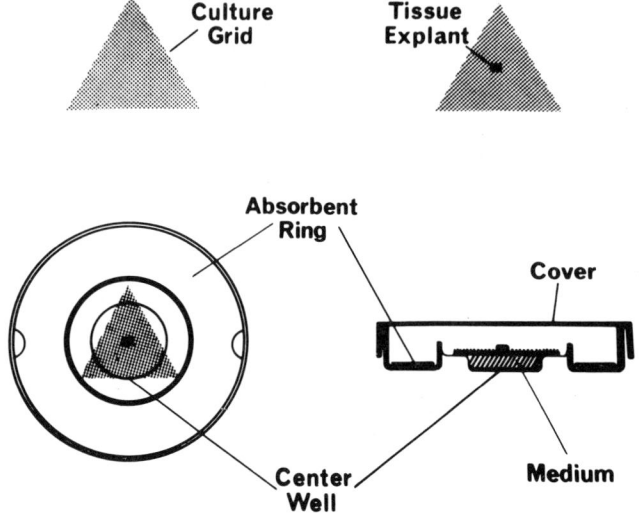

Fig. 1. Diagram of an organ culture (from Vernadakis, 1971).

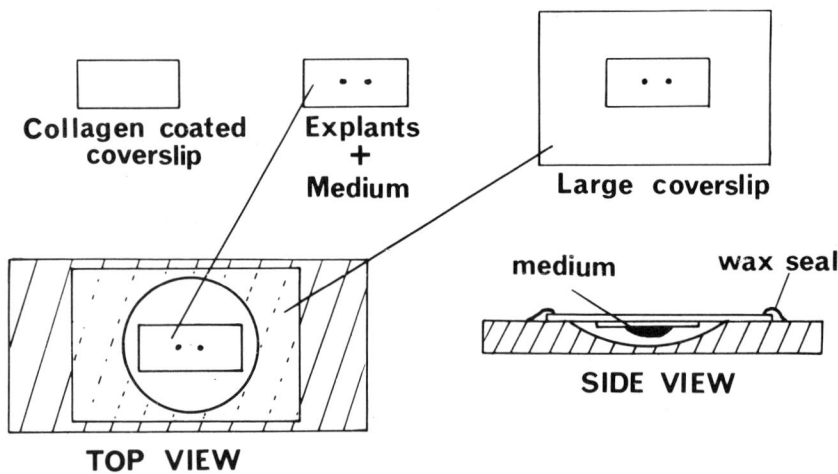

MAXIMOW DOUBLE COVERSLIP ASSEMBLY

Fig. 2. Diagram of a Maximow Double Coverslip Assembly (from Vernadakis, 1971).

Table 1

EFFECTS OF CORTISOL, CORTICOSTERONE, AND ESTRADIOL ON
DNA CONCENTRATION OF CEREBELLAR EXPLANTS REMOVED FROM
16-DAY-OLD CHICK EMBRYOS AND MAINTAINED
AS ORGAN CULTURES FOR 24 HOURS

Culture Medium[a]	DNA Concentration (μg/mg wet tissue)
Noncultured	5.04 ± 0.23[b]
Basal	6.61 ± 0.27 (<0.001)[c]
Cortisol	8.07 ± 0.49 (<0.02)[d]
Corticosterone	6.30 ± 0.40
Estradiol	7.87 ± 0.66 (0.05-0.1)

[a] The medium was Eagle's basal medium with Earle's salts. Hormones, when added, were in the following concentrations: cortisol 2.76×10^{-5} M; corticosterone 2.89×10^{-5} M; estradiol dipropionate 2.65×10^{-5} M.
[b] Each value represents the mean ±SE of 12-16 explants.
[c] Numbers in parentheses are values for comparison with noncultured control group.
[d] Numbers in parentheses are P values for comparison with basal medium group.

after 14 days of embryonic age (Hanaway, 1967). This view is further supported by data obtained from organotypic culture studies (Table 2). In cerebellar explants cultured in the presence of cortisol, corticosterone, estradiol, or progesterone, but not in the presence of testosterone, the migration rate of glial cells from explants was markedly higher than that in control cultures.

Recently several investigators have been using C-6 glial cells, a rat astrocytoma line, as a cell model with which to study glial cell properties and functions. These undifferentiated cells, when cultured in the presence of dibutyryl cyclic AMP (DBcAMP), develop astrocytic-like processes and resemble mature astrocytes (Vernadakis and Nidess, 1976). Thus, one has the opportunity to study these cells at an immature glioblast stage and as mature glial cells. We have investigated the effects of cortisol on ^{3}H-uridine incorporation into RNA, and ^{3}H-leucine incorporation into protein in C-6 glial cells, both at the immature glioblast state and the mature astrocytic-like stage. Cortisol treatment of glioblasts only slightly increased ^{3}H-leucine incorporation into protein but significantly decreased the uptake (unincorporated ^{3}H-leucine) of ^{3}H-leucine. In contrast to the effects of cortisol on glioblast cells, when mature astrocytic cells (following DBcAMP-induced differentiation) were exposed to cortisol, both ^{3}H-leucine

Table 2

EFFECTS OF HORMONES ON THE MIGRATION RATE
OF CELLS IN CULTURED CEREBELLAR EXPLANTS[a]

Culture Medium[b]	Number of Explants	Migration Rate[c] (mean ratio ± SE)
Control for cortisol	67	4.07 ± 0.12[d]
Cortisol (2.76×10^{-8} M)	50	5.06 ± 0.23 (<0.001)
Control for corticosterone	16	3.55 ± 0.16
Corticosterone (2.89×10^{-8} M)	21	4.56 ± 0.28 (<0.01)
Control for estradiol	21	4.36 ± 0.24
Estradiol dipropionate (2.65×10^{-8} M)	24	5.14 ± 0.25 (<0.05)
Control for progesterone	36	3.78 ± 0.15
Progesterone (3.18×10^{-8} M)	45	4.41 ± 0.21 (<0.02)
Control for testosterone	21	3.88 ± 0.26
Testosterone (3.47×10^{-8} M)	19	3.76 ± 0.21

[a]Cerebellar explants were removed from 15-day-old chick embryos and maintained in culture (Maximow double coverslip assembly) for five days.
[b]Medium consisted of 45% Gey's BSS, 5% ascitic fluid, 5% chick embryo (nine-day-old) extract and glucose.
[c]Migration rate was calculated during the growth period between the first and fifth days in culture. Migration rate was calculated as follows: Outlines of the explants were traced at one and five days by focusing the image of the explant on translucent paper placed on the translucent back of a camera attached to the microscope. The images were cut from the paper and the papers were weighed to determine relative surface areas. The ratio between the surface of one-day and that of five-day culture was calculated as "migration" rate.
[d]Mean ± SE.
[e]Numbers in parentheses are P values for comparison with appropriate control group.

incorporation into protein and ^{3}H-leucine uptake were decreased, and the net effect was no change in protein synthesis. These findings demonstrate that glial cells respond to their microenvironment differently at different maturational stages. This differential responsiveness is further demonstrated by an increase of RNA synthesis (Vernadakis, 1973b). Schwartz (1972) has also reported an increase in a large RNA species (450,000 molecular weight) in retinas removed from 12-day chick embryos and cultured for three hours in medium containing cortisol; this RNA species was assumed to be the steroid-induced messenger RNA coding for the enzyme glutamine synthetase. de Vellis and associates (de Vellis and Brooker, 1973) have reported that in C-6 glial cells the activity of glycerol phosphate dehydrogenase is controlled by glucortocoids and is dependent on RNA synthesis.

The possibility that steroid hormones may influence glial cell function via an intracellular mechanism is supported by recent findings that these hormones accumulate in glial cells. We have used both C-6 glial cells and dissociated brain-cell cultures prepared from chick embryos to study the uptake of steroid hormones into glial cells (Vernadakis et al., 1978). We found that corticosterone accumulates in both neuronal and glial cultures (Fig. 3). Moreover, the retention of corticosterone appears to be specific in the maturing neurons but not in glial cells (Fig. 4).

DeVellis and associates have used the C-6 glial cell to explore the role of glial cell in neural growth and differentiation (deVellis and Brooker, 1973; de Vellis and Kukes, 1973; de Vellis et al., 1971a,b). They have used enzyme induction by hormones as a cellular index of cell differentiation. Their work has been extensively reviewed (recent review deVellis et al. 1977) and will be briefly summarized here. These investigators found that the induction of glycerol phosphate dehydrogenase (GPDH) by hydrocortisone in C-6 glial cells is due to an increased rate of synthesis without alteration of the rate of degradation of the enzyme, thus resulting in a greater number of molecules. Similarly, they have shown that GPDH in the brains of normal (induced) and hypophysectomized (uninduced) rats is identical (McGinnis and de Vellis, 1977). Thus in vivo as in cell culture, the hormonal regulation of GPDH activity is brought about by a change in the number of molecules, not in their catalytic efficiency. Using immunoperoxidase methods deVellis and associates have demonstrated that only one type of neural cell is GPDH positive and have defined by light and electron microscopic criteria as oligodendrocytic. This finding supports two novel conclusions: (a) in nervous tissue, GPDH is a biochemical marker for oligodendrocytes, and (b) oligodendrocytes are target cells for glucocorticoids.

Steroid hormones have shown to increase adhesion of cells to growth substrates in culture (Wrenn and Wessels, 1970; Berliner and Gerschenson, 1975; Bonney et al., 1974; Ballard and Tomkins, 1970). More recently, de Vellis and associates (Berliner et al., 1978) studied the effects of cortisol on cell morphology using the glioma cell line, C6-2B glial cells. When cultures were in log phase of growth (three days) or stationary phase (ten days), 5.5×10^{-7} M hydrocortisone was added to the medium for various periods. Cytochalasin B, colcemid, actinomycin, and cycloheximide were used to study the mechanism of shape changes. Hydrocortisone was found to induce cell spreading by 24 hours after its addition. This spreading phenomenon was correlated with an increase in the fraction of the peripheral cytoplasm occupied by microfilaments. Cytochalasin B caused disorganization of microfilaments in the peripheral cytoplasm of the cells. Additionally, it also prevented cell spreading in response to hormonal stimulation. Both spreading of cells and increases in peripheral microfilaments produced by hydrocortisone were shown to be dependent on RNA and protein synthesis. The foregoing discussion establishes the fact that steroid hormones may influence neural growth by their actions on glial cell function.

Thyroid Hormones. As discussed earlier in this chapter one of the best characterized effects of thyroid hormones is on activity and levels of the sodium pump, $Na^+ - K^+$ ATPase (Valcana and Timiras, 1969; Edelman, 1976). Recently, Timiras and associates (Draves and Timiras, 1980) reported differential effects of thyroid hormones on neurons and glial cells using neuroblastoma and C-6 glial cells respectively as cell models. They found that ATPase activity was negligible in neuroblastoma cells

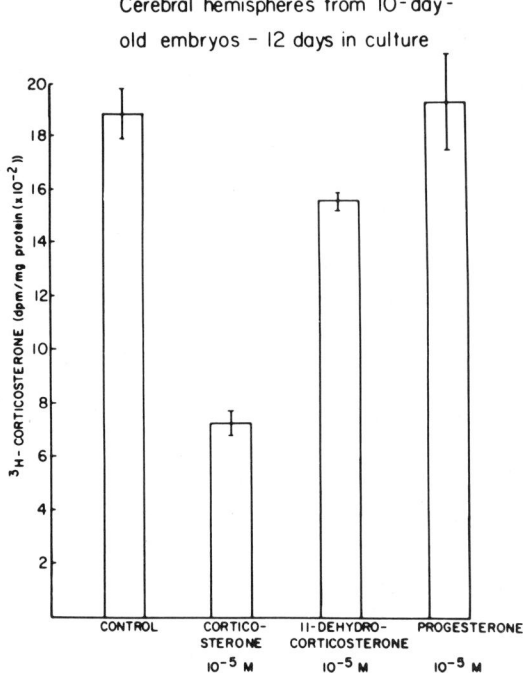

Fig. 3. Retention of 1,2-^{3}H-corticosterone (10-9 M0 in 11-day cultures of dissociated cerebral hemispheres from 10-day-old chick embryos (E10c11). Cultures were preincubated with unlabeled hormones (10-5M) for 60 min. Bars represent means ± S.E. (from Vernadakis et al., 1978).

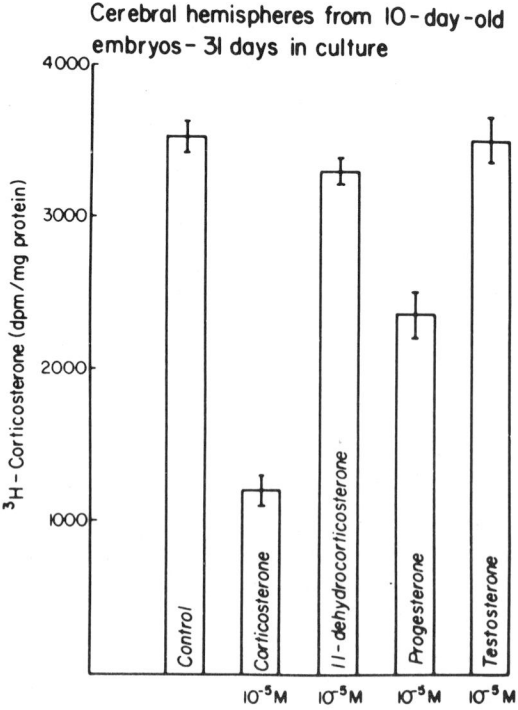

Fig. 4. As in Figure 3, except in E10c31 cultures.

(N_{1g} and N_aA) cultured in medium containing 10% fetal calf serum which has low T_4 or T_3 concentration (hypothyroid) (Rockland Farms: T_4, 0.01 μg g%; T_3, negligible). The effect of low T_4 or T_3 concentration in the medium on cell morphology is demonstrated in Figs. 5(a) and (b). In contrast, ATPase activity was not significantly different in C-6 glial cells cultured in low or T_3 medium as compared with high T_4 or T_3 medium. Moreover, all neuroblastma cell showed alterations in intracellular electrolyte concentrations whereas C-6 glial cells seem better able to withstand thyroid hormone deficiency. Thus, the greater adaptability of glial cells is consistent with their proposed nurse cell and K^+ environment regulatory functions.

SECTION II: HORMONAL INFLUCENCE ON BIOCHEMICAL ASPECTS OF BRAIN MATURATION

MYELINATION

Myelination can be measured by isolating myelin by differential centrifugation or by estimating certain relatively specific constituents of myelin. It is generally agreed that cerebrosides and sulfatides are good markers for myelin; cholesterol, which is also used as an index, is less specific. Sulfatides contain almost all the ^{35}S which is incorporated into the brain lipids after the administration of inorganic (^{35}S) sulfate (Davison and Gregson, 1962), and in some studies myelination has been followed by the determination of (^{35}S) lipids. The rapid phase of myelination begins in the rat brain during the second week after birth and continues for

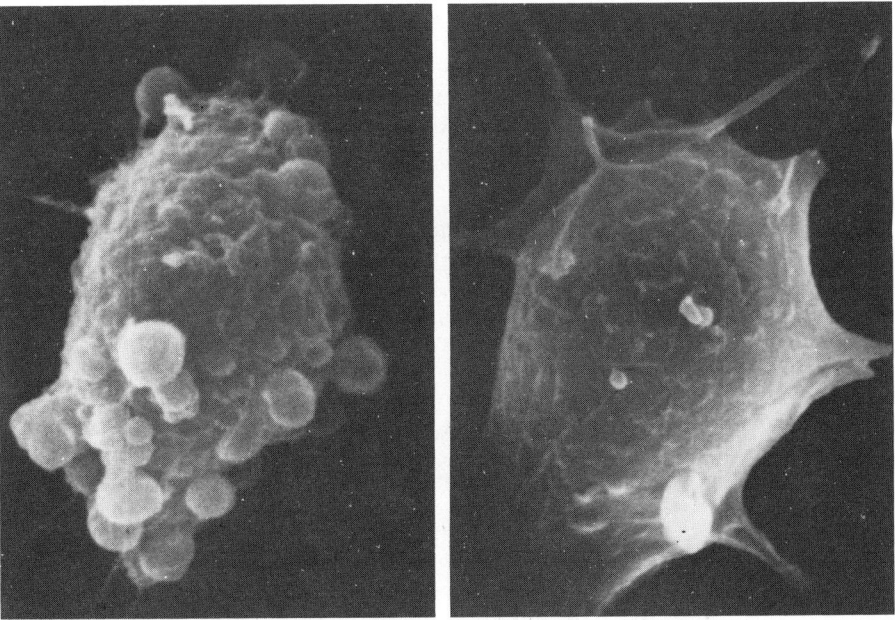

Fig. 5. Neuroblastoma N_2A cultured in medium containing 10% fetal calf serum (euthyroid) (GIBCO: T_3, 186 ng%; T_4, 12.4 μg%) (A) or in medium containing 10% fetal calf serum with low T_3 or T_4 (hypothyroid Rockland Farms: T_4, 0.01 μg%; T_3, negligible). (B) Scanning 600X (Courtesy of Maria Letitia Timiras).

another 3 - 4 weeks, although a slow rate of myelin deposition can be detected up to six months.

Thyroid deficiency in infancy cuases significant reduction in myelin formation in the rat brain (Balazs et al., 1969; Walravens and Chase 1969). Dalal et al. (1971) have reported a decrease of 32% in myelin cerebrosides, 40% in sulfatides and 23% in cholesterol of hypothyroid rats. Of importance is the finding that thyroxine restored these lipids to near control values. Neonatal thyroxine treatment of intact rats accelerated myelin formation early postnatally (10 days). Acceleration of myelinogenesis by thyroxine has also been reported by Hamburgh (1968) in cerebral explants maintained in culture. Schapiro (1966) has reported a moderate increase in the levels of cerebral cholesterol in the rat treated with thyroxine and a similar increase in brain phospholipids has been observed by Myant and Cole (1966). In view of the effects of hypothyrodism on myelin lipids Valcana et al. (1975) examined the interrelationship between thyroid hormone activity and myelination using as an index of myelinogenesis the detection of myelin proteins with special emphasis on the sequential appearance of the basic protein(s) and proteolipids as well as in both fast- and slow- moving proteins. The activity of $2',3'$-cyclic nucleotide phosphohydrolase, and enzyme associated with myelin (Kurihara and Tsukada, 1967; Norton, 1972) was consistently lower in myelin of the hypothyroid animals than controls at all ages studied (10, 24, and 37 days after birth). Mathieu et al. (1975) reported a similar decrease. Valcana and associates suggest that thyroid hormones may influence myelination through their effects on oligodendrocytes, throught to be involved in the formation of myelin (Bunge et al., 1962; Mickel and Gilles, 1970).

More recently, Rosman and Malone (1977) reported the effects of hypothyroidism on myelin purified according to the method of Poduslo and Norton (1973) with slight modifications. Two fractions were obtained, one designated light myelin and the other heavy myelin. Electron microscopic examination of these fractions showed no evidence of contamination. Biochemical analyses were then done on the light and heavy myelin fractions obtained from the hypothyroid and control animals. In agreement with Valcana et al. (1975) the most striking biochemical alteration in the hypothyroid brains was seen in myelin basic protein. In control animals heavy myelin basic protein appeared at 12 days, light myelin basic protein did not appear until 16 days, and by 24 days, there still was no basic protein in the heavy myelin fraction. When individual myelin proteins were examined in the hypothyroid animals after 24 days myelin yields increased sharply and by 40 days the myelin yields from hypothyroid and control animals were similar. Hypothyroid animals showed a persistence of high molecular-weight protein. Thus although early hypothyroidism causes a reduction in total myelin yield the hypothyroid animal has the capacity during later development to increase its synthesis of myelin and eventually to produce myelin at the rate comparable to that in controls.

The effects of steroid hormones on myelination have not been studied in detail, however in an early study Casper et al. (1967) investigated the effects of estradiol and cortisol on lipids and cerebrosides in the developing central nervous system of the rat. Lipids of the cerebrum were higher in the estradiol-treated animals than in controls, whereas in the cerebellum and spinal cord the concentration of lipids was the same in both groups. No differences were found in lipids of the cortisol-treated animals and those of controls. When lipids were fractionated there was a significantly higher amount of cerebrosides in the spinal cord and cerebrum of the estradiol-treated than in controls and cerebrosides were higher in the spinal cord of the cortisol-treated animals. In a later study Granich and Timiras (1971) examined the effects of cortisol on brain lipid patterns in chick embryos and young chicks. They found that in embryos treated on the 15th day of incubation and sacrificed at day 19, cortisol produced a significant increase over control values in total lipid content relative to brain weight. Similarly, the percentages of total sphingolipid and of cerebroside in brain lipid extracts from treated embryos were greater than in controls. Thus, some steroid hormones appear to influence the process of myelination and support histological studies (Curry and Heim, 1966) showing that myelin appears earlier in rat brain after estradiol treatment, during postnatal periods of development. Recently, Dawson and Kernes (1979) using oligodendroglioma clonal lines obtained evidence that cortisol may be involved in the process of myelination via their influence on glial cells.

NEUROTRANSMITTER METABOLISM

Uptake and Synthesis

Developmental Changes in the Rat. The actions of a number of neurotransmitters are partly terminated by a reuptake of the transmitter across the presynaptic membrane and reincorporation of the transmitter into synaptic vesicles. In the adult brain it is possible to demonstrate neuronal uptake of norepinephrine (NE) in vitro in slices (Dengler et al., 1962; Rutledge and Jonason, 1967; Rutledge, 1970), in isolated synaptosomes (Davis et al., 1967), and in homogenates (Snyder and Coyle, 1969).

Various organs of the rat develop the ability to take up and store ^{3}H-norepinephrine (^{3}H-NE) by the time of birth or early in the prenatal period (Glowinski et al., 1964; Iversen et al., 1967). The development of the ability to accumulate ^{3}H-NE parallels the outgrowth of sympathetic adrenergic neurons, as demonstrated by the histochemical fluorescent methods (DeChamplain et al., 1970). The fact that immunosympathectomy with antiserum to nerve growth factor (Iversen et al., 1966) or chemical sympathectomy with 6-hydroxydopamine (Sachs et al., 1970) markedly reduces the accumulation of ^{3}H-NE indicates that the uptake occurs mainly into adrenergic neurons. Thus, the ability of tissues to accumulate ^{3}H-NE is a sensitive index of the extent of sympathetic innervation (Iversen et al., 1966).

In the brain of the adult rat, it has been possible to demonstrate the neuronal uptake of NE in vitro in slices (Dengler, et al., 1962), in isolated synaptosomes (Davis et al., 1967), and in homogenates (Synder and Coyle, 1969). The neuronal mechanism for uptake in the rat brain has distinct anatomical, kinetic, and pharmacological characteristics (Snyder and Coyle, 1969; Coyle and Snyder, 1969). The uptake mechanism has also been used to label the norepinephrine-containing nerve endings in the rat brain with radioactive NE in order to separate them on sucrose gradients from nerve endings that take up other putative neurotransmitters (Kuhar et al., 1970).

The uptake of ^{3}H-NE into slices and synaptosomes prepared from brain is mediated by a number of different mechanisms. A major proportion of the total uptake at 37°C is related to energy-requiring processes, since the uptake is markedly reduced by incubation in the absence of glucose (Shaskan and Snyder, 1970) or in the presence of metabolic inhibitors (White and Keen, 1971). Incurbation with high concentration of ouabain inhibits uptake of ^{3}H-NE to levels near that observed at 0°C (Shaskan and Snyder, 1970). The inhibition of uptake by ouabain appears to be related to inhibition of the neuronal Na^+-K^+ATPase (Tissari, et al., 1969), although other factors may be involved (White and Keen, 1971).

Coyle and Axelrod (1971) have studied the uptake properties of ^{3}H-NE in the developing rat brain. In homogenates prepared from the rat brain at 14 days of gestation, there is no significant difference between 0° and 37°C for the initial uptake of ^{3}H-NE; thus, at this stage of development nonspecific uptake is the sole factor responsible for uptake. At 17 days of gestation, homogenates of rat brain exhibit a temperature-dependent, initial uptake of ^{3}H-NE; however, the uptake mechanism does not show saturability. The brain homogenates first exhibit the high affinity, saturable uptake of ^{3}H-NE at 18 days of gestation.

Studies by Snyder and associates (Snyder et al., 1968; Snyder and Coyle, 1969) indicate that there are regional differences in the uptake of NE in the rat brain. Since the peripheral noradrenergic neurons innervate organs at different stages of development, the various regions of the brain may exhibit similar differences. At birth, the cerebral cortex, cerebellum, brain stem, and hypothalamus-midbrain all exhibit the specific NE uptake mechanism. Between birth and adulthood, the cerebral cortex shows the largest increase in uptake, whereas the cerebellum shows the least change. The hypothalamus-midbrain and brain stem show intermediate increases in uptake between birth and adulthood.

Chick Brain as a Developmental Model. The chick embryo has been a very useful animal model to study factors influencing brain development and thus this animal has been extensively explored biochemically, morphologically and electrophysiologically. We have also investigated the development of the uptake of ^{3}H-NE in chick from embryonic age up to 3 years of hatching. In our studies (Kellogg et al., 1971; Vernadakis, 1973c) of ^{3}H-NE uptake and storage we found that in both the cerebral hemispheres and cerebellum of chicks ^{3}H-NE accumulation progressively increased during brain maturation. In the cerebral hemispheres ^{3}H-NE accumulation increased up to three months after hatching, leveled off up to one year, and markedly declined during aging (Figs. 6,7). In the cerebellum ^{3}H-NE accumulation had already reached maximum level at 20 days of embryonic age and did not significantly change with age (Figs. 8,9). The increase in ^{3}H-NE accumulation during brain maturation is attributed to maturation of the uptake and storage processes which appear to develop earlier in the cerebellum than in the cerebral hemispheres.

Other studies from this laboratory have shown that endogenous levels of NE in the cerebral hemispheres and cerebellum of chicks are high during early embryonic periods, decrease slightly at hatching and do not change significantly during aging (Vernadakis, 1973d). Although the endogenous level of NE is high in the cerebral hemispheres at 10 days of embryonic age, accumulations of ^{3}H-NE is barely detectable in 10-day check embryos (Fig. 6). Thus, it is speculated that the high levels of NE during early embryonic development indicate a functional role other than the classical proposed neurotransmission role. Since functional activity does not develop until 17 days in chick brain (Corner et al., 1967), monoamines have been proposed to be involved in biochemical cellular differentiation (Renson, 1971), and the presence of high levels of NE in the cerebral hemispheres prior to electrical activity is supporting evidence for this role. Also, it is speculated that the relatively high levels of endogenous NE in the aging chick brain, when the uptake and possibly storage processes have declined may reflect that NE has a neurohumoral role in addition to neurotransmission in the aging brain.

The decline in ^{3}H-NE accumulation in the cerebral hemispheres of chicks with aging (Fig. 7) represents neuronal loss as well as changes in uptake and storage processes with age. The continued high ^{3}H-NE accumulation

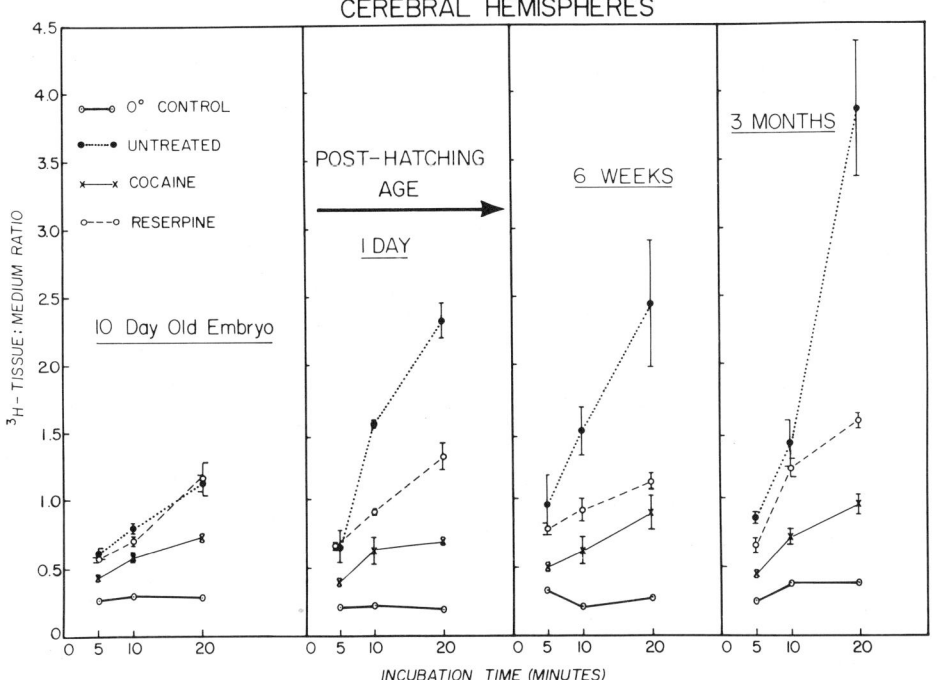

Fig. 6. Uptake of ^{3}H-NE into slices of cerebral hemispheres taken from chick embryos at 10 days of age and from chicks at one day, six weeks, and three months after hatching. The effects of cocaine (10 µg/ml) and reserpine (10^{-6}M) are illustrated. Results represent mean ± S.E.M. for five or six determinations. (From Vernadakis, 1973c.)

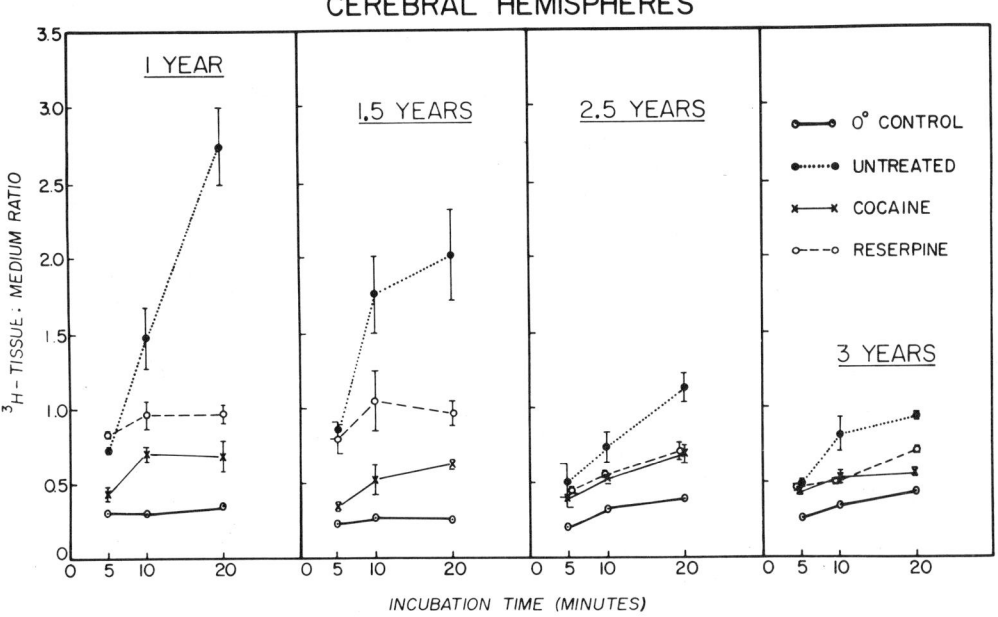

Fig. 7. As in Fig. 6 except that tissues were taken from hens at 1.0, 1.5, 2.5, and 3 years of age. (From Vernadakis, 1973c.)

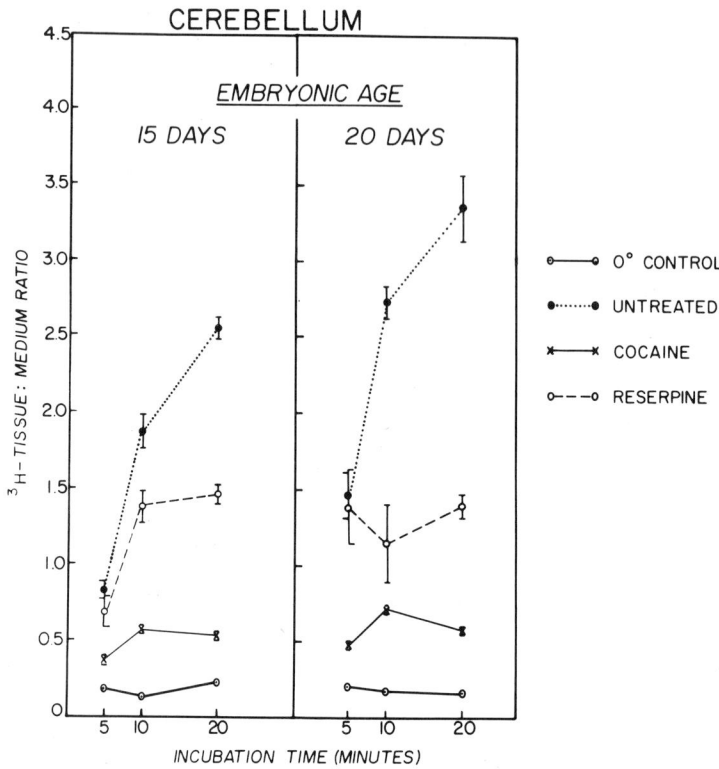

Fig. 8. Uptake of ^{3}H-NE into slices of cerebellum taken from chick embryos at 15 and 20 days of embryonic age. Points as in Fig. 6 (From Vernadakis, 1973c).

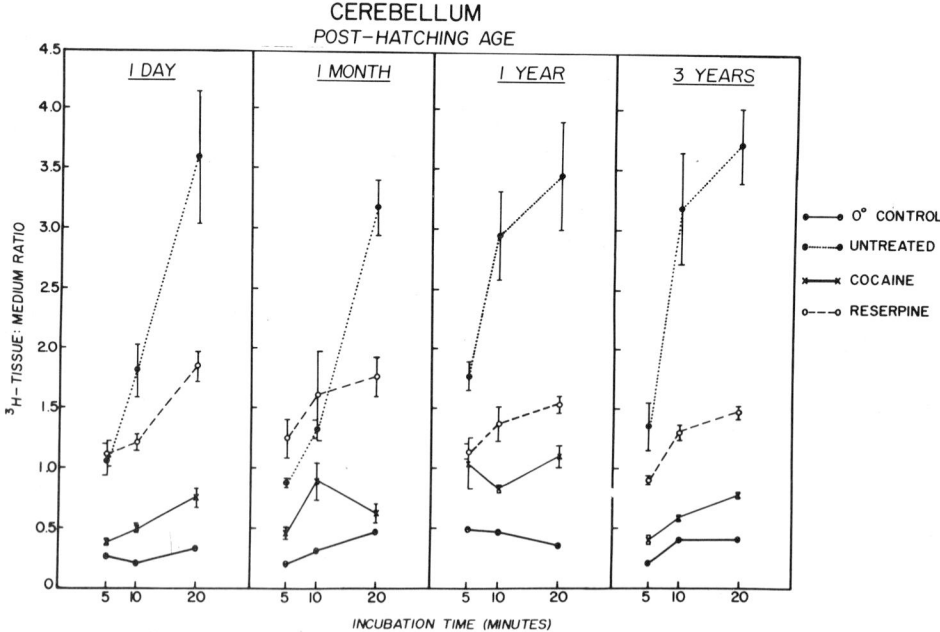

Fig. 9. As in Fig. 8 except that tissues were taken from hens one day after hatching, one month, one year and three years.

in the cerebellum with aging reflects active uptake and storage processes and also maintenance of adrenergic neurons with age. It is of importance to note here that the activity of choline acetyltransferase, the synthesizing enzyme of acetylcholine, markedly declines with age in the cerebral hemispheres but not in the cerebellum of chicks (Vernadakis, 1973d). Thus, both adrenergic and cholinergic function, presumably excitatory, appears intact in the aging cerebellum. However, it is speculated that the cerebellar inhibitory function may decline with age since it is the Purkinje cells that contain δ-aminobutyric acid (Obata and Takeda, 1969; Otsuka et al., 1971), an inhibitory neurotransmitter, that are lost.

Hormonal Effects

Glucocorticoids: Certain steroids have been shown to inhibit extraneuronal uptake of NE, uptake$_2$, in the peripheral nervous system (Ivensen and Salt, 1970). We propose that extraneuronal uptake of NE in the central nervous system may be represented by glial cells. Glial cells accumulate NE at a low affinity (Henn and Hamberger, 1971; Vernadakis and Nidess, 1976) and thus could represent some of the cells which may regulate the extraneuronal concentration of NE.

Recent studies by our laboratory have shown that cortisol at 2.76×10^{-5} M inhibits the uptake of NE 10^{-6} M in cerebellar explants removed from a 16-day-old chick embryo and maintained as organ culture in the presence of cortisol for four hours (Vernadakis, 1974); uptake of NE 10^{-7} M was not affected by cortisol. Thus, it appears from these findings that low-affinity NE uptake in embryonic neural tissue is influenced by cortisol. Since during this period of brain embryonic development in the chick there is active proliferation of glial cells (Vernadakis, 1973a), the inhibition of NE uptake by cortisol has been interpreted to reflect inhibition in glial cells. More recent studies using C-6 glial cells, a rat astrocytoma cell line, have also shown that cortisol, 5.5×10^{-5} M, inhibits uptake of NE (Vernadakis and Nidess, 1976). Thus, the findings of Iversen and Salt (1970) that cortisol inhibits extraneuronal NE uptake in the peripheral nervous system can also be extended in the CNS that cortisol inhibits NE uptake in glial cells.

We have proposed, along with other investigators, that neurotransmitter substances are involved in growth processes during early neural growth and differentiation (Vernadakis and Gibson, 1974; Lauder and Krebs, 1978). Moreover, glial cells play an important role in regulating the metabolic environment of neurons by their responses to neurohumor substances and to hormones. If neurohumors are important during early neural growth and differentiation, including maturation of the synapse, then availability of these substances at the synaptic cleft is of importance. Hormones such as cortisol by inhibiting the uptake of NE into glial cells may enhance the availability of this neurohumor to be taken up by the presynaptic neuron or to act postsynaptically. The following scheme has been proposed for the metabolic fate of NE at the nerve ending (Fig. 10) (Vernadakis et al., 1979a).

Thyroid Hormones. As discussed earlier in this chapter, neonatal treatment of rats with thyroxine (T_4) accelerates the development of synaptic structures such as spine density (Schapiro et al., 1970). The influence therefore of this hormone on neurotransmitter mechanisms continues to be investigated. Geller and associates (see review Geller, 1977) reported changes in in vitro uptake of NE in synaptosomes of cerebral cortex after T_4 and cortisol treatment of rats early postnatally. These authors found that the number of NE binding sites was somewhat lower in animals treated with either T_4 or cortisol than controls at all ages studied. Since these

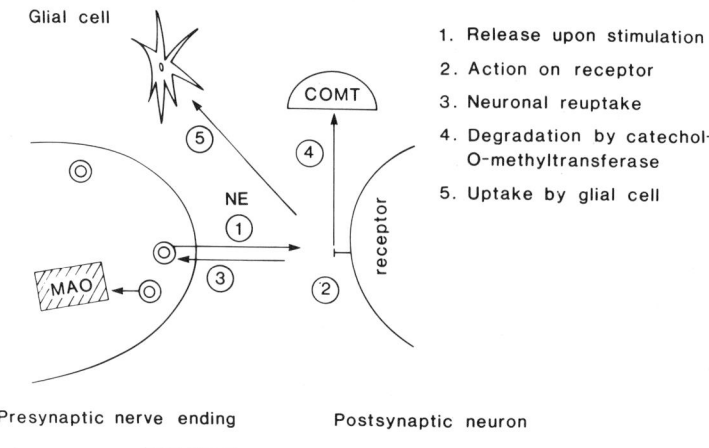

Fig. 10. Schematic representation of the fate of norepinephrine in the central nervous system (From Vernadakis et al., 1979a).

findings were not expected, perhaps binding sites in isolated synaptosomes may not be a good measure of presynaptic development. Moreover, as Geller and associates suggest the binding sites in the treated animals may be a result of different processes in the two hormonal conditions.

Hyperthyroidism appears to increase synthesis and utilization of catecholamine and 5-HT in the maturing brain. Rastogi and Singhal (1976) reported that treatment of neonatal rats with L-triiodothyronine (T_3) increased the endogenous concentration of striatal tyrosine and the activity of tyrosine hydroxylase as well as the levels of dopamine in several brain regions. The concentration of striatal homovanillic acid and 3,4-dihydroxy-phenylacetic acid, the chief metabolites of dopamine, was also increased and the magnitude of change was greater than the rise in dopamine. However, despite increase in the activity of tyrosine hydroxylase and the availability of the substrate tyrosine, the steady-state levels of NE remain unaltered in various regions of brain except in cerebellum. Furthermore, neonatal hyperthyroidism significantly increased the levels of midbrain tryptophan and tryptophan hydroxylase activity but produced no change in 5-HT levels of several discrete brain regions, except hypothalamus and cerebellum where 5-HT level was slightly decreased.

More, recently, Ito et al. (1977) reported changes in monoamine metabolism in mature thyroidectomized rats. The content but not accumulation of 5-HT was significantly decreased in the cerebral hemispheres and mesencephalon; NE content but not accumulation was decreased in the cerebellum and dopamine content was decreased in all brain regions studied whereas synthesis was increased. When animals were made hyperthyroid, the changes in monoamines were again brain area specific: 5-HT content but not accumulation increased in the cerebral hemispheres; NE was not affected; and dopamine accumulation was increased in the mesencephalon.

Gonadal Hormones. Functional, morphological, and biochemical evidence suggests that maturation of the nervous system follows different timetables in the male and female (see reviews in Timiras, 1972). Gonadal hormones in addition to their effects on differentiation of sex organs (discussed in

Chapter 2), may also influence the development of neural pathways by acting on such neurogenic events as cell division and synaptogenesis.

Sex-related differences in neurotransmitter levels have been reported by several investigators. For example, 5-HT and NE levels in whole brain and cerebral cortex are similar before puberty but become higher in female compared to males in adulthood (Hardin, 1973a,b; Kato, 1960). However, whole brain NE levels appear to be already higher in female than males at two days of age (Hardin, 1973a,b) even though these neonatal differences disappear by 5-10 days (Hardin, 1973a,b) to become manifested again after puberty (Hyyppë and Rinne, 1971).

Giulian et al. (1973) reported higher levels of 5-HT in the forebrain-midbrain of female as compared with male rats on 10, 12, and 14 days after birth but not on 2, 4, and 8 or on day 16 after birth. Further experiments using gonadectomy and hormones injection support the conclusion that the transient increases of 5-HT levels in the female brain around 10 days after birth seem to reflect an increase in ovarian activity. Additionally, both testosterone propionate and 5-HT dihydrotestosterone given on day one prevent the normal rise in female brain 5-HT. It is possible that these androgens exert their effects by blocking ovarian function or even by decreasing brain sensitivity to estrogen. In contrast, pharmacological doses of estrogens given on day one resulted in abnormal facilitation of the developing serotonergic system. As discussed later (see neural enzymes), these hormones may in some way alter levels of synthetic or degradative enzymes involved in serotonin metabolism, thus altering the steady state brain level of 5-HT.

The finding that androgens and estrogens have their greatest effects on 5-HT levels only when administered on day one is consistent with the notion of a sensitive period for testosterone action on the developing rat brain, as established by a number of investigators for gonadotropin secretion and female sexual behavior. The serotonin change reported is only one of several perhaps interrelated effects of neonatal estrogens in the brain which were reviewed elsewhere and include altering GABA levels (Hudson et al., 1970), promoting myelination (Curry and Heim, 1966) and enhancing permanently the electrical activity of the brain (Heim and Timiras, 1963).

Neurotransmitter Enzymes

Cholinergic Enzymes. Valcana and associates (see review Valcana, 1971) have examined the changes of acetycholinesterase (AChE) the hydrolysing enzyme of acetylcholine (ACh), and choline acetyltransferase (ChA), the synthesizing enzyme of ACh with age and the influence of thyroid function thereon. Acetylcholinesterase activity showed a marked increase in both cerebral cortex and cerebellum structures from 11 to 29 days of postnatal development in both control and hypothyroid animals. In the cerebral cortex of hypothyroid animals the AChE acitivty was lower than controls at all age intervals studied although not significantly so until 22 days of age. This decrease is in agreement with previous studies by these investigators (Geel and Timiras, 1967). The AChE activity in the cerebellum of hypothyroid animals was similarly lower than controls in early development (11-15 days of age) but not higher in later stages of development (22-29 days of age). In the cerebral cortex of both groups, ChA activity increased from 11 to 22 days. In the cerebellum, however, ChA activity, while high early in development (11 days), decreased from 11 to 15 days with no significant changes thereafter.

The changes in AChE and ChA activity in the cerebral cortex and cerebellum of the developing rat correspond to those described in other species (Hebb, 1956; McCaman and Aprison, 1964; Vernadakis, 1973d). The increased activity of the cerebral cortex may be associated with the formation of enzymes in the cholinergic synapses between the endings of cholinergic neurons in the reticular activating system and of cortical neurons in the cerebrun (Lewis and Shute, 1966; Whittaker, 1969). The general decrease in ChA activity in the developing cerebellum is thought to be due to the development of noncholinergic elements (Hebb, 1956; McCaman and Aprison, 1964). Its increase in hypothyroid animals at 15 days and its subsequent return to normal levels may indicate that maturation of these noncholinergic elements has been delayed as a result of hypothyroidism. Such an interpretation is supported by the work of Balazs et al. (1968) who have shown a reversible decrease in the number of cells (as indicated by DNA content) in the developing cerebullum, with hypothyroidism.

The decrease in AChE and ChA activity in the cerebral cortex of hypothyroid rats represent a depression in the development of cholinergic synapses and/or a depression in the synthesis and activity of these enzymes and therefore, in overall ACh metabolism. Subcellular fractionation showed that the depression in AChE and ChA activities was associated mainly with the synaptosomal fraction (Valcana, 1971). A decrease in the specific activity may indicate a decrease in the synthesis of these proteins relative to the rest of proteins of the fraction. It is also possible that, although the total mass of synaptosomes remains the same, as indicated by the protein content of this fraction, the relative proportion of cholinergic and noncholinergic synaptosomes has been altered by neonatal hypothyroidism, and is manifested by a decrease in AChE and ChA activities.

Enzymes Involved in GABA Metabolism. Growing evidence suggests that GABA functions as an inhibitory neurotransmitter in the central nervous system of vertebrates (Krnjevic, 1974). There is also considerable concern regarding the manner in which extracellular GABA is removed from the synaptic environment and inactivated following performance of its inhibitory function. It is known that free GABA is taken up by a variety of nerve endings and presynaptic inhibitory interneurons (Krnjevic, 1974), but the capacity of glial cells to accumulate GABA is considerably greater (Hamberger and Sellstrom, 1974). Exogenous 3H-GABA is taken up by glioma cells (Schrier and Thompson, 1974), by satellite cells in sensory ganglia (Gottesfeld et al., 1973), by Muller cells in retina (Neal and Hokfelt and Ljungdahl, 1972), and by glial cells in cerebral cortex (Henn and Hamberger, 1971). Hokfelt and Ljungdahl (1973) have reported uptake of 3H-GABA by Bergman glial cells in rat cerebellum. Purkinje cells comprise the only output from the cerebellar cortex. Basket and stellate cells make inhibitory, GABA-ergic synapses on Purkinje cell somata and dendrites (Roberts, 1974), and the sheaths of Bergmann glial cell cytoplasm that enclose Purkinje cells are intimately associated with these synaptic sites.

Hypothyroidism results in permanent reduction in cerebellar levels of δ-aminobutyric acid transaminase (GABA-T), the enzyme which catalyzes the oxidation of GABA to succinic semialdehyde (SSA) (Garcia Argiz et al., 1967). A subsequent investigation by Krawiec et al. (1969) demonstrated that GABA-T concentration could be restored to normal by administering thyroid hormone to hypothyroid rats prior to 15 days of age but not thereafter. Pesetsky and Burkart (1977) recently summarized their histochemical studies on GABA-degradative enzymes in cerebellar cortical cells of animals rendered chronically hypothyroid and in normal controls. GABA-T and succinic semialdehyde dehydrogenase (SSADH) were demonstrable in Purkinje cell bodies and their processes, in the neuropil of the molecular layer and in glomeruli of the granule cell layer. High levels of enzyme activity were also found in Bergmann glial fibers, in perikarya of their cells or origin, as well as in astrocytes dispersed throughout the white matter. GABA-T activity in the cerebellum was detectable as early as postnatal day 3, and confirm the findings of Garcia-Argiz et al. (1967). Until day 6, comparison of control and hypothyroid animals revealed no differences between the two groups. On day 6 and more obviously on day 5, 8 and 10 activity in conttols was more intense than in hypothyroid young. From day 6 through 23 activity of GABA-T and SSADH was consistently more intense in the Purkinje dendrites molecular layer neuropil, glomeruli and glial of controls than in hypothyroid rats. After 26 days of age such differences were no longer apparent in most cells nor in the neuropil of the molecular layer. With respect to the Bergmann glial cells, the situation was quite different. In the hypothyroid rats, as in controls, Bergmann fibers and cells displayed considerable GABA-T SSADH activity between days 10 and 23. At 40 days of age enzyme activity was nearly invisible in the hypothyroid animals.

Enzymes Involved in Monoamine Metabolism. Coyle and Axelrod (1972) studied the development of the activity of tyrosine hydroxylase (TH), the enzyme which catalyses the hydroxylation of dihydroxyphenylalanine (DOPA). The activity was detected in the fetal rat brain at 15 days of gestation and increased four-fold between day 15 and 17. Overall, between the 15th day of gestation and adulthood, TH activity exhibited a 15-fold increase in specific activity.

We have reported the maturational profiles of TH, monoamineoxidase (MAO) and aromatic δ-amino-acid decarboxylase in the cerebral hemispheres of the developing chick embryo (Waymire et al., 1974; Arnold and Vernadakis, 1979). Tyrosine hydroxylase activity in the cerebral hemispheres was undetectable prior to 14 days of embryonic age. From this time until immediately prior to hatching TH activity markedly increased. This increase in TH activity parallels the time course of synaptogenesis in the embryonic chick brain (Corner et al., 1974, 1977). The activities of MAO and aromatic δ-amino acid decarboxylase were measurable at day 14 in all brain regions studied, and progressively during embryonic development. In addition, we have studied the development of TH activity in dissociated cultures of chick embryo brain (Arnold and Vernadakis, 1979). Moreover, several studies by Prasad and us (see review Prasad, 1975; Arnold and Vernadakis, 1979) have shown that TH activity is a characteristic feature of biochemical differentiation in mature neurons.

Various stressors have been found to affect central catecholamine metabolism and TH activity appears to be inducible in sympathetic ganglia and other peripheral organs, Diez et al. (1974) have reported the effects of glucocorticoids and stress on brain TH activity. They found that a postnatal developmental increase in whole brain TH activity was temporally correlated with that of plasma and brain corticosterone levels in mice. Hormone levels and TH activity rose markedly after 12 days of age. Whole brain TH was increased 34% in nine-day-old mice which had received five injections of hydrocortisone acetate (20 mg/kg, i.p., at 12-hr interval).

Tryptophan hydroxylase (TPH) is the first of two enzymes responsible for the biosynthesis of 5-hydroxytryptamine from tryptophan. Azmitia and McEwen (1969) have shown that TPH activity in adult rat midbrain decreased after adrenalectomy and was restored by treatment with corticosterone. Cycloheximide, administered intracisternally, prevented the restoration of the enzyme activity by corticosterone. Cycloheximide administration to adrenalectomized rats resulted in further decrease in the enzyme activity, an indication that the enzyme has a rapid turnover even in the absence of corticosterone.

More recently, Sze et al. (1976) also reported that the normal developmental rise in TPH levels in developing rat brain was blocked by adrenalectomy and this effect was reversed by replacement injections of corticosterone. Adrenalectomy was perfomed at nine days of age and corticosterone (5 mg/kg) was given to rats starting at nine days of age. Brain TPH activity was determined after three days and after six days. Also adrenalectomy prevented the reserpine-induced elevation of TPH activity in the brain stem of adult mice. However, replacement with large doses of corticosterone did not enhance the reserpine effect but merely restored the drug-induced increase of TPH activity to that normally achieved in intact animals. On the basis of these observations it seems that glucocorticoids act in these events mainly as a permissive factor in modulating cellular responses to specific regulatory signals that lead to an elevation of enzymatic activity.

Sex-related differences in the activity of two of the enzymes involved in the metabolism of catecholamines and 5-HT have also been reported. Monoamine oxidase activity in whole-brain is similar in male and female rats during the neonatal period (Hardin, 1973a,b) but becomes higher in whole-brain and hypothalamus (Kamberi and Kobayashi, 1970; Skillen et al., 1961) and lower in the cerebral cortex of adult female rats (Kamberi and Kobayashi, 1970) as compared with males. The acitvity of 5-hydroxytryptophan decarboxylase (5-HTPdC) was also found to be higher in whole-brain of adult female than in male rats (Skillen et al., 1961), a difference present in the newborn as well (Hardin, 1973a,b).

Recently, Vaccari et al. (1977) examined sex-related differences in brain monoamine synthesizing and catabolizing enzymes. They found that TPH was constantly more active in the cerebral cortex and mesodiencephalon of adult females. In general, sex-related differences for catabolizing enzymes were less marked than those for the synthesizing. For example, females showed a higher MAO specific activity than males in the cerebral cortex, starting from the 12th day of age to the 60th day; however, earlier in development (at seven days of age), MAO acitvity was less in females than in males. Even though low MAO activity has been reported in female hypothalamus (Kamberi and Kobayashi, 1970), the majority of reports indicate that MAO activity is higher in whole-brain of adult female than of male rats (Skillen et al., 1961) and humans (Murphy, 1976; Robinson et al., 1971). Differences in the ratio of water to solids per unit of fresh tissue have been reported in the rat brain, depending on sex and stage of development (Soriero and Ford, 1971); inasmuch as MAO is the only particle-bound enzyme among those studied, changes in the density of subcellular components with age and sex could explain the discrepancy between the various reports in the literature and between total and specific activity. Sporadic sex differences were observed in catechol-O-methyl transferase (COMT) activity, such as a 40% higher activity in adult female cerebella and a 50% higher activity in pons medulla of 20-day-old females. In contrast, 60-day-old female adrenals displayed 21% less active COMT than did male adrenals.

It is conceivable that hormonal influences affect differentially monoamine synthesis, transport, storage, and release and the respective sites where these functions occur. In addition, enzymatic maturational timetables in the central nervous system of male and female rats (Gregory, 1975; Soriero and Ford, 1971) may be related to qualitative and quantitative differences in hormonal development in the two sexes (Dorner, 1974; Harris, 1964);

and, even though interactions between sex hormones and sex differentiation on enzymes during development are extremely complex, many aspects of neurotransmitter metabolism appear to be modulated by sex hormones or other hormones (Coppola, 1971; Luine et al., 1975; Terasawa et al., 1975). For example, female sex-hormone levels have been related to MAO acitivty; thus, in females, progesterone levels follow a maturational timetable similar to that of MAO, low levels during the first two weeks of life and high levels later in development (Dohler and Wuttke, 1975). During the estrous cycle, MAO in cerebral cortex (Kamberi and Kobayashi, 1970), adrenals, and other tissues (Holzbauer and Youdim, 1973) is less active when progesterone levels are low. Moreover, the inversion of this tendency of MAO acitvity in the female cerebral cortex at 12 days coincides with very high levels of LH in females (Dohler and Wuttke, 1975) and high levels of LH correspond to high levels of MAO in brain (Kamberi and Kobayashi, 1970; Ramirez and McCann, 1963). Although it cannot be concluded whether these differences reflect the differential effects of androgenic and estrogenic steroids, rather than inborn sex differences, the former possibility seems to be pre-eminent, since in the rat measurable hormone levels early in development have been shown (Dohler and Wuttke, 1975).

GENERAL METABOLIC PROCESSES

Amino-Acid Content

The amino acids provide the substrate for protein synthesis which is fundamental to cellular growth. They may also be concerned with the regulation of protein synthesis. These substances and their metabolites are also involved in many of the specific functions of the central nervous system. Several amino acids are themselves neurotransmitters; for example, the concentration of the substance tryptophan is thought to be a determining factor in the formation of serotonin; tyrosine is a precursor of catecholamines; and methionine and choline may serve as methyl donors for the synthesis of catecholamines.

In early studies we have investigated changes in several amino acids in the developing rat brain (Vernadakis and Woodbury, 1962). In general, the concentrations of the amino acids, except for threonine, progressively increase during maturation. Hudson et al. (1970) have studied in more detail regional changes in amino acid concentration in the developing rat brain. The normal developmental pattern showed glutamic acid to be present in highest concentration in the hemispheres and lowest in the pons-medulla, with no significant change in either region throughout the 21-day experimental period. In the cerebellum and mesodiencephalon, glutamic acid increased significantly between 12 and 16 days and continued to increase to 21 days, but the concentration was still lower than in the cerebral hemispheres.

Aspartic acid was highest in the hemispheres and lowest in the cerebellum at 12 days (Hudson et al., 1970). However, while the concentration of aspartic acid remained unchanged in the hemispheres, cerebellar concentrations continued to increase significantly up to 21 days. In the mesodiencephalon and pons-medulla, aspartic acid increased significantly between 12 and 16 days of age. By 21 days, differences among the four structures disappeared.

GABA concentrations in the hemispheres and pons-medulla were constant while that of the mesodiencephalon showed a barely significant increase between 12 and 16 days (Hudson et al., 1970). The greater proportional change of GABA concentration occurred in the cerebellum between 12 and

16 days of age, coinciding with the period of most rapid growth of this structure. Regional comparison of GABA concentration showed it to be highest in the mosodiencephalon and lowest in the cerebellum.

In contrast to the general increase observed with development in the foregoing amino acids, glutamine and alanine tended to decrease (Hudson et al., 1970). Regional comparisons of these two amino acids indicate that both decreased in the hemispheres and pons-medulla from 12 to 16 days, and that, while glutamine also decreased in the cerebellum, alanine decreased in the mesodiencephalon during this time. Glutamine was highest in the cerebellum whereas alanine was highest in the hemispheres; both were lowest in the mesodiencephalon. Whereas the glutamate group of amino acids showed variability in its pattern during development, the remaining amino acids studied essentially demonstrated a decrease with age, most striking in taurine and phosphoethanolamine. Some variability appeared in glycine and cystathionine with respect to age and brain area. Glycine decreased during the 12-16-day period in the hemispheres and mesodiencephalon only. Cystathionine concentration, on the other hand, decreased in the hemispheres throughout the experimental period and in the cerebellum from the 16th to 21st day. In contrast, in the mesodiencephalon, cystathionine increased from the 12th to the 16th day and showed no changes in the pons-medulla.

The changes in amino acid concentrations with development are of interest with respect to the maturational timetables specific to each brain structure studied. The levels and constancy of glutamic acid, aspartic acid and GABA in the hemispheres of control rats after 12 days may be viewed as an index of the degree of maturity of this structure, an interpretation that is consistent with the morphological findings in the rat cerebral cortex at this time. It is of interest that the attainment of constant levels of various amino acids also seems to be strain specific. For example, in the Long-Evans rat glutamic acid levels in the hemispheres increase by one-third from 9 to 12 days of age and thereafter the level appears to be constant whereas in the Sprague-Dawley rat this increase occurs later and relatively constant levels are not reached until 17 days (Hudson et al., 1970). In the slower-maturing cerebellum for example, these amino acids continued to increase until the 21st day. The mesodiencephalon little studied because of its complexity, has been assumed to be relatively mature by the 12th postnatal day; however, the increase in amino acid concentrations in this structure, from the 12th to the 16th day, would indicate that the structure may be still developing.

Estradiol treatment during early postnatal development (from the 6th to the 10th day after birth) significantly accelerated the maturation increase of excitatory amino acid group, glutamate and aspartic acid, in the slower developing cerebellum. Moreover, estradiol treatment accelerated the maturation decrease of the inhibitory amino acid group, taurine, glycine and cystathionine. It is suggested that the accelerated decline in these amino acids amy explain the increased sensitivity of estradiol-treated rats to electrical stimulation (Heim and Timiras, 1963).

Bachmann et al. (1975) have reported changes in amino acid concentrations during normal fetal development of primates and alterations due to prenatal hormonal and nutritional imbalance. In normal animals, the concentrations of most free amino acids were higher in the cerebrum than in the cerebellum and both had levels higher than in the plasma. There was some variation in the concentrations of amino acids in the cerebrum and cerebellum. In hypothyroid fetuses most of the amino acids remained unchanged in the cerebrum. However, the levels per unit tissue of GABA, ornithine, lysine, histidine, and cystathionine, showed a trend toward an increase in the experimental group. There was a trend toward a reduction of nonessential amino acids. Both aspartic and glutamic acid showed some reduction in cerebrum and cerebellum. In both the cerebrum and cerebellum there was a slight but significant increase in tryptophan and in the cerebellum there was in addition an increase in serine. The significance of the change in tryptophan is not known but the synthesis of serotinin in the brain may well be substrate-dependent.

Protein Synthesis

During brain development the synthesis of protein and nucleic acid declines, however not as rapidly as originally was accepted. In studies of the fate ^{14}C-leucine in brain tissue during the period from birth to 35 days, total ^{14}C and ^{14}C combined in protein and specific radioactivity free leucine are low up to 9 days of age, rise up to 20 days then decline as the free leucine concentration declines reflecting free leucine pool (see review Balazs and Richter, 1973). Oja (1967) has examined the fluxes of ^{3}H-tyrosine between blood and brain, as well as between the free amino acid pool and proteins in both directions at different stages of brain development. Protein synthesis does not change much during the first 30 days after birth, but it is highly relative to that in adults. On the other hand, the rate of protein catabolism increases during the first week of life, and the rate is lower in the adult than in the brain of 30-day-old rats. These results are in variance with some observations obtained with brain preparations in vitro; for example a striking fall has been reported from birth to the end of the 2nd week in the rate of both protein and, to a smaller extent, RNA synthesis in mouse brain cell suspensions and rat cortex slices (Orrego, 1967; Johnson, 1971). There are many controversial observations concerning the developmental changes in the activity of cerebral protein synthesizing systems which may relate to factors such as the relative purity and stability of the isolated preparations (See review Balazs and Richter, 1973).

Thyroid Hormones. Early studies (see Gelber et al., 1964) have shown that amino acid incorporation into protein is approximately 3-4 times more rapid in cell-free preparations from infant rat brain than in preparations from adult rat brain. The addition of L-thyroxine to the reaction mixture inhibits amino acid incorporation into protein in adult brain preparations, regardless of the oxidizable substrate used. In contrast, in infant brain L-thyroxine stimulates amino acid incorporation when α-hydrorybutyrate is the oxidiazable substrate; with d-oxoglutarate, the effects of thyroxine in infant brain preparations are similar to those in adult brain. Since L-thyroxine stimulates amino acid incorporation into protein in both adult and infant liver these authors suggest that these differences in the effects of thyroxine on amino acid incorporation into protein may explain the differences in the responses of metabolic rate of adult and infant brain to thyroxine.

Thyroid deficiency in early life leads to a decrease in the rate of protein synthesis in the brain (Geel et al., 1967). Following the injection of uniformly labeled ^{14}C-leucine in proportion to body weight, the rate of disappearance of the label from plasma and cerebral free amino acid pool is depressed in thyroid-deficient animals. The specific activity of cerebral protein-bound leucine expressed as a function of the specific acitvity of the label in the free amino acid pool was significantly reduced in hypothyroid rats, indicating a decrease in protein synthesis. Geel et al. (1967)

have inferred that changes in cerebral Na^+, K^+, and Cl^- ions in hypothyroid rats may partially mediate the effects on brain protein synthesis. At 22 days of age the percent water in the cerebral cortex was significantly higher in the hypothyroid group indicating a delay in solids differentiation. The K^+ concentration of the cerebral cortex was significantly lower and the Na^+ concentration significantly higher in the hypothyroid animals compared to the other groups. The Cl^- concentration on the other hand was the same for all groups. No differences were observed in cerebral water and ionic concentrations between the hypothyroid rats receiving physiological levels of L-thyroxine from the sixth day of age and the normal control animals. It has been inferred from the changes in cerebral Na^+, K^+, and Cl^- ions in hypothyroid rats that the $Na+-K+$ pump is altered and the transport of amino acids is adversely affected. Thyroxine therapy reverses the changes in cerebral protein synthesis and ionic concentration. These findings suggest that the relationship between thyroid hormone action and brain protein synthesis is partially mediated through the effect of this hormone on ionic distribution and amino acid transport.

In a later study, Valcana and associates used the in vitro brain-slice system to further examine aspects of amino acid transport and incorporation into protein (see review Valcana and Eberhardt, 1977). The greatest difficulty in determining protein synthesis has been a lack of knowledge of the specific activity and size of the amino acid precursor pool and how these may be altered by hypothyroidism. Changes in plasma amino acids, alterations in cerebral blood flow, and their consequent differential equilibration in the brain amino acid pools are factors which have to be considered in the evaluation of protein synthesis. The amino acid incorporation into protein was higher in the brain slices obtained from hypothyroid animals than those from controls. This was observed in protein of all subcellular compartments, at early (13 days) as well as at later (42 days) stages of hypothyroidism (Table 3). This effect was noted when leucine, tyrosine, or proline were used as precursors. These findings agree with results obtained by others in vivo in which both the activity incorporated into cerebral protein and the activity of the acid-soluble fraction are higher in thypothyroid tissue (see review Valcana and Eberhardt, 1977). In view of the overally hypoplasia that characterizes the developing hypothyroid brain and the depression of the rate of protein synthesis in the hypothyroid liver, the observed rate of incorporation of amino acids into brain protein could not reflect a higher rate of protein systhesis, but, rather some other changes, such as a difference in transport or changes in the amino amino acid pool participating in protein synthesis (Valcana and Eberhard, 1977). Cycloheximide depressed the incorporation of amino acid into both the protein and the acid soluble fraction in control tissue (Table 4). In tissue slices from hypothyroid animals, however, the incorporation into protein was depressed, but the incorporation into acid soluble fraction was enhanced. Thus, the higher activity in this fraction in hypothyroidism in the absence of cycloheximide does not represent a response to a higher rate of synthesis, but rather may reflect enhanced transport. Unlike leucine, the uptake of both glutamic and lysine into the acid-soluble fraction was not significantly higher. It sould be concluded that this higher incorporation of glutamic acid into protein arises from the known depression in the glutamic acid concentration with hypothyroidism. Hypothyroidism induced a marked depression in the free concentration of leucine, tyrosine, and lysine in the cerebral cortex of 29-day-old rats. Leucine was depressed by 65%, lysine by 50%, and tyrosine was not detectable in the hypothyroid cerebral cortical tissue.

Table 3

IN VITRO AMINO ACID INCORPORATION INTO PROTEIN OF
VARIOUS SUBCELLULAR FRACTIONS FROM
CONTROL AND HYPOTHYROID RAT BRAIN

Precursor	Age (days)	Specific Activity (cpm/mg protein)							
		Homogenate		Cytoplasm		Microsomes		Myelin	
		C	H	C	H	C	H	C	H
Leucine	13	8,561	10,528 (23%)	22,521	20,584	12,985	16,269 (25%)	9,143	13,306 (45%)
	43	1,102	1,347 (13%)	2,887	3,876 (34%)	6,740	7,460 (11%)	---	----
Tyrosine	13	4,609	6,064 (31%)	9,476	8,691	9,680	11,363 (17%)	4,705	5,553 (18%)
	43	1,108	3,465 (212%)	4,345	6,035 (39%)	8,080	11,580 (43%)	---	----
Proline	13	3,073	4,643 (51%)	6,183	7,597 (23%)	12,079	15,828 (31%)	4,409	8,287 (88%)
	43	399	1,890 (374%)	1,916	2,867 (49%)	10,000	11,720 (17%)	---	----

Brain tissue slices, 0.5 mm thick coronal sections, from whole brain of control (C) and hypothyroid (H) rats were incubated at 37°C for 2 hr in Krebs-Ringer medium (100 mg/ml). Final amino acid concentrations and specific activity in incubation medium were leucine: 1.6 μm, 312 μCi/μM; tyrosine: 12 μM, 460μCi/μM; proline: 1.0 μM, 260 μCi/μM. Numbers represent means from two determinations, 4 animals per determination. Numbers in parentheses represent percent change from control values. (From Valcana and Eberhardt, 1977).

These foregoing findings show that determination of the rate of protein synthesis in brain tissue in vivo and in vitro is complicated by such factors as changes in amino acid pools, compartmentation, transport and protein degradation. The findings by Valcana and associates indicate that such differences do exist between control and hypothyroid tissue.

More recently Verity et al. (1976, 1977) investigated the effects of thyroid hormone on amino acid uptake in synpatosomes. Current evidence indicates that two, presumably independent systems, are responsible for protein synthesis in synaptosome fractions. The first is a mitochondrial system sensitive to chloramphenicol and the second is a classical eukaryotic ribosomal system sensitive to cycloheximide. This second system is distributed between synaptosomal particiles and small resiculated contaminating particles of unknown origin. Thyroid hormone inhibits the incorporation of ^{14}C-leucine into TCA-precipitable fraction of cerebral synaptosomes and this effect appears to be independent of age.

Sex Hormones. Cavallotti and Bisanti (1972) have reported changes in protein content in rats after chronic postnatal administration of estradiol, ovariectomy at the 14th day of age and replacement therapy with estradiol.

Table 4

EFFECTS OF CYCLOHEXIMIDE ON THE IN VITRO INCORPORATION
OF AMINO ACIDS INTO PROTEIN BY CONTROL
AND HYPOTHYROID BRAIN TISSUE

Precursor ± inhibitor	Specific activity (cpm/mg protein)		Acid soluble (cpm/mg wet tissue)	
	Control	Hypothyroid	Control	Hypothyroid
Leucine (−)	1,102	1,347	1,382	1,480
Leucine (+)	751 (32%)	174 (84%)	1,187	1,749
Proline (−)	399	1,890	2,144	1,711
Proline (+)	59 (85%)	153 (92%)	1,816	2,144
Leucine and Tyrosine (−)	2,870	6,472	2,125	2,849
Leucine and Tyrosine (+)	1,097 (62%)	1,490 (77%)	2,859	3,113

Whole brain from 35-42-day-old control (C) and hypothyroid (H) rats was sliced into 9.5-mm coronal sections and incubated (100 mg/ml) for 2 hr under the conditions described in Table 1. Numbers represent means of 2-4 determinations. Numbers in parentheses indicate the percent inhibition by cycloneximide (40 mg/mi). (From Valcana and Eberhardt, 1977).

In estradiol-treated animals the protein content of cerebral hemispheres was slightly but significantly increased at 7 and 14 days of age whereas in ovariectomized rats it was reduced. Replacement therapy with estradiol restored normal content of proteins in brain tissue. The accelerated increase in protein-content in the estradiol-treated animals may reflect the general increase in brain weight observed by these investigators. It is of interest that these biochemical findings correlate with the increased brain electrical activity observed by Heim and Timiras (1963) and Bisanti and Cavallotti (1972).

Kartzinel et al. (1971) reported neonatal castration inhibits amino acid accumulation in the rat brain at five weeks. Other workers have found that castration decreases and testosterone propionate administration increases the accumulation of several amino acids in seminal vesicle, ventral prostate, kidney, liver, and skeletal muscle in both in vivo and in vitro preparations (Frieden, 1964; Williams-Ashman, 1965). This however, appears to be the first report of the effect of castration on protein metabolism in the central nervous system. It is reasonable to assume that androgen withdrawal is responsible for the inhibition of amino acid accumulation in the brain. It also seems likely that the sex difference in ^{3}H-lysine accumulation in the brain is androgen-dependent, since it is eliminated by castration. Further experimentation is needed, however, to prove an androgen effect on brain protein synthesis. While the significance of this finding is not clear, it is possible that the observed depression in protein synthesis may be in part responsible for the behavioral changes produced by neonatal castration (Chapter 8).

An extensive series of autoradiographic studies by Litteria and associates (Litteria, 1973a,b,c; 1977a,b,c; Litteria and Thorner, 1974a,b,c,d) has shown that the administration of estradiol to neonatal rats during the critical period of brain differentiation is followed postpuberally by an inhibition in the incorporation of ^{3}H-lysine into proteins of specific neurons of the hypothalamus, cerebellum and cerebral cortex. In addition, these investigators have determined that neonatally administered estradiol has long-term effects on the incorporation of ^{3}H-lysine into neurons of selected limbic and paralimbic structures. They propose several possible mechanisms to account for the decreased incorporation of ^{3}H-lysine into specific neurons of the limbic and paralimbic structures examined. (1) The most probable explanation is the immediate occurrence of permanent metabolic changes in these neurons of the undifferentiated brain following the administration of estradiol. The demonstration of putative estrogen receptors in limbic tissue from neonatal rat brain provides a possible biochemical basis for these results. (2) Decreased incorporation into proteins could be secondary to other undefined biochemical changes induced in the steroid responsive neurons. (3) Primary steroid induced changes in "extralimbic" brain structures could secondarily influence the metabolism of selected limbic and paralimbic areas via multiple synaptic interconnections. (4) Different hormonal profiles between the postpuberal control and estrogenized rats could be responsible for inducing (directly or indirectly) metabolic changes in certain neurons.

Nucleic Acid Metabolism

Thyroid Hormones. Timiras and associates (Timiras, 1972) and Balazs and associates (Balazs et al., 1968) have extensively investigated in the rat changes in nucleic acid content with age and the role of thyroid function there on. The content of both ribonucleic acid (RNA) and deoxyribonucleic acid (DNA) is high during early brain development but with continued growth, it gradually decreases, DNA at a more rapid rate than RNA. In hypothyroid rats DNA content is significantly higher and RNA content is lower at 22 days (Geel and Timiras, 1967) and 35 days (Balazs et al., 1968) of age as compared with control.

The major defects in the composition of the brain attributable to neonatal thyroidectomy have proven to be manifested mainly after the first two weeks of postnatal life. As also discussed earlier, Balazs and associates have shown that, during the period studied (up to 35 days after birth), thyroid deficiency affects the size of the cells but not their number. The majority of the new cells in the brain after birth are probably glial cells and the DNA contents of the cerebrum and cerebellum indicate their formation is reduced up to the age of 35 days by which time these structures may be considered already completely mature. The increase of the DNA content/unit wt. and the unaltered content of DNA per organ as a result of thyroidectomy indicate a reduction in the size of the cells of both cerebrum and cerebellum.

Gourdon et al. (1973) further explored the effects of neonatal hypothyroidism on the development of the protein and nucleic acid contents of the cerebellum in young rats ranging in age from 6 to 35 days. They also found that the cerebella of the hyperthyroid and normal animals contain about the same number of cells, but the organ from the hyperthyroid animal is on the average less rich in proteins.

Although the changes in rat cerebellar DNA content in hyperthyroid states are now well documented, there is little known concerning the mechanisms involved in such manipulation of the developmental process by hormones. Such changes in cerebellar DNA content have led to studies involving relationships between thyroxine, DNA, and activities of enzymes involved in pyrimidine biosynthesis during cerebellar development. Enzymes

from the de novo pyrimidine biosynthetic pathway have been shown to be
potential markers for cerebellar cell replication (Weichsel et al., 1972)
and more recently thymidine kinase, a salvage pathway enzyme for
pyrimidine biosynthesis, has been shown to peak in activity at age 6 days
(Sung, 1971; Yamagami et al., 1972) just prior to the period of most rapid
cerebellar cell replication as determined by incorporation of radioactivity
labeled thymidine (Altman, 1966; Altman, 1969). Because of the inhibition
of that incorporation by thyroxine (Nicholson and Altman, 1972b),
Weichsel and associates (1974, 1976) examined whether a relationship
might exist between thyroxine, DNA synthesis, and the activity of thymidine
kinase during cerebellar development. In hypothyroid rats the activity
of cerebellar thymidine kinase was suppressed at ages two and five days
and was in excess of control values on days 15 and 22, thus resulting
in a delay in the developmental spectrum for thymidine kinase, and
extending the time span of activity beyond that of controls. The profound
delay in maximal activity and subsequent prolonged activity of cerebellar
thymidine kinase correlates with the persistence of the fetal external
granular layer of the cerebellum in the hypothyroid state (Altman, 1966,
1969; Gourdon et al., 1973; Hamburgh et al., 1971; Nicholson and Altman,
1972a) and suggests that there may be an important relationship between
the activity of thymidine kinase and the synthesis of cerebellar DNA.
In thyroxine-treated rat pups an acceleration of cerebellar DNA synthesis
from age two to six days of life was noted with a concomitant acceleration
in the activity of thymidine kinase which became measurable one day prior
to the measurable increase in DNA synthesis. Thus, the delay in the
time course of cerebellar thymidine kinase activity in the hypothyroid state
further supports the possibility that thymidine kinase may be an essential
regulatory enzyme in DNA biosynthesis during cerebellar development
(Weichsel, 1974).

Glucocorticoids. Studies by Howard (1965, 1973a,b) have shown that
intensive treatment with corticoserone in infancy interferes with brain
growth as well as with body growth in mice. After termination of the
corticosterone treatment, growth is resumed, but both body size and the
weight and DNA content of the cerebra remain below the values in control.
A similar interference with the accumulation of DNA in the brain follows
the injection of cortisol acetate during the first 5 days after birth in rats
(Balazs and Cotterrell, 1972; Cotterrell et al., 1972). Incorporation of
$(2-^{14}C)$-thymidine into brain DNA was strongly inhibited by the cortisol
treatment. Balazs and associates concluded that cortisol acted primarily
by inhibiting DNA synthesis and cell formation, rather than by causing
cell destruction. However, thymidine uptake rates compared with DNA
accumulation rates suggested a considerable amount of cell death in postnatal cerebra of both control and treated rats.

The occurrence of cell death in the nervous system during normal
development has been recognized by histologists (Glucksmann, 1965;
Prestige, 1970). The reason for this cell death is not well understood:
neurons that fail to make adequate connections might be discarded, and
perhaps glial cells that have functioned as developmental scaffolding may
be eliminated. During embryonic day 14, cell division in the cerebrum
is considered mainly, though not exclusively, in neuronal precursors
(Altman, 1970; Angevine, 1970); hence, a loss of thymidine label
subsequently would suggest loss of neurons. On the other hand, after
birth in rodents glial cell formation greatly predominates over neuronal
in the cerebrum, so that postnatal inhibition of DNA synthesis by corticosterone would involve, on a quantitative basis, chiefly the glia, although

relatively small numbers of neurons of the area dentate might also be suppressed, as well as olfactory neurons (Altman, 1970).

To obtain further evidence regarding the mechanisms that are responsible for the lasting deficit in cerebral DNA after corticosterone treatment in infancy in rodents, Howard and Benjamins (1975) examined whether corticosterone causes loss of cells that have been formed prior to the initiation of treatment. Their findings suggest that no more than a small fraction of the DNA deficit after corticosterone treatment may be due to loss of preformed cells, and that most of the deficit is apparently due to suppression of postnatal DNA synthesis by the steroid as also shown by Cotterrell et al. (1972) with cortisol.

As is also discussed later in this chapter, the biochemical effects of corticosterone on the developing brain are not without functional consequences. In studies of operant behavior of mice given the steroid at two days, and examined as adults, it has been found that the treated mice overreact, in comparison with the controls, in several different test situations. They also show a greater difficulty in making an adaptive transition from a schedule requiring rapid responding to obtain food to a schedule requiring a slow rate of responding rats given steroid have difficulty in acquiring a conditioned active avoidance response, relative to controls (Olton et al., 1974) and show what appears to be an emotionally accentuated polydipsia (Howard et al., 1974). Both mice and rats appear to act as if they have some difficulty in processing new information and are less able to direct emotion into adaptive channels that are the controls.

Enzymes Involved in General Metabolic Processes

Early studies by Hamburgh and Flexner (1957) showed that the activity of succinic dehydrogenase in the frontal cortex is at a relatively low level until the 10th day after birth, when it begins to rapidly increase to a level characteristic of more mature animals. Much the same pattern is followed by cytochrome oxidase, the 10th day again marking the time when enzyme activity begins sharply to increase. The activity of aldolase follows this pattern less clearly, since there is a substantial rate of increase in activity from birth to the 10th day. The rate of change from the 10th to the 15th day, however, is approximately five time the maximum rate observed before the 10th day.

Thyroidectomy at birth leads to a statistically significant decrease of succinic dehydrogenase activity. Hypothyroidism was without significant effect on the activities of both cytochrome oxidase and aldolase. It is uncertain whether absence of thyroid hormone leads to a decrease in rate of synthesis of succinic dehydrogenase or whether it simply lowers the activity of a numerically normal population of enzyme molecules. Drabkin (1950) found that the concentration of cytochrome c in several tissues is reduced in hypothyroidism and increased in hyperthyroidism. The activity of succinoxidase and cytochrome oxidase has also been shown to increase in hyperthyroidism (Cohen and Gerard 1937; Tipton, 1950). Unlike cytochrome c, this effect is not necessarily the result of an increase in concentration of the enzymes, but may be due to an increase in their activities mediated by thyroid hormone, since it has been found that the activity of succinoxidase in an in vitro system is increased by addition of thyroxine (Gemmill, 1952).

The administration of thyroxine to Rana pipiens larvae results in precocious maturation of many elements of the central nervous system. Pesetsky (1965) examined histochemically the activities of dinucleotide phosphate diaphorase (NADPH or TPNH diaphorase) and glucose 6-phosphate dehydrogenase (G6-PD) in the brain of thyroxine-stimulated Rana pipiens larvae. Fifteen animals, ranging from Stage III to VII were treated with L-thyroxine, 222 µg/l of spring water for 4-7 days. The activities of NADPH diaphorase and G6-PD were more intense in thyroxine-treated larvae than in untreated animals. The microanatomical localization of activity was identical for both enzymes. Increased activity was evident in most tissue elements of the hindbrain, but the greatest changes occurred in the choroid plexus, the ependyma, and in the many small neurons of the mantle. In general, the intensity of histochemical staining was greater in small neurons than in large motor cells, and activity in the giant Mauthner neuron is especially diffuse. It proved difficult to ascertain just how much, if any, of the stain was deposited in glial cells or in pericellular neuropil. The most prominent thyroxine-stimulated increase in enzyme activity was found in the central ependymal cells. This increase is especially interesting. In recent years, the interests of a number of investigators has been drawn to the ependymal lining of the brain. Ving et al. (1962) and Leveque and Stern (1964) have suggested that some neuroendocrine cells discharge their products into the ventricular fluid rather than directly into the blood.

Ornithine decarboxylase (ODC) catalyzes the first step in polyamine biosynthesis, which is the conversion of L-ornithine to putrescine (Tabor and Tabor, 1964). The demonstration that the activity of this enzyme changes markedly during perinatal development of the rat brain (Anderson and Schanberg, 1972; Parker et al., 1978) and that these changes correlate with those of the polyamine spermidine (Pearce and Schanberg, 1966) is in agreement with considerable evidence showing a dramatic increase in the activity of the enzyme very early in the transition of tissues from a nongrowing to a growing state (Kremzner, 1970). Much evidence in the literature suggests that polyamines influence nucleic acid metabolism (Tabor and Tabor, 1964), and that they accumulate in parallel with RNA in many mammalian systems (Raina and Janne, 1970). In normally developing rat brain changes in spermidine concentration with time are related directly to net RNA and DNA synthesis (Kremzner, 1970).

Anderson and Schanberg (2975) examined the effects of thyroxine and cortisol on brain ODC in the developing rat and correlated the changes with swimming behavior. In control rats, ODC activity was high initially and declined rapidly to low adult values by day 11. In general, thyroxine was effective in all brain regions studied if given during the first 10 postnatal days. The overall effect of thyroxine treatment was to compress the time course in which ODC activity passed through its ontogenetic pattern; both the peak in activity and the drop in the characteristic low adult activity level occurred earlier and were accelerated. The overall effect of cortisol treatment was to delay and spread out the normal ontogenetic pattern of ODC activity.

As cell proliferation is high in the cerebral hemispheres at birth and low in the cerebellum, it is reasonable that early treatment with cortisol would have a more pronounced suppressive effect on the forebrain. Whereas the initial action of thyroxine to increase ODC is somewhat delayed, it is not surprising that the cerebellum, which proliferates and differentiates primarily after birth, is more sensitive to the action of thyroxine than the cerebral hemispheres which already have undergone a major portion of their development before birth. Since cell proliferation in the brain stem at this time is low, these hormones have a correspondingly less significant effect on this brain region. The actions of the hormones are not necessarily specific to ODC. Berlin and Schimke (1965) have shown that, in a situation where an inducing agent stimulates general protein synthesis, enzymes with a fast turnover increase much more rapidly than those with a slow turnover. Hence, since ODC has an extremely short half-life, a general stimulation of all protein synthesis could lead to a rapid increase in ODC. However, as indicated by Tabor and Tabor (1972), although such an increase in an enzyme may be nonspecific, it still could be physiologically important. The fast turnover of ODC would enable it to increase quickly preparatory to rapid growth without the need of specific induction.

McEwen and associates have extensively investigated the effects of estrogens on brain metabolic enzymes (Luine et al., 1974). Specifically, they have been interested in determining whether estrogen interaction with putative hormone receptor sites within the pituitary, hypothalamus, preoptic area, and amygdala can be causally related to changes in levels of specific enzymes in those tissues as appears to be the case in the uterus (Baquer and McLean, 1972). Enzyme activities were examined in ovariectomized rats (70-80 days of age) and in ovariectomized rats that received injections of estradiol daily for one week. Pituitary showed estrogen-dependent increases in glucose-6-phosphate dehydrogenase (G6PHD), 6-phosphogluconate dehydrogenase (6-PGDH) and lactic dehydrogenase (LDH), and no change in $NADP^+$-dependent isocitric dehydrogenase (ICDH), $NADP^+$-dependent malic dehydrogenase (MDH) or hexokinase (HK). MDH and ICDH were elevated in whole hypothalamus. Enzyme activities did not change significantly in whole amygdala, cerebral cortex, or hippocampus. Increases in enzyme activities were observed in subregions of the preoptic area, hypothalamus and amygdala, and were related to the total in vivo dose of estradiol given. Since adrenalectomy did not prevent increases in enzyme activity by estrogen in the whole hypothalamus and pituitary, the possibility that endogenous adrenal hormones mediating the estrogen effects is excluded.

The foregoing findings show that estradiol may act directly to cause changes in activities of some brain and pituitary enzymes associated with oxidative metabolism of glucose. Moreover, the findings that the response to estrogen was greater in the pituitary, a brain area with a large amount of putative estrogen receptor sites, suggests a hormone-receptor interaction. This hormone-receptor interaction in enzyme activity is demonstrated by another study by Luine et al. (1975). They found that testosterone did not increase the activity of these enzymes in the brain or pituitary of female rats. Testosterone does not bind to putative estradiol receptors. In addition the estrogen antagonist MER-25, which binds to putative estrogen receptors (Clark et al., 1973; Lerner et al., 1958) and blocks estrogen-dependent female sex behavior (Arai and Gorski, 1968; Bickel et al., 1972) blocks increases in pituitary G6PHD and 6PGDH found when estrogen is given alone. In contrast to the female, testosterone was effective in males and elevated G6PHD and ICDH in the hypothalamus; MDH in the amygdala; and G6PDH, 6PGDH, LDH, MDH, and ICDH in the pituitary. These findings further support the view that hormonal effects on brain enzyme levels provide a biochemical endpoint for defining the consequences of sexual differentiation on the functioning of the brain and pituitary.

Ionic Composition

One parameter of special significance to normal CNS growth and function is that of its ionic composition. Maturational disturbances in the levels of various ions would be expected to have their repercussions on the development of specialized CNS functions.

During normal maturation Cl, Na and Ca concentrations in plasma increase whereas K concentration decreases (Vernadakis and Woodbury, 1962;

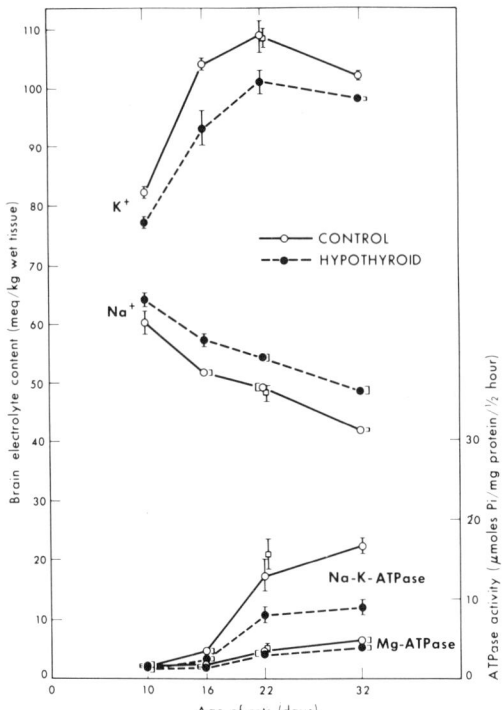

Fig. 11. Effects of neonatal hypothyroidism on Na and K contents, Mg-ATPase and Na+-K+ ATPase activity in the DOC-extractable fraction (pH 6.5) from the cerebral cortex of the rat brain during development. Na-K ATPase activity is the difference between the total ATPase activity determined in the presence of NA, K, and Mg ions and that of Mg-ATPase determined in the presence of Mg ions only. The Mg-ATPase activity was determined in a 2-cc volume of incubation mixture of the following composition: 100 mM-histidine, pH 7.2, 3 mM-tris-ATP, and 3 mM-Mg; the incubation mixture for determination of total ATPase activity contained in addition 100 mM-Na and 20 mK-K. Incubation 30 min at 37°. For ATPase activity, each point represents mean ± S.E. of duplicate determinations from 16 enzyme preparations and for Na and K contents, each point represents mean ± S.E. of eight or more samples. The squares in all parameters at day 22 represent mean values obtained from hypothyroid animals treated daily with thyroxine (day 6 to day 22, 10 μg/100 g body weight). (From Valcana and Timiras (1969).

Valcana and Timiras, 1969; Valcana and Timiras, 1971; review Valcana, 1974). In the cerebral cortex K and Mg content increase and Na and Cl content decrease with development (Vernadakis and Woodbury, 1962; review Valcana, 1974). These ionic changes have been correlated with changes in the specific activity Mg-ATPase and Na+-K+-activated ATPase which increases with age and the latter shows a significant increase during that period of brain development characterized by marked redistribution of Na and K content (Fig. 9). The decrease in Na concentration and the increase in K concentration between the 10th and 22nd postnatal dyas implies an exchange of these ions during neonatal development (Fig. 11). This ionic redistribution in the cerebral cortex has been previously ascribed to acid-base changes that may occur during maturation according to Woodbury and associates (Woodbury, 1954; 1958), for example, K concentration is regulated by total anion content of the cell, which may vary considerably, being dependent on cellular pH as well as on the metabolic processes affecting the concentrations of free acidic and basic amino acids and on the cellular concentration of bicarbonate. In addition, however, K concentration depends on the coupling of potassium movement to active sodium transport. The results by Valcana and associates suggest that Na and K redistribution may be consequent to the onset of active transport, inasmuch as it occurs concomitantly with a marked increase in the Na+-K+-activated ATPase activity in this tissue. This interpretation is consistent with the role ascribed to ATPase in brain and other tissues (Deul and McIlwain, 1961; Skou, 1972, 1965; Bonting et al., 1962).

Early studies by Timiras and Woodbury (1956) have shown that thyroidectomy of adult rats leads to an increase in plasma Na concentration with no change in plasma K or Cl concentrations. Furthermore, these investigators observed a decrease in the intracellular Na concentration in the brain, when using the Cl space as a measure of cerebral extracellular space, whereas brain K and Cl concentrations were not changes. On the other hand, thyroxine and triiodothyronine treatment of normal adult rats significantly decreased Na and increased K of plasma; intracellular Na concentrations and Na space were increased by hormonal treatment whereas brain Cl space and Cl concentration in plasma and brain were not modified.

Hypothyroidism increases Na (and Cl, not shown here) and decreases K levels. Hypothyroidism also induces a decrease in the specific activity of Na+-K+-activated ATPase in the cerebral cortex but does not appear to significantly affect the specific activity of Mg-ATPase (review Valcana, 1974). When hypothyroid rats are treated with thyroxine, enzyme activity reverts to control values. That thyroxine may affect ionic levels through interactions at the membrane level is supported by the findings of Matty and Green (1962) and Green and Matty (1964) which indicate that thyroxine accelerates active sodium transport across isolated toad skin and bladder membranes, and by many studies showing that thyroxine alters mitochondrial membrane properties (Pitt-Rivers and Tata, 1959; Peachey and Grief, 1965). More recently, support has come from the work of Ismail-Beigi and Edelman (1970, 1971) who have shown that thyroid hormones activate Na extrusion and K accumulation by stimulating the Na pump in all tissues in which thyroxine is found to have a stimulatory effect on oxygen consumption.

Valcana et al. (1967) also studied the effects of estradiol and cortisol on early ionic development in the rat. Hormones were administered from the 6th to the 10th day after birth. They found that the Na and Cl content of the cerebellum at 12 days was higher in estradiol-treated rats than in controls; and in the cortisol-treated animals Na and Cl contents were higher in the spinal cort but not in the cerebellum. This differential brain region response to steroid action is also demonstrated in another study where estradiol administered from 6th to the 10th day after birth increases the glycolipid content in the cerebrum whereas cortisol increases glycolipid content in the spinal cord (Casper et al., 1967).

HORMONAL INFLUENCE ON PHYSIOLOGICAL ASPECTS OF BRAIN MATURATION

Seizure Activity

Normal Development. Responses to electroshock stimulation and to direct stimulation of the spinal cord are used as indices of gross regional sequence to CNS development and activity. The developmental pattern of electrically induced brain seizures provides indirect evidence of gross maturation of higher CNS systems involved in minimal and maximal seizures, whereas the developmental pattern of spinal cord convulsions provides evidence of the maturation of lower CNS centers. There are two types of electroshock seizure threshold (EST) tests: the threshold for 60 cps alternating current stimulation (ac EST) and the threshold for low frequency stimulation with unidirectional pulses at 6 cps for 3 s (lf EST). The EST tests are measures of the threshold for evoking clonic discharge characterized by facial clonus and thythmic movements of the vibrissae, jaws, and ears. Seizure threshold involves more than simply the threshold for initial excitation of the individual neurons involved. In order for the full minimal seizure (for example, 5 s of sustained clonus) to occur, it is necessary for a substantial number of neurons to discharge for a considerable period of time. This collection of neurons, the discharge of which maintains minimal seizure activity, has been termed the <u>oscillator</u> to distinguish it from the seizure focus which ordinarily serves to trigger the oscillator. The lf stimulus excites the oscillator with maximum efficiency. The discharge elicited by the ac EST method is more intense and spreads over a wider area. Evidence indicates that the anatomical substratum of the oscillator may be that portion of the upper brain stem designated as the <u>centrencephalic system</u> by Penfield and Jasper (1954). Numerous investigations (summarized by Woodbury and Esplin, 1959) clearly show the role of specific components of this region (reticular activating system and thalamus) in the initiation and maintenance of minimal seizure discharge, for example, petit mal. Thus, manifestations of minimal seizure activity, namely, loss of consciousness and slight clonus, can be related to activity in the centrencephalic system, including some involvement in the cortex.

In maturing rats and mice, the ac EST for clonus exhibited only after eight days of age progressively and markedly decreases until 16 days of age in rats (Vernadakis and Woodbury, 1965) and 18 days in mice and slowly increases somewhat thereafter. Similarly, the lf threshold for clonus decreases markedly until 30 days of age in mice and remains relatively constant thereafter (Vernadakis and Wood bury, 1965). The progressive decreases in thresholds are interpreted to be a result of development of the oscillator and discharge of more excitatory neuronal elements representing the oscillator have not fully matured and a self-sustained discharge cannot occur. As the oscillator develops and more neuronal elements respond to stimulation, a lower voltage can elicit repetitive neuronal elements respond to stimulation, a lower voltage can elicit repetitive neuronal discharge. With ac stimulation, both the oscillator and other developing excitatory and inhibitory systems are stimulated. If there is merit to the proposal that the decrease in the lf threshold during maturation is probably associated with development of the oscillator, then it is likely that the progressive decrease in ac threshold is a measure of the extent to which neuronal elements other than the oscillator participate in the seizure discharge.

With electrical stimuli substantially above threshold, a tonic-clonic seizure (maximal electroshock seizure, MES) replaces the purely clonic minimal seizure. This convulsion has a characteristic stereo-typed pattern consisting of tonic hindlimb flexion, tonic hindlimb extension, and clonus. A distinctive qualitative difference between this maximal seizure and a threshold convulsion concerns the extent of spread of seizure discharge, which in the maximal seizure involves the entire cerebrospinal axis. In the maturing rat the maximal electroshock pattern develops in phases which appear in the following sequence first described by Millichap (1958): 1-8 days of age, hyperkinesia (paddling and running movement, shaking, and hyperextension of the head); 13-15 days of age, forelimb flexion followed by forelimb extension and hindlimb flexion; 16 days of age and older, full tonic-clonic seizure pattern, including hindlimb extension (reviews Vernadakis and Woodbury, 1969a,b; Vernadakis and Timiras, 1963a).

Spinal reflex systems play a substantial role in integrating maximal seizure discharge into the motor pattern. Stimulation of the cervical spinal cord in adult spinal animals can duplicate all the hindlimb motor patterns seen during generalized seizure activity in intact animals (Esplin, 1959). Flexor-extensor spinal cord convulsions can be elicited by electrical stimulation of the cord in one-day-old rats (Vernadakis, 1962).

Glucocorticoids. As also discussed in Chapter 12, Woodbury and Timiras (Woodbury, 1954; Timiras et al., 1954) have demonstrated that cortisol administration in adult rats increases brain excitability as measured by a decrease in electroshock seizure threshold and by facilitation of the tonic phase of the maximal electroshock seizure.

In early studies Vernadakis and Woodbury (1963) examined the effects of cortisol on the electroshock seizure thresholds in developing rats. The threshold for eliciting forelimb flexion decreased with age in both control and cortisol-treated groups (Fig. 12). However, the thresholds of the

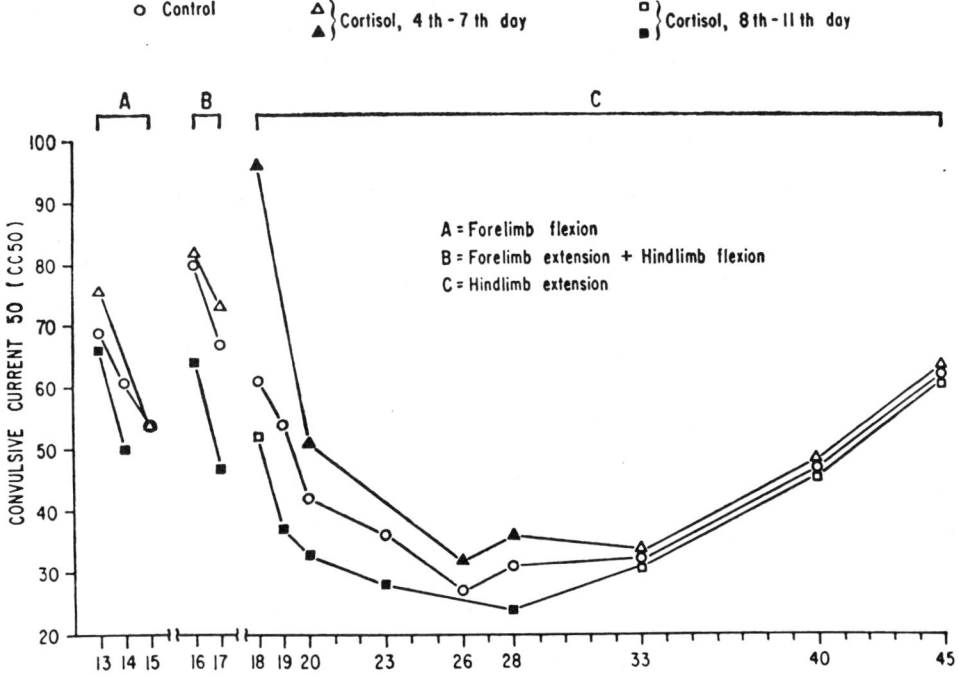

Fig. 12. Effect of cortisol on the electroshock threshold for sequentially developing seizure patterns in rats during maturation. Each point represents the convulsive current 50 (CC50). Filled-in symbols represent values significantly different (P <.05) from the appropriate control for each day. (From Vernadakis and Woodbury, 1962).

two cortisol-treated groups differed from those of the controls; in the animals treated with cortisol early postnatally (days 4-7) the threshold for forelimb flexion was higher than that of the controls whereas in the animals treated with cortisol at later postnatal periods (days 8-15) the thresholds were significantly lower than those of the controls. One of the possible mechanisms by which cortisol, given at a critical period of development, may influence brain excitability is by enhancing myelination (Casper et al., 1967). It is possible that cortisol given during the age period between 8 and 15 days enchances metabolic systems concerned with myelination (Casper et al., 1967), and in this manner accelerates maturation of the nervous system. Since increase in brain excitability appears to correlate with maturation, the decrease in threshold induced by cortisol indicates that this steroid, given at an appropriate period, accelerates maturation of the nervous system.

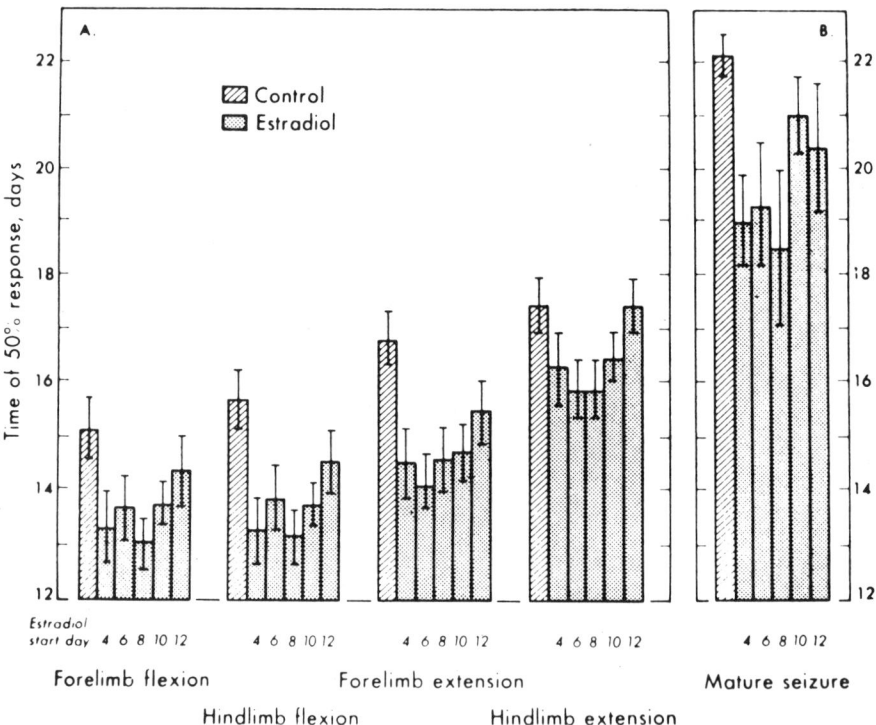

Fig. 13. Effect of estradiol on brain maturation as measured by electroshock convulsive responses in infant rats. (A) Appearance of early flexor and extensor tonic phases. (B) Appearance of fully mature tonic-clonic convulsion. Ordinate is time in days and represents the day of appearance of each phase (A) or of the entire adult convulsive pattern (B) in 50% of animals. Abscissa indicates the days at which estradiol injections were started. The bracketed vertical lines represent 95% confidence limits. (From Heim and Timiras, 1963).

Heim and Timiras (1963) investigated the effects of estradiol in developing rats using responses to maximal electroshock seizure as an index of brain maturation. Estradiol induced precocious appearance of the adult pattern of maximal seizures (Fig. 13). Estradiol was most effective in accelerating brain maturation when administered between days 6-12 of postnatal development (Fig. 13). This observation is of particular interest when correlated with other studies which show that this age coincides with a period of rapid CNS growth. As discussed earlier, this age also appears to be cirtical for the acceleration of CNS maturation after cortisol (Vernadakis and Woodbury, 1963).

An increase in convulsability was also observed when spinal cord convulsions were used as indices (Vernadakis and Timiras, 1963b). The duration of flexion has been utilized as an index of seizure intensity; the shorter the duration of flexion the more severe is the convulsion. In the estradiol-treated animals, the duration of flexion is significantly decreased when the animals were tested at either 12 or 21 days of age. Thus, it appears that estradiol given at a critical period of nervous system development increases the activity of spinal reflex systems.

Thyroid Hormones. Meisami et al. (1970) studied the changes in brain excitability with age in control and neonatally to thyroidectomized female rats by assessing responses to electroshock stimulation. Although the overall pattern of EST development in normal animals, characterized by an initial fall and subsequent rise, was essentially similar in hypothyroid rats, EST remained significantly lower in the latter throughout the experimental period. Moreover, the duration of flexion was significantly shorter in the thypothyroid animals at 25 days of age. It would appear, then, that the developing brain of the hypothyroid rat is more excitable and more sensitive to electrically-induced seizures than the brain of normal animals. Two interpretations, possible complementary, may explain these findings: first, inasmuch as the inhibitory influence of cortical neurons on seizure mechanisms has long been recognized and since most of the available evidence indicates that it is the cerebral cortical development that is impaired in young hypothyroid rats, it can be assumed that the lowering of seizure threshold and the shortening of the duration of flexion reflect impaired maturation of these inhibitory mechanisms. Second, it is assumed that hypothyroidism has impaired the development of the subcortical "centrencephalic system" as well as the cortex, an interpretation that is supported by the finding that certain thalamic lesions as well as cortical lesions can lower the convulsive threshold to chemically-induced seizures in the adult rat (Adler, 1969). The lack of proper cortical influence and/or an improperly developed "oscillator" may also explain the sustained excitation observed in some of the hypothroid animals after minimal stimulation.

As discussed earlier, the critical period for the action of thyroid hormones on the developing brain of the rat is usually assumed to be from the 10th to the 15th postnatal day. However, the significant difference between the EST of control and hypothyroid animals at 10 days of age suggests that thyroid hormones may affect the developing nervous system even before this time. This suggestion is supported by other studied by these authors (discussed below) who found a significant difference in the threshold and amplitudes of the transcallosal response in hypothyroid and normal rats by the 10th postnatal day (Hatotani and Timiras, 1967).

Evoked Cortical Activity

Normal Development. The evoked cortical potential is dependent upon the intensity and quality of the afferent sensory input. It is suggested that the evoked cortical response represents the combined effect of two asynchronic ascending impulse discharges upon cortical structures (Anokhin, 1964). The nerve impulses producing the primary cortical responses are conducted through a short pathway, the specific thalamic radiation, whereas apparently impulses eliciting the secondary repetitive slow waves that follow the primary response are conducted by an extralemniscal nonspecific route. The latter repetitive slow waves are assumed to be related to attentional and perceptual processes (Callaway, 1966; Thompson and Shaw 1965; Wilkinson and Morlock, 1967).

The dendritic spine is a distinctive feature of the postsynaptic apparatus of the cerebral cortex, the cerebellum and other subcortical structures (Globus and Scheibel, 1967). Evidence exists that the development of the dendritic spines is influenced by sensory input (Globus and Scheibel, 1967; Rosenweig et al., 1969). In the rat, dendritic spines are absent at birth. They increase progressively in number until they, at about three weeks of age, seem to have achieved adult appearance.

As discussed in other sections in this book, evidence has been reported that axodendritic synapses involved in the generation of electrical activity of excitatory neurons mature relatively early (Pappas and Purpura, 1973; Voeller et al., 1963). By contrast, the axosomatic and juxtasomatic synapses subserving inhibitory activity, are sparsely distributed and do not increase markedly in number until the second and third postnatal week (Eccles, 1964; Purpura et al., 1965). Present results indicate the occurrence of changes in evoked primary cortical potentials and in the appearance of evoked late repetitive waves. The slowing down of the development of powerful cortical synaptic systems due to reduction in number of spines and afferent thalamic terminals may, in turn, produce a differential retardation of cortical neuron development. Cells responsible for excitatory and inhibitory processes mature at different ages and consequently thyroxine as other environmental influences may permanently disturb the normal balance between excitatory and inhibitory actions.

Thyroid Hormones. Hatotani and Timiras (1967) have described in detail the development of transcallosal responses in rats and the effects of thyroid hormones thereon. The evoked transcallosal responses of normally developing rats consist of a major surface-negative phase preceded by a brief shallow positive phase. In some cases, the initial positive phase. In the course of development the evoked transcallosal response of the rat undergoes characteristic changes with age with peak latency, threshold and duration decreasing and amplitude increasing with age (Figs. 14-17). Latency, threshold and duration are higher and amplitude is lower in hypothyroid rats as compared to controls at all ages studied (10 to 60 days). In hyperthyroid rats latency and duration were markedly lower than controls at most ages studied with no significant changes observed in amplitude.

The formation of myelin sheaths most likely underlies the increase with age in conduction velocity and electrical excitability of the callosal fibers. The histological data obtained by Hatotani and Timiras (1967) and supported by the extensive work of Jacobson (1963) show that myelination of the callosal fibers in the rat starts around 10 days and continues until about

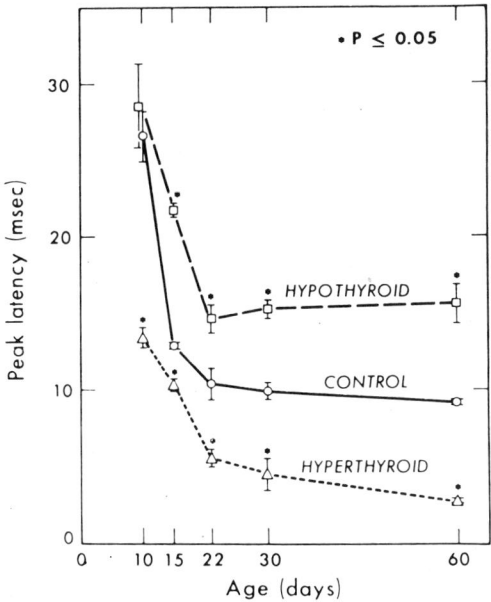

Fig. 14. Mean values for peak latency in controls, hypo- and hyperthyroid rats during development: bracketed lines represent standard errors of the mean. Asterisks indicate statistical significance (by t test) of differences between control and experimental animals (From Hatatani and Timiras, 1967)

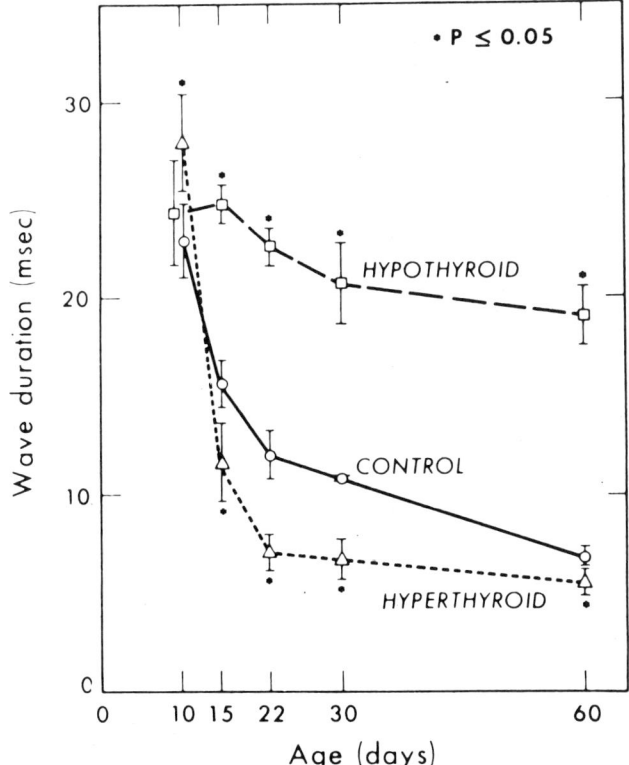

Fig. 15. Mean values for wave duration in controls, hypo- and hyperthyroid rats during development. Bracketed lines represent standard errors of the means. Asterisks indicate statistical significance (by t test) of differences between control and experimental animals. (From Hatatani and Timiras, 1967).

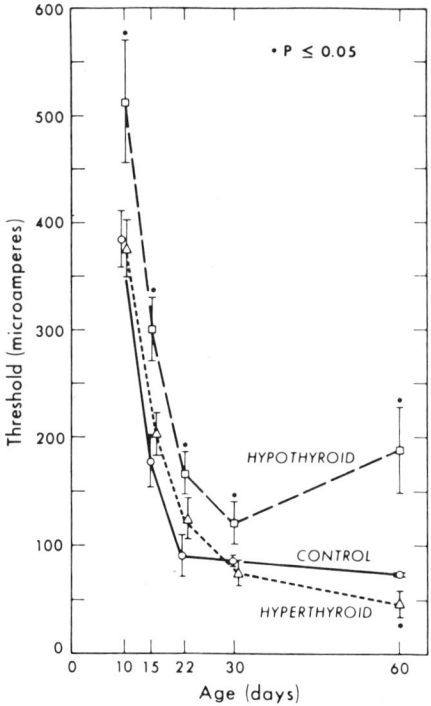

Fig. 16. Mean values for threshold in controls, hypo- and hyperthyroid rats during development. Bracketed lines represent standard errors of the means. Asterisks indicate statitsitcal significane (by t test) of differences between control and experimental animals. (From Hatotani and Timiras, 1967)

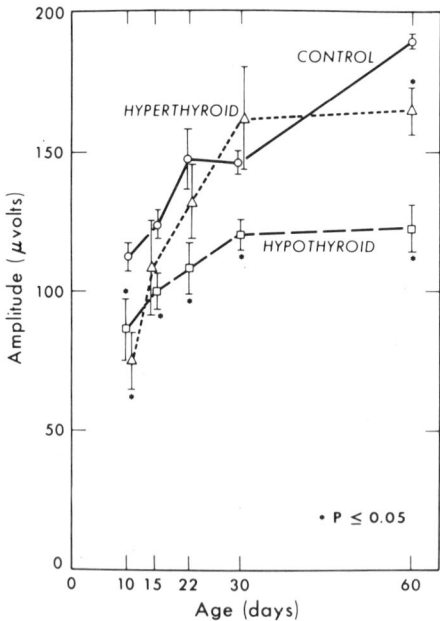

Fig. 17. Mean values for amplitude in controls, hypo- and hyperthyroid rats during development. Bracketed lines represent standard errors of the means. Asterisks indicate statistical significance (by t test) of differences between control and experimental animals. (From Hatotani and Timiras, 1967)

30 days. The time course of latency in control rats appears to change with that of myelination. Impairment of myelin growth might be the cause for the retardation in shortening of latency in the hypothyroid animals. Conversely, acceleration of myelinogenesis could account for the early appearance of shorter latency in the thyroxine-injected rats. As discussed earlier in this chapter Hamburgh (1966) reported that myelinogenesis was accelerated in tissue culture of the cerebella of newborn rats in the presence of thyroxine. These findings strongly suggest that changes in the peak latency are closely related to the myelination process of the callosal fiber. This does not necessarily preclude the possibility that changes in latency may be related to thyroid-dependent metabolic processes, e.g., through changes in nerve membrane properties or in electrical resistance.

Glucocorticoids. Curry and Timiras (1972) examined the development of evoked potentials in specific brain systems after neonatal administration of estradiol. The development of the transcallosal response (TCR) and the thalamically evoked relayed pyramidal response (RPR) were studied by analyzing evoked potentials at several ages. Estradiol-treated animals generally showed a lower threshold, shorter peak latency, and greater peak amplitude of the TCR than controls, indicating accelerated development at these ages. The shortening of peak latency in estradiol-treated rats is ascribed to an increase in axonal conduction velocity, implicating precocious myelination, whereas the increase in peak amplitude reflects the progressive establishment of synaptic connections. Lowering of the threshold in the hormone-treated animals reflects changes in either the electrical excitability of the neurons involved or in development of synaptic processes, or both. Inasmuch as the effects of estradiol administration on threshold of the TCR did not coincide with an alteration in peak amplitude, the changes in threshold in this case are more likely to represent alterations in neuronal excitability consequent to disturbances in ionic balance rather than precocious development of synaptic connections. Based on the specific findings that estradiol differentially affects peak latency and peak amplitude in the TCR depending upon age, that these effects are independent from those on threshold, and that only peak latency is affected in the RPR, the authors suggest that estradiol exerts a three-fold effect on functional development of specific brain system, acting independently on myelination, the development of synaptic processes and neuronal excitability.

Spontaneous Electrical Activity

The effects of hormones on spontaneous electrical activity have been sporadically investigated. Bisanti and Davallotti (1972) studied the effects of β-estradiol on spontaneous electrical activity. The experimental groups consisted of controls, neonatally estradiol-treated female rats, animals ovariectomized 14 days after birth, control sham-operated females, and ovariectomized rats receiving hormone replacement therapy. During the first week after birth, in normal rats the electrical activity detectable from the cerebral cortex was irregular, intermittent and characterized by a theta rhythm of a relatively low amplitude; during the second week a marked trend towards regularity, rhythmicity and continuous activity was observed, and by the fourth week of age the patterns observed were basically similar to those of the adult rat. Treatment with β-estradiol modified the spontaneous EEG recorded during rat brain development. In general, spontaneous activity was accelerated after hormone treatment, a finding which is in accordance with other studied discussed earlier in this chapter that estradiol enhances brain maturation when given at critical periods of neural growth. Early studies by Bradley et al. (1961) have shown several abnormalities in the EEG of the rat as a result of thyroidectomy performed at birth. A marked reduction in the amplitude in the potentials recorded from the surface of the skull was observed in the thyroidectomized rats tested from 15 to 48 days. Sensory stimuli, which in normal animals cause "blocking" of these potentials accompanied by behavioral alerting produced little or no change in the records of hypothyroid rats; and finally photic driving, which can be elicited in most normal animals from the age of 24 days does not appear in the hypothyroid animals. When animals thyroidectomized at birth were treated with thyroxine the differences between control and thyroidectomized treated rats were not marked. If the slow components in the EEG are attributal to fluctuations in dendritic potentials it would seem reasonable to postulate that the low amplitude and absence of slow activity in the EEG of the hypothyroid animal is due to reduced growth of dendrites. As described elsewhere in this book, a reduction in the length and number of dendrites developed by each neuron is observed in hypothyroid animals.

BEHAVIORAL EFFECTS OF HORMONES

As discussed throught this chapter, at birth, the rat is a very immature organism, and many of the coordinated physiological mechanisms so essential for adult survival do not function effectively (see review, Timiras, 1972). In addition to perinatal homeostatic immaturity, numerous adult behavioral adaptive mechanisms also develop slowly (see review, Timiras, 1972).

Thyroid Hormones

Despite both clinical and experimental evidence that infantily hypothyroidism results in permanent impairment of learning capacities in mammalian species, our knowledge is incomplete regarding the age boundaries of the critical period of early thyroid deficiency for inducing irreversible learning deficits and with respect to the detailed nature of deficits themselves.

Eayrs (1961a) induced thyroid deficiency by radiothyroidectomy, surgical thyroidectomy, or by daily injections of methyl thiouracil on postnatal days one, 4, 10, or 24, and tested the performance of rats at 50 and 120 days of age in a closed-field maze series. Error scores on the maze problems showed maximal impairment in the day-one treated groups, with the degree of deficit progressively declining to no impairment in day-24 animals. Restoration to a euthyroid condition in the day-one thyroidectomized group resulted in essentially normal maze performance whereas delay of replacement therapy to the 24th day resulted in learning deficits as in unmedicated rats thyroidectomized at birth. These findings establish a critical period for the induction of learning deficits by thyroid deficiency.

Evidence of enduring deficit in learning capacities induced by early thyroid deficiency has also been obtained by Davenport (1970) in studies involving the antithyroid drugs tricyanoaminopropene (TCAP) and thiouracil. Rats exposed to these drugs for 16 prenatal days and for 130 or 212 postnatal days displayed deficits most clearly in post drug tests in the symmetrical maze, a semiautomated close-field problem series. To what extent the induced deficits may have been due to the prenatal exposure;

i.e., the transplacental passage which is known to occur for both drugs, was not determined in these studies. No methods were employed for separating the prenatal effects from the effects of transmammary passage of the drugs during the early postnatal period or from influences arising after the lactation period. The authors noted that the long postnatal periods of drug administration, during which behavioral tests requiring food deprivation were conducted, produced extreme growth retardation and other physical signs of cretinism and may have introduced contaminating factors into studies, such as undernutrition and drug toxicity. In another study Davenport and Dorcey (1972) obtained thiouracil-induced learning deficits in rats without the concomitant, irreversible deficits in physical development. The exact nature of the learning abilities which are tapped by the symmetrical maze series is not yet clarified, however, it does appear that the thiouracil-induced deficits found with this task were more reflective of deficiencies in associative capacities than of "motivational" inadequacies in the thiouracil groups, considering the absence of group differences in body weights, consummatory behavior, activity, lever-pressing rates and latencies, and locomotion speeds. These authors concluded that the deficits may be attributed to thyroid deficient conditions near birth rather than to conditions in later stages of development.

As discussed in Chapter 2, thyrotropin releasing hormone activity is found in hypothalamic extracts of the rat brain early prenatally (Conklin et al., 1973). In addition to the hypothalamus, a substantial amounts of TRH are found in the thalamus, cerebrum, and the brain stem of adult rats (Plotnikoff et al., 1975; Winokur and Utiger, 1974). The appearance of TRH activity early in life and its pervasiveness in brain tissue suggests a role in neuronal, synaptic or metabolic events which would affect behavioral and developmental processes. Recently, Stratton et al. (1976) examined the effects of neonatal treatment with TRH on learning and emotionality in the rat. Rats treated with TRH from two to seven days after birth were less emotional and more active in an open field both as pups and as adults.

Schapiro (1968a) found that neonatal thyroxine administration for several days immediately after birth produced earlier opening of the eyes and the animals exhibited greater spontaneous locomotor activity. In addition, the maturation of the pituitary-adrenal response to stress appeared earlier, the development of the cortical EEG was advanced, as was its response to novel stimuli. Young thyroxine-treated rats, learned to avoid an electric shock faster than controls. However, when the experimental animals are adults they make more errors on two different maze-learning tasks (Schapiro, 1968a). These results suggest that: (1) The presence of specific hormones in excess of ontogenetically inappropriate times during development may compromise the capability of certain adaptive or survival value mechanisms at later stages of the life cycle; (2) The immature status of some neuroendocrine systems during such hormone sensitive periods may therefore provide for the later full expression of these adaptive mechanisms.

In a later study Schapiro et al. (1970) examined the maturation of swimming behavior and the evoked cortical response to sciatic stimulation in newborn rats receiving thyroxine or cortisol. Compared to that of controls the maturation of swimming is accelerated or delayed 2-3 days by thyroxine or cortisol treatment, respectively, and this corresponds to ontogenetic shifts in the characteristics of evoked potential. Front leg movement during swimming normally diminishes at about 16 days of age and is inhibited by day 22. Thyroxine also advances, whereas cortisol delays, the age at which this inhibitory mechanisms become evident.

Schapiro et al. suggest that swimming and its development may be a productive model to use in studying functional integration of the multitudinous component reflexes usually studied in isolation that enables the organism to effectively adapt to its environment.

Glucocorticoids

Howard and associates (Howard and Granoff, 1968; Howard, 1973b; Olton et al., 1974) have extensively explored the influence of glucocorticoids on innately organized behavior. They have shown that mice given corticosterone at two days postnatally were hyperactive when given access to revolving wheels at 14-26 days of age but after two months, this form of hyperactivity was not present. As adults, mice given corticosterone as infants were tested for their ability to swim to an escape ladder in a water maze of the Lashley Type III pattern. The experimental animals made more errors than the controls on day 2 of testing, but thereafter their performance was not inferior. They showed no gross tremor or ataxia; but in a test of ability to maintain stance on a slowly rotating bar, the mice given corticosterone were distinctly inferior. This would be consistent with some decrease in motor coordination related to cerebellar dysfunction but other factors might have been contributing. The treated mice showed decreased exploratory activity in an open field.

In operant conditions studies (Howard, 1973b) mice treated neonatally with corticosterone bar pressed on continuous reinforcement at the same rate as controls but left more food uneaten; they responded at higher rates than controls during extinction and on a fixed-ratio-64 (FR-64) schedule. On 20-s differential reinforcement of low rate (DRL), experimental animals again responded at higher rates and were less successful than controls in making a transition from FR-99 to 20 sec DRL. However, naive experimental animals were as efficient as control animals on 10-s DRL. Thus, after hypercoticism adult mice were hyperresponsive when working for food and showed an impaired ability to adapt to a schedule change. The same group of investigators (Olton et al., 1974) observed hyperresponsive on conditioned, active avoidance test in adult treated neonatally with corticosteron rats.

It is well established that the pituitary-adrenal system play an important role in the defense mechanisms of the organism in response to noxious stimuli. It has therefore been suggested that if the functioning of the neuroendocrine system is maintained at a higher level during infancy than under normal conditions, the resulting exposure to the action of adrenocortical hormones will influence the organization of the CNS in such a way as to modify certain aspects of neuroendocrine regulation and the behavior of the adult animal (Levine and Mullins, 1968; Levine, 1970). In experiments designed to test this hypothesis rats were exposed to different types of daily stresses ("handling") during the period of relative adrenocortical rest (Denenberg, 1968). It was found that when these handled animals grew up, their responses to the demands of the environment were significantly more appropriate than those of the unhandled controls (Levine and Mullins, 1968). Production of adrenocortical hormones increased even during the period of relative adrenal quiescence after the animals had been subjected to stress, although the increase was much less than in adults. However, the response of the handled animals was not significantly different from controls, except perhaps before and after the period of adrenal quiescence, at 3 and 21 days, respectively.

The mode of action of the early "handling" experience is not yet known. However, the increase in pituitary-adrenal activity in response to handling may be involved in the changes in the development of the CNS, which result in the observed changes in behavior. Although the increase in the production and secretion of adrenocortical hormones is small, it is possible that infant animals are more sensitive to corticosteroids than older rats. Schapiro (1968b) observed that the atrophy of the adrenal glands was more pronounced in the infant than in the adult rat after injection of small doses of cortisol. The low plasma concentraion of corticosteroid-binding globulin in infants may play a role in the increased sensitivity (Mills et al., 1959). Furthermore, it is known that under certain conditions very low concentrations of adrenocortical hormones may have important effects on biological systems.

Recent investigations have shown that certain polypeptides synthesized in the pituitary and secreted in response to stressful stimuli are required for the acquisition and retention of conditioned avoidance responses. The "neurogenic" property is exhibited by ACTH, and it is confined to a a heptapeptide section of the N terminal part of the molecule which has no corticotrophic activity (de Wied et al., 1968). Other polypeptides with neurogenic properties have also been isolated from the pituitary (de Wied et al., 1970). The possibility that the effects of early handling on the development of the CNS are mediated by secretory products of the pituitary should therefore be considered.

Gonadal Hormones

Extensive research is focusing on the role of sex hormones on various behaviors and a complete review will not be attempted in this chapter. The rat displays marked sex differences on a number of measures of emotional behavior, the male usually behaving as though it were the more fearful or emotional (Harris, 1964; Gray and Keynes, 1969) and more aggressive (Edwards, 1969; Bronson and Desjardins, 1970). Spontaneous aggression occurs more frequently among male mice than it does among females of almost all strains (Scott, 1966). Several studies have shown that postpuberal castration of males is accompanied by a decrease in fighting behavior and that testosterone replacement results in the return of at least a degree of aggressiveness. Additionally, at least some of the neural control systems mediating aggression in adult males are known to be organized neonatally by testicular androgens (Bronson and Desjardins, 1970; Edwards, 1969). Bronson and Desjardins (1970) and Whitsetti et al. (1972) report that single injections of testosterone were maximally effective in enhancing adult aggressiveness when administered to the females on the day of birth, less effective when given after that time, and becoming ineffective some time between days 12 and 24 of life. Additionally, uptake of radiolabeled testosterone by several brain areas declined significantly from day zero to day 12 after birth. The implication, then, is that early androgen treatment sensitizes appropriate neural elements to androgen encountered in adulthood.

A recent report by Meyer-Bahlburg et al. (1977) summarizes the role of hormones on human behavior. They show that clinical research on the effects of prenatal sex hormone exposure in man on sex-dimorphic behavior are compatible with the androgen organization hypothesis. However, studies of prenatal progestogens in males suggest, but do not substantiate, emasculinizing effect.

ACKNOWLEDGEMENTS

The writing of this chapter was supported in part by a Research Scientist Career Development Award to Dr. Vernadakis K02 MH from the National Institute of Mental Health. The author is grateful to Ms. Gina Carter for her invaluable assistance in library research.

REFERENCES

Adler, M.W. Laboratory evaluation of antiepileptic drugs; the use of chronic lesions. *Epilepsia, Amst.* 10, 263-280 (1969).

Aghajanian, G.K. and Bloom, F.E. The formation of synaptic junctions in developing rat brain: a quantitative electron microscopic study. *Brain Res.* 6, 716-727 (1967).

Altman, J. Autoradiographic and histologic studies of postnatal neurogenesis. II. A longitudinal investigation of the kinetics, migration, and transformation of cells incorporating tritiated thymidine in infant rats, with special reference to postnatal neurogenesis in some brain regions. *J. Comp. Neurol.* 128, 431-474 (1966).

Altman, J. Autoradiographic and histological studies of postnatal neurogenesis, III. Dating the time of production and onset of differentiation of cerebellar microneurons in rats. *J. Comp. Neurol.* 136, 269-294 (1969).

Altman, J. Postnatal neurogenesis and the problem of neural plasticity, in *Development Neurobiology*. W.A. Himwich, ed. Charles C. Thomas, Pub., Springfield, Ill. (1970) pp. 197-237.

Anderson, T.R. and Schanberg, S.M. Ornithine decarboxylase activity in developing rat brain. *J. Neurochem.* 19, 1471-1481 (1972).

Anderson, T.R. and Schanberg, S.M. Effect of thyroxine and cortisol on brain ornithine decarboxylase activity and swimming behavior in developing rat. *Biochem. Pharmacol.* 24, 495-501 (1975).

Angevine, J.B., Jr. Critical cellular events in the shaping of neural centers, in *The Neurosciences, II*. F.O. Schmitt, ed. Rockefeller University Press, New York (1970) pp. 62-72.

Anokhin, P.K. The electroencephalogram as a resultant of ascending influences on the cells of cortex. *Neurophysiol.* 16, 27-43 (1964).

Arai, Y. and Gorski, R.A. Effect of anti-estrogen on steroid induced sexual receptivity in ovariectomized rats. *Physiol. Behav.* 3, 351-353 (1968).

Arnold, E.B. and Vernadakis, A. Development of tyrosine hydroxylase activity in dissociated cerebral cell cultures. *Dev. Neurosci.* 2, 46-50 (1979).

Azmitia, E. and McEwen, B. Corticosterone regulation of tryptophan hydroxylase in midbrain of the rat. *Science* 166, 1274-1276 (1969).

Bachman, C., Nyhan, W.L., Kulovich, S., and Hornback, M.E. Amino acids in the brain during fetal growth and the effects of prenatal hormonal and nutritional imbalance. Fetal and postnatal cellular growth, in *Hormones and Nutrition*. Donald B. Cheek, ed. Wiley Biomedical-Health Publication, John Wiley & Sons Pubs. New York (1975) pp. 169-205.

Balazs, R. Biochemical effects of thyroid hormones in the developing brain, in *Cellular Aspects of Neural Growth and Differentiation*. D.C. Pease, ed. University of California Press, Los Angeles (1971a) pp. 273-320.

Balazs, R. Effects of hormones on the biochemical maturation of the brain, in *Influence of Hormones on the Nervous System*. D.H. Ford, ed. S. Karger, Basel (1971b) pp. 150-164.

Balazs, R. Effect of thyroid hormone and undernutrition on cell acquisition in the rat brain, in *Thyroid Hormones and Brain Development*. D. Gilman-Grave, ed. Raven Press, New York (1977) pp. 287-302.

Balazs, R. and Cotterrell, M. Effect of hormonal state on cell number and functional maturation of the brain. *Nature* 236, 348-350 (1972).

Balazs, R. and Richter, D. Effects of hormones on the biochemical maturation of the brain, in *Biochemistry of the Developing Brain*.

W. Himwich, ed. Marcel Dekker, Inc., New York (1973) pp. 253-356.

Balazs, R., Lewis, P.D., and Patel, A.J. Effects of metabolic factors on brain development, in *Growth and Development of the Brain*. M.A.B. Brazier, ed. Raven Press, New York (1975a) pp. 83-115.

Balazs, R., Patel, A.J., and Hajos, F. Factors affecting the biochemical maturation of the brain: effects of hormones during early life. *Psychoneuroendocrin.* 1, 25-36 (1975b).

Balazs, R., Brooksbank, B.W.L., Davison, A.N., Eayrs, J.T., and Wilson, D.A. The effect of neonatal thyroidectomy on myelination in the rat brain. *Brain Res.* 15, 219-232 (1969).

Balazs, R., Kovacs, S., Cocks, W.A., Johnson, A.L., and Eayrs, J.T. Effect of thyroid hormone on the biochemical maturation of rat brain: postnatal cell formation. *Brain Res.* 25, 555-570 (1971).

Balazs, R., Kovacs, S., Teichgraber, P., Cocks, W.A., and Eayrs, J.T. Biochemical effects of thyroid deficiency on the developing brain. *J. Neurochem.* 15, 1335-1345 (1968).

Ballard, P.J. and Tomkins, G.M. Glucocorticoid induced alterations in the surface membrane of cultured hepatoma cells. *J. Cell Biol.* 47, 222-234 (1970).

Baquer, N.Z. and McLean, P. The effect of oestradiol on the profile of constant and specific proportion groups of enzymes in rat uterus. *Biochem. Biophys. Res. Commun.* 48, 729-734 (1972).

Berlin, C.M. and Schimke, R.T. Influence of turnover rate on the responses of enzymes to cortisone. *Molec. Pharmac.* 1, 149-156 (1965).

Berliner, J.A. and Gerschenson, L.E. The effects of a glucocorticoid on the cell surface of RLC-GAI cells. *J. Cell Physiol.* 86, 523-532 (1975).

Berliner, J.A., Bennett, K., and deVellis, J. Effect of hydrocortisone on cell morphology in C-6 cells. *J. Cell Biol.* 94, 321-334 (1978).

Bickel, M., Kuhn, H., Eng Tan, J.S., and Taubert, H.D. Evidence for a sex-specific effect of testosterone and progesterone upon L-cystine-amino peptidase activity in the hypothalamus and paleopallium of the rat. *Neuroendocrinology* 9, 321-331 (1972).

Bisanti, L. and Cavallotti, C. Hormonal regulation of rat brain development: III Effect of β-estradiol on electrical activity and behavior, in *Progress in Brain Research; Topics in Neuroendocrinology*. J.A. Kappers and J.P. Schade, eds. Elsevier, Amsterdam (1972) pp. 327-335.

Bohn, C.M. and Lauder, J.M. The effects of neonatal hydrocortisone on rat cerebellar development. *Dev. Neurosci.* 1, 250-265 (1978).

Booher, J. and Sensenbrenner, M. Growth and cultivation of dissociated neurons and glial cells from embryonic chick, rat, and human brain in flask cultures. *Neurobiology* 2, 97-105 (1972).

Bonney, R.J., Becker, J.E., Walder, P.R., and Potter, V.R. Primary monolayer cultures of adult liver parenchymal cells suitable of the regulation of enzyme synthesis. *In Vitro* 10, 130-141 (1974).

Bonting, S.L., Caravaggio, L.L., and Hawkins, N.M. Studies on sodium-potassium-activated adenosine triphosphatase. IV. Correlation with cation transport sensitive to cardiac glycosides, *Arch. Biochem. Biophys.* 98, 413-419 (1962).

Bornstein, M.B. Reconstituted rat-tail collagen used as substitute for tissue culture on coverslips in Maximow slides and roller tubes. *Lab. Invest.* 7, 134-137 (1958).

Bradley, P.B., Eayrs, J.T., Glass, A., and Health, R.W. The maturational and metabolic consequences of neonatal thyroidectomy upon the recruiting response in the rat. *Electroenceph. Clin. Neurophysiol.* 13, 577-586 (1961).

Bronson, F.A. and Desjardins, C. Neonatal androgen administration and adult aggressiveness in female mice. *Gen. and Comp. Endocrinol.* 15,

320-325 (1970).

Bunge, M.B., Bunge, R.P., and Pappas, G.D. Electron microscopic demonstration of connections between glia and myelin sheaths in the developing mammalian central nervous system. *J. Cell Bio.* 12, 448-453 (1962).

Callaway, E. Average evoked responses in psychiatry. *J. Nerv. Ment. Dis.* 143, 80-94 (1966).

Cavallotti, C. and Bisanti, L. Hormonal regulation of rat brain development. II. Biochemical changes induced by β-estradiol. *Prog. Brain Res.* 38, 69-83 (1972).

Casper, R., Vernadakis, A., and Timiras, P.S. Influence of estradiol and cortisol on lipids and cerebrosides in the developing brain and spinal cord of the rat. *Brain Res.* 5, 524-526 (1967).

Clark, J.H., Anderson, J.N., and Peck, E.J. Estrogen receptor-antiestrogen complex: a typical binding by uterine nuclei and effects on uterine growth. *Steroids* 22, 707-718 (1973).

Cohen, R.A. and Gerard, R.W. Hypothyroidism and brain oxidations. *J. Cell Comp. Physiol.* 10, 223-240, (1937).

Conklin, P.M., Schindler, W.J., and Huff, S.F. Hypothalamis thyrotropin-releasing factor. *Neuroendocrinology* 11, 197-211 (1973).

Coppola, J.A. Brain catecholamines and gonadotropin secretion, in *Frontiers in Neuroendocrinology*. W. Ganong and L. Martini, eds. Oxford University Press, New York (1971) pp. 129-143.

Corner, M.A., Romijin, H.J., and Richter, A.P. Synaptogenesis in the cerebral hemisphere (accessory hyperstriatum) of the chick embryo. *J. Neurosci. Lett.* 4, 15-19 (1977).

Corner, M.A., Smith, J., and Romijin, H.J. Maturation of cerebral bioelectric activity in the chick embryo in relation to morphological and biochemical factors, in *Ontogenesis of the Brain*, 2, L.Jilek and S.Trojan, eds. Charles University Press, Prague (1974) pp. 31-42.

Corner, M.A., Schade, J.P. Sedlacek, J., Stoeckart, R., and Bot, A.P.C. Developmental patterns in the control nervous system of birds. I. Electrical activity in the cerebral hemisphere, optic lobe and cerebellum. *Prog. Brain Res.* 26, 145-192 (1967).

Cotterrell, M., Balazs, R., and Johnson, A.L. Effects of corticosteroids on the biochemical maturation of rat brain: postnatal cell formation. *J. Neurochem.* 19, 2151-2167 (1972).

Coyle, J.T. and Axelrod, J. Development of the uptake and storage of L-^{3}H-norepinephrine in the rat brain. *J. Neurochem.* 18, 2061-2075 (1971).

Coyle, J.T. and Axelrod, J. Tyrosine hydroxylase in rat brain: developmental characteristics. *J. Neurochem.* 19, 1117-1123 (1972).

Coyle, J.T. and Snyder, S.H. Catecholamine uptake by synaptosomes in homogenates of rat brain: stereospecificity in different areas. *J. Pharmac. Exp. Ther.* 170, 221-231 (1969).

Curry, J.J. and Heim, L.M. Brain myelination after neonatal administration of estradiol. *Nature* 209, 915-916 (1966).

Curry, J.J. and Timiras, P.S. Development of evoked potentials in specific brain systems after neonatal administration of estradiol. *Exper. Neurol.* 34, 129-139 (1972).

Dalal, K.B., Valcana, T., Timiras, P.S., and Einstein, E.R. Regulatory role of thyroxine on myelinogenesis in the developing rat. *Neurobiology* 1, 211-224 (1971).

Davenport, J.W. Cretinism in rats: enduring behavioral deficit induced by tricyanoaminopropene. *Science* 167, 1007-1009 (1970).

Davenport, J.W. and Dorcey, T.P. Hypothyroidism: learning deficit induced in rats by early exposure to thiouracil. *Hormones and Behavior* 3, 97-

112, (1972).

Davenport, J.W., Hagquist, W.W., and Rankin, G.R. The symmetrical maze: an automated closed-field test series for rats. *Behav. Res. Methods Instrum.* 2, 112-118 (1970).

Davis, J.M., Goodwin, F.K., Bunney, W.E., Murphy, D.L., and Colburn, R.W. Effects of ions in uptake of norepinephrine by synaptosomes. *Pharmacologist* 9, 184 (1967).

Davison, A.N. and Gregson, N.A. The physiological role of cerebron sulphuric acid (sulphatide) in the brain. *Biochem. J.* 85, 558-568 (1962).

Dawson, G. and Kernes, S.M. Mechanism of action of hydrocortisone potentiation of sulfogalactosylceramide synthesis in mouse oligodendroglioma clonas cell lines. *J. Bio. Chem.* 254, 163-167 (1979).

DeChamplain, J., Malmfors, T., Olson, L., and Sachs, Ch. Ontogenesis of peripheral adrenergic neurons in the rat: pre- and postnatal observations. *Acta. Physiol. Scand.* 80, 276-288 (1970).

DeKirmenjian, H. and Brunngraber, E.G. Distribution of protein bound N-acetylneuraminic acid in subcellular particulate fractions prepared from whole rat brain. *Biochem. Biophys. Acta.* 177, 1-10 (1969).

Denenberg, V.H. A consideration of the usefulness of the critical period hypothesis as applied to the stimulation of rodents in infancy, in *Early Experience and Behavior*. G. Newton and S. Levine, eds. Charles C. Thomas, Springfield, Ill. (1968) pp. 142-167.

Dengler, H.J., Michaelson, I.A., Spiegel, H.E., and Titus, E. The uptake of labeled norepinephrine by isolated brain and other tissues of the cat. *Int. J. Neuro-pharmacol.* 1, 23-28 (1962).

Deul, D.H. and McIlwain, H. Activation and inhibition of adenosine triphosphatases of subcellular particles from the brain. *J. Neurochem.* 8, 246-256 (1961).

deVellis, J. and Brooker, G. Induction of enzymes by glucocorticoids and catecholamines in a rat glial cell line, in *Tissue Culture of the Nervous System*. G. Sato, ed., Plenum Press, New York (1973) pp. 231-245.

deVellis, J. and Kukes, G. Regulation of glial cell functions hormones and ions: a review. *Texas Rep. Biol. and Med.* 31, 271-293 (1973).

deVellis, J., Inglish, D., and Galey, F. Effects of cortisol and epinephrine on glial cells in culture, in *Cellular Aspects of Growth and Differentiation in Nervous Tissue*. D. Pease, ed. UCLA Forum on Medical Sciences, No. 14 (1971a) pp. 23-32.

deVellis, J., Inglish, D., Cole, R., and Molson, J. Effects of hormones on the differentiation of cloned lines of neurons and glial cells, in *Influence of Hormones on the Central Nervous System*. D. Ford, ed. S. Karger, Basel (1971b) pp. 25-39.

deVellis, J., McGinnis, J.F., Breen, G.A.M., LeVeille, P., Bennett, K., and McCarthy, K. Hormonal effects on differentiation in neural cultures, in *Cell, Tissue, and Organ Culture in Neurobiology*. S. Fedoroff and L. Hertz, eds. Academic Press, New York (1977) pp. 485-511.

DeWied, D., Bohus, B., and Grever, H.M. Influence of pituitary and adrenocortical hormones on conditioned avoidance behavior in rats, in *Endocrinology and Human Behaviour*. R.P. Michael, ed. Oxford University Press, London (1968) pp. 188-199.

DeWied, D., Willer, A., and Lande, S. Anterior pituitary peptides on avoidance acquisition of hypophysectomized rats, in *Progress in Brain Research, Vol. 32, Pituitary, Adrenal and Brain*. DeWied and J.A.M. Weijnen, eds. Elsevier, Amsterdam (1970) pp. 213-218.

Diez, J.A., Sze, P.Y., and Ginsburg, B.E. Effect of glucocorticoid and stress on brain tyrosine hydroxylase activity. *Trans. Amer. Soc. Neurochem.* 5, 79 (1974).

Dohler, D. and Wuttke, W. Changes with age in levels of serum gonadotropins, prolactin, and gonadal steroids in prepubertal male and female rats. *Endocrinology* 97, 898-907 (1975).

Dorner, G. Environment-dependent brain differentiation and fundamental processes of life. *Acta. Biol. Med. Ger.* 33, 129-148 (1974).

Drabkin, D.L. Cytochrome c metabolism and liver regeneration: influence gland and thyroxine. *J. Biol. Chem.* 182, 335-357 (1950).

Draves, D.J. and Timiras, P.S. Differential effects of altered thyroid hormone states on nervous tumor cells, in *A Multidisciplinary Approach to Brain Development*. C. DiBenedetta, R. Balazs, G. Gombos and G. Porcellati, eds. International Meeting, Selva di Fasano, April 16-21, 1979, Elsevier, (198) pp. 313-315.

Eayrs, J.T. Age as a factor determining the severity and reversibility of the effects of thyroid deprivation in the rat. *J. Endocrinol.* 22, 409-419 (1961a).

Eayrs, J.T. and Goodhead, B. Postnatal development of the cerebral cortex in the rat. *J. Anat.* 93, 385-402 (1959).

Eccles, J.C. *The Physiology of Synapses*, Springer-Verlag, Berlin, 1964.

Edelman, I.S. Transition from poikilotherm to homeotherm: possible role of sodium transport and thyroid hormone. *Fed. Proc.* 35, 2180-2184 (1976).

Edwards, D.A. Early androgen stimulation and aggressive behavior in male and female mice. *Physiology and Behavior* 4, 333-338 (1969).

Esplin, D.W. Spinal cord convulsions. *Arch. Neurol.* 1, 485-490 (1959).

Frieden, E.H. Sex hormones and the metabolism of amino acids and proteins, in *Actions of Hormones on Molecular Processes*. G. Litwack and D. Kritchevsky, eds. John Wiley & Sons, New York (1964) pp. 509-559.

Garcia Argiz, C.A., Rasquini, J.M., Kaplun, B., and Gomez, C.J. Hormonal regulation of brain development: II. Effect of neonatal thyroidectomy on succinate dehydrogenase and other enzymes in developing cerebral cortex and cerebellum of the rat. *Brain Res.* 6, 635-646 (1967).

Geel, S.E. and Gonzales, L.W. Cerebral cortical ganglioside and glycoprotein metabolism in immature hypothyroidism. *Brain Res.* 128, 515-525 (1977).

Geel, S.E. and Timiras, P.S. The influence of neonatal hypothyroidism and of thyroxine on the ribonucleic acid and deoxyribonucleic acid concentrations of rat cerebral cortex. *Brain Res.* 4, 135-142 (1967).

Geel, S.E., Valcana, T., and Timiras, P.S. Effect of neonatal hypothyroidism and thyroxine on L-(^{14}C) leucine incorporation in protein *in vivo* and the relationship to ionic levels in the developing brain of the rat. *Brain Res.* 4, 143-150, (1967).

Gelber, S., Campbell, P.L., Deibler, G.E., and Sokoloff, L. Effects of L-thyroxine on amino acid incorporation into protein in mature and immature rat brain. *J. Neurochem.* 11, 221-229 (1964).

Geller, E. Effect of thyroxine on neurotransmitter uptake into rat cerebral synaptosomes during development, in *Thyroid Hormones and Brain Development*. G.D. Grave, ed. Raven Press, New York (1977) pp. 215-234.

Gemmill, C.L. Enzymatic mechanisms of thyroxine. *J. Clin. Endocrinol.* 12, 1300-1305 (1952).

Giulian, D., Phorecky, L.A., and McEwen, B. Effects of gonadal steroids upon brain 5-hydroxytryptamine levels in the neonatal rat. *Endocrinology* 93, 1329-1335 (1973).

Globus, A. and Scheibel, A.B. Synaptic loci on visual cortical neurons of the rabbit: the specific afferent radiation. *Exp. Neurol.* 18, 116-131 (1967).

Glowinski, J., Axelrod, J., Kopin, I.J., and Wurtman, R.J. Physiological disposition of ^{3}H-norepinephrine in the developing rat. *J. Pharmac. Exp. Ther.* 146, 48-53 (1964).

Glucksmann, A. Cell death in normal development. *Arch. Biol.* (Liege) 76, 419-437 (1965).

Gottesfeld, Z., Kelly, J.S., and Schon, F. Uptake of γ-aminobutyric acid (GABA) by sensory root ganglia. *Br. J. Pharmacol.* 47, 640 (1973).

Gourdon, J., Clos, J., Coste, C., Daninat, J., and Legrand, J. Comparative effects of hypothyroidism, hyperthyroidism, and undernutrition on the protein and nucleic acid contents of the cerebellum in the young rat. *J. Neurochem.* 21, 861-871 (1973).

Granich, M. and Timiras, P.S. Mechanism of cortisol in maturation of brain lipid patterns in embryonal and young chicks, in *Hormones in Development.* M. Hamburgh and E.J.W. Barrington, eds. Appleton-Century-Crofts, New York (1971) pp. 213-219.

Gray, J.A. and Keynes, L.A. Infant androgen treatment and adult open-field behavior: direct effects and effects of injections to siblings. *Physiology and Behavior* 4, 177-181 (1969).

Green, K. and Matty, A.J. The effects of thyroid hormones on water permeability of the isolated bladder of the toad *Bufo bufo. J. Endocrinol.* 28, 205-211 (1964).

Gregory, E. Comparison of postnatal CNS development between male and female rats. *Brain Res.* 99, 152-156 (1975).

Hajos, F., Patel, A.J., and Balazs, R. Effect of thyroid deficiency on the synaptic organization of the rat cerebellar cortex. *Brain Res.* 50, 387-401 (1973).

Hamberger, A. and Sellstrom, A. Techniques for separation of neurons and glia and their application to metabolic studies, in *Metabolic Compartmentation and Neurotransmission, NATO Advanced Study Institute on Metabolic Compartmentation in Relation to Structure and Function of the Brain.* S. Berl, D.D. Clarke, and D. Schneider, eds. Plenum Press, New York (1974) pp. 145-166.

Hamburgh, M. Evidence of a direct effect on temperature and thyroid hormone on myelinogenesis *in vitro. Develop. Biol.* 13, 15-30 (1966).

Hamburgh, M. An analysis of the action of thyroid hormone on development based on *in vivo* and *in vitro* studies. *Gen. Comp. Endocr.* 10, 198-213 (1968).

Hamburgh, M. and Flexner, L.B. Biochemical and physiological differentiation during morphogenesis. XXI. Effect of hypothyroidism and hormone therapy on enzyme activities of the developing cerebral cortex of the rat. *J. Neurochem.* 1, 279-288 (1957).

Hamburgh, M., Mendoza, L.A., Burkhart, J.F., and Weil, F. The thyroid as a time clock in the developing nervous system, in *Cellular Aspects of Neural Growth and Differentiation.* D.C. Pease, ed. UCLA Forum Med. Sci., University of California Press, Los Angeles (1971) pp. 321-328.

Hanaway, J. Formation and differentiation of the external granular layer of the chick cerebellum. *J. Comp. Neurol.* 131, 1-14 (1967).

Hardin, C.M. Sex differences and the effects of testosterone injections on biogenic amino levels of neonatal rat brain. *Brain Res.* 62, 286-290 (1973a).

Hardin, C.M. Sex differences in serotonin synthesis from 5-hydroxytryptophan in neonatal rat brain. *Brain Res.* 59, 437-439 (1973b).

Harris, G.W. Sex hormones, brain development and brain function. The Upjohn Lecture of the Endocrine Society. *Endocrinology* 75, 627-648 (1964).

Hatotani, N. and Timiras, P.S. Influence of thyroid function on the postnatal development of the transcallosal response in the rat. *Neuro-*

endocrinology 2, 147-156 (1967).

Hebb, C.O. Choline acetylase in the developing nervous system of the rabbit and guinea pig. *J. Physiol.* Lond. 133, 566-570 (1956).

Heim, L.M. and Timiras, P.S. Gonad-brain relationship: precocious brain maturation after estradiol in rats. *Endocrinology* 72, 598-606 (1963).

Henn, F.A. and Hamberger, A. Glial cell function: uptake of transmitter substances. *Proc. Natl. Acad. Sci. U.S.A.* 68, 2686-2690 (1971).

Hokfelt, T. and Ljungdahl, A. Histochemical determination of neurotransmitter distribution, in *Neurotransmitters*. I.J. Kopin, ed. Res. Nerv. Dis. Proc. Vol. 50, Williams and Wilkins, Baltimore, Md. (1972) pp. 1-24.

Holian, O., Dill, D., and Brunngraber, E.G. Incorporation of radioactivity of D-glucogamine-1-^{14}C into heteropolysaccharide chains of glycoproteins in adult and developing rat brain. *Arch. Biochem. Biophys.* 142, 111-121 (1971).

Holzbauer, M. and Youdim, M.B.H. The oestrus cycle and monoamine oxidase activity. *Brit. J. Pharmacol.* 48, 600-608 (1973).

Howard, E. Effects of corticosterone and food restriction on growth and on DNA, RNA, and cholesterol contents of the brain and liver of infant mice. *J. Neurochem.* 12, 181-191 (1965).

Howard, E. Reductions in size and total DNA of cerebrum and cerebellum in adult mice after corticosterone treatment in infancy. *J. Comp. Physiol. Psychol.* 85, 211-220 (1973a).

Howard, E. Increased reactivity and impaired adaptability in operant behavior of adult mice given corticosterone in infancy. *J. Comp. Physiol. Psychol.* 85, 211-220 (1973b).

Howard, E. Hormonal effects on the growth and DNA content of the developing brain, in *Biochemistry of the Developing Brain, II*. W.A. Himwich, ed. Marcel Dekker Inc., New York (1974) pp. 1-68.

Howard, E. and Benjamins, J.A. DNA, ganglioside and sulfatide in brains of rats given corticosterone in infancy, with an estimate of cell loss during development. *Brain Res.* 92, 73-87 (1975).

Howard, E. and Granoff, D.M. Increased voluntary running and decreased motor coordination in mice after neonatal corticosterone implantation *Exper. Neurol.* 22, 661-673 (1968).

Howard, E., Olton, D.S., and Taylor, M.H. Polydipsia in mice and rats given corticosterone in infancy: accentuation by variable internal food reinforcement. *J. Comp. Physiol. Psychol.* 87, 120-125 (1974).

Hudson, D.B., Vernadakis, A., and Timiras, P.S. Regional changes in amino acid concentration in the developing brain and the effects of neonatal administration of estradiol. *Brain Res.* 23, 213-222 (1970).

Hughes, R.C. Glycoproteins as components of cellular membranes, in *Progress in Biophysics and Molecular Biology*, Vol. 26. J.A.V. Butler and D. Nobel, eds. Pergamon Press, New York (1973) pp. 191-268.

Hyyppe, M. and Rinne, U.K. Hypothalamic monoamines after neonatal androgenization, castration, or reserpine treatment of the rat. *Acta. Endocr.* 66, 317-324 (1971).

Ismail-Beigi, F. and Edelman, I.S. Mechanism of thyroid calorigenesis: role of active sodium transport. *Proc. Nat. Acad. Sci.* 67, 1071-1078 (1970).

Ismail-Beigi, F. and Edelman, I.S. The mechanism of the calorigenic action of thyroid hormone. *J. Gen. Physiol.* 57, 710-722, (1971).

Ito, J.M., Valcana, T., and Timiras, P.S. Effect of hypo- and hyperthyroidism on regional monoamine metabolism in the adult rat brain. *Neuroendocrinology* 24, 55-64 (1977).

Iversen, L.L. and Salt, P.J. Inhibition of catecholamine uptake$_2$

by steroids in the isolated heart. *Brit. J. Pharmacol.* 40, 528-530 (1970).

Iversen, L.L., DeChamplain, J., Glowinski, J., and Axelrod, J. Uptake storage and metabolism of norepinephrine in tissues of the developing rat, *J. Pharmac. Exp. Ther.* 157, 509-516 (1967).

Iversen, L.L., Glowinski, J., and Axelrod, J. The physiologic disposition and metabolism of norepinephrine in immunosympathectomized animals. *J. Pharmac. Exp. Ther.* 151, 273-284 (1966).

Jacobson, G. Sequence of myelination in the brain of the albino rat. *J. Comp. Neurol.* 121, 5-29 (1963).

Johnson, T.C. Regulatory mechanisms responsible for alterations in protein and nucleic acid synthesis in developing brain tissue, in *Cellular Aspects of Neural Growth and Differentiation.* D.C. Pease, ed. U.C.L.A. Forum of Medical Science No. 14, University of California Press, Los Angeles (1971), pp. 473-478.

Kamberi, I.A. and Kobayashi, Y. Monoamine oxidase activity in the hypothalamus and various other brain areas and in some endocrine glands of the rat during the estrous cycle. *J. Neurochem.* 17, 261-268 (1970).

Kartzinel, R., Ford, D.H., and Rhines, R.K. Lysine accumulation in the protein-containing fraction of the rat brain: the effect of age, sex, and neonatal castration, in *Influence of Hormones on the Nervous System, Proc. Int. Soc. Psychoneuroendocrinol.* D.H. Ford, ed. S. Karger, Basel (1971) pp. 296-305.

Kato, R. Serotonin content of rat brain in relation to sex and age. *J. Neurochem.* 5, 202 (1960).

Kellogg, C., Vernadakis, A., and Rutledge, C.O. Uptake and metabolism of ^{3}H-norepinephrine in the cerebral hemisphere of chick embryos. *J. Neurochem.* 18, 1931-1938 (1971).

Krawiec, L., Garcia Argiz, C.A., Gomez, C.J., and Pasquini, J.M. Hormonal regulation of brain development. III. Effects of triiodothyronine and growth hormone on the biochemical changes in the cerebral cortex and cerebellum of neonatally thyroidectomized rats. *Brain Res.* 15, 209-218 (1969).

Kremzner, L.T. Metabolism of polyamine in the nervous system. *Fedn. Proc.* 29, 1583-1588 (1970).

Krnjevic, K. Chemical nature of synapatic transmission in vertebrates. *Physiol. Rev.* 54, 418-540 (1974).

Kuhar, M.J., Green, A.I., Snyder, S.H., and Gfeller, E. Separation of synaptosomes storing catecholamines and gamma-aminobutyric acid in rat corpus striatum. *Brain Res.* 21, 405-417 (1970).

Kurihara, T. and Tsukada, Y. The regional and subcellular distribution of 2', 3' -cyclic nucleotide 3'-phosphohydrolase in the central nervous system. *J. Neurochem.* 14, 1167-1174 (1967).

Lapetina, E.G., Soto, E.F., and DeRobertis, E. Gangliosides and acetylcholinesterase in isolated membranes of the rat brain cortex. *Biochem. Biophys. Acta.* 135, 33-43 (1967).

Lauder, J.M. and Krebs, H. Serotonin and early neurogenesis, in *Maturation of Neurotransmission: Biochemical Aspects.* A. Vernadakis E. Giacobini, and G. Filogamo, eds. S. Karger, Basel (1978) pp. 171-180.

Legrand, J. La maturation du cervelet chez la rat blanc hypothyroidien, in *Regional Development of the Brain in Early Life.* A. Minkowski, ed. Blackwell, Oxford (1967a) pp. 485-493.

Legrand, J. Analyse de l'action morphogenetique des hormones thyroidiennes sur le cervelet du jeune rat. (Analysis of the morphogenetic action of thyroid hormones on the cerebellum of young rats).

Archs. Ant. Microsc. Morph. Esp. 56, 205-244 (1967b).

Lerner, L.L., Holthaus, F.J., and Thompson, D.R. A non-steroidal estrogen antagonist, 1-C p-2-diethyl-aminothoxyphenyl-1-phenyl-2-p-methoxyphenol ethanol. *Endocrinology* 63, 295-318 (1958).

Leveque, T.F. and Stern, J.T. A periventricular PAS reactive site in the frog hypothalamus. *Anat. Record* 148, 306 (1964).

Levine, S. The pituitary-adrenal system and the developing brain, in *Progress in Brain Research*. D. deWied and J.A.W.M. Weijnen, eds. Elsevier, Amsterdam (1970), pp. 79-85.

Levine, S. and Mullins, R.J., Jr. Hormones in infancy, in *Early Experience and Behavior*. G. Newton and S. Levine, eds. Charles C. Thomas, Springfield, Ill. (1968), pp. 168-197.

Lewis, P.R. and Shute, C.C.D. The distribution of cholinesterase in cholinergic neurons demonstrated with the electron microscope. *J. Cell Sci.* 1, 381-390 (1966).

Litteria, M. Increased incorporation of ^{3}H-lysine in specific hypothalamic nuclei following castration in the male rat. *Exp. Neurol.* 40, 309-315 (1973a).

Litteria, M. Inhibitory action of neonatal androgenization on the incorporation of [^{3}H] lysine in specific hypothalamic nuclei of the adult female rat. *Exp. Neurol.* 41 no. 2, 395-401 (1973b).

Litteria, M. *In vivo* alterations in the incorporation of [^{3}H] lysine into the medial preoptic nucleus and specific hypothalamic nuclei during the estrous cycle of the rat. *Brain Res.* 55, 234-237 (1973c).

Litteria, M. Inhibitory action of neonatal estrogenization on the incorporation of [^{3}H] lysine into protein of specific limbic and paralimbic neurons of the adult rat. *Brain Res.* 127, 164-167 (1977a).

Litteria, M. The effects of neonatal androgenization on the *in vivo* transport of alpha-aminoisobutyric acid into specific regions of the rat brain. *Brain Res.* 132, 287-299 (1977b).

Litteria, M. Effects of neonatal estrogen on *in vivo* transport of α-aminoisobutyric acid into cat brain. *Exp. Neurol.* 57, 817-827 (1977c).

Litteria, M. and Thorner, M.W. Inhibition in the incorporation of [^{3}H] lysine in the Purkinje cells of the adult female rat after neonatal endrogenization. *Brain Res.* 69, 170-173 (1974a).

Litteria, M. and Thorner, M.W. Alterations in the incorporation of [^{3}H] lysine into proteins of the medial pre-optic area and specific hypothalamia nuclei after ovariectomy in the adult female rat. *J. Endocrinology* 60, 377-378 (1974b).

Litteria, M. and Thorner, M.W. Inhibitory effect of neonatal estrogenization on the incorporation of [^{3}H] lysine in the Purkinje cells of the adult male and female rat. *Brain Res.* 80, 152-154 (1974c).

Litteria, M. and Thorner, M.W. Increased incorporation of [^{3}H] lysine into proteins of the Purkinje cells of the adult male and female rat after castration. *Experientia* 30, 904 (1974d).

Litteria, M. and Thorner, M.W. Inhibitory action of neonatal estrogenization on the incorporation of [^{3}H] lysine into proteins of specific hypothalamic nuclei in the adult, male rat. *Brain Res.* 90, 175-180 (1975a).

Litteria, M. and Thorner, M.W. Inhibition in the incorporation of [^{3}H] lysine into proteins of specific hypothalamic nuclei of the adult female rat after neonatal estrogenization. *Exp. Neurol.* 49, 592-595 (1975b).

Luine, V.N., Khylcherskaya, R.I., and McEwen, B.S. Estrogen effects on brain and pituitary enzyme activities. *J. Neurochem.* 23, 925-934 (1974).

Luine, V.N., Khylcherskaya, R.I., and McEwen, B.S. Effect of gonadal hormones on enzyme activities in brain and pituitary of male and female

rats. *Brain Res.* 86, 283-292 (1975).
Margolis, R.V., Margolis, R.K., Chang, L.B., and Petri, C. Glycosaminoglycans of brain during development. *Biochemistry* 14, 85-88 (1975).
Matthieu, J., Reir, P., and Sawchak, J. Proteins of rat brain myelin in neonatal hypothyroidism. *Brain Res.* 84, 443-451 (1975).
Matthieu, J.M., Quarles, R.H., Brady, R.O., and Webster, H. deF. Variation of proteins, enzyme markers and gangliosides in myelin subfractions. *Biochem. Biophys. Acta.* (Amst.) 329, 305-317 (1973).
Matty, A.J. and Green, K. Active sodium transport in response to thyroxine. *Life Sci.* No. 9, 487-489 (1962).
McCaman, R.E. and Aprison, M.H. The synthetic and catabolic enzyme systems for acetylcholine and serotinin in several discrete areas of the developing cerebral cortex and cerebellum of the rat. *Brain Res.* 6, 621-634 (1967).
McGinnis, J.F. and deVellis, J. Differential hormonal regulation of L-glycerol-3-phosphate dehydrogenase in rat brain and skeletal muscle. *Arch. Biophys.* 179, 682-691 (1977).
Meisami, E., Valcana, T., and Timiras, P.S. Effects of neonatal hypothyroidism on the development of brain excitability in the rat. *Neuroendocrinology* 6, 160-167 (1970).
Meyer-Bahlburg, H.F.L., Grisanti, G.C., and Ehrhardt, A.A. Prenatal effects of sex hormones on human male behavior: medroxyprogesterone acetate (MPA). *Psychoneuroendocrinology* 2, 383-390 (1977).
Mickel, H.S. and Gilles, F.H. Changes in glial cells during human telencephalic myelinogenesis. *Brain* 93, 337-346 (1970).
Mills, I.H., Chan, P.S., and Bartter, F.C. The protein-binding steroids as studied by ultrafiltration. *J. Endocrinol.* 18, 30-31 (1959).
Millichap, J.G. Seizure patterns in young animals: significance of brain carbonic anhydrase. II, *Proc. Soc. Exp. Biol. Med.* 97, 606-611 (1958).
Murphy, D.L. Clinical, genetic, hormonal and drug influences on the activity of human platelet monoamine oxidase, in *Monoamine Oxidase and its Inhibition,* G. Wolstenholme and J. Knight, eds. CIBA Foundation Symposium No. 39 (new series). Elsevier, Amsterdam (1976), pp. 345-346.
Murray, M.R. Nervous tissues *in vitro,* in *Cells and Tissues in Culture Methods, Biology and Physiology.* E.N. Willmer, ed. Academic Press, New York (1965), pp. 373-455.
Myant, N.B. and Cole, L.A. Effects of thyroxine on the deposition of phospholipids in the brain *in vivo* and on the synthesis of phospholipids by brain slices. *J. Neurochem.* 13, 1299-1307 (1966).
Neal, M.J. and Iversen, L.L. Autoradiographic localization of ^{3}H-GABA in rat retina. *Nature* (New Biol.) 235, 217-218 (1972).
Nicholson, J.L. and Altman, J. Synaptogenesis in the rat cerebellum: effects of early hypo- and hyperthyroidism. *Science* 176, 530-531 (1972a).
Nicholson, J.L. and Altman, J. The effects of early hypo- and hyperthyroidism on the development of rat cerebellar cortex. I. Cell proliferation and differentiation. *Brain Res.* 44, 13-23 (1972b).
Norton, W.T. Myelin, in *Basic Neurochemistry.* R.W. Albers, G.J. Siegel, R. Katzman, and B.W. Agranoff, eds. Little, Brown and Co., Boston, Mass. (1972), pp. 365-386.
Norton, W.T. and Pudoslo, S.E. Myelination in rat brain: changes in myelin composition during brain maturation. *J. Neurochem.* 21, 759-773 (1973).
Obata, K. and Takeda, K. Release of γ-aminobutyric acid into the fourth ventricle induced by stimulation of the cat's cerebellum. *J. Neurochem.* 16, 1043-1047 (1969).

Oja, S.S. Studies on protein metabolism in developing rat brain. *Ann. Acad. Sci. Fenn.* 131, 7-81 (1967).

Olton, D.S., Johnson, C.T., and Howard, E. Impairment of conditioned active avoidance in adult rats given corticosterone in infancy. *Develop. Psychobiol.* 8, 55-61 (1974).

Orrego, F. Synthesis of RNA in normal and electrically stimulated brain cortex slices *in vitro*. *J. Neurochem.* 14, 851-858 (1967).

Otsuka, K., Obata, K., Migata, Y., and Tanaka, Y. Measurement of γ-aminobutryic acid in isolated nerve cells of cat central nervous system. *J. Neurochem.* 18, 287-295 (1971).

Pappas, G.D. and Purpura, D.P. Fine structure of dendrites in the superficial neocortical neuropil. *Exp. Neurol.* 29, 145-150 (1973).

Parker, K., Arnold, E., Strong, S., and Vernadakis, A. Stimulation of ornithine decarboxylase activity in culture. *Trans. Amer. Soc. Neurochem.* 10, 130 (1978).

Peachy, C.D. and Greif, R.L. Alterations of mitochondrial structure induced by thyroid hormones *in vivo* and *in vitro*. *Endocrinology* 77, 61-77 (1965).

Pearce, L. and Schanberg, S.M. Hormonal influence on brain organization in infant rats. *Science* 152, 1585-1592 (1966).

Penfield, W. and Jasper, H. *Epilepsy and the Functional Anatomy of the Human Brain*. Little, Brown and Co., Boston, 1954.

Pesetsky, I. Thyroxine-stimulated oxidative enzyme activity associated with precocious brain maturation in anurans: a histochemical study. *Gen. Comp. Endocrinol.* 5, 411-417 (1965).

Pesetsky, I. and Burkart, J.F. A role of thyroid hormones in development of GABA-metabolic enzymes in cerebellar neurons and glia: loss of enzymatic activity in Bergmann cells of hypothyroid rats, in *Thyroid Hormones and Brain Development*. G.D. Grave, ed. Raven Press, New York (1977), pp. 99-106.

Pitt-Rivers, R. and Tata, J.R. *The Thyroid Hormones*. Pergamon Press, London (1959), pp. 99-123.

Plotnikoff, N.P., Breese, G.R., and Prange, A.J. Thyrotropin releasing hormone (TRH): DOPA potentiation and biogenic amine studies. *Pharmac. Biochem. Behav.* 3, 665-670 (1975).

Poduslo, S.E. and Norton, W.T. The characterization of the plasma membrane and myelin fractions obtained from isolated oligodendroglia. *Trans. Amer. Soc. for Neurochem.* 4, 123 (1973).

Prasad, K.N. Differentiation of neuroblastoma cells in culture. *Biol. Rev.* 50, 129-165 (1975).

Prestige, M.C. Differentiation, degeneration and the role of the periphery: quantitative considerations, in *The Neurosciences II*. F.O. Schmill, ed. Rockefeller University Press, New York (1970), pp. 73-82.

Purpura, D.P., Shofer, R.J., and Scarff, T. Properties of synaptic activities and spike potentials of neurons in immature neocrotex. *J. Neurophysiol.* 28, 925-942 (1965).

Quarles, R.H. and Brady, R.O. Synthesis of glycoproteins and gangliosides in developing rat brain. *J. Neurochem.* 18, 1809-1820 (1971).

Rabie, A. and Legrand, J. Effects of thyroid hormone and undernourishment on the amount of synaptosomal fraction in the cerebellum of the young rat. *Brain Res.* 61, 267-278 (1973).

Raina, A. and Janne, J. Polyamines and the accumulation of RNA in mammalian systems. *Fed. Proc.* 29, 1568-1574 (1970).

Ramirez, D.V. and McCann, S.M. Comparison of the regulation of luteinizing hormone (LH) secretion in immature and adult rats. *Endocrinology* 72, 452-464 (1963).

Rastogi, R.B. and Singhal, R.L. Influence of neonatal and adult hyperthyroidism on behavior and biosynthetic capacity for norepinephrine, dopamine, and 5-hydroxytryptamine in rat brain. *J. Pharmacol. Exp. Ther.* 198, 609-618 (1976).

Renson, J. Development of monoaminergic transmission in the rat brain in *Chemistry and Brain Development.* R. Paoletti and A.N. Davison, eds. Plenum Press, New York (1971), pp. 175-184.

Roberts, E. γ-aminobutyric acid and nervous system function: perspective. *Biochem. Pharmacol.* 23, 2637-2649 (1974).

Robinson, D.S., Davis, J.M., Nies, A., Ravaris, C.L., and Sylwester, D. Relation of sex and aging to monoamine oxidase activity of human brain, plasma and platelets. *Arch. Gen. Psychiat.* 24, 536-539 (1971).

Rosenweig, M.R., Bennett, E.L., Diamond, M.C., Yuwu, S., Slagle, R.W., and Saffran, E. Influences of environmental complexity and visual stimulation on development of occipital cortex in rat. *Brain Res.* 14, 427-445 (1969).

Rosman, N.R. and Malone, M.J. Brain myelination in experimental hypothyroidism: morphological and biochemical observations, in *Thyroid Hormones and Brain Development.* G.D. Grave, ed. Raven Press, New York (1977), pp. 169-198.

Rutledge, C.D. The mechanisms by which amphetamine inhibits oxidative deamination of norepinephrine in brain. *J. Pharmacol. Exp. Ther.* 157, 493-502 (1970).

Rutledge, C.D. and Jonason, J. Metabolic pathways of dopamine and norepinephrine in rabbit brain *in vitro.* *J. Pharmacol. Exp. Ther.* 157, 493-502 (1967).

Sachs, Ch., de Champlain, J., Malmfors, T., and Olson, L. The postnatal development of noradrenaline uptake in the adrenergic nerves of different tissues from the rat. *Eur. J. Pharmac.* 9, 67-69 (1970).

Schapiro, S.C. Metabolic and maturational effects of thyroxine on the infant rat. *Endocrinology* 78, 527-532 (1966).

Schapiro, S. Some physiological, biochemical and behavioral consequences of neonatal hormone administration: cortisol and thyroxine. *Gen. and Comp. Endocrinol.* 10, 214-228 (1968a).

Schapiro, S. Maturation of the neuroendocrine response to stress in the rat, in *Early Experience and Behavior: Psychobiology of Development.* G. Newton and S. Levine, eds. Charles C. Thomas, Springfield, Ill. (1968b), pp. 198-257.

Schapiro, S., Salas, M., and Vukovich, K. Hormonal effects on ontogeny of swimming ability in the rat: assessment of central nervous system development. *Science* 168, 147-151 (1970).

Schapiro, S., Vukovich, K., and Globus, A. Effects of neonatal thyroxine and hydrocortisone administration on the development of dendritic spines in the visual cortex of rats. *Exp. Neurol.* 40, 286-296 (1973).

Schrier, B.K. and Thompson, E.J. On the role of glial cells in the mammalian nervous system: uptake, excretion and metabolism of putative neurotransmitters by cultured glial tumor cells. *J. Biol. Chem.* 249, 1769-1780 (1974).

Schwartz, R.J. Steroid control of genomic expression in embryonic chick retina. *Nature New Biol.* 237, 121-125 (1972).

Scott, J.P. Agonistic behavior of mice and rats: a review. *Amer. Zool.* 6, 683-701 (1966).

Sensenbrenner, M., Booher, J., and Mandel, P. Culturation and growth of dissociated neurons from chick embryo cerebral cortex in the presence of different substrates. *Z. Zellforsch.* 117, 559-569 (1971).

Shaskan, E.G. and Snyder, S.H. Kinetics of serotonin accumulation into slices from rat brain: relationship to catecholamine uptake. *J. Pharmac. Exp. Ther.* 175, 404-418 (1970).

Skillen, R.G., Thienes, C.H., and Strain, L. Brain 5-hydroxytryptamine, decarboxylase, and monoamine oxidase of normal, thyroid-fed, and propylthiouracil-fed male and female rats. *Endocrinology* 69, 1099-1102 (1961).

Skou, J.C. Preparation from mammalian brain and kidney of the enzyme system involved in active transport of Na^+ and K^+. *Biochim. Biophys. Acta.* 58, 314-325 (1962).

Skou, J.C. Enzymatic basis for active transport of Na^+ and K^+ across cell membrane. *Physiol. Rev.* 45, 596-617 (1965).

Snyder, S.H. and Coyle, J.T. Regional differences in ^3H-norepinephrine and ^3H-dopamine uptake into rat brain homogenates. *J. Pharmac. Exp. Ther.* 165, 78-86 (1969).

Snyder, S.H., Green, A.I., and Hendley, E.D. Kinetics of ^3H-norepinephrine accumulation into slices from different regions of the rat brain. *J. Pharmac. Exp. Ther.* 164, 90-102 (1968).

Soriero, O. and Ford, D.H. Age and sex: the effect on the composition of different regions of the neonatal rat brain, in *Influence of Hormones on the Nervous System*, D.H. Ford, ed. S. Karger, Basel (1971), pp. 322-333.

Stratton, L.O., Gibson, C.A., Kolar, K.G., and Kastin, A. Neonatal treatment with TRH affects development, learning and emotionality in the rat. *The Neuropeptides: Pharmacol., Biochem. and Behav.* 5, 65-67 (1976).

Sung, S.C. Thymidine kinase in the developing rat brain. *Brain Res.* 35, 268-271 (1971).

Suzuki, K. The pattern of mammalian brain gangliosides. III. Regional and developmental differences. *J. Neurochem.* 12, 969-979 (1965).

Sze, P.Y., Neckers, L., and Tawle, A.C. Glucocorticoids as a regulatory factor for brain tryptophan hydroxylase. *J. Neurochem.* 26, 169-173 (1976).

Tabor, H. and Tabor, C.W. Spermidine, spermine, and related amines. *Pharmac. Rev.* 16, 245-300 (1964).

Tabor, H. and Tabor, C.W. Biosynthesis and metabolism of 1,4-diaminobutane, spermidine, spermine, and related amines. *Adv. Enzymol.* 36, 203-268 (1972).

Terasawa, E., Bridson, W.E., Davenport, J.W., and Goy, R.W. Role of monoamines in release of gonadotropin before proestrus in the cyclic rat. *Neuroendocrinology* 18, 345-358 (1975).

Thompson, R.F. and Shaw, J.A. Behavioral correlates of evoked activity recorded from association areas of the cerebral cortex. *J. Comp. Physiol. Psychol.* 60, 329-339 (1965).

Timiras, P.S. *Developmental Physiology and Aging.* The MacMillian Company, New York (1972).

Timiras, P.S. and Vernadakis, A. Structural, biochemical and functional aging of the nervous system, in *Developmental Physiology and Aging,* Chap. 26. P.S. Timiras, ed. The MacMillian Company, New York (1972), pp. 502-526.

Timiras, P.S., Vernadakis, A., and Sherwood, N. Development of plasticity of the nervous system, in *Biology of Gestation.* N. Assali, ed. Academic Press, New York (1968), pp. 261-319.

Timiras, P.S., Woodbury, D.M., and Baker, P.H. Effect of hydrocortisone acetate, desoxycorticosterone acetate, insulin, glucagon, and dextrose, alone or in combination, on experimental convulsions and carbohydrate metabolism. *Arch. Int. Pharmacodyn.* 105, 450-467 (1956).

Timiras, P.S., Woodbury, D.M., and Goodman, L.S. Effect of adrenalectomy hydrocortisone acetate and desoxycorticosterone acetate on brain excitability and electrolyte distribution in mice. *J. Pharmacol. Exp. Ther.* 112, 80-93 (1954).

Tipton, S.R. Relationship between certain vitamine B factors and response to thyroid of succinoxidase and cytochrome oxidase of rat liver. *Amer. J. Physiol.* 161, 29-34 (1950).

Tissari, A.H., Schonhofer, P.S., Bogdanski, D.F., and Brodie, B.B. Mechanism of biogenic amine transport. II. Relationship between sodium and the mechanism of ovabain blockade of the accumulation of serotonin and norepinephrine by synaptosomes. *Mol. Pharmac.* 5, 593-604 (1969).

Vaccari, A., Brotman, S., Cimino, J., and Timiras, P.S. Sex differentiation of neurotransmitter enzymes in central and peripheral nervous systems. *Brain Res.* 132, 176-185 (1977).

Valcana, T. Effect of neonatal hypothyroidism on the development of acetylcholinesterase and choline acetyltransferase activities in the rat brain, in *Influence of Hormones on the Nervous System.* D.H. Ford, ed. Proc. Int. Soc. Psychoneuroendocrinology, Brooklyn, S. Karger, Basel (1971), pp. 174-184.

Valcana, T. Developmental changes in ionic composition of the brain in hypo and hyperthyroidism, in *Advances in Behavioral Biology. Vol. 9: Drugs and the Developing Brain.* A. Vernadakis and N. Weiner, eds. Plenum Press, New York (1974), pp. 289-304.

Valcana, T. and Eberhardt, N.L. Effects of neonatal hypothyroidism on protein synthesis in the developing rat brain: an open question, in *Thyroid Hormones and Brain Development.* G.D. Grave, ed. Raven Press, New York (1977), pp. 271-186.

Valcana, T. and Timiras, P.S. Effect of hypothyroidism on ionic metabolism and Na-K-activated ATP phosphohydrolase activity in the developing rat brain. *J. Neurochem.* 16, 935-943 (1969).

Valcana, T. and Timiras, P.S. Effect of thyroid hormones on ionic metabolism of the developing rat brain, in *Hormones in Development.* M. Hamburgh and E.J.W. Barrington, eds. Appleton-Century-Crofts, New York (1971), pp. 453-463.

Valcana, T., Vernadakis, A., and Timiras, P.S. Influence of estradiol and cortisol on electrolytes in the central nervous system of developing rats. *Neuroendocrinology* 2, 326-239, (1967).

Valcana, T., Einstein, E.M., Csegtey, J., Dalal, K.R., and Timiras, P.S. Influence of thyroid hormones on myelin proteins in the developing rat brain. *J. Neurol. Sci.* 25, 19-27 (1975).

Vanier, M.T., Holm, M., Ohman, R., and Svennerholm, L. Developmental profiles of gangliosides in human and rat brain. *J. Neurochem.* 18, 581-592 (1971).

Verity, M.A., Brown, W.J., Cheung, M.K., and Czer, G.T. Thyroid hormone inhibition of synaptosome amino acid uptake and protein synthesis. *J. Neurochem.* 29, 853-858 (1977).

Verity, M.A., Brown, W.J., Gheung, M., Huntsman, H., and Smith, R. Effects of neonatal hypothyroidism on cerebral and cerebellar synaptosome development. *J. Neurosci. Res.* 2, 323-335 (1976).

Vernadakis, A. Spinal cord convulsions in developing rats. *Science* 137, 532 (1962).

Vernadakis, A. Hormonal factors in the proliferation of glial cells in culture, in *Influence of Hormones on the Nervous System.* D.H. Ford, ed. Proc. Int. Soc. Psychoneuroendocrinology, Brooklyn, S. Karger, Basel (1971), pp. 42-55.

Vernadakis, A. Changes in nucleic acid content and butyrylcholinesterase activity in CNS structures during the lifespan of chick. *J. Gerontol.* 28, 281-286 (1973a).

Vernadakis, A. RNA synthesis in embryonic cerebellar explants cultured with estradiol, in *Hormones and Brain Function*. K. Lissak, ed. Plenum Press, New York (1973b), pp. 78-79.

Vernadakis, A. Uptake of ^{3}H-norepinephrine in the cerebral hemispheres and cerebellum of the chicken throughout the lifespan. *Mechanisms of Aging and Development* 2, 371-379 (1973c).

Vernadakis, A. Comparative studies of neurotransmitter substances in the maturing and aging nervous system of the chicken, in *Neurobiological Aspects of Maturation and Aging*. D.H. Ford, ed. Elsevier, Scientific Pub. Co., Amsterdam (1973d), pp. 231-243.

Vernadakis, A. Neurotransmission: a proposed mechanism of steroid hormones in the regulation of brain function, in *Proceedings of the Mie Conference of the International Society for Psychoneuroendocrinology*. N. Hatotani, ed. S. Karger, Basel (1974), pp. 251-258.

Vernadakis, A. and Berni, A. Changes in the resting membrane potential of glial cells in culture. *Brain Res.* 57, 223-228 (1973).

Vernadakis, A. and Culver, B. Neural tissue culture: a biochemical tool, in *The Biochemistry of Brain*. S. Kumar, ed. Pergamon Press, Ltd. (1980) pp. 407-477.

Vernadakis, A. and Gibson, D.A. Role of neurotransmitter substances in neural growth, in *Perinatal Pharmacology: Problems and Priorities*. J. Dancis and J.C. Hwang, eds. Raven Press, New York (1974) pp. 251-258.

Vernadakis, A. and Nidess, R. Biochemical characteristics of C-6 glial cells. *Neurochem. Res.* 1, 385-402 (1976).

Vernadakis, A. and Timiras, P.S. Effects of whole-body x-irradiation on electroshock seizure responses in developing rats. *Am. J. Physiol.* 205, 177-180 (1963a).

Vernadakis, A. and Timiras, P.S. Effect of oestradiol on spinal cord convulsions in developing rats. *Nature* (London) 197, 906 (1963b).

Vernadakis, A. and Timiras, P.S. Effects of estradiol and cortisol in neural tissue in culture. *Experentia* 23, 467-468 (1967).

Vernadakis, A. and Timiras, P.S. Pathophysiology of the nervous system disorders, in *Pathophysiology of Gestation*, N. Assali, ed. Vol. 3, Chap. 5. Academic Press, New York (1972), pp. 233-304.

Vernadakis, A. and Woodbury, D.M. Electrolyte and amino acid changes in rat brain during maturation. *Amer. J. Physiol.* 203, 748-752 (1962).

Vernadakis, A. and Woodbury, D.M. Effect of cortisol on the electroshock seizure thresholds in developing rats. *J. Pharmacol. Exp. Ther.* 139, 110-113 (1963).

Vernadakis, A. and Woodbury, D.M. Effects of diphenylhydantoin on electroshock seizure thresholds in developing rats. *J. Pharmacol. Exp. Ther.* 148, 144-150 (1965).

Vernadakis, A. and Woodbury, D.M. The developing animal as a model. *Epilepsia* (Amst.) 10, 163-178 (1969a).

Vernadakis, A. and Woodbury, D.M. Maturational factors in development of seizures, in *Basic Mechanisms of the Epilepsies*. H.H. Jasper, A.A. Ward, and A. Pope, eds. Little, Brown and Co. (1969b), pp. 535-541.

Vernadakis, A., Nidess, R., and Bragg, E.A. Role of glial cells in neural growth, in *Neural Growth and Differentiation*, Vol. 5. E. Meisami and M.A.B. Brazier, eds. International Brain Research Organization, Raven Press, New York (1979) pp. 27-38.

Vernadakis, A., Culver, B., and Nidess, R. Actions of steroid hormones on neural growth in culture: role of glial cells. *Psychoneuroendocrinology* 3, 47-64 (1978).

Vernadakis, A., Nidess, R., Culver, B., and Bragg, E.A. Glial cells: modulators of neuronal environment. *Mechanisms of Ageing and Development* 9, 553-556 (1979a).

Ving, B., Aros, B., Zarand, P., Tork, I., and Wenger, T. Ependymal neurosecretion. II. Gomori-positive secretion in the paraventricular organ and the ventricular ependymal of different vertebrates. *Acta. Morphol. Hung.* 11, 336-350 (1962).

Voeller, K., Pappas, G.D., and Purpura, D.P. Electron microscope study of development of cat superficial neocortex. *Exp. Neurol.* 7, 107-130 (1963).

Von Hahn, H.P. Distribution of DNA and RNA in the brain during the life span of the albino rat. *Gerontologia* 12, 18-29 (1966).

Walravens, P. and Chase, H.P. Influence of thyroid on formation of myelin lipids. *J. Neurochem.* 16, 1477-1484 (1969).

Wardell, W.M. Electrical and pharmacological properties of mammalian neuroglial cells in tissue culture. *Proc. Reg. Soc. B.* 165, 326-361 (1966).

Waymire, J.C., Vernadakis, A., and Weiner, N. Studies on the development of tyrosine hydroxylase, monoamine oxidase and aromatic-L-amino acid decarboxylase in several regions of the chick brain, in *Drugs and the Developing Brain*. A. Vernadakis and N. Weiner, eds. Plenum Press, New York (1974), pp. 149-170.

Weichsel, M.E., Jr. Effect of thyroxine on DNA synthesis and thymidine kinase activity during cerebellar development. *Brain Res.* 78, 455-465 (1974).

Weichsel, M.E., Jr. and Dawson, L. Effects of hypothyroidism and under-nutrition on DNA content and thymidine kinase activity during cerebellar development in the rat. *J. Neurochem.* 26, 675-681 (1976).

Weichsel, M.E., Jr., Hoogenraad, N.J., Levine, R.L., and Kretchmer, N. Pyrimidine biosynthesis during development of the rat cerebellum. *Pediat. Res.* 6, 682-686 (1972).

White, T.D. and Keen, P. Effects of inhibitors of $(Na^+,\ K^+)$-dependent adenosine triphosphalase on the uptake of norepinephrine by synaptosomes. *Mol. Pharmac.* 7, 40-45 (1971).

Whitsett, J.M., Bronson, F.H., Peters, P.J., and Hamilton, T.H. Neonatal organization on aggression in mice: correlation of critical period with uptake of hormone. *Hormones and Behavior* 3, 11-21 (1972).

Whittaker, V.P. The synaptosome, in *Handbook of Neurochemistry*, Vol. 2. A. Lajtha, ed. Plenum Press, New York (1969), pp. 327-364.

Wilkinson, R.T. and Morlock, H.C. Auditory evoked response and reaction time. *Electroenceph. Clin. Neurophysiol.* 23, 50-56 (1967).

Winokur, A. and Utiger, R.D. Thyrotropin-releasing hormone: regional distribution in rat brain. *Science* 185, 265-266 (1974).

Williams-Ashman, H.G. Androgenic control of nucleic acid and protein metabolism in male accessory genital organs. *J. Cell. Comp. Physiol.*, Supp. 1, Vol. 66, 111-124 (1965).

Woodbury, D.M. Effects of hormones on brain excitability and electrolytes. *Rec. Prog. Horm. Res.* 10, 65-107 (1954).

Woodbury, D.M. Relation between the adrenal cortex and the central nervous system. *Pharmacol. Rev.* 10, 275-357 (1958).

Woodbury, D.M. and Esplin, D.W. Neuropharmacology and neurochemistry of anticonvulsant drugs. *Res. Publ. Ass. Res. Nerv. Ment. Dis.* 27, 1-56 (1959).

Wrenn, J.T. and Wessells, N.K. Cytochalasin B: effects upon microfilaments involved in morphogenesis of estrogen-induced glands of oviduct *Proc. Nat. Acad. Sci.* 66, 904-912 (1970).

Yamagami, S., Mori, K., and Kawakita, Y. Changes of thymidine kinase in the developing rat brain. *J. Neurochem.* 19, 369-376 (1972).

Copyright©1982, Spectrum Publications, Inc.
Hormones in Development and Aging

Chapter 8
Current Issues in the Neurobiology of Sexual Differentiation
Richard E. Whalen

INTRODUCTION

When female guinea pigs are treated with testosterone propionate during pregnancy the sexual behavior of their female offspring is permanently altered. The prenatally "androgenized" daughters are less likely than controls to display female-type sexual responses in adulthood, even when administered exogenous estrogen and progesterone, and they are more likely than controls to show male-type mounting responses when administered testosterone. Phoenix et al. (1959) interpreted these findings in the following way: "The embryonic and fetal periods are periods of organization or differentiation in the direction of masculinization or feminization."

While many studies, utilizing several different mammalian species, have confirmed the basic observation of Phoenix and associates that hormonal stimulation during limited, sensitive periods of early development can alter the organism's potential to respond to hormones during adulthood, the interpretation of the observations has changed and advances have been made toward an understanding of the biology of the differentiation process.

In this chapter, we will review three areas of current active research on the psychobiology of sexual differentiation. The first area we discuss is concerned with the conceptual aspects of the problem. It is suggested that masculinization and defeminization are independent, sexually differentiated, neurobehavioral processes and that these two can and should be distinguished from the neural control of gonadotropin secretion. It is further noted that the sexual differentiation of these systems depends upon a very precise pattern of hormonal stimulation during sensitive periods of development and that the expression of sex differences may be dependent or independent of hormonal stimulation after the sensitive period.

The second area covered is concerned with the morphological and biochemical correlates of sex differences. It will be shown that hormonal stimulation during development can influence the characteristics of individual neurons as well as their synaptic connectivity. The possible differential binding of hormones by neural tissues, and their cytoplasm, nuclei, and chromatin will also be examined. It has been noted that neural tissues or their elements may bind hormones more, less, or equally in males, females, and perinatally hormone-treated animals depending upon the study selected. Consensus has not yet been reached. Possible reasons for these differences are suggested but no resolution to this problem is yet evident.

The third area concerns the nature of the differentiating agent. Early work in this area assumed that testicular androgens were the active agents. However, more recent work has suggested that estrogens produced by the aromatization of androgens may indeed be the active agents, at least in some species.

The issue of species differences in differentiation becomes obvious and the literature suggests that the molecular biology of differentiation may differ across species. This presents difficulties for drawing generalizations, yet it was suggested that the differences which exist between species may provide an opportunity for gaining a better understanding of the differentiation process.

It is hoped that this summary of the current research literature on sexual differentiation will help medically oriented investigators and practitioners to become more sensitive to the idea that the brain, as well as the genitalia, is a target tissue for gonadal hormones both during development and in adulthood. Newborns with genital anomalies resulting either from genetic anomalies or atypical hormonal stimulation during gestation may also have atypical neural development. An awareness of this factor may become a consideration in the therapeutic program. Phenotypic females, detected at puberty as being XY genotypically, but insensitive to androgens, may secrete gonadotropins in the male-typical tonic fashion rather than in the female-typical cyclic fashion. This factor might be a consideration for therapy. Finally, the practitioner may be asked to participate in decision making relative to sex-change therapy. An awareness of the theoretical issues raised and experimental findings provided by laboratory scientists should broaden the decision maker's perspective about this controversial procedure.

CONCEPTUAL ASPECTS

The interpretation of the behavioral effects of early hormonal stimulation advanced by Phoenix and associates held that differentiation occurs in the direction of "masculinization or feminization" (italics added). This view, which I termed the linear model (Whalen, 1974), states that masculinity and femininity are opposite ends of the same continuum. As stated by Goy (1970), this model has important implications: "Clearly implied, however, is the corollary that an inverse relationship between the capacities to display masculine and feminine behaviors will normally exist in the adult. That is, feminine characteristics will always be suppressed to an extent which corresponds to the degree of augmentation of masculine characteristics." An early hint that this model may not be fully adequate came from a study in which male and female rats were gonadectomized or gonadectomized and treated with steroids at birth and tested for their potential to display masculine and feminine behavior in adulthood (Whalen and Edwards, 1967). We found that both genetic males and females, regardless of hormonal manipulation at birth, would show male-typical mounting behavior when administered testosterone in adulthood. When administered estrogen and progesterone in adulthood, however, only females ovariectomized at birth or in adulthood and males castrated at birth would show the female-typical response of lordosis which is normally indicative of sexual receptivity. Males castrated when adult, or males and females castrated and steroid-treated at birth failed to display lordosis in adulthood. Thus, both males and females could show feminine behavior in adulthood if gonadal hormonal stimulation was absent during postnatal development and would fail to show such behavior if hormonal stimulation was present shortly after birth. We concluded: "The present data lead us to believe that the essential characteristic of sexual behavioral differentiation in the rat lies in the suppression of the development of the female neurobehavioral control system rather than in the enhancement of the masculine control system." This inhibition of the development of feminine potential we termed defeminization.

The implication of this interpretation is that feminization (and defeminiation) and masculinization (and demasculinization) can occur independently. A number of more recent studies have provided data which are consistent with this hypothesis and which suggest that masculinization and feminization are independent processes (Beach et al., 1972; Clemens, 1974; DeBold and Whalen, 1975). Because of these observations we proposed an "orthogonal" model (see below) of sexual differentiation in which masculinity and femininity are viewed as independent rather than as dependent dimensions of sexuality (Whalen, 1974).

The orthogonal model of sexuality was based upon studies which showed that the experimental manipulation of hormonal levels during various periods of development could result in animals which could display both masculine and feminine sexual behaviors, masculine but not feminine behaviors or vice versa, or neither masculine nor feminine behaviors. Reinisch (1976), however, has noted that such demonstrations do not reflect normal physiological development and that indeed masculinization and defeminization are usually correlated. Reinisch, therefore, proposed what she has termed the "oblique" model: "The oblique model states that masculinization and defeminization, as well as feminization and demasculinization, are correlated, but that one is not necessarily determined, under all circumstance, by the other." Indeed, the point is well taken—normally, the adult male rat displays mounting behavior because he is masculinized by gonadal hormones prenatally (Clemens, 1974) and is unlikely to show feminine responses because he is defeminized by gonadal hormones postnatally. The male hamster is masculinized, but is only partially defeminized by gonadal hormones postnatally. The male hamster is masculinized, but is only partially defeminized, because of the particular pattern of hormonal stimulation which is found in that species during the perinatal period (Tiefer, 1970).

The point is that for a given species and for a given behavior an individual may be capable of displaying the male-typical, the female-typical, or both responses depending upon the precise pattern of hormonal stimulation during a particular period of fetal or neonatal development. Two aspects of this conclusion are important. The first concerns capacity. The normal female rat does not often display male-type mounting responses, nor does the male hamster often display the lordosis response. They may, however, possess the capacity to display these responses lacking only the endogenous hormonal milieu which is required. Consider, for example, normal male and female mice and female mice treated with testosterone at birth. All three groups are gonadectomized at the time of puberty. If these animals are socially isolated for a few weeks and then paired with another mouse, social relations are peaceful. If all three are then treated with testosterone, the male and neonatally androgenized female will begin to attack their new companion, while the normal female will not (Edwards, 1969). The expression of the aggressive behavior, therefore, requires both the capacity to respond to testosterone in adulthood (and this is determined by the presence of androgen during the perinatal period) and the presence of the hormone during adulthood.

Of course not all hormonally differentiated behavior patterns require the presence of hormones for their expression. Several social behaviors displayed by rhesus monkeys fall into this category. For example, male rhesus show facial threat responses more frequently than females, and females androgenized prenatally show levels of this behavior which approach those characteristic of males. Gonadectomy at birth, however, has no influence upon the frequency of display of this behavior (Goy, 1970).

Thus, the capacity to display this behavior is determined by hormonal stimulation during early development, but its expression is hormone independent.

The second facet of our conclusion concerns the "precise pattern of hormonal stimulation" during sensitive periods of development. As noted earlier, in general, female rats ovariectomized at birth, develop the capacity to display male-like mounting responses when administered testosterone in adulthood. Clemens (1974) and Clemens et al. (1978) have shown that not all female rats develop this capacity. Females which develop in all female litters do not respond to testosterone in adulthood. However, females which develop in utero between two males, are capable of displaying high frequencies of mounting. The farther away a female is from a male during uterine growth the less likely she is to show masculine behavior in adulthood. Similarly, female mice which develop in utero between two males are almost as responsive as males in tests of testosterone stimulated fighting; females with no contiguous males in the uterus are quite unresponsive to testosterone in adulthood (Gandelman et al., 1977). Thus, even the microenvironment of hormonal stimulation can critically determine the organism's later behavioral potential.

The fact that male neighbors in utero can masculinize females while not at the same time defeminizing them, provides further evidence that masculinization and defeminization can indeed be independent events. This concept becomes important which one wishes to examine the neural substrate for sex differences. For example, Gregory (1975) reported that a sex difference in nuclear cell volume develops in the somatosensory cortex, but not in the hippocampus, of rats between 25 and 35 days after birth. This neuronal difference could underlie some aspect of sex differences in behavior. However, a simple comparison of the sexes provides no insight into the possible relation of this finding to the sexual differentiation process. The observed difference could reflect the masculinization and defeminization which is characteristic of the male, or the fact that only some females are masculinized normally; the difference could also be quite independent of a hormonally controlled differentiation process. Any assumption that "more male" is equivalent to "less female" would be quite misleading. Masculinization and defeminization are independent, although often correlated.

NEURAL CORRELATES OF SEX DIFFERENCES

Studies of the neural correlates of sexual differentiation have taken two primary directions; morphological and biochemical. The former studies are based on the assumption that "anatomy is destiny," while the latter reflect the development in the last two decades in our understanding of the molecular biology of hormone action (see also Chapters 2, 6, 7).

Morphology

Some of the earliest work on the morphological correlates of neuroendocrine events involved the measurement of nuclear and nucleolar size. Ifft (1964) examined the size of nucleoli in several hypothalamic areas of female rats following hypophysectomy, ovariectomy, thyroidectomy, adrenalectomy, and adrenal demedullation. The latter two operations increased nucleolar size in all 16 areas studied, while hypophysectomy increased nucleolar size in only ten sites, having no effect in the others. Ovariectomy increased nucleolar size in the paraventricular and supraoptic nuclei and in the premammillary and supramammillary areas, and decreased nucleolar size in the arcuate and medial mammillary nuclei. Ovariectomy had no effect in other areas including the preoptic area, anterior hypothalamic area, and ventromedial nucleus, areas which are known to selectively accumulate estradiol (Pfaff and Keiner, 1973). These findings, it may also be noted, are not consistent with the findings of Lisk and Newlon (1963) who reported that implants of estradiol into the arcuate nucleus significantly reduce nucleolar size.

Nuclear and Nucleolar Size

More directly related to the issue of sexual differentiation are the studies of Docke and Smollich (1968), Pfaff (1966), and Staudt and Dorner (1976). Docke and Smollich reported that the size of medial preoptic nuclei is smaller in females than in males, but that this relationship reverses when the rats are castrated 18 days after birth, that is, after the sensitive period of differentiation. Females administered testosterone three days after birth have preoptic nuclei which are intermediate in size between gonadally intact males and females. Nuclei from males castrated on the day of birth are somewhat smaller than those of males castrasted on day 18 and much smaller than those of females ovariectomized on day 18.

Pfaff also found the nuclei to be smaller in gonadally intact females than in males in the ventromedial hypothalamus, dentate gyrus, reticular formation and medial habenular nucleus. He found no sex difference in nuclear area in the lateral preoptic area, neocortex and lateral habenula nucleus. Males castrated seven days after birth, that is after testes-induced defeminization (Feder and Whalen, 1965), had smaller nuclei than intact males in all sites examined except the lateral preoptic area.

Finally, Staudt and Dorner (1976) noted that the nuclear volumes of cells in the central and medial amygdala were smaller in males than in females gonadectomized after the sensitive period of differentiation. Males castrated on the day of birth had nuclei which did not differ from those of females, while males castrated at birth and administered testosterone three days later had nuclei which did not differ in volume from those of males castrated 2-3 weeks after birth. In this study, all animals received the same testosterone treatment before sacrifice.

These studies show that nuclear and nucleolar size are responsive to the hormonal status of the animal. However, since these studies do not for the most part clearly distinguish between the differentiating and concurrent effects of the gonadal hormones upon the morphological measure, it is difficult to evaluate the relevance of the findings to the differential sensitivities of males and females to gonadal hormones. The one exception is the comparison between nuclear size in the preoptic nucleus as reported by Docke and Smollich (1968). Nuclear sizes in the three groups which were sexually differentiated, but which were not being stimulated by gonadal hormones at the time of sacrifice were: females castrated at day 18: 323.3 mu^3; males castrated at day 18: 278.9 mu^3; males castrated at birth: 259.7 mu^3. The males castrated at birth, and which would be expected to show female behavior if administered estrogen (Whalen et al., 1971), possessed nuclei which were similar in size to those found in males castrated after the sensitive period for differentiation and which would not show lordosis behavior if given estrogen. Medial preoptic nuclear size, therefore, appears unrelated to the capacity of the rat to show female-type sexual behavior. It should also be noted that the male rat castrated at birth, unlike the male castrated on day 18, can maintain the female pattern of cyclic gonadotropin secretion. Preoptic nuclear size is therefore also unrelated to the neural control of cyclic gonadotropin secretion.

Possibly more closely related to physiologically important sex differences are the findings of Raisman and Field (1971, 1973). These workers first reported that the morphology of synaptic connections within the preoptic region, but not in the ventromedial nucleus, are sexually differentiated. The number of nonamygdaloid stria terminalis synapses on dendritic shafts relative to the number of synapses on dendritic spines within the preoptic region was found to be higher in males than in females. These authors later reported that preoptic spine synapses per unit area were lower in gonadally intact males, males castrated seven days after birth and in females androgenized four days after birth than in intact females, females androgen-treated 16 days after birth, and males castrated on the day of birth. They concluded that the animals which were potentially cyclic (that is, which should be able to demonstrate cyclic secretion of gonadotropins) had more frequent spine synapses than animals which presumably exhibit a tonic secretion of gonadotropins.

Greenough et al. (1977), using hamsters, have also found morphological differences between the sexes in the preoptic area. Males tended to have longer dendrites than females, dendritic density was greater in males and the dendrites showed a more regular distribution in males than females. Since both males and females were gonadectomized at the time of testing and had received the same prior treatment with exogenous hormones these differences in preoptic morphology presumably reflect "true" sex differences in neural differentiation. However, since the hormonal conditions during the sensitive periods were not manipulated it is not possible to relate these differences to the hormonally controlled sexual differentiation process.

The most recent evidence of morphological differentiation of the brain comes from Gorski and colleagues (1978). These workers have reported that the volume of the medial preoptic nucleus is substantially larger in males than females. This difference was uninfluenced by gonadectomy and by hormone treatment in adulthood. However, the volume of the preoptic area (POA) was reduced in males castrated neonatally and enlarged in females which were androgenized a few days after birth.

These studies make clear that mature males and females differ in both neural and neuronal morphology and in the pattern of neural connections. The extent of these differences has not yet been explored adequately to any degree nor has the relationship of these differences to hormonal sensitivity been determined.

Biochemistry

Studies of the biochemistry of hormone action had their advent in the late 1950s with the production of radiolabeled steroids of high specific activity. These studies have led to the currently held notion that steroids readily enter both target and nontarget tissue cells. Target tissue cells contain specific cytoplasmic proteins, or receptors, which selectively bind particular hormones. This hormone-receptor interaction induces a conformational change in the receptor, termed receptor activation, which allows the complex to enter the cell nucleus. Within the nucleus, the hormone, receptor and/or hormone-receptor complex binds to the nuclear chromatin at specific "acceptor" sites. It is thought that the acceptor sites are acidic, nonhistone proteins which comprise part of the chromatin material. Target and nontarget tissue cell nuclei appear to differ in the degree of "masking" of acceptor sites by other acidic proteins (Spelsberg et al., 1976). Following acceptor site binding, RNA and, later, protein synthesis are initiated. The sequence of molecular events outlined here for the action of steroid hormones has been established in studies of peripheral tissues of the reproductive system such as the uterus, oviducts and prostate. All steps of this sequence have not yet been established for the brain, although it is clear that hypothalamic cells of adult female rats, mice, and guinea pigs contain cytoplasmic receptors for estradiol, that the estradiol can be translocated to the nucleus and that within the nucleus chromatin binding of estradiol does occur (see Chapter 12).

Sex differences could occur at one or more steps in the sequence: (1) cytoplasmic receptor binding (2) receptor activation (3) the translocation process (4) chromatin binding, or (5) activation of the genome.

Whole Tissue Analyses

To obtain information as to whether sex differences do indeed occur in the interaction of gonadal steroids with neural tissues investigators first examined whole tissue accumulation and retention of a radiolabeled steroid. Typically adult animals were administered labeled hormone, presumed target (hypothalamus) and nontarget (cortex) tissues were removed and prepared for autoradiography or liquid scintillation counting either after digestion of the tissue or extraction of the hormone. In some of these studies males were compared to females, but many of the studies compared normal females with females which had been made anovulatory by neonatal treatment with either testosterone or estradiol. In some studies the animals were gonadectomized; in others they were not. These differences could be important for interpreting the results obtained.

Probably the first study to compare males and females was that of Eisenfeld and Axelrod in 1966. These workers found no sex differences in the retention of estradiol in the preoptic region, septum, cerebrum or cerebellum of gonadally intact rats one hour after hormone administration. However, they also found no differential concentration in the preoptic region and "cerebrum," a difference which is commonly found in studies of this type.

The first reports of a sex difference in estradiol retention by the brain appeared in 1968. Pfaff, using autoradiographic procedures, found a greater concentration of radioactivity over cell bodies in limbic structures of female than male rats following estradiol injection. In a parallel autoradiographic study Anderson and Greenwald (1969) reported that female rats, treated with testosterone shortly after birth and ovariectomized when adult, retained less radioactivity in the preoptic area and ventromedial nucleus than did control rats.

In 1969 several reports appeared comparing estradiol binding in the brain of males and females and/or neonatally androgenized females using liquid scintillation techniques. Flerko et al. found significantly less retention of estradiol in the anterior and middle hypothalamus, but not the posterior hypothalamus or cortex of female rats androgenized two days after birth as compared with controls. McGuire and Lisk, however, failed to find an inhibition of estradiol binding in the anterior, middle or posterior hypothalamus (except for the two-hour time point in the middle hypothalamus) in female rats androgenized at birth. The dose of testosterone, 500 µg, and time of treatment, the day of birth, differed from the values used by Flerko and colleagues (1.25 mg TP, two days after birth). McGuire and Lisk, however, did find that neonatal estrogen treatment did inhibit estradiol binding in the anterior hypothalamus and to a lesser extent in the middle and posterior hypothalamus.

In our own laboratory (Green et al., 1969), we found that a relatively high dose of testosterone propionate (500 µg TP) given four days after birth lowered the retention of estradiol by the hypothalamus; a lower dose (100 µg TP) given neonatally had no effect. In addition we found that males, castrated when adult, retained less estradiol in the hypothalamus than either normal females or females which were androgenized neonatally. In this study, however, the animals were given a constant dose of radiolabeled hormone, even though they differed in body weight. When we compared the retention of radioactivity in adult males and females of equal body weight which differed by three weeks in age, we found that males and females did not differ in hypothalamic or preoptic radioactivity levels suggesting that sex differences in body weight could account for differences in target tissue accumulation for the hormone.

Factors Influencing Gonadal Steroid Retention in Brain

The few foregoing studies raised most of the basic issues regarding sexual differentiation and the selective accumulation and retention of gonadal steroids by brain tissues. The first question was whether differences existed among males, females, and neonatally hormone treated females. The initial studies were equivocal on this point and continued research failed to resolve this issue. Some studies reported finding sex differences (e.g., McEwen and Pfaff, 1970; Flerko et al., 1971; McEwen et al., 1970; Maurer and Woolley, 1971, 1974, 1975; Tuohimaa and Johnsson, 1971; Tuohimaa, 1971; Tuohimaa and Niemi, 1972) while others did not (Aguilar et al., 1971; Attramadal, 1970a,b; Attramadal and Aakvaag, 1970; Maurer and Woolley, 1974; Whalen and Luttge, 1970).

The lack of consistency in these findings has several possible causes. Major differences in these studies are the degree, nature, and timing of the hormonal stimulation received by the experimental subjects. In the studies comparing males and females, the animals were gonadectomized at different ages and they were administered radiolabeled steroid after different periods of endogenous hormone deprivation. Even within a given study, where animals are usually sacrificed after the same period of hormone deprivation, it remains an assumption that the target tissues of males and females are in equivalent states after gonadectomy. If this assumption is incorrect, comparisons between studies become tenuous indeed.

In studies comparing females and neonatally hormone-treated females there are large differences in the steroid given neonatally (both testosterone and estradiol were administered as various esters) and in the time and dosage of hormone treatment. Our own work (Green et al., 1969) indicated that dosage can be important and there is no question that the timing of treatment is crucial (Clemens et al., 1969; Gorski, 1968).

Comparisons between studies of males versus females and androgenized females may also be inappropriate. Androgenized females are not males, even though, like males, they may be acyclic and fail to show lordosis behavior. The pattern of hormone stimulation received by the neonatally androgenized animal does not mimic that received by the male during normal development. Androgenization should, there, be considered a pharmacological manipulation. Of course, this does not mean that such procedures are not useful for gaining insight into normal developmental processes.

A second possible basis for the differences between the findings of different studies lies in the procedures used to assess the neural accumulation and retention of hormones. In the studies reviewed, both autoradiographic and liquid scintillation procedures have been employed. The labeled hormones have been administered subcutaneously, intraperitoneally and intravenously. The animals were sacrificed at different times after treatment and the tissues have been processed in different ways. These procedural differences can be important since in most studies the kinetics of binding were not examined. The importance of these details is revealed by the study of McGuire and Lisk (1969) who reported that anterior hypothalamic retention of estradiol did not differ between females and neonatally androgenized females when the tissue was examined either two or four hours after tritium administration. Six hours after hormone treatment more radioactivity was retained by the anterior hypothalamus of the androgenized animals than by the tissue in the controls. Thus, if a single time point of two hours after tritium administration had been selected for analysis, the conclusion would have been that neonatal androgenization has no effect on the binding of estradiol by the anterior hypothalamus. If a six-hour time point had been selected, the conclusion would have been that androgenization facilitates estradiol retention by the anterior hypothalamus. The kinetic analysis, however, revealed that neonatal androgenization has little overall effect on estradiol retention by the brain. Clearly, the particular techniques and parameters of hormone administration used in the reported studies could account for the variance in the results obtained.

Finally, rather trivial factors, such as body weight differences, may also have contributed to the lack of consistency in the findings reported by different laboratories.

Cell Fractionation Studies

Cytoplasmic Binding. As was noted earlier, the cytoplasm of target tissue cells contains proteins, called receptors, which bind the hormone. In early work, Ginsburg and colleagues reported that receptors appeared in the hypothalamus of female rats at about 20 days of age (Ginsburg et al., 1972a) and after puberty showed cyclic variations (Ginsburg et al., 1972b). These workers, however, failed to find any high-affinity cytoplasmic receptors in males (Ginsburg et al., 1971).

While other research did indicate a cytoplasmic accumulation in male and female (Maurer and Woolley, 1974; Whalen and Massicci, 1975) and in neonatally androgenized females (Vertes and King, 1971) it was the latter study which first indicated that protein bound hormone could be found in the defeminized animal. The findings, however, were not simple. Vertes and King, using a protamine precipitate of the estradiol-receptor complex, found *enhanced* binding in both anterior and posterior hypothalamus, but not the cortex, of *immature* androgenized female rats as compared with controls. In *mature* animals, cytoplasmic binding was lower in the anterior hypothalamus and cortex, but not in the posterior hypothalamus of the androgenized animals.

Later work provided even more convincing evidence for the presence of cytoplasmic receptors in males and androgenized females. Maurer (1974), for example, using sucrose gradient procedures, showed selective cytoplasmic binding of estradiol in the preoptic region and median eminence, but not cortex, of male, female, and androgenized female rats. There were no qualitative differences in binding among these groups. That same year, Korach and Muldoon (1974a) also reported finding no quantitative differences in the estradiol binding properties of hypothalamic cytosols from adult gonadectomized male and female rats. This finding was replicated in other work by the same authors using a variety of techniques to assess binding (Korach and Muldoon, 1974b).

It is no longer a question as to whether cytoplasmic estradiol receptors are present in hypothalamic cytosols from males and androgenized females. Questions remain as to whether they are quantitatively the same and whether they demonstrate the same physiological properties as the receptors found in the brains of females. Concerning the first question, the evidence is again equivocal. Korach and Muldoon (1974a,b) found no quantitative differences. Ginsburg et al. (1975) reported that the number of binding sites in the hypothalamus of gonadally intact males was at the level found in proestrous females and approximately one-third the level found in metestrus. Davies et al. (1975) found cytosol binding to be 15% lower in the anterior hypothalamus of immature male as compared with immature female rats. Similarly, Kniewald et al. (1976) using a calorimetric analysis also found receptor levels to be lower in immature male than female rats. However, in the Ginsburg et al. (1975), Kniewald et al. (1976), and Davies et al. (1975) studies, comparisons were made between nongonadectomized males and females. Their values therefore represent net receptor levels existing under unknown conditions of endogenous hormone stimulation. This may be an important consideration since, even in males, castration can alter estradiol binding by hypothalamic cytosols (Vreeburg et al., 1975).

One study at least does suggest a "true" sex differences in hypothalamic cytosol binding. Poppe et al. (1975) found specific binding in the middle hypothalamus of males to be 50% that of female levels when both groups had been castrated for 14 days. These workers found no difference in binding either in the preoptic-anterior hypothalamic area or in the cortex.

Thus, as with total tissue accumulation of estradiol, the currently available studies allow no firm conclusion about baseline levels of estradiol receptors in neural target tissues. Individual studies indicate that the levels in males and androgenized females are higher than or equal to those of normal females, or as low as 50% of females levels. Precise, reliable quantitation remains to be accomplished.

Receptor dynamics. The dynamics of receptors have as yet received even less attention than receptor levels. As noted, following estrogen stimulation, the hormone-receptor complex is translocated to the cell nucleus. This process depletes cytoplasmic receptors; receptor replenishment begins within a few hours and returns to initial levels 6-24 hours later (Cidlowski and Muldoon, 1974; Sarff and Gorski, 1971; Whalen et al., 1975). Cidlowski and Muldoon (1976a) have compared the depletion-replenishment process in gonadally intact and in castrated male and female rats. Estradiol depletes hypothalamic receptors in both sexes regardless of gonadal state. In gonadally intact animals receptor replenishment occurred much more rapidly in males than in females. No sex difference in receptor replenishment was found in the gonadectomized animals. Similarly the sexes did not differ when immature animals were used. Thus, neural sex differentiation as reflected by patterns of gonadotropin secretion and sexual behavior does not seem to be paralleled by differences in receptor depletion and replenishment. However, receptor replenishment does seem influenced by endogenous hormone levels — the gonadally intact, but neither the castrated nor immature males, differed from females.

Cidlowski and Muldoon (1976a) may also have provided further evidence that neonatally androgenized females are not functional males. They produced persistent estrus in females by an intraperitoneal (IP) injection of 15 μg TP three days after birth. When mature these females showed less receptor depletion in hypothalamic cytosols than did control females. However, since these animals were not castrated at the time of treatment, their different patterns of endogenous hormone secretion could account for the results.

Cidlowski and Muldoon (1976b) have also compared the effects of antiestrogens on receptor depletion and replenishment in castrated males and females. They found that the antiestrogen CI-628 failed to deplete receptors in the hypothalamus of females, but caused 50% depletion in males. MER-25 did not cause a receptor depletion in either sex, while dimethylstilbestrol caused dramatic depletion in the female and only modest depletion in the male. While a definitive interpretation of these findings is, of course, not possible at this time, the data suggest that the interaction of estradiol with the protein receptor differs between male and female. Possibly the male synthesizes a cytoplasmic protein with a different conformation or amino acid composition than does the female. Yu (1975) reported finding no such differences, however, using a polyacrylamide gel electrophoretic analysis of forebrain cytoplasmic proteins, although it is possible that subtle differences would not have been detected by the techniques used.

The available data thus suggest that in the male as in the female, cytoplasmic estradiol-receptor complexes are formed, that they are then depleted from the cytoplasm and that receptors are subsequently replenished. The capacity for depletion and replenishment following estradiol stimulation seems qualitatively similar in the two sexes. Reliable quantitative measures are still needed. The validity of such measures, however, will depend upon several variables such as the molecular identity and baseline levels of the receptors in the two sexes. If estradiol binding proteins of different character exist in different amounts in the cytoplasm of neural target cells of males and females, interpretation of observed similarities or differences in depletion-replenishment dynamics will be problematical.

Nuclear Binding. Following translocation from the cytoplasm, estradiol, receptor, or the estradiol-receptor complex is bound within the nucleus, presumably to the chromatin material of target tissues. Vertes and King (1971) reported that under *in vivo* conditions less binding occurred in anterior hypothalamic nuclei of both immature and mature androgenized rats than of controls. In the mature animals they also found reduced binding in the KCl soluble fraction of posterior hypothalamic nuclei. In this study, the mature animals were ovariectomized at the time of sacrifice so the group differences cannot be explained in terms of different patterns of endogenous hormone secretion.

In contrast to Vertes and King (1971), Clark et al. (1972) reported finding no saturable nuclear binding in immature male hypothalamus; such binding was found in the hypothalamus of females. They suggested that any estradiol found in the hypothalamus of males was bound in a nonspecific manner. The difference between these findings could have been due to the difference in the comparison made-males vs. females and females vs. androgenized females. Maurer and Woolley (1974b) attempted to resolve this issue by comparing the in vivo nuclear binding in all three groups following gonadectomy in adulthood. They found no difference between males and females in nuclear binding in the preoptic-anterior hypothalamic area, median eminence region or amygdala, tissues which bound 16, 14, and 7 times as much estradiol as the cortex. Diethystilbestrol (DES) competed for binding in the three active sites in both sexes. Androgenized females, however, showed less binding of estradiol in the preoptic and median eminence regions, but not in the amygdala, when compared with normal females, although binding did occur to a much greater extent in these regions than in the cortex. (It should be noted that in this experiment the cytosols of the androgenized females also accumulated less estradiol than they did in normal females which might account for the difference in nuclear binding.)

These findings are, of course, quite the opposite of what one would have predicted from the Vertes and King (1969, 1971) and Clark et al., 1972 studies which suggested that males differ from females in the nuclear binding process more than do androgenized females. However, the procedures used by Maurer and Woolley (1974b) were more similar to those used by Vertes and King (1971) (in vivo; gonadectomized adult animals) than to those used by Clark et al. (in vitro; gonadally intact immature animals) and these differences could account for the findings. Nonetheless, Maurer and Woolley did confirm the earlier finding that estradiol can be bound to certain neural nuclei of males and androgenized females as well as of females.

The finding that under in vivo conditions estradiol-receptor complexes are translocated to the nucleus of hypothalamic cells of males as well as of females has been confirmed in several studies. A question remains, however, about quantitative differences in nuclear binding which might exist between the sexes. Maurer and Woolley (1974a,b) found no sex differences. Anderson et al. (1973) also found no quantitative differences between males and females in hypothalamic nuclear binding, a finding replicated by Ogren et al. (1976). In our laboratory, however, we did find a male-female difference in nuclear binding using a kinetic analysis (Whalen and Massicci, 1975). During the first hour after the IP administration of labeled estradiol, there was a progressive increase in the accumulation of radioactivity by the hypothalamic but not cortical nuclei of both males and females. Between one and two hours after injection radioactivity levels increased in female hypothalamic nuclei, but declined in male nuclei, and at two hours, female levels were more than twice male levels. In both sexes this binding was of limited capacity as indicated by competition experiments. More recently, Marrone and Feder (1977) have reported a similar finding in guinea pigs, namely, male-female differences in nuclear binding at some time points, but not at others, following the administration of hormone. These later studies suggest that in males certain neural nuclei are capable of binding estradiol, but that the dynamics of the binding system differ between the sexes. To date, a kinetic analysis of nuclear binding in androgenized females has not been reported.

Chromatin Binding. Within the nucleus of peripheral target tissues (e.g., uterus) some aspect of the estradiol-receptor complex binds to the chromatin material. We (Whalen and Olsen, 1978) have recently found that chromatin in hypothalamic, but not cortical nuclei binds estradiol as well. The time course of binding is illustrated in Fig. 1. This binding is of limited capacity and appears to be specific. As with nuclear binding, the kinetics of chromatin binding differ between the sexes. Under the conditions used, chromatin binding was significantly lower in male than in female hypothalamic nuclei 30 and 60 minutes after an IV injection of labeled estradiol. This sex difference disappeared 2-4 hours after treatment (Fig. 1). To date, this is the only report of in vivo chromatin binding in the brain. The findings certainly need to be replicated if the studies of cytoplasmic and nuclear binding provide an appropriate example of the variability of findings in studies of this nature.

THE NEURAL BINDING OF ESTRADIOL AND THE SEXUAL DIFFERENTIATION PROCESS

Strategies

As indicated in this chapter, a great deal of research effort has been directed toward an assessment of the relationship between the cellular binding of estradiol and the sexual differentiation process. Two major strategies have evolved; the direct comparison of binding in males and females, and the comparison of binding in normal females with that in neonatally androgenized (or estrogenized) females. Few investigators have specified the reason for choosing one strategy over the other, however. This choice, nonetheless, is important since the androgenized animal may provide evidence of little relevance to questions of normal male differentiation, and studies of the male may provide little insight into the basis of the clearly pharmacological, persistent estrus syndrome.

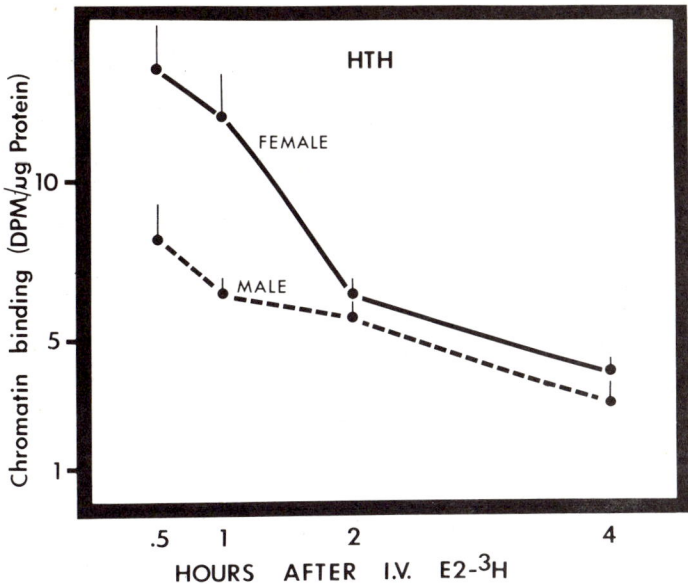

Fig. 1. Chromatin binding of estradiol in hypothalamic (HTH) cells of adult gonadectomized rats 30 minutes to four hours after the intravenous administration of labeled hormone. (From Whalen and Olsen, 1978).

A clarification of the question(s) to be answered might prove useful for future studies of sexual differentiation. For example, if one is interested in neurocellular events mediating cyclic gonadotropin secretion, but not those mediating mating responses, it might be best to compare normal females with females administered a low dose of androgen (10 μg TP) at birth. This treatment will produce acyclicity but it will not inhibit estrogen-induced mating responses. The most common treatment reported in the literature is with doses ranging between 500 μg and 1250 μg TP, doses which alter both gonadotropin secretion and sexual behavior. Similarly, if one is interested in the neural basis of lordotic behavior, it might be best to compare males castrated on the day of birth and administered 10 μg TP with males castrated at birth and given 500 μg TP. Both sets of males will be "masculinized" by endogenous androgens prenatally, neither will support a cyclic secretion of gonadotropin, but the former and not the latter will show lordosis if administered estrogen. Thus by the selective use of gonadectomy plus androgen treatment, one should be able to dissect hormone induced differentiation of the neural systems which control behavior and gonadotropin secretion. These possibilities are outlined in Table 1. Table 1 assumes that all animals are gonadectomized at the time of testing, since endogenous hormones can interact with cellular binding in ways which are not yet fully known.

As can be seen in Table 1, it is possible to prepare genetic males and females which are "acyclic" in terms of gonadotropin release, but which are either capable or incapable of showing lordosis behavior. And it is possible to prepare genetic males and females which are cyclic and are capable of showing lordosis behavior. Absent are males and females which are cyclic but which are incapable of showing lordosis. Rats with these capabilities have not been described probably because of the overlap in

Table 1. Effects of Perinatal Hormonal Manipulation on the Development of the Potential to Display Cyclic Gonadotropin Secretion and Masculine and Feminine Sexual Behavior in Adulthood.

Condition or Treatment	Genetic Sex	Pattern of Gonadotropin Secretion		Masculine Behavior with Testosterone	Feminine Behavior with Estrogen
		Cyclic	Acyclic		

ANIMAL SPECIES: RAT

Condition or Treatment	Genetic Sex	Cyclic	Acyclic	Masculine Behavior with Testosterone	Feminine Behavior with Estrogen
Intact					
Male	M		X[a]	++[b]	-
Female	F	X		+	++
Male Castrated at Birth	M	X		+	++
Low Dose Testosterone	M		X	+	+ to ++
High Dose Testosterone	M		X	++	-
Female Ovariectomized at Birth	F	X		+	++
Low Dose Testosterone	F		X	+	+ to ++
High Dose Testosterone	F		X	+	-

ANIMAL SPECIES: HAMSTER

Intact					
Male	M		X	++	+/-
Female	F	X		-	++
Male Castrated at Birth	M	?		-	++

Table 1. (Continued)

Condition or Treatment	Genetic Sex	Pattern of Gonadotropin Secretion		Masculine Behavior with Testosterone	Feminine Behavior with Estrogen
		Cyclic	Acyclic		
Low Dose Testosterone					
High Dose Testosterone					
Female Ovariectomized at Birth					
Low Dose Testosterone	F	X		+	++
High Dose Testosterone	F		X	+	-

[a] "X" indicates the type of gonadotropin pattern.
[b] "-" indicates absence of the specific behavior. "+" indicates presence of the behavior and the number of + the magnitude of the response.

the times at which the differentiation of gonadotropin secretion and feminine behavior occur and because the inhibition of cyclicity occurs at a lower dose of androgen than does the inhibition of feminine behavior. Possibly with the careful application of steroids systemically or by localized intracerebral implants of androgen (Nadler, 1973) it might be possible to inhibit those systems which control behavior while leaving the gonadotropin control system unaltered.

Table 1 also indicates that all groups are "masculinized" to some degree, that is, they are capable of showing male-type mounting responses when given androgen in adulthood. One could obtain nonmasculinized females by selecting females from all female litters (Clemens, 1974) and possibly one could obtain nonmasculine males and females by treating pregnant rats with an antiandrogen such as cyproterone acetate (Neumann, 1977), by the selective application of steroids and gonadotropin antiserum (Goldman et al. 1972), or by immunizing males against testosterone (Bidlingmaier et al., 1977).

Another alternative solution to the problem of the prenatally masculinized animal is the use of hamsters rather than rats. Normally the adult female hamster will not show male-like mounting responses when administered testosterone, even if given testosterone prenatally (Nucci and Beach, 1971). Low doses of TP given postnatally will facilitate later mounting behavior, but will not inhibit lordosis, while high doses will both increase mounting and inhibit lordosis (Carter et al., 1972; DeBold and Whalen, 1975).

Less is known about the induction of anovulatory sterility in the hamster than rat; however, Alleva et al. (1969), did report that 100 µg TP administered two days after birth would produce anovulatory sterility. The same dose (a total of 500-730 µg TP) given between days 4 and 6 did not interrupt vaginal cyclicity, but in these animals mating did not disrupt cyclicity nor induce pregnancy. Animals treated on day 12 were normal. We found that a single injection of 10 µg TP on the day of birth would inhibit lordosis behavior; possibly that dose would not inhibit ovulation. Were this the case, one could generate an animal which is cyclic but which has an impaired capacity to show female behavior. This condition is difficult to produce in the rat.

This analysis could be extended to other species in which the sensitive periods for the differentiation of the neuroendocrine and masculine and feminine control systems are even more different than in rat or hamster. However, it should be clear from these examples that it is possible to dissociate male from female on a variety of dimensions which probably have different underlying neural systems. Such distinctions would be valuable in studies of the cellular binding of the sex steroids.

Assumptions

It remains an assumption that steroid hormones activate behavior only following genomic activation and the production or inhibition of synthesis of particular proteins. It also remains an assumption that sex differences in behavioral capacity are reflected in differences in cytoplasmic, nuclear, and chromatin binding. These assumptions do not seem overly unreasonable. Agents, such as antiestrogens, which can compete for cellular binding sites can block the expression of behavior following estrogen treatment (Arai and Gorski, 1968; Morin et al., 1976; Powers, 1975; Whalen and Gorzalka, 1973) as do agents such as actinomycin-D and cyclohexamide which can block protein synthesis (Quadagno et al., 1971; Quadagno and Ho, 1975; Whalen et al., 1974). Thus, it is reasonable to assume that estrogen facilitates receptive behavior *because* it is bound to the nuclear chromatin and induces protein synthesis. However, *is* it reasonable to assume that the reduced chromatin binding of estradiol which is found in the hypothalamus of males underlies the reduced effectiveness of estradiol on behavior which is also found in males? As noted earlier, males differ from females not only in their behavior, but also in the nature of their release of gonadotropins.

It is possible that sex differences in chromatin binding are more related to the neural control of gonadotropin secretion than to the neural control of behavior. To test this hypothesis we (Olsen and Whalen, unpublished) have recently studied hypothalamic chromatin binding of estradiol in male and female rats which we hormonally manipulated at birth in such a way as to dissociate these two systems. Some females were administered 10 μg of testosterone propionate at birth. These females became anovulatory, but did display sexual receptivity when hormone treated in adulthood. Some females were given 500 μg of TP at birth. These females were both anovulatory and behaviorally inhibited. Some males were castrated at birth and others 10 days after birth. The former would show lordosis in adulthood; the latter would not. The behavioral effects of these treatments are illustrated in Fig. 2. Following behavioral testing we examined hypothalamic chromatin binding in these animals. As is shown in Fig. 3, the animals which were behaviorally responsive to estrogen in adulthood showed the highest levels of chromatin binding regardless of their genetic sex. The females given a high dose of testosterone at birth showed less binding, while males castrated after the sensitive period showed the least binding. Thus a correlation exists between the behavioral sensitivity of estrogen and hypothalamic chromatin binding. The presence or absence of cyclic gonadotropin secretion (females, day 1 castrate males vs. androgenized females, and day 10 castrate males) did not correlate with chromatin binding. A comparison of Figures 2 and 3 also indicates that the magnitude of behavioral inhibition produced by early postnatal hormone stimulation is greater than the inhibition of chromatin binding.

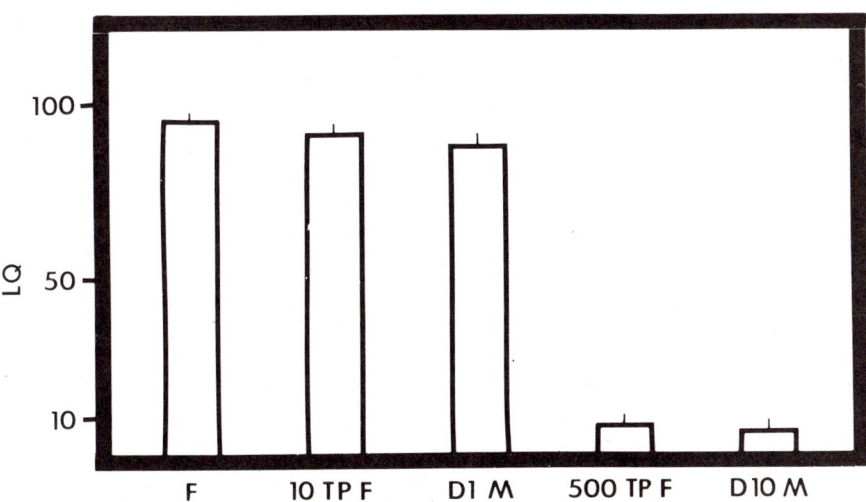

Fig. 2. Lordosis behavior (LQ = lordosis responses/mounts x 100) in adult gonadectomized rats administered estradiol benzoate and progesterone. The animals were hormonally manipulated on the day of birth. F = oil-treated females; 10 TP F = females given 10 μg of testosterone propionate at birth; D1 M = males castrated at birth; 500 TP F = females given 500 μg of testosterone propionate at birth; D10 M = males castrated 10 days after birth.

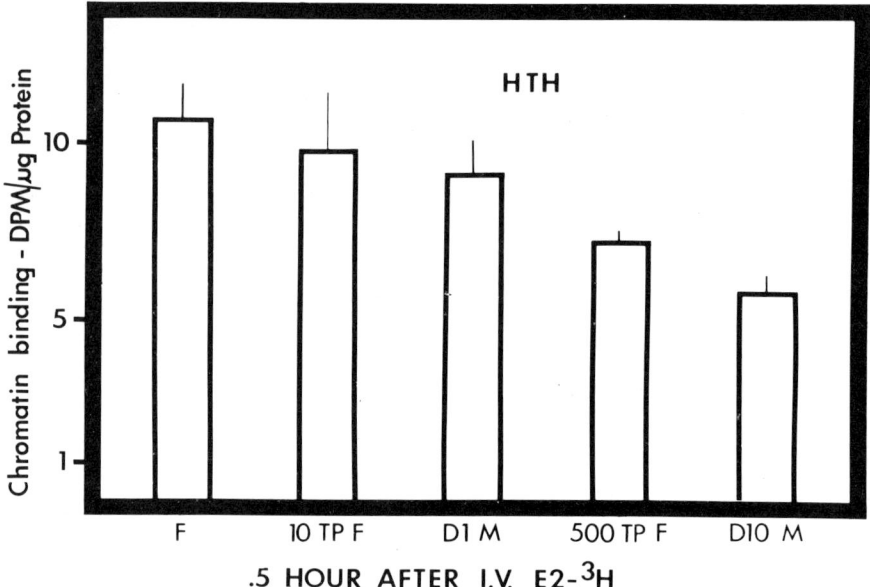

Fig. 3. Chromatin binding of estradiol in hypothalamic (HTH) cells of adult gonadectomized rats hormonally manipulated at birth. The animals were sacrificed 30 minutes after the intravenous administration of tritiated estradiol. Group designations are listed under Figure 2.

It is important to note that all estrogen-sensitive systems are not sexually differentiated or are not sexually differentiated to the degree that gonadotropin regulation and lordosis behavior are differentiated. As mentioned both males and females show estrogen induced inhibition of gondotropin secretion (negative feedback). Male and female rats both show male-type mounting responses when given estrogen. Estrogen stimulates running activity and causes a reduction of weight in both sexes, although the latter function is partially sex differentiated. Thus the fact that chromatin binding occurs in all groups, including males, should not be surprising.

If we assume that the chromatin binding found in males is attributable to nondifferentiated functions and that in the females is attributable to both nondifferentiated and differentiated functions we are led to the conclusion that 57% of maximal binding is related to nondifferentiated functions, 6% is related to cyclic secretion of gonadotropin and 37% is related to lordosis behavior (Fig. 4). While these percentages are certainly not precise, this analysis may prove to be a useful way of conceptualizing chromatin binding of estrogen. Estrogen certainly subserves multiple functions and it is reasonable to think that different numbers of cells are devoted to these different functions. By altering functions selectively as in this study we altered chromatin binding. At least in theory one could subdivide all estrogen binding into different functional systems.

At a superficial level, the present findings suggest that chromatin binding of estradiol is essential for the elicitation of receptive behavior. This interpretation, however, requires that some assumptions be made.

The essential problem for the interpretation of the chromatin (or nuclear) binding studies is that they require the assumption that there are no deficits in cytoplasmic binding or in the activation and translocation

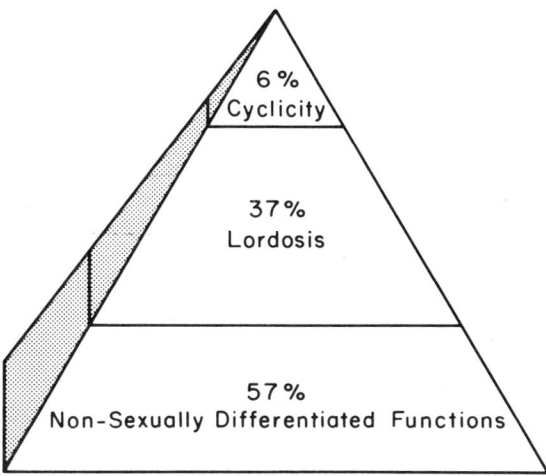

Fig. 4. Partition of estradiol binding by hypothalamic nuclear chromatin into functional groups. Both males and females possess estrogen regulated (nonsexually differentiated) functions such as weight control and over half (57%) of chromatin binding can be attributed to such functions. Estrogen also controls sexual behavior (lordosis in rats) (37%) and is involved in the stimulation of the cyclic release of gonadotropins (6%). Lordosis and the positive feedback action of estrogen are seen only in females.

of the hormone-receptor complex in males and androgenized females relative to normal females or neonatally castrated males. Unfortunately, the studies reviewed above provide no compelling reason to believe that the number or nature of the binding sites in hypothalamic cytoplasm are indeed the same in males, females and androgenized females. If, as suggested by Davies et al. (1975) and are 15 percent fewer sites in males than females, this difference could account for the reported sex differences in nuclear and chromatin binding. Even if the number of binding sites were the same (Korach and Muldoon, 1973) the sexes might differ in receptor dynamics (Cidlowski and Muldoon, 1976), a factor which could also lead to sex differences in nuclear binding. Clearly more must be known about the initial stages of hormone-cell interaction before later events can be readily interpreted.

Finally one other (of several possible) assumption must be identified, namely the assumption that the locus of the neural basis of sex differences lies in hormone sensitive elements in the hypothalamus. A great deal of data indicate that hypothalamic and limbic system structures, but not the cortex, of adult rats contain cytoplasmic estrogen receptors and can display a nuclear binding of estradiol. Moreover, numerous studies have shown that these structures respond to the local administration of estradiol with changes in behavior and gonadotropin secretion. However, it need not be these cells that are sexually differentiated. The neuronal systems which control sexual behavior in the adult animal presumably contain both hormone dependent and hormone independent elements, and it could be the latter which are altered by hormones during the sensitive period of differentiation. This possibility is suggested by recent studies which show the presence of numerous cytoplasmic estradiol binding sites and of nuclear binding in the cortex of newborn rats. Cytoplasmic and nuclear binding of estradiol disappear, however, in the cortex, but not in hypothalamic and other limbic sites as the animal matures (Barley et al., 1974; MacLusky et al., 1976; McEwen et al., 1975; Westley and Salaman, 1976). Thus, particular neurons could possess cytoplasmic receptors and nuclear acceptor sites at one stage of the life cycle and not at another. Hormonal stimulation during the sensitive (receptor-acceptor present) period could determine neuronal functioning during a later hormone-independent (receptor-acceptor absent) period. During adulthood, estradiol could interact with hormone responsive cells in the hypothalamus and limbic system which are not sexually differentiated and yet have different behavioral and physiological effects because certain cells which are not estradiol targets, but which are sexually differentiated, react differently to inputs from hormone-responsive cells.

Two examples might illustrate this possibility. First, as noted earlier, not all behaviors which are sexually differentiated as a result of hormonal stimulation during a sensitive period require hormonal stimulation for their later expression. Prenatally androgenized female rhesus monkeys show socially assertive behaviors at male-like frequencies even when castrated at birth (Goy, 1970). Presumably, therefore, these behaviors do not require a cytoplasmic receptor-nuclear acceptor system for their display. Second, Goldman et al. (1974) have shown that cortical lesions in year-old rhesus monkeys have different effects on the sexes. These lesions disrupt delayed-response performance in males but not in females. This sex difference in cortical function is also probably not mediated by a steroid receptor system at the time of the lesion, but probably reflects sex differences in brain maturation which could have been sensitive to hormones at some earlier stage of development. Thus, there appear to be examples of sexually differentiated brain-behavior relationships which are independent of hormonally activated steroid receptor systems in the hypothalamus. Our assumption, therefore, that the neural basis of sexually differentiated behaviors "resides" in the hypothalamus must be considered tentative.

NATURE OF THE DIFFERENTIATING SUBSTANCE

Defeminization

In his recent book, Dorner (1976) reviews the work in his laboratory on the effects of hormonal stimulation during the perinatal period upon the development of sexual behavior in rats. He notes that early investigators (e.g. Wilson, 1943) found that estrogen as well as androgen treatment of neonatal female rats would lead to anovulatory sterility and that such rats do not mate when given estrogen and progesterone. These findings have been confirmed in various laboratories (Whalen and Nadler, 1963; Kincl et al., 1965; Dorner, 1976). Dorner felt that this effect was "paradoxical," testosterone being the normal differentiating agent, and due to doses of estrogen which were "unphysiologically high." Indeed, the estrogen-induced inhibition of feminine behavior may be pharmacological. However, the theory has developed that estrogen *is* the active agent in the diffentiation of the brain. This hypothesis derives both from the observation that perinatal estrogen treatment can mimic the effects of testosterone and from the observation that neural tissues can aromatize androgens to estrogens (Naftolin et al., 1971, 1972). Plapinger and McEwen (1978) have recently provided an excellent review of this problem.

The initial suggestion that testosterone and androstenedione would defeminize only following their metabolism to estradiol and estrone, respectively, received support from studies indicating that androgens such as dihydrotestosterone which are not aromatized (in a placental microsome system, Ryan, 1960) do not defeminize the rat (McDonald and Doughty, 1972a) even when administered chronically in subcutaneously implanted silastic capsules (Whalen and Rezek, 1974). Two studies, however, dispute this conclusion (Paup et al., 1975; Gerall et al., 1976).

Studies utilizing agents that antagonize the actions of estrogen have also been used to study the hormonal control of the differentiation process. In the first study of this type, McDonald and Doughty (1972b) showed that the antiestrogen MER-25 could block testosterone-induced persistent estrus; the behavioral effects of the treatment were not analyzed. Later, these workers *failed* to demonstrate a protective effect of MER-25 on receptivity (McDonald and Doughty, 1973/1974; Doughty and McDonald, 1974); however, the dose of testosterone used in these studies failed by itself to inhibit later lordosis behavior. Similar findings were reported by Brown-Grant (1974).

In 1975, Doughty et al. (1975a) demonstrated that the synthetic estrogen RU-2858 would inhibit later receptivity when given neonatally, much as does estradiol benzoate; indeed, RU-2858 was found to be 100 times more potent than EB. In a companion paper (Doughty et al., 1975b) these workers did show that the antagonist MER-25 could block the behavioral inhibition produced by RU-2858.

While this last study, along with the studies using dihydrotestosterone, is consistent with the aromatization hypothesis, it in no way proves that aromatization is a necessary prerequisite for the defeminizing actions of androgens as suggested by Booth (1977) who stated: "Testosterone must be converted to oestrogen in order to influence sexual differentiation of the neonatal brain." Booth did however, report that male rats castrated at birth and administered testosterone plus the aromatization inhibitor androst-4-ene-3,6,17-trione (ADT) were protected from the behavioral defeminization induced in males treated with testosterone alone (1978). Unfortunately, no quantified data were presented so the magnitude of the "protection" is not known.

More compelling evidence for the aromatization hypothesis using androst-1,4,6-triene-3,17-dione (ATD) as an inhibitor has recently been provided by McEwen et al. (1977). These workers implanted newborn rats with silastic capsules containing testosterone, ATD, T and ATD, or cholesterol. The implants were removed 10 days after birth. When adult the ATD treated males displayed lordosis when given estrogen and progesterone. Cholesterol-treated males showed no lordotic responding. Cholesterol-treated females did show lordosis, those treated with testosterone showed infrequent lordosis and those given T and ATD neonatally were indistinguishable from control females. The aromatization inhibitor clearly prevented the defeminizing action of exogenous testosterone in the females and of endogenous testicular hormones in the males. In the females ATD by itself had no defeminizing action. (These investigators also reported finding a partial protective effect of CI-628, an anti-estrogen which is more potent than MER-25.) These data provided support for the hypothesis that the defeminizing actions of testosterone are due to the aromatization of that steroid to estradiol.

The use of nonaromatizing androgens, estrogen antagonists, and aromatase inhibitors supplement the early studies showing that estrogen given neonatally could defeminize the rat. Indeed, these later studies may provide even more convincing evidence for the theory than do the estrogen-treatment studies themselves. One of the apparent problems with the estrogen-treatment studies was with the dose needed to alter normal differentiation. For example, Brown-Grant (1973) reported that estradiol benzoate was only three times more potent than testosterone propionate in inducing defeminization, yet Reddy et al. (1974) reported that the percentage of androstenedione converted to estrone by neonatal limbic system structures was only 0.1-0.37%. One would, therefore, expect estradiol to be over 100 times more potent then testosterone in inducing defeminization rather than the three times greater potency which was observed.

This problem appears to have been resolved by the finding that the neonatal brain contains large amounts of an estrogen-binding macromolecule which is similar to or identical with plasma α-fetoprotein (Plapinger et al., 1973). This fetoneonatal estrogen binding protein may well protect the newborn rat from the defeminizing actions of estrogen. Consistent with this hypothesis is the finding that RU-2858, which does not readily bind to the fetoneonatal protein, defeminizes the female rat at a dose which is 1/100 of that needed for testosterone propionate to be effective (Doughty et al., 1975a). In our own laboratory, we have found that in hamsters neonatal treatment with 5 μg, but not 1 μg of TP would significantly reduce lordosis duration (DeBold and Whalen, 1975). Only 0.1 ng of RU-2858 was needed to inhibit lordosis behavior to a similar degree (Whalen and Etgen, 1978). Thus RU-2858 was found to be 50,000 times more potent than TP in inducing defeminization in this species. Clearly estrogens can defeminize at doses which are consistent with magnitude of the intracellular conversion of androgens to estrogens.

Species Differences. The studies of the defeminizing effects of RU-2858 cited above indicate that the hamster is some 500 times more responsive to this synthetic than is the rat. This may or may not be true, as we have found that in hamsters the dose-response relationship between RU-2858 and the inhibition of lordotic behavior is not linear (Fig. 5) and to date no one has studied the defeminizing effects of nanogram-level doses of estradiol in rats. Nonetheless, the possibility exists that there may be important species differences in the hormonal control of the differentiation process. Indeed, it would be surprising if such differences were not found, since the rodents most commonly studied (rat, mouse, guinea pig, and hamster) differ in their responsiveness to hormones in adulthood. For example, estradiol benzoate will induce the complete pattern of masculine behavior in castrated male rats (Gorzalka et al., 1975) and mice (Edwards and Burge, 1971) but this treatment is ineffective in male guinea pigs (Alsum and Goy, 1974) and only partially effective in male hamsters (Noble and Alsum, 1975). Dihydrotestosterone (DHT) is almost completely ineffective in most strains of rat (Whalen and Luttge, 1971; Feder, 1971; Paup et al., 1975), is weakly active in hamsters (Whalen and DeBold, 1974), is active in SW mice but not in males of the CD-1 strain (Luttge and Hall, 1973), and is as effective as testosterone in male guinea pigs (Alsum and Goy, 1974). Species differences in the response to androgens and estrogens are indeed great.

With respect to the hamster, it has already been noted that defeminization can be induced by testosterone propionate, estradiol benzoate, and RU-2858. Gerall et al. (1975) have recently reported that both androsterone and dihydrotestosterone can defeminize the female hamster. Neither steroid was as potent as testosterone even though the animals received chronic stimulation for nine days, beginning the day after birth. Payne (1976) has also reported decreased lordosis duration in adult hamsters following brief exposure to DHT at birth. These findings contrast with those or Paup et al. (1972) and Coniglio et al. (1973) however, who failed to find reduced receptivity following neonatal treatment with andro-

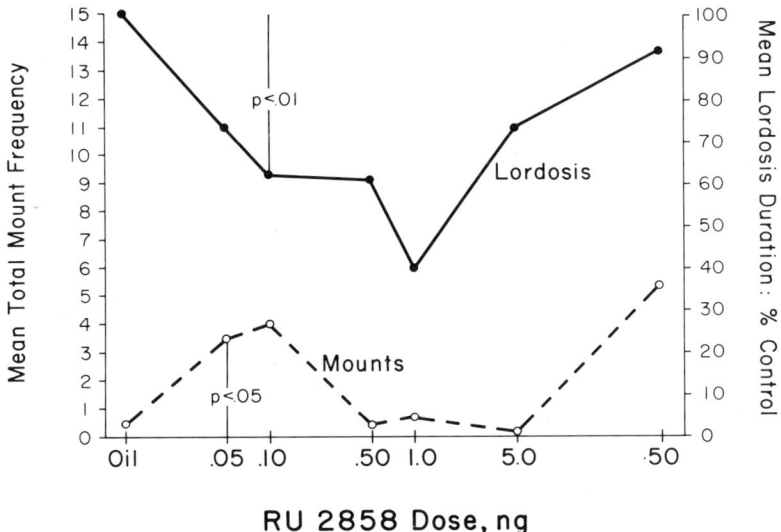

Fig. 5. Mounting and lordosis behavior in adult ovariectomized hamsters given oil or 0.05-50 ng of the synthetic estrogen RU-2858 at birth. When adult, the animals were treated with testosterone propionate or estradiol benzoate and progesterone before testing. (From Whalen and Etgen, 1978).

sterone and DHT in hamsters. These latter two studies did, however, report that defeminization was induced by the synthetic estrogen DES. Thus, in the hamster, testosterone propionate and estrogens defeminize and nonaromatizable androgens are weakly effective in this regard. This parallels the effects of these steroids in adult hamsters.

The findings with the hamster are important in showing that the aromatization hypothesis, even if correct for the rat, may not have general applicability. In hamsters, two androgens (androsterone and DHT) which are not thought to undergo aromatization, can suppress the female's potential to display lordosis in adulthood. Of course, the possibility remains that these androgens can be aromatized by neural tissue in the hamster. The potential for aromatization of these steroids has only been tested in the placental microsome system! There is no question that neural tissue can aromatize testosterone and androstenedione; the failure of the *brain* to aromatize androsterone and DHT, however, has not been tested. It should be noted that Sholl et al. (1975) failed to find evidence of aromatization in adult guinea pigs indicating that even the aromatization of testosterone is not a universal process.

Masculinization

Most of the studies directed toward identifying the active agent in the differentiation process have focused upon the hormone-induced defeminization of the neural systems underlying receptive behavior. Possibly this reflects the fact that in the rat masculinization occurs in the most part prior to birth, and the rat has been the most actively studied species. Some authors might dispute this statement. However, it is clear that castration of the male rat at birth does not prevent the later display of mounting behavior; intromission and ejaculatory behaviors are less frequent (Whalen and Edwards, 1967). Controversy remains whether changes in intromission and ejaculation behavior can be accounted for solely in terms of inadequate penile development (Whalen, 1968; Beach et al., 1969) or whether they also reflect incomplete central nervous system (CNS) differentiation (Hart, 1977). Because of this difficulty in distinguishing the central neural and peripheral hormone effects in the rat, this species is less than ideal for studying masculinization. The masculinizing actions of hormones, however, may better be studied in the hamster, since in this species, the female does not respond to testosterone in adulthood with the display of mounting responses unless she is stimulated by hormones shortly after birth (Nucci and Beach, 1971; Swanson and Crossley, 1971; DeBold and Whalen, 1975). In 1972, Paup et al., demonstrated that masculinization could be induced in the hamster by neonatal treatment with DES as well as testosterone and TP but not by androsterone, a nonaromatizable androgen. In a later paper Coniglio et al. (1973) extended this research to show that masculinization could also be induced by estradiol and estradiol benzoate but not by DHT. Payne (1976) has since replicated the finding that DHT is inactive.

The evidence thus strongly suggests that, in the hamster at least, both masculinization and defeminization may be controlled by estrogens. However, in earlier work (DeBold and Whalen, 1975) we had found that masculinization could be induced in the female hamster with a dose of TP (1 μg) which did not defeminize. Assuming that the rate of aromatization was similar to that in the rat (that is less than 1%) the data suggested that if estradiol were the active masculinizing agent it should be so at a dose of approximately 5-10 ng. It seems unlikely that such a dose of estradiol would masculinize. We, therefore, treated newborn hamsters with 500-2000 ng of estradiol benzoate (or with one μg TP) at birth. We found to our surprise that the lowest dose administered significantly facilitated mounting behavior in adulthood and that 500 ng induced a more intense masculinization than did one μg TP (Fig. 6); also see Whalen and Etgen, 1978.

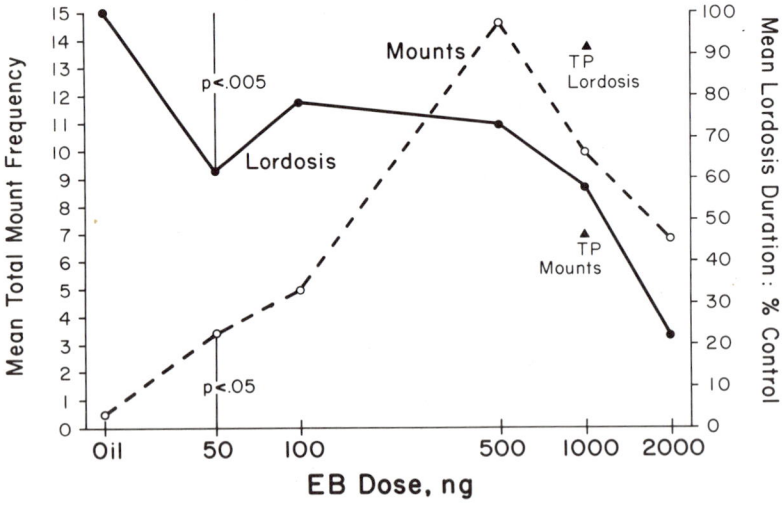

Fig. 6. Mounting and lordosis behavior in adult ovariectomized hamsters given oil, one μg of testosterone propionate (TP) or 50-2000 ng of estradiol benzoate (EB) at birth. When adult, the animals were treated with testosterone propionate or estradiol benzoate before testing.

Treatment of newborn female hamsters with the synthetic estrogen RU-2858, which apparently does not bind to plasma α-fetoprotein, also masculinized and did so at a dose of 0.05 ng. The dose-response relationship between RU-2858 and mounting behavior was not linear (Fig. 5) which raises questions about the nature of the mechanism of action of this synthetic. Nonetheless, that fact that 50 pg of an estrogen can masculinize is certainly consistent with the hypothesis that estrogens can be involved in neural differentiation under normal physiological conditions.

In spite of the increasing number of observations which suggest validity of the aromatization hypothesis, a note of caution must be expressed. Androgens as well as estrogens are bound in the brain of developing rats and mice (Fox, 1975; Kato, 1976). Lieberburg et al. (1977) for example, found that following the administration of testosterone to neonatal female rats the brain nuclear concentration of testosterone was 1.54 fmoles/mg protein; the concentration of estradiol was 1.78 fmoles and that of DHT was 0.65 fmoles. Thus, while estradiol is indeed found in brain nuclei, this does not occur at the expense of testosterone and DHT binding. There is no compelling reason to assume that the androgens which are found play no role in the differentiation process.

A second reason for caution relates to the species studied. Even if it becomes evident that aromatization is critical for differentiation of the rat or hamster brain, this metabolic process need not be involved in all mammalian species. As noted, the adult guinea pig does not appear to possess neural aromatizing enzymes (Sholl et al., 1975). Possibly this species achieves differentiation solely through the actions of androgens. This caution is expressed because it has become popular to assume that the rat provides an adequate model for the study of differentiation. It may not.

A clear understanding of the molecular events underlying the sexual differentiation process in lower animals should help us to understand the human sexual differentiation process. There is no question that differentiation in our own species is also regulated by hormonal events prenatally and possibly postnatally as well. We need to know how these processes are regulated and we need to know the behavioral and neuroendocrine outcomes of these processes. Our studies of animals are generating what should prove to be fruitful hyoptheses about human development.

ACKNOWLEDGMENTS

The research of the author was supported by grant HD-00893 from the National Institute of Child Health and Human Development.

REFERENCES

Aguilar, E., Schiaffini, O., and Oriol-Bosch, A. Estradiol-H^3 uptake by the hypothalamus and the lymbic structures of the androgenized female rats. *J. Neuro-Visc. Rel.* Suppl X., 112-116 (1971).

Alleva, F.R., Alleva, J.J., and Unberger, E.J. Effect of a single prepubertal injection of testosterone propionate on later reproductive functions of the female gold hamster. *Endocrinology* 85, 312-318 (1969).

Alsum, P. and Goy, R.W. Action of esters of testosterone, dihydrotestosterone or estradiol on sexual behavior in castrated male guinea pigs. *Horm. Behav.* 5, 207-217 (1974).

Anderson, C.H. and Greenwald, G.S. Autoradiographic analysis of estradiol uptake in the brain and pituitary of the female rat. *Endocrinology* 85, 1160-1165 (1969).

Anderson, J.N., Peck, Jr., E.J., and Clark, J.H. Nuclear receptor estrogen complex: Accumulation, retention and localization in the hypothalamus and pituitary. *Endocrinology* 93, 711-717 (1973).

Arai, Y. and Gorski, R.A. Effect of anti-estrogen on steroid induced sexual receptivity in ovariectomized rat. *Physiol. Behav.* 3, 351-353 (1968).

Attramadal, A. Cellular localization of ^{3}H-oestradiol in the hypophysis. *Z. Zellforsch* 104, 572-581 (1970a).

Attramadal, A. Cellular localization of ^{3}H-oestradiol in the hypophysis. *Z. Zellforsch* 104, 597-614 (1970b).

Attramadal, A., and Aakvaag, A. The uptake of ^{3}H-oestradiol by the anterior hypophysis and hypothalamus of male and female rats. *Z. Zellforsch* 104, 582-596 (1970).

Barley, J., Ginsburg, M., Greenstein, B.D., MacLusky, N.J., and Thomas, P.J. A receptor mediating sexual differentiation? *Nature* 252, 259-260 (1974).

Beach, F.A., Kuehn, R.E., Sprague, R.H., and Anisko, J.J. Coital behavior in dogs. XI. Effects of androgenic stimulation during development on masculine mating responses in females. *Horm. Behav.* 3, 143-168 (1972).

Beach, F.A., Nobel, R.G., and Orndoff, R.K. Effects of perinatal androgen treatment on responses of male rats to gonadal hormones in adulthood. *J. Comp. Physiol. Psychol.* 68, 490-497 (1969).

Bidlingmaier, F., Knorr, D., and Neumann, F. Inhibition of masculine differentiation in male offspring of rabbits actively immunised against testosterone before pregnancy. *Nature* 266, 647-648 (1977).

Booth, J.E. Effects of the aromatization inhibitor, androst-4-ene-3,6,17-trione on sexual differentiation induced by testosterone in the neonatally castrated rat. *J. Endocr.* 72, 53P-54P (1977).

Brown-Grant, K. Recent studies on the sexual differentiation of the brain, in *Foetal and Neonatal Physiology*. K.S. Comline, K.W. Corss, G.S. Dawes, and P.W. Nathanielsz, eds. Cambridge University Press, Cambridge (1973), pp. 527-545.

Brown-Grant, K. Failure of ovulation after administration of steroid hormones and hormone antagonists to female rats during the neonatal period. *J. Endocr.* 62, 683-684 (1974).

Carter, C.S., Clemens, L.G., and Hoekema, D.J. Neonatal androgen and adult sexual behavior in the golden hamster. *Physiol. Behav.* 9, 89-95 (1972).

Cidlowski, J.A. and Muldoon, T.G. Estrogenic regulation of cytoplasmic receptor populations in estrogen-responsive tissues of the rat. *Endocrinology* 95, 1621-1629 (1974).

Cidlowski, J.A. and Muldoon, T.G. Sex-related difference in the regulation of cytoplasmic estrogen receptor levels in responsive tissues of the rat. *Endocrinology* 98, 833-841 (1976a).

Cidlowski, J.A. and Muldoon, T.G. Dissimilar effects of antiestrogens upon estrogen receptors in responsive tissues of male and female rats. *Biol. Reprod.* 15, 381-389 (1976b).

Clark, J.H., Campbell, P.S., and Peck, Jr., E.J. Receptor-estrogen complex in the nuclear fraction of the pituitary and hypothalamus of male and female immature rats. *Neuroendocrinology* 10, 218-228 (1972).

Clemens, L.G. Neurohormonal control of male sexual behavior, in *Reproductive Behavior*. W. Montagna and W. Sadler, eds. Plenum Press, New York (1974), pp. 23-53.

Clemens, L.G. Gladue, B.A. and Coniglio, L.P. Prenatal endogenous androgens influences on masculine sexual behavior and genital morphology in male and female rats. *Horm. Behav.* 10, 40-53 (1978).

Clemens, L.G., Hiroi, M., and Gorski, R.A. Induction and facilitation of female mating behavior in rats treated neonatally with low doses of testosterone propionate. *Endocrinology* 84, 1430-1438 (1968).

Coniglio, L.P., Paup, D.C., and Clemens, L.G. Hormonal factors controlling the development of sexual behavior in the male golden hamster. *Physiol. Behav.* 10, 1087-1094 (1973).

Davies, J., Naftolin, F., Ryan, K.J., and Siu, J. Estradiol receptors in the pituitary and anterior hypothalamus of the rat: measurement by agar gel electrophoresis. *Steroids* 25, 591-609 (1975).

DeBold, J.F. and Whalen, R.E. Differential sensitivity of mounting and lordosis control systems to early androgen treatment in male and female hamsters. *Horm. Behav.* 6, 197-209 (1975).

Docke, F. and Smollich, A. Morphologic effect of a single postnatal administration of androgen on medial preoptic nucleus in prepubertal female rats. *Endocrinolgia Experimentalis* 3, 107-112 (1968).

Dorner, G. *Hormones and Brain Differentiation.* Elsevier, Amsterdam (1976).

Doughty, C. and McDonald, P.G. Hormonal control of sexual differentiation of the hypothalamus in the neonatal female rat. *Differentiation* 2, 275-285 (1974).

Doughty, C., Booth, J.E., McDonald, P.G., and Parrott, R.F. Effects of oestradiol-17β, oestradiol benzoate and the synthetic oestrogen RU-2858 on sexual differentiation in the neonatal female rat. *J. Endocr.* 67, 419-424 (1975a).

Doughty, C., Booth, J.E., McDonald, P.G., and Parrott, R.F. Inhibition, by antioestrogen MER-25, of defeminization induced by the synthetic oestrogen RU-2858. *J. Endocr.* 67, 459-460 (1975b).

Edwards, D.A. Early androgen stimulation and aggressive behavior in male and female mice. *Physiol. Behav.* 4, 333-338 (1969).

Edwards, D.A. and Burge, K.G. Estrogenic arousal of aggression behavior and masculine sexual behavior in male and female mice. *Horm. Behav.* 2, 239-245 (1971).

Eisenfeld, A.J. and Axelrod, J. Effect of steroid hormones, ovariectomy estrogen pretreatment, sex and immaturity on the distribution of 3H-estradiol. *Endocrinology* 79, 38-42 (1966).

Feder, H.H. The comparative actions of testosterone propionate and 5α-androstan-17β-ol-3-one propionate on the reproductive behavior, physiology and morphology of male rats. *J. Endocr.* 51, 241-252 (1971).

Feder, H.H. and Whalen, R.E. Feminine behavior in neonatally castrated and estrogen-treated male rats. *Science* 147, 306-307 (1965).

Flerko, B., Illei-Donhoffer, A., and Mess, B. Oestradiol-binding capacity in neural and non-neural target tissues of neonatally androgenized female rats. *Acta Biol. Acad. Sci. Hung.* 22, 125-130 (1971).

Flerko, B., Mess, B., and Illei-Donhoffer, A. On the mechanism of androgen sterilization. *Neuroendocrinology* 4, 164-169 (1969).

Fox, T.O. Androgen- and estrogen-binding macromolecules in developing mouse brain; biochemical and genetic evidence. *Proc. Nat. Acad. Sci. USA* 72, 4303-4307 (1975).

Gandleman, R., vomSaal, F.S., and Reinisch, J.M. Contiguity to male foetuses affects morphology and behavior of female mice. *Nature* 266, 722-724 (1977).

Gerall, A.A., Dunlap, J.L., and Wagner, R.A. Effects of dihydrotestosterone and gonadotropins on the development of female behavior. *Physiol. Behav.* 17, 121-126 (1976).

Gerall, A.A., McMurray, M.M., and Farrell, A. Suppression of the development of female hamster behaviour by implants of testosterone and non-aromatizable androgens administered neonatally. *J. Endocr.* 67, 439-455 (1975).

Ginsburg, M., MacLusky, N.J., Morris, I.D., and Thomas, P.J. Oestradiol binding in hypothalamic cytosol. *Brit. J. Pharmac.* 43, 422P-423P (1971).
Ginsburg, M., MacLusky, N.J., Morris, I.D., and Thomas, P.J. Ontogenesis of high affinity binding of oestradiol in hypothalamic and pituitary cytosol. *J. Endocr.* 55, xx-xxi (1972a).
Ginsburg, M., MacLusky, N.J., Morris, I.D., and Thomas, P.J. Cyclic fluctuation of oestradiol receptors in hypothalamus and pituitary. *J. Physiol.* 224, 72P-74P (1972b).
Ginsburg, M., MacLusky, N.J., Morris, I.D., and Thomas, P.J. Physiological variation in abundance of oestrogen specific high-affinity binding sites in hypothalamus, pituitary and uterus of the rat. *J. Endocr.* 64, 443-449 (1975).
Goldman, B.D., Quadagno, D.M., Shryne, J., and Gorski, R.A. Modification of phallus development and sexual behavior in rats treated with gonadotropin antiserum neonatally. *Endocrinology* 90, 1025-1031 (1972).
Goldman, P.S., Crawford, H.J., Stokes, L.P., Golkin, T.W., and Rosvold, H.E. Sex-dependent behavioral effects of cerebral cortical lesions in the developing rhesus monkey. *Science* 186, 540-542 (1974).
Gorski, R.A. Influence of age on the response to paranatal administration of low dose of androgen. *Endocrinology* 82, 1001-1004 (1968).
Gorski, R.A., Gordon, J.H., Shryne, J.E., and Southam, A.M. Evidence for a morphological sex difference within the medial preoptic area of the rat brain. *Brain Res.* 148, 333-346 (1978).
Gorzalka, B.B., Rezek, D.L., and Whalen, R.E. Adrenal mediation of estrogen-induced ejaculatory behavior in the male rat. *Physiol. Behav.* 14, 373-376 (1975).
Goy, R.W. Experimental control of psychosexuality. *Phil. Trans. Roy. Soc. Lond. (Biol.)* 259, 149-162 (1970).
Green, R., Luttge, W.G., and Whalen, R.E. Uptake and retention of tritiated estradiol in brain and peripheral tissues of male female and neonatally androgenized female rats. *Endocrinology* 85, 373-378 (1969).
Greenough, W.T., Carter, C.S., Steerman, C., and DeVoogd, T.J. Sex differences in dendritic patterns in hamster preoptic area. *Brain Res.* 126, 63-72 (1977).
Gregory, E. Comparison of postnatal CNS development between male and female rats. *Brain Res.* 99, 152-156 (1975).
Hart, B.L. Neonatal dihydrotestosterone and estrogen stimulation: Effects on sexual behavior of male rats. *Horm. Behav.* 8, 193-200 (1977).
Ifft, J.D. The effect of endocrine gland extirpations on the size of nucleoli in rat hypothalamic neurons. *Anat. Rec.* 148, 549-604 (1964).
Kato, J. Cytosol and nuclear receptors for 5α-dihydrotestosterone and testosterone in the hypothalamus and hypophysis and testosterone receptors isolated from neonatal female rat hypothalamus. *J. Steroid Biochem.* 7, 1179-1187 (1976).
Kincl, F.A., Folch Pi, A., Maqueo, M., Herrera Lasso, L., Oriol, A., and Dorfman, R.I. Inhibition of sexual development in male and female rats treated with various steroids at the age of five days. *Acta Endocr.* 49, 193-206 (1965).
Kniewald, J., Cala, J., Mildner, P. and Kniewald, Z. Calorimetric approach to the study of 5α-dihydrotestosterone and estradiol receptors in rat hypothalamus cytosol. *J. Steroid Biochem.* 7, 1077-1081 (1976).
Korach, K.S. and Muldoon, T.G. Comparison of specific 17β-estradiol-receptor interactions in the anterior pituitary of male and female rats. *Endocrinology* 92, 322-326 (1973).
Korach, K.S. and Muldoon, T.G. Studies on the nature of the hypothalamic estradiol-concentrating mechanism in the male and female rat. *Endocri-*

nology 94, 785-793 (1974a).

Korach, K.S. and Muldoon, T.G. Characterization of the interaction between 17β-estradiol and its cytoplasmic receptor in the rat anterior pituitary gland. *Biochem.* 13, 1932-1938 (1974b).

Lieberburg, I., Wallach, G., and McEwen, B.S. The effects of an inhibitor of aromatization (1,4,6-androstatriene-3,17,dione) and an anti-estrogen (CI-628) on in vivo formed testosterone metabolities recovered from neonatal rat brain tissues and purified cell nuclei. Implications for sexual differentiation of the rat brain. *Brain Res.* 128, 176-181 (1977).

Lisk, R.D. and Newlon, M. Estradiol: evidence for its direct effect on hypothalamic neurons. *Science* 139, 223-224 (1963).

Luttge, W.G. and Hall, N.R. Differential effectiveness of testosterone and its metabolites in the induction of male sexual behavior in two strains of albino mice. *Horm. Behav.* 4, 31-43 (1973).

MacLusky, N.J., Chaptal, C., Lieberburg, I., and McEwen, B.S. Properties and subcellular interrelationships of presumptive estrogen receptor macromolecules in the brain of neonatal and prepubertal female rats. *Brain Res.* 114, 158-165 (1976).

Marrone, B.L. and Feder, H.H. Characteristics of (^{3}H) estrogen and (^{3}H) progestin uptake and effects of progesterone on (^{3}H) estrogen uptake in brain anterior pituitary and peripheral tissues of male and female guinea pig. *Biol. Reprod.* 17, 42-57 (1977).

Maurer, R.A. ^{3}H-estradiol binding macromolecules in the hypothalamus and anterior pituitary of normal female, androgenized female and male rats. *Brain Res.* 67, 175-177 (1974)..

Maurer, R., and Woolley, D. Distribution of ^{3}H-estradiol in clomiphene-treated and neonatally androgenized rats. *Endocrinology* 88, 1281-1287 (1971).

Maurer, R.A. and Woolley, D.E. ^{3}H-estradiol distribution in normal and androgenized female rats using an improved hypothalamic dissection procedure. *Neuroendocrinology* 14, 87-94 (1974a).

Maurer, R.A. and Woolley, D.E. Demonstration of nuclear ^{3}H-estradiol binding in hypothalamus and amygdala of female, androgenized-female, and male rats. *Neuroendocrinology* 16, 137-147 (1974b).

Maurer, R.A. and Woolley, D.E. ^{3}H-estradiol distribution in female, androgenized female and male rats at 100 and 200 days of age. *Endocrinology* 96, 755-765 (1975).

McDonald, P.G. and Doughty, C. Comparison of the effect of neonatal administration of testosterone and dihydrotestosterone in the female rat. *J. Reprod. Fert.* 30, 55-62 (1972a).

McDonald, P.G. and Doughty, C. Inhibition of androgen-sterilization in the female rat by administration of an antioestrogen. *J. Endocr.* 55, 455-456 (1972b).

McDonald, P.G. and Doughty, C. Androgen sterilization in neonatal female rat, and its inhibition by an estrogen antagonist. *Neuroendocrinology* 13, 182-188 (1973/1974).

McEwen, B.S., Lieberburg, I., Chaptal, C., and Krey, L.C. Aromatization: important for sexual differentiation of the neonatal rat brain. *Horm. Behav.* 9, 249-263 (1977).

McEwen, B.S. and Pfaff, D.W. Factors influencing sex hormone uptake by rat brain regions. I.. Effects of neonatal treatment, hypophysectomy, and competing steroid on estradiol uptake. *Brain Res.* 21, 1-16 (1970).

McEwen, B.S., Pfaff, D.W., and Zigmond, R.E. Factors influencing sex hormone uptake by rat brain regions. II. Effects of neonatal treatment and hypophysectomy on testosterone uptake. *Brain Res.* 21, 17-28 (1970).

McEwen, B.S., Plapinger, L., Chaptel, C., Gerlach, J., and Wallach, G. Role of fetoneonatal estrogen binding proteins in the associations of estrogen with neonatal brain cell nuclear receptors. *Brain Res.* 96, 400-406 (1975).

McGuire, J.L. and Lisk, R.D. Oestrogen receptors in androgen or oestrogen sterilized female rats. *Nature* 221, 1068-1069 (1969).

Morin, L.P., Powers, J.B., and White, M. Effects of the anti-estrogens MER-25 and CI-628 on rat and hamster lordosis. *Horm. Behav.* 7, 283-292 (1976).

Nadler, R.D. Further evidence on the intrahypothalamic locus for androgenization of female rats. *Neuroendocrinology* 12, 110-119 (1973).

Naftolin, F., Ryan, K.J., and Petro, Z. Aromatization of androstenedione by limbic system tissue from human foetuses. *J. Endocr.* 51, 795-796 (1971).

Naftolin, F., Ryan, K.J., and Petro, Z. Aromatization of androstenedione by the anterior hypothalamus of adult male and female rats. *Endocrinology* 90, 295-297 (1972).

Neumann, F. Pharmacology and potential use of cyproterone acetate. *Horm. Metab. Res.* 9, 1-13 (1977).

Noble, R.G., and Alsum, P.B. Hormone dependent sex dimorphisms in the golden hamster *(Mesocricetus auratus)*. *Physiol. Behav.* 14, 567-574 (1975).

Nucci, L.P., and Beach, F.A. Effects of prenatal androgen treatment on mating behavior in female hamsters. *Endocrinology* 88, 1514-1515 (1971).

Ogren, L., Vertes, M., and Woolley, D. In vivo nuclear ^{3}H-estradiol binding in brain areas of the rat: reduction of endogenous and exogenous androgens. *Neuroendocrinology* 21, 350-365 (1976).

Paup, D.C., Coniglio, L.P., and Clemens, L.G. Masculinization of the female golden hamster by neonatal treatment with androgen or estrogen. *Horm. Behav.* 3, 121-131 (1972).

Paup, D.C. Mennin, J.P., and Gorski, R.A. Androgen- and estrogen-induced copulatory behavior and inhibition of luteinizing hormone (LH) secretion in the male rat. *Horm. Behav.* 6, 35-46 (1975).

Payne, A.P. A comparison of the effects of the neonatally administered testosterone, testosterone propionate and dihydrotestosterone on aggressive and sexual behaviour in the female golden hamster. *J. Endocr.* 69, 23-31 (1976).

Pfaff, D.W. Morphological changes in the brains of adult male rats after neonatal castration. *J. Endocr.* 36, 415-416 (1966).

Pfaff, D.W. Autoradiographic localization of radioactivity in rat brain after injection of tritiated sex hormones. *Science* 161, 1355-1356 (1968).

Pfaff, D., and Keiner, M. Atlas of estradiol-concentrating cells in the central nervous system of the female rat. *J. Comp. Neurol.* 151, 121-158 (1973).

Phoenix, C.H., Goy, R.W., Gerall, A.A., and Young, W.C. Organizing action of prenatally administered testosterone propionate on the tissues mediating mating behavior in the female guinea pig. *Endocrinology* 65, 369-382 (1959).

Plapinger, L. and McEwen, B.S. Gonadal steroid-brain interactions in sexual differentiation, in *Biological Determinants of Sexual Behaviour*. J.B. Hutchison, ed. Wiley, Chichester (1978), pp. 153-218.

Plapinger, L., McEwen, B.S., and Clemens, L.E. Ontogeny of estradiol binding sites in rat brain. II. Characteristics of a neonatal binding macromolecule. *Endocrinology* 93, 1129-1139 (1973).

Poppe, L. Stahl, F., Gotz, F., and Dorner, G. Sex specific oestrogen binding in the middle hypothalamus of rats. *Endokrinologie* 65, 227-228 (1975).

Powers, J.B. Anti-estrogenic suppression of the lordosis response in female rats. *Horm. Behav.* 6, 379-392 (1975).

Quadagno, D.M. and Ho, G.K.W. The reversible inhibition of steroid-induced sexual behavior by intracranial cycloheximide. *Horm. Behav.* 6, 19-26 (1975).

Quadagno, D.M., Shryne, J., and Gorski, R.A. The inhibition of steroid-induced sexual behavior by intrahypothalamic antinomycin-D. *Horm. Behav.* 2, 1-10 (1971).

Raisman, G., and Field, P.M. Sexual dimorphism in the preoptic area of the rat. *Science* 197, 731-733 (1971).

Raisman, G. and Field, P.M. Sexual dimorphism in the neuropil of the preoptic area of the rat and its dependence on neonatal androgen. *Brain Res.* 54, 1-29 (1973).

Reddy, V.V.R., Naftolin, F., and Ryan, K.J. Conversion of androstenedione to estrone by neural tissues from fetal and neonatal rats. *Endocrinology* 94, 117-121 (1974).

Reinisch, J.M. Effects of paranatal hormone exposure on physical and psychological development in humans and animals: with a note on the state of the field, in *Hormones, Behavior and Psychopathology*. E.J. Sacher, ed. Raven Press, New York (1976) pp. 69-93.

Ryan, K.J. Estrogen formation by the human placenta: studies on the mechanisms of steroid aromatization by mammalian tissue. *Acta Endocr.* 35, 697-698 (1960).

Sarff, M., and Gorski, J. Control of estrogen binding protein concentration under basal conditions and after estrogen administration. *Biochem.* 10, 2557-2563 (1971).

Sholl, S.A., Robinson, J.A., Goy, R.W. Neural uptake and metabolism of testosterone and dihydrotestosterone in the guinea pig. *Steroids* 25, 203-215 (1975).

Spelsberg, T.C., Webster, R., Pickler, G., Thrall, C. and Wells, D. Role of nuclear proteins as high affinity sites ("acceptor") for progesterone in the avian oviduct. *J. Steroid Biochem.* 7, 1091-1101 (1976).

Staudt, J. and Dorner, G. Structural changes in the medial and central amygdala of the male rat following castration and androgen treatment. *Endokrinologie* 67, 296-300 (1976).

Swanson, H.H. and Crossley, D.A. Sexual behaviour in the golden hamster and its modification by neonatal administration of testosterone propionate, in *Hormones and Development*. M. Hamburgh and J.W. Barrington, eds. Appleton-Century-Crofts, New York (1971), pp. 677-687.

Tiefer, L. Gonadal hormones and mating behavior in the adult golden hamster. *Horm. Behav.* 1, 189-202 (1970).

Tuohimaa, P. The radioautographic localization of exogenous tritiated dihydrotestosterone, testosterone, and oestradiol in the target organs of female and male rats, in *Basic Actions of Sex Steroids on Target Organs*. P.O. Hubinot, F. Leroy, and P. Goland, eds. S. Krager, Basal (1971), pp. 208-214.

Tuohimaa, P. and Johansson, R. Decreased estradiol binding in the uterus and anterior hypothalamus of androgenized female rats. *Endocrinology* 88, 1159-1164 (1971).

Tuohimaa, P. and Niemi, M. *In vitro* uptake of sex steroids by the hypothalamus of adult male rats treated neonatally with an antiandrogen (cyproterone). *Acta Endocr.* 71, 45-54 (1972).

Vertes, M. and King, R.J.B. The influence of androgen on oestradiol binding by rat hypothalamus. *J. Endocr.* 45, xxii-xxiii (1969).

Vertes, M. and King, R.J.B. The mechanism of oestradiol binding in rat hypothalamus: effect of androgenization. *J. Endocr.* 51, 271-282 (1971).

Vreeburg, J.T.M., Schretlen, P.J.M., and Baum, M.J. Specific, high-affinity binding of 17β-estradiol in cytosols from several brain regions and pituitary of intact and castrated adult male rats. *Endocrinology* 97, 969-977 (1975).

Westley, B.R. and Salaman, D.F. Role of oestrogen receptor in adrogen-induced sexual differentiation of the brain. *Nature* 262, 407-408 (1976).

Whalen, R.E. Differentiation of the neural mechanisms which control gonadotropin secretion and sexual behavior, in *Perspectives in Reproduction and Sexual Behavior*. M. Diamond, ed. Indiana University Press, Bloomington (1968), pp. 303-340.

Whalen, R.E. Sexual differentiation: models, methods and mechanisms, in *Sex Differences in Behavior*. R.C. Friedman, R.M. Richart, and R.L. Vande Wiele, eds. John Wiley, New York (1974), pp. 467-481.

Whalen, R.E. and DeBold, J.F. Comparative effectiveness of testosterone androstenedione and dihydrotestosterone in maintaining mating behavior in castrated male hamster. *Endocrinology* 95, 1674-1679 (1974).

Whalen, R.E. and Edwards, D.A. Hormonal determinants of the development of masculine and feminine behavior in male and female rats. *Anat. Rec.* 157, 173-180 (1967).

Whalen, R.E. and Etgen, A.M. Masculinization and defeminization induced in female hamsters by neonatal treatment with estradiol benzoate and RU-2858. *Horm. Behav.* 10, 170-177 (1978).

Whalen, R.E. and Gorzalka, B.B. Effects of an estrogen antagonist on behavior and on estrogen retention in neural and peripheral target tissues. *Physiol. Behav.* 10, 35-40 (1973).

Whalen, R.E. and Luttge, W.G. Long-term retention of tritiated estradiol in brain and peripheral tissues of male and female rats. *Neuroendocrinology* 6, 255-263 (1970).

Whalen, R.E., and Luttge, W.G. Testosterone, androstenedione and dihydrotestosterone: effects on mating behavior of male rats. *Horm. Behav.* 2, 117-125 (1971).

Whalen, R.E. and Massicci, J. Subcellular analysis of the accumulation of estrogen by the brain of male and female rats. *Brain Res.* 89, 225-264 (1975).

Whalen, R.E. and Nadler, R.D. Suppression of the development of female mating behavior by estrogen administered in infancy. *Science* 141, 273-275 (1963).

Whalen, R.E. and Olsen, K.L. Chromatin binding of estradiol in the hypothalamus and cortex of male and female rats. *Brain Res.* 152, 121-131 (1978).

Whalen, R.E. and Rezek, D.L. Inhibition of lordosis in female rats by subcutaneous implants of testosterone, androstenedione or dihydrotestosterone in infancy. *Horm. Behav.* 5, 125-128 (1974).

Whalen, R.E., Luttge, W.G., and Gorzalka, B.B. Neonatal androgenization and the development of estrogen responsivity in male and female rats. *Horm. Behav.* 2, 83-90 (1971).

Whalen, R.E., Martin, J.V., and Olsen, K.L. Effect of oestrogen antagonist on hypothalamic oestrogen receptors. *Nature* 258, 742-743 (1975).

Whalen, R.E., Gorzalka, B.B., DeBold, J.F., Quadagno, D.M., Ho, G., and Hough, J.C. Studies on the effects of intracerebral antinomycin-D implants on estrogen-induced receptivity in rats. *Horm. Behav.* 5, 337-343 (1974).

Wilson, J.G. Reproductive capacity of adult female rats treated prepuberally with estrogenic hormone. *Anat. Rec.* 86, 341-359 (1943).

Yu, J. Y-L. Regional and sexual distribution patterns of chromatin proteins and of soluble cytoplasmic proteins in the rat brain. *J. Neurochem.* 24, 1111-1116 (1975).

Copyright©1982, Spectrum Publications, Inc.
Hormones in Development and Aging

Chapter 9
The Neuroendocrinology of Puberty
Judith A. Ramaley

INTRODUCTION

Definition of Puberty

The word puberty itself derives from a reference to the bodily changes that it produces (from the Latin root <u>pubes</u> meaning hair). In discussing the mechanisms of puberty onset we will really be considering the initiation of gametogenesis: follicle development in females and spermatogenesis in males (see also Chapter 1). It should be kept in mind that the events controlling puberty in the two sexes may be very similar. Gametogenesis occurs in intimate association with granulosa cells in the ovary and Sertoli cells in the testis. These two cell types are related to each other embryologically (see also Chapter 2) and have many properties in common, including the capacity to secrete steroids (Armstrong and Dorrington, 1976) and factors that may control gametogenesis and follicle-stimulating hormone (FSH) secretion (Steinberger and Steinberger, 1976; Marder et al., 1977).

In the course of this chapter we will review three classes of theories—the missing link hypothesis and the inhibitory and gonadostat theories—that attempt to explain puberty onset. As the field stands now, it is safest to say that puberty probably is the end result of a whole closely nested sequence of events and that no single theory as currently set forth has a chance of explaining the whole process. Rather, it is best to read this chapter with the notion that many things influence the time at which mature gametogenesis begins and the mechanisms underlying these timing events probably lie both in the central nervous system and in peripheral tissues such as the gonads themselves, the liver and the adrenal cortex.

The Rat as a Model for Puberty Onset

We will begin our discussion of puberty by spending some time describing the process of sexual maturation in rats even though the main purpose of this text is to describe maturation and aging in humans. The underlying assumption is that the solutions to basic biological problems such as how to control the development of ova within follicles will be similar across species lines. It is, therefore, more efficient to develop explanations for these events by testing hypotheses on an animal model first since experiments can be done on animals that often cannot be done in humans. It will be much easier to deal with what is known about the control of human puberty onset once this background has been covered.

Although the rat estrous cycle (the interval between ovulations) is only four or five days long, the basic process of follicle development appears to be the same as in animals with longer cycles, such as the human; and it has been suggested that these longer cycles are made up of several shorter units, equivalent in length to a rat cycle strung together to make a long follicular phase (Baird et al., 1975). Spermatogenesis appears to take about the same length of time in a variety of species and may also represent a process controlled in similar fashion across species lines (Moller et al., 1977). A recent report commissioned by the Ford Foundation compared a number of features of the reproductive cycles of several animals (rat, sheep, monkey, human) and concluded that most aspects of the control system examined were similar in all of the animals (Greep et al., 1976). It has become clear that, for most issues in reproductive biology, the rat is a good model; and the answers obtained in the rat apply to the human. Major differences that should be kept in mind are (a) rodents such as the rat and hamster have a greater dependence upon time-of-day information in the running of their reproductive cycles than humans appear to have, (b) the corpus luteum of the rat cycle is much more short-lived than that of the sheep, monkey, or human cycle (Nalbandov, 1973), and (c) there is a clear sex difference in the development of gonadotrophin regulation that can be found in rodents and sheep (Karsch and Foster, 1975), but not in monkeys (Karsch et al., 1973) or humans (Forest et al., 1976).

DESCRIPTION OF REPRODUCTIVE DEVELOPMENT IN THE RAT

Some aspects of gametogenesis have already been discussed with respect to fertilization (see Chapter 1) and the embryonal development of gonads has been summarized in Chapter 2. We will consider here primarily postnatal development of the gonads.

Males

For the purpose of description, I will divide the early development of the male rat into four stages: (a) the neonatal period, extending from birth until around the end of the first week during which testosterone secretion is predominant and the first crop of Leydig cells degenerages (b) the early juvenile period which lasts until around day 20 during which the predominant androgen is androstenedione (c) the late juvenile period which begins at weaning age (21 days of age) with the closure of the blood-testis barrier and the cessation of Sertoli cell proliferation and terminates at the onset of mature Leydig cell function and spermatogenesis after about day 30, and (d) the peripubertal period just at the time of the first release of sperm from the germinal epithelium around day 40-45.

The testes contain gonadotropin receptors prior to birth in males (Frowein and Engel, 1975) and are capable of steroidogenesis prenatally and neonatally (Ficher and Steinberger, 1971; Lording and DeKretser, 1972). Shortly after birth there is an FSH peak in the circulation that quickly declines (Dohler et al., 1977) until a resurgence beginning around day 30 (see Fig. 1). FHS then slowly declines to adult levels. This fall has been attributed to the increased secretion of a material called inhibin, whose levels may rise as spermatogenesis reaches adult magnitude (Rager et al., 1975). Luteinizing hormone (LH) levels rise neonatally, fall again around weaning age (21 days) and then rise again to a steady adult level. The role that gonadotrophins may play in the initiation of steroidogenesis and spermatogenesis is only partially understood, as we shall see.

Androgen secretion changes during the period from birth to puberty

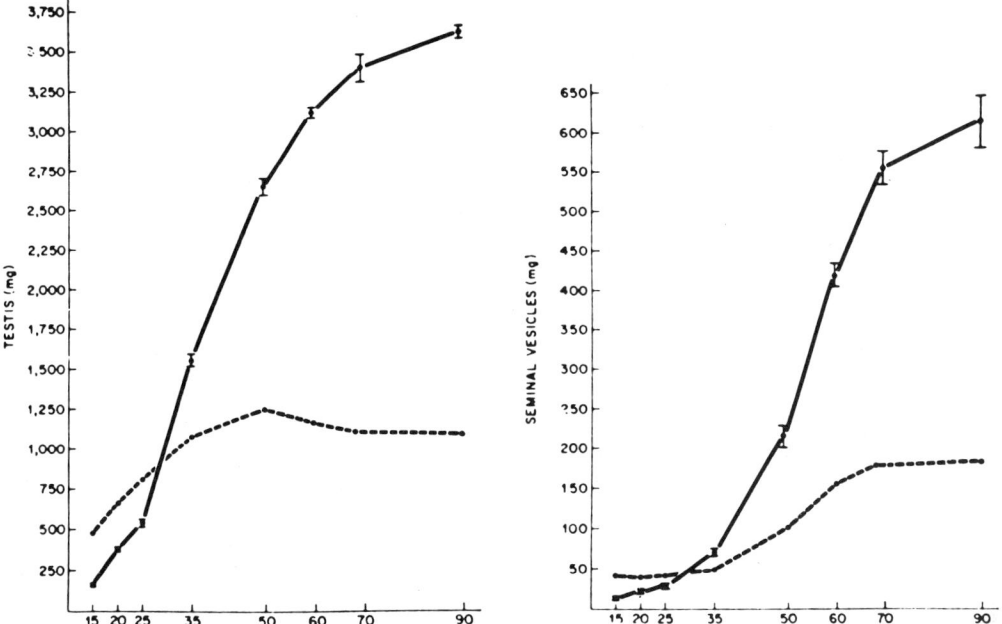

Fig. 1(A), (B). Development of Reproductive System in Males. Testis and seminal vesicle weight of rats from 15 to 90 days of age—absolute organ weights—organ weights per 100 gm of body weight. (From Negro-Vilar et al., 1973a).

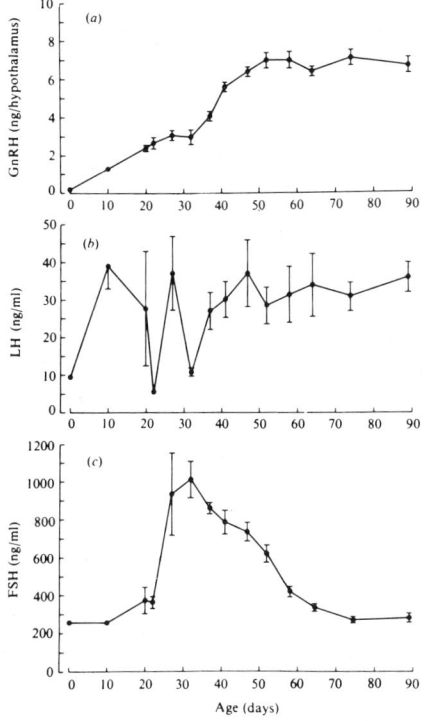

Fig. 1 (C) Changes in hypothalamic GnRH and serum LH and FSH in males during maturation. (From Payne et al., 1977.)

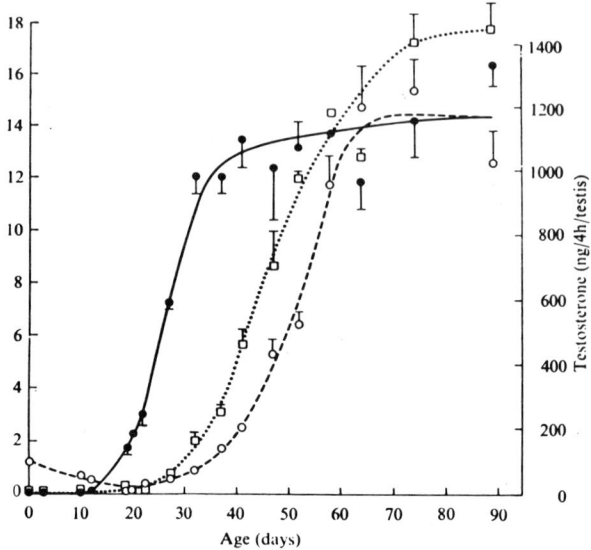

Fig. 1 (D) Changes in testicular function with age. Testosterone synthesis per testis per 4 hr (0 - - 0), 17 β hydroxysteroid dehydrogenase (- -) measured as testosterone produced/min/testis as marker for rate limiting step in testosterone production and 5-ene-3β-progesterone produced/min/testis as marker for Leydig cell development. (From Payne et al., 1977.)

All data in Figures (A)-(D) expressed as mean ±S.E.M.

in the rat. There appear to be two populations of Leydig cells, one that peaks prenatally and undergoes rapid dissolution in the neonatal period and a second crop that emerges just prior to the pubertal rise in androgen secretion (Lording and DeKretser, 1972). Associated with this biphasic pattern of Leydig cells are two spectra of androgen secretion, one in the neonate and adult and the other in the interim between these two stages. In the neonate, testosterone secretion is dominant. Testosterone production then falls off and androstenedione output rises until the pubertal resurgence of testosterone secretion (Ficher and Steinberger, 1971; Moger, 1977).

One explanation that has been offered for the switchover to testosterone production at puberty is that the immature Sertoli cell is sensitive to FSH and is capable of producing estrogens in response to FSH stimulation. The estrogens, in turn, suppress testosterone secretion by Leydig cells (Steinberger and Steinberger, 1972; Chen et al., 1977). After about 20 days of age, Sertoli cell sensitivity to FSH declines (Steinberger and Walther, 1977), estrogen production falls off (Armstrong and Dorrington, 1976), and the Leydig cells are released from suppression.

Another suggestion has been that either FSH (Chen et al., 1976; Odell et al., 1973) or prolactin (Arogona et al., 1977) directly stimulates Leydig-cell steroidogenesis, either by generating a crop of LH receptors or by some effect on steroidogenesis itself (Charreau et al., 1977). FSH has only a transitory effect on the testis and appears to be required for the initiation of spermatogenesis in the immature rat or the reinitiation of this process in hypophysectomized adults, but not for ongoing sperm production once it begins (Steinberger and Steinberger, 1972). The FSH effect may disappear at puberty due to the loss of FSH receptors on Leydig cells (Lipsett, 1977) or the development of a factor that inhibits FSH binding (Reichert and Abou-Issa, 1977).

A final possibility that has been offered to explain the onset of mature testicular function has been proposed by Frowein and Engel (1975). According to their hypothesis, the immature testis has a full complement of LH receptors and LH stimulation is ineffective because the cyclic AMP generated is rapidly inactivated by phosphodiesterase (Engel and Frowein, 1974). As corticosterone secretion increases in the late juvenile period (Ramaley, 1973), the glucocorticoid restrains the activity of phosphodiesterase and testosterone secretion rises. Alcohol dehydrogenase activity (a marker for LH responsivity) can be increased precociously by either ACTH or glucocorticoids. It should be kept in mind, however, that excess glucocorticoids suppress androgen production, perhaps by depleting LH receptors (Saez et al., 1977).

Once testosterone begins to rise, it stimulates the growth and secretory function of the accessory organs (Fugii, 1976; also see Fig. 1). About the same time, the process of spermatogenesis undergoes significant changes. Prior to 40 days of age, the rate at which spermatogonia pass through reduction division and enter spermiogenesis occurs much faster than in the adult, although the process of spermiogenesis itself proceeds at adult rates from the earliest appearance of spermatids (C. Huckins, 1978; Clermont and Perey, 1957). As spermiogenesis begins, the definitive blood-testis barrier and the adult vascular pattern are formed (Vitale et al., 1973). The newborn testis has no blood-testis barrier; that is, there is very little restriction of the passage of molecules into the seminiferous tubules. As spermiogenesis beings (Mills et al., 1977), Sertoli cell proliferation ceases (Steinberger and Steinberger, 1971), and a protective sheath of Sertoli cell cytoplasm is completed which divides the germinal epithelium into a basal and an adluminal compartment (Setchell and Waites, 1973). The role of gonadotrophins or steroids in this structural process remains uncertain, although it has been shown that estrogen can retard the formation of the blood-testis barrier (Vitale et al., 1973).

Females

The first developmental period in the female occurs around the time of birth. During this neonatal phase, the ovaries are insensitive to gonadotrophin stimulation. Beginning around the end of the first week, the ovaries enter the early juvenile phase during which they begin to show follicular maturation and gonadotrophin sensitivity. Around weaning age, the late juvenile phase begins during which it becomes possible to induce ovulation by special stimuli. Finally, around day 26 to day 28 the control system enters a new functional phase and the animals can be said to be in the peripubertal phase during which first ovulation may begin spontaneously. In the neonatal period, gonadotrophin secretion is high and then falls during the first week (Eneroth et al., 1975; Dohler et al., 1977). This activation of the brain pituitary axis appears general, since other hormones are also elevated neonatally including thyrotropic hormone, or TSH (Florsheim et al., 1966), adrenocorticotropic hormone, or ACTH (Chiappa and Fink, 1977), and growth hormone, or GH (Walker et al., 1977). Despite the high levels of gonadotrophin, the newborn ovary shows minimal follicle development and steroidogenesis (Dawson and McCabe, 1951) and during the neonatal period, the ovaries do not respond to exogenous gonadotrophin stimulation (Kolena, 1976).

At the end of the first week, FSH levels begin to rise, receptors for gonadotrophin can be demonstrated in the ovary (Peluso et al., 1976; Siebers et al., 1977) and the ovary begins to show both a growth and steroidogenic response to gonadotrophic stimulation (Rennels, 1951;

Kolena, 1976). Although changes in the ovary coincide with an FSH rise, attempts to block ovarian development with antisera to FSH have been relatively unsuccessful (Schwartz, 1974). In the prepubertal rat, FSH appears identical in its properties to adult FSH (Ojeda and Jameson, 1977). The FSH rise does not seem to be generated by light cues, although its peak usually coincides with eyeopening (Cons and Timiras, 1975), but light and dark can modulate both FSH and LH levels in the late juvenile period. It is possible that the peak is generated by steroid action, although only LH appears to be under positive feedback regulation prepubertally as indicated by the ability of antisera to estradiol to reduce circulating LH levels (Döhler et al., 1975; Gelato and Wuttke, 1975) (Fig. 2).

The first organized steroid-secreting tissue to appear in the early juvenile period is the interstitial compartment rather than the follicles. By about 8-10 days of age, follicles with antra can be seen (Dawson and McCabe, 1951). At the end of the second week of life, the late juvenile stage begins. Note that the basic developmental intervals are very similar in timing to those described for males. From this point on, the ovary will

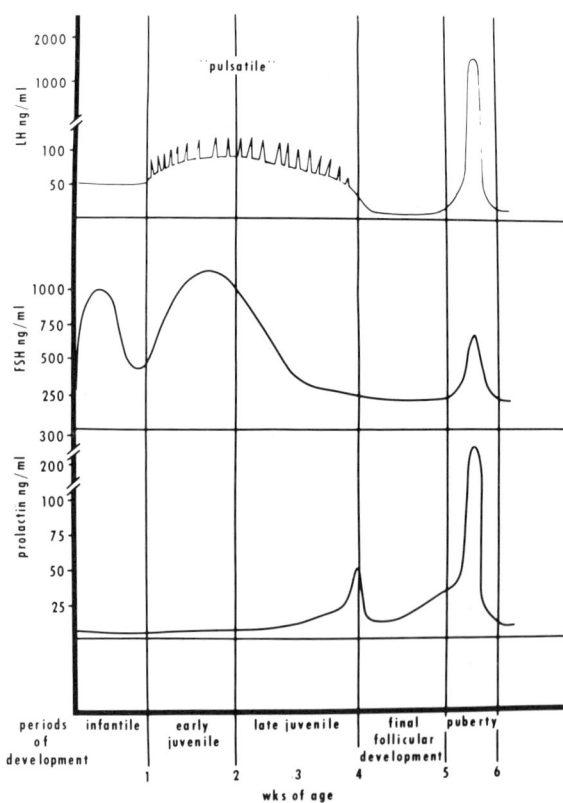

Fig. 2. Gonadotrophin profiles in the immature female rat. (From Döhler and Wuttke, 1975; Meijs-Roelofs et al., 1975; Campbell and Ramaley, 1977, Osmon, 1975.)

respond to gonadotrophic stimulation by showing increased follicular growth, estrogen synthesis and ovulation. The shift from the early to the late juvenile stage may be due to estrogen which can induce FSH receptors in the follicle wall (Reiter et al., 1972).

As the late juvenile phase begins, the FSH peak begins to fall. This could be due to either steroid feedback (Ramirez, 1973) or to the secretion of an FSH-inhibiting substance by the increasing numbers of large follicles present in the maturing ovary. Such a material has recently been isolated from porcine follicular fluid and can be used to reduce FSH secretion selectively in both adult and immature rats (Schwartz and Channing, 1977; Marder et al., 1977).

The prepubertal secretory pattern of LH is quite different from FSH. There are no major peaks and troughs in the prepubertal period but LH secretion may be pulsatile until around 26 days of age (Döhler and Wuttke, 1975). These pulses are estrogen dependent since they are abolished by antiserum to estradiol (Dohler et al., 1975). Both sheep (Foster et al., 1975) and humans (Boyer et al., 1972) show similar prepubertal gonadotrophin pulses which may be involved in the initiation of adult steroidogenesis and gametogenesis.

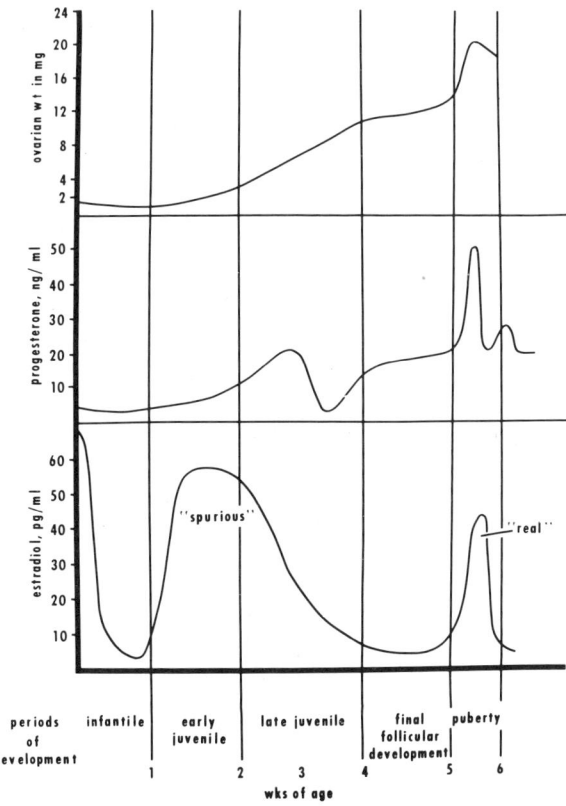

Fig. 3. Steroid profiles ovarian weights in the immature female rat. (From Dohler and Wuttke, 1975; Meijs-Roelofs et al., 1975; Ramaley and Schwartz (1980); Osmon (1975).

By around 15 days of age, the ovary is producing enough steroid to affect uterine weight (Cierciorowska and Russfield, 1968). Estrogen levels in the blood rise during the late juvenile period, followed shortly by a progesterone rise (Fig. 3) (Meijs-Roelofs et al., 1975). Both rises are brief. Beginning a few days before puberty onset, estrogen and progestin begin to drift upward again (Parker and Mahesh, 1976). As this happens, ovarian and uterine weights show a steplike increase (Ramaley and Schwartz, 1980; Sehgal and Hoffmann, 1977) and uterine sensitivity to estrogen is enhanced (Advis and Alvarez, 1977). These changes can be activated precociously by brain lesions (Advis and Alvarez, 1977) or brain stimulation (YoungLai et al., 1976).

The hormonal events of the first ovulatory cycle are difficult to distinguish from those of any adult cycle except for minor differences in the timing of the estradiol rise and ova release (Osmon and Meijs-Roelofs, 1976).

The adrenals also appear to play a role in sex steroid secretion in the immature female rat. The prepubertal adrenals are the source of progesterone (Ramaley and Bartosik, 1975) and of estrogenic materials. These estrogens behave chemically like estradiol and estrone during column chromatography and in some radioimmunoassays (RIAs) but appear to lack biological activity when assayed using uterine cytosol estrogen receptors (Weisz and Gunsalus, 1973).*

The secretion of "spurious estrogens" begins to fall off around 21 days of age (Rabii and Ganong, 1976; Meijs-Roelofs et al., 1975) and the adrenal may convert to an adult steroid pattern (Campbell and Schwartz, 1977).

What role might spurious estrogens play in gonadotrophin regulation? They may be estrogenic, antiestrogenic, or of no biological importance. In one study, combined ovariectomy-adrenalectomy resulted in a substantial reduction in pituitary sensitivity to a (GRH) challenge test when compared with ovariectomy alone (Ojeda et al., 1977) but other steroids such as corticosterone might be responsible for the effect. It has been suggested that 25-day-old rats respond more rapidly to ovariectomy than do nine-day-olds because they lack adrenal spurious estrogens (Ojeda et al., 1975). Döhler et al. (1975) found that either ovariectomy or adrenalectomy in rats during the early juvenile period caused an elevation in LH. It is tempting to speculate that adrenal spurious estrogen secretion can delay puberty onset and that the reduction in spurious estrogen production after weaning age can trigger ovarian stimulation that leads a few days later to a rise in "true" estrogen secretion (Fig. 3).

THE MECHANISM OF PUBERTY ONSET

Unlike other body systems, the reproductive system does not mature rapidly. After an initial flurry of activity in the prenatal or neonatal period, an interval of relative calm sets in which is succeeded by the initiation of mature gametogenesis and a rising hormone titer that stimulates the development of secondary sexual characteristics. Regardless of the species examined, the crucial question is: "What controls the timing of puberty onset?"

---

*These compounds, which have been called "spurious estrogens" are to be distinguished from another class of estrogens, the catechol estrogens, which bind avidly to estrogen receptors in brain and pituitary but lack biological activity in uterine weight bioassays. It has been suggested that catechol estrogens may be natural antiestrogens (Paul and Skolnick, 1977). If so, they might also play a role in puberty onset.

The Missing Link Hypothesis

When the issue of what initiates sexual maturation was first raised in the 1930s, the hypothesis was advanced that puberty occurred when the necessary elements had matured sufficiently to support mature gametogenesis and steroidogenesis. To test this, attempts were made to stimulate the ovaries precociously with exogenous gonadotrophin, derived either from pituitary extracts or from the serum of pregnant mares (PMSG, pregnant mare's serum gonadotrophin) or delivered to an animal via a parabiotically linked vascular connection to an older animal or an animal in which pituitary gonadotrophin secretion had been elevated by castration (reviewed in Critchlow and Bar-Sela, 1967; Ramirez, 1973). After the infantile period, the immature ovary was able to respond. In short order, it was also established that the pituitary contained sufficient gonadotrophin to stimulate the ovaries (Clark, 1935), that the secretion of this gonadotrophin could be activated precociously either by prepubertal castration (Hohlweg and Junkman, 1932) or by signals that can initiate precocious puberty such as PMSG (Stauss and Meyer, 1962), certain patterns of estrogen (Caligaris et al., 1972; Ramirez and Sawyer, 1965; Ying and Greep, 1971) or direct brain stimulation (Ruf et al., 1974; Kawakami and Terasawa, 1972; Meijs-Roelofs, 1972). After the identification of a structure for GRH and the development of assays capable of detecting GRH-reactive material, it was shown that the infant hypothalamus contained releasing hormone which can be depleted by castration (Gross and Baker, 1977); and, morphological studies demonstrated the presence of an organized hypophyseal portal system capable of delivering brain signals to the pituitary via the drainage from the median eminence (Halasz et al., 1972).

On the basis of these observations, it is clear that puberty does not await forging of a missing link. All of the elements of the brain-pituitary-gonadal axis that are accessible to manipulation in the prepubertal rat are capable of functioning in an adult fashion at least two or three weeks before they normally do so. This is, of course, not to say that more subtle developmental events may not underlie the time of puberty.

Inhibitory Theories: Neural Inputs

Another way of approaching the question of puberty onset is to assume that the circuitry is ready to function but is held in check by inhibitory pathways, either neural or hormonal, until the time of puberty onset. This interpretation is based in part on the effects of lesions in the central nervous system and in part on the response of immature and adult rats to castration and steroid replacement therapy.

Lesions of the anterior hypothalamus can induce precocious vaginal opening and ovulation in rats (Ruf et al., 1976; Donovan and Van der Werff ten, Bosch, 1959). Extrahypothalamic connections are also important since lesions in the amygdala (Elwers and Critchlow, 1961; Velasco, 1972) or the use of knife cuts that sever the connections between the hypothalamus and other portions of the brain (Ramaley and Sieck, 1975) can advance vaginal opening while lesions in the hippocampus can delay puberty (Riss et al., 1963). In cases where the necessary connections for an ovulatory discharge remain intact, precocious ovulation can also occur after brain lesions. The usual interpretation given to such experiments is that the lesions or knife cuts have destroyed or interrupted inhibitory pathways passing from extrahypothalamic sites to the essential gonadotrophin regulatory centers in the hypothalamus and that, once released from this check, gonadotrophin secretion increases. Enhanced gonadotrophin secretion then stimulates the gonads, follicles are prepared for ovulation and the necessary hormonal signal that activates the LH surge is delivered into the bloodstream by the ripening follicles.

In the past several years, new approaches have been taken to the intrepretation of lesion studies. First, the question has been raised whether lesions remove a pathway or structure and thus eliminate the effects of that connection or whether lesions are actually sources of chronic irritation and stimulation that increase impulse traffic over the pathways that lie in the lesion site (Reynolds, 1965). To test this, direct current irritative lesions have been made with stainless steel electrodes that deposit iron in the lesion site or nonirritative lesions with platinum electrodes to prevent iron deposition. In the former case, iron can provide an irritative or "epileptic" focus while in the latter cases, iron is not deposited at the lesion site and the actual degree of trauma is less. In one study, both irritative and nonirritative lesions placed in the rostral hypothalamus produced an advancement in vaginal opening and ovulation (Sherwood and Timiras, 1974). The authors concluded that there was no evidence for the irritative focus hypothesis and that puberty onset resulted from removal of inhibitory pathways. In another study, irritative lesions placed in the amygdala stimulated gonadotrophin secretion while nonirritative ones did not (Velasco, 1972). This may represent a difference due to the more complex circuitry of the amygdala.

Using another approach, Ruf and his colleagues have placed small unilateral lesions just above the pituitary stalk and argue that the advancement of puberty produced by such lesions is due to direct activation of the pituitary followed by a rapid rise in ovarian steroidogenesis and puberty onset since these lesions spare most of the recognized feedback sites of sex steroids on gonadotrophin activation (Ruf et al., 1976; Young-Lai et al., 1976). Lesions also may induce prolactin secretion which might initiate puberty (Advis and Alvarez, 1977; Alvarez et al., 1977).

A second approach to the interpretation of lesion studies has been to ask whether puberty onset is initiated by a release from suppression or by a change in the balance between suppression and facilitation. The balance of input to the hypothalamus appears to change with age, with the prepubertal animal seemingly in a more "inhibited" state. In adults, stimulation of the basolateral nuclei of the amygdala reduced the magnitude of the LH surge in proestrus rats (Carrillo et al., 1977) while stimulation of the corticomedial nuclei of the amygdala could produce early LH surges in similar animals (Velasco and Taleisnik, 1969a). This study is representative of the findings that suggest that in adults the amygdala contains two clusters of nuclei, one facilitory to GTH secretion, the other inhibitory. There is some evidence that amygdalar influence is initially inhibitory until a facilitatory circuit develops around weaning age. Döcke (1976) has argued, using nonirritative lesions, that the anteromedial nucleus is inhibitory to gonadotrophin secretion in three week old females. Coinciding with the development of the capacity of rats to release an ovulatory discharge of LH around four week of age, the cortical amygdala develops a site of positive feedback. Another area in the cortical amygdalar complex may develop into a negative feedback site around the same time. Various brain areas, including the amygdala, also show changes in protein content and in the rate of incorporation of precursors such as ^{35}S-methionine into protein in vivo during sexual development (MacKinnon, 1973). These biochemical changes could reflect changes in circuitry. Terasawa and Timiras (1968) also have evidence for a shift in amygdalar function with age. In their study there was a change at 27 days of age in the threshold required for seizure activity induced by local stimulation of the amygdala, which could be induced early by the administration of PMSG. A similar maturation of function may occur in the hippocampus which appears to be inhibitory to gonadotrophin secretion in adults but possibly stimulatory before puberty. Electrical stimulation of the hippocampus can block ovulation in adult proestrus rats (Carrillo et al., 1977) and immature rats (Velasco and Taleisnik, 1969b), while lesions of the hippocampus have been reported to delay puberty (Riss et al., 1963).

Another approach to neural changes in development has been to define the developmental history of the hypothalamus in terms of morphological and chemical criteria. There are increases in the number of synapses in the arcuate nucleus that correlate with the functional maturity of these neurons (Matsumoto and Arai, 1976). The induction of precocious puberty with PMSG can lead to establishment of the adult synaptic pattern early, suggesting that the synaptic changes follow the activation of the brain-gonadal axis rather than precipitate it (Matsumoto and Arai, 1977), a situation reminiscent of the changes in rest-activity cycle at puberty which can also be induced precociously with PMSG (Sieck et al., 1976). The monoamine patterns of the median eminence have been reported to reach adult configuration by 21 days of age, about the time that rats can first be induced to secrete an ovulatory surge of LH by appropriate stimuli (Smith and Simpson, 1970). Whether the emergence of an adult compliment of neurotransmitters is involved in puberty onset is a moot point since even drastic alterations in catecholamine turnover do not alter puberty onset unless they are accompanied by anorexia and a retardation of weight gain (Weiner and Capagnini, 1974).

Inhibitory Theories: Steroid Feedback

Another source of potential inhibition of the brain-pituitary axis before puberty is the steroids secreted by the gonads. It has been proposed that the sensitivity of the gonadal axis to sex steroid feedback changes during the process of puberty and that this shift in setpoint of the control system is responsible for the escape of the pituitary from inhibition and the initiation of mature gametogenesis by increased gonadotrophin secretion.

A set of observations first made in the late 1930s by Hohlweg and Junkmann (1932) and later confirmed by Ramirez and McCann (1963) support the concept of hormonal feedback as an inhibitor of the immature gonadal axis. To test the functioning of the brain-pituitary axis before and after puberty, early investigators chose the route of opening the feedback loop between the brain-pituitary axis and the gonads by castration and then determining how much steroid therapy would be necessary to turn off a long established castration response. It was necessary to wait some time after castration in order to see changes in pituitary morphology (the development of castration cells) or in blood levels of gonadotrophin (using bioassays which are less sensitive, on the whole, than RIAs). In this situation, immature animals required smaller daily doses of estrogen or androgen to suppress an established castration response than did adults. From these observations developed the "gonadostat" concept (Ramirez, 1973).

There are several problems with these early experiments. First, later studies using RIA have permitted the measurement of gonadotrophins before puberty in intact rats and the complex patterns that have been found are hard to explain using a simple feedback model or gonadostat. Second, using RIA, Ramirez and Ojeda (1973-74) established that feedback sensitivity to steroids in the gonadectomized rat is not uniform throughout the prepubertal period. In the female, for example, steroid replacement is ineffective until the late juvenile period (Ramirez, 1973), although steroids can suppress gonadotrophins in the intact rat (Morishita et al., 1975) and antisera to estradiol can cause an elevation in FSH levels

(Gelato and Wuttke, 1975). In males during the peripubertal period steroid sensitivity cannot be demonstrated at all (Negro-Vilar et al., 1973a,b; Block et al., 1974; Ojeda and Ramirez, 1973-74). One difficulty in explaining why different doses of steroid are needed to suppress gonadotrophin in rats castrated before and after puberty is that immature rats may not be comparable with adults since the interval of steroid therapy often spans a period of insensitivity to steroid replacement.

A third problem is that the sensitivity of both immature and mature rats to steroid replacement may change at different rates as the duration of the castration interval increases. In adults, acutely castrated rats are more sensitive to steroids than are long term castrates (Swerdloff and Walsh, 1973). When Swerdloff et al. (1971, 1972) initiated replacement therapy immediately after castration, they found no age differences in steroid sensitivity. One way to explain the difference between the behavior of a recently opened feedback loop and one left open for several days (as in Ramirez and McCann, 1963) might be that immature animals and adults differ in the stability of their steroid receptors and have a different fall off rate in sensitivity to steroid replacement as the castration interval increases. There is very little information on the dynamics of receptor turnover and replenishment as a function of age although one report suggests there may be a difference (Ferguson and Katzenellenbogen, 1977). If this should turn out to be so, then either the gonadostat hypothesis is based on a behavior of the system that is irrelevant to puberty onset since the loop is never opened or else the process of puberty may involve an opening of the loop perhaps due to the secretion of natural antiestrogens or antiandrogens during the peripubertal period.*

In a novel approach to the elusive gonadostat issue, Steele and Weisz (1974) began continuous infusion of estradiol in 26-day-old females that had been castrated a week before. LH began to rise in the castrates at about the age of puberty in intact rats, suggesting that the gonadostat had reset and that the rats had "escaped" from suppression. The samples were collected in the afternoon and thus the high LH might have reflected the onset of daily LH surges due to continuous estradiol. To control for this, Steele (1977) repeated the study and obtained morning samples which also showed an escape of LH from estradiol suppression. It would be interesting to find out whether LH would rise at puberty age in rats given estradiol immediately after castration. In other words, does the behavior defined as a gonadostat exist only after the loop has been kept open for a minimum length of time?

Although the research emphasis has been on feedback sensitivity, the rate of initiation of a gonadotrophin rise after castration might be different in young rats (Goldman et al., 1971). According to one study, the LH rise after castration does not change at puberty (Sehgal and Hoffmann, 1977) while other laboratories have reported that in both males and females the castration response is more sluggish after puberty (Swerdloff et al., 1972; Eldridge and Mahesh, 1974; Campbell and Schwartz, 1977).

To confuse the issue further, the whole question of a gonadostat may have to be reexamined in the light of recent observations that suggest that the castration response may not be initiated by falling steroid levels as has been traditionally thought (reviewed in Campbell and Schwartz, 1977). If the whole issue of feedback regulation is unsettled in adults, its role in puberty initiation is even less well defined.

---

*Since this review was written in 1977, several labs have reported evidence that the prepubertal rat is androgen-dominated not estrogen dominated. See, for example, Andrews and Ojeda, 1981.

If There Is a Gonadostat, What Is It?

A number of attempts have been made to find physiological and chemical correlates that reflect changes in feedback sensitivity and gonadotrophin secretion in the prepubertal period. For example, early studies suggesting that the preweanling rat had no specific receptors for estradiol (Plapinger and McEwen, 1973) could be used to explain the absence of steroid suppression of gonadotrophin secretion in castrates before weaning age and changes in receptors might be used to explain changes in feedback sensitivity. Subsequent studies have revealed that the neonatal brain does have specific estradiol receptors but that their presence is masked by high affinity steroid binding proteins (the α-fetoproteins) that circulate in the blood of rats before weaning age. When steroids are used that do not bind to α-fetoproteins (McEwen et al., 1975) or when autoradiographic techniques are used (Sheridan et al., 1974), specific uptake of steroid by brain tissue can be demonstrated in the neonate. Hypothalamic content of receptors increases with age (Kato, 1975). It seems likely that the blood proteins do prevent some endogenous steroid from reaching brain tissue in newborns (Farquahar et al., 1976; Andrews and Ojeda, 1977), although androgens, which do not bind to α-fetoproteins, may have freer access to the central nervous system (CNS) (Raynaud et al., 1971).

The greater steroid sensitivity of castrated immature rats may be based on the metabolic conversion of secreted steroids to more potent compounds in the brain and pituitary. It has been shown that testosterone can be converted to 5-α-dihydrotestosterone (DHT) more effectively in the neonatal period than in the late prepubertal period as a result of higher 5-α-reductase activity (Denef et al., 1974). DHT is a more potent inhibitor of gonadotrophin secretion than testosterone. The immature brain is also very efficient in converting testosterone to estrogens, and according to one view, androgens exert some of their biological effects after such conversions (Ryan et al., 1972). In adults, each region of the hypothalamus and limbic system has its own metabolic profile as defined by the products into which testosterone can be converted (Lieberburg and McEwen, 1977), and it may be that immature animals gradually develop these regional differences. If the immature brain processes estrogen and androgen differently or produces more potent compounds, a smaller daily replacement dose would be required to suppress a castration response.

Another possibility is that the greater sensitivity of young animals may not reside in the properties of the CNS at all. The delivery of hormone signals to the brain may change as a function of age and the brain may simply respond to the signals that reach it. There is some evidence for a pubertal change in steroid output by the ovaries (Eckstein, 1975) and testes (see earlier). Another possible explanation for the sensitivity of younger rats could be that steroids secreted into the blood stream might circulate for different periods of time before and after puberty. It has been reported that the clearance rates of estradiol decrease after puberty (Ojeda et al., 1975; deHertogh et al., 1970). If so, a given dosage of steroid given to a prepubertal rat would be more effective. There is a gradual increase in the binding capacity of serum proteins for sex steroids at puberty (Ramaley, 1971) but these changes can be induced precociously with estrogen and do not seem causal in puberty onset (Ramaley, 1970). Finally, steroid delivery might change because of peripheral conversions to other compounds produced as a result of the maturation of steroid metabolism in the liver or skin (Gustafsson, 1974). If such were the case, the original gonadal signal might be converted to a different message as puberty approached.

The Role of Clock Mechanisms

The clock theory shares some thought in common with the inhibitory concepts in the sense that it postulates the development of new neuronal circuitry. Puberty may be activated by the development of the capacity to tell time of day and to link gonadotrophin secretion to photoperiod. There is now some evidence that rats set their biological clocks to the adult mode at about 26 days of age, about the same time that spurious estrogen levels fall in the blood stream. Beginning at 26 days, the peripheral rhythm of corticosterone acquires its adult time relationship to light and dark (Ramaley, 1973) and the rat develops the capacity to reset its rhythms rapidly in response to photoperiod challenges (Ramaley, 1975). It first becomes possible to elicit daily LH (Caligaris et al., 1972; and prolactin surges (Smith and Ramaley, 1978) about this time as well. In addition to hormonal rhythms, rats also begin to show a variety of behavioral rhythms such as diurnal patterns in food intake (Sieck et al., 1977).

How might biological rhythms be linked to follicle maturation and ovulation? It has been estimated that the follicles that rupture during a given ovulation began their preparation about 10-17 days before (Pedersen, 1969). Counting forward from 26 days of age, we arrive at 36-43 days, the usual time of first ovulation in most rats. It appears possible that some signal associated with the maturing of clock rhythms may set into motion a train of follicular development. Indirect evidence for such an activation can be found in the presence of four day cycles of food intake and body weight gain (Sieck et al., 1977) in the prepubertal period. These cycles are ovarian dependent since they disappear in the ovariectomized animal. It seems unlikely that the adrenal rhythm itself is the signal since only a minimal background level of corticosterone seems to be necessary to permit normal puberty onset (Ramaley, 1976a,b). Recently Campbell and I (Ramaley and Campbell, 1977) proposed that the signal may be prolactin. Prolactin has emerged as a possible modulator of hormone action (Hafiez et al., 1972) and a regulator of the gonadotrophin binding capacity of the ovaries and testes (Richards and Williams, 1976; Aragona et al., 1977).

Does puberty in males also require a trigger process that might be activated by the development of clock mechanisms? In males, the development of the blood testis barrier, the cessation of Sertoli-cell proliferation and the onset of mature rates of spermatogenesis and steroidogenesis emerge fairly abruptly. A trigger may be required for these events.

Environment and Puberty

The most obvious environmental cue that might alter puberty onset would be food availability which could be monitored as an optimal body weight, body composition, or metabolic rate at which puberty might be triggered. Frisch (1972) has proposed that puberty occurs in human females at a critical body wieght or body composition. A similar suggestion has been made in rats (Kennedy and Mitra, 1963). Glass and Swerdloff (1977) found a good correlation between growth rates (defined as grams gained per day) and puberty onset, but no apparent relationship with body weight. The mechanism by which nutritional status or growth rate might alter puberty onset is now known. It may act to alter the sensitivity of the gonadal axis to feedback (Howland and Ibrahim, 1973). It is interesting in this regard to compare these findings to what happens in patients with anorexia nervosa while they remain below a critical body weight. In these patients, the capacity to release LH in response to clomiphene is lost and the pubertal pattern of gonadotrophin secretion reemerges (Katz et al., 1977). It may be that passage through a critical body-weight zone, whether due to weight gain in the prepubertal period or weight loss in adults, can trigger a change in gonadotrophin secretion that is important in priming the gonads for gametogenesis.

Other environmental cues such as noise, crowding or extreme temperatures can also affect puberty onset (reviewed in Moltz, 1975). It is possible that these changes are mediated by the adrenal axis since ACTH (Hagino et al., 1969; Macfarland and Mann, 1977) or excess glucocorticoids (Ramaley, 1976a,b) can delay puberty. It should be kept in mind, however, that stressful stimuli can also activate other pituitary hormones including prolactin (Neill, 1979), LH (Dunn et al., 1972), and TSH (Ducommon et al., 1966). Since stressful stimuli can affect food intake and metabolic status, there may also be a link between stress and the metabolic theory of puberty onset.

CORRELATIONS WITH THE HUMAN

Theories of puberty onset in the rat can be groups into three categories: the missing link, inhibitory and clock concepts. To what extent do these ideas apply to humans? First, it is evident that the gonadal axis in humans can be activated precociously, as witnessed by clinical evidence of puberty onset in young children (Bierich, 1975). This situation is comparable with the induction of precocious puberty in rats by a variety of means. Clearly, the circuitry is ready to go and a developmental theory which ties puberty to the final closure of the circuit cannot apply.

What evidence is there for an active inhibition of the gonadal axis before puberty in humans? Information about neural inhibitory circuits is difficult to come by in humans; although there are cases in which hypothalamic damage has been associated with precocious puberty, a situation similar to the effects of hypothalamic or limbic system lesions in rats. There is, however, some information about the existence of a gonadostat. Children appear to be more sensitive to the feedback action of sex steroids on gonadotrophin secretion than are adults (Kulin et al., 1972).

There is also some suggestion that humans may undergo a triggering process, similar to that seen in rats, which involves a daily clock. In both rats and humans, the adrenal cortex releases considerable sex steroid and the amount of sex steroid released changes as a function of age. In the human, the activation ("adrenarche") of the adrenal may play a role in early pubertal growth and in gonadotrophin secretion (Gupta, 1975; Root, 1973). As in the rat, there may be a time-of-day signal involved in puberty onset. During the onset of puberty, pulsatile gonadotrophin secretion increases (Boyer et al., 1973) as it does in the early juvenile phase female rat. This daily rhythm is lost in adulthood in humans. A special feature of the human at puberty is that these gonadotrophin pulses are linked to sleep and wakefulness with maximal gonadotrophin secretion occurring during sleep. Unlike the rat, the human appears to use social cues rather than photoperiod as a timer, since the pattern of gonadotrophin secretion shifts immediately if a pubertal age child sleeps during the day rather than at night (Boyer et al., 1972).

Human females appear to have a number of "silent" ovarian cycles prior to the first menstrual period. This is suggested by the presence of monthly cycles of gonadotrophin excretion in the urine of premenarcheal girls (Hansen et al., 1975). In contrast to rats, the human female does not ovulate at the first external sign of ovarian cycles. First ovulation is generally delayed until several months after menarche (Gupta, 1975).

In both rats and humans, an LH surge can be elicited prior to the onset of ovarian cyclicity (Kulin et al., 1972). In contrast to the rat, the human male retains the capacity to secrete an LH surge if presented with the appropriate hormonal signals (Kulin and Reiter, 1976). Apparently the human brain does not undergo sexual differentiation in the way that the rat brain does, although the difference may be one of degree only, since the male rat retains the LH surge circuitry which can only be activated by direct electrical stimulation of the hypothalamus (Parker and Makesh, 1977).

Finally, in both species, metabolism or growth rates appear to play a role in sexual maturation (Frisch, 1972).

REFERENCES

Advis, J.P. and Alvarex, E.O. Changes in uterine responsiveness to estradiol in maturing female rats with precocious puberty induced by hypothalamic lesions. *Biol. Reprod.* 17, 321-326 (1977).

Alvarez, E.O., Hancke, J.L., and Advis, J.P. Indirect evidences of prolactin involvement in precocious puberty induced by hypothalamic lesions in female rats. *Acta Endocrin.* 85, 11-17 (1977).

Andrews, W.W. and Ojeda, S.R. On the feedback actions of estrogen on gonadotrophin and prolactin release in infantile female rats. *Endocrin.* 101, 1517-1523 (1977).

Andrews, W.W. and Ojeda, S.R. A quantitative analysis of the maturation of steroid negative feedbacks controlling gonadotropin release in the female rat. *Endocrin.* 108, 1313-1320 (1981).

Aragona, C., Bohnet, H.G., and Friesen, H.C. Localization of prolactin binding in prostate and testis: the role of serum prolactin concentration on the testicular LH receptor. *Acta Endocrin.* 84, 402-409 (1977).

Armstrong, D.T. and Dorrington, J.H. Estrogen biosynthesis in the ovaries and testes. *Advances in Sex Hormone Res.* 3, 217-258 (1976).

Baird, D.T., Baker, T.G., McNatty, K.P., and Neal, P. Relationship between the secretion of the corpus luteum and the length of the follicular phase of the ovarian cycle. *J. Reprod. Fert.* 45, 611-619 (1975).

Bierich, J.R. (ed.). *Disorders of Puberty.* Clinics in Endo. & Metab. 4(1). W.B. Saunders Co., London (1975).

Bloch, G.J., Masken, J., Kragt, C.L., and Ganong, W.F. Effect of testosterone on plasma LH in male rats of various ages. *Endocrin.* 94, 947-951 (1974).

Boyar, R.M., Finkelstein, J.W., Roffwarg, H.P., Kapen, S., and Weitzman, E.D.: Synchronization of augmented luteinizing hormone secretion with sleep during puberty. *New Engl. J. Med.* 287, 582-587 (1972).

Boyar, R.M., Finkeltstein, J.W., David, R., Roffwarg, E., Weitzman, E., and Hellman, L.: Twenty-four hour patterns of plasma luteinizing hormone and follicle stimulating hormone in sexual precocity. *New Engl. J. Med.* 289, 282-289 (1973).

Brown-Grant, K. and Raisman, G. Reproductive function in the rat following selective destruction of afferent fibers to the hypothalamus from the limbic system. *Brain Res.* 46, 23-42 (1972).

Caligaris, L., Astrada, J.J., and Taleisnik, S. Influence of age on the release of luteinizing hormone induced by oestrogen and progesterone in immature rats. *J. Endocrinol.* 55, 97-103 (1972).

Campbell, C.S. and Schwartz, N.B. Steroid feedback regulation of LH and FSH secretion rates in male and female rats. *J. Toxicol. Envir. Health* 3, 61-96 (1977).

Campbell, G.T. and Ramaley, J.A. Maturation of the FSH control system:

effects of early androgenization. Tenth annual meeting of the Society for the Study of Reproduction (abstract) (1977).

Carrillo, A.J., Rabil, J., Carrer, H.F., and Sawyer, C.H. Modulation of of the proestrous surge of luteinizing hormone by electrochemical stimulation of the amygdala and hippocampus in the unanesthetized rat. *Brain Res.* 128, 81-92 (1977).

Charreau, E.H., Attramadal, A., Jorgensen, P.A., Purvis, K., Calandra, R., and Hansson, V. Prolactin binding in rat testis: specific receptors in interstitial cells. *Molecular and Cellular Endocrin.* 6, 303-307 (1977).

Chen, Y.D., Payne, A.H., and Kelch, R.P. FSH stimulation of Leydig cell function in the hypophysectomized immature rat. *Proc. Soc. Exp. Biol. Med.* 153, 473-475 (1976).

Chen, Y.D., Shaw, M.J., and Payne, A.H. Steroid and FSH action on LH receptors and LH sensitive testicular responsiveness during sexual maturation of the rat. *Molecular and Cellular Endocrin.* 8, 291-299 (1977).

Chiappa, S.A. and Fink, G. Releasing factor and hormonal changes in the hypothalamic-pituitary-gonadotrophin and adrenocorticotrophic systems before and after birth and puberty in male, female and androgenized female rats. *J. Endocrin.* 72, 211-224 (1977).

Cierciorowska, A. and Russfield, A.B. Determination of estrogenic activity of the immature rat ovary. *Arch. Pathol.* 85, 658-662 (1968).

Clark, H.M. A prepubertal reversal of the sex difference in the gonadotrophic hormone content of the pituitary gland of the rat. *Anat. Res.* 61, 175-202 (1935).

Clermont, Y. and Perey, B. Quantitative study of the cell population of the seminiferous tubules in immature rats. *Am. J. Anat.* 100, 241-267 (1957).

Cons, J.M. and Timiras, P.S. Developmental patterns of FSH and LH in female rats deprived of light before puberty. *Environ. Physiol. Biochem.* 5, 355-360 (1975).

Critchlow, V. and Bar-Sela, M.E.: Control of the onset of puberty, in *Neuroendocrinology*, L. Martini and W.F. Ganong, eds. Academic Press, New York (1967).

Dawson, A.B. and McCabe, __. The interstitial tissue of the ovary in infantile and juvenile rats. *J. Morphol.* 88, 543-564 (1951).

deHertogh, R., Ekka, E., Vanderheyden, I., and Hoet, J.J. Metabolic clearance rates and the interconversion factors of estrone and estradiol 17β in the immature and adult female rat. *Endocrinol.* 87, 874-880 (1970).

Denef, C., Magnus, C., and McEwen, B.S. Sex dependent changes in pituitary 5α dihydrotestosterone and 3α androstanediole formation during postnatal development and puberty in the rat. *Endocrin.* 94, 1265-1274 (1974).

Döcke, F. Age-dependent changes in the puberty controlling function of the medial and cortical amygdaloid nuclei. *Ann. Biol. Anim. Bioch. Biophys.* 16, 423-432 (1976).

Döhler, K.D. and Wuttke, W. Changes with age in levels of serum gonadotrophins, prolactin and gonadal steroids in prepubertal male and female rats. *Endocrin.* 97, 899-907 (1975).

Döhler, K.D., von zur Muhler, A., and Dohler, U. Effects of ovariectomy, adrenalectomy, or antiestrogen treatment on serum LH in postnatal female rats. *Acta Endocrin.* Suppl. 199, 93 (1975).

Döhler, K.D., von zur Muhler, A., and Dohler, U. Pituitary luteinizing hormone (LH), follicle stimulation hormone (FSH) and prolactin from birth to puberty in female and male rats. *Acta Endocrin.* 85, 718-728 (1977).

Donovan, B.T. and van der Werff ten Bosch, J.J. The hypothalamus and sexual maturation in the rat. *J. Physiol.* 147, 78-92 (1959).

Ducommon, P., Sakiz, E., and Guillemin, R. Lability of plasma TSH levels in the rat in response to nonspecific exteroceptive stimuli. *Proc. Soc. Exp. Biol. Med.* 121, 921-923 (1966).

Dunn, J.D., Arimura, A., and Scheving, L.E.: Effects of stress on circadian periodicity in serum LH and prolactin concentration. *Endocrin.* 90, 29-33 (1972).

Eckstein, B. Studies on the mechanism of the onset of puberty in the female rat. *J. Steroid Biochem.* 6, 873-878 (1975).

Eldridge, J.C. and Mahesh, V.B. Pituitary-gonadal axis before puberty: evaluation of testicular steroids in the male rat. *Biol. Reprod.* 11, 385-397 (1974).

Elwers, M. and Critchlow, V. Precocious ovarian stimulation following interruption of stria terminalis. *Am. J. Physiol.* 201, 281-284 (1961).

Eneroth, P., Gustafsson, J.A., Skett, P., and Stenberg, A.L.: Sex-dependent prepubertal gonadotrophin surges in the rat. *J. Endocrin.* 65, 91-98 (1975).

Engel, W. and Frowein, J. Glucocorticoids and hCG sensititivy of rat testicular Leydig cell. *Nature* 251, 146-148 (1974).

Farquhar, M.N., Namiki, H., and Gorbman, A. Accumulation of ^{3}H-testosterone in nuclear and cytoplasmic fractions of rat brain during postnatal development. *Neuroendocrin.* 20, 136-150 (1976).

Ferguson, E.R. and Katzenellenbogen, B.S. A comparative study of antiestrogen action: temporal patterns of antagonism of estrogen stimulated uterine growth and effects on estrogen receptor levels. *Endocrin.* 100, 1242-1251 (1977).

Ficher, M. and Steinberger, E. In vitro progesterone metabolism by rat testicular tissue at different stages of development. *Acta Endocrin.* 68, 285-292 (1971).

Florsheim, N.H., Faircloth, M.A., Corcorran, N.L., and Rudko, P. Perinatal thyroid function in the rat. *Acta Endocrin.* 52, 375-382 (1966).

Forest, M.G., dePeretii, E., and Bertrand, J. Hypothalamic-pituitary gonadal relationships in man from birth to puberty. *Clinical Endocrin.* 5, 551-569 (1976).

Foster, D.L., Lemons, J.A., Jaffe, R.B., and Niswender, G.D. Sequential patterns of circulating luteinizing hormone and follicle-stimulating hormone in female sheep from early postnatal life through the first estrous cycles. *Endocrin.* 97, 985-994 (1975).

Frisch, R.E.: Weight at menarche: similarity for well-nourished and undernourished girl at different ages and evidence for historical constancy. *Pediatrics* 50, 445-461 (1972).

Frowein, J. and Gengel, W. Inhibition of gonadotrophin-stimulated cAMP synthesis in rat testis: a model for the onset of male puberty. *Acta Endocrin.* suppl., 193, 83 (1975).

Fugii, R. Roles of age and androgen in the regulation of sex accessory organs. *Advances in Sex Hormone Res.* 3, 103-137 (1976).

Gelato, M. and Wuttke, W. Evidence for a positive feedback action of estrogen on LH secretion in immature female rats. *IRCS Medical Science* 5, 260 (1975).

Gelato, M., Dibber, J., Marshal, S., Meites, J., and Wuttke, W. Prolactin-adrenal interactions in the immature female rat. *Ann. Biol. Anim. Bioch. Biophys.* 16, 395-397 (1976).

Glass, A.R. and Swerdloff, R.S. Serum gonadotrophins in rats fed a low-valine diet. *Endocrin.* 101, 702-707 (1977).

Goldman, B.D., Grazia, Y.R., Kamberi, I.A., and Cooper, J.C. Serum gonadotrophin concentrations in intact and castrated neonatal rats. *Endocrin.* 88, 771-776 (1971).

Greep, R.O., Koblinksy, M.A., and Jaffe, F.S. *Reproduction and human welfare: a challenge to research.* The MIT Press, Cambridge, Mass. (1976).

Gross, D.S. and Baker, B.L. Immunohistochemical localization of gonadotrophin-releasing hormone (GnRH) in the fetal and early postnatal mouse brain. *Am. J. Anat.* 148, 195-216 (1977).

Gupta, D. Changes in the gonadal and adrenal steroid patterns during puberty. *Clinics in Endocrin. and Metabolism* 4, 27-56 (1975).

Gustafsson, J.A. Androgen responsiveness of the liver of the developing rat. *Biochem. J.* 144, 225-229 (1974).

Hafiez, A.A., Bartke, A., and Lloyd, C.W. The role of prolactin in the regulation of testis function: the synergistic effects of prolactin and LH on the incorporation of 1-^{14}C acetate into testosterone and cholesterol by testes from hypophysectomized rats in vitro. *J. Endocrin.* 53, 223-230 (1972).

Hagino, N., Watanabe, M., and Goldzieher, J.W. Inhibition by adrenocorticotropon of gonadotrophin-induced ovulation in immature female rats. *Endocrin.* 84, 308-314 (1969).

Halasz, B., Kosaras, B., and Lengvari, I. Ontogenesis of the neurovascular link between the hypothalamus and the anterior pituitary in the rat, in *Brain-Endocrine Enteraction. Median Eminence: Structure and Function.* K.M. Knigge, D.E. Scott, and A. Weindl, eds. S. Karger, Basel (1972), pp. 27-34.

Hansen, J.W., Hoffmann, H.J., and Ross, G.T. Monthly gonadotrophin cycles in premenarcheal girls. *Science* 190, 161-163 (1975).

Hohlweg, W. and Junkaman, K. Die hormonal-nervose Regulierung der Funktion des hypophysenvorderlappens. *Klin. Wochenschr.* 11, 321-323 (1932).

Howland, B.E. and Ibrahim, E.A. Increased LH-suppressing effect of oestrogen in ovariectomized rats as a result of underfeeding. *J. Reprod. Fertl.* 35, 545-548 (1973).

Huckins, C. Personal communication (19XX).

Karsch, F.J. and Foster, D.L. Sexual differentiation of the mechanism controlling the preovulatory discharge of luteinizing hormone in sheep. *Endocrin.* 97, 373-379 (1975).

Karsch, F.J., Weick, R.F., Hotchkiss, J., Dierschke, D.J., and Knobil, E. An analysis of the negative feedback control of gonadotropin secretion utilizing chronic implantation of ovarian steroids in ovariectomized rhesus monkeys. *Endocrin.* 93, 478-486 (1973).

Kato, J. The role of hypothalamic and hypophyseal 5α dihydrotestosterone, estradiol and progesterone receptors in the mechanism of feedback action. *J Steroid Biochem.* 6, 979-987 (1975).

Katz, J.L., Boyer, R.M., Roffwarg, H., Hellman, L., and Weiner, H. LHRH responsiveness in anorexia nervosa: intactness despite prepubertal circadian LH pattern. *Psychosomatic Med.* 39, 241-251 (1977).

Kawakami, M. and Terasawa, E. Electrical stimulation of the brain on gonadotrophin secretion in the female prepubertal rat. *Endocrinol. Japon.* 19, 335-347 (1972).

Kennedy, G.C. and Mitra, J. Body weight and food intake as initiating factors for puberty in the rat. *J. Physiol.* 16, 408-418 (1963).

Kolena, J. Reversal of the unresponsiveness of neonatal rat ovary to LH in cAMP synthesis by estrogen. *Hormone Res.* 7, 152-157 (1976).

Kulin, H.E., Grumbach, M.M., and Kaplan, S.L. Gonadal-hypothalamic interaction in prepubertal and pubertal man: effect of clomiphene citrate on urinary FSH and LH and plasma testosterone. *Pediatric Res.* 6, 162-171 (1972).

Kulin, H.E. and Reiter, E.O. Gonadotrophin and testosterone measurements after estrogen administration to adult men, prepubertal and pubertal boys and men with hypogoadotropism: evidence for maturation of positive feedback in the male. *Pediatric Res.* 10, 46-51 (1976).

Lieberburg, I. and McEwen, B.S. Brain cell nuclear retention of testosterone metabolites, 5α-dihydrotestosterone and estradiol 17β in adult rats. *Endocrin.* 100, 588-597 (1977).

Lipsett, M.B. Regulation of androgen secretion, in *Androgens and Antiandrogens*, L. Martini and M. Motta, eds. Raven Press, New York (1977), pp. 11-17.

Lording, D.W. and DeKretser, D.M. Comparative ultrastructural and histochemical studies of the interstitial cells of the rat testis during fetal and postnatal development. *J. Reprod. Fert.* 29, 261-269 (1972).

Macfarland, L.A. and Mann, D.R. The inhibitory effects of ACTH and adrenalectomy on reproductive maturation in female rats. *Biol. Reprod.* 16, 306-314 (1977).

MacKinnon, P.C.B. Changes in volume and incorporation of 35S or 3H methionine in the amygdala before and after puberty in male and female mice. *Brain Res.* 50, 115-123 (1973).

McEwen, B.S., Plapinger, L., Chaptal, C., Gerlach, J., and Wallach, G. Role of fetoneonatal estrogen binding proteins in the associations of estrogen with neonatal rain cell nuclear receptors. *Brain Res.* 96, 400-406 (1975).

Marder, M.L., Channing, C.P., and Schwartz, N.B. Suppression of serum follicle stimulating hormone in intact and acurely ovariectomized rats by procine follicular fluid. *Endocrin.* 101, 1639-1642 (1977).

Matsumoto, A. and Arai, Y.: Developmental changesin synaptic formation in the hypothalamic arcuate nucleus of female rats. *Cell Tissue Res.* 169, 143-156 (1976).

Matsumoto, A. and Arai, Y. Precocious puberty and synaptogenesis in the hypothalamic arcuate nucleus in pregnant mare serum gonadotrophin (PMSG) treated immature female rats. *Brain Res.* 129, 375-378 (1977).

Means, A.R., Fakunding, J.L., Huckins, C., Tindall, D.J., and Vitale, R.: Follicle stimulating hormone, the Sertoli cell and spermatogenesis. *Recent Prog. Hormone Res.* 32, 477-528 (1976).

Meijs-Roelofs, H.M.A. Effect of electrical stimulation of the hypothalamus on gonadotrophin release and the onset of puberty. *J. Endocrin.* 54, 277-284 (1972).

Meijs-Roelofs, H.M.A., Uilenbroek, J.T., deGreef, W.J., deJong, F.H., and Kramer, P. Gonadotrophin and steroid levels around the time of first ovulation in the rat. *J. Endocrin.* 67, 275-282 (1975).

Mills, N.C., Mills, T.M., and Means, A.R. Morphological and biochemical changes which occur during postnatal development and maturation of the rat testis. *Biol. Reprod.* 17, 124-130 (1977).

Moger, W.H. Serum 5α androstane-3α-17β-diol, androsterone and testosterone concentrations in the male rat: influence of age and gonadotrophin stimulation. *Endocrin.* 100, 1027-1032 (1977).

Moller, D.W., Flickinger, C.J., and Howards, S.S. Duration of the cycle of the seminiferous epithelium in the guinea pig determined by tritiated thymidine autoradiography. *Biol. Reprod.* 17, 532-534 (1977).

Moltz, H. The search for the determinants of puberty in the rag, in *Hormonal Correlates of Behavior, Vol. I, A lifespan View*. B.E. Eleftherious and R.L. Sprott, eds. Plenum Press, New York (1975) pp. 35-154.

Morishita, G.H., Naftolin, F., Todd, R.B., Wilen, R., Davies, I.J., and Ryan, K.J. Lack of an effect of dihydrotestosterone in serum LH in neonatal female rats. *J. Endocrin.* 67, 139-140 (1975).

Nalbandov, A.V. Control of luteal function in mammals, in *Handbook of Physiology, Section 7, Endocrinology, Volume II. Female Reproductive System, Part I.* R.O. Greep and E.B. Astwood, eds. Am. Physiol. Soc., Washington, D.C. (1973), pp. 153-168.

Negro-Villar, A., Krulich, A., and McCann, S.M.: Changes in serum prolactin and gonadotrophins during sexual development of the male rat. *Endocrin.* 93, 660-664 (1973a).

Negro-Vilar, A., Ojeda, S.R., and McCann, S.M. Evidence for changes in sensitivity to testosterone negative feedback on gonadotrophin release during sexual development in the male rat. *Endocrin.* 93, 729-735 (1973b).

Neill, J.D. Effect of stress on serum prolactin and LH levels during the estrous cycle of the rat. *Endocrin.* 87, 1192-1197 (1970).

Odell, W.D., Swerdloff, R.S., Jacobs, H.S., and Hescox, M.A.: FSH induction of sensitivity to LH: one cause of sexual maturation in the male rat. *Endocrin.* 92, 160-165 (1973).

Ojeda, S.R. and Jameson, H.E. Studies on the biological properties and gel filtration behavior of pituitary and serum follicle stimulating hormone of infantile female rats. *Endocrin.* 101, 475-484 (1977).

Ojeda, S.R. and Ramirez, V.D. Short term steroid treatment on plasma LH and FSH in castrated rats from birth to puberty. *Neuroendocrin.* 13, 100-114 (1973/74).

Ojeda, S.R., Jameson, H.E., and McCann, S.M. Developmental changes in pituitary responsiveness to LHRH in the female rat: ovarian-adrenal influence during the infantile period. *Endocrin.* 100, 440-451 (1977).

Ojeda, S.R., Kalra, P.S., and McCann, S.M. Further studies on the maturation of the estrogen negative feedback on gonadotrophin release in the female rat. *Neuroendocrin.* 18, 242-255 (1975).

Osmon, P. Preovulatory changes in the ovaries during the first spontaneous proestrus in the rat. *J. Endocrin.* 67, 259-265 (1975).

Osmon, P. and Meijs-Roelofs, H.M.A. Effects of sodium pentobarbitone administration on gonadotrophin release, first ovulation and ovarian morphology in pubertal rats. *J. Endocrin.* 68, 431-437 (1976).

Parker, C.R. and Mahesh, V.B. Hormonal events surrounding the natural onset of puberty in female rats. *Biol. Reprod.* 14, 347-353 (1976).

Paul, S.M. and Skolnick, P. Catechol oestrogens inhibit estrogen elicited accumulation of hypothalamic cyclic AMP suggesting role as endogenous antiestrogen. *Nature* 266, 559-560 (1977).

Payne, A.H., Kelch, R.P., Murone, E.P., and Kerlan, J.T. Hypothalamic, pituitary and testicular function during sexual maturation of the male rat. *J. Endocrin.* 72, 17-26 (1977).

Pedersen, T.: Follicle development in the immature mouse ovary. *Acta Endocrin.* 62, 117-132 (1969).

Peluso, J.J., Steger, R.W., and Hafiez, E.S.E. Development of gonadotrophin-binding sites in the immature rat ovary. *J. Reprod. Fert.* 47, 55-58 (1976).

Plapinger, L. and McEwen, B.S. Ontogeny of estradiol-binding sites in rat brain. I. Appearance of presumptive adult receptors in cytosol and nuclei. *Endocrin.* 93, 1119-1128 (1973).

Rabii, J. and Ganong, W.F. Responses of plasma estradiol and plasma LH to ovariectomy, ovariectomy plus adrenalectomy and estrogen injection at various ages. *Neuroendocrin.* 20, 270-281 (1976).

Ragar, J., Zarzycki, J., Eichner, M., and Gupta, D. Effect of experimental biolateral cryptochidism and castration on the plasma gonadotrophins in male rats during sexual maturation. *Res. Exp. Med.* 165, 55-59 (1975).

Ramaley, J.A. Effect of gonadal hormones upon serum proteins and steroid binding in maturing female rats. *Life Sciences* 9, 673-682 (1970).

Ramaley, J.A. Steroid binding to serum proteins in maturing male and female rats. *Endocrin.* 89, 545-552 (1971).

Ramaley, J.A. The development of daily changes in serum corticosterone in pre-weanling rats. *Steroids* 21, 433-442 (1973).

Ramaley, J.A. The effect of an acute light cycle change on adrenal rhythmicity in prepubertal rats. *Neuroendocrin.* 19, 126-136 (1975).

Ramaley, J.A. The role of corticosterone rhythmicity in puberty. *Biol. Reprod.* 14, 151-157 (1976a).

Ramaley, J.A. Effects of corticosterone treatment on puberty in female rats. *Proc. Soc. Exp. Biol. Med.* 153, 514-517 (1976b).

Ramaley, J.A. and Bartosik, D. Precocious puberty: the effect of adrenalectomy on PMS-induced ovulation and progesterone secretion. *Endocrin.* 96, 269-274 (1975).

Ramaley, J.A. and Campbell, G.T. Serum prolactin concentrations in the adrenalectomized rat: relationship to puberty onset. *Endocrin.* 101, 890-898 (1977).

Ramaley, J.A. and Sieck, G.C. Adrenal-gonadal function in rats with frontal hypothalamic transections. *Neuroendocrin.* 18, 55-64 (1975).

Ramaley, J.A. and Schwartz, N.B. The pubertal process in the rat. Effect of chronic corticosterone treatment. *Neuroend.* 30, 213-219 (1980).

Ramirez, V.D. Endocrinology of puberty, in *Handbook of Physiology*. Greep, R.O. (ed.), Section 7, Endocrinology, Vol. II, Female Reproductive System, Part 1. Am. Physiol. Soc., Washington, D.C. (1973).

Ramirez, V.D. and McCann, S.M. Comparison of the regulation of luteinizing hormone (LH) secretion in immature and adult rats. *Endocrin.* 72, 452-464 (1963).

Ramirez, V.D. and Sawyer, C.H. Advancement of puberty in the female rat by estrogen. *Endocrin.* 76, 1158-1168 (1965).

Raynaud, J.P., Mercier-Bodard, C., and Baulieu, E.E. Rat estradiol binding plasma protein (EBP). *Steroids* 18, 767-788 (1971).

Reichert, L.E. and Abou-Issa, H. Students on a low molecular weight testicular factoe which inhibits binding of FSH receptor. *Biol. Reprod.* 17, 614-621 (1977).

Reynolds, R.W. An irritative hypothesis concerning the hypothalamic regulation of food intake. *Psych. Review* 72, 105-116 (1965).

Richards, J. and Williams, J.J. Luteal cell receptor content for prolactin and luteinizing hormone: regulation by LH and PRL. *Endocrin.* 99, 1571-1581 (1976).

Riss, W., Burstein, S.D., and Johnson, R.W. Hippocampal or pyriform lobe damage in infancy and endocrine development of rats. *Am. J. Physiol.* 204, 861-866 (1963).

Root, A.W. Endocrinology of puberty. I. Normal sexual maturation. *J. Pediatrics* 83, 1-19 (1973).

Ruf, K.B., Wilkinson, M., deZiegler, D., and Cassard, D. Ovarian control of gonadotropin secretion during induced precocious sexual maturation in the rat. *Neuroendocrin.* 22, 226-230 (1976).

Ryan, K.J., Naftolin, F., Reddy, V., Flores, F., and Petro, Z. Estrogen formation in the brain. *Am. J. Ob. Gyn.* 114, 454-460 (1972).

Saez, J.M., Morera, A.M., Haour, F., and Evain, D. Effects in vivo administration of dexamethasone, corticotropin and human chorionic gonadotropin in steroidogenesis and protein and DNA synthesis of testicular and interstitial cells in prepubertal rats. *Endocrin.* 101, 1256-1263 (1977).

Schwartz, N.B. The role of FSH and LH and their antibodies on follicle

growth and ovulation. *Biol. Reprod.* 10, 236-272 (1974).
Schwartz, N.B. and Channing, C.P. Suppression of the secondary rise in serum FSH at proestrus by porcine follicular fluid (PFF). *Fed. Proc.* 36, 322 (1977) (abstract).
Sehgal, A. and Hoffmann, J.C. Serum LH levels in intact and ovariectomized female rats during puberty. *Indian J. Exp. Biol.* 15, 229-231 (1977).
Setchell, B.P. and Waites, G.M.H. The blood-testis barrier, in *The Handbook of Physiology*, Section 7: Endocrinology, Vol. V., Male Reproductive System, D.W. Hamilton and R.O. Greep, eds. Am. Physiol. Soc., Washington, D.C. (1973) pp. 143-172.
Sheridan, P.J., Sar, M., and Stumpf, W.E. Autoradiographic localization of ^{3}H-estradiol or its metabolites in the central nervous system of the developing rat. *Endocrin.* 94, 1386-1390 (1974).
Sherwood, N.M. and Timiras, P.S. Comparison of direct-current and radio frequency current lesions in the rostral hypothalamus with respect to sexual maturation in the female rat. *Endocrin.* 94, 1275-1285 (1974).
Siebers, J.W., Peters, F., Zenzes, M.Y., Schmidtke, J., and Engel, W. Binding of human chorionic gonadotrophin in rat ovary during development. *J. Endocrin.* 73, 491-496 (1977).
Sieck, G.C., Nance, D.M., Ramaley, J.A., Newman-Taylor, A., and Gorski, R.A. Prepubertal cyclicity in feeding behavior and body weight regulation in the female rat. *Physiol. Behav.* 18, 299-305 (1977).
Sieck, G.C., Ramaley, J.A., Harper, R.M., and Newman-Taylor, A. Sleep wakefulness changes at the time of puberty in the female rat. *Brain Res.* 116, 346-352 (1976).
Smith, M.S. and Ramaley, J.A. Development of ability to initiate and maintain prolactin surges induced by uterine cervical stimulation in immature rats. *Endocrin.* 102, 351-354 (1978).
Smith, G.C. and Simpson, R.W. Monoamine and fluorescence in the median eminence of foetal, neonatal and adult rats. *Z. Zellforsch* 104, 541-556, (1970).
Steele, R.E. Role of the ovaries in maturation of the estradiol-luteinizing hormone negative feedback system of the pubertal rat. *Endocrin.* 101, 587-597 (1977).
Steele, R.E. and Weisz, J. Changes in sensitivity of the estradiol and LH feedback system with puberty in the female rat. *Endocrinol.* 95, 513-520 (1974).
Steinberger, A. and Steinberger, E. Replication pattern of Sertoli cells in maturing testis in vivo and in organ culture. *Biol. Reprod.* 4, 84-87 (1971).
Steinberger, A. and Steinberger, E. Secretion of an FSH-inhibiting factor by cultured Sertoli cells. *Endocrin.* 98, 918-921 (1979).
Steinberger, A. and Walther, K. Age dependent responses of isolated Sertoli cells to FSH. Tenth Annual Meeting, Society for the Study of Reproduction (1977) (abstract).
Steinberger, E. and Steinberger, A. The testis: growth versus function, in *Regulation of Organ and Tissue Growth*, R.J. Goss, ed. Academic Press, New York (1972), pp. 299-314.
Strauss, W.F. and Meyer, R.K. Neural timing of ovulation in immature rats treated with gonadotropin. *Science* 137, 860-861 (1962).
Swerdloff, R.S. and Walsh, P.C. Testosterone and oestradiol suppression of LH and FSH in adult male rats: duration of castration, duration of treatment and combined treatment. *Acta Endocrin.* 73, 11-21 (1973).
Swerdloff, R.S., Jacobs, H.S., and Odell, W.D. Hypothalamic-pitiutary-

gonadal interrelationships in the rat during sexual maturation, in *The Gonadotrophins*, B.B. Saxena, C.G. Beling, and H.M. Gandy, eds. Wiley, New York (1972), pp. 546-561.

Swerdloff, R.S., Walsh, P.C., Jacobs, H.S., and Odell, W.D. Serum LH and FSH during sexual maturation in the male rat: effect of castration and cryptochidism. *Endocrin.* 88, 120-128 (1971).

Terasawa, E. and Timiras, P.S. Electrophysiological study of the limbic system in the rat at onset of puberty. *Am. J. Physiol.* 215, 1462-1467 (1968).

Velasco, M.E. Opposite effects of platinum and stainless steel lesions of the amygdala on gonadotropin secretion. *Neuroendocrin.* 10, 301-308 (1972).

Velasco, M.E. and Taleisnik, S. Effect of hippocampal stimulation on the release of gonadotropin. *Endocrin.* 85, 1154-1159 (1969a).

Velasco, M.E. and Taleisnik, S. Release of gonadotrophins induced by amygdaloid stimulation in the rat. *Endocrin.* 84, 132-139 (1969b).

Vitale, R., Fawcell, D.W., and Dym, M. The normal development of the blood testis barrier and the effects of compliphene and estrogen treatment. *Anat. Rec.* 176, 333-344 (1973).

Walker, P., Dussault, J.H., Alvarado-Urbina, G., and Dupont, A. The development of the hypothalamo-pituitary axis in the neonatal rat: hypothalamic somatostatin and pituitary and serum growth hormone concentrations. *Endocrin.* 101, 782-787 (1977).

Weiner, R.I. and Scapagnini, U. Effect of central acting drugs on the onset of puberty, in: *Narcotics and the Hypothalamus*, E. Zimmermann and R. George, eds. Raven Press, New York (1974), pp. 175-182.

Weisz, J. and Gunsalus, P.: Estrogen levels in immature rats: true or spurious-ovarian or adrenal? *Endocrin.* 93, 1057-1965 (1973).

Ying, S.Y. and Greep, R.O. Effect of age of rat and dose of a single injection of estradiol benzoate on ovulation and the facilitation of ovulation by progesterone. *Endocrin.* 89, 785-790 (1971).

YoungLai, E.V., Holmes, M.J., and Ruf, K.B. Changes in the concentration of LH, FSH and estrogen in the immature female rat during precocious sexual maturation induced by electrochemical stimulation of the brain. *Hormone Res.* 7, 34-42 (1976).

BIBLIOGRAPHY

Bierich, J.R., ed. *Disorders of Puberty*. Clinics in Endocrinology and Metabolism, vol. 4(1). W.B. Saunders Co. Ltd., London (1975), pp. 225.

Critchlow, V. and Bar-Sela, M.E. Control of the onset of puberty, in *Neuroendocrinology*, vol. I. L. Martini and W.F. Ganong, eds. Academic Press, New York (1967).

Donovan, B.T. and van der Werff ten Bosch, J.J.: *Physiology of Puberty*. The Williams and Wilkins Co., Baltimore (1976), pp. 216.

Forest, M.G., DePeretti, Ed., and Bertrand, J. Hypothalamic-pituitary-gonadal relationships in man from birth to puberty. *Clinical Endocrin.* 5, 551-569 (1976).

Girard, J. and Nars, P.W. Some aspects of the activity of the hypothalamo-pituitary gonadal system in children, in *The Endocrine Function of the Human Ovary*, V.H.T. James, M. Serio, and G. Giusti, eds. Academic Press, London (1976).

Greep, R.O., ed. MTP International Review of Science, vol. 8, *Reproductive Physiology*, Butterworths, London (1974), p. 323.

Greep, R.O., ed. International Review of Physiology, vol. 13, *Reproductive Physiology II*, University Park Press, Baltimore (1977), p. 248.

Grumbach, M.M., Grave, G.D., and Mayer, F.E., eds. *The control of the onset of puberty*. John Wiley and Sons, New York (1974) pp. 484.

Gupta, D., Rager, K., Attamasio, A., Klemm, W., and Eichner, M. Sex steroid hormones during multiphase pubertal developments. *J. Steroid Biochem.* 6, 859-868 (1975).

Moltz, H. The search for the determinants of puberty in the rat, in *Hormonal Correlates of Behavior, Vol. I. A Lifespan View.* B.E. Eleftherious and R.L. Sprott, eds. Plenum Press, New York (1975) pp. 35-154.

Ramaley, J.A. Adrenal-gonadal interactions at puberty. *Life Sciences*, 14, 1623-1633 (1974).

Ramirez, V.D. Endocrinology of puberty, in *Handbook of Physiology*, Section 7, Endocrinology, vol. II, Female Reproductive System, part 1, R.O. Greep, ed. Am. Physiol. Soc., Washington, D.C. (1973), pp. 1-28.

Root, A.W. Endocrinology of puberty. I. Normal sexual maturation. *J. Pediatrics* 83, 1-19 (1973).

Root, A.W. and Reiter, E.O. Evaluation and management of the child with delayed pubertal development. *Fertility and Sterility* 27, 745-755 (1976).

Copyright©1982, Spectrum Publications, Inc.
Hormones in Development and Aging

Chapter 10
Clinical Aspects of Puberty and Adolescence
Arthur F. Kohrman

INTRODUCTION

The physician who cares for children and adolescents is frequently challenged with the recognition and appropriate management of a variety of interesting and important chromosomal, endocrinologic, or developmental abnormalities which manifest as aberrances of pubertal development. The psychological problems posed by entities such as true precocious puberty, virilizing tumors, or various forms of gonadal dysgenesis require astute diagnosis, complicated decisions about management, and a sensitive and humane approach to the patient and family. In turn they explicate important issues of human biology, yet in by far the greatest number of patients seeking medical advice concerning development in puberty no such specific abnormality is present. Because of the wide variation from individual to individual in the timing and patterns of pubertal development, large differences between individuals in the same peer group at the same time may exist within the range of normality. The generalized ignorance of concepts of biological variation and deviation from the mean of a series of developmental and somatic markers contributes to the puzzlement of the patient and primary care physician, and difficulty with the recognition of the true pathological state. Contemporary culture through the media projects a series of ideals or norms of appearance, stature, and sexual development, which both consciously and unconsciously have conditioned our expectations.

These issues are really problems of discrepancy from or failure to meet expectations which are held unconsciously by the child, the parents, the school and social cultures in which the child grows up. Of particular importance are the imaginary "ideals" for development held by the child's peer culture. The pressures on a child who is discrepant from the expectations of peers in any significant measurement or appearance during development can be monstrous. The idealization of beauty, athletic ability (and with it, usually, extraordinary size) perpetuated by the mass media which penetrate to the deepest reaches of our culture has led to a progressively narrowed conception of "normality." As a consequence, children who are seen in their development to be deviant from that concept of normality are immediately assumed to have some variety of "medical problem" and frequently present to the medical care system because of a variety of anxieties in the child, parent, or teachers. In a sense, these <u>are</u> medical problems in that the continued unrelieved anxiety and compensatory behaviors can seriously affect the development of the individual child in terms of a misshaped self-image and sense of worth. Appropriate management of these problems by the physician who is so often the first line of consultation should be seen as effective preventive medicine and prophylaxis against serious future dysfunction.

The present chapter will deal with the recognition of these sorts of problems and with some analysis of their roots, with suggestions for general differential diagnosis and brief discussion of approaches to counseling the affected individuals and families. There are many texts which extensively delineate the differential diagnosis and specific management of endocrinopathies, chromosomal disorders, and developmental anomalies which specifically interfere with pubertal development, or otherwise manifest as disorders of puberty; I will not attempt to duplicate them here. In addition, there are within this text extensive discussions of the mechanisms and sequence of normal puberty and the psychological development and specific abnormalities during puberty in adolescents. The reader is referred to the meticulous study of the stages of puberty by Tanner (1975) for details concerning biological, somatic, and psychological events which provide the background for the following discussions (see Table 1 and Figs. 1 and 2).

It is particularly important that those caring for children in puberty recognize the wide variation in time of onset and the duration of pubertal changes which exist in a given population. These variations result in marked differences in size and state of sexual development and psychosexual outlook in children of the same age. Thus it is apparent in a school culture in which children are primarily stratified by age that there will be times in which a given age-cohort of children will have members who are at both extremes of stages of sexual development. This simple but essential fact is in itself responsible for a great number of the concerns about abnormality of development which present to the physician.

It is important for the ensuing discussion also to make clear the distinction between puberty and adolescence. In my terminology, puberty includes the somatic and endocrinologic changes which complete and accompany genital development and preparation for reproduction in the human male and female. It is useful to distinguish those biologic and somatic changes from the process of adolescence - a cultural and psychological phenomenon with great differences between cultures and subcultures, which changes in its definition and imperatives with the changing history of our culture. Adolescence is clearly related to the endocrinologic and somatic changes; that is, the cultural imperatives are directed at the individuals undergoing those changes. In fact, the culture of the adolescent is one of response to and interaction with social, familial, and cultural phenomena which are directed at the emerging sexuality and ultimate sexual function of the young adult. While I shall discuss both puberty and adolescence in this chapter, the distinction is made in order to permit analysis of what would otherwise be a hopelessly complex interweaving of psychological, cultural, and physical phenomena and the individual's response to them.

In a sense it is my intent to establish a perspective on normal biological variation and its consequences that will allow the physician and medical care personnel to "demedicalize" a series of normal variations about the mean, with the goal of helping children and their families to return to a more balanced perspective concerning development and to lessen anxiety. I hope also that this perspective will permit human biologists of all sorts to approach these problems with appropriate precision on one hand and breadth of inquiry and of intervention on the other.

Table 1.

Sequence of Events in Puberty, adapted from Tanner, 1975
(also see Figs. 1 and 2)

1. The pubic hair stages (in both charts in Figs. 1 and 2) are:

    Stage 1. Predolescent: the fine hair over the pubic area is the same as that over the abdominal wall; i.e., no pubic hair.

    Stage 2. There is sparse growth of long, slightly pigmented, downy hair, straight, or slightly curled, chiefly at the base of the penis or along the labia.

    Stage 3. The hair is considerably darker, coarser, and more curled. It spreads sparsely over the pubic bones.

    Stage 4. Hair is now adult in type, but the area covered is still considerably smaller than in the adult. There is no spread to the medial surface of the thighs.

    Stage 5. The hair is adult in quantity and type. Spread is to the medial surface of the thighs. The horizontal or classically "feminine" pattern is complete.

    In about 80 percent of men and 10 percent of women, pubic hair spreads further upward towards the umbilicus (classical "masculine" pattern).

2. Axillary hair appears on the average, some two years after the beginning of pubic hair growth, i.e., when pubic hair is reaching Stage 4. There is much variability in and disassociation of these events. In a few children axillary hair actually appears first.

3. The male genital development stages (as shown by penis and testis growth) are:

    Stage 1. Preadolescent: testes, scrotum, and penis are about the same size and shape as in early childhood.

    Stage 2. Scrotum and testes are slightly enlarged. The skin of the scrotum is reddened and changed in texture. There is little or no enlargement of the penis at this stage.

    Stage 3. Penis is slightly enlarged, at first mainly in length. Testes and scrotum are further enlarged than in Stage 2.

    Stage 4. Penis is further enlarged, with growth in breadth and development of glans. Testes and scrotum are further enlarged than in Stage 3; scrotal skin is darker than in earlier stages.

    Stage 5. Genitalia are adult in size and shape.

4. The breast development stages are:

    Stage 1. Preadolescent: there is elevation of the papilla only.

Stage 2. Breast bud stage. There is elevation of the breast and the papilla as a small mound. Areolar diameter is enlarged over Stage 1.

Stage 3. Breast and areola are both enlarged and elevated more than in Stage 2, but with no separation of their contours.

Stage 4. The areola and papilla form a secondary mound projecting above the contour of the breast.

Stage 5. Mature stage. The papilla only projects, with the areola recessed to the general contour of the breast.
   The Stage 4 development of the areolar mound does not ever occur in some girls; in probably a quarter it is absent and in a further quarter slight. Furthermore, when it does occur, it may persist well into adulthood. Thus Stages 4 and 5 are not distinct in all girls.

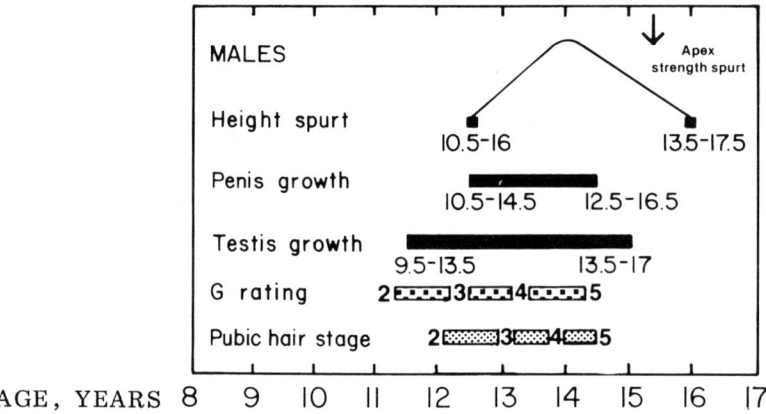

Fig. 1. Average Ages of Onset and Duration of Changes in Secondary Sex Characters in Males (European and North American White Children). (From Tanner, 1975.)

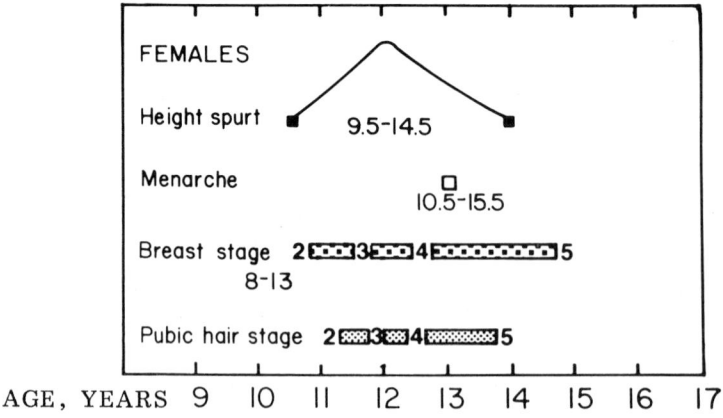

Fig. 2. Average Ages of Onset and Duration of Changes in Secondary Sex Characters in Females (European and North American White Children). (From Tanner, 1975.)

PROBLEMS IN STATURE

Certainly the most common problems in both males and females presented to the physician around puberty are those of short stature. In many cases the child's perceived short stature is simply a concomitant of the delay of onset of puberty and its accompanying growth spurt. Often in this circumstance children and families will be adequately relieved with simple reassurance of the inevitability of puberty in the child and a short education in the normal variation. It is also useful in these situations to develop a careful family history of the adult height of several generations of family members and both parents, and to make an estimate of the growth patterns and age of onset of puberty of the parents and their siblings. Not only are distributions of ultimate adult height clearly familial but the time of onset and sequences of pubertal development tend to replicate within families. It is useful in this context to note that the standard growth curves used for plotting normal childhood development are derived from cross-sectional data. The actual longitudinal patterns of growth for any individual child may not necessarily conform to these charts. Many children enter puberty late and have a growth period extended beyond the average, which results in a relatively short stature at the mean pubertal age, but with ultimate attainment of height in the normal adult range at a time later than that of their peers. At the opposite end of the spectrum some children achieve nearly all of their pubertal height increment with the initial growth spurt early in puberty and thus will pass through a period during which they are significantly taller than their peers. Again, the differences in size become less obvious with the passage of time.

SHORT STATURE

The analysis of the problem of short stature around puberty in both sexes must include a careful neonatal and childhood history and a search for concomitant systemic illness which might explain the short stature. For example, it is becoming clear that a certain subset of those children who appear to have undergone intrauterine growth retardation will continue in their somatic development below the normal range for their sex and age through puberty and into adulthood; some may predictably become short adults. Thus careful perinatal history and measurements can be helpful in identifying such an individual. It also appears likely that children who undergo severe environmental or psychosocial deprivation during early phases of prepubertal development may not totally recover normal growth velocity. They may be permanently small even if restored to a more supportive and nurturant environment. For those children the normal growth spurt at puberty may be blunted and their ultimate adult height less than that predicted from familial patterns.

A variety of systemic illnesses may first be identified when a child presents with delayed puberty or short stature. Among the group of subtle but important entities which can profoundly affect growth and development are chronic inflammatory bowel disease (regional ileitis and colitis) and renal disease (both congenital dysplasias of the kidneys as well as acquired renal problems such as chronic pyelonephritis). Occasionally, various forms of hypopituitarism will not be suspected until failure of the child to enter into puberty or to initiate a pubertal growth spurt is seen. Craniopharyngiomas may first be recognized at this time. However, these organic causes make up by far the smallest proportion of all those children seen by the physician and endocrinologist for short stature and delayed puberty.

Short Stature in the Male

It is not uncommon for boys with serious concerns about short stature to claim a variety of alleged illnesses or other problems. Interviewing these children and their parents may require some degree of cleverness and insight to discover real sources of concern. When the boy himself seeks care, it is usually because of the disadvantages of his size he perceives compared to that of his peers. This discrepancy is especially difficult for 11 to 13 year-old males when they may even be significantly shorter than the girls in their classes who have already begun or completed pubertal growth spurts. An obvious source of concern is the disadvantage in the youngster's ability to compete in the dominant athletic activities of his age and peer group. This concern may vary depending upon the particular sports which are in high regard in that school, region or culture. In those areas where team sports dependent on size such as basketball or football have the highest social value, small males, of course, are at the greatest disadvantage. In other areas, where individual sports such as swimming and wrestling are valued, the short male may find an appropriate athletic role. Another large group of short males present as "school problems," sometimes as academic failure or indifferent performance, in other cases because of poor relationships with their peers. These are often manifest by frequent fights and complaints of disruptiveness. In this last group further inquiry will reveal that the small youngster is often venting great anger in inappropriate situations, and suffering the consequences in terms of physical abuse or rejection by his colleagues who tire of his constant contentiousness and provocation.

When parents bring the major complaints and the child appears to be less concerned, the interviewer must immediately identify the parents' expectations for the child's size, appearance, ability, or performance. We have been astonished to find parents with deep concern about boys with perfectly normal development, because the parents had excessive expectations or because they wished their son to be somehow larger or stronger or different-looking in order to fulfill some imagined ideal. With the short male in whom the family history clearly predicts for reasonably short stature, the parents may harbor unrealistic goals of growth for their son perhaps, in fact, hoping, equally unrealistically, that he somehow will grow larger than they and avoid the social and psychological penalties that they have suffered. Sometimes the parents present the child as a "behavior problem," particularly when there is significant acting-out behavior or anger on the part of the child with siblings or in school.

Another concern of parents, not illogical, is to take care of potentially treatable problems "early enough" to avoid penalizing the youngster permanently with short stature. A great deal of such anxiety is generated by materials in the media which sensationalize announcements of various therapies for short stature without appropriately noting that these apply only to very specific small groups of children with identified abnormalities. A common problem which needs correction if detected is the parents' concern that if their son would only "eat better" he would grow more. It may be that the short stature is used by these families as a demonstration of the penalty that the child is paying for noncompliance with the expected eating patterns. Finally, discovery of a certain concern may take exceptional insight on the part of the interviewer; some parents feel that their child's failure of growth or, particularly, of sexual development is a sign of impending lack of sexual capability or even a precursor of homosexuality. These fears are rarely stated as such, but careful interaction will often allow the parents to admit their concerns about ultimate sexual performance and preference in their son.

Occasionally, referrals of youngsters with short stature are made to the physician by the schools; any of the reasons already mentioned may be the motivation of the school authorities. The clinician in this circumstance should be extremely thorough in ascertaining what the attitudes are about children with variance in stature or sexual development in that particular school. It has been our frequent experience that individuals in the school itself, particularly physical education instructors, have been guilty of fostering an environment which heightens the concerns about size and development in the small male; in a misguided attempt to be jocular, the teacher can reinforce the child's sense of deviance and inadequacy.

Management of the Problems of the Small Male

It is, of course, first necessary to ascertain that the small male does not exhibit one of the rare primary organic causes of growth failure outlined earlier. A thorough history, analysis of the child's growth patterns and those of the family, and careful physical examination will usually serve to make those distinctions. It is important that the examiner well understand the normal sequence of pubertal events; if there is disunity in these, a thorough search for organic disease must be made. The detection of early changes in testicular size and/or scrotal elongation and thinning can be useful predictors of the onset of full puberty. The knowledge of the parents and the child concerning the normal and expected changes of puberty should be thoroughly probed and the expectations for this particular child's development should be brought out into the open by interview of the parents and the child in each other's presence. Usually this procedure will bring to the surface the true concerns and the genesis of the reason for the consultation. Once these have been determined and the state of the child's development ascertained by physical examination, the entire family unit should be informed of the physician's impression, and the expected progress should be outlined in some detail. The medical team should offer continuing support for the child and the family of the child with short stature. One of the most important elements in that support is giving both the children and the parents permission to express their feelings. The child must have a "safe" place to ventilate the anger and sadness which size discrepancy from his peers creates in order that inappropriate and possibly self-destructive acting-out behaviors can be moderated. Usually, when the appropriate information is given to the family and child, and offers of continuing observation and support are made, there is marked relief of anxiety. Efforts then can be made to construct a more positive environment to help the child deal with the consequences of short stature. The physician must help the family and the short boy to set realistic goals for his ultimate development and social performance. The use of judicious estimation of adult height with the tables of Bayer and Bayley (1959) based upon bone-age determination can be a great source of relief to families and children and give the physician an optimistic base for future management.

Finally, when it appears from familial or other data that the child's ultimate height will be significantly shorter than that of his peers, long-term supportive help may be necessary to bring the child into acceptance of his stature and development of a realistic life plan. Failure to deal with the roots of anger and self-hatred can result in significant worsening of acting-out behavior, with important delays in psychosocial maturation and loss of self-esteem and social mobility. Of great help in dealing with youngsters in whom short stature appears to be their ultimate fate, are organizations such as the Little People of America and groups sponsored

Short Stature in the Female

The social consequences of shortness in females in our society are generally much less severe than those for males. It is considered "cute" or "pert" for a girl to be small, and short females have fit better into traditional sex roles of the American culture. Nonetheless, concerns about height in the pubertal female are not infrequent.

It is much more common for girls with short stature to present not because of size alone but because of the accompanying delays of sexual development, specifically secondary sex characteristics and menses, which they see their peers acquiring around them. In girls as in boys, "school problems" may be the initial complaint and the girl who has failed to keep up with her peers in size or sexual development may present a variety of behavioral responses to her sense of deviance and alienation from her peer group. Not infrequently such girls are noted to be extremely shy and have not evidenced interest in heterosexual activities. This may be a greater source of concern to the parent than to the child. The parents may fear their daughter's ultimate unmarriageability or noncompetitiveness in the sexual arena. On the other hand, the child herself may, with encouragement, report her own fears of rejection by males.

Parents of the short preadolescent or adolescent girl may express some of the same concerns about size as those described of parents with short boys. In addition, there are some special concerns that the parents of girls harbor which may require subtle and experienced inquiry. Among these are questions of the girl's ultimate fertility and the ability to provide grandchildren for the parents. I have found this concern about reproductive capability to be present even in parents who have apparently clear understandings of the normal variations in onset and rate of pubertal development. They express concerns that lateness of puberty somehow predicts for impaired fertility. Another frequent anxiety of parents of the small girl relates to diet and appetitite. This is particularly prominent when the child herself has expressed desire to appear slimmer and has placed herself on a very restrictive or fad diet in an attempt to lose weight or to remain slim. The parents may seek medical consultation to enlist support in their advocacy of a more rational diet or because of their fears of anorexia nervosa or organic illness resulting from severe prolonged dietary restriction.

School authorities, teachers, or counselors may refer a short girl because of concerns similar to those noted for boys. However, in my experience, it is more frequent that girls deal with their short stature and discrepancy from their peer size and sexual development by withdrawal and alienation than by active acting-out or by evidences of overt hostility.

Management of the Problems of the Short Female

Again, the cornerstone of management is the establishment of a good historical base with careful delineation of the child's previous growth pattern and of the familial growth patterns. The age of menarche of the mother and of older sisters can be of help in establishing a familial pattern of delay. In that group of short girls who present with primary amenorrhea gonadal dysgenesis is by far the most frequent organic finding. The particular clue to this diagnosis on physical examination is the presence of some secondary sexual hair (presumably stimulated by adrenal androgen secretion) with complete absence of any estrogen effect (i.e., breast development or keratinization of the vaginal mucosa) confirmed by very low serum estrogen levels. In children with gonadal dysgenesis there is wide variation in the presentation of associated signs and stigmata. The diagnosis is particularly complicated by the existence of a number of mosiac forms of gonadal dysgenesis. It is not uncommon in my experience that girls with one of the various forms of gonadal dysgenesis have as their only presenting findings short stature and primary amenorrhea. Thus, it is imperative that the physician consider the diagnosis and make appropriate laboratory investigations. Early diagnosis of gonadal dysgenesis will permit initiation of a realistic program for the child and family concerning their expectations of ultimate size, physical development, and most important in my experience, helping the family and child to understand and accept the almost inevitable infertility which is part of the syndrome.

Similar considerations apply to the much less common forms of hypogonadism, either primary or secondary, which also may be associated with short stature and failure of secondary sexual development in the female. When one of these forms of hypogonadism is identified, careful decisions concerning hormone replacement therapy should be made in consultation with the child and the family, grounded on a firm knowledge base of the physiological variables and the risks and benefits of various therapies. Communication of these facts and choices may be difficult, and the physician should be prepared to work intensively and extensively with children and their families with the diagnosis of hypogonadism.

In that much larger group of small girls in whom the only diagnosis is normal variation of time of onset or rate of pubertal development and its accompanying growth spurt, the true concerns of the child and the family should be carefully elicited and directly addressed. In many of our patients these discussions provide an excellent opportunity to present or reinforce education about menstruation; in some cases these discussions lead quite naturally and effectively into discussions of sexual function, reproduction, and contraception. Thus, a concern about a nonexistant "medical" problem can be capitalized upon as an opportunity to provide useful patient education and counseling in important aspects of impending adulthood.

In both males and females who present with short stature, there is excellent opportunity to help children and their parents with an understanding of normal physiology and reproductive function. The physician and staff should be prepared to vigorously and effectively capitalize on the possibilities for health education in areas which are so frequently neglected or inadequately dealt with by parents and schools.

Finally, I must note the persistent tendency of parents, teachers, and physicians to address and treat short adolescents as if they were younger than their chronological ages. This tendency must be consciously identified and countered; continued use of diminutive and childlike language and terminology by adults will serve to reinforce the child's sense of inadequacy and deviance. It may, as well, encourage dependence and help to impede maturation appropriate to the child's age and social peers, thus fostering further alienation.

EXCESSIVE GROWTH

The problems of excessive growth or stature are much less frequently seen in contemporary practice in North America where the size of children has progressively exceeded that of their parents for several generations. Excessive height is becoming less and less of a social or developmental concern. Here again, for both males and females, analysis of familial growth patterns and previous growth patterns for the individual will often give reassurance that a particular child is growing within reasonable expectations. There are relatively few organic conditions in which excessive height occurs around the adolescent period. These include the rare pituitary tumors with giantism, late-onset forms of virilizing adrenal hyperplasia and androgen-secreting tumors of the adrenal or gonad and the rare chromosomal or genetic abnormalities such as Klinefelter's or Marfan's syndromes. In the preadolescent or adolescent youngster with thyrotoxicosis there is often a noticeable acceleration of growth; however, other signs of increased thyroid function usually dominate the clinical picture and the growth spurt becomes only a confirming sign.

The Tall Male

The social consequences of being tall for a male in our culture are much fewer than those for a short individual. Nonetheless there are still potential problems for the male who grows to excessive height and particularly for one who begins his growth spurt much earlier than his peers.

The most obvious and common problem for these young boys is the teasing and derision they receive because of their size, often from other males who are envious of the presumed advantage in the competition for adult privileges and heterosexual activities. There are frequent demands for athletic performance on males who undergo exuberant growth spurts early in puberty. For the individual who is athletically inclined, of course, his size may be an advantage. However, for the large number of boys for whom athletics are only an avocation or amusement, or for whom they hold no attraction at all, the pressures to perform can be severe; their inability or unwillingness to do so can result in perplexity on the parts of their peers and parents, and alienation from their peer group. Often teachers and athletic instructors are particularly guilty of holding expectations for athletic performance which are not congruent with the child's own interests. The situation can be worsened by the clumsiness which is often present in young pubertal males who have had rapid growth. Adults in authority positions frequently fail to recognize that the peak of muscular development and fine motor coordination occurs many months or even years after the peak velocity of growth. Thus, these youngsters are expected to perform in a variety of tasks to which their neuromuscular development is not yet suited. They then become the constant victims of failure of the expectations of the adults in their world and consequently may become severely critical of themselves as well. It is not uncommon in careful interviewing to find that the young male who grows rapidly has a series of expectations more appropriate to his size than to his age placed upon him by his parents. Failure to understand this situation can lead to progressive deterioration of both the parent-child relationship and of the child's sense of his own competence.

Parents who bring large or rapid-growing pubertal males to the physician are often concerned that he will be some sort of a "freak" and often urgently inquire as to what needs to be done to prevent his growing to a height which will be in the clearly deviant range. Another frequent complaint by parents or school personnel is that the large male is guilty of bullying the smaller children or is a source of disorder in the school and play setting. In many cases these behaviors are in reaction to constant teasing and a sense of frustration against which the large boy acts out in inappropriate and ultimately antisocial manners.

Management of the Tall Male

It is essential that the expectations of the child's growth and development be carefully elicited both from the child and his parents, preferably in their presence. In many cases the parents and the child may not identify the same problems and a joint interview will often bring out the discrepancies.

Concerns about ultimate height can often be met by prediction of adult height from bone age films and the use of the Bayer and Bayley tables. Once a reasonable final adult height is assured, there is frequently relief on the parts of all the concerned parties. Then it is much easier to deal with actual issues of concern and to initiate a program of reassurance and support. It can be very helpful to reassure the child that size alone does not determine athletic ability or motor skill and that the development of his own interests and personality can take other directions than those which the dominant culture would expect from a person of his size. This kind of support from an authority figure such as the physician or other health-care professional can be of great importance to the youngster who is concerned about meeting the expectations of the other adults around him.

The Tall Female

Nowhere in the spectrum of problems arising from normal variations in puberty is the changing cultural pattern better seen than in the issues around excessively tall girls. Early in this century in the United States and until the present in Europe, the tall girl has been one of the most frequently referred patients to the pediatric endocrinologist. Concerns around the possibility of her becoming a woman of excessive height derive almost entirely from anxieties about her ultimate marriageability and competitiveness in heterosexual relationships. It has been long felt that the excessively tall woman (i.e., one who is taller than the majority of men) will be at a distinct disadvantage in terms of her attractiveness to males and in her opportunities for "marrying well." The emotional burden carried by these girls, especially those who grow rapidly in early adolescence, is immense. Many of them have historically sought seclusion, have had a severe sense of their own inferiority and freakishness, and have developed a variety of compensatory behaviors.

It is of great interest that because of the sequential increases in height seen in succeeding generations of children in the last 50 years in the United States and because of the greater social acceptability of tall, active women, the referrals for excessive height have dramatically diminished in number. However, these children are still seen and in most cases their concerns are those mentioned. There may be also a certain curiosity on the child's part as to her ultimate height and whether or not she will "ever stop growing." This may be a particular problem for 11 to 14-year-old girls where it is not uncommon to find many girls taller than the boys in their classes and peer groups. The girls may undergo a great deal of teasing with predictable secondary effects of either withdrawal or acting-out.

Here again, medical personnel must be alert to the source of the concern. It is my experience that parents are often concerned when the girls themselves are not and the identification of this situation can relieve a great deal of tension between the child and parent and allow the child to proceed with her own adaptation to her physical size and appearance. Certain parents and children are aware that there are possible hormonal interventions, i.e., estrogen therapy, which have been recommended in an attempt to slow down or stop excessive growth in females. It is not uncommon that those parents come to the physician demanding such therapy.

Approach to Management of the Tall Female

After elimination of the relatively rare organic causes of excessive height, a bone-age and subsequent height-projection estimate can be of great value in relieving anxiety about the girl's unlimited growth to the point of freakishness. For the parent or child who is insistent upon an attempt of estrogen therapy to inhibit or slow growth, careful discussion of the questionable value and possible side effects of this approach must be undertaken. Of value in this discussion is a recent review of estrogen therapy in childhood (New, 1978), the result of a recent conference of European and American pediatric endocrinologists in which the relative values, the possible hazards, and the unknown variables are thoroughly discussed. It is my practice not to acquiesce to such parental demands. However, estrogen treatment for girls with excessive growth is accepted therapy in the European community. Those investigators with large experience with estrogens report a significant reduction in ultimate height from the anticipated height estimated by bone age studies before the initiation of therapy. There is some concern about the effects of large doses of estrogen in terms of carcinogenic potential and possible effects on the maturation of the hypothalamic-pituitary-gonadal axis in young girls in whom the normal menstrual cycle has not yet been well established. The actual risks of these are unknown.

PROBLEMS WITH SEXUAL DEVELOPMENT

The important role of sexuality in our culture and especially in the culture of the adolescent places a high value on variations in sexual development and, as in so many situations in which variance is presumed to represent some form of organic abnormality, these issues are frequently presented as medical problems.

Thorough review of history and physical examination can usually quickly identify the uncommon individual with an organic disorder of sexual development. As with problems of short stature, the patient in whom there is an organic abnormality is uncommon.

Early Sexual Development in the Male

Early sexual development in the male is rarely a clinical problem, possibly because the changes of early puberty in the male are gradual and not as clearly noted by the child or parents as are the more dramatic events of breast development and menarche in the female. When there is the onset of pubic hair, penile enlargement or other signs of virilization in the male below the age of about nine years (North American children) it is extremely likely that the boy has true precocious puberty, which in males has frequent association with intracranial lesions, or another virilizing disorder such as adrenal hyperplasia or tumor. An important differential point is the presence or absence of significant testicular enlargement, which only occurs under stimulation from the gonadotropins present in true puberty.

It is true for males as for females that the average age of onset of puberty is earlier in the last several decades. Parents who are concerned about early puberty in a normal male may need to be reminded of this phenomenon. The male who begins pubertal development with a growth spurt may be the victim of social isolation or secondary acting-out behavior if he is much bigger and more sexually developed than his peers. Particularly if progressive sexual development is accompanied by concomitant psycho-sexual interests, the boy may be seen both as a nuisance in the classroom and as a source of distraction in his pursuit of sexual games in a group of children whose general somatic, sexual, and psychosexual development are behind his.

A frequent and often worrisome problem which is the source of many medical referrals in the adolescent male is the occurrence of breast enlargement (gynecomastia) during pubertal development. The transient enlargements of the breast are often associated with hypersensitivity and even, at times, pain. It is often a source of great anxiety for the parents and embarrassment for the boy. Careful surveys of populations of pubertal males reveal that the majority of males undergo some period of transient breast hypertrophy during puberty, especially during the time of peak velocity in growth. The boy is extremely concerned about the change in his appearance and the "feminine" qualities which it imparts. He is often the subject of teasing in locker-room and athletic situations. He may take elaborate measures to avoid the necessity for public exposure of his bared chest; these youngsters often present with minor or undetectable "injuries" for which they are seeking excuses for absence from school activities. Because of the acute embarrassment of having a temporarily "female" habitus the true cause for their concern is rarely the presenting complaint. Sensitivity from parents and medical personnel is required both to identify and to gently confront the cause of the child's concern.

Parents' anxieties around gynecomastia can be extreme and their motives may also be elaborately concealed under a variety of euphemistic and elliptical statements and presenting complaints. There is, of course, a general concern in a cancer-conscious population that the hypertrophy of breast tissue represents a malignant process; this is heightened in the not infrequent situation in which the gynecomastia is asymmetrical. It is our experience that the fathers of pubertal males with gynecomastia may fear for their son's ultimate sexual identity and behavior; fears of homosexuality, while not immediately evident, very frequently are part of the motivation for bringing the child to medical attention. These kinds of parental concerns, whether explicit or more subtle, can be a source of great confusion to the pubertal male's own emerging psychosocial identity in a highly-charged sexual environment. The parental fears of homosexuality can be easily transferred to such a boy; it may take several hours of delicate inquiry and careful counseling to bring out and address the true issues.

The parents can be assured by the fact that the great majority of boys with adolescent gynecomastia have complete resolution of the enlarged breast tissue within 12-18 months of its appearance. While there are scattered reports of malignancies occurring in the pubertal male breast, the only group in which biopsy is indicated are those in which markedly asymmetrical breast development persists well beyond this period of most rapid growth or in which one breast shows progressive enlargement in the absence of any palpable enlargement of the other. Extremely rarely, it may be necessary to consider plastic reduction of the breasts in the adolescent male with gynecomastia, i.e., if the enlargement persists and is of a sufficient degree to create sex-role confusion or continued embarrassment for the male.

Among the conditions in the male which are associated with gynecomastia are Klinefelter's syndrome and the extremely uncommon estrogen-producing tumors of the testis. The ingestion of estrogens by adolescent and young adult males with sex-role confusion in a deliberate attempt to become feminized has been reported.

The management of gynecomastia in the pubertal male requires careful sympathetic inquiry and reassurance. Family counseling where severe confusion or anxiety about sex roles or gender identification exists may be of great value.

Delayed Sexual Development in the Male

Many of the concerns about delayed sexual development in the pubertal male are inseparable from those around short stature which have been discussed. However, there are some special problems in sexual development which may specifically be the cause for concern in boys and their parents.

The boy who has delayed onset of puberty and consequently small genitalia may suffer severe embarrassment in the inevitable comparison games which take place in locker-rooms and other places where he is required to undress in a public setting. Teasing about genital size and pubic hair development is a common activity in a group of adolescent males many of whom are concerned about their own adequacy and anxious to demonstrate superiority. The late-developing boy can become the brunt of a great deal of derision with potentially serious consequences in terms of his own confidence and self-concept. Teasing may also contain some allusions to potential homosexuality. These may feed into already existing anxieties about gender development in the child or his parents and trigger a medical referral which is often cloaked in indirectness and euphemism. Permanent serious effects on self-confidence and sexual identity and role can result from these undetected and unaddressed concerns. Education concerning variations in development for parents and teachers as well as for the affected youngster is often required.

A curious situation seen not infrequently is that of the parent who brings a normally-developed male to the physician because of concerns about growth and development. The presentation of such children may be extremely confusing and after careful inquiry it is usually determined that the family holds hopes that the child will develop in a super-normal manner to become more competitive in athletic or other highly rewarded activities. Through these kinds of expectations parents can make normal individuals feel quite insecure and inadequate, and the detection of this sequence of events is important to protect the child's identity and sense of adequacy. Such parents must be strongly admonished as to the inappropriateness of their expectations and behavior; the boy may need repeated reassurance of his normality to restore what is an entirely preventable and potentially destructive situation. These youngsters are completely the victims of a cultural phenomenon and a set of familial expectations unrelated to the biology of development. Their failure to meet those expectations and cultural norms can become the root of serious personality disorders and degradation of self-image.

Early Sexual Development in the Female

With the steadily declining age of onset of puberty in both sexes, the early appearance of pubertal changes in the female may be a puzzle for the child, her parents, teachers, and her physicians. The onset of puberty appears to be more quickly detected in females in our culture, probably because the first sign of puberty is breast development which has very high significance with both male and female peers of the pubertal girl. This is particularly evident in those situations in which significant breast development (so-called premature thelarche) occurs as the only pubertal manifestation present for several months or years. A flurry of concern by parents and teachers (often greater than that of the child herself) about emerging sexuality and the disruption which the girl with early breast development can cause in a classroom situation may lead to a medical referral. Parental concerns in girls with early sexual development within the normal range of pubertal onset may include anxiety about cancer in asymmetrically growing breasts (a common normal finding). Especially prominent in some families of girls with early puberty are the parents' fears that the onset of secondary sexual characteristics is accompanied by a similar psychosexual precocity. It has been our experience that these parents project upon the child expectations of sexual behavior for which the girl herself in most cases is by no means ready. In this way the parents may create a self-fulfilling prophecy; these girls, because of their parents constant questioning and cautioning about sexual behavior, may be led to experiment at an age earlier than they would have ordinarily. Similarly, teachers become extremely concerned that the early-developing female in a classroom of still prepubertal boys and girls will become disruptive; at times, those fears are borne out in reality. It is not uncommon for parents of the early-developing female to seek advice concerning contraception, a situation which is somewhat perplexing to the girl who has no particular desires for sexual activity. These girls most often are psychologically at the same stages as their age-peers and often find the parents' and teachers' concerns confusing.

The increasingly earlier age of onset of puberty makes a distinct definition of precocious puberty in the female difficult to state with certainty. In addition it is not uncommon that girls have the development of pubic hair alone (so-called premature adrenarche) or of breasts alone (premature thelarche) occurring several months or even years prior to the continuation of the normal pubertal sequence. While the very early appearance of these findings requires a careful differential diagnosis, in the majority of cases no specific pathologies will be found. True precocious puberty in the female is much more common than in the male and is extremely rarely the result of a diagnosable abnormality. After sorting out these variations, the physician must then deal with the concerns of the child and the parents. In this circumstance it is often useful to interview them separately which will usually permit identification of the source of the concerns.

Delayed Sexual Development in the Female

The clinical problems of delayed sexual development in the female have been discussed in concert with those of short stature. The special concerns which arise around the girl with late onset of puberty or in whom breast development is less than desired, center around issues of competition in the sexual arena. Both the child and parent may be extremely concerned about her ability to participate in the peer "dating" culture and ultimately to capture successfully a mate. An additional concern, more often voiced by parents than by the girls themselves, is that of fertility. The irregular menses which are part of the onset of puberty in most girls may feed those concerns of parents and may result in demands for medical intervention to "make her periods regular." That same request from parents may also be an indirect way of their seeking contraceptive protection for their daughter because they believe or fear that she is or will become sexually active. Here again appropriate identification of the real concerns, confrontation of the responsible parties, and factual identification of the normal events of puberty can be very effective in reducing anxiety and conflict between the parents and the child and within the child herself.

ENDOCRINE PATHOLOGIES WHICH MAY PRESENT AS PROBLEMS IN PUBERTY

There is a large number of pathologic conditions which may first present during puberty or as possible abnormalities of pubertal development; several of these have been alluded to in the preceding text. Almost all of these conditions can be recognized as true pathological states when the examiner or interviewer has a firm understanding of the normal sequences of puberty in both the male and female, and has taken a careful history and performed an appropriate physical examination. While the generalist may not be able to make the final definitive diagnosis of some of these unusual conditions, the differentiation from the variations of normal already discussed is usually made without difficulty. It is not our intent to discuss their physiology nor differential diagnosis; it seems useful, however, to mention the general categories of these abnormalities to guide further inquiry.

True precocious puberty, particularly in the female, is the most common problem that must be sorted out from normal variations of pubertal onset and course. Not only is this entity likely to occur more frequently in females but the vast majority of those girls in whom the diagnosis is made will be found to have no discernible pathologic lesions; that is, true precocious puberty appears to occur as the result of an unexplained exceptionally early initiation of the normal mechanisms of hypothalamic-pituitary maturation which initiate normal puberty. True precocious puberty in the female also occurs with some regularity in various forms of mental retardation, and following encephalitides. Long-standing undiagnosed hypothyroidism sometimes presents as the precocious onset of puberty. This presumably occurs as a result of elevated hypothalamic gonadotropin-releasing hormone which occurs along with elevated thyroid-releasing hormone released in the absence of negative feedback resulting from primary hypothyroidism. Finally, there is a series of very rare disorders such as McCune-Albright syndrome and neurofibromatosis in which precocious puberty is one of the somatic manifestations of multi-system disease.

True precocious puberty in the male, while much rarer than in the female, is also a greater cause for concern about the presence of an organic lesion. A high percentage of males with true precocious puberty will be found to have primary tumors of the brain in one of several deep-seated locations in the hypothalamus or pineal areas. Extremely rare causes of precocious puberty are gonadotropin-secreting tumors; the liver has been the primary site of most of those described.

A variety of adrenal lesions may cause the stimulation of primarily virilizing manifestations which may be mistaken for components of the pubertal sequence. In the various enzymatic forms of virilizing congenital adrenal hyperplasia pubic hair and facial hair growth in either sex may cause confusion with some of the changes of puberty. Androgen-secreting tumors of the adrenal, testis, or ovary likewise will be recognized either by the occurrence of inappropriately virilizing manifestations in the female or by the presence of androgen effects without evidence of the usual pubertal sequence of symmetrical testicular enlargement which proceeds virilization in the normal male entering puberty.

The true pathologies which may present as delays in the onset of puberty, incomplete pubertal development, or complete absence of puberty have been discussed in some detail. In the female, chromosomal disorders and the various types of gonadal dysgenesis are the most common forms of primary hypogonadism. Various degrees of hypothalamic-pituitary hypogonadism also exist in the female; these may be seen either as isolated disorders of gonadotropin or releasing-hormone secretion, or in conjunction with other hypothalamic or pituitary dysfunctions in the chromosomally normal female. The group of disorders which appear to be related to abnormalities of end-organ responsiveness, i.e., the syndromes of testicular feminization, usually will present as cases of primary amenorrhea in an individual who has undergone otherwise normal female development.

In the male, hypogonadism may be either primary, manifested as a disorder of testicular development or maturation, or secondary to disorders of the hypothalamus or pituitary, possibly in conjunction with other deficiencies of hypothalamic releasing or pituitary trophic hormones. In these the affected males almost always fail to enter into the earliest sequences of pubertal development; the absence of gonadal enlargement or the presence of abnormal gonads by physical examination will lead into the appropriate differential diagnostic studies.

Psychological and Educational Management

The professional who is consulted for and deals with problems around the physical development of puberty must be an inquisitive, humane, well-informed human biologist and student of human behavior and culture. The vast majority of these problems are in the realm of normal variations around the mean of development and require not classic medical intervention but a careful sorting of the child's and parents' concerns and the initiation of educational and supportive activities. These must be directed not only at the child and immediate family; complete management may also require educational interventions with the schools and community at large which make up the child's environment and so significantly contribute to the narrowed definitions of normality which lead to the initial perceptions of abnormality in the child (also see Chapter 11).

REFERENCES

Bayer, L.M. and Bayley, N. *Growth Diagnosis*. University of Chicago Press, Chicago (1959).
New, M. (Ed.) Estrogen treatment of the young - (Supplement) *Pediatrics* 62. 1087-1217 (1978).
Tanner, J.M. Growth and endocrinology of the adolescent, in *Endocrine and Genetic Diseases of Childhood and Adolescence* (Second Ed.). L.I. Gardner, ed. W.B. Saunders Co., Philadelphia (1975), pp. 14-64.

Copyright©1982, Spectrum Publications, Inc.
Hormones in Development and Aging

Chapter 11
Psychology and Psychopathology of Adolescence
John Griffin
Richard P. Michael

INTRODUCTION

Although the biological phenomena of adolescence in the human vary within broad limits, the behavioral manifestations of these phenomena are often very different from one culture to another. In primitive cultures, progression from childhood to adulthood is often a relatively simple affair. When a youth has developed sufficient physical maturity to work, engage in warfare, and participate in sexual activity, he or she does so. Since little formal training is required for these adult activities, adolescents can begin to participate when their bodies first become capable of adult performance. In primitive cultures, restrictions upon sexual activities, although variable, are sometimes less conspicuous than in more advanced societies.

Western culture has developed an elaborate set of restrictions upon the sexual and work activities of the adolescent. It would seem that this stems from a need to delay the establishment of a long-lasting sexual relationship until the adolescent has completed preparation for complex work activities. The culturally enforced delay between biological maturity and the assumption of complete adult privileges has led to a sequence of characteristic emotional and social adjustments in adolescents. Thus, adolescence in the United States involves a complicated set of compromises in which the teenager and society handle the problems presented by the lag between the ages of physical and social maturity. These adjustments are most marked in adolescents from the middle and upper social classes who frequently wait until the early or mid-twenties before marrying, but they are seen in abbreviated form in the lower social class in which marriage tends to occur earlier.

Growth from dependent child to independent adult involves spectacular changes: physical, cognitive, emotional, and social. The normal adolescent at times reacts to these changes with insecurity. Since decisions made in adolescence can have lifelong consequences, there is cultural pressure to delay decisions about marriage and final career choice until the early adult years. Obviously, the adolescent is in transition, and parents continue to have the responsibility of protecting the teenager against decisions that would have long-term detrimental effects. They need to allow sufficient experimentation for the adolescent to learn from mistakes as well as successes. A careful balance is needed in which parents encourage self-reliance but are available to place restrictions upon experimentation in directions that might be irreversibly damaging. Thus, most parents would strongly intervene to prevent heroin use but would allow the teenager to have great freedom in matters of dress.

PHYSICAL DEVELOPMENT

Teenagers are intensely preoccupied with their bodies; not only with its physical changes but also with the body's fresh insistence on sexual gratification. This need is different from those, such as hunger, thirst, and the need for sleep, with which most children have previously had to cope. The adolescent is confronted with fresh social restrictions that prevent the gratification of sexual needs. Since economic self-sufficency often comes five to eight years after the onset of puberty, the adolescent has a considerable period in which the satisfaction of sexual urges is a complicated process fraught with conflicting emotions.

In both sexes, there are major increases in height, weight, and muscle strength. By the eleventh year, the majority of girls have begun the period of rapid adolescent growth. At the end of the twelfth year, the average girl has attained over 95% of her adult height. Most boys experience the beginning of the adolescent growth spurt during the thirteenth year. The period of most rapid increase in height occurs in boys during the fourteenth year. By age 15, the average boy has reached about 95% of his mature height. Since girls begin their period of rapid growth approximately two years before boys, girls of eleven and twelve are often taller than many of their male counterparts. Sexual changes are prominent. Boys experience growth of axillary, facial, pubic and body hair, deepening of the voice and enlargement of the penis and testicles. By the end of the fourteenth year, the majority of males have developed ejaculatory capacity (Gesell, 1956). Girls have growth of axillary, pubic, and extremity hair, enlargement of the breasts and genitalia, and onset of menstrual cycles. The average age of menarche for American girls is just under 13 years with a range between $10\frac{1}{2}$ and 15 years. Strong concern is felt by those whose growth patterns fall outside the usual limits (see Chapter 10). Adolescents desperately desire to be like each other. To be too tall, too short, too fat, or too thin can be devastating. Delayed development of sexual function can also produce much anxiety. Adolescents test the new capacities of their bodies over and over again through athletics, work, and sexual experimentation.

COGNITIVE DEVELOPMENT

Cognitive potential reaches adult levels during adolescence. The extensive studies of Piaget indicate that the capacity for cognitive operations at the highest level of intellectual function is first developed at age 11-12 and reaches an equilibrium at 14-15 years. He characterized this as the stage of formal operations. The adolescent "comes to control not only hypothetico-deductive reasoning and experimental proof based on the variation of a single factor with the others held constant (all other things being equal) but also a number of operational schemata which he will use repeatedly in experimental and logicomathematical thinking" (Inhelder and Piaget, 1958). Males and females acquire this ability at approximately the same age.

Although the potential for cognitive performance at the highest level of intellectual achievement exists, it must be emphasized that most adolescents do not consistently function at adult levels. This discrepancy is due to the lack of knowledge and changing priorities of motivation. Most adolescents are still in the process of acquiring fundamental principles; they cannot consistently deal with complex areas of learning until basic material is mastered. The area of highest priority in a teenager's life changes many times. At one stage it may be sports; at another, a girl or boy friend; at another, making money to buy a car. It is rare that an adolescent consistently sees intellectual achievement as the area of first importance.

EMOTIONAL DEVELOPMENT

Sigmund Freud (1920) has emphasized the recurrence in adolescence of Oedipal feelings. The unconscious resurgence of sexual feelings toward the parent of the opposite sex is often handled by reaction formation; thus, the teenager may defend against an unacceptable feeling of sexual attraction by rebelliousness and overt rejection of one or both parents. Erikson (1950) sees the primary emotional task of this stage of life as the achievement of personal identity. The adolescent tries to obtain a clear sense of self and a sense of his place in the world. Without such concepts the adolescent's attempts to find a clear direction in life will fail.

Anna Freud has pointed out that the efforts of the adolescent to master the intensified instinctual impulses of this period frequently involve two attitudes: asceticism and intellectualization. Asceticism involves so complete a repudiation of instinctual impulses that not only is the instinctual wish itself avoided but this may be extended to include ordinary physical needs such as eating or sleeping. Unlike neurosis, asceticism does not allow for compromise formations. Often the adolescent changes at some point from asceticism to instinctual excess, suddenly embracing everything that was formerly prohibited. In contrast to ascetism, by using intellectualization the adolescent turns towards the disturbing impulse. It is not, however, dealt with directly but only in an abstract intellectual form. The highest ideals and most lofty aims may be expressed but this is without any effect on actual behavior. A beautiful concept of love may be verbalized but, at the same time, there may be considerable callousness in romantic relationships. Anna Freud summarizes the tasks and aims of ascetism and intellectualization as follows: "The task which ascetism sets itself is to keep the id within limits by simply imposing prohibitions; the aim of intellectualization is to link up instinctual processes closely with ideational content and so to render them accessible to consciousness and amenable to control" (Freud, 1966). In behavioral terms, the adolescent can be seen as moving by virtue of age and development into a new set of social contingencies. From the reinforcing stimuli of their environment, adolescents learn behavioral patterns that gradually become more and more like those of the adult.

One of the characteristic findings in the emotional life of this period is the high intensity and marked instability of affects. The frequent surges of strong emotion are at least in part responsible for the difficulties that occur in the following general areas: (1) low self esteem (2) tempestuous interpersonal relationships, and (3) lack of precision in reality perception. Adolescents attempt many unfamiliar emotional tasks, and their increasing independence is both welcome and frightening. Small reversals cause a severe decline in self-esteem. The plaintive comment "I can't do anything right" is frequently heard after minor blunders. Low self-esteem is related to the struggle for a firm concept of identity.

Their insecurity tends to create considerable strain in all interpersonal relationships. Adolescents are so concerned about their own needs that it becomes difficult to make the emotional commitments that are necessary for more mature relationships. Those in which each gives emotional support to the other usually occur when emotional reserves have developed from satisfaction of one's own needs. For the adolescent this is exactly the problem. Many adolescents in therapy frequently wonder why they quarrel so constantly with their romantic partner. They do not know how the quarrels begin but they are often over some "little thing." Close inspection reveals that each member of the adolescent pair feels a need for constant reassurance concerning their own desirability. With such a pattern, sooner or later one member of the pair may tire of the quarreling and break the relationship. While having the superficial appearance of mature emotional closeness, the relationships actually involve a process of strengthening the adolescent's sense of self. Thus, the adolescent boy may see himself as someone's boyfriend and then pattern his life in a way that is appropriate to that role.

The lack of precision in the perception of reality can be quite striking. In therapy, an adolescent who is in the midst of having one short-lived affair after another can understand intellectually that the previous ones were not realistic. This knowledge is invariably of little help with the current romance because it is perceived as being quite different from those preceding it. When emotional needs are high, quite serious threats of danger are ignored; the propensity of adolescents for accidents with motor vehicles is reflected in the higher insurance rates for this high-risk group. When danger (self-preservation) conflicts with emotional gratification, the teenager tends to respond to the emotional need. Thus, teenagers tend to take chances based on poor reality perception and exceed the limits of safety in many areas.

As the mentally healthy teenager progresses through adolescence, a gradual awareness of his or her capacities for meeting adult levels of performance develops. The self-esteem which had earlier been so easily shaken is now stronger, and a clearer sense of purpose usually becomes evident in late adolescence. The troublesome surges of emotion decrease. Emotional responses become more appropriate to the external stimuli. The extravagant orientation to the immediate present, and difficulty in postponing gratification, is replaced by long-term goals. In late adolescence there is an integration of emotional responses with intellectual awareness and increasing success in many areas. Independence becomes possible without the need for constant rebellion. The young adult has mastery in many areas, approval from others is still sought, although less often at the expense of severe compromises.

SEXUAL DEVELOPMENT

Much of the observational data on infrahuman primates point to behavioral differences between the sexes becoming manifest very early in life. Male primates exhibit erections from within a few weeks of birth and, particularly in the great apes, the interest and attention directed towards the genitals by the mother is greater for the male than for the female. Well before puberty, male rhesus monkeys indulge in more rough-and-tumble play than females, and mounting behavior by males both on other males and females occurs with increasing frequency as puberty approaches (Goy, 1968). Furthermore, the pattern of mounting gradually assumes a more adult form. Genital self-manipulation is observed but, unlike in the human, ejaculation does not seem to take place until the first or second intromission in the life of the male (Michael and Wilson, 1973). Thus, infantile and prepubertal sexuality is the rule in most infrahuman primate species, and we know that these cognitive rehearsals are necessary for normal sexual development. This is particularly true for the male and, if he is deprived of these opportunities experimentally, adult sexual performance is impaired (Harlow and Harlow, 1965). Naturalistic observations are, of course, quite difficult to make, but we have no evidence at this time of anything resembling the human latency period in other higher primates.

One matter which deserves some consideration here is the question of incest. There is no evidence for a strong barrier against incestuous behavior in lower mammals where in-line breeding is the easiest way to develop pure strains. However, some interesting findings have begun to emerge from long-term studies of known social groups of primates in which the kinships and genealogies of individuals are known. Incestuous matings (mother-son) in rhesus monkeys and Japanese macaques turn out to be quite rare (Missakian, 1973). Recent data from the wild chimpanzees studied by Jane Goodall and her colleagues indicate that sexual interactions between siblings are common before puberty but are actively avoided thereafter. Individual immature female chimpanzees maintained company with males that were either known or judged to be their brothers, and these females copulated with the male companions. However, after the females went through puberty and began to show sexual swellings, they no longer associated with males from their natal group nor permitted them to copulate. Instead, they associated with males from other groups. For the human, there is recent data from Israeli kibbutzim where small groups of unrelated boys and girls are reared together from infancy in quite intimate groups. They are permitted extensive interactions with each other in the interests, so we are told, of developing repression-free adult sexuality. Remarkably, the incidence of adult sexual relationships between group members is extremely low, and marriage is totally absent: in more than 2,700 second-generation kibbutz-dwellers' marriages, none involved those reared together (Shepher, 1971). This was despite the absence of incest taboo since group members were not consanguinous, and there would have been no objection to matrimony.

Since humans ordinarily seek conditions of privacy for sexual activity, the development of accurate normative data has presented problems. Masters and Johnson (1966) have made direct observations on human sexual behavior under laboratory conditions, and have produced valuable information concerning the physiology of human sexual responses. It is not possible to make such direct observations on large populations in their usual setting, and frequency data must be derived from questionnaires or interviews. The information that emerges is what people say they do, and this is not always representative of what actually happens. Even the most widely quoted studies have serious methological problems, and an excellent review of the difficulties has been given by Katchadourian and Lunde (1975). Kinsey et al. (1965) stated that they always assumed "that everyone has engaged in every type of activity"; the rationale behind this being that subjects were more likely to deny behavior in which they had engaged than to claim behavior in which they had not engaged. Although this assumption may be true, the contention can be made that it might produce false positive responses. Studies such as that sponsored by the Playboy Foundation and carried out by "The Research Guild, Incorporated" (Hunt, 1974), as quoted by Katchadourian and Lunde, have problems related to the self-selection of volunteers; about 80% of those originally chosen in the sample refused to participate. The Sorensen (1973) survey which related specifically to adolescent sexuality also suffered from the fact that fewer than half of the original randomly selected group actually participated. One might anticipate that the respondents in such a study would tend to be more sexually permissive than those who refused to participate, and this would bias the results. Although data from these studies must be viewed with some reserve, areas of general agreement can be found and these help to identify trends in behavior. When trends seem congruent with the clinical impressions of professionals such as psychiatrists, psychologists, ministers, and social workers, greater confidence in the validity of the data seems warranted. For example, one can say with some certainty that there has recently been a trend toward increased sexual permissiveness among adolescents.

There is a consensus that masturbation constitutes a significant portion of adolescent sexual activity. Kinsey and his coworkers found it to be the source of the first ejaculation in about 66% of males and of the first orgasm in approximately 37% of females. Other sources of first ejaculation in boys were nocturnal emissions (22.2%), coitus (6.2%), homosexual contacts (2.9%), spontaneous ejaculation (1.5%), petting (0.4%), and animal coitus (0.2%). For girls in the Kinsey study, 13% reached first orgasm in premarital petting, and 30% in coitus after marriage. Only small percentages of females reached first orgasm in nocturnal dreams, premarital coitus, homosexual relations, animal contacts, or psychologic stimulation (Kinsey et al., 1965). Thus, for a majority of boys and a significant number of girls, masturbation provides an entry into sexual experience.

Most boys and girls begin extensive heterosexual social contacts in the early to mid-adolescent period. For most adolescents these social contacts lead to various forms of physical contact known as "petting." Petting consists of the same type of behavior that constitutes the sexual foreplay of normal adults. However, in petting the aim is not coitus but that of providing and obtaining extensive physical and emotional closeness. For the majority of adolescents, petting constitutes the first sociosexual activity. Kinsey has stated that petting "provides most females with their first real understanding of a heterosexual experience" (Kinsey et al., 1953). This is probably equally true for the male. Light petting, also referred to as "necking," begins with simple touching, usually involving the hands. Touching is typically followed by prolonged kissing. Touching and oral contact with the girl's breast by the male often constitute the next stage of petting. Heavy petting involves manipulation below the waist and mutual caressing of the genitalia. Oral stimulation of the genitalia may also be a part of heavy petting. Direct apposition of the sexual organs without penetration also may occur at times. Light petting is quite common in the adolescent age group, and heavy petting is somewhat less common. Sorensen's study (1973) produced the following data: 98% of all adolescent girls have felt the sex organs of boys; 32% of all adolescent boys have had their sex organs felt by girls; 56% of all adolescent boys have felt the sex organs of girls; 95% of all adolescent boys have felt girls' breasts. Many factors contribute to the large individual variations, one being social class. For example, Kinsey data indicate that petting forms a larger part of the sexual experience of higher socioeconomic group males than of lower socioeconomic group males, and it is not rare for petting to be carried to the point of orgasm for one or both partners. Petting to climax is a more significant portion of the total orgastic outlet of adolescent females than it is for adolescent males. However, even among girls it forms less than 20% of the total orgastic outlet, and petting to climax formed less than 5% of the sources of orgasm for the total U.S. adolescent male population.

Nocturnal emissions are a part of the normal sexual experience of males. Approximately 83% of males experience them some time in their life, but they never account for a large portion of the total number of orgasms experienced in the male population. Nocturnal sex dreams in females apparently occur somewhat less frequently, with an accumulative incidence curve indicating that 37% of females experience dreams leading to orgasm by the age of 45 (Kinsey et al., 1953). As for males, orgasm during nocturnal dreams is never more than a minor outlet for females.

Sexual contact with animals is extremely rare among adolescent males and females. Kinsey found that 17% of boys reared on farms had experienced orgasm as a result of animal contacts, but he pointed out that "there is no other type of sexual activity which accounts for a smaller portion of the total outlet of the total population."

Homosexual behavior occurs during adolescence, but for most adolescents, it is rather sporadic. It occurs most usually in early adolescence as an extension of exploratory sexual activities. Kinsey's data indicated that 37% of the total male population had at least some homosexual experience to the point of orgasm between adolescence and old age. However, only 4% of white males were exclusively homosexual throughout their lives. The incidences for females were even lower. "By age 40, 19% of the females in the total sample had had some physical contact with other females which was deliberately and consciously, at least on the part of one of the partners, intended to be sexual" (Kinsey et al., 1953). However, of females who had had sexual contacts with other females, only one-half to two-thirds reached orgasm during the contacts. In the total sample, only 4% of females had, by the age of 20 years, experienced orgasm in homosexual relations. Recently, the increased public visibility of homosexuals in the United States and elsewhere has been construed as an indication of increased incidence. However, this does not appear to be the case for the early and middle years of adolescence. Sorensen (1973) found that 9% of all adolescents reported having had one or more homosexual experiences. Only 2% of all adolescent boys and virtually no adolescent girls reported having one or more homosexual experience during the preceding month. The picture is somewhat less clear for adolescents of college age. The activity of gay liberation groups on American campuses may have attracted more attention than previously, but data are not yet clear as to whether or not there has been an increased incidence recently in this group.

Sexual intercourse with prostitutes appears to constitute a very minor portion of the total sexual experience of adolescents. In the Kinsey study between 3.5 and 4.0% of the total outlet of the total male population (single and married) occurred during relations with female prostitutes. Younger single males show a considerably lower percent of their total outlet (3.7% in late teens) than older single males (almost 10% by age 30).

Heterosexual intercourse is the preferred sexual activity for the vast majority of both males and females during adolescence as well as during other times of life. When freely available, heterosexual intercourse constitutes by far the largest source of sexual outlet for both males and females. Teenagers are significantly discouraged from early marriage. This can be rationalized on a number of grounds. Marriage in early and middle adolescence occurs at a time of intense emotional flux; consequently, the choice of mate may be based on insecure grounds. An ideal partner at 15 may be totally unacceptable at 21. The high divorce rate among married teenagers (50% in the first five years) is therefore to be expected. Early marriage may tend to limit the opportunity for advanced training, and economic success is hampered. The lower socioeconomic group marries, in general, considerably earlier than the higher socioeconomic group. Strong arguments have been raised against premarital coitus. Among these is the risk of producing offspring without a family unit, and the spread of venereal disease that accompanies promiscuity. Modern technology has mitigated these hazards somewhat.

Premarital sexual intercourse appears now to be increasing among teenagers. Sorensen (1973) found that 21% of the adolescents in his group were serial monogamists. This was defined as a "relationship between two people of uncertain duration and to which partners generally intend to be true. Either partner, however, may depart when he or she desires, often to participate in another such relationship." On the other hand, 15% of American adolescents were sexual adventurers who sought many different sexual mates with little interest in any continuous or monogamous relationship. It seems likely that along with the increase in premarital sexual activity there has been an increase in extramarital sexual activity among married teenagers. At this point it does not appear that any significant portion of teenagers are rejecting the concept of marriage since 85% of all boys and 92% of all girls indicated that they agreed with the statement "Someday I will probably want to get married and have children."

There are two rather distinct issues here which cannot be considered in detail. One concerns the role of consummatory sexual experiences in reinforcing the pair-bond, and hence contributing to its stability and maintenance. The other concerns the stability of the nuclear family which is necessitated, even in sophisticated Western societies, by the prolonged period of infantile dependence in the human. It is well-known that copulation in our species is concerned as much with providing gratification as with producing offspring: in this, we do not differ substantially from other mammalian forms. This does not eliminate romantic love, but it does provide an infrastructure on which it can be based. The stability of the nuclear family, however, is clearly threatened to some extent since the divorce rate in the United States has increased from a rate of 2.2 per 1,000 population in 1960 to a rate of 5.0 per 1,000 population in 1976 (Kreps, 1977).

SOCIAL DEVELOPMENT

In a series of carefully developed steps, the American teenager is given increasing social responsibilities. Ceremonies such as Confirmation and Bar Mitzvah signal adult privileges and responsibilities in the area of religion. In cases of divorce, the adolescent can, at around the age of 14 (which varies with the State), decide whether to reside with the mother or the father. Scholastic achievement is recognized by yearly progression in grade level and by special exercises at times of graduation. Options increase dramatically at about the age of 16: a driver's license can be obtained; attendance at school is no longer required by law; parental permission is no longer necessary for employment. Establishment of close heterosexual relationships progresses from the stage of group parties to one-to-one pairings. Couples pass from regular dating to a variety of closer stages of bonding, ranging from "going steady" (restricting dating activities to each other) to being "pinned" (the pre-engagement ritual in which a girl wears a boy's fraternity pin), and to formal engagement and marriage. At about the age of 17 the adolescent can legally choose to live independently from the parents. At 18, full adult privileges and responsibilities are accorded. The teenager of 18 can legally consume alcoholic beverages, attend "X-rated" (sexually explicit) movies, serve in the armed forces, enter contracts including marriage, and vote in political elections.

During the past 50 years a variety of cultural changes in our society have had an impact upon adolescent behavior. The overall result has been an increase in permissiveness in child rearing. Adolescents have become more aware of their rights, protests against injustices are more frequent, and remedies have been obtained; for example, recent U.S. court decisions have given adolescents the right of due process in many court proceedings. Adolescents in several states can now obtain contraceptive information and abortions without parental consent. The development of the counterculture movement and the establishment of hippie communes in different parts of the country have offered dissatisfied teenagers places to go. Previously, teenagers who ran away from home had to maintain themselves or return home for help. Counterculture groups offer alternatives where teenagers can find food, shelter, and help.

The women's rights movement appears to have its effect. Increasing numbers of adolescent girls do not see themselves as moving only toward the role of housewife and mother. Many feel a need to have a career in addition to marriage, and plan their lives with this in mind.

Antiwar protests attracted many adolescent, and many continue to feel a cynicism and disenchantment towards government. There has been a definite increase in the use of drugs by adolescents. This phenomenon appears to be related in part to increased acceptance of drug usage by many cultures during the past 18 years. Alcohol, marihuana, and tobacco are the substances most widely abused. Because of the emotional quality of the phase of adolescent development, teenagers are more easily swayed by social changes than adults.

The prolonged period of economic dependence during adolescence creates severe emotional strains. While some of their peers achieve independence rapidly, others must still answer to parents who pay their expenses. Chafing under these restrictions, so much anger may be generated that a promising career is abandoned in disgust. Parental flexibility can do much to ease these strains, and increasing participation in their own economic support through part-time jobs enhances the adolescents' self-esteem and often smoothes the negotiating process between adolescent and parent. The extent to which the phenomena of adolescence are genetically or environmentally determined cannot be resolved at this time. However, as Scharfman has pointed out (1977), in some more primitive societies the period of adolescence can hardly be separately delineated; a finding which suggests that many of the emotional responses of adolescence seen in civilized societies may be culturally determined.

PSYCHOPATHOLOGY

Normal adolescent emotional turmoil is at times difficult to distinguish from true psychopathology since the behavior exhibited may be quite similar in both. In these circumstances a clear understanding of the underlying feelings is crucial. A teenage girl who runs away from home to escape an alcoholic father who threatens sexual abuse is in a different diagnostic category from a schizophrenic girl who runs away to escape her auditory hallucinations. Adolescents experiencing emotional conflict tend to respond to stress with action. The actions are a way of venting in some indirect manner the primary emotion that is felt. Such behavior is called "acting out." For example, a teenager in hospital might act out by destroying ward property as a way of expressing anger towards a nurse. Adults also act out, but it is a type of behavior that is most characteristic of adolescents. Because emotions are strongly felt and changes in mood are more rapid, impulse control presents special problems even for the healthy teenager.

The classification of psychiatric disorders has long been a troublesome issue. In their most characteristic form, most clinical entities are relatively easy to identify. In many patients, however, there may be partial presentation where some but not all of the clinical features of a particular syndrome are present. Furthermore, the clinical picture can change quite rapidly, and the same symptoms may occur in different clinical conditions. For example, immobility closely resembling the waxy flexibility seen in catatonic schizophrenia may occur in cases of phencyclidine (PCP) intoxication. Many psychiatrists dealing with adolescents have felt that the diagnostic nomenclature was not sufficiently detailed for describing adolescent problems. The Committee on Child Psychiatry of the Group for the Advancement of Psychiatry has published an extensive nomenclature which a allows detailed diagnosis of psychopathological conditions in children and adolescents (Committee on Child Psychiatry, Group for the Advancement of Psychiatry, 1966). However, this nomenclature has not been widely used. The American Psychiatric Association has recently published the third Edition of the "Diagnostic and Statistical Manual of Mental Disorders" which includes a more complete description of disorders of adolescence than was present in earlier versions of the manual. The present discussion of psychopathological conditions in adolescence will use terms which have sufficiently wide acceptance in the mental health field to allow easy understanding even with the changes in terminology which are occurring. The following discussion is not an exhaustive description of all the emotional issues of adolescence. The aim is to include enough information about the major psychiatric syndromes to allow accurate clinical diagnosis and adequate treatment. For more detailed descriptions see: Shaw and Lucas, 1970; Freedman et al., 1975; Kolb, 1977; Nicholi, 1978.

Situational Disturbances

Situational disturbances in adolescents have frequently been referred to as adjustment reactions. These basically relate to problems arising from difficulties with the external environment. Frequently, these reactions are related to emotional issues of normal adolescence. The situation and the reaction to it may merely be an intensification of the same responses seen in milder form in normal teenagers. For example, a 16-year-old girl whose parents are rigid and restrictive may become quite angry about their refusal to allow her to date. She may then act out her anger toward them by poor school performance. If the problem falls into the category of a situational disturbance, some increased flexibility on the part of the parents will resolve the conflict, and the girl's scholastic performance will improve. Thus, by definition, the patient with a situational disturbance has not internalized his or her unconscious conflicts into deeper neurotic or psychotic patterns and when the environmental stress is corrected, there is a prompt return to normal emotional function.

In situational disturbances, the reactions of the adolescent are easily understandable. There is nothing bizarre about what the adolescent feels. They may represent an overreaction, but they are not qualitatively different from what most normal adolescents would feel in a similar situation. Delusions and hallucinations are absent, and in most cases the prognostic outlook is excellent.

Antisocial Behavior

The term "antisocial behavior" often carries the connotation of illegality, but as used here it refers in a broader sense not only to unlawful activities but also to other self-detrimental behavior in which the adolescent refuses to follow the acceptable norms of society. For example, running away from home for older adolescent may not be illegal, but in many cases it is symptomatic of psychopathology.

Delinquency

Although at times delinquent behavior occurs as a part of a major sensory distortion, as when a teenager obeys an auditory hallucination telling him or her to set fire to a building, most cases of delinquent behavior by teenagers do not occur as a part of a psychosis. At times, activities such as mugging or stealing are clearly goal-directed, much like similar behavior in many adult criminals. However, in many cases of vandalism the behavior is more difficult to explain. Vandalism involving property belonging to persons with whom the adolescent is not acquainted frequently falls into this category. The causes for delinquent behavior vary from case to case. In some instances it is a response to a need for peer acceptance when the peer group is also engaged in such behavior. At other times the delinquent behavior amounts to an expression of anger directed at parents or other authority figures. This is the situation in some cases of shoplifting and vandalism. Johnson and Szurek (1962) have pointed out that some teenagers appear to have parents who show "super ego lacunae"; the parents unconsciously give tacit encouragement to antisocial behavior from which they derive vicarious pleasure.

The most serious forms of delinquent behavior are aggravated assault and homicide. The number of homicides by teenagers appears to be increasing in the United States. Those who commit murder are typically easily angered, explosive individuals who show little or no remorse for what they have done. Their restraint mechanisms are weak and laws are held in low regard. King (1975) has pointed out that these youths often have marked difficulty with mastering reading, social symbols, and with comprehension in general. They are limited in their capacity to cope in society, and many show characteristics of the antisocial personality seen in certain adult criminals. Rehabilitation is difficult but, when successful, usually requires long-term residence in a carefully controlled environment.

Runaway Reactions

When faced with psychological stresses, adolescents may make an effort to escape from their difficulties by running away. At times, this represents a plea for help. At other times, it is an expression of hostility directed towards those with whom the teenager is in conflict. The runaway reaction of adolescents is not new, but the situation has changed greatly since World War II. Formerly, teenagers knew that they must, through their own efforts, provide themselves with the necessities of life. If they failed in this, they were forced to return home. In every large city there are now significant collections of counterculture groups, frequently heavily drug-oriented, who help runaways by providing food and shelter. Some of these groups teach runaways effective methods for utilizing the welfare systems. Life in such groups is often seductively satisfying to teenagers who feel alienated, alone, and unable to succeed in normal society. Female runaways are frequently enticed into prostitution. Other groups offer help to runaway adolescents within a religious or pseudo-religious context. Runaways in general have a less favorable self-image than nonrunaways (Wolk and Brandon, 1977). Although some runaways "find themselves" and eventually develop successfully, for many of them the action of running away represents a significant distortion of reality and they fail to come to grips both with the dangers and the long-range consequences of their actions.

School Refusal

School refusal refers to persistent resistance to attending school, and there are five main reasons for this. (1) Family background of limited educational achievement: many teenagers from lower socioeconomic groups are reared by parents who did not themselves finish high school. Often these parents very much want their children to complete their education. However, many children in such families love their parents and identify strongly with them, they do not want to compete, and thus are satisfied with the family lifestyle and are not motivated to continue school. (2) Difficulties in learning: children who are slow learners, frequently those with I.Q. scores in the range of 70-85, often find it very difficult to achieve academic success in high school. The sense of frustration and failure which they often experience may lead eventually to refusal to attend school. Shifting such students to vocational programs, if done early enough, may preserve their interest in continued training. (3) School refusal related to alienation: a number of teenagers identify with deviant groups who reject the values of conventional society. These students frequently develop apathy or antipathy toward school. Many of these students use drugs of abuse. Heavy use of drugs frequently adds to their dislike for school since intellectual performance may be impaired. (4) School refusal associated with acting out behavior: adolescents who are very angry with their parents may express this anger passively through poor school performance or truancy. (5) School phobia: this phenomenon is described under phobic neurosis. Although it occurs in adolescents, it is much more frequently encountered in prepubertal children.

Substance Abuse

Teenagers are especially vulnerable to loss of control when using abusable substances. One of the major psychological characteristics of this period of development is a limitation of the ability to foresee the consequences of actions. Although teenagers may be intellectually aware of the dangers of using drugs, they frequently behave as if they have an absolute conviction that such difficulties could never befall them personally. The increased use of drugs by teenagers in the United States since 1960 has been documented (Wald and Hutt, 1972; LeDain, 1972). Many social factors have contributed to this use. The Vietnamese War brought many American soliders in close contact with a culture where drugs were easily accessible (The Domestic Council Drug Abuse Task Force, 1975). The war itself was unpopular with many young people and use of drugs was one of many expressions of anger toward civil authority. Rock music has also been very influential with teenagers. Many of the lyrics of popular rock melodies openly support drug use. Some of the most famous rock artists have made it clear that they are themselves users. The enormous financial gains from rock concerts have led some communities to make little effort to enforce laws against drug abuse. Many teenagers experiment with drugs but never use them heavily and show little evidence of long-term psychological or physical damage. However, heavy users of barbiturates, amphetamines, LSD, or marihuana, find it increasingly difficult to control their intake, and they develop a lifestyle that revolves almost entirely around rock music and drug use. It is then only a short step to illegal activities as a means of obtaining money. The use of heroin is particularly dangerous since its extremely addictive properties make it unlikely that the teenager can experiment without progressing to a marked physical addiction. Among teenagers, alcohol, tobacco and marihuana are by far the most widely used substances of abuse. The easy accessibility of alcohol accounts in part for the increasing incidence of alcoholism among adolescents. Barnes (1977), in an extensive review of adolescent drinking behavior, offers the hypothesis that problem drinking is a manifestation of incomplete and inadequate socialization within the family. The pattern of alcohol use exhibited by the teenager tends, in general, to be like that of the parents.

Most teenagers are introduced to substance abuse by their friends. Although they may initially use drugs primarily to achieve peer acceptance, and for the pleasurable pharmacologic effects, those who develop severe problems frequently use drugs to relieve psychological distress. When the teenager feels sad, fearful, or angry, drugs give immediate relief, and this gradually results in a psychological state in which the teenager becomes unable to tolerate almost any stress-provoking situation. This can become quite incapacitating. Teenagers who find their job or school boring, exhausting, or otherwise distasteful are constantly tempted to withdraw from the situation and use drugs. There does not appear to be a specifically addictive personality type among teenagers, and even those who initially show no evidence of emotional difficulties may become involved in drug abuse. However, teenagers experiencing high levels of psychological stress are the ones that are more prone to develop serious addictions.

Neurosis

A number of neuroses occur in adolescents, and they are characterized primarily by conflict and anxiety. The anxiety arises largely from the unconscious conflicts. The symptoms themselves are ways of providing at least partial relief from the underlying fear and anxiety. Patients with these conditions do not ordinarily have delusions, hallucinations, or illusions. Typically, the adolescent is torn by desires to behave and act simultaneously in opposite directions. Although symptoms may serve partially to relieve anxiety, it is at the cost of considerable capacity to make behavioral adaptations.

Depressive Neurosis

Depression is an extremely frequent reaction to internal emotional conflict among adolescents. The adolescent experiences some actual or perceived loss which results in decreased self-esteem. Usually, there is anger concerning the loss, but this cannot be adequately expressed. Rado (1956, 1962) and others have described this rage as being turned inward against the self. Subjectively, it is then experienced as depression. Signs of depression in adolescents include episodes of weeping, poor appetite, weight loss, difficulty sleeping, and amenorrhea. Sexual interest may also decline but this is not as marked as in adults. Sadness, irritability, and feelings of hopelessness are common, as is a general withdrawal of interest from all matters outside the self. Suicide ranks fourth as a cause of death in the 15-19 year age group (Gallagher, 1966). The rate of suicide among adolescents has shown a greater rise than that for any other age group (Toolan, 1975). The labile moods and impulsiveness of adolescents make it imperative to take adequate precautions against suicide. The following factors tend to increase the risk of suicide in teenagers: (1) a change in circumstances that results in loneliness, especially if there is a concomitant decrease in contact with friends and relatives (2) a family history of suicide especially in a close relative (3) stated intention to commit suicide (4) careful plans concerning the methods to be used (5) easy access to instruments of suicide such as guns or sedatives, and (6) a disorganized home situation, frequently where one parent is absent (Toolan, 1975).

A history of previous suicidal attempts must be evaluated in individual cases, and many teenagers who commit suicide have previously made an unsuccessful attempt. However, some teenagers clearly make suicidal gestures with no real intention of harming themselves. The method for an attempt gives some information concerning its seriousness. Those involving bodily mutilation are almost always serious while those that do not, such as taking "sleeping" tablets may or may not actually be intended to cause death. The family should be advised to remove dangerous drugs from the medicine cabinet and all weapons from the house. Although those sufficiently determined to kill themselves will do so, the danger is lessened if extensive plans are need to acomplish it.

Anxiety Neurosis

As the name suggests, the patient's primary complaints are pervaded by the constant presence of anxiety. These teenagers are described by lay persons as very "nervous" which means that the patient is a constant worrier who shows many nonverbal manifestations of anxiety such as a tremulous voice, shakiness, twisting of the hands, involuntary movements and motor restlessness such as tapping the fingers and so forth. Some of these teenagers are burdened by guilt and unconsciously fear punishment and retribution. These conflicts often concern masturbation and the control of sexual impulses.

A particularly frequent form of anxiety attack in adolescents is the hyperventilation episode. This does not itself indicate the presence of an anxiety neurosis, but is not uncommonly associated with it. During a hyperventilation episode, the patient begins feeling anxious and breathes with increasing rapidity. Eventually the increased respiratory rate results in an alkalosis by lowering the carbon dioxide content of the blood. The patient then experiences a number of physical symptoms which can be quite frightening. These include numbness and tingling of the hands, feet, and lips, shortness of breath, carpopedal spasm, and dizziness. These symptoms may cause the patient to fear that death is imminent. Although the hyperventilation episode looks quite serious, it is in fact self-limiting because syncope will occur, whereupon the respiratory rate slows and the patient soon recovers. Therapy, whether by reassurance, sedation, or breathing into a paper bag, is directed at slowing the respiratory rate.

Hysterical Neurosis

Although not seen with great frequency in teenagers, hysterical neuroses, particularly the conversion reaction and the dissociative reaction, can be quite significant. As with adults, conversion reactions characteristically involve loss of nervous system function expressed as loss of voluntary activity or sensory perception. In psychodynamic terms an unconscious conflict is converted into a failure of function which symbolizes the original conflict and alleviates some of the anxiety resulting from it. For example, a teenage boy who became very angry with his father had the repressed, unconscious desire to kick him. The repression of these feelings was expressed by an hysterical paralysis of his legs. It is typical for the disability to symbolize the repressed desire, in this case: to kick. The anxiety associated with an actual confrontation with father is completely relieved, and it is this that accounts for "la belle indifference," a lack of concern about the paralysis, that is so characteristic of the condition. In cases of conversion hysteria there are often considerable secondary gains from the symptoms (family concern, etc.), and this may result in the symptoms persisting long after the original conflict has been resolved. Conversion reactions are much less frequent than depression among teenagers. Bernstein (1969) described three cases of psychogenic seizures in adolescent girls who were reported to be struggling with sexual pressures. He pointed out that such cases still occur despite reports that the current incidence of conversion reactions is lower than was the case fifty years ago.

Dissociative episodes also occur in adolescents but are not frequent. Like adults, the teenager solves a severe intrapsychic conflict by dissociating from one aspect of the problem. If the teenager is facing a situation of intense unconscious conflict, amnesia or a fugue state may develop, thus avoiding, at least temporarily, the reality of some aspect of the situation. Cases of multiple personality represent a type of dissociative reaction in which the patient changes from one personality to another depending upon the circumstances being faced at a particular time. For example, a teenage girl who feels tremendous guilt related to sexual activity might shift into a licentious, guilt-free personality that allows her to act out her desires. Multiple personality syndromes have received much publicity in recent years through widely read books such as *Sybil* by Flora Rheta Schreiber (1973) and *The Three Faces of Eve* by Hervey Cleckley and Crobett Thigpen (1974).

Phobias

Phobias in teenagers represent the use of the mental mechanism of displacement to avoid an anxiety-provoking situation. When a phobia develops, the teenager typically displaces the anxiety from the actual situation that causes fear to another situation which is continguous in time or circumstance. Avoidance of the second situation must also provide escape from the primary anxiety-provoking situation. A typical example is that of a 15-year-old high-school student who became phobic about attending school. She resisted all efforts to force her to attend. Psychological investigation indicated that her phobia actually represented a displacement of her actual fears which were of being separated from her mother. Shortly before the onset of the school phobia, her mother had developed a severe depression. The girl was very frightened about possible dangers to her mother. She worried that her mother might commit suicide and repressed these fears, but relieved her anxiety by displacing them to the situation of attending school. By avoiding school she was able to stay at home and be reassured that nothing serious would befall her mother.

Adolescent school phobia constitutes a serious situation that warrants intensive effort. Some teenagers with this syndrome require inpatient care. In a follow-up study on 100 adolescents treated for school phobia in a psychiatric inpatient service (Berg et al., 1976), it was found that three years after discharge about one-third of the group had persistent symptoms of severe emotional disturbance and social impairment. Another third had improved appreciably and had primarily neurotic symptoms rather than social impairment, and the remaining third were markedly improved or completely well. A wide variety of other phobias such as fear of crowds, fear of heights, and fear of various animals involve similar use of the mental mechanism of displacement.

Psychosis

Psychoses are characterized by a loss of reality contact often manifested by delusions, hallucinations, and illusions. In addition, there is some disorganization of the personality. This disorganization usually is sensed by the patient as a loss of control of mental functions. The patient is plagued by intrusive thoughts which he cannot control. There are difficulties in maintaining any concentrated mental effort, confused thought processes, looseness of associations, and often much anxiety and panic. The usual difficulties which teenagers have in judging the consequences of their actions are compounded during a psychosis by the major sensory distortions. The two psychotic conditions which will be discussed in this chapter are schizophrenia and manic-depressive psychosis.

Schizophrenia

With the exception of organic psychoses such as those due to drug abuse, schizophrenia is the most frequently seen psychotic condition in teenagers. In keeping with the limitations of schizophrenic patients in tolerating stress, it is understandable that the tensions of adolescence might precipitate a breakdown. Schizophrenic adolescents typically feel isolated and different. Their thought processes are not logical, associations are loose, there is flatness of affect, and marked ambivalence. The tendency to autistic withdrawal and a disorganized state of mind make interpersonal relationships difficult or impossible. They often perceive, quite accurately, that they are disliked by other teenagers. Many struggle almost constantly with feelings of depression, anger, and inferiority. One reaction to such stress is an increasing withdrawal into fantasy and a loss of contact with the real world. As the disorganization increases, some schizophrenic adolescents begin hearing voices which are typically threatening or critical. Although less common than auditory hallucinations, visual hallucinations may occur but they lack the vivid distinctness of those occurring with the use of psychedelic drugs. Persecutory delusions are not infrequent and, during a period of decompensation, the schizophrenic adolescent shows essentially the same symptomatology as the acutely psychotic adult (Annesley, 1961).

It is important to remember that the lability and intensity of adolescent emotions may cause them to appear more ill than they really are. A teenager who is anxious or depressed may appear so disorganized that the diagnosis of a psychosis is considered. Extreme care must be taken to avoid mis-labeling patients in this age group. Toxic psychoses associated with use of LSD, phencyclidine, amphetamines, or other drugs of abuse may mimic schizophrenia, but a correct differential diagnosis can usually be made on the basis of history, physical examination, and laboratory findings. The potentially psychotic teenager is a sensitive, vulnerable person for whom major preventive efforts are needed, and psychotic decompensation can often be avoided if early signs are correctly interpreted.

In many cases, an impending psychotic break may be heralded by the intensification of abnormalities that have been present for many years. Such adolescents have often been rather isolated and feel disliked by peers. As children, they were prone to blame others for their difficulties. Signs of impending psychosis in the teenager include marked anxiety, restlessness, inordinate worrying, and motor restlessness indicative of great inner tension. Not infrequently they will have outbursts of uncontrollable rage and members of their families become fearful of them. Schoolwork deteriorates, and episodes of bizarre talk or strange behavior may occur. Not infrequently such youngsters threaten harm to themselves. Bender (1959) has described patterns of behavior in which teenagers first become known to the psychiatrist because of delinquent, antisocial or other disordered behavior that was basically part of the schizophrenic process. The clinician must be alert for signs of hallucinations even if the patient denies them. Episodes during which the youngster looks away with an attitude of listening and is not aware of what is being said may indicate their presence. Other evidence of impending psychosis is an increased looseness of association and thought processes that are not logical but which jump rapidly and tangentially from one area to another with little apparent connection between topics. Such patients may lose their train of thought entirely (thought block), look bemused, and wonder where they are.

Manic-Depressive Psychosis

Bipolar depressive illness (manic-depressive psychosis) is much less frequent in teenagers than schizophrenia. When it occurs in the bipolar form, the patient alternates between periods of excited and periods of depressed behavior. There is sometimes a family history of the condition. During the manic episodes, the adolescent appears energetic and elated, speaks rapidly, shows great pressure of thought and flights of ideas. The mood is unrealistically euphoric and judgment is severely impaired. He or she tends to make optimistic and grandiose plans that have no means of being fulfilled. During the depressed phases, the patient presents quite an opposite picture. The mood is depressed, energy is lacking, there may be motor retardation, and the outlook is gloomy with hopelessness, emptiness, and suicidal thought predominating. There may be well-organized somatic delusions and a conviction that some incurable disease has taken hold. Hallucinations and delusions may occur in both the manic and depressed phase. In some patients only the depressed phase or the manic phase may be seen. In contrast to the schizophrenic youngster, who has characteristically shown abnormalities throughout childhood, manic-depressive teenagers are often thought by their families to have been emotionally normal during childhood.

Anorexia Nervosa

This is an unusual and serious emotional disturbance which is typically seen in adolescent girls (Bruch, 1965). The major symptom is a very marked reduction in food intake amounting, in some cases, to a virtual refusal to eat. Typically, there is a conviction of being overweight so that the girl starts dieting and purging herself. Weight loss can be very rapid and, despite encouragement from parents to take more food, the weight loss may become life-threatening. Mortality rates as high as 15%-21% have been reported (Halmi, 1977). There is an associated endocrine disorder with an early and persistent cessation of menstruation, and there is a decline in sexual interest. It has recently been reported (Russell et al., 1975) that there is a distortion of the body image. Clinically, although patients become extremely emaciated they continue to show surprising amounts of activity and energy. Although 50% of the original body weight may have been lost, the patient still may be seen vigorously exercising herself. Hospitalization may be a matter of urgency, and re-feeding has to be instituted cautiously. Even in hospital on controlled diets, self-induced vomiting in secret may continue to restrict the intake of calories. When re-feeding has brought the body weight up to a safe level, psychotherapy for the patient and counseling for the family can commence. Although the physiological result is the same, the psychological and behavioral manifestations of anorexia nervosa are quite different from those seen in starvation by other causes and to the weight loss in certain endocrinopathies (Casper and Davis, 1977). A long-term follow-up study indicates that patients whose illness is associated with hysterical personality features tend to do best, those with primarily obsessive personality features do next best, and the patients with the worst prognosis are those with schizoid personalities.

Sexual Dysfunction

Complaints about the physical aspects of the sexual act (i.e., impotence, frigidity, premature ejaculation, and failure of ejaculation) are much less frequent in teenagers than in older adults. Teenage girls not infrequently experience difficulty achieving orgasm but almost never come for psychiatric treatment with this as a primary complaint. Orgastic difficulty, when present, is often found to be related to guilt feelings, anxiety about discovery, fears of pregnancy, or the inconvenience of the setting (e.g., the back seat of a car). Teenagers complain much more frequently of distress related to the interpersonal aspects of sex. Developmentally, teenagers feel great insecurity about their own attractiveness. This often leads to a need for frequent reassurance in sexual relationships. When sexual activity is a part of the relationship, one member of the pair, typically the female, often feels concern that the interest of the partner is in obtaining sexual satisfaction rather than in making a lasting emotional commitment. Many teenagers, despite increased sexual permissiveness, continue to feel guilty about their sexual activities. They often express the feeling that love is good but sex is bad. This leads to the idea that sex is permissible if it occurs as a natural part of feelings of love but that planning for its occurrence is somehow immoral. This may make it difficult to take precautions against pregnancy.

The small percentage of individuals who establish a homosexual orientation during adolescence often encounter severe opposition from their families. Although a family constellation for male homosexuality is one in which the mother is a dominating woman who dislikes males and discourages masculine behavior in her son (Biever et al., 1962), it is clear that many homosexuals do not come from such a background. Other factors such as overwhelming insecurity, rejection by girls, or identification with a homosexual role model during adolescence may be of significance in the development of this condition. Many of these teenagers come for treatment, and chances for success are enhanced if the teenager has a strong desire not to become a part of the homosexual culture. Unusual sexual behavior such as voyeurism, exhibitionism, fetishism, pedophilia, cross-dressing in secret, transvestism, and transsexualism often begins in adolescence. However, the number coming for treatment of these conditions is small.

The possibility of a physical etiology for sexual dysfunction in teenagers must not be overlooked. It is clear that organic conditions can either increase or decrease sexual performance. Goodman (1976) reported five cases of hypersexuality in girls with hyperadrenalism. Use of dopamine in adults with Parkinson's Disease has resulted in increased sexual activity (Bowers and Van Woert, 1972). Use of thioridazine (Mellaril) in male adolescents can cause inability to ejaculate. There is also evidence to suggest that prenatal influences may at times affect gender identity. A careful genetic and neuroendocrine study should be a part of the evaluation of adolescents showing marked deviation from the usual patterns of sexual behavior.

Brief Considerations for Treatment

Treatment for emotional difficulties in teenagers is rarely a unitary process. A combination of several methods of treatment is needed. Perhaps that most widely used is psychotherapy. Psychotherapy (or other forms of counseling) whether done on a one-to-one basis or as group therapy, is an effort to help the teenager understand the causes of his or her behavior and then, through interaction with the therapist, make changes that will improve adaptation in life. Many counseling techniques also rely upon modeling. This unconscious process, whereby the teenager changes his or her behavior in the direction of that which is perceived as being shown by the therapist, is usually based on identification or role modeling. Properly utilized, this can be a powerful force for changing attitudes and behavior. Although many practitioners do not regard behavior modification as a counseling procedure, it does bear some similarity in that the therapist and client work together in an interpersonal endeavor directed at resolving the problem. Behavior modification does not usually attempt to find reasons for behavior, but rather focuses on methods for changing problem behavior, using techniques derived from operant conditioning.

Pharmacological means for modifying behavior are widely employed. Specific therapy in the form of lithium carbonate has resulted in dramatic improvement in many manic-depressive patients, but blood lithium levels must be carefully monitored. The use of lithium has allowed many patients to lead normal lives.

Major tranquilizers such as chlorpromazine, haloperidol, and thioridazine have an antipsychotic action which effectively decreases many of the symptoms of schizophrenia. Delusions and hallucinations can be stopped by the use of these medications. Patients become less disorganized and function at school or work again becomes possible. Antidepressant medications such as imipramine and amitriptyline are effective in many cases. Minor tranquilizers such as diazepam and chlordiazepoxide can be of use in reducing anxiety, particularly in nonpsychotic individuals. All pharmacological agents have side effects which can be dangerous, and their use is restricted to experienced, properly trained personnel. The criticism is sometimes made that medications do not solve problems and are at best palliative. This is not strictly true. In many instances, patients are so unable to function as a result of their emotional illness that they no longer have the capacity to make any adaptive maneuvers in their lives. Frequently, medication raises the level of adaptive capacity to the point at which patients can take constructive steps to change the situations in which they find themselves.

A third major area of therapeutic intervention involves environmental manipulation. In cases of extreme illness this requires complete care in a hospital or residential treatment setting. In less severe cases, the environmental manipulation may involve boarding schools, foster homes, moving to a new area to provide a different peer group, and various forms of increased supervision carried out by parental or other authority figures. Behavior patterns in adolescence are, in general, less firmly fixed than those in adulthood. This makes for more rapid and more extensive changes than are possible in many adult patients. Resolution of difficulties in adolescence often serves a long-term preventive function by helping to avoid harmful decisions in such areas as marriage, education, and work. The cost may be considerable, but the impressive benefit for thousands of teenagers is worth the investment.

REFERENCES

Annesley, P.T. Psychiatric illness in adolescence: presentation and prognosis. *J. Mental Sci.* 107, 268-278 (1961).

Barnes, G.M. The development of adolescent drinking behavior: an evaluative review of the impact of the socialization process within the family. *Adolescence* 12, 571-591 (1977).

Berg, J., Butler, A., and Gabrielle, H. The outcome of adolescent school phobia. *Brit. J. Psy.* 128, 80-85 (1976).

Bender, L. The concept of pseudopsychopathic schizophrenia in adolescents. *Am. J. Orthopsy.* 29, 491-512 (1959).

Bernstein, N.R. Psychogenic seizures in adolescent girls. *Beh. Neuropsy.* 1, 31-34 (1969).

Bieber, I. *Homosexuality, A Psychoanalytic Study.* Basic Books, Inc., New York (1962).

Bowers, B. and Van Woert, M.H. Sexual behavior during L-dopa treatment of Parkinson's disease. *Medical Aspects of Human Sexuality* 6, 88-98 (1972).

Bruch, H. The psychiatric differential diagnosis of anorexia nervosa, in *Anorexia Nervosa, Symposium in Gottingen.* J.E. Meyer and H. Feldman, eds. Thieme Verlag, Stuttgart (1965), p. 70.

Casper, R.C. and Davis, J.M. On the course of anorexia nervosa. *Am. J. Psy.* 134, 974-978 (1977).

Cleckley, H. and Thigpen, C. *The Three Faces of Eve.* Popular Library (1974).

Committee on Child Psychiatry, Group for the Advancement of Psychiatry. *Psychopathological Disorders in Childhood: Theoretical Considerations and a Proposed Classification,* New York, 6, Report No. 62 (1966).

Domestic Council Drug Abuse Task Force. R.D. Parsons, Chairman. White Paper on Drug Abuse, U.S. Government Printing Office, Washington, D.C. (1975), pp. 1-34.

Erikson, E. *Childhood and Society.* W.W. Norton and Co., New York (1950).

Freedman, A.M., Kaplan, H.I., and Sadock, B.J. *Modern Synopsis of the Comprehensive Textbook of Psychiatry, II.* Williams and Wilkins, Baltimore (1977).

Freud, A. *The Ego and the Mechanisms of Defence.* International Universities Press, Inc., New York (1966).

Freud, S. Beyond the pleasure principle, group psychology, and other works, in *The Standard Edition of the Complete Psychological Works of Sigmund Freud.* J. Strachey, ed. (1920), p. 18.

Gallagher, J.R. *Medical Care of the Adolescent, Second Edition.* Appleton-Century-Crofts, New York (1966).

Gesell, A., Ilg, F., and Ames, L.B. *Youth: The Years From Ten to Sixteen.* Harper and Row, New York (1956).

Goodman, J.D. The behavior of hypersexual delinquent girls. *Am. J. Psy.* 133, 662-668 (1976).

Goy, R.W. Organizing effects of androgen on the behaviour of rhesus monkeys, in *Endocrinology and Human Behaviour.* R.P. Michael, ed. Oxford University Press, London (1968), pp. 12-31.

Halmi, K.A. Effectiveness of behavioral therapy in anorexia nervosa. *Psychiatry Digest* 38, 19-24 (1977).

Harlow, H.F. and Harlow, M.K. The affectional systems, in *Behavior of Nonhuman Primates.* A.M. Schrier, H.F. Harlow, and F. Stollnitz, eds. Academic Press, New York (1965), pp. 287-334.

Hunt, M. *Sexual Behavior in the 1970's.* Playboy Press, Chicago (1974).

Inhelder, B. and Piaget, J. *The Growth of Logical Thinking from Childhood to Adolescence.* Basic Books, Inc., New York (1958).

Johnson, A. and Szurek, S. The genesis of antisocial acting out in children and adults. *Psychoan. Quar.* 21, 323 (1952).

Katchadourian, H.A. and Lunde, D.T. *Fundamentals of Human Sexuality, Second Edition.* Holt, Rinehart and Winston, New York (1975).

King, C. The ego and the integration of violence in homicidal youth. *Am. J. Orthopsy.* 45, 134-135 (1975).

Kinsey, A.C., Pomeroy, W.B., and Martin, C.E. *Sexual Behavior in the Human Male.* W.B. Saunders Co., Philadelphia (1965).

Kinsey, A.C., Pomeroy, W.B., Martin, C.E., and Gebhard, P.H. *Sexual Behavior in the Human Female.* W.B. Saunders Co., Philadelphia (1953).

Kolb, L.C. *Modern Clinical Psychiatry.* W.B. Saunders Co., Philadelphia (1977).

Kreps, J.M. *Statistical Abstract of the United States, 1977.* U.S. Department of Commerce, 98th edition, U.S. Bureau of the Census, Washington, D.C. (1977).

LeDain, G., Chairman. *Cannabis, A Report of the Commission of Inquiry into the Non-Medical Use of Drugs.* Crown Copyrights, Ottawa, Canada (1972), pp. 184-204.

Masters, W. and Johnson, V. *Human Sexual Response.* Little, Brown and Co., Boston (1966).

Michael, R.P. and Wilson, M.I. Changes in the sexual behaviour of male rhesus monkeys (*M. mulatta*) at puberty. *Folia Primatologica* 19, 384-403 (1973).

Missakian, E.A. Genealogical mating activity in free-ranging groups of rhesus monkeys (*Macaca mulatta*) on Cayo Santiago. *Behaviour* 45, 225-241 (1973).

Nicholi, A.M., ed. *The Harvard Guide to Modern Psychiatry.* The Belnay Press of Harvard University Press, Cambridge, Massachusetts (1978).

Rado, S. *Psychoanalysis of Behavior,* Vol. 1. Grune and Stratton, New York (1956).

Rado, S. *Psychoanalysis of Behavior,* Vol. 2. Grune and Stratton, New York (1962).

Russell, G.F.M., Campbell, P.G., and Slade, P.D. Experimental studies on the nature of the psychological disorder in anorexia nervosa. *Psychoneuroendocrinology* 1, 45-56 (1975).

Scharfman, M.A. Preadolescence, puberty, and early adolescence, in *Understanding Human Behavior in Health and Illness.* R.C. Simons and H. Pardes, eds. Williams and Wilkins, Baltimore (1977).

Schreiber, F.R. *Sybil.* Henry Regnery Co., Chicago (1973).

Shaw, C.R. and Lucas, A.R. *The Psychiatric Disorders of Childhood, Second Edition.* Appleton-Century-Crofts, New York (1970).

Shepher, J. Mate selection among second generation kibbutz adolescents and adults: incest avoidance and negative imprinting. *Arch. Sex Behavior* 1, 293-307 (1971).

Sorensen, R.C. *Adolescent Sexuality in Contemporary America.* World Publishing Co., New York (1973).

Task Force on Nomenclature and Statistics of the American Psychiatric Association. *Diagnostic and Statistical Manual of Mental Disorders, Third Edition,* American Psychiatric Association, Washington, D.C. (1980).

Toolan, J.M. Suicide in children and adolescents. *Am. J. Psychother.* 29, 339-344 (1975).

Wald, P. and Hutt, P.B. The drug abuse survey project: summary of findings, conclusions and recommendations, in *Dealing with Drug Abuse: A Report to the Ford Foundation.* Praeger Publishers, New York (1972), pp. 3-61.

Wolk, S. and Brandon, J. Runaway adolescents' perceptions of parents and self. *Adolescence* 12, No. 46 (1977).

Chapter 12
Neuroendocrine Control Systems in the Adult
Antonia Vernadakis
Bruce Culver

INTRODUCTION

The influence of hormones in regulating whole body growth, sex differentiation, development of reproductive function, and maturation of the nervous system has been considered at several levels of organization, from the molecular to the organismic and psychologic in previous chapters in this book. The role of hormonal and neural interactions in some aspects of aging and in the etiopathology of somatic disease and affective disorders often associated with development and aging has been discussed in subsequent chapters. In this chapter, the ubiquitous effects of hormones in controlling normal physiologic competence in the adult organism are emphasized.

It is known that during adulthood hormones either act directly on target organs and tissues (primary actions of hormones) or are necessary for the optimal expression of certain functions (permissive actions of hormones) or condition the target tissue to actions of hormones (conditioning actions of hormones). All these endocrine actions are the subject of a great number of textbooks to which the reader is referred for an overview of endocrine function. Within the classical framework of the metabolic and regulatory actions of hormones increasing evidence also shows that hormones in their relationship to the nervous system may be responsible for the timetable of some of the periods of the lifespan. Indeed it has been suggested the some "neuroendocrine pacemakers" possibly situated in the brain may signal the progression from one critical period of the lifespan to another. If that were the case, of particular importance for our understanding of the role of hormones of specific ages would be knowledge of neuroendocrine interrelations in the adult, that is, the age of optimal health. These interrelations, therefore represent the topic of this chapter. Our discussion may clarify mechanisms by which the two control systems, brain and endocrine, interact, and will serve as a background to better understand the changes which occur at old age. This chapter will, therefore, review some aspects of neurosecretory regulation, hormone uptake, and distribution and biochemical substrates of hormonal actions on normal CNS activity.

HORMONAL NEUROSECRETION MECHANISMS

Hypothalamic-Pituitary Stimulating and Inhibiting Hormones

As a result of the work of many investigators during the last three decades, it is now clear that the final common pathway between hypothalamus and anterior pituitary gland is bridged by a family of neurohormones. These substances are released into the hypophyseal portal vessels and act directly and specifically on the adenohypophysis to increase or decrease the release of each pituitary tropic hormone.

The origin within the central nervous system of a neural signal that could influence the function of the anterior pituitary was noted by Hinsey and Markee in 1933. In 1937, Hinsey proposed that hypothalamic elements release a neurohumoral substance(s) that on entering the blood in the primary capillary plexus of the portal vessels would be carried to the sinusoids of the anterior pituitary. In 1952, Harris and Jacobsohn reported the results of an experiment that gave much support to the notion of neurohumoral control of the adenohypophysis. Since 1952, our understanding of the hypothalamo-hypophyseal relationship has been greatly expanded. In this chapter, only a brief account will be presented of the hypothalamo-hypophyseal relationships.

Endocrine neurons are effector cells whose main secretory product is a hormone which is released at the terminal membrane in response to the depolarization caused by action potentials arriving from the perikaryon. In mammalian brains endocrine neurons are largely restricted to the hypothalamus, where they occur as two distinct populations. One population includes the large neurons of the supraoptic and paraventricular neclei which release the hormones oxytocin or vasopressin. The other population comprises neurons in the arcuate nucleus and in the paraventricular zone that synthesize and secrete the releasing hormones responsible for the regulation of the adenohypophysis.

Verney in 1947 summarized the evidence for the existence of hormone-secreting cells within the CNS of mammals and showed that one such hormone, the antidiuretic hormone (ADH) could be released both by emotional stress and increasing osmotic pressure of carotid blood. The neurons in the supraoptic nucleus of the hypothalamus were subsequently localized and identified. A good review has been presented by Cross et al. (1975).

Besides the hormones, oxytocin, and vasopressin, extracts of the neurohypophysis contain cysteine-rich proteins - the neurophysins. Each vertebrate species elaborates a family of neurophysins (see Pickering et al., 1974). The neurophysins are found within the same secretory granules as the hormones (Ginsburg and Ireland, 1963, 1966; Dean and Hope, 1966, 1967) and are thus packaged along with them in the perikarya. Because of the readiness with which hormone and protein associate, it has been suggested the neurophysin functions as an intracellular carrier molecule to keep the hormones within the granules as they pass along axons and are stored in the nerve terminals (see Ginsburg, 1968). Others (Pickering et al., 1971, 1974; Sachs et al., 1969, 1971) suggest that the same biosynthetic process is concerned in the formation of the neurophysins and their associated hormones. A review of the biosynthesis of the neurohypophyseal hormones is presented by Cross et al. (1975) in a hypothetical scheme shown in Fig. 1.

Many different types of tissues with a secretory capability, including neurons releasing chemotransmitters, endocrine and exocrine cells, have been shown to sequester their product within discrete intracellular organelles and to share a similar mechanisms for release of the stored product. The complex of events which results in a physiological stimulus causing the release of secretory product into the extracellular space has been termed stimulus-secretion coupling by Douglas and others (Douglas and Rubin, 1963) and calcium plays a central, obligatory, role in this process. However, for the endocrine neurons the storage and release of hormones is not well understood.

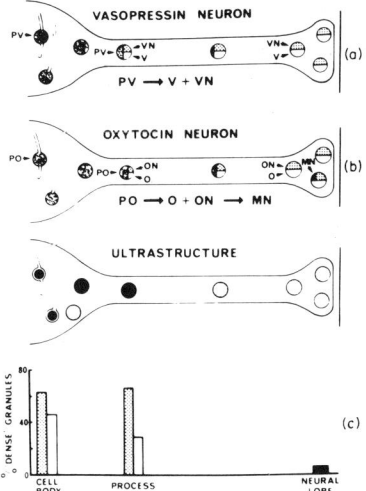

Fig. 1. (a) and (b) Hypothetical scheme for the maturation of neurosecretory granules in vasopressin and oxytocin neurons. PV = provasopressin; V = vasopressin; VN = vasopressin-neurophysin; PO = pro-oxytocin; O = oxytocin; ON = oxytocin-neurophysin; MN = minor neurophysin. (c) Proportion of granules that are resistant to alkaline fixatives (? immature) in the paraventricular nucleus (PVN), supraoptic nucleus (SON), and neural lobe of the rate. (From Cross et al., 1975).

Cross et al. (1975) have reviewed the similarities and differences between the release mechanisms of cholinergic, adrenergic and endocrine neurons. Table 1, summarizes the evidence (Cross et al., 1975). The cholinergic neuron does not have a reuptake process and recent evidence indicates that the endocrine neuron does not possess this process either (Edwards, 1971; Pliska et al., 1971). The peptidergic neuron synthesizes and releases in absolute terms much less than the cholinergic neuron and probably also the adrenergic neuron. In the neurotransmitter neuron synthesis of product is localized within the nerve terminals utilizing enzymes which have been formed in and transported from, the cell body. The endocrine neuron synthesizes the active product in the cell body using enzyme systems of this region.

Corticotropin-Releasing Factor (CRF). After it became apparent that variation in the concentration of adrenocortical steroids (Ingle and Higgins, 1938; Ingle and Kendall, 1937), and hence variation in the negative feedback of these steroids, was not adequate to explain many of the observed alterations in the rate of release of adrenocorticotropin hormone (ACTH) by the pituitary, investigators began to seek an alternate unifying explanation for the responsiveness of the anterior pituitary to a large array of stimuli. Thus a search was begun for the presumed hormonal substances that might stimulate the release of this pituitary polypeptide.

The results obtained by means of tissue culture provided more convincing data in support of the concept of hypothalamic control of the release of ACTH (Guillemin and Rosenberg, 1955). The release of ACTH by cultured pituitaries was potentiated significantly by the addition of median eminence tissue or posterior hypothalamus to the cultures. Addition of anterior hypothalamus, brain cortex, spleen, or liver to the anterior pituitary cultures was ineffective.

Table 1.

Quantitative Comparison of Endocrine with other Neurons[a]

Parameter	Endocrine Neuron (hypothalamoneuropypophysial peptidergic neuron)	Adrenergic Neuron (postganglionic fibers: cat skeletal muscle)	Cholinergic neuron (preganglionic fibers: cat superior cervical ganglion)
A. Total store[b]	2000 mg/neural lobe = mU/neural lobe	8000 ng/100 gm muscle = 80 ng/mg	255 ng/sup. cervical ganglion = 133 ng/mg
B. Portion releasable by nerve impulses	10%	25%	85%
C. Maximum rate of synthesis	250 mU/day = 0.5 µg/day = 0.35 ng/min	20 mg/min/100 gm muscle	28 ng/min
D. Time necessary for total replacement of store	5700 min	400 min/100 gm muscle	10 min
E. Basal release rates	50 mU/day = 0.1 µg/day = 1 x 10^4 granules/sec	3-13 ng/min	0.3 ng/min
F. Maximum release rates	a. Chronic stimulation 250-500 mU/day = 0.5-1 µg/day ∿ 1 x 10^5 granules/sec b. Acute stimulation 10 mU/min = 22 µg/min 4 x 10^6 granules/sec	283 ng/min	33 ng/min
G. Reuptake process	No	Yes	No

Table 1. (cont'd.)

Quantitative Comparison of Endocrine with other Neurons[a]

Parameter	Endocrine Neuron (hypothalamoneuropyhypophysial peptidergic neuron)	Adrenergic Neuron (postganglionic fibers: cat skeletal muscle)	Cholinergic neuron (preganglionic fibers: cat superior cervical ganglion)
H. Quantal content	?	Not yet estimated	5.0×10^4 molecules = 1.21×10^{-5} pg
I. Content of storage organelle	2 µg in 2×10^{10} granules = 1×10^{-4} pg/granule = 6×10^4 molecules/granule	1.6-5.0×10^{-6} pg/granule = 1-5.7×10^3 molecules/granule	0.06-1.21×10^{-5} pg/vesicle 2.3×10^3-5×10^4 molecules/vesicle
J. Amount released per impulse	1 mU = 540,000 NAPs = 0.005 pg/NAP = 3×10^6 molecules/NAP = 50 granules/NAP		0.25 pmoles/NAP = 36.5 pg/NAP = 1.5×10^{11} molecules/NAP 10^7-10^8 vesicles/NAP
K. Proportion of total store release per NAP	0.005 pg of 2 µg = 0.0000003%	1×10^{-4} of store = 0.01%	36.5 pg of 266 ng = 0.014%

[a] Calculations for transmitter neurons are based on values collected in Hubbard (1970).

[b] Stores have been calculated on a milligram basis assuming the neural lobe and cat superior cervical ganglion have the same weight (2 mg) and that neural tissue in the cat skeletal muscle is 0.1% of muscle weight.

Source: Cross et al. (1975)

If a hormonal substance has a significant role in the regulation of function of the anterior pituitary, it is evident that the substance must be present in the effluent blood of the gland. The first evidence indicating the pituitary stalk blood of dogs did in fact possess an activity that stimulated the release of ACTH was reported in 1955 (Porter et al., 1955). The activity was due to a substance acquired by the blood on its passage through the primary capillary bed of hypophyseal portal vessels, i.e., the pituitary stalk and median eminence of the hypothalamus.

The ability of crude hypothalamic extracts to stimulate the secretion of ACTH in vivo in rats with hypothalamic lesions or in rats in which the nonspecific response to stress had been inhibited by adrenocortical steroids, was documented in the late 1950s. In the case of hypothalamic extracts, in contrast to those from the posterior lobe, it was apparent that the activity could not be accounted for by vasopressin. Furthermore, cerebral cortical extracts possessed little or no activity, so that CRF appeared to be localized to the hypothalamus, in contrast to the findings with the early in vitro assay (Guillemin, et al., 1957; Saffran and Schally, 1955), that indicated an unibquitous distribution for CRF.

The first purification of CRF from hypothalamic tissue was reported by two groups of workers in 1959 and 1960 (Royce and Sayers, 1960; Rumsfeld and Porter, 1959). The fraction containing the ACTH-releasing activity (i.e., CRF) was prepared by ion-exchange chromatography. Subsequently other purifications have been reported (Dhariwal et al., 1965-1966; Dhariwal et al., 1966; Schally et al., 1962).

Thyrotropin-Releasing Hormone (TRH). Early evidence suggesting the importance of the hypothalamus in maintenance of pituitary-thyroid function was obtained by examining the effects of hypothalamic lesions on thyroid responses to propylthiouracil (Martin et al., 1970). These studies indicated that the region of the hypothalamus most important in TSH control extends from the paraventricular nucleus through the anterior hypothalamus to the anterior portion of the ventromedial (VMN) and arcuate (ARC) nuclei. Bilateral electrolytic lesions in this zone resulted in a striking reduction in resting pituitary-thyroid function (Martin et al., 1970).

The first unequivocal demonstration of the presence of a substance in hypothalamic tissue with thyrotropin-releasing-activity using a specific assay was reported in 1962 (Guillemin et al., 1962). TRH has been purified from hypothalamic tissue from a variety of sources including the pig (Schally, et al., 1966a; Schally et al., 1966c), sheep (guillemin et al., 1966), ox (Schally et al., 1966b) and man (Bowers et al., 1965). The structure of TRH was determined to be pyroglutamyl-histadyl-prolineamide (Bassiri and Utiger, 1972). Synthetic TRH has been produced and a highly sensitive and specific radioimmunoassay has been developed using the synthetic hormone (Bassiri and Utiger, 1972). In 1974, Oliver et al. (1974), Winokur and Utiger (1974), and Brownstein et al. (1974) reported the distribution and concentration of TRH in adult rat brain, including hypothalamus, thalamus several areas of forebrain, brainstem, posterior diencephalon, posterior cortex, cerebellum, and anterior and posterior pituitary.

Recently, Jackson and Reichlin (1977) examined the question whether extrahypothalamic brain TRH arises from the hypothalamus or is synthesized in situ. They made electrolytic lesions of the "thyrotropic area" of the rat hypothalamus, and examined the effect on TRH content in the hypothalamus itself, the rest of the brain and the anterior and posterior pituitary. They found that such lesions markedly reduce the hypothalamic content of TRH, deplete the pituitary gland of TRH, and result in severe chemical hypothyroidism, but do not affect the large amount of TRH from

the posterior pituitary in the lesioned animals supports the concept of a
third neurosecretory hypothalamo-hypophyseal system extending into the
neurohypophysis. This neurosecretory system has been suggested by
Hokfelt et al. (1975a-c) on the basis of immunohistochemical staining of
TRH and somatostatin positive fibers reaching into the posterior pituitary
from the median eminence. Whether TRH has a role in posterior pituitary
function is uncertain, but it may be relevant that TRH is found in high
concentrations in the pituitary complex of lower vertebrates (Jackson and
Reichlin, 1977), and there is evidence that in bony fish the neurohypophysis may be an homoloque of the median eminence in higher animals.

Gonadotropin-Releasing Hormones (or Factors). With the realization
that a CRF controled the release of ACTH, it was logical to search for
other such factors that might regulate the secretion of other hypophyseal
tropic hormones. The hypothalamus is now known to modulate the secretion of gonadotropins (Everett, 1964; McCann and Ramirez, 1964), and
it was reasonable to postulate the existence of gonadotropin releasing
factors to bridge the gap between hypothalamus and pituitary.

The introduction of a very sensitive and specific bioassay for luteanizing
hormone (LH), the ovarian ascorbic acid depletion test (Parlow, 1961), led
to the demonstration of an LRF in crude acid extracts of rat stalk-median
eminence tissue (McCann et al., 1960). These extracts caused a depletion
of ovarian ascorbic acid and a dose-response relationship was obtained in
immature gonadotropin-pretreated rats. The results could not be accounted for by contamination of the extracts with LH or by their content
of other physiologically active substances, such as oxytocin, vasopressin,
epinephrine, serotonin, histamine, or substance P. There was not any
activity obtained with cerebral cortical extracts, which indicated that the
effect was not simply a nonspecific response to the injection of tissue
extracts. As determined by both in vivo and in vitro assay, the LRF is
localized to the basal hypothalamus in a rather broad zone rostrocaudally
that extends from the suprachiasmatic region to include the median eminence and pituitary stalk (McCann, 1962). There is little or no LH-releasing activity in the posterior lobe of the pituitary itself. Further
evidence for this localization of LRF comes from elegant experiments in
which the pituitary was transplanted to the hypothalamus of hypophysectomized rats (Flament-Durand, 1965; Halasz et al., 1965).

The LH-releasing activity has been purified and separated from other
hypothalamic factors (Dhariwal et al., 1965) and recently a decapeptide
has been synthesized based on structural studies of the nature product
which has a potent effect in releasing LH (Burgus et al., 1971; Matsuo
et al., 1971).

A follicle stimulating hormone-releasing action of hypothalamic extracts
was demonstrated shortly after the discovery of LRF. These extracts were
found to elevate plasma FSH in spayed female rats in which the release of
FSH had been inhibited either by the injection of estrogen and progesterone
or by lesions in the median eminence, and a dose-response relationship
was obtained (Dhariwal et al., 1965; Igarashi and McCann, 1964; Kuroshima
et al., 1965). Cerebral cortical extracts and basopressin or oxytocin were
ineffective.

In addition to increasing FSH release in vivo, FRF increases FSH release into the medium of pituitaries incubated in vitro in either short-
term (Kuroshima et al., 1965; Mittler and Meites, 1966; Watanabe and
McCann, 1968) or long-term experiments (Mittler and Meites, 1964). Implantation of FSH into hypothalamus has been reported to lower both FRF
and pituitary FSH stores (Corbin and Story, 1967) which suggests that
FSH feeds back to alter its secretion via alterations in FRF release.

Prolactin-Inhibiting Factor (PIF). In sharp contrast to other pituitary hormones, the hypothalamus exercises an inhibitory influence on the secretion of prolactin in both the male and female as illustrated by the increased release of prolactin that occurs after hypothalamic lesions or from the grafted pituitary (Everett, 1974; McCann et al., 1968; Meites and Nicoll, 1966). When the gland is incubated in vitro, enhanced prolactin released also occurs in the face of curtailed secretion of other pituitary hormones (Gala and Reace, 1965; McCann et al., 1968; Meites and Nicoll, 1966). PIF has been purified and separated from most other releasing factors; however, it has proved difficult to separate from LRF (Dhariwal et al., 1968).

Growth Hormone-Releasing Factor (GRF). The hypothalamus exerts a stimulatory influence on the secretion of growth hormone (GH) by the adenohypophysis, since hypothalamic lesions led to stunting of growth in some mammals (Han et al., 1965; Han and Liu, 1966) that is accompanied by diminished stores of GH (Reichlin, 1961) and a blockade in the release of GH in response to stimuli that trigger release of the hormone in the normal animal, such as insulin-induced hypoglycemia (Abrams et al., 1966). Hypothalamic extracts have been found to possess a GH-releasing activity as determined by both in vivo and in vitro assays. However, GRF has not yet been isolated.

Growth Hormone-Inhibiting Factor (GIF, Somatostatin). In the course of routine screening of fractions from Sephadex columns for GRF by in vitro assay, certain fractions were found to diminish the output of bioassayable GH by about 50% (Krulich et al., 1968). This inhibitory effect was obtained in uniform position in four fractionations of ovine hypothalamic extract and was also recovered from a fractionation of rat hypothalami. The inhibitor, termed GIF, also prevented the increased GH release that followed addition of GRF to the incubated glands, but it failed to alter the release of other pituitary hormones such as FSH, LH or ACTH. It thus appeared to have a specific effect on somatotrophs. In vivo GIF has no detectable effect on pituitary GH stores when injected alone, but it abolishes the depletion of GH that follows the administration of GRF.

It is now recognized that somatostatin has many biological effects other than as an inhibitor of GH secretion (Brazeau et al., 1973). In addition to being found in the hypothalamus, somatostatin also occurs in neuronal elements and axonal fibers in multiple locations in the CNS, including the spinal cord (Hokfelt et al., 1975c), and in discrete secretory cells of classical epithelial appearance in all the parts of the stomach, gut, and pancreas (Dubois, 1975; Luft et al., 1974). An excellent review on the biological activity of somatostatin is by Guillemin (1978).

Melanocyte Stimulating Hormone-Releasing and -Inhibiting Factors (MSH,MRF,MIF). The hypothalamus inhibits the release of MSH in amphibians (Etkin, 1962; Ito, 1968; Kastin and Ross, 1965) but the nature of its control over secretion of this hormone in mammals is not clear (Bal and Smelik, 1967; Taleisnik et al., 2966). Although current evidence is quite conflicting, it appears likely that there is a dual control of MSH release in both amphibians and mammals that is mediated by an MRF and MIF.

MRF has been localized to the region of the paraventricular nuclei and to the median eminence with lesser amounts found in the posterior lobe (Taleisnik et al., 1966). The secretory neurons are thought to have their cells bodies in the vicinity of the paraventricular nuclei with axons that project to the median eminence, stalk, and neural lobe.

Pituitary Hormones

The hormones of the pituitary regulate many important processes in the body; not only are they the mediators of various disturbances in the endocrine system but they are themselves sensitive to the aberrations of systemic disease. Their secretion is profoundly influenced by many hormones of the peripheral endocrine glands as well as by stimulatory and inhibitory hormones of hypothalamic origin. Similarly striking effects are exerted by many drugs, including natural hormones, hormonal analogs, and inhibitors of hormone synthesis and action.

Among the vertebrates, ten adenohypophyseal hormones are recognized: growth hormone, prolactin, two gonadotropins (luteinizing hormone, follicle-stimulating hormone), thyrotropin, corticotropin, two melanocyte-stimulating hormones (α and β), and two lipoproteins (β and α). Of these, the first six are demonstrably important in humans. However, it would be premature to deny the possibility that there may be others.

Recent elucidation of the amino acid sequences of the recognized hormones of the adenohypophysis has shed light on the amazing diversity of the organ (see Li, 1972). An excellent review of the properties of the hormones of the adenohypophysis is presented by Gilman and Murad (1975).

The posterior lobe of the pituitary is mainly a storage organ for hormones produced in the hypothalamus and various somatic conditions cause their release. Neurohypophyseal extracts of vertebrates have yielded seven octapeptide hormones whose structures have been defined and whose biological activities have been extensively investigated. The neurohypophyseal hormones fall into two categories: vasopressor-antidiuretic and oxytocic-milk ejecting. In the first category there are the three dibasic peptides: arginine basopressin, lysine vasopressin, and arginine vasotocin; the monobasic oxytocin-like peptides include oxytocin, mesotocin, isotocin, and glumitocin. The pituitaries of mammals contain only oxytocin and either arginine or lysine vasopressin. For reviews in this field see Walter et al. (1967), Sawyer (1947), and Sachs (1970).

The most important physiological function of the vasopressin peptides stems from their ability to act as an antidiuretic agent. It has been amply demonstrated that the neurons of the supraoptic-hypophyseal tract are essential components of the interlocking systems controling the volume and composition of the body fluids. The role of oxytocin in parturition and milk ejection in the female is reasonably well-established. It is beyond the scope of this chapter to discuss the physiology of the mammalian neurohypophysis and/or those studies concerned with the mechanisms of action and metabolism of vasopressin and oxytocin; these subjects have been reviewed in a number of publication (for refs see Sachs, 1970).

Endorphins

The findings by Pert and Snyder (1973) of synaptosomal opiate receptors in the brains of mammals has led to the search of what have been termed the endogenous-ligands of these opiate receptors now called endorphins. It is not the scope of this chapter to review this new and rapidly expanding field. An excellent brief review of the subject is presented by Guillemin (1978).

Other Peptides: Peripheral division of the neuroendocrine system

The concept that the neuroendocrine system is diffused and is constituted by the cells of the central and peripheral divisions of the amine precursor uptake and decarboxylation (APUD) series has been advocated by Pearse and associates (review Pearse and Takor, 1979). The central division contains the neuroendocrine and endocrine cells of the hypothalamo-pituitary axis and pineal gland, while the peripheral division contains all the APUD cells outside these regions. The majority of the APUD cells are situated in the gastrointestinal tract and pancreas where they comprise the gastroenteropancreatic (GEP) endocrine cells. In the fetus these cells are found also in the respiratory and urogenital tracts, the thyroid and parathyroid glands, the thymus, the adrenal medulla and accessory chromaffin tissue, and in the sympathetic system itself.

The APUD cells produce more than 35 physiologically active peptides and a small number of equally active amines. Recently, 17 of these peptides have been identified jointly in endocrine cells and in neuronal cells bodies or processes and thus have been defined as "common peptides." Pearse and associates (1979) propose that the neuroendocrine system is to be regarded as a third division of the nervous system whose products suppress, amplify or modulate the activities of the other two divisions.

Hormonal Inputs and Feedback Sites

Adrenocorticotropic Hormone Secretion. Feedback in the adrenocortical system has been the subject of numerous research which has resulted in an almost unlimited possible interactions of the brain sites involved, i.e., pituitary, hypothalamus, limbic system. Some of the possible feedback relationships are depicted in Figures 2 and 3 (Kendall et al., 1975). The classical so-called long-loop corticoid feedback system includes steroid feedback sites on the anterior pituitary; hypothalamus, and other brain areas. Table 2 is taken from Stumpf and Sar (1973) and summarizes the effective feedback sites as defined by hormone implantation and autoradiographic data. More recently, short loops has been conceptualized which includes ACTH-anterior pituitary and ACTH-hypothalamic feedback sites. There is evidence for nearly all these feedback sites. However, the relative importance of the various possibilities and for that matter, the absolute physiological importance of feedback in this system remain unknown. Comprehensive reviews of the general topic of corticosteroid feedback have been provided in recent years and include arguments about pituitary and non-pituitary sites (Gann and Cryer, 1972; Jones et al., 1974; Kendall, 1971; Papaikonomou, 1974; Yates et al., 1971).

Several methods have been used to study feedback sites. Evidence favoring a pituitary feedback site has been provided from studies on (a) in vitro monolayer culture (Fleischer and Rawls, 1970; Tang and Spies, 1974); (b) dexamethasone suppression of adrenal corticosterone secretion in rats with pituitary islands (Matsuda et al., 1963); (c) suppression of ACTH secretion in the pituitary-transplanted rat (Kendall and Allen, 1968); (d) suppression of human ACTH-secreting tumors (Liddle et al., 1969; Strott et al., 1968). Evidence of an intrapituitary site of feedback has been reported by Kracier et al. (1973) and more recently by Kendall et al. (1975). It has found that various physiological manipulations of the adrenocortical system produced alterations in ACTH content of the anterior lobe of the rat pituitary but not of the intermediate or posterior lobes. However, a corticotropin-like intermediate lobe peptide (CLIP) indistinguishable from the 18-39 amino acid sequence of the ACTH molecule, has been reported to be present in the rat pars intermedia (Scott et al., 1973). From these studies the relationship of CLIP to the adrenocortical axis is not clear, since neither corticosteroid deficiency or excess alter the ACTH content of parts intermedia.

Exogenous ACTH has been reported to suppress endogenous ACTH secretion. Evidence for such direct feedback are: (a) exogenous ACTH

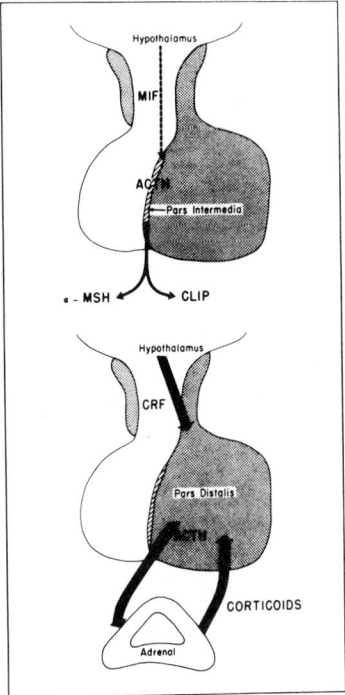

Fig. 2. Schematic representation of a hypothesis for control of anterior lobe ACTH as contrasted with intermediate lobe ACTH. (From Kendall et al., 1975).

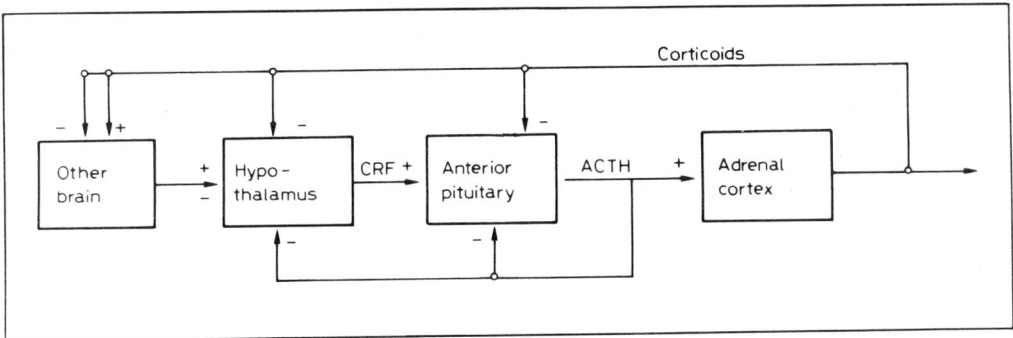

Fig. 3. Controls in the pituitary-adrenocortical system. Long-loop feedback by glucocorticoids is shown on top and short-loop and ultrashort-loop feedback at the level of the brain and pituitary are shown on the bottom. (From Kendall et al., 1975).

Table 2

EFFECTS OF ADRENOCORTICOID IMPLANTS IN THE BRAIN ON ADRENAL FUNCTION[a]

Species	Hormone	Site of Implants	Effect	Reference
Rat, cat	Cortisone	Basomedial hypothalamus	Inhibits adrenal secretion (at rest) and prevents the increase caused by operative stress	Endröczi et al., 1961
Rabbit	Corticosterone	Basal hypothalamus, anterior portion of median eminence	Inhibits stress response to rise in plasma corticoids	Smelik and Sawyer, 1962
		Anteromedial hypothalamus	Less inhibition	
		Post. hypothalamus, mesencephalon	No effect	
		Adenohypophysis	No effect on blood corticoid level	
Rat	Cortisol, hydro-cortisone acetate	Median eminence, anteromedial hypothalamus	Abolish CAH and the AAAD in the remaining gland following unilateral adrenalectomy	Davidson and Feldman, 1963
		Midbrain, anterior forebrain	Slight or no inhibition of CAH	
		Posterior diencephalon, cerebellum	Normal CAH	
		Pituitary	No effect	
Rat	Cortisol acetate	Median eminence	Adrenal atrophy, inhibition of AAAD	Chowers et al., 1963
		Pituitary	No effect	

Table 2 (cont'd.)

Species	Hormone	Site of Implants	Effect	Reference
Rat	Corticosterone	Anterobasal hypothalamus	Decreased stress-induced ACTH secretion, plasma corticosterone level. Small decrease in adrenal weight	Smelik, 1965
Rat	Dexamethasone, cortisol	Median eminence	Reduced plasma and adrenal corticosterone levels. Decreased adrenal weight. (Cortisol did not reduce adrenal weight.)	Corbin et al., 1965
	Dexamethasone	Pituitary	No effect	
		Cerebral cortex	No effect	
		Midbrain, lateral reticular formation	Lower plasma and adrenal corticosterone levels. Reduced adrenal weight.	
Rat	Cortisol, corticosterone	Median eminence	Block of stress response to increased adrenal corticoids	Davidson et al., 1965
Rat	Cortisol	Median eminence	Decreased adrenal weight. Depression of adrenal corticosterone	Slusher, 1966
		Midbrain reticular formation, ventral hippocampus	Decreased level of adrenal corticosterones. No change in adrenal weight	
Rat	Cortisol	Median eminence	Substantial suppression of the stress-induced ACTH release	Bohus, 1968

Table 2 (cont'd.)

Species	Hormone	Site of Implants	Effect	Reference
Rat	Cortisol, corticosterone	Mesencephalic reticular formation	Moderate suppression of the stress-induced ACTH release	Grimm and Kendall, 1968
Rat	Cortisol, corticosterone	Median eminence	Decrease in rise of plasma corticosterone following either stress	Grimm and Kendall, 1968
Rat	Cortisol acetate	Median eminence	Depressed adrenal function, adrenal atrophy	Kendall and Allen, 1968
Rabbit	Corticosterone	Hippocampus: CA2 and CA3 and part of fascia dentata	Adrenocortical biosynthetic activity increased as measured by $[1-^{14}C]$ acetate incorporation into corticosterone	Kawakami et al., 1968
		CA1	No influence	
		Amygdala: nucleus centralis nucleus basolateralis nucleus medialis nucleus corticalis	Decreased corticosterone synthesis	
		Hypothalamus	No change	
Rat	Dexamethasone	Hypothalamus	Marked suppression of diurnal peak in non-stress plasma corticosterone levels	Zimmerman and Critchlow, 1969
		Ventral diencephalon, rostral midbrain	Suppressed non-stressed corticosteroid level	
		Amygdala	No suppression	

Table 2 (cont'd.)

Species	Hormone	Site of Implants	Effect	Reference
Rat		Pituitary	No suppression	Bohus and Strashimirov, 1970
	Dexamethasone, cortisol, corticosterone, 11-dehydrocorticosterone, 11-deoxycorticosterone	Anterior median eminence region	Suppression of ACTH release in response to stress	
	11-deoxycortisol, tetrahydrocortisol, pregnenolene, progesterone, testosterone	Anteromedian eminence region	No effect	
	Dexamethasone, cortisol, 11-deoxycorticosterone	Infundibular region	Suppressed ACTH release	
	Dexamethasone, 11-deoxycorticosterone	Anterior pituitary (bilateral implants)	Suppressed ACTH release	

[a] Animals other than rat are included in this table, because only a few studies on extrahypothalamic implants have been reported.

AAAD, adrenal ascorbic acid depletion. ACTH, corticotrophin. CAH, compensatory adrenal hypertrophy.

Source: Stumpf and Sar (1973)

increases pituitary stores of ACTH in adrenalectomized rats and prevents pituitary ACTH depletion that follows stress in these rats (Hodges and Vernikos, 1959; Kitay et al., 1959); (b) ACTH-secreting transplanted pituitary tumors prevent pituitary and plasma ACTH rises following adrenalectomy (Vernikos-Danellis and Trigg, 1967); (c) the CRF content of the median eminence is greater in adrenalectomized-hypophysectomized rats than in the same rats treated with ACTH (Motta et al., 1968b). In all these studies it is presumed that ACTH causes direct suppression of CRF secretion. Alternatively, there could be an ultrashort loop, termed "mini-feedback" by Kastin et al. (1971) in which ACTH suppresses its own formation or release by product inhibition. The physiological significance of these "short loops" is not understood.

Thyrotropin Stimulating Hormone Secretion

Thyroid-stimulating hormone (TSH) secretion is regulated by the interaction of a neural component mediated by the hypothalamic thyrotropin-releasing hormone (TRH) discussed earlier and a feedback component mediated by circulating levels of thyroid hormones (See review Martin and Jackson, 1975).

Although there was evidence that thyroid hormone or lack of it causes cellular changes in the pituitary, the control of secretion of TSH by the negative-feedback action of thyroid hormone was not appreciated fully until its central role in the pathogenesis of goiter was elucidated in the early 1940s. It is now recognized that the rate of secretion of TSH is delicately controlled by the quantity of thyroid hormone in the circulation. If extra hormone is given, the secretion of thyrotropin is suppressed and the thyroid becomes inactive and regresses, whereas any decrease in the normal rate of secretion of the thyroid evokes an enhanced secretion of thyrotropin and the thyroid is stimulated to increased growth and function.

It is now believed that several phasic changes in TSH secretion, including diurnal rhythm, stress-related and sleep associated release are probably mediated by higher limbic system structures, such as the hippocampus and amygdala, and the midbrain reticular formation. Evidence exists that the hippocampus and amygdala may have exictatory influences on TSH secretion (Dupont et al., 1972; Eleftheriou and Zolovick, 1968; Martin, 1974; Shizume et al., 1974).

Gonadotropin Secretion

A variety of evidence has clearly established the important role of the hypothalamus in controlling the release of both gonadotropins, follicle-stimulating hormone (FSH), and luteinizing hormone (LH). The current concept is that LH release is controlled by a hypothalamic mechanism, possibly of two types (Barraclough, 1966). The rostral center located in the anterior hypothalamic area is concerned with the cyclic discharge of LH and FSH which is triggered by estrogen and/or progesterone. The more caudal region in the arcuate nucleus-median eminence area is concerned with basal gonadotropin release and the so-called negative feedback of gonadal steroids may be mediated primarily here. Both implantation of gonadal steroids in various hypothalamic loss and suprachiasmatic lesions support this concept (Davidson, 1969; Halasz, 1969).

Since the introduction of oral contraceptives, there has been much interest in the inhibitory feedback effects of estrogen and progesterone. Davidson et al. (1968) showed that the effective dose of estradiol benzoate for inhibition of pituitary castration cell formation in female rats three weeks after ovariectomy was roughly similar to that required to restore sexual behavior and uterine weight to normal levels. This experiment could provide no more than an approximation of the estrogen-gonadotropin relationship. However, the effective doses were considerably above those used by Ramirez and McCann (1963), suggesting that gonadectomy is followed by a decrease in sensitivity to gonadal steroids of various target organ responses including that of feedback mechanism.

Estrogens have been investigated more than other steroids from the point of view of feedback receptor localization (see reviews, Flerko, 1966; Stumpf and Sar, 1973). The findings are summarized as follows: prevention by anterior hypothalamic lesions of the gonadotropin-inhibitory effects of estrogen administration; prevention of "compensatory" ovarian changes following hemicastration by atropine or by anterior hypothalamic lesions; failure of systemic administration of estrogen to block the effect of exogenous gonadotropin releasing factors; changes in gonadotropin releasing factor activity of the hypothalamus following gonadectomy; cytological changes in hypothalamic nuclei following changes in circulating estrogen levels; failure of estrogen to inhibit gonadotropin secretion in incubated pituitaries. In summary, the evidence is generally in favor of the medial basal hypothalamus as the main site for negative feedback inhibition of gonadotropin secretion by estrogen in females, while the pituitary may be a secondary site of estrogen sensitivity (see Table 3, from Stumpf and Sar, 1973). Despite the well-known effect of progesterone feedback action, studies on the localization of this effect using intracranial administration of progesterone have shown only transient effects in inhibition of ovulation (Davidson, 1969).

The negative feedback inhibition of testosterone on LH secretion has been shown (see review, Davidson, 1969), but testosterone-FSH relationship has not been thoroughly investigated. Although considerably less work has been done on localization of the negative feedback receptor for androgen, the results are clearer than they are for estrogen (see review Davidson, 1969; also Table 4, from Stumpf and Sar, 1973).

Growth Hormone Secretion

A number of stimuli primarily metabolic are involved in the regulation of GH secretion: fasting after a meal or drinking glucose, hypoglycemia induced by insulin, or other causes, physical exertion and emotional excitement are normal stimuli to enhance GH secretion. After sectioning the pituitary stalk, there is no change in the plasma concentration of GH in response to hypoglycemia or to glucose, although the basal concentration of GH is normal. Obesity causes reduction or absence of responses of GH to fasting and other stimuli (Beck et al., 1964). Inhibitory influences on secretion of GH are exerted by free fatty acids (Blackard et al., 1971) and perhaps, by way of a negative-feedback loop, by GH itself (Abrams et al., 1971).

Several provocative tests have been devised to evaluate the capacity of the pituitary to secrete GH. The intravenous infusion of arginine (in a dose of 30 g in 30 minutes in adults or 0.5 g/kg in children) is safer and just as useful as the induction of hypoglycemia with insulin (Parker et al., 1967). Three to five hours after a dose of glucose, as used in the glucose tolerance test, there is normally a rise in the concentration of plasma GH. In this test, excessively obese subjects often do not respond (Theodoridis et al., 1969). The oral administration of L-Dopa can also be used to evoke secretion of GH (Weldon et al., 1973).

A consistent finding and probably a most important one is the rise in the concentration of GH in the plasma shortly after the onset of deep sleep. This is not just a reflection of a circadian rhythm; if the subject

Table 3

EFFECTS OF ESTROGEN IMPLANTS IN THE FEMALE RAT BRAIN ON REPRODUCTIVE FUNCTION

Hormone	Site of implants	Effect	References
Estradiol	Arcuate nucleus mammillary bodies	Atrophy of the genital tract and gonads	Lisk, 1960
	Preoptic regions anterior hypothalamus, medial hypothalamus	No significant effect on genital tract and gonads	
Estradiol	Basal tuberal hypothalamus	Prevented post castration elevation in plasma and hypophysial LH and evoked adenohypophysial enlargement	Ramirez et al., 1963
	Suprachiasmatic region, posterior hypothalamus, globi palladi	No effect	
Estradiol	Preoptic area, lateral hypothalamus, basolateral amygdala, fornix-hippocampus, caudate-putamen, subarachnoidal space	No effect on COH	Littlejohn and DeGroot, 1963
	Mammillary complex (peduncle)	Inhibited COH	
	Anterior or antermedial amygdala	Greater than normal COH	
Estradiol	Median eminence or pituitary	Increased release and synthesis of prolactin by the adenohypopsis	Ramirez and McCann, 1964
Estradiol benzoate (acute)	Anterior hypothalamic preoptic region	Advance vaginal opening; ovaries, uteri, and cycles normal	Smith and Davidson, 1968
	Median eminence	No apparent effect	
Estradiol benzoate (chronic)	Median eminence	Inhibited uterine and ovarian development	

Table 3 (cont'd.)

Hormone	Site of implants	Effect	References
Estradiol	Median eminence	Precocious opening of the vagina, ovulation, reduction in pituitary LH, increase of plasma LH levels, increase in uterine weight	Motta et al., 1968a
	Habenular region	Retardation of puberty, reduction of the weight of ovaries and uteri, increased pituitary LH stores, plasma LH undetectable	
Estradiol benzoate	Amygdaal	No effect on the vaginal opening	Davidson, 1969
Estradiol	Cortical amygdaloid nucleus or hypothalamic preoptic area	Tendency to stimulate LH secretion	Lawton and Sawyer, 1970
	Arcuate-ventromedial nuclear region	Depressed LH secretion	
Estradiol benzoate (unilateral)	Preoptic anterior hypothalamic area	Elevated plasma LH at 6 h	Kalra and McCann, 1972
Estradiol benzoate (bilateral)	Median eminence, arcuate region, amygdala	Increased plasma level of LH and FSH at 30 h	
Estradiol benzoate	Dorsal and ventral hippocampus	Significant decrease of only plasma FSH at 30 h	

[a] Intact 26-day immature rats were used. COH, compensatory ovarian hypertrophy. FSH, follicle-stimulating hormone. LH, luteinizing hormone. Source: Stumpf and Sar (1973)

Table 4

EFFECTS OF ANDROGEN IMPLANTS IN THE RAT BRAIN ON REPRODUCTIVE FUNCTION

Hormone	Site of implants	Effect	References
Testosterone	Basal tuberal median eminence region, arcuate nucleus	Atrophy of the ventral prostate and seminal vesicles	Lisk, 1962
	Lateral hypothalamus, anterior pituitary	No effect	
	Arcuate region[a]	Some atrophies of the ovaries	
Testosterone propionate[b]	Hypothalamus	Androgen sterilization	Wagner et al., 1966
Testosterone propionate[c]	Median eminence	Atrophy of the testis, seminal vesicle and prostate	Smith and Davidson, 1967a
Testosterone propionate[d]	Median eminence	Retardation of puberal development of the testis and accessory sex glands	Smith and Davidson, 1967b
Testosterone propionate	Arcuate nucleus	Decreased testicular biosynthesis of testosterone, atrophy of the seminal vesicles and urethral bulbs	Matsuyama, 1970
	Preoptic area	Slight increase in the weight of the accessory sex organs	
Testosterone propionate[e]	Ventromedial-arcuate nucleus area	More effective in androgen sterilization	Nadler, 1971
	Basal-preoptic suprachiasmatic region	Less effective in androgen sterilization	

[a]Female rats, mature. [b]Female rats, 4 days old; Hypophysectomized rat with renal pituitary graft; [d]Prepuberal; [e]Female rats, 5 days old. Except for [a], [b], and [e], all experiments listed in this table were done with male rats.

is kept awake all night or fitfully naps often, the rise does not take place until after he falls fast asleep the next day (see Parker et al., 1967). In fact, prepubertal children may secrete GH primarily during sleep, while secretion of GH during waking hours becomes more significant in adolescents (Finkelstein et al., 1972). Both prepubertal and pubertal boys have higher concentrations of GH than do adult males (Thompson et al., 1972). The role of hypothalamic stimulatory and inhibitory hormones in the regulation of GH secretion was discussed earlier.

Prolactin Secretion

Prolactin release is believed to be predominantly under inhibitory control by the mammalian hypothalamus through the secretion of a prolactin release-inhibiting factor (PIF) (Meites et al., 1972). TRH has been shown to induce release of prolactin as well as TSH in animals (Convey et al., 1973; Mueller et al., 1973; Tashijian et al., 1971) and in man (Bowers et al., 1971; Jacobs et al., 1971) and it has been proposed that TRH is an important physiological stimulator of prolactin secretion (Bowers et al., 1971). As also discussed earlier the presence of a prolactin-releasing factor (PRF) in the hypothalamus distinct from TRH has been reported (Dular et al., 1974; Grimm and Reichlin, 1973; Valverde et al., 1972).

The existence of prolactin in human blood has been established (Frantz and Kleinberg, 1970; Loewenstein et al., 1971). It has been now demonstrated by bioassay systems sensitive enough to be applied to unextracted human plasma that the isolated material is immunologically distinct from growth hormone. Isolation of the human hormone from pituitary glands (Hwang et al., 1972; Lewis et al., 1971) has permitted the development of a homologous radioimmunoassay. There is also a heterologous radioimmunoassay in which porcine prolactin is incubated with antiovine prolactin can be used to measure the hormone in human plasma (Jacobs et al., 1972).

Some of the physiological and pharmacological factors which have been shown to affect prolactin release in human are not well understood. The act of suckling in postpartum women, is one of the most powerful and specific of all stimuli to prolactin release. Plasma samples obtained at frequent intervals before, during, and after nursing indicate the prolactin begins to rise within 5 to 10 minutes after the onset of suckling, continues to rise throughout its duration, and begins to fall almost immediately after the cessation of nursing (Frantz et al., 1972a; Noel et al., 1972). A diurnal rhythm of prolactin secretion in humans has recently been documented (Sassin et al., 1972). A progressive rise in prolactin occurs throughout pregnancy in humans. It has been speculated that this phenomenon, which does not occur to any comparable degree in monkeys, may be linked with the rise in estrogen production that takes place in human pregnancy (Hwang et al., 1971. A rise in baseline plasma prolactin has been observed in males receiving large doses of estrogens (Frantz et al., 1972a). It has been suggested that estrogens may sensitize the hypothalamus and/or pituitary to prolactin-releasing stimuli of various kinds.

Regulation of Hormonal Secretion by Neurohumors

The role of central neural systems on endocrine function continues to be investigated on many fronts. Considerable evidence has been accumulating for varying degrees of catecholaminergic influence in the control of virtually every CNS pituitary-target organ axis (Ganong, 1972; McCann and Ojeda, 1976; Muller, 1973; Porter et al., 1972; Reichline et al., 1974). The role of serotonergic neural systems in neuroendocrine processes has been less thoroughly explored. The presence of serotonergic terminals in hypothalamus and their origin from cell bodies in the midbrain has been established by fluorescence histochemistry (Dahlstrom and Fuxe, 1964; Fuxe, 1965; Loizou, 1972; Ungerstedt, 1971). Hypothalamic structures shown to receive serotonergic input include the supraoptic, suprachiasmatic and arcuate nuclei. Although the presence and localization of dopamine and norepinephrine have been carefully documented (Björklund et al., 1973; Hökfelt and Fuxe, 1972; Jonsson et al., 1972), the presence of 5-HT nerve terminals are absent in the median eminence. Later reports, however (Baumgarten et al., 1973; Saavedra et al., 1974) point to the possibility that the apperance of 5-HT nerve terminals characterized with fluorescence histochemistry may be masked by the high percentage of dopamine and norepinephrine present in the median eminence. Recently, Knigge et al. (1975) used acute in vitro incubates or longer term organ cultures of rat and mink median eminences to study the uptake and localization of ^{14}C-5-HT. Their studies showed that organ-cultured median eminence retains approximately 80% of the 5-HT uptake capacity of fresh tissue, characteristics of the uptake process appear to be the same as in fresh tissue. Since ependyma were the major component of median eminence remaining in organ-cultured tissue, it appears likely that these cells are responsible for the uptake of 5-HT. Further studies are required to clarify indolamine metabolism in median eminence.

Hypothalamic-Hypophyseal-Adrenal Axis

As discussed earlier, the hypothalamus is the focal point at which neural stimuli converge to influence the secretion of ACTH and the median eminence is regarded as the final common path through which information is transmitted to the anterior pituitary. Since hypothalamic neurosecretory cells are located in a post-synaptic position with regard to the terminals of the central nervous pathways converging on the hypothalamus (Kobayashi and Matsui, 1969; Zambrano, 1968) and since many of the terminals contain monoamine granules (Fuxe and Hökfelt, 1969) one might assume that there is at that level a regulation by naturally occurring synaptic transmitters.

A possible role of brain catecholamines in the regulation of ACTH secretion has been considered by numerous investigators. The data up to the present show that the secretion of ACTH is inhibited by a catecholaminergic system and that the catecholamine neurons exert their effects via α-adrenergic receptors on the cells that secrete CRF. Only a brief review of the evidence for the noradrenergic-inhibitory hypothesis will be presented here. Early studies showed that reserpine produced depletion of brain catecholamines in association with increased secretion of ACTH (Maickel et al. 1961; Munson, 1963) and CRF (Bhattacharya and Marks, 1969). Moreover, drugs interfering with the synthesis (Bhattacharya and Marks, 1969; Carr and Moore, 1968; Vernikos-Danellis, 1968) and with the site of action (Bhattacharya and Marks, 1969; DeWied, 1967) of brain catecholamines increase the activity of the hypothalamic-hypophyseal-adrenal axis. On the other hand, drugs such as amphetamine, which releases catecholamines from nerve endings and also inhibits neuronal uptake, and monoamine oxidase (MAO) inhibitors, which decrease intraneuronal catabolism of catecholamines, decrease ACTH secretion (Bhattacharya and Marks, 1969; Hirsch and Moore, 1968; Lorenzen and Ganong, 1967). Stresses such as electroshock, hemorrhage, and hypoglycemia are associated with decreased brain norepinephrine content (Bliss et al., 1968; Thierry et al., 1968) and increased ACTH secretion (Ganong and Lorenzen, 1967).

More recently, Scapagnini and Preziosi in a review (1973a,b) have re-evaluated the hypothesis that brain adrenergic system inhibits ACTH secretion. Evidence is presented in favor of a central noradrenergic system that tonically inhibits ACTH secretion via stimulation of α-receptors. In dogs, the injection into the third ventricle of catecholamine precursors, catecholamines, and drugs that release or protect catecholamines from inactivation inhibits stress-induced adrenocortical activation (Van Loon et al., 1971). In rats, drugs that decrease brain catecholamine levels provoke adrenocortical activation. For example, intraperitoneal administration of α-methyl-p-tyrosine (α-MPT), a drug that inhibits catecholamine synthesis, regardless of time of day caused an increase in adrenocortical activity. Moreover, L-Dopa inhibited the mean increase in plasma corticosterone produced by α-MPT if administered simultaneously, whereas NE had no such effect since NE does not cross the blood-brain barrier (Van Loon et al., 1971). Further evidence that the inhibitory system is a central one is supplied by experiments using guanethidine, a drug which depletes catecholamines and does not cross the blood-brain barrier. Intraventricular administration of guanethedine produced an elevation of plasma corticosterone and a depletion of hypothalamic NE and dopamine. Inhibition of dopamine synthesis, resulted in a decreased NE level, and increased plasma corticosterone with no change in dopamine. These findings suggest that NE may be primarily involved in the regulation of the hypothalamic-hypophyseal-adrenocortical axis.

Role of Biogenic Amines in the Control of Gonadotropin Secretion

The first evidence that catecholamines participate in regulation of gonadotropin secretion was obtained by Sawyer and associates (Sawyer, 1947; Sawyer et al., 1947), who showed that ovulation can be blocked by phenoxybenzamine an agent which blocks a-adrenergic receptors and reserpine, an agent which depletes brain catecholamines, would block ovulation. More recently, Coppola (1971) reported that pseudopregnancy could also be induced by other catecholamine depletors, such as L-methyl-Dopa or tetrabenazine and that this effect could be counteracted by treatment with L-Dopa or monoamine inhibitors. Fuxe and Hökfelt (1969; also review Hökfelt and Fuxe, 1972) using fluorescence microscopy demonstrated the presence of catecholaminergic and serotoninergic neurons in the hypothalamus. A role for these particular neurons in control of gonadotropins secretion is suggested by the alterations in their content of dopamine in situations associated with altered gonadotropin secretion.

Unfortunately, due to conflicting reports from different laboratories of the specific role of each of the monoamines in control of gonadotropin secretion, a clear picture has not yet emerged. Two opposing views exist regarding the specific role of dopamine in controling release of gonadotrophins. Evidence, derived from variations in intensity of fluorescence and turnover rates of dopamine in neurons of medial basal hypothalamus of the rate during different phases of the reproductive cycle, led Fuxe and associates to believe that increased activity in dopamine neurons inhibits the release and/or synthesis of the gonadotropin-releasing factors. On the other hand, Schneider and McCann (1969) and Kamberi and associates (1970, 1971a,b) have provided evidence suggesting a stimulating effect of dopamine on LRF and FRF both in vitro and in vivo. Kamberi in a recent review article (1973) presents evidence that catecholamines have a stimulatory effect on gonadotropin release. Kamberi and McCann (1969) found that a variety of amines and even basic peptides, if used in sufficient amounts can alter FSH release (as measured by bioassay) from anterior pituitary incubated in vitro. Dopamine in small doses (0.2-1 μg/ml) was without effect, whereas in large doses (10-100 μg/ml) had an inhibitory effect. However, in anterior pituitary co-incubated with ventral hypothalamic tissue, dopamine in relatively small doses (1-5 μg/ml) had a stimulatory effect on FSH release. In addition, phentolamine and phenoxybenzamine, α-adrenergic blocking agents, prevent the effect of dopamine to stimulate discharge of gonadotropins from the hypothalamus.

Early and more recent reports have shown that indoleaminergic and catecholaminergic central nervous pathways exert a mutually antagonistic influence on the secretion of gonadotrophins. Early studies showed that the administration of p-chlorophenylalanine, a specific inhibitor of the synthesis of serotonin (Koe et al., 1966) stimulates sex behavior in male rats and rabbits (Tagliamonte et al., 1969). Pinealectomy, which suppresses an abundant source of idoleamines, also liberates the reproductive system from an inhibitory influence, inducing in female rats in acceleration of puberty and ovarian hypertrophy (Kitay and Altchule, 1954); in males an increase in accessory organ weights as well as increment in pituitary content of gonadotropins has been reported (Fraschini and Martini, 1970). Other studies, however, have shown that the administration of indoleamines, either by direct implantation into the median eminence (Fraschini, 1969) or by intraventricular administration, inhibits gonadotropin secretion (Kamberi et al., 1970, 1971a,b). Kamberi and associates (see Kamberi 1973) found that intraventricular injection of serotonin, or its metabolic product melatonin, in anesthetized male rats, appears to have an opposite effect than that seen with small doses of dopamine or large doses of norepinephrine or epinephrine. Either agent, serotonin or melatonin, suppresses the release of LH or FSH with simultaneous increase in prolactin release, and these effects were dose-related. These authors also observed a suppression of LH and FSH release with concomitant increase in prolactin release in castrated male and female rats after intraperitoneal injection of 5-hydroxytryptophan, a precursor of serotonin (see Kamberi, 1973). Moreover, intraventricular injection of melatonin or serotonin in unanesthetized cyclic female rats suppressed proestrous surge of LH and FSH and inhibited ovulation (see Kamberi, 1973).

The influence of the pineal gland in sexual maturation and gonadal function has been clearly evidenced by several studies (Collu et al., 1973; Mess et al., 1973; Reiter and Fraschini, 1969; Wurtman et al., 1968). More recently the effects of pineal estract on sexual maturation and of indoles on ovulation and gonadotrophin release have been reported (see Kamberi, 1973). Pineal extract delays precocious puberty as a result of inhibition of LH and FSH release and pineal indoles have an inhibitory effect on the release of plasma LH and FSH. Intraventricular injection of melatonin on the day of proestrous inhibits ovulation in rats examined on the day of estrous.

The role of the cholinergic system being involved in neural control of gonadotrophin secretion is not well documented. Everett et al. (1949) have reported that atropine, a blocker of muscarinic cholinergic receptors, was capable of blocking ovulation. Kato and Minaguchi (1964) reported changes into choline acetyltransferase activity in states of altered gonadotrophin secretion. Kamberi (1972) found that subcutaneous or intraventricular injection of atropine in female rats suppresses the proestrous surge of LH and FSH.

The foregoing studies demonstrate rather convincingly the involvement of catecholaminergic and indolaminergic pathways in the control of gonadotropin secretion. The precise origin of aminergic pathways relevant for the neuroendocrine regulation is not well defined. It has been suggested (Kamberi, 1973) that nomoaminergic neurons may have axons which run parallel to those of hypothalamic-hypophyseal hormone(HH)-containing neurons and may establish axo-axonal contacts with them. Thus, release of monoamines may depolarize the axon of HHH-containing neurons and produce their discharge. Alternatively, HHH-containing vesicles may be present in the axons of monoaminergic neurons. Indeed, vesicles of the two different sizes have been described in neurons in this locus. Release of monoamines would then provoke release of HHH from the same neurons. Presumbalby, specificity of different HHH is obtained by the fact that any given neuron synthesizes and releases only one HHH and is affected by monoaminergic neurons linked to the controling mechanism(s) for the specific hormone.

Hypothalamic-Hypophyseal-Thyroid Axis

The elucidation of the chemical structure of TRH has made it possible to study the synthesis of endogenous TRH by incubating hypothalamic fragments with radioactive amino acid precursors and measuring TRH synthetase activity (Reichlin et al., 1972). With this technique Reichlin and associates found that reserpine-treated animals had a reduced concentration of TRH synthetase activity compared with nontreated animals. Furthermore, the addition of NE or dopamine to hypothalamic in vitro enchanced the rate of discharge of labeled TRH. They proposed that the total function of the TRH peptidergic neuron is subject to control by noradrenergic neurons.

As reported earlier by Fuxe and Hökfelt (1966) dopamine granules have been found in proximity to the pituitary portal system, and dopamine levels are increased in the hyopthalamus and median eminence (Kleinberg et al., 1971). Burrow et al. (1974) examined the effect of L-Dopa therapy of Parkinsonian patients on the response of the pituitary to TRH injection. They measured TSH and thyroxine levels in the blood. In this study, TRH did not elevate TSH levels or thyroxine in patients receiving L-Dopa. From these findings a direct effect of L-Dopa or the thyroid gland to release thyroxine which in turn through the negative feedback mechanism would decrease TSH responsiveness to TRH is not apparent. Dopamine has been reported to have a direct effect on thyroid hormone release (Maayan et al., 1973). It has been proposed that perhaps catecholamines might stimulate the release of a hypothalamic TSH inhibitory factor in a manner similar to the stimulation of TRH. L-Dopa might thus inhibit TRH-induced release of TSH by mechanisms similar to those regulating TRH-induced prolactin release (Kamberi et al., 1971a). Such an inhibitory factor has been reported in teleosts where it is apparently the major regulator (Peter, 1971).

Prolactin Secretion

Prolactin release has been reported to be inhibited by catecholamines, mainly dopamine, and stimulated by serotonin in the hypothalamus (Kamberi and McCann, 1969; Lu and Meites, 1973; MacIndoe and Turkington, 1973; Meites et al., 1972). Dopamine injected into the third ventricle (Kamberi, 1969) and L-Dopa injected systematically (Meites et al., 1972) were found to promote PIF discharge from the hypothalamus, and dopamine also was found to act directly on the pituitary to inhibit prolactin release (MacLeod and Lehmeyer, 1972; Birge et al., 1970).

It has been known that phenothiazine derivatives and other drugs of the neuroleptic class, such as reserpine and butyrophenones, which have the property of depleting or antagonizing cerebral catecholamines, can cause prolactin release in animals (Sulman, 1970). Early in vitro studies involving pituitary glands co-cultured with hypothalamus from normal and perphenazine-treated rats suggest that the drug suppresses prolactin-inhibiting factor (PIF) liberated from the hypothalamus (Danon et al., 1963). In normal humans, prolactin levels rise sharply one hour after acute chlorpromazine treatment. In addition, elevated prolactin concentrations have been reported in psychiatric patients receiving high doses of chlorpromazine (Frantz et al., 1972b). Pretreatment with oral L-Dopa can effectively block the prolactin rise that ordinarily follows chlorpromazine in normal individuals (Frantz, 1973). Moreover, L-Dopa depresses baseline prolactin concentrations in endocrinologically normal individuals (Frantz et al., 1972b; Friesen et al., 1972). Although it appears that dopaminergic pathways act as inhibitors of prolactin secretion in humans, the exact loci of their action is still unsettled. Also uncertain is the extent of participation, if any, of noradrenergic pathways in prolactin regulation.

Growth Hormone Secretion

Animal studies have shown that growth hormone secretion is regulated by both peptidergic and monoaminergic systems, but the interaction of these systems is not yet understood. Data in rats and baboons suggest that hypothalamic norepinephrine and serotonin stimulate and dopamine inhibits growth hormone secretion (Collu et al., 1972; Toivola and Gale, 1970). In view of the animals studied on the possible interrelation of biogenic amines and growth hormone several investigators became interested on the effects of L-Dopa on growth hormone secretion in man. Lebowitz et al. (review, 1974) in their early observations on the effects of L-Dopa administration to patients with Parkinson's disease found that doses as small as 0.5 g orally caused a dramatic and rapid rise in plasma growth hormone levels. The most striking characteristic of L-Dopa-stimulated growth hormone secretion in Parkinsonian patients is the failure of oral or intravenous glucose to suppress it. The effect of L-Dopa on growth hormone stimulation occurs also in patients who have been on chronic L-Dopa therapy. Many subsequent studies have shown that L-Dopa stimulates the secretion of growth hormone in normal individuals (Eddy et al., 1971; Kansal et al., 1972). One study (Sachar et al., 1972) has suggested that the growth hormone response to L-Dopa is decreased in elder individuals (ages 48 to 68) as compared with younger individuals (ages 20 to 32). The implication that a growth hormone deficiency may be associated with Parkinson's disease is supported by the findings of Boyd and associates (1971) who have reported that patients with Parkinson's disease exhibit borderline growth hormone secretion in response to a stimulus such as insulin-induced hypoglycemia. Further studies are required before a more clear concept on growth hormone in Parkinson patients can be put forward.

UPTAKE, DISTRIBUTION, AND METABOLISM OF HORMONES IN THE CENTRAL NERVOUS TISSUE

Extensive research is focusing on elucidating the mechanisms of actions of hormones on cellular processes. Their direct cellular actions have been supported by extensive evidence that hormones accumulate in neural cells, are bound to specific cytosol or nuclear sites (receptors binding sites) and also some hormones are metabolized in nervous tissue. Moreover the demonstration of hormone-binding macromolecules in brain cell nuclei has led to the proposal that the actions of some hormones, for example, steroid hormones, on many target tissues in the body involves the activation of genes leading to increased RNA and protein synthesis.

The presence of hormones in specific nervous cells or specific brain regions has been correlated with specific physiological and behavioral hormonal effects.

Hormone uptake and hormone receptors are discussed in several chapters in this book as they relate to sex differentiation and to neural growth and differentiation (Chapters 2, 4, 7, 8). Therefore, a brief account of uptake, distribution and metabolism of some hormones in the CNS tissue in the adult organism will be presented in this chapter.

Glucocorticoids

As also discussed in Chapter 5, early studies of endogenous glucocorticoid levels in plasma and brain of experimental animals (Henkin et al., 1968; Butte et al., 1972) and humans (Touchstone et al., 1966) suggested that the CNS possesses some mechanism of glucocorticoid accumulation and that this mechanism is responsive to physiological fluctuations of plasma levels of these steroid hormones. Subsequently, it has been shown that circulating glucocorticoids rapidly enter both the brain and cerebrospinal fluid (McEwen et al., 1969; 1970 a-c; 1972a,b). High concentrations of labeled corticosterone have been reported in hippocampus, septum, and amygdala, with lower concentrations in hypothalamus, midbrain and preoptic area (McEwen et al., 1970c, 1972a, 1976). A more detailed description of the localization of labeled glucocorticoids in brain has been provided by the autoradiographic studies of Stumpf and Sar (1975a-c). The intensity of nuclear labeling of neurons in rat brain is highest in the hippocampus, including the spracallosal and precommissural hippocampus, followed by the septum, the cortical nucleus of the amygdala, and the entorhinal and piriform cortex. A similar regional distribution of glucocorticoids has been reported in hamster and mouse (McEwen, 1976a,b), peking duck (Rhees et al., 1972), pig (Stith et al., 1976), and rhesus monkey (Gerlach et al., 1976, Pfaff et al., 1976). It has been suggested that these findings point to an ancestral origin in evolution for the brain glucocorticoid-receptor system (McEwen, 1976a).

Soluble glucocorticoid-binding macromolecules have been extracted from the cytosols of rat brain (Grosser et al., 1971; McEwen et al., 1972a; Stevens et al., 1973). The macromolecular protein from rat brain cytosol also exhibits several properties characteristic of a "receptor," including selectivity for corticosterone, stereospecificity, and a high affinity and limited capacity for this steroid (Grosser et al., 1973; McEwen et al., 1972a; Stevens et al., 1975). Moreover, in vitro binding of [^{3}H] B to cytosol receptors is reduced by progesterone but dexamethasone produces even less competition than progesterone. Subsequent studies indicate that there may be different corticoid-binding proteins in brain cytosol (Anderson and Fanestil, 1976; DeKloet and McEwen, 1976; Maclusky et al., 1977).

Sex Steroids

Estrogen Accumulation and Metabolism

Significant accumulation and retention of radioactive estradiol occurs in certain brain areas, particularly in hypothalamic, preoptic, and limbic areas, as well as in pituitary, uterus, and vagina (Anderson and Greenwald, 1969; Eisenfeld and Axelrod, 1965; McEwen and Pfaff, 1970; McGuire and Lisk, 1968; Kato, 1970; Kato and Villee, 1967; Pfaff, 1968; Stumpf, 1968). Although the role of estrogens in neuroendocrine tissue has been studied more extensively in regard to estradiol than other agents, attention is directed to the metabolism of estrogens in neuroendocrine tissue with the possibility that differential biotransformations may help to rationalize the variety of effects elicited by the estrogens (Ryan et al., 1973; Fishman, 1976). Transformations of estrogen in neuroendocrine tissue include hydrolysis of estrone sulfate, interconversion of estradiol and estrone and 2-hydroxylation to catechol estrogens (Ball et al., 1978; Fishman, 1976; Jenkin and Heap, 1976; Paul and Axelrod, 1977). There has been some speculation on the roles of these various biotransformations although their actual significance is not yet clear. For example, the estrone sulfate, present in fetal circulation, may act as a prohormone since it is readily converted in vitro into the more biologically active estrogens, estrone and estradiol, especially in the fetal pituitary (Jenkin and Heap, 1976). Also, 2-hydroxylation of estrogens results in catecholestrogens which have been proposed to have an important role in neuroendocrine regulation including actions on estrogen control of gonadotrophin release and other estrogenic effects (Paul and Axelrod, 1977; Fishman, 1976). Catecholestrogens have been shown to have profound effects on the circulating levels of plasma gonadotrophins (Naftolin et al., 1975) and to effectively compete with estrogens for estrogen receptors of the pituitary and brain (Davies et al., 1975). It has also been shown that catecholestrogens are excellent competitive inhibitors of catechol-0-methyl transferase mediated 0-methylation of catecholamines (Ball et al., 1972) which are themselves implicated in gonadotropin control mechanisms. Thus, it appears that catecholestrogens are not only metabolic end products, but possess potent biological and endocrine activities of their own.

Estrogen Binding Sites

Autoradiographic studies have suggested that brain binding sites for estradiol are, like those of the uterus, found in cell nuclei (Pfaff, 1968; Zigmond and McEwen, 1970) and the soluble (cytosol) fraction (Eisenfeld, 1970; Kahwango et al., 1969; Pfapinger and McEwen, 1973). Furthermore, there are specific nuclear receptors for estradiol and that estradiol is translocated to the nucleus after binding to a cytosol macromolecule.

Isolation and physiochemical characteristics of estrogen receptors and their organ specificity has been reviewed (Davies et al., 1976; Kato, 1977; McEwen, 1976a, 1978; Vernikos-Danellis, 1972). Similar molecular characteristics of rat uterine and anterior pituitary estrogen receptors suggest that the differential effects are not related to different properties of the receptors but are the result of subsequent molecular events evoked by the receptors (Notides, 1970). However differences in rate of uptake and retention time of estrogens in rat hypothalamus and pituitary compared with uterine tissue have been reported (Mowles et al., 1971) suggesting that different types of estrogen-receptor complexes may be responsible for different actions of estrogen in different target organs. Thus, information regarding possible organ specificity of estradiol receptors is still inconclusive.

Androgen Accumulation and Metabolism

Several investigators have demonstrated uptake of radioactivity by brain and pituitary tissues following administration of (^{3}H) testosterone ((^{3}H) T) to experimental animals (McEwen et al., 1970a,b; Resko et al., 1967; Stern and Eisenfeld, 1971; Whalen et al., 1969). The pattern of uptake of (^{3}H)T appears to parallel endogenous testosterone concentrations (Robel et al., 1973; Challis et al., 1976) with pituitary generally highest followed by hypothalamus and cerebral cortex. However, it is important to note that unlike estradiol, which is retained in its target organs largely unmetabolized, radioactive testosterone can be metabolized both to dihydrotestosterone (DHT) and other ring A reduced androgens (Massa et al., 1974) or be aromatized to estrogen (Naftolin et al., 1976). Thus, it becomes necessary to consider metabolites derived from testosterone in studies of testosterone accumulation. (For detailed studies, see McEwen, 1978; McEwen et al., 1970a,b; Stern and Eisenfeld, 1971).

Transformations of testosterone into more or less active metabolites is considered important in the mechanism of action of this hormone and has led to the concept that testosterone may function, at least in part, as a prehormone (Robel et al., 1973; Naftolin et al., 1975; 1976; McEwen, 1976b, 1978). Of the active metabolites of testosterone, considerable attention has been focused on dihydortestosterone (DHT) and estradiol. These metabolites are known to be produced from testosterone by brain tissue and both have been suggested as mediators of androgen action. Also, it has been demonstrated that brain and pituitary tissues possess enzymes that produce these and other metabolites from testosterone, although it appears that there are regional differences in these conversions (Denef et al., 1973; Farquhar et al., 1976; Martini, 1976; McEwen, 1976b, 1978; Verhoeven et al., 1974). It has been suggested (Farquhar et al., 1976) that the enzymes involved in reduction of testosterone and DHT may function in the metabolism of progesterone or other steroids. Although functional correlates of DHT, DIOL and other 5α-reduced metabolites have not yet been established in neuroendocrine tissue, a relationship between DHT formation and gonadotrophin secretion has been proposed (Denef et al., 1973). It has also been suggested that transformation of testosterone into 5α-reduced metabolites is a necessary step for regulation of androgen-induced feedback and behavioral responsiveness (Martini, 1976).

It has been proposed that aromatization of androgens by neuroendocrine tissue is involved in sexual differentiation of the brain, initiation and maintenance of sexual behavior, control of gonadotropin secretion and timing of puberty (Naftolin et al., 1975, 1976). Aromatization of androgens to estrogens is the result of a multiple reaction process which is represented by formation of estradiol and estrone, respectively, from testosterone and androstendione (Naftolin et al., 1975, 1976). The hypothalamus is a prominent region of aromatization reactions but that the pituitary has low aromatizing activity. These studies have also indicated that limbic brain regions such as the amygdala have high aromatizing activity. This is in contrast to the very low 5α-reductase activity formed in limbic brain regions (Martini, 1976). Also in contrast to androgen reduction reactions, sex differences have been reported for aromatization, with higher activity in male brains than female brains (see McEwen, 1976b).

Androgen Binding Sites

Subcellular localization of androgens and characterization of androgen binding sites is complex and confounded even further by androgen metabolism (McEwen et al., 1970a,b). Lieberburg and McEwen (1975) and Lieberburg et al., (1977) have indicated that the presence or absence DHT receptors, and not the rate of DHT formation is the limiting factor in cell nuclear retention of (^{3}H) DHT.

Several laboratories have described binding of (^{3}H)T or (^{3}H) DHT to macromolecules in the cytosol of pituitary and brain regions of male rats (Ginsberg et al., 1974; Gustafason et al., 1976; Jouan et al., 1973; Kato and Onouchi, 1973; Naess and Attramadal, 1974). Characteristics of these macromolecules have been extensively reviewed (Davies et al., 1976; McEwen, 1976a, 1978; Kato, 1977). Although castrated male rats have been used in most of these studies, binding components in female rat brains have also been described (Ginsberg et al., 1974). Gustafason et al. (1976), however, were unable to detect a macromolecular receptor protein in 8-week-old female rats of a similar nature to those they found in male rats of the same age. However, they did detect testosterone receptor protein in 28-day-old female rats. It is suggested (Gustafsson et al., 1976) that the absence of receptor proteins in hypothalamus and pituitary of the older female rats is responsible for the unresponsiveness of these animals to a central androgen-mediated regulation of hepatic enzyme activity. Studies of the ontogeny of androgen receptors in rat brain cytosol indicate that they are present at a low level in neonatal CNS by postnatal day 1-5 and increase gradually to adult levels over the first four weeks of life (Kato, 1977; Lieberburg et al., 1978). These studies indicate that the neonatal androgen receptors exhibit properties typical of receptors found in adult animals, including: sedimentation in low ionic strength glycerol gradients at approximately 8S; binding to DNA cellulose with the same affinity; high affinity for the antiandrogen, cyproterone, and low affinity for the synthetic estrogen, diethylstilbestrol.

Thyroid Hormone Accumulation

Accumulation of thyroid hormones, both thyroxine (T_4) and triiodothyronine (T_3) has been demonstrated in early studies by Ford and associates (see Chapter 5) and will not be discussed here in detail. The distribution of these hormones in the CNS is dependent upon the specific brain area and the age and sex of the animal (Eberhardt et al., 1976; Ford and Rhines, 1967; 1970; Ford et al., 1971; Timiras and Luchoch, 1974). Subcellular fractionation studies of rat brain following intravenous administration of (^{125}I) T_3 have indicated that the hormone is rapidly and selectively taken up by the nerve ending fraction (Dratman et al., 1976). Findings of selective uptake, concentration, and retention of T_3 in nerve terminals of rat brain led Dratman et al. (1976) to suggest that these processes might be related to sympathomimetic actions and behavioral effects produced by thyroid hormones.

Thyroid Hormone Binding

More recently developmental studies have significantly contributed to an understanding of binding properties of cellular receptors in the brain for thyroid hormones (Eberhardt et al., 1976, 1978; Geel, 1977; Naidoo et al., 1978; Valcana and Timiras, 1978a,b). The rationale of the developmental approach has generally been based on the hypothesis that the differential sensitivity of the brain during early development compared with later ages might be the result of age-related differences in thyroid hormone-binding sites. Indeed, Geel (1977) reported that the affinity of rat brain cytosol for T_3 was high during an early age period, but declined dramatically with age with no apparent change in the number of receptor sites. The decline in the affinity constant of brain tissue (cerebellum) in this study appeared to correlate with the limited period of responsiveness of the brain to functional differentiation influenced by thyroid hormone. On the other hand, it was also shown that liver, which apparently does not demonstrate an age-dependent tissue sensitivity to thyroid hormones, shows a marked increase in the number of binding sites with little change in the affinity constant (Geel, 1977).

The significance of thyroid binding proteins in the cytosol of brain and other tissues (Davis et al., 1974; Dillman et al., 1974) is not completely understood, but Naidoo et al. (1978) have described a hypothetical model for T_3 interaction with its target cell. The sequence of events, assumed to be similar for T_3 and T_4, is different than those described in previous sections of this chapter for the interaction of steroid hormones with their target sites. In the model described by Naidoo et al. (1978), T_3 enters the cell and is reversibly bound to a cytosol binding protein. The hormone-protein complex then exists in reversible equilibrium with a minute pool of free T_3 in the cytoplasm which can interact reversibly with nuclear T_3 receptors and perhaps also with receptors in mitochondria. Studies showing that nuclei from thyroid hormone target tissues are capable of specific T_3 binding in vitro (Samuels et al., 1974a,b; DeGroot and Torresani, 1975) have suggested that it is unlikely that a special mechanism exists for the translocation of T_3 from cytoplasm to nucleus as proposed for steroid hormones. Indeed, it has been proposed that the characteristics of cytosol T_3 binding proteins are consistent with a cytoplasmic function of modulating the intracellular levels of free T_3 in a manner similar to plasma thyroxine binding proteins.

Correlation of Hormone Binding Sites and Biological Activity

A considerable body of evidence indicates that thyroid hormones may mediate biological effects by binding to nuclear receptor proteins (see Eberhardt et al., 1978; Naidoo et al., 1978; Valcana and Timiras, 1978a,b). Specific high-affinity and low-capacity binding sites for T_3 and T_4 have been demonstrated in cell nuclei from anterior pituitary, pituitary tumor cells, cerebral hemispheres and other brain regions, as well as from liver, kidney and other nonneural tissues (Eberhardt et al., 1976, 1978; Oppenheimer et al., 1972, 1974; Samuels et al., 1974a,b). The nuclear receptors appear to be salt extractable, acidic chromatic proteins of approximately 50,000 daltons and bind to DNA (Latham et al., 1976; MacLeod and Baxter, 1976; Samuels et al., 1974b; Surks et al., 1973).

In an in vivo isotopic displacement study of the binding of T_3 to nuclei from various rat organs, Oppenheimer et al. (1974) reported that the highest receptor concentration in the tissues studied was present in the anterior pituitary followed by the liver and brain. However, more recent in vitro studies of T_3 binding to nuclei from rat tissues have reported that although the anterior pituitary contains 2.5 times as many binding sites as the cerebral hemispheres, the cerebral hemispheres contain 2-4 times the number of binding sites as in liver (Eberhardt et al., 1978). Eberhardt et al. have suggested that the discrepancy between in vivo and in vitro results may be due to a combination of the following factors: differential receptor stability of liver brain tissues and different methods of preparing nuclei; regional differences in concentration of nuclear receptors in brain (whole brain vs. cerebral hemispheres; and differences in the equilibrium of the exchange reaction at the level of the hormone-receptor interaction between liver and brain. Although these studies were conducted using adult rats, other studies have reported an age-related dependence in the density of nuclear thyroid hormone receptors. Studies of rat brain indicate that nuclei from cerebral hemispheres have a high density of T_3 receptors at birth which declines to adult levels by the end of the second postnatal week (Naidoo et al., 1978; Valcana and Timiras, 1978a). In contrast, developing liver appears to show an increase in nuclear T_3 receptor density.

The high density of nuclear T_3 receptors found in brain during the first week of brain development might be correlated with the period of critical sensitivity of the brain during early development whereas the decline in nuclear receptor density might explain the decline in brain sensitivity to thyroid hormones (Naidoo et al., 1978; Valcana and Timiras, 1978a,b). However, data showing that the density and affinity of nuclear T_3 receptors in brain remain relatively high in the mature brain, relative to liver or other thyroid-hormone-responsive tissues, suggest that these hormones may have a continued functional role in the mature brain (Eberhart et al., 1978, Valcana and Timiras, 1978b). In the adult brain, in contrast to the neonatal brain, thyroid hormones may regulate an entirely different set of biochemical reactions which may underly the neurological and psychological deficits seen in the absence of these hormones (Naidoo et al., 1978).

Hypothalamic Peptides

The development of highly specific radioimmunoassays for LHRH, TRH, and somatostatin revealed the presence of relatively large quantities of these "hypothalamic" peptides in extrahypothalamic brain regions. The wide distribution of these peptides in brain, together with the findings that systemic administration of these peptides produce behavioral modifications (Cohn and Cohn, 1977; Moss et al., 1972; Prange et al., 1975) led to speculation that peptide hormones subserve an important role in brain function, possibly as neurotransmitters or neuromodulators, quite apart from their role in anterior pituitary regulation (Bloom, 1977; Brownstein, 1977; Jackson, 1978; Moss, 1977; Vale et al., 1977).

Luteinizing Hormone Releasing Hormone (LHRH).

Immunohistochemical and radioimmuno-assay techniques have led to descriptions of the localization of LHRH-containing elements within the CNS of a variety of mammals, including rat (Baker et al., 1975; Palkovits et al., 1974), mouse (Zimmerman et al., 1974), guinea pig (Silverman, 1976), rabbit (Barry, 1976), cat and dog (Barry and Dubois, 1975), several species of monkeys (Barry and Carette, 1975; Silverman et al., 1977), and human (Barry, 1977; Bugnon et al., 1977). In general, the distribution of LHRH-containing neurons ranges from the septal-preoptic region anteriorly to the premammillary nucleus posteriorly. The highest concentration of LHRH detected in rat brain by radioimmunoassay is found in the median eminence with lower concentrations in the arcuate (infundibular) and ventromedial nuclei of the hypothalamus (Palkovits et al., 1974). LHRH is also concentrated in the preoptic and suprachiasmatic areas. A dual central influence of LHRH on pituitary LH has been proposed, whereby LHRH from the basal hypothalamus controls the tonic, and LHRH from preoptic-suprachiasmatic areas the cyclic, secretion of LH (Barry, 1977; Jackson, 1978). Interestingly, Silverman et al. (1977) noted that in their study of rhesus monkey, areas which contained high concentrations of LHRH neurons, especially the bed nucleus of the stria terminalis, infundibular nucleus, and the medial preoptic nucleus, corresponded to regions containing estradiol-concentrating neurons in this same species (Gerlach et al., 1976). However, it was not determined whether or not estradiol was accumulated within cells that synthesize LHRH.

Ultrastructural detection of LHRH has been hampered by technical problems and by dispersion of the peptide and the small number of immunoreactive elements. However, localization of LHRH in perikarya granules, axons, dendrites, and nerve endings of neurons and in tanycytes has been reported (Barry, 1977; Bugnon et al., 1977; Joseph and Knigge, 1978; Vale et al., 1977; Zimmerman, 1976, 1977). LHRH-positive granules are observed in nerve endings close to the basement membrane of the pericapillary space, and are occasionally seen in the subependymal layer (Goldsmith, 1977). The finding of LHRH in hypothalamic synaptosomal particles (Barnea et al., 1975) with biochemical characteristics similar to those of the catecholamines has led to the suggestion of a neurotransmitter role as well as neuroendocrine function for this decapeptide (Goldsmith, 1977).

Thyrotropin Releasing Hormone

The highest concentration of TRH is in the hypothalamus, but over 70% of the total brain TRH is found in the rat extrahypothalamic brain (see Vale et al., 1977; Jackson, 1978). Although this tripeptide was initially identified in relation to its ability to induce the release of thyroid stimulating hormone (TSH) from the pituitary, its extrahypothalamic distribution suggested a broader functional role than regulation of pituitary function (Bloom, 1977; Elde and Hökfelt, 1978; Moss, 1977).

Thyrotropin releasing hormone is localized in a broad band of tissue beginning with the bed nucleus of the stria terminalis, extending caudally to the dorsomedial nucleus and ventrally through the ventromedial and arcuate nuclei to the median eminence where the bulk of the activity is stored (see McCann and Ojeda, 1976). Immunofluorescence studies of TRH localization (Elde and Hökfelt, 1978) have shown that the external layer of the median eminence contains a high density of TRH-positive terminals that appear to abut on the capillary loops of the ortal plexus. Several hypothalamic areas, including the dorsomedial, paraventricular, periventricular (arcuate), and ventromedial nuclei, the perifornical region and the zona incerta, also contain TRH-positive nerve terminals. Occasional fibers are found in the suprachiasmatic and magnocellular paraventricular nuclei, ventral preoptic area, medial forebrain bundle and the organum vasculosum of the lamina terminalis. Extrahypothalamic TRH terminals have been described in the nucleus accumbens, nucleus interstitialis stria terminalis and in nuclei of the septum, brainstem and spinal cord. It has been reported that these concentrations of TRH are maintained independently of the hypothalamus (Brownstein et al., 1975a,b; Jackson and Reichlin, 1977).

The broad distribution of TRH is consistent with the concept that, in addition to its hypophysiotropic function, this peptide has direct actions within the CNS, possibly through a role in neurotransmission. Studies of TRH in nerve endings have provided additional support for a neurotransmitter role of this peptide. Concentrations of TRH have been demonstrated in hypothalamic synaptosomes (Barnea et al., 1975). Also, TRH release has been reported following exposure of synaptosomes to depolarizing concentrations of K^+ in the presence of 1-2$\underline{mM}$ Ca^{2+} (Warberg et al., 1977).

Somatostatin

The tetradecapeptide somatostatin is distributed in several tissues and possess a wide range of physiological effects. Somatostatin has been shown to inhibit release of growth hormone, insulin, glucagon, gastrin and gastric acid (for review, see Vale et al., 1977; Goldsmith, 1977). Somatostatin, like TRH, is widely distributed in the brain and is also present in mammalian stomach and pancreas where it is localized in argyrophilic delta or D (A_1) cells (Arimura et al., 1975; Hökfelt et al., 1975a,c) and guinea pig (Dube et al., 1975). Immunofluroescent somatostatin-positive nerve fibers have been seen in the substantia gelatinosa of the spinal cord (Hökfelt et al., 1975a,c) consistent with the significant extrahypothalamic distribution of this peptide throughout the central nervous system

(Brownstein et al., 1975a,b; Brownstein, 1977; Goldsmith, 1977; Jackson, 1978). The finding that a certain population of primary sensory nerves contain somatostatin suggests that this substance may act as a depressant neurotransmitter in sensory neurons (Hökfelt et al., 1975a,c).

Somatostatin has been localized in perikarya in rat hypothalamus by light microscopy (Hökfelt et al., 1975a,c). These cell bodies were distributed in the periventricular region anteriorly between the anterior commisure and optic chiasm and posteriorly to the region of the ventromedial nucleus. Pelletier and associates (1975) localized somatostatin without 900 to 1,100-A° granules in 30% of the nerve endings on portal capillaries in the zona externa of the rat. Subcellular fractionation studies of medial basal hypothalamus, preoptic area and amygdata of rat brain have shown that over 70% of somatostatin detectable by radioimmunoassay was localized to the synaptosome fraction (Epelbaum et al., 1977). It was proposed that somatostatin, in addition to being released into blood vessels of the median eminence, may also be liberated from nerve terminals in other brain regions.

EFFECTS OF HORMONES ON BRAIN METABOLIC FUNCTIONS

Water and Electrolytes

The effects of hormones on regulation of water and electrolyte metabolism range from the well-known role of antidiuretic hormone, and adrenal and gonads on Na, K, and water balance, to those of the thyroid and parathyroid glands on calcium metabolism. The effect of steroid hormones on brain electrolyte metabolism and their relationship to brain excitability have been extensively studied by Woodbury and associates and have been frequently reviewed (Withrow and Woodbury, 1972; Woodbury, 1954; Woodbury et al., 1957; Woodbury, 1958; Woodbury, 1972). Thus, only a brief account will be presented here.

Deoxycorticosterone (DOC) has been extensively studied with respect to tissue electrolytes, including brain. In tissues other than brain, it has been shown that DOC causes a rise in intracellular Na concentration and a concomitant loss of intracellular K. The effects of DOC on brain electrolytes have been studied by Woodbury and associates (see reviews by Woodbury) who found that total brain Na and K concentration, calculated on the assumption that Cl space is a measure of extracellular fluid volume, is markedly decreased. Despite the fact that intracellular K concentration remains unchanged, the ratio of intracellular to extracellular K in brain is increased by DOC treatment because plasma K is decreased. The decrease in brain intracellular Na has been associated with an increase in electroshock seizure threshold (EST)–a decrease in brain excitability. The increase in EST is ascribed to possible membrane hyperpolarization caused by enhanced Na transport from brain cells. That a fundamental difference exists in intact rats between the effects of DOC on brain and other tissues is indicated by the fact that only in brain does DOC decrease intracellular Na concentration. Intracellular concentration in Na is increased while that of K is decreased in muscle, heart, liver, and skin of DOC-treated animals.

The changes induced by aldosterone in brain and skeletal muscle electrolytes have been studied in mice (see reviews by Woodbury). This steroid increases the ratio of plasma to intracellular Na concentration and intracellular to plasma K concentration in both brain and muscle. Thus, aldosterone and DOC produce similar effects on brain electrolytes but opposite effects on muscle electrolytes. The discrepancy was resolved by Withrow and Woodbury (1972) who found that DOC had the same influence as aldosterone on Na and K concentrations in muscle in nephrectomized, partially eviscerated rats. In this preparation, the marked renal and gastrointestinal losses of K usually caused by DOC were minimized. Thus, the stimulatory effects of DOC on muscle Na-K transport were separated from the inhibitory effects of low extracellular K on Na+-K+ ATPase.

The influence of acute administration of cortisol and DOC on brain electrolyte metabolism has also been studied by Woodbury and associates (see reviews by Woodbury). The results for DOC are the same as those obtained with chronic injection of this steroid but relatively greater since sufficient time to allow of lowering of plasma K does not occur. However, the acute administration of cortisol differs from the chronic in that it increases brain intracellular Na concentration, decreases the Na ratio between brain and plasma, and increases brain excitability. Thus the changes in brain electrrolytes induced acutely by cortisol correlate with the observed changes in exictability. The results differ from those for chronic cortisol treatment in which no measurable changes in intracellular brain electrolytes have been noted, perhaps because cortisol, like DOC, also increases K excretion. In this particular situation CSF electrolytes have not been measured, hence further conclusions as to the lack of effect of chronic cortisol administration on electrolytes are unwarranted.

Woodbury and associates have also examined the influence of other adrenocortical steroids and ACTH on brain electrolytes. The results may be summarized as follows: In rats, chronic administration of ACTH, cortisone, cortisol, corticosterone, dehydrocorticosterone, and 11-desoxy-17-hydroxycorticosterone has no effect on brain electrolytes in the doses used and for the period administered (28 days). However, they markedly affect brain excitability, and markedly increase brain Na and Cl spaces. The increase in both Na and Cl spaces suggests an effect of cortisol on the permeability of brain cells, such that both Na and Cl enter the cells without a net increase in intracellular Na concentration, or an effect on the glial cells. The increase in brain Cl space induced by cortisol in mice suggests that extracellular space is influenced by this steroid; and increased permeability may occur, that is, the ground substance and glial cells may become more fluid. On the other hand, recent studies using neural explants in organotypic culture have shown that cortisol enhances proliferation of glial cells (Vernadakis 1971). Since Cl space mostly reflects glial space (Vernadakis and Woodbury, 1965), the changes in extracellular space by cortisol may reflect the effects of this hormone on glial cells.

The glucocorticoids have been extensively used for the clinical treatment of cerebral edema and have been shown to be effective for the prevention and/or reduction of cerebral edema in several experimental animal models (Bakay, 1965; Johnson and Assam, 1966; Maxwell et al., 1971). In cases in which steroids have been found to reduce brain edema and in which electrolytes have been measured, the diminution of brain edema has been correlated with a reversal of the pathological water and electrolyte changes. Whether these agents exert their effects by decreasing abnormal permeability to ion movement in disturbed membranes seems reasonable but remains to be clarified.

Millichap (1969) has reviewed the interrelationship between electrolyte and neuroendocrine mechanisms with respect to its role in seizures disorders in humans. He concludes that systemic electrolyte disorders particularly hyponatremia, hypernatremia, hypocalcemia, and magnesium deficiency may be associated with an increased susceptibility to seizures, but abnormalities of electrolyte and acid-base metabolism show no constant correlation with the exacerbation or remission of seizures in patients with epilepsy. Moreover, the influence of hormones on seizures and brain excitability may be correlated with changes in electrolyte and water balance by a direct depressant effect on neuronal cell membrane and inhibitory or facilitatory cerebral mechanisms must also be postulated.

The thyroid gland has also been shown to affect brain electrolytes and brain excitability by a mechanism distinct from its general stimulatory action on general metabolism. As discussed elsewhere, both thyroidectomy and treatment with propylthiouracil decrease brain excitability and treatment with thyroxine and triiodothyronine increase it (see reviews by Woodbury). Triiodothyronine is more potent in this respect than thyroxine, although its action is faster and more transient. The effects of thyroxine are still evident after 24 hours. The effects of the thyroid appear to bear a direct relationship to brain Na concentrations.

Also, as discussed elsewhere in this chapter Woolley and Timiras (1962a,b) have shown that administration of estradiol has a stimulatory effect on brain excitability as opposed to the depressant action of progesterone. They also refer to a sex influence as shown by the greater depressant effect of progesterone in the female than in male rats. However, their findings indicate that estradiol affects brain excitability by mechanisms other than electrolyte changes since brain excitability was increased in spite of elevated plasma Na concentration and an increased extracellular/intracellular Na ratio.

Protein Metabolism

The effects of hormones on brain protein metabolism are reviewed more extensively in Chapter 7 since considerable more evidence is available in protein metabolism in the developing brain. It is generally agreed that the stimulant action of the thyroid hormones is not concerned with the amino acid-activating enzymes. Klee and Sokoloff (1964) conclude that thyroxine stimulates the rate of uptake by the ribosomes of bound amino acids attached to transfer RNA. Tata and Windnell (1964) report evidence that the point of action of the thyroid hormones is the DNA-independent RNA polymerase, since stimulation of this system by thyroid hormones can be shown to precede the increase incorporation of amino acids into protein: they conclude that the thyroid hormones act primarily on the genetically linked regulatory mechanisms of protein synthesis. It would therefore be expected that the central effects of drugs such as methylthiouracil, which reduce the level of thyroid hormones, are also due to their influence on portein metabolism.

The influence of ACTH and adrenocortical steroids on protein and amino acid metabolism of brain has not been adequately studied, although their effects on other tissues, particularly the liver, have received considerable attention (Tomkins and Maxwell, 1963). The evidence indicates that steroid hormones can influence the protein metabolism of nervous tissue in a number of different ways: it may act on the enzymes directly concerned in protein synthesis; it may act by affecting the transport or metabolism of amino acids, or it may act by influencing the energy metabolism of the cell. Nurberger (1953) has investigated the concentrations of nucleic acids and proteins in the cytoplasm, nucleus, and nucleolus of cells of the liver and of the supraoptic neurons in the hypothalamus of rats. Changes of these cellular constituents following fasting and exposure to cold stress were found and correlated with changes in adrenal ascorbic acid. Adrenalectomy approximately doubled the protein concentration of both types of cells. For example, liver cell protein increased from 26% to 51% and that of the supraoptic nucleus from 14% to 32%. Changes in RNA concentration were not significant following adrenalectomy. When intact rats were exposed to one hour of cold, cellular protein content was increased from 26% to 40% in liver and RNA fell from 1.8% to 1% with comparable changes occurring in the supraoptic nucleus cells. The most pronounced changes, however, were found in adrenalectomized rats exposed to cold where liver protein fell from 51% to 17% and supraoptic nucleus protein fell from 32% to 11%. These changes in protein concentrations were not accompanied by changes in RNA. Cell protein changes seem to be very sensitive to the level of adrenocortical function and it is important to note that the effect of the adrenal on brain protein metabolism may be similar to that on liver cells.

Neurotransmitter Metabolism

Intercommunication between neurons involves chemical substances, neurotransmitters. An excellent review on neurotransmitter systems in the central nervous system is that by Weiner (1974). For a substance to qualify as a respectable candidate neurotransmitter several criteria should be satisfied: (1) The substance should be synthesized and stored in specific neurons. (2) It should not be uniformly distributed throughout the central nervous system, but should be present in certain specific neuronal pathways. (3) The enzymes involved in the synthesis of the transmitter should be present in the neurons. (4) Neurons operate intermittently and may be required to fire very intensively for a variable duration or they may remain quiescent for different intervals. Neurons which fire very frequently and intensively must have the potential for restoring the neurotransmitter substance which they release and which, at least to some extent, is metabolized or degraded upon release. Thus, the synthesis of neurotransmitter should be regulated and should be sensitive to neural activity. (5) The substance must be released on nerve stimulation. (6) When the substance is applied to the brain tissue, it must act upon specific postsynaptic receptor sites to produce the appropriate neurophysiological effect. (7) The action of the applied putative neurotransmitter, as is true for effects of nerve stimulation, should be rapidly terminated in some manner. There are at least a half dozen substances in the brain which are respectable candidates as neurotransmitters; including acetylcholine (ACh), norepinephrine (NE), dopamine, 5-hydroxytryptamine (5HT, serotonin), gamma-aminobutyric acid (GABA) and glycine (see review Weiner, 1974). However, the influence of hormones on neurotransmitter metabolism is only beginning to be investigated and the data are very sporadic.

Uptake

The idea that catecholamines might be taken up into tissue binding sites was suggested by Burn (1932), but it is only in the past decade that the properties and the physiological and pharmacological significance of such tissue uptake processes have been appreciated. The studies of Whitby et al. (1961) demonstrated that the uptake and binding of circulating NE in tissues represented an important mechanism for the inactivation of this substance. Subsequent studies have defined the properties of the uptake processes involved and the importance of these mechanisms, not only for terminating the actions of NE at adrenergic synaptic junction, but also in explaining the mechanism of action of many adrenergic drugs.

Several different uptake processes exist for the catecholamines in animal tissues. Norepinephrine and related amines are known to be transported from the extracellular space across the axonal membranes of adrenergic nerves, a process which is termed "uptake"; a further mechanism exists to promote the transfer of free catecholamines from the axoplasm into the membrane-bound storage vesicle in adrenergic nerves; finally, catecholamines are transported across the membranes of smooth muscle and various other postsynaptic cells by another process known as "uptake." The properties of the catecholamine uptake processes have been reviewed extensively (Iversen 1971, 1973).

Early studies in the peripheral nervous system have shown that cortisol potentiates the vascular responses of catecholamines in vivo (Besse and Bass, 1966) and in vitro (Besse and Bass, 1966; Hapke and Green, 1970; Kalsner, 1969a,b). Besse and Bass (1966) suggested that cortisol increases the affinity of adrenergic receptors for catecholamines due to alternation in the relationship between the receptor and the p-hydroxyl group for NE. However, Hapke and Green (1970) found that cortisol had no effect on the affinity of NE for its receptor in isolated aortic strips from rabbits. Kalsner (1969a,b) has demonstrated that cortisol (and other steroid hormones) potentiate contractile responses to sympathomimetic amines and reduce the rate of relaxation of rabbit aorta contracted with epinephrine.

More recently, Williams and Hudgins (1973) reported the effects of cortisol (2.5×10^{-8} M) on accumulation and metabolism of ^{14}C-NE by isolated rabbit aorta over a two-hour period. Cortisol significantly reduced uptake of ^{14}C-NE after 40 min. In addition O-methylated ^{14}C-NE metabolites were significantly reduced. Since monamine oxidase (MAO) and catecholamine-O-methyltransferase (COMT) are located intramuscularly in the microsomal fraction of rabbit aorta (Verity et al., 1972), in order for enzymatic metabolism to occur, it is assumed that ^{14}C-NE must penetrate into the tissue. It may be possible, therefore, to explain the actions of steroid hormones on the basis of their membrane effects. Steroids may induce some conformational change in membranes which reduces their permeability to NE. Luch (1968) has presented evidence for steroid-induced alterations in membrane structure and permeability.

Iversen and Salt (1970) reported that steroids inhibit the uptake$_2$ mechanism in the rat heart with potencies comparable with the most active inhibitors previously described (Iversen, 1971). The concentrations of free corticosterone in the plasma of the rat are normally at least ten times lower than the inhibitor concentration value of steroids for uptake$_2$. In certain conditions, however, plasma concentrations may rise towards levels at which an effective block of uptake$_2$ could occur, for instance during stress. In such circumstances the physiological effects of catecholamines might be elevated.

Using neural tissue culture of the experimental model we have investigated the effects of cortisol on the uptake of NE in CNS tissue (Vernidakis, 1974). We found that the accumulation of ^3H-NE at 10^{-6} M was lower in explants cultured in the presence of cortisol, 2.76×10^{-5} M. This inhibitory effect of cortisol on NE uptake was observed at a low concentration of ^3H-NE, 10^{-6} M. We interpret these findings to mean that cortisol inhibits the extraneuronal uptake of NE in neural tissue as it does in the periphery (Iversen and Salt, 1970), and we propose that this uptake · represents glial cells. That glial cells accumulate NE was first shown by Henn and Hamberger (1971) using glial-enriched brain fractions. Similar findings we have found using neural cultures of glial cells (Vernadakis and Culver, 1976; Vernadakis et al., 1978). The proposed role of glial cells in neurotransmission mechanisms is discussed in more detail in Chapter 7. It has been proposed that a possible role of glial cells in neurotransmission is to provide a safety valve, i.e., to limit the possible build up of neurotransmitter substances extracellularly. Thus, inhibition of uptake of neurotransmitter substances in glial cells by hormones could lead to an intracellular-extracellular imbalance and may result in deleterious cellular effects. For example, excessive amounts of NE in the synaptic cleft would make more NE available to stimulate the CNS and would result in CNS hyperexcitability known to occur with cortisol treatment as discussed later in this chapter.

The evidence on the role of glucocorticoids on the metabolism of brain 5-hydroxytryptamine (5-HT) is controversial. Early studies have shown that adrenalectomy depresses the rate of formation of 5-^{3}HT from intracisternally injected ^{3}H-tryptophan (Azmitia et al., 1970). Tryptophan dydroxylase (TPH) activity in brain has also been reported to be reduced by adrenalectomy and to be restored by corticosterone (Azmitia and McEwen, 1969). However, Curzon and Green (1971) reported a decrease in absolute levels of brain 5-HT in rats following hydrocortisone treatment. In addition, several laboratories failed to confirm the effects of corticosterone and adrenalectomy on brain TPH activity (Lovenberg et al., 1973; Renson, 1973). More recently Neckers and Sze (1975) reported that intraperitoneal cortisol treatment in mice accelerates the accumulation of 5-HT in whole brain after inhibition of MAO activity by pargyline without affecting tryptophan hydroxylase activity. These authors further found that in in vitro synaptosomal preparations, cortisol at 10^{-5}-10^{-7} M or corticosterone at 10^{-5} M stimulates the uptake of L-^{3}H-trytophan by the synaptosomes while androgenic and progesterone-like steroids are ineffective. These findings support the view that glucocosteroids may act on presynaptic sites.

Several studies show that sex steroids affect brain catecholamine metabolism. Janowsky and Davis (1970) have reported that estradiol and progesterone exhibit significant inhibitory effects on uptake of ^{3}H-NE by rat brain synaptosomes. In a later study this same group (Nixon et al. 1974) found that uptake of ^{3}H-NE and ^{3}H-metaraminol by synaptosomes was decreased by progesterone and estradiol and that testosterone inhibited 5-^{3}HT uptake. The steroids tested did not affect ^{3}H-dopamine uptake. These authors further suggest that these steroids diminish the synaptosomal accumulation of monamines by affecting the membrane pump directly, since they inhibit the net uptake of a non-metabolized amine, metaraminol, and fail to appreciably alter monoamine metabolism. Wirz-Justice et al. (1974) studied uptake of monoamines in brain slices of rats treated with various hormones in vivo for two hours. They found that in estradiol-treated animals uptake of dopamine was decreased in the thalamus but in no other brain region examined; dopamine uptake was decreased in cortex of animals treated with estradiol plus progesterone; uptake of 5-HT was increased in the preoptic and septal region of animals.

Synthesis, Turnover, and Enzyme Levels

As discussed also in other chapters in this book some of the physiological manifestations of altered thyroid states may be mediated through peripheral and central adrenergic systems. For example, thyroxine treatment produces a state resembling hyperthyroidism; this state is characterized by hyperthermia, tachycardia, sweating, and elevated blood pressure. These symptoms are all manifestations of increased adrenergic activity. Conversely hypothyroid animals generally show signs of decreased peripheral adrenergic function. In the heart and other adrenergically innervated peripheral tissues, turnover of NE is decreased in thyroxine-treated animals, while an accelerated turnover is associated with hypothyroid state (see references in Emlen et al., 1972). A similar relationship between thyroxine and catecholamines has been proposed to exist in the central nervous system. Similarities between the symptomatology of thyroid deficiency and certain psychiatric illnesses particularly endogenous depressions are well-known and is discussed elsewhere in this book (see Chapter 18).

Thyroid state influences the synthesis and metabolism of NE in the CNS. The rate of conversion of isotopically labeled tyrosine to NE is decreased in the brains of hyperthyroid rats and is increased in hypothyroid rats (Lipton et al., 1968; Prange et al., 1969, 1970a,b, 1975). To further explore the possibility that thyroid hormone may exert some of its action by facilitating catecholamine action Emlen et al. (1972) studied the effects of thyroid state on the specific activity of tyrosine hydroxylase (TH) from the midbrain in rats correlated to spontaneous motor activity. They found that in hypothyroid rats, spontaneous motor activity was less than that in matched normal controls, and the specific activity of TH in the midbrain was significantly greater than in controls. Rats made hyperthyroid, by treating them with thyroxine became hyperactive and showed increased sensitivity to the behaviorally activating effects of NE administered intraventricularly. In hyperthyroid rats the specific activity of TH in the midbrain remained within the normal range. The increase in TH activity may be responsible for the increase in catecholamine synthesis reported by Prange and associates. However, since TH was not changed in the hyperthyroid rats, the decreased in NE turnover reported by Prange and associates may result from factors other than the activity of the rate-limiting enzyme for NE biosynthesis. One possible explanation for these disparities is that changes in amount of TH are secondary to the effect of thyroid state on sensitivity of adrenergic receptors. Emlen et al. (1972) found that motor activity of thyroxine-treated rats was significantly increased during intraventricular infusion of NE. This increased response to NE seen in thyroxine-treated rats could be due to an increase in the amount or accessibility of NE available for release, or to an increase in the sensitivity of adrenergic receptors.

Changes in catecholamine enzyme activity have been also observed with steroid hormones. As also discussed in Chapter 7, Sze and associates (Diez et al., 1974) found that a postnatal developmental increase in whole brain TH activity was temporally correlated with that of plasma and brain corticosterone levels in mice. Whole brain TH increased 34% in nine-day-old mice which had received 5 injections of hydrocortisone acetate (20 mg/kg). In contrast to the effects of hydrocortisone in the brain, dexamethasone or ACTH markedly stimulate the activity of phenylethanolamine N-methyl transferase but decrease TH activity in the superior cervical ganglion in newborn rats (one up to eight days after birth) (Ciaranello and Axelrod, 1975). It is suggested that dexamethasone administration to newborn rats before 5 days of age impairs development of the preganglionic neuron. This is based on the findings that choline acetyltransferase activity is also depressed in the superior cervical ganglion. It has been repeatedly demonstrated that surgical or pharmacological manipulations which impair the development of the preganglionic nerve also retard the normal maturation of TH in ganglion cells (Black et al., 1971; Thoenen, 1972). As discussed also in Chapter 7, the foregoing findings emphasize that the effects of hormones on neurotransmitter processes are age dependent.

It has been reported that $ACTH_{4-10}$ causes an increase in the turnover rate of NE in the brain of intact rats (Leonard, 1974; Versteeg and Wurtman, 1975) as calculated from the rate of decline of NE following inhibition of its synthesis. It appears that the effect of ACTH $4-10$ on catecholamine metabolism in the brain is indirect. Since neither in adreneloctomized rats, in corticosterone-treated adrenalectomized rats nor hypophysectomized rats ACTH $4-10$ affected the accumulation of brain ^3H-NE or ^3H-dopamine, it appears that the effects of ACTH on catecholamine in the brain is indirect.

Luine et al. (1975) have examined the effects of gonadal steroids on some neural enzymes. They examined specifically changes in MAO, and choline acetyltransferase (ChA). Several laboratories have shown that levels of MAO in the hypothalamus and amygdala fluctuate during the estrus cycle when levels of ovarian and pituitary hormones are changing (see references in Luine et al., 1975). In this study gonadectomized male and female rats were treated with estradiol benzoate or testosterone propionate daily for three or seven days. Changes in enzyme activity were found in those brain regions where gonadal hormones are known to affect sexual behavior and for gonadotropin release and which contain putative hormone receptor sites, namely, corticomedial-amygdala, basomedial-hypothalamus, and medial preoptic area. More specifically, estradiol administration to females resulted in decreased activity of MAO in the corticomedial amygdala and basomedial hypothalamus and an elevation in ChA activity in the medial preoptic and corticomedial amygdala. While testosterone did not alter enzyme activity in any brain region. In contrast, estradiol administration to castrated males was without significant effect on enzyme activities while testosterone administration resulted in increased activity of MAO and ChA in the medial-preoptic area. Gonadal hormone-dependent changes in the activity of MAO may be important in regulation of behavioral and neuroendocrine aspects of the gonadol-hypothalamic-pituitary axis. As discussed earlier in this chapter adrenergic synapses within the preoptic-hypothalamic-amygdaloid areas may be related to the control of gonadotropin and prolactin secretion.

The involvement of the cholinergic system in neuroendocrine events has received less study than the adrenergic system. Libertun et al. (1973) found that ChA activity was higher in the female than in the male preoptic-suprachiasmatic area and that this sex difference was dependent on the early hormonal environment. No differences were found between males and females in the arcuate-mammillary area of the hypothalamus or in cerebral cortex. The effects of estradiol on ChA activity of the amygdala may be of special importance since this brain structure is known to have putative estradiol receptors (Zigmond and McEwen, 1970; Pfaff and Keiner 1973) and has been implicated in the regulation of gonadotropin secretion. McEwen and associates (review, McEwen et al., 1972b) have also reported changes in ChA activity in the hippocampus is much lower than in the hypothalamus. These authors suggest that the increased ChA activity in the hippocampus after estradiol may be a result of the action of this hormone in the cell bodies in the medial septal area where estrogen-labeled cells are found and where enzyme proteins would be synthesized and then transported along the neuron into the hippocampus.

Bernard and Paolino (1974) studied changes in brain biogenic amine dynamics following castration in 40-day-old male rats. Significant increases were observed in whole-brain (minus hypothalamus) 5-hydroxyindole acetate acid (5-HIAA) levels and hypothalamic dopamine levels, fractional rate constants and utilization rates at three but not a six weeks following castration. Whole brain NE turnover rates in castrated animals did not differ significantly from those of sham-castrate control animals. However, a tendency toward increased hypothalamic NE turnover rates was seen in the castrated animals. Inferences about the possible functional significance of these biochemical changes resulting after castration were made by comparing these data with the behavioral effects observed by these authors (Bernard and Paolino, 1973) using the identical experimental design. They found that gonadectomy resulted in increased open field activity at three weeks but not six weeks after the operation. However, more detailed studies are needed before definite correlations can be made between hypothalamic biogenic amine changes and open field behavior.

To further explore the interrelation between biogenic amines and behavior Bernard (1976) investigated the effects of testosterone on ranacide (frog killing) aggression and brain monoamine metabolism in the adult female rat. Aggressors were defined as animals which attacked or killed during this a 30-min testing session, while nonaggressors failed to do so. Using either aggressors or nonaggressors, testosterone and sesame oil (the vehicle) equally increased aggressive behavior. Testosterone treated animals had significantly higher brain NE and NE/5-HT levels than sesame treated animals. Aggressors treated with either testosterone or sesame had higher NE/5-HT ratios than nonaggressors. It is concluded from these findings that ranacide behivor in the adult female is not androgen dependent nor is this behavior functionally related to the observed differences in biogenic amine metabolism. These findings support the hypothesis of these authors that alterations in testosterone levels could not be employed to alter ranacide behavior in the adult male rat (Bernard, 1974).

Great interest has been focusing on the physiological role of melanocyte stimulating hormone (MSH) in mammals besides its role in pigmentation. As described elsewhere in this book (Chapter 18), it has been suggested that MSH produces anxiety, motor restlessness, and alterations in EEG in humans, and exacerbation of symptomatology in patients suffering from Parkinsonism (see references in Friedman et al., 1973). Friedman et al. (1973) have studied the effects of an inhibitory factor of MSH release (MIF) 1-prolyl-1-leucylgylcinamide, on brain dopamine metabolism. It was found that MIF exerts a stimulatory influence on striatal dopamine synthesis while not affecting hypothalamic NE synthesis in the rat. The influence of MIF on dopamine synthesis appears to be mediated by the pituitary because prior hypophysectomy prevents this MIF effect. However, it is not possible to unequivocally ascribe this action of MIS solely to its influence on MSH release because other pituitary hormones whose influence is eliminated by hypophysectomy may also interact with brain dopamine and so mask the MIF effect.

Hormones and Cyclic AMP Interactions

Adenosine 3',5'-monophosphate (Cyclic 3',5'-AMP) has now been established as an intracellular second messenger mediating many of the actions of a variety of different hormones. Several reviews dealing with selected aspects of this subject are available (see monograph Ribison et al., 1968). In view of the tremendous progress in this area it is not possible to review here all recent findings. Only a brief account of the history of cyclic AMP will be presented and the role of cyclic AMP in hormonal actions in relation to CNS function will be briefly reviewed.

Cyclic 3',5'-AMP was discovered in the course of investigations into the mechanism of the hyperglycemic action of epinephrine and glucagon by E.W. Sutherland and T.W. Rall (1960). The level of cyclic AMP at any given instant depends upon the activities of at least two enzymes. Adenyl cyclase catalyzes the formation of cyclic AMP from adenosine triphosphate (ATP) in a reaction which requires magnesium. Cyclic 3',5'-AMP is inactivated by a specific phosphodiesterase which catalyzes its hydrolysis to 5'-AMP. It is now known that many of the actions of a number of hormones in addition to epinephrine and glucagon are mediated by cyclic 3',5'-AMP (see review Robison et al., 1968).

The concept has been developed that many hormones act by way of a two-messenger system. A schematic representation of the two-messenger system is shown in Fig. 4 (from Sutherland et al., 1965). Following the release of the hormone from the endocrine gland the hormone interacts with the adenyl cyclase systems of the effector cell membrane producing an increased rate of synthesis of cyclic 3',5'-AMP from ATP. Detailed studies for the establishment of cyclic 3',5'-AMP as second messenger are elegantly described by Haynes et al. (1960), Sutherland et al. (1965) and Robison et al. (1968), and will not be reviewed here.

Rall and colleagues have demonstrated that several endogenous chemical agents such as catecholamines, histamine, 5-hydroxytryptamine, adenosine and adenine nucleotides have significant effects of the intracellular levels of 3',5' cyclic AMP of the mixed population of cells of a brain slice (Kakiuchi and Rall, 1968a,b; Kaikuchi et al., 1969; Sattin and Rall, 1970). It has been clearly documented from these studies that there are species differences, brain regional differences, potentiative interactions between chemical agents, anomalous effects of cyclic neucleotide phosphodiesterase inhibitors, loss of neurohormonal responses, that enter in the regulation of cyclic 3',5' cyclic AMP levels.

The experimental model which has provided a large volume of information on the possible functions of cyclic AMP and its intracellular regulation is the neural cultures. Several culture systems have been used and include several clonal lines of neuronal and glial tumor cells and reaggregated

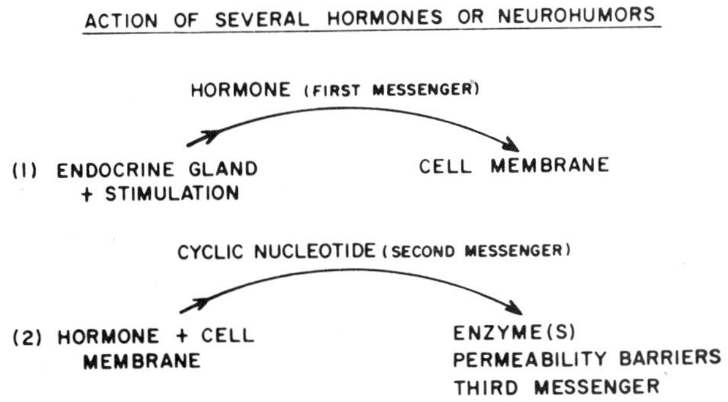

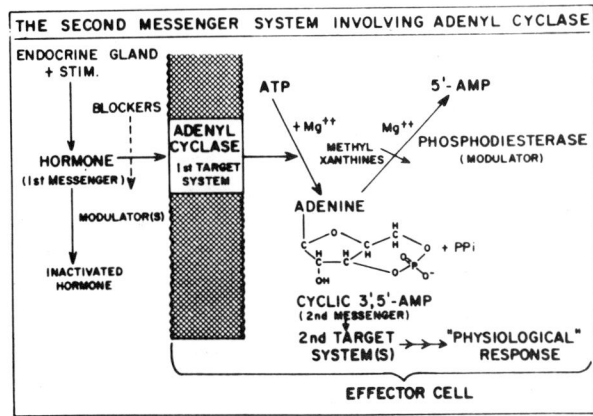

Fig. 1. The two-messenger system.

Fig. 4. Upper portion: The two messenger system is shown. Upon stimulation of the endocrine gland a first messenger is released, interact at the cell membrane; a second messenger may be formed which functions within the cell to modify enzyme activity and permeability barriers; a third messenger may be formed.

Lower portion: The two-messenger system involving adenyl cyclase. (From Sutherland et al., 1965).

brain cell cultures derived from embryonic mouse brain. The early findings
using these systems are well reviewed by Gilman (1972). The evidence
derived using clonal lines of glial tumors strongly demonstrates that glial
cells specifically respond to catecholamines to increase intracellular levels
of cyclic AMP and this effect is a result of interaction with β-receptor.
The evidence that the β-receptor is a regulatory component of the membrane
bound adenylate cyclase system has been reviewed by Robison et al.
(1971). More recently, Perkins (review 1975) has reviewed the work from
his laboratory and others which demonstrates that under certain circumstances hormonal responsiveness may be regulated as a result of the
variable expression of one or more of the macromolecular components of
the second messenger system. Changes in the responsiveness of cells
to hormones or neurotransmitters has been detected: (1) in steroid and
thyroid hormone-deficient animals, (2) during ontogenic development (3)
as a result of malignant transformation (4) during the cell cycle of cells
in culture (5) during the growth cycle of cells in culture, and (6) as a
result of overexposure or underexposure of cells or tissues to hormones
or neurotransmitters. Studies suggest (see review by Perkins, 1975)
that cells that receive adrenergic innervation regulate the responsiveness
of their norepinephrine-sensitive adenylate cyclase in a compensatory
manner with respect to their recent history of exposure to norepinephrine;
that is to say, excess adrenergic nerve activity results in a decrease in
the responsiveness of adenylate cyclase to norepinephrine while prolonged
periods of limited nerve activity result in an increase of such responsiveness.

Steroid hormones modulate many cell functions which are mediated
primarily by cyclic 3',5'-AMP. For example, glucocorticoids modulate the
glycogenolytic response of liver to epinephrine by controlling the sensitivity
of the system to cyclic AMP (Exton et al., 1972); they increase the concentration of cyclic AMP generated in response to enpinephrine in hepatoma
cells (Manganiello and Vaughan, 1972) and lymphocytes (Parker et al.,
1973). Adenylate cyclase activity is increased by progesterone action on
the oviduct (Kissel et al., 1970) and by testosterone action on seminal
vesicles (Thomas and Singhal, 1973), but estradiol abolishes norepinephrine
activation of adenylate cyclase of the pineal gland (Weiss and Crayton,
1970).

Evidence exists for steroid hormone (Sawyer and Gorski, 1971) and
cyclic nucleotide (Brostrom et al., 1974) participation in specialized functions
of the CNS, as well as in basic controls over cell growth and differentiation
(Abell and Monahan, 1973). However, understanding of interactions
between the two classes of compounds of the cellular level in brain is extremely limited. DeVellis and associates using the C-6 glioma cell line
have extensively explored the regulation of glycerol phosphate dehydrogenase (GPDH) by glucocorticoids and the role of cyclic nucleotides (see
reviews deVellis et al, 1971; deVellis and Brooker, 1973; deVellis et al.,
1977). They found that glucorticoid treatment of the cells results in
doubling of GPDP activity apparently through a mechanism independent
of cyclic AMP (see also Chapter 7). Brostrom et al. (1974) have further
explored the possibility of a hormone cyclic nucleotide interaction in C-6
glial cells that might pertain to brain function. They found that intact
C-6 glial cells treated with corticosterone elevated basal concentrations of
cyclic AMP and augmented cyclic AMP response to NE. The authors suggest
that the effect of the hormone is on adenylate cyclase, since the steroid
incubation time required to increase adenylate cyclase activity coincided
with that for increased cyclic AMP response of intact cells to NE. Only
glucocorticoids were effective in both intact and broken cell, and the
half-maximal effective concentration of corticosterone was the same for

augmenting the cyclic AMP response of intact cells and for increasing adenylate cyclase. The time course of the cyclic AMP response to NE for glucocorticoid-treated cultures did not differ from that of control cultures. In addition, cyclic nucleotide phosphodiesterase activity of cell lysates measured at several concentration of cyclic AMP was unchanged in glucocorticoid-treated cells. Kinetic studies also supported the hypothesis that more adenylate cyclase is synthesized in the presence of glucocorticoids. The apparent K_m for ATP was unaltered after treatment with corticosterone, but the V_{max} was doubled. The possibility exists that most adenylate cyclase synthesis in C-6 cells is under the control of the glucocorticoids and that the presence of these substances in serum in the medium is responsible for activity in the absence of added steroid.

In the brain glial cells are in intimate contact with norepinephrine-containing axon terminals and in some locations exhibit specialized contacts with them (Guldner, 1973). Astrocytes are abundantly represented in those brain regions which selectively accumulate corticosterone (McEwen et al., 1971), so that it seems certain that these cells are exposed in situ to both catecholamines and glucocorticoids. The foregoing discussion supports the view proposed by us and others that glial cells modulate the neuronal microenvironment by their response to steroid hormones and neurohumor substances (Vernadakis et al., 1979) (see also Chapter 7).

EFFECTS OF HORMONES ON BRAIN ACTIVITY

Brain Excitability

The pharmacology and biochemistry of adrenal steroids as they relate to the central nervous system (CNS) have been extensively reviewed by Withrow and Woodbury (1972), Woodbury (1958, 1972), and Vernikos-Danellis (1972).

Early studies by Selye (1941) have demonstrated that several steroid hormones including deoxycorticosterone (DOC) progesterone and other adrenal steroids have anesthetic effects in a number of animals. Large doses of DOC and progesterone produced deep anesthesia in rats undergoing prolonged abdominal operations. In man, progesterone has been reported to have soporofic effect in minimal anesthetic doses (Merryman et al., 1954).

Woodbury and associates in early studies showed that adrenocortical steroids influence brain excitability as assessed by responses to electroshock stimulation (Woodbury, 1958). The electroshock seizure threshold (EST) has been used as a measure of brain excitability. EST may be defined as the amount of alternating current in milliamperes delivered for 0.2 s through corneal electrodes which will just elicit a barely detectable minimal convulsion characterized by facial clonus and rhythmic movements of the vibrissae, jaws and ears and lasting for several seconds. The technique for measuring EST in rats has been described by L.A. Woodbury and Davenport (1952). The properties of electroshock seizures produced in animals have been described extensively (reviews Woodbury, 1958; Woodbury and Esplin, 1959). Toman et al. (1948) state, "...that some properties of neuronal interaction in seizures elicited in animals are qualitatively similar to the properties of individual neurons in isolation, although the time-seal differs by several orders of magnitude at these two different levels of organization." By the technique of determining

EST it is possible to measure at the organismic level the effects of various factors on the excitability of the brain. Using this technique, Woodbury and associates (review 1958) showed that adrenadectomy in rats markedly lowers EST and thus increases brain excitability. Adrenocortical steroids given chronically to rats for several days (28 days) produced quantitatively different effects on brain excitability: deoxycorticosterone acetate (DCA) elevates the EST in an expotential fashion; 11-desoxy-17-hydroxycorticosterone acetate (Substance S acetate), reaches a maximum elevation of about 6% in nine days and stays at this level for the duration of the 28-day experiment; corticosterone acetate (Compound B acetate) produces a transient decrease during the first two days then EST returns to the control level; 11-deoxycorticosterone acetate (Compound A acetate) decreases EST progressively up to 16 days of treatment; 17-hydroxycorticosterone acetate (hydrocortisone, Compound F acetate) steadily lowers the EST throughout the 28-day experimental period; the effect of 11-dehydro-17-hydroxycorticosterone acetate (cortisone acetate, Compound E acetate) on EST is more pronounced than that of hydrocortisone acetate. ACTH produces a gradual increase in EST. Moreover, ACTH given simultaneously with DCA over a 28-day period completely prevents the increased EST induced by DCA alone. In addition, ACTH and Compound A antagonize the EST-lowering action of cortisone. It has been suggested (Woodbury, 1954) that some product(s) of the adrenal cortex, released as a result of ACTH stimulation and similar to Compound A, regulate(s) CNS excitability in such a way as to restore an abnormally elevated (DCA) EST as well as a lowered (cortisone) EST to normal.

As discussed earlier the changes on EST by these steroids have been associated with changes in electrolyte concentrations in plasma, brain and muscle (reviews, Withrow and Woodbury, 1972; Woodbury, 1954). Of the adrenocortical steroids tested, only in the case of DCA can changes in electrolyte metabolism be correlated with changes in EST. The other steroids must then act directly on the brain in some as yet undisclosed manner to produce their striking effect on brain excitability.

The effects of thyroxine on the electroshock seizure threshold of adult rats have been studied by Woodbury (1954) and Timiras et al. (1955). Thyroxine markedly decreases EST when given in a dose of 100 μg per day. Moreover thyrodectomized animals treated with half this dose of thyroxine showed an equal decrease in EST.

The role of sex hormones on brain activity has been extensively explored by Timiras and associates. Early studies by Woolley and Timiras (1962a,b) showed that the EST of the female rat is lower than that of the male (Woolley and Timiras, 1962a) and fluctuates with phases of the estrous cycle, reaching its lower level at the time of ovulation. In the adult rat, testosterone is capable of both convulsant and anticonvulsant action depending on the age of the animal, the dose used, and the type of seizure pattern under study (Woolley and Timiras, 1962a,b). When administered to newborn rats testosterone does not affect seizure responses. Yet lack of physiological levels of the hormone, as in rats that have been castrated neonatally, results in delayed appearance of the maximal seizure pattern (see review Timiras, 1969). Thus, it has been suggested that certain hormones may influence CNS development by a "permissive" rather than a regulatory action.

Studies in awake, unrestrained rats chronically implanted with electrodes have demonstrated that hormones are also capable of influencing localized seizure activity. For example, changes in seizure threshold were observed in discrete strucutres of the limbic system during the estrous cycle, after ovariectomy, and following administration of estrogen or progesteron

(Terasawa and Timiras, 1968a). Seizure threshold in the dorsal hippocampus and the medial part of the amygdala was significantly decreased during proestrus and estrus, whereas that of the lateral part of amygdala was significantly increased during the same period and decreased during diestrus (Fig. 5). Ovariectomy of young adult rats eliminated the cyclic aspects of activity in the limbic system, but a dampened cyclicity was

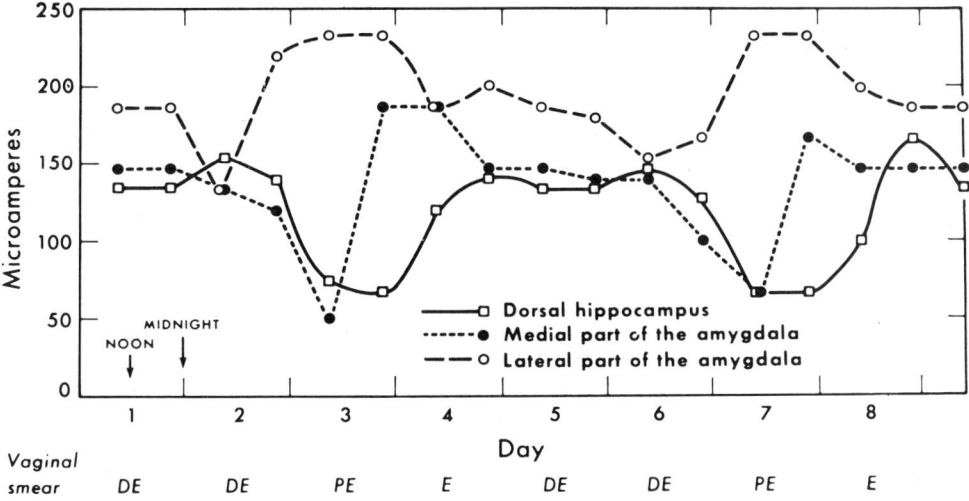

Fig. 5. Comparison of localized seizure threshold curves a three structures of limbic system during 2 estrous cycles. Data on hippocampus and lateral part of amygdala from one rat; data for medial part of amygdala from another rat. DE-diestrous day; E-estrous day; PE-proestrous day. (From Timiras, 1969).

restored by a single injection of estradiol (Fig. 6). The same dose of estrogen was somewhat less effective in older ovariectomized rats and failed to establish cyclicity in rats that had been ovariectomized at birth (Fig. 7). Progesterone in relatively high doses affected seizure threshold only slightly and, generally, its effects were opposite to those of estradiol. These findings indicate that estrogen plays a role not only in the regulation of seizure activity of the hippocampus and the amygdala in the adult animal, but also on its development.

Based on the observations cited above Timiras and associates further investigated the development of seizure activity in specific CNS areas from birth to sexual maturity in the rat (Terasawa and Timiras, 1968b). Specifically they studied the development of seizure activity in the dorsal hippocampus and the medial part of the amygdala in intact female rats, rats ovariectomized at birth, and rats in which precocious puberty was induced by injection of gonadotropin. Seizure activity significantly changes with age in both structures; in the amygdala, seizure threshold dropped sharply around 30 days of age, a few days before the onset of puberty (as evidenced by vaginal and ovarian signs), whereas no such changes occurred in the ovariectomized animals. When puberty was accelerated by administration of gonadotropin, amygdaloid threshold

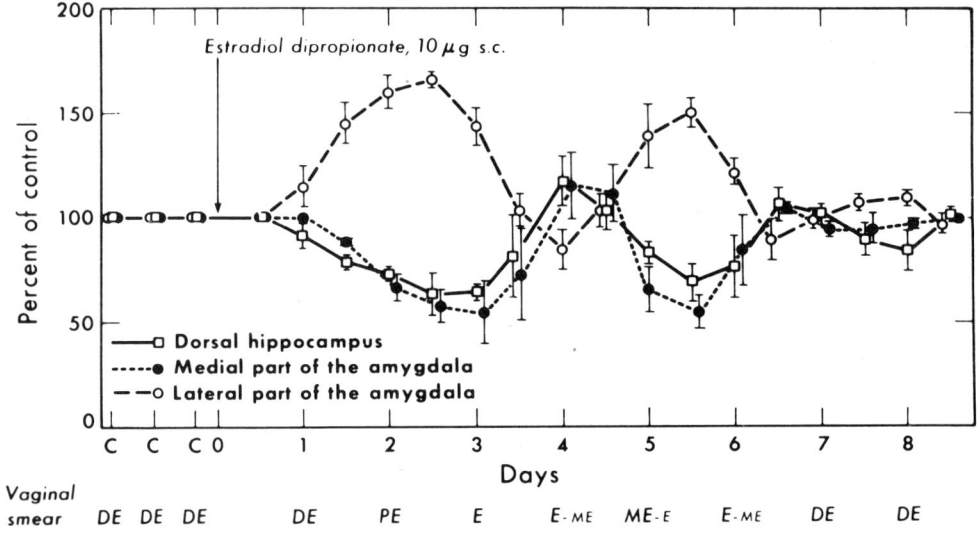

Fig. 6. Effect of 10 μg of estradiol dipropionate on localized seizure thresholds in limbic areas of ovariectomized rats. (Values are espressed with mean and standard errors as a percentage of control values.) E-me: Nearly all the animals show cornified vaginal smears, but some animals show metestrous smears. ME-e represents the reverse situation. (In time legend, C indicates control. Other abbreviations are the same as in Fig. 5). (From Timiras, 1969).

similarly dropped preceding the onset of sexual function (Fig. 7).

From the foregoing findings several important physiological conclusions can be drawn: The development of seizure activity in the limbic system and especially in the amygdala is influenced by gonadal hormones. Indeed, it may be postulated that estrogens have an organizing action in the amygdala comparable to that of androgens in the hypothalamus; that is, these hormones are indispensable to normal functional differentiation at a specific neonatal age. It can also be inferred from these studies that maturation of specific CNS structures and their changing sensitivity to gonadal hormones may be key factors in the onset of puberty.

Spontaneous Electrical Activity

Adrenal Steroids

The direct effects of cortisol on the electroencephalogram (EEG) on evoked potentials and on single cells have been extensively studied by Feldman and associates (see references Feldman, 1971). They have shown that intraventricular injection of cortisol into various subcortical regions in cats enhances electrical activity ranging from the appearance of localized spikes in the hippocampus and lateral hypothalamus to generalized convulsive activity. The opposite effect on brain electrical activity is seen in the hypoadrenal state. This is characterized by a generalized slowing in electrical activity in the cortex as well as in subcortical structures (Bergen, 1951). The conduction time of somatosensory impulses and neuronal recovery are prolonged in the polysynaptic pathways of the

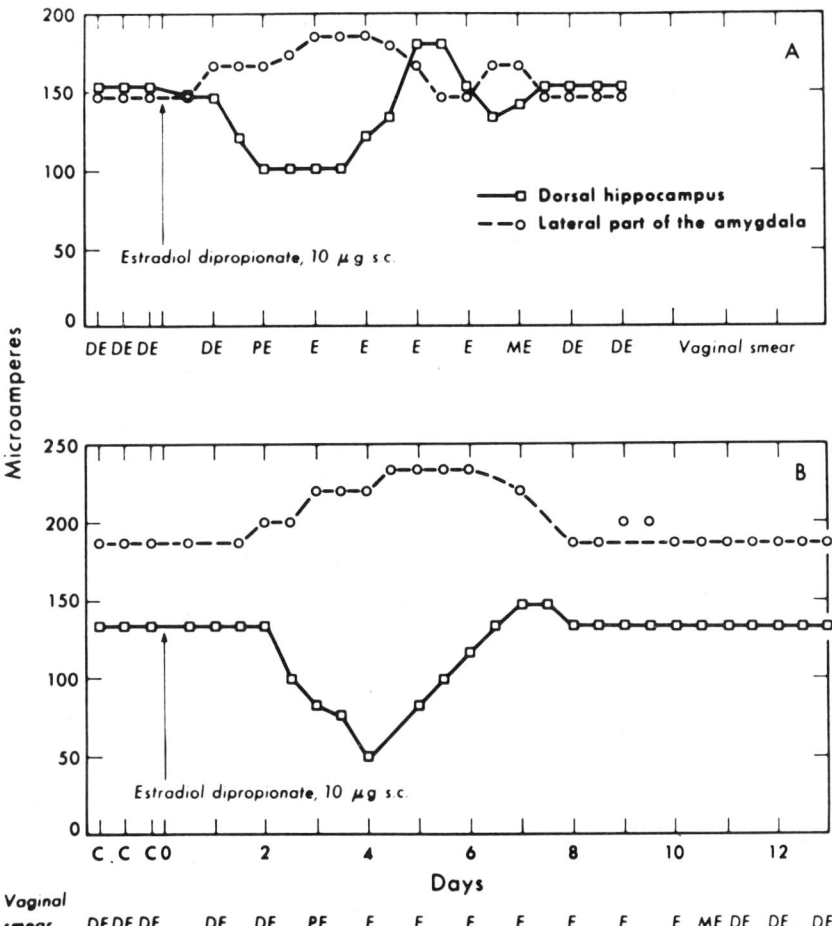

Fig. 7. (A) Effect of 10 µg of estradiol dipropionate on localized seizure thresholds in rats of advanced age ovariectomized in adulthood. (B) Effect of 10 µg of estradiol dipropionate on localized seizure thresholds in adult rats ovariectomized at birth. (From Timiras, 1969).

midbrain reticular formation and the anterior hypothalamus, but not in the medial lemniscus (Feldman and Robinson, 1968). The involvement of polysynaptic pathways in hypoadrenalism is also indicated by the experiments of Cook et al. (1960) and Chambers et al. (1963) studying blockage of the EEG arousal response and the conduction from the midbrain to the cortex in adrenalectomized cats. The electrical activity of the brain of rats implanted chronically with cortical and subcortical electrodes decrease in wave frequency with time after adrenalectomy. The slowing in wave frequency appears to correspond to the changes in circulating ACTH and the increase coricotropin-releasing factor content of the median eminence following adrenalectomy (Vernikos-Danellis, 1965).

Gonadal Steroids

Perhaps some of the most significant findings in this field are those of Kawakami and Sawyer (1959) who showed that the behavioral "after-reaction" akin to paradoxical sleep, which follows coitus in the female rabbit is associated with changes in electrical activity which can also be elicited by giving LH but not FSH. Since the stimulus of coitus in the female rabbit operating through a neurohumoral pathway originating in the hypothalamus is known to induce the release of LH by the pituitary, these findings indicate that LH, in addition to causing ovulation, reacts back upon the CNS to cause behavioral and EEG changes which block further release of "ovulating hormones." Further work along similar lines (Kawakami and Sawyer, 1959) has shown that progesterone markedly elevates the threshold for the EEG manifestations of the after-reaction, is associated with a reluctance to accept the male and causes inhibition of the gonadotropic activity of the pituitary.

Progesterone was found not only to suppress the activity level of the posterior hypothalamus, but the limbic midbrain area as well (Kobayashi et al., 1962). During the estrous cycle, for instance, the activity of the hippocampus decreases during estrous (when estrogen levels are higher than progesterone) and increases during the postcoital stage and pregnancy (when the reverse hormonal pattern occurs), while the excitability of the amygdala increases at estrous and decreases in postcoital and pregnant stages. The same relationship is reported (Kawakami et al., 1966) in alterations of the amplitude and latency of evoked potentials recorded from the periventricular arcuate nucleus in the hypothalamus of rabbits by stimulation of the hippocampus and amygdala. The authors point out that an inverse relationship of excitability exists between the hippocampus and the amygdala at the estrous and postestrous stages. After injection of progesterone the arcuate nucleus potential evoked by stimulation of the hippocampus was facilitated, whereas the arcuate potential evoked by stimulation of the amygdala was inhibited. LH injection facilitated both negative and positive components of this potential evoked by hippocampal stimulation and inhibited both components evoked by amygdaloid stimulation. Since electrical stimulation of the hippocampus resulted in increased progesterone synthesis and secretion, Kawakami et al. (1967) suggested that the hippocampus is the critical area of positive feedback control and the medial amygdaloid complex of negative feedback control of progesterone.

Considerably less is known about the mechanism of action of androgens on neural tissue. Although it has been generally assumed that gonadal hormones have little direct influence on central neural tissue outside of certain areas of the forebrain, Hart (1967) found that testosterone administration influenced quantitatively the sexual reflexes of castrated spinal male rats. Hart and Haugen (1968) further reported that testosterone implanted into the spinal canal near spinal neurons which mediate sexual reflexes had a facilitory influence on such reflexes suggesting a direct effect on spinal neural tissue.

Thyroid Hormones

Slowing of the α frequency has been reported in the EEG of hypothyroid subjects. In rats, thyroidectomy was first thought not to affect spontaneous brain electrical activity despite decreased basal metabolic rate (Lee and Van Buskirk, 1928). Later investigations showed that thyroidectomy depressed both spontaneous activity and the metabolic rate; dinitrophenol restored the basal metabolic rate to normal but did not correct the spontaneous activity (Hall and Lindsay, 1938) suggesting that the effects of thyroid hormone on the CNS are independent of its effect on the general metabolism of the animal. In an experiment in which EEGs were taken from animals which had been thyroidectomized neonatally, there was a reduction in frequency and voltage (Bradley et al., 1960) and the capacity of the cortex to "block" in response to auditory stimuli or to "follow" the frequency of photic stimulation was either delayed in appearance or absent. Since axonal and dendritic hypoplasia in the sensorimotor cortex occurs (Eayrs, 1955) under these conditions, the failure of sensory input to synchronize or desynchronize cortical rhythms may be equated with afferent axonal hypoplasia. Other work prompted the suggestion that the prolonged duration of electrocortical potentials was related primarily to metabolic factors and the decreased amplitude primarily to the processes of growth and maturation. This hypothesis was based on the observation that alterations in the former could be rectified by thyroid hormone administration, but the latter could not (Bradley et al., 1964). Since it appears that the origin of the slow potentials associated with electrocortical activity depends on the summation of postsynaptic dendritic potentials (Clare and Bishop, 1955), Eayrs (1964) postulated that the reduced amplitude of recordings in animals having had neonatal thyroidectomy, but not in those having had thyroidectomy when mature, may be attributable to a comparable reduction in the probability of axodendritic interaction as orginating and maintaining these potentials.

Other Hormones

Melatonin, a principle substance of the pineal gland, has been reported to have acute effects on sleep mechanisms (Barchas et al., 1967). It increased by 50% the hexobarbital sleeping time in mice and induced sleep for about 45 min following intravenous administration in four-day-old chicks in which the blood-brain barrier is relatively undeveloped. Similarly, Marczynski et al. (1964) found that crystalline melatonin injected through steel cannulae implanted directly into areas of the hypothalamus of cats caused sleep lasting about 2 hours.

Nir et al. (1969) have also shown that pinealectomy in female rats results in changes in cerebrocortical electrical activity characterized by intermittent general paroxysmal outbursts of slow waves with high amplitudes of centrocephalic origin. These seizure-like discharges occur against a background of basic electrical activity of permanent symmetrical monomorphic and monorhythmic waves of 9-12 Hz. The authors suggest that the cortical hyperactivity of the pinealectomized rat may be due to an indirect effect mediated by the gonads (high estrogen levels of permanent estrus) rather than a direct effect of the pineal hormones on the CNS. MSH has also been implicated in producing hyper-excitability in mice (Sakamoto, 1966).

Cotzias et al. (1967) found that larger doses of MSH aggravated Parkinson's disease, that a result from an abnormality in the mechanism controling the release of MSH from the pituitary gland (Kastin et al., 1968). This is supported by the observation in animals that tranquilizers do indeed release pituitary MSH (Kastin and Schally, 1966). It is also of interest that Kastin et al. (1968) found a significant decrease in serum calcium following MSH infusion. Patients with hypocalcemia seem to be more sensitive to extrapyramidal symptoms induced by tranquilizing drugs (Schaaf and Payne, 1966; Lichtigfeld and Simpson, 1967). Kastin et al. (1968) suggest the intriguing concept that the decline in serum calcium induced by MSH is responsible for the increase in responsiveness of the CNS to MSH. Hypocalcemia is indeed generally considered to be responsible for the psychic and neurological alterations in hypoparathyroidism, which include emotional liability, anxiety, irritability, and delirium.

HORMONES AND BEHAVIOR

In view of the numerous reviews and books that already exist concerning the influence of hormones on behavior, especially animal behavior, it is beyond the scope of this chapter to attempt to discuss behavioral aspects of hormonal action. The interested reader is advised to look into books by Beach (1948), Whalen (1967), Levine (1972), and Eleftheriou (1975). Moreover, information concerning influences of hormones on human behavior is presented in Chapters 18 and 19.

REFERENCES

Abell, C.W. and Monahan, T.M. The role of adenosine 3',5'-cyclic monophosphate in the regulation of mammalian cell division. *J. Cell Biol.* 59, 549-558 (1973).

Abrams, R.L., Grumbach, M.M., and Kaplan, S.L. The effect of administration of human growth hormone on the plasma growth hormone, cortisol, glucose, and free fatty acid response to insulin: evidence for growth hormone autoregulation in man. *J. Clin. Invest.* 50, 940-950 (1971).

Abrams, R.L., Parker, M.L., Blanco, S., Reichlin, S., and Daughaday, W.H. Hypothalamic regulation of growth hormone secretion. *Endocrinology* 78, 605-613 (1966).

Anderson, C.H. and Greenwald, G.S. Autoradiographic analysis of estradiol uptake in the brain and pituitary of the female rat. *Endocrinology* 85, 1160-1165 (1969).

Anderson, N.S., III and Fanestil, D.D. Corticoid receptors in rat brain: evidence for an aldosterone receptor. *Endocrinology* 98, 676-684 (1976).

Arimura, A., Sato, H., Dupont, A., Nishi, N., and Schally, A.V. Somatostatin: abundance of immunoreactive hormone in rat stomach and pancreas. *Science* 189, 1007-1009 (1975).

Azmitia, E.C. and McEwen, B.S. Corticosterone regulation of tryptophan hydroxylase in midbrain of the rat. *Science* 166, 1274-1276 (1969).

Azmitia, E.L., Algeri, S., and Costa, E. *In vivo* conversion of 3H-L-trytophaninto ^{3}H-serotonin in brain areas of adrenolectomized rats. *Science* 169, 201-203 (1970).

Bakay, L. and Lee, J.C. *Cerebral Edema.* Charles C. Thomas Co., Springfield, Illinois (1965).

Bal, H. and Smelik, P.G. Effect of hypothalamic lesions on MSH content of the intermedia lobe of the pituitary gland in the rat. *Experentia* 23, 759-760 (1967).

Ball, P., Haupt, M., and Knuppen, R. Comparative studies on the metabolism of oestradiol in the brain, the pituitary and the liver of the rat. *Acta Endocrinol.* 82, 1-11 (1978).

Ball, P., Knuppen, R., Haupt, M., and Breuer, H. Interactions between estrogens and catecholamines. *J. Clin. Endocrinol.* 34, 736-746 (1972).

Baker, B.L., Dermody, W.C., and Reel, J.R. Distribution of gonadotropin releasing hormone in the rat brain as observed with immunocytochemistry. *Endocrinology* 97, 125-135 (1975).

Barchas, J., DaCosta, F., and Spector, S. Acute pharmacology of melatonin. *Nature* 214, 919-920 (1967).

Barnea, A., Ben-Jonathon, N., Colston, C., Johnston, J.M., and Porter, J.C. Differential subcellular compartmentalization of thyrotropin releasing hormone (TRH) and gonadotropin releasing hormone (LH-RH) in hypothalamic tissue. *Proc. Natl. Acad. Sci.* (Wash.) 72, 3153-3157 (1975).

Barraclough, C.E. Modifications in the CNS regulation of reproduction after exposure of prepubertal rats to steroid hormones. *Rec. Prog. Horm. Res.* 22, 503-539 (1966).

Barry, J. Characterization and topography of LH-RH neurons in the rabbit. *Neurosci. Lett.* 2, 201-205 (1976).

Barry, J. Immunofluorescence study of LRF neurons in man. *Cell Tiss. Res.* 181, 1-14 (1977).

Barry, J. and Carette, B. Immunofluorescence study of LRF neurons in primates. *Cell Tiss. Res.* 164, 163-178 (1975).

Barry, J. and Dubois, M.P. Immunofluorescence study of LRF producing neurons in the cat and the dog. *Neuroendocrinology* 18, 290-298 (1975).

Bassiri, R.M. and Utiger, R.D. The preparation and specificity of antibody to thyrotropin releasing hormone. *Endocrinology* 90, 722-727 (1972).

Baumgarten, H.S., Bjorklund, A., Lachenmayer, L., and Nobin, A. Evaluation of the effects of 5' dihydroxytryptamine on serotonin and catecholamine neurons in the rat CNS. *ACTA Physiol. Scand.* (suppl.) 3-19 (1973).

Beach, F.A. *Hormones and Behavior.* Paul B. Hoeber, New York (1948).

Beck, P., Koumans, J.H.T., Winterling, C.A., Stein, M.F., Daughaday, W.H., and Kipnis, D.M. Studies of insulin and growth hormone secretion in human obesity. *J. Lab. Clin. Med.* 64, 654-667 (1964).

Beese, J. and Bass, A. Potentiation by hydrocortisone of responses to catecholamines in vascular smooth tissue. *J. Pharmacol. Exp. Ther.* 154, 224-238 (1966).

Bergen, J.R. Rat electrocorticogram in relation to adrenal cortical function. *Am. J. Physiol.* 164, 16-22 (1951).

Bernard, B.K. Testosterone manipulations: effects on ranacide aggression and brain monoamines in the adult female rat. *Pharmacol. Bio. Behav.* 4, 59-65 (1976).

Bernard, B.K. and Paolino, R.M. Brain norepinephrine levels and turnover rates in castrated mice isolated for 13 months. *Separatum Experientia* 29, 221-223 (1973).

Bernard, B.K. and Paolino, R.M. Time-dependent changes in brain biogenic amine dynamics following castration in male rats. *J. Neurochemistry*, 22, 951-956 (1974).

Bhattacharya, A.N. and Marks, B.H. Effects of pargyline and amphetamine upon acute stress response in rats. *Proc. Soc. Exp. Biol. Med.* 130, 1194-1198 (1969).

Birge, C.A., Jacobs, L.S., Hammer, C.T., and Daughaday, W.H. Catecholamine inhibition of prolactin secretion by isolated rat adenohypophyses. *Endocrinology* 86, 120-130 (1970).

Björklund, A., Nobin, A., and Stenevi, V. Effects of 5,6-dihydroxytryptamine on nerve terminal serotonin and serotonin uptake in the rat brain. *Brain Res.* 53, 117-127 (1973).

Black, I.B., Hendry, I., and Iversen, L.L. Trans-synaptic regulation of growth and development of adrenergic neurones in a mouse sympathetic ganglion. *Brain Res.* 34, 229-240 (1971).

Blackard, W.G., Hull, E.W., and Lopez-S. A. Effect of lipids on growth hormone secretion in humans. *J. Clin. Invest.* 50, 1439-1443 (1971).

Bliss, E.L., Ailion, J., and Zwanziger, J. Metabolism of norepinephrine, serotonin and dopamine in rat brain with stress. *J. Pharmacol. Exp. Ther.* 164, 122-134 (1968).

Bloom, F.E. Peptide transmitters: clues to the chemical cryptogram of interneuronal communication. *Bio. Systems*, 179-183 (1977).

Bohus, B. Pituitary ACTH release and avoidance behavior of rats with cortisol implants in mesencephalic reticular formation and median eminence *Neuroendocrinology* 3, 355-365 (1968).

Bohus, B. and Strashimirov, D. Localization and specificity of corticosteroid "feedback receptors" at the hypothalamo-hypophyseal-level. Comparative effects of various steroids implanted in the median eminence or the pituitary of the rat. *Neuroendocrinology* 6, 197-209 (1970).

Bowers, C.Y., Guyda, H.J., and Folkers, K. Prolactin and thyrotropin release in man by synthetic pyroglutamyl-histidyl-prolinamide. *Biochem. Biophys. Res. Commun.* 45, 1033-1041 (1971).

Bowers, C.Y., Redding, T.W., and Schally, A.V. Effect of thyrotropin releasing factor (TRF) of ovine, bovine, porcine, and human origin on thyrotropin release *in vitro* and *in vivo*. *Endocrinology* 77, 609-616 (1965).

Boyd, A.E., III, Lebovitz, H.E., and Feldman, J.M. Endocrine function and glucose metabolism in patients with Parkinson's disease and their alteration by L-DOPA. *J. Clin. Endocrinol. Metab.* 33, 829-837 (1971).

Bradley, P.B., Eayrs, J.T., and Richards, N.M. Factors influencing potentials in normal and cretinous rats. *Electroencephalogr. Clin. Neurophysiol.* 17, 308-313 (1964).

Bradley, P.B., Eayrs, J.T., and Schmalbach, K. The electroencephalogram of normal and hypothyroid rats, in *Electroencephalography and Clinical Neurophysiology* 12, 467-477 (1960).

Brazeau, P., Vale, W., Burgus, R., Ling, N., Butcher, M., Rivier, J., and Guillemin, R. Hypothalamic polypeptide that inhibits the secretion of immunoreactive pituitary growth hormone. *Science* 179, 77-79 (1973).

Brostrom, M.A., Kon, C., Olson, D.R., and Breckenridge, B.M. Adenosine 3',5'-monophosphate in glial tumor cells treated with glucocorticoids. *Mol. Pharmacol.* 10, 711-720 (1974).

Brownstein, M. Neurotransmitters and hypothalamic hormones in the central nervous system. *Federation Proc.* 36, 1960-1963 (1977).

Brownstein, M.J., Utiger, R.D., Palkovits, M., and Kizer, J.S. Effect of hypothalamic differentiation on thyrotropin releasing hormone levels in rat brain. *Proc. Natl. Acad. Sci. (U.S.A.)* 72, 4177-4179 (1975a).

Brownstein, M., Arimura, A., Sato, H., Schally, A.V., and Kizer, J.S. The regional distribution of somatostatin in the rat brain. *Endocrinology* 96, 1456-1461 (1975b).

Brownstein, M.J., Palkovits, M., Saavedra, J.M., Bassiri, R.M., and Utiger, R.D. Thyrotropin-releasing hormone in specific nuclei of rat brain. *Science* 185, 267-269 (1974).

Bugnon, C., Bloch, B., Lenys, D., and Fellmann, D. Ultrastructural study of the LH-RH containing neurons in the human fetus. *Brain Res.*, 137, 175-180 (1977).

Burgus, R., Butcher, M., Ling, N., Monohan, M., Rivier, J., Fellows, R., Amoss, M., Blackwell, R., Vale, W., and Guillemin, R. Structure moleculaire du facteur hypothalaminque (LRF) d'ougine avine controlant la secretion de l'hormone gonadotrope hypophysaise de luteinisation (LH). *C.R. Acad. Sci.* 273, 1611-1613 (1971).

Burn, J.H. The action of tyramine and ephedrine. *J. Pharmacol. Exp. Ther.* 46, 75-95 (1932).

Burrow, G.N., Spaulding, S.W., Donabedian, R., VanWoert, M., and Ambani, L. The effect of L-dopa on the hypothalamic-pituitary-thyroid axis. *Advances in Neurol.* 5, 489-493 (1974).

Butte, J.C., Kakihana, R., and Noble, E.P. Rat and mouse brain corticosterone. *Endocrinology* 90, 1091-1100 (1972).

Carr, L.A. and Moore, K.E. Effects of reserpine and α-methyl-tyrosine on brain catecholamines and the pituitary adrenal response to stress. *Neuroendocrinology* 3, 285-302 (1968).

Challis, J.R.G., Naftolin, F., Davies, I.J., Ryan, K.J., and Lanman, T. Endogenous steroids in neuroendocrine tissues, in *Subcellular Mechanisms in Reproductive Endocrinology*, F. Naftolin, K.J. Ryan, and J. Davies, eds. Elsevier, Amsterdam (1976), pp. 247-261.

Chambers, W.F., Freeman, S.L., and Sawyer, C.H. The effect of adrenal steroids on evoked reticular responses. *Exp. Neurol.* 8, 458-469 (1963).

Chowers, I., Feldman, S., and Davidson, J.M. Effects of intrahypothalamic crystalline steroid on acute ACTH secretion. *Am. J. Physiol.* 205, 671-673 (1963).

Ciaranello, R.D. and Axelrod, J. Effects of dexamethasone on neurotransmitter enzymes in chromaffin tissue of the newborn rat. *J. Neurochem.* 24, 775-778 (1975).

Clare, M.C. and Bishop, G.H. Dendritic circuits: the properties of cortical paths involving dendrites. *Am. J. Psychiatry* 111, 818-825 (1955).

Cohn, M.L. and Cohn, M. Comparison of the regulation of rotational behavior by hypothalamic oligopeptides. *Psychoneuroendocrinology* 2, 197-202 (1977).

Collu, R., Fraschini, F., and Martini, L. Role of indoleamines and catecholamines in the control of gonadotropin and growth hormone secretion, in *Drug Effects on Neuroendocrine Regulation Progress in Brain Research*, Vol. 29. C. Zimmerman, W.H. Sispen, B.H. Marks, and O. Delvied, eds. Elsevier Scientific Publishing Co., Amsterdam (1973).

Collu, R., Fraschini, F., Visconti, P., and Martini, L. Adrenergic and serotoninergic control of growth hormone secretion in adult male rats. *Endocrinology* 90, 1231-1237 (1972).

Convey, E.M., Tucker, H.A., Smith, V.G., and Zolman, J. Bovine prolactin, growth hormone, thyroxine, and corticoid response to thyrotropin-releasing hormone. *Endocrinology* 92, 471-476 (1973).

Cook, S., Mavor, H., and Chambers, W.F. Effects of reticular stimulation in altered adrenal states. *Electroencephalogr. Clin. Neurophysiol.* 12, 601-608 (1960).

Coppola, J.A. Brain catecholamines and gonadotropin secretion, in *Frontiers in Neuroendocrinology*. L. Martini and W.F. Ganon, eds. Oxford University Press, New York (1971), pp. 129-143.

Corbin, A. and Story, J.C. Internal feedback mechanism: response of pituitary FSH and of stalk-median eminence follicle stimulating hormone-releasing factor to median eminence implants of FSH. *Endocrinology* 80, 1006-1012 (1967).

Corbin, A., Mangili, C., Motta, M., and Martini, L. Effects of hypothalamic and mesencephalic steroid implantation on ACTH feedback mechanism. *Endocrinology* 76, 811-818 (1965).

Cotzias, G.C., Van Woert, M.H., and Schiffer, L.M. Aromatic amino acids and modification of parkinsonism. *New Engl. J. Med.* 276, 374-379 (1967).

Cross, B.A., Dyball, R.E.J., Dyer, R.G., Jones, C.W., Lincoln, D.W., Morris, J.F., and Pickering, B.T. Endocrine neurons. *Recent Prog. Horm. Res.* 31, 243-294 (1975).

Curzon, G. and Green, A.R. Regional and subcellular changes in the concentration of 5-hydroxytryptamine and 5-hydroxyindoleacetic acid in the rat brain caused by hydrocortisone, DL-α-methyl-tryptophan, L-kyninenine and immobilization. *J. Pharm. Pharmacol.* 43, 39-52 (1971).

Dahlstrom, A. and Fuxe, K. Evidence for the existence of monoamine-containing neurons in the central nervous system. I. Demonstration of monoamines in the cell bodies of brain stem neurons. *Acta Physiol. Scand.* 62 (suppl.) 232, 1-55 (1964).

Danon, A., Dikstein, S., and Sulman, F.G. Stimulation of prolactin secretion by perphenazine in pituitary-hypothalamus organ culture. *Proc. Soc. Exp. Biol. Med.* 114, 366-368 (1963).

Davidson, J.M. Feedback control of gonadotropin secretion, in *Frontiers in Neuroendocrinology*, Vol. I. W.F. Ganong and L. Martini, eds. Oxford University Press, New York (1969), pp. 343-388.

Davidson, J.M. and Feldman, S. Cerebral involvement in the inhibition of ACTH secretion by hydrocortisone. *Endocrinology* 72, 936-964 (1963).

Davidson, J.M., Jones, L.E., and Levine, S. Effects of hypothalamic implantation of steroids on plasma corticosterone. *Fed. Proc.* 24, 191 (Abstract) (1965).

Davidson, J.M., Smith, E.R., Rogers, C.H., and Bloch, G.J. Relative thresholds of behavioral and somatic responses to estrogen. *Physiol. Behav.* 3, 227-229 (1968).

Davies, I.J., Naftolin, F., Ryan, K.J., and Siu, J. Specific binding of steroids by neuroendocrine tissues, in *Subcellular Mechanisms in Reproductive Neuroendocrinology*. F. Naftolin, K.J. Ryan, and J. Davies, eds. Elsevier, Amsterdam (1976), pp. 263-275.

Davies, I.J., Naftolin, F., Ryan, K.J., Fishman, J., and Siu, J. The affinity of catechol estrogens for estrogen receptors in the pituitary and anterior hypothalamus of the rat. *Endocrinology* 97, 554-557 (1975).

Davis, P.J., Handwerger, B.S., and Glaser, F. Physical properties of a dog liver and kidney cytosol protein that binds thyroid hormones. *J. Biol. Chem.* 249, 6208-6217 (1974).

Dean, C.R. and Hope, D.B. Protein constituents of neurosecretory granules isolated from the posterior lobes of bovine granular glands. *Biochem. J.* 101, 17-18 (1966).

Dean, C.R. and Hope, D.B. The isolation of purified neurosecretory granules from bovine pituitary posterior lobes: comparison of granule protein constituents with those of neurophysin. *Biochem. J.* 104, 1082-1088 (1967).

DeGroot, L.J. and Torresani, J. Triiodothyronine binding to isolated liver cell nuclei. *Endocrinology* 96, 357-369 (1975).

DeKloet, E.R. and McEwen, B.S. A putative glucocorticoid receptor and a transcortin-like macromolecule in pituitary cytosol. *Biochim. Biophys. Acta* 421, 115-123 (1976).

Denef, C., Magnus, C., and McEwen, B.S. Sex differences and hormonal control of testosterone metabolism in rat pituitary and brain. *J. Endocrinol.* 59, 605-621 (1973).

DeVellis, J. and Brooker, G. Induction of enzymes by glucocorticoids and catecholamines in a rat glial cell line, in *Tissue Culture of the Nervous System*. G. Sato, ed. Plenum Press, New York (1973), pp. 231-245.

DeVellis, J., Inglish, D., and Galey, F. Effects of cortisol and epinephrine on glial cells in culture, in *Cellular Aspects of Growth and Differentiation in Nervous Tissue*. D. Pease, ed. UCLA Forum in Medical Sciences, No. 14, University of California Press, Berkeley (1971), pp. 23-32.

DeVellis, J., McGinnis, J.E., Breen, G.A.M., Le Veille, P., Bennett, K., and McCarthy, K. Hormonal effects on differentiation in neural cultures, in *Cell, Tissue, and Organ Culture in Neurobiology*. S. Fedoroff and L. Hertz, eds. Academic Press, New York (1977), pp. 485-511.

DeWied, D. Chlorpromazine and endocrine function. *Pharmacol. Rev.* 19, 251-288 (1967).

Dhariwal, A.P.S., Antunes-Rodrigues, J., and McCann, S.M. Purification of ovine luteinizing hormone-releasing factor by gel filtration and ion exchange chromatography. *Proc. Soc. Exp. Biol. Med.* 118, 999-1003 (1965).

Dhariwal, A.P.S., Antunes-Rodrigues, J., and McCann, S.M. Separation of growth hormone-releasing factor (GHRF) from corticotrophin-releasing factor (CRF). *Neuroendocrinology* 1, 341-349 (1965-1966).

Dhariwal, A.P.S., Grosvenor, C.E., Antunes-Rodrigues, J., and McCann, S.M. Studies on the purification of ovine prolactin-inhibiting factor (PIF). *Endocrinology* 82, 1236-1240 (1968).

Dhariwal, A.P.S., Antunes-Rodrigues, J., Reeser, F., Chowers, I., and McCann, S.M. Purification of hypothalamic corticotrophin-releasing factor (CRF) of ovine origin. *Proc. Soc. Exp. Biol. Med.* 121, 8-12 (1966).

Diez, J.A., Sze, P.Y., and Ginsburg, B.E. Effect of glucorticoid and stress on brain tyrosine hydroxylase activity. *Trans. Amer. Soc. Neurochem.* 5, 79 (1974).

Dillman, W., Surks, M.I., and Oppenheimer, J.H. Quantitative aspects of iodothyronine binding by cytosol proteins of rat liver and kidney. *Endocrinology* 95, 492-498 (1974).

Douglas, W.W. and Rubin, R.P. The mechanisms of catecholamine release from the adrenal medulla and the role of calcium in stimulus-secretion coupling. *J. Physiol.* (London) 167, 288-310 (1963).

Dratman, M.B., Crutchfield, F.L., Axelrod, J., Colburn, R.W., and Thoa, N. Localization of triiodothyronine in nerve ending fractions of rat brain. *Proc. Nat. Acad. Sci.* (U.S.A.) 73, 941-944 (1976).

Dube, D., LeClerc, R., Pelletier, G., Arimura, A., and Schally, A.V. Immunohistochemical detection of growth hormone release inhibiting hormone (somatostatin) in the guinea-pig brain. *Cell Tiss. Res.* 161, 385-392 (1975).

Dubois, M. Immunoreactive somatostatin is present in discrete cells of the endocrine pancreas. *Proc. Natl. Acad. Sci.* (U.S.A.) 72, 1340-1343 (1975).

Dular, R., Labella, F., Vivian, S., and Eddie, L. Purification of prolactin-releasing and inhibiting factors from beef. *Endocrinology* 94, 563-567 (1974).

Dupont, A., Bastarache, E., Endroczi, E., and Fortier, C. Effect of hippocampal stimulation on the plasma thyrotropin (TSH) and corticosterone responses to acute cold exposure in the rat. *Can. J. Physiol. Pharm.* 50, 364-367 (1972).

Eayrs, J.T. The cerebral cortex of normal and hypothyroid rats. *Acta Anat.* 25, 160-183 (1955).

Eayrs, J.T. Endocrine influence on cerebral development. *Arch. Biol.* 75, 529-565 (1964).

Eberhardt, N.L., Valcana, T., and Timiras, P.S. Hormone-receptor interactions in brain: uptake and binding of thyroid hormones. *Psychoneuroendocrinology* 1, 399-409 (1976).

Eberhardt, N.L., Valcana, T., and Timiras, P.S. Triiodothyronine nuclear receptors: an *in vitro* comparison of the binding of triiodothyronine to nuclei of adult rat liver, cerebral hemisphere and anterior pituitary. *Endocrinology* 102, 556-561 (1978).

Eddy, R.L., Jones, A.L., Chakmakjian, Z.H., and Silverthorne, M.C. Effect of levodopa (L-DOPA) on human hypophyseal trophic hormone release. *J. Clin. Endocrinol. Metab.* 33, 709-712 (1971).

Edwards, B.A. The uptake of tritiated lysine vasopressin by rat and pig pituitary neural lobes *in vitro*. *J. Endocrinol.* 50, 669-677 (1971).

Eisenfeld, A.J. ^{3}H-Estradiol: *in vitro* binding to macromolecules from the rat hypothalamus, anterior pituitary and uterus. *Endocrinology* 86, 1313-1318 (1970).

Eisenfeld, A.J. and Axelrod, J. Selectivity of estrogen distribution in tissues. *J. Pharmacol. Exp. Ther.* 150, 469-475 (1965).

Elde, R. and Hokfelt, T. Distribution of hypothalamic hormones and other peptides in the brain, in *Frontiers in Neuroendocrinology*, Vol. 5. W.F. Ganong and L. Martini, eds. Raven Press, New York (1978), pp. 1-33.

Eleftheriou, B.E. and Sprott, R.L. *Hormonal Correlates of Behavior*. Plenum Press, New York (1975).

Eleftheriou, B.E. and Zolorick, A.J. Effect of amygdaloid lesions on plasma and pituitary thyrotropin levels in the deer-mouse. *Proc. Soc. Exp. Biol. Med.* 127, 671-674 (1968).

Emlen, W., Segal, D., and Mandell, A.T. Thyroid state: effects on pre- and postsynaptic central noradrenergic mechanisms. *Science* 175, 79-82 (1972).

Endröczi, E., Lissak, K., and Tekeres, M. Hormonal feedback regulation of pituitary adrenocortical activity. *Acta Physiol. Acad. Sci. Hung.* 18, 291-299 (1961).

Epelbaum, J., Brazeau, P., Tsang, D., Brawer, J., and Martin, J.B. Subcellular distribution of radioimmuno-assayable somatostatin in rat brain. *Brain Res.* 126, 309-323 (1977).

Etkin, W. Hypothalamic inhibition of pars intermedia activity in the frog. *Gen. Comp. Endocrinol.* Suppl. 1, 148-159 (1962).

Everett, J.W. Central neural control of reproductive functions of the adenohypophysis. *Physiol. Rev.* 44, 373-431 (1964).

Everett, J.W., Sawyer, C.H., and Markee, J.E. A neurogenic timing factor in control of the ovulatory discharge of luteinizing hormone in the cyclic rat. *Endocrinology* 44, 234-250 (1949).

Exton, J.H., Friedmann, N., Wong, E.H.A., Brineaux, J.P., Corbin, J.D., and Park, C.R. Interaction of glucocorticoids with glucagon and epinephrine in the control of gluconeogenesis and gluco-gendysis in liver and of lipolysis in adipose tissue. *J. Biol. Chem.* 247, 3579-3588 (1972).

Farquhar, M.N., Namiki, H., and Gorbman, A. Cytoplasmic and nuclear metabolism of testosterone in the brains of neonatal and prepubertal rats. *Neuroendocrinology* 20, 358-372 (1976).

Feldman, S. Electrical activity of the brain following cerebral microinfusion of cortisol. *Epilepsia* 12, 249-262 (1971).

Feldman, S. and Robinson, S. Electrical activity of the brain in adrenalectomized rats with implanted electrodes. *J. Neurol. Sci.* 6, 1-8 (1968).

Finkelstein, J.W., Roffwarg, H.P., Boyar, R.M., Kream, J., and Hellman, L. Age-related change in the twenty-four-hour spontaneous secretion of growth hormone. *J. Clin. Endocr. Metab.* 35, 665-670 (1972).

Fishman, J. Estrogen metabolism by neuroendocrine tissues, in *Subcellular Mechanisms in Reproductive Neuroendocrinology*. F. Naftolin, K.J. Ryan, and I.J. Davies, eds. Elsevier, Amsterdam (1976), pp. 357-362.

Flament-Durand, J. Observations on pituitary transplants into the hypothalamus of the rat. *Endocrinology* 77, 446-454, 1965.

Fleischer, N. and Rawls, W. ACTH synthesis and release in pituitary monolayer culture: effect of desamethasone. *Am. J. Physiol.* 219, 445-448 (1970).

Flerko, B. Control of gonadotropin secretion in the female, in *Neuroendocrinology*, Vol. 1. L. Martini and W.F. Ganong, eds. Academic Press, New York (1966), pp. 613-653.

Ford, D.H. and Rhines, R.K. Accumulation of [^{131}I] triiodothyronine in neurons and other tissues following intravenous injection of the labeled hormone. *Brain Res.* 6, 481-488 (1967).

Ford, D.H. and Rhines, R.K. Effect of age on the accumulation of [^{131}I] triiodothyronine in male and female rat brains and other tissues. *Brain Res.* 21, 265-274 (1970).

Ford, D.H., Rhines, R.K., and Stieg, C. Hormone localization in the nervous system, in *Influence of Hormones on the Nervous System.* D.H. Ford, ed. S. Karger, Basel (1971), pp. 2-16.

Frantz, A.G. Catecholamines and the control of prolactin secretion in humans, in *Drug Effects on Neuroendocrine Regulation, Progress in Brain Research,* Vol. 39. E. Zimmerman, W.H. Gispen, B.H. Marks, and D. DeWied, eds. Elsevier, Amsterdam (1973), pp. 311-322.

Frantz, A.G. and Kleinberg, D.L. Prolactin: evidence that it is separate from growth hormone in human blood. *Science* 170, 745-747 (1970).

Frantz, A.G., Kleinberg, D.L., and Noel, G.L. Studies on prolactin in man. *Rec. Prog. Horm. Res.* 28, 527-573 (1972).

Frantz, A.G., Habif, D.V., Hyman, G.A., and Sich, H.K. Remission of metastatic breast cancer after reduction of circulating prolactin in patients treated with L-DOPA. *Clin. Res.* 20, 864 (1972).

Fraschini, F. The pineal gland and the control of LH and FSH secretion, in *Progress in Endocrinology.* G. Gual, ed. Excerpta Medica, Amsterdam (1969), pp. 637-644.

Fraschini, F. and Martini, L. Rhythmic phenomena and pineal principles in *The Hypothalamus.* C.L. Martini, M. Motta, and F. Fraschini, eds. Academic Press, New York (1970), pp. 529-549.

Friedman, E., Friedman, J., and Gershon, S. Dopamine synthesis: stimulation by a hypothalamic factor. *Science* 182, 831-832 (1973).

Friesen, H., Webster, B.R., Hwang, P., Guyda, H., Munroe, R.E., and Read, L. Prolactin synthesis and secretion in a patient with the Forbes Albright syndrome. *J. Clin. Endocrinol.* 34, 192-199 (1972).

Fuxe, K. Evidence for the existence of mono-amine neurons in the central nervous system. IV. Distribution of monoamine nerve terminals in the central nervous system. *Acta. Physiol. Scand.* (suppl.) 247, 36-85 (1965).

Fuxe, K. and Hokfelt, T. Further evidence for the existence of tubero-infundibular dopamine neurons. *Acta Physiol. Scand.* 66, 245-246 (1966).

Fuxe, K. and Hokfelt, T. Catecholamines in the hypothalamus and the pituitary gland, in *Frontiers in Neuroendocrinology.* L. Martini and W.F. Ganong, eds. Oxford University Press, New York (1969), pp. 47-96.

Gala, R.R. and Reace, R.P. *In vitro* lactogen production by anterior pituitaries from various species. *Proc. Soc. Exp. Biol. Med.* 120, 263-264 (1965).

Gann, D.S. and Cryer, G.L. Models of adrenal cortical control. *Adv. Biol. Med. Engl.* 2, 1-60 (1972).

Ganong, W.F. Evidence for a central noradrenergic system that inhibits ACTH secretion, in *Brain-Endocrine Interaction. Median Eminence: Structure and Function Int. Symp. Munich, 1971.* K.M. Knigge, D.E. Scott, and A. Weindl, eds. S. Karger, Basel (1972), pp. 254-266.

Ganong, W.F. and Lorenzen, L.C. Brain neurohumors and endocrine function, in *Neuroendocrinology,* Vol. 2. L. Martini and W.F. Ganong, eds. Academic Press, New York (1967), pp. 583-640.

Geel, S.E. Development-related changes of triiodothyronine binding to brain cytosol receptors. *Nature* 269, 428-430 (1977).

Gerlach, J.L., McEwen, B.S., Pfaff, D.W., Moskowitz, S., Ferin, M., Carmel, P.N., and Zimmerman, E.A. Cells in regions of rhesus monkey brain and pituitary retain radioactive estradiol, corticosterone and cortisol differentially. *Brain Res.* 103, 603-612 (1976).

Gilman, A.G. Regulation of cyclic AMP metabolism in cultured cells of the nervous system. *Advances in Cyclic Nucleotide Res.* 1, 389-410 (1972).

Gilman, A.G. and Murad, F. Adenohypophyseal hormones and related substances, in *The Pharmacological Basis of Therapeutics*. L.S. Goodman and A. Gilman, eds., and A.G. Gilman and G.B. Koelle, assoc. eds. MacMillan Publishing Co., New York (1975), pp. 1372-1397.

Ginsburg, M. Production, release, transportation and elimination of the neurohypophysial QV_{34} hormones, in *Handbuch der experimentellen Phardakologie*, Vol. 23. B. Berde, ed. Springer-Verlag, Berlin and New York (1968), pp. 286-371.

Ginsburg, M. and Ireland, M. Isolation hormone binding capacity and subcellular distribution of neurophysin. *J. Physiol.* (London) 169, 114-115 (1963).

Ginsburg, M. and Ireland, M. Role of neurophysin in the transport and release of neurohypophysial hormones. *J. Endocrinol.* 35, 289-298 (1966).

Ginsburg, M., Greenstein, B.D., MacLusky, N.J., Morris, I.D., and Thomas, P.J. Dihydrotestosterone binding in brain and pituitary cytosol of rats. *J. Endocrinol.* 61, XXIV (1974).

Goldsmith, P.C. Ultra-structural localization of some hypothalamic hormones. *Federation Proc.* 36, 1968-1972 (1977).

Grimm, Y. and Kendall, J.W. A study of feedback suppression of ACTH secretion utilizing glucocorticoid implants in the hypothalamus. The comparative effects of cortisol, corticosterone and their 21-acetates *Neuroendocrinology* 3, 55-63 (1968).

Grimm, Y., and Reichlin, S. Thyrotropin-releasing hormone (TRH): neurotransmitter regulation of secretion by mouse hypothalamic tissue in vitro. *Endocrinology* 93, 626-631 (1973).

Grosser, B.I., Stevens, W., and Reed, D.J. Properties of corticosterone-binding macromolecules from rat brain cytosol. *Brain Res.* 57, 387-395 (1973).

Grosser, B.I., Stevens, W., Brnenger, F.W., and Reed, D.J. Corticosterone binding in rat brain cytosol. *J. Neurochem.* 18, 1725-1732 (1971).

Guillemin, R., Burgus, R., Sakiz, E., and Ward, D.N. Nourelles donne'es sur la purification de l' hormone hypothalamique TSH-hypophysiotrope, TRF. *Compt. Rend.* 262, 2278-2280 (1966).

Guillemin, R., Hearn, W.R., Cheek, W.R., and Householder, D.E. Control of corticotrophin release: further studies with in vitro methods. *Endocrinology* 60, 488-506 (1957).

Guillemin, R., Yamazaki, E., Jutisz, M., and Sakiz, E. Presence dans un extrait de tissue hypothalamiques d'une substance stimulant la secretion de l' hormone hypophysaire thyreotrope (TSH). Premie're purification par filtration sur gel Sephadex. *Compt. Rend.* 255, 1018-1020 (1962).

Guldner, F.H. and Wolff, J.M. Neurono-glial synaptoid contacts in the median eminence of the rat: ultrastructure, staining properties and distribution on tanycytes. *Brain Res.* 61, 217-234 (1973).

Gustafsson, J.A., Pousette, A., and Svensson, E. Sex-specific occurrence of androgen receptors in rat brain. *J. Biol. Chem.* 251, 4047-4054 (1976).

Halazs, B. The endocrine effects of prolation of the hypothalamus from the rest of the brain, in *Frontiers in Neuroendocrinology*. L. Martini and W.F. Ganong, eds. Oxford University Press, New York (1969), pp. 307-342.

Halazs, B., Pupp, L., Uhlarik, S., and Tima, L. Further studies on the hormone secretion of the anterior pituitary transplanted into the hypophysiotrophic area of the rat hypothalamus. *Endocrinology* 77, 343-355 (1965).

Hall, V.E. and Lindsay, M. The relation of the thyroid gland to the spontaneous activity of the rat. *Endocrinology* 22, 66-72 (1938).

Han, P.W., and Liu, A.C. Obesity and impaired growth of rats force fed 40 days after hypothalamic lesions. *Am. J. Physiol.* 211, 229-231 (1966).

Han, P.W., Liu, C.-H., Chu, K.C., Liu, J.-V., and Liu, A.-C. Hypothalamic obesity in weanling rats. *Am. J. Physiol.* 209, 627-631 (1965).

Hapke, M. and Green, R. Studies on the mechanism of epinephrine potentiation by hydrocortisone in aortic strips of the rabbit. *Fed. Proc.* 29, 613 (1970).

Harris, G.W. and Jacobsohn, D. Functional grafts of the anterior pituitary gland. *Proc. Roy. Soc.* (London), Ser. B., 139, 263-279 (1952).

Hart, B.L. Testosterone regulation of sexual reflexes in sinal male rats. *Science* 155, 1283-1284 (1967).

Hart, B.L. and Haugen, C.M. Activation of sexual reflexes in male rats by spinal implantation of testosterone. *Physiol. Behav.* 3, 735-738 (1968).

Haynes, R.C., Sutherland, E.W., and Rall, T.W. The role of cyclic andenylic acid in hormone action. *Rec. Prog. Horm. Res.* 16, 121-133 (1960).

Henkin, R.I., Casper, A.G.T., Brown, R., Harlan, A.B., and Bartter, F.C. Presence of corticosterone and cortisol in the central and peripheral nervous system of the cat. *Endocrinology* 82, 1058-1061 (1968).

Henn, F.A. and Hamberger, A. Glial cell function: uptake of transmitter substances. *Proc. Nat. Acad. Sci. (Wash.)* 68, 2686-2690 (1971).

Hinsey, J.C. The relation of the nervous system to ovulation and other phenomena of the female reproductive tract. *Colo. Springs Harbour Symp. Quant. Biol.* 5, 269-279 (1937).

Hinsey, J.C. and Markee, J.E. Pregnancy following bilateral section of the cervical sympathetic trunks in the rabbit. *Proc. Soc. Exp. Biol. Med.* 31, 270-271 (1933).

Hirsch, G.H. and Moore, K.E. Brain catecholamines and the reserpine-induced stimulation of the pituitary-adrenal system. *Neuroendocrinology* 3, 398-405 (1968).

Hodges, J.R. and Vernikos, J. Circulating corticotrophin in normal and adrenalectomized rats after stress. *Acta. Endocrin., Kbh.* 30, 188-196 (1959).

Hökfelt, T. and Fuxe, K. On the morphology and the neuroendocrine role of the hypothalamic catecholamine neurons, in *Brain-Endocrine Interaction. Median Eminence: Structure and Function, Int. Symp. Munich 1971.* K.M. Knigge, D.E. Scott and A. Weindl, eds. S. Karger, Basel (1972), pp. 181-223.

Hökfelt, T., Elde, R., Johansson, D., Luft, R., and Arimura, A. Immunohistochemical evidence for the presence of somatostatin, a powerful inhibitory peptide in some primary sensory neurons. *Neurosci. Lett.* 1, 231-235 (1975a).

Hökfelt, T., Fuxe, K., Johansson, I., Jeffcoate, S., and White, N. Distribution of thyrotropin-releasing hormone (TRH) in the central nervous system as revealed with immunohistochemistry. *Eur. J. Pharmac.* 34, 389-392 (1975b).

Hökfelt, T., Effendic, S., Hellerstrom, C., Johansson, O., Luft, R., and Arimura, A. Cellular localization of somatostatin in endocrine like cells and neurons of the rat with special references to the A, cells of the pancreatic islets and to the hypothalamus. *Acta Endocrinol.* 80, (suppl. 200) 1-41 (1975c).

Hubbard, J.I. Mechanism of transmitter release. *Progr. Biophys. Mol. Biol.* 21, 33-124 (1970).

Hwang, P., Guyda, H., and Friesen, H. A radioimmunoassay for human prolactin. *Proc. Natl. Acad. Sci.* U.S.A. 68, 1902-1906 (1971).

Hwang, P., Guyda, H., and Friesen, H. Purification of human prolactin. *J. Biol. Chem.* 247, 1955-1958 (1972).

Igarashi, M. and McCann, S.M. A hypothalamic follicle stimulating hormone-releasing factor. *Endocrinology* 74, 446-452 (1964).

Ingle, D.J. and Higgins, G.M. Atrophy of the adrenal cortex in the rat produced by administration of large amounts of cortin. *Anat. Record* 71, 363-372 (1938).

Ingle, D.J. and Kendall, E.C. Atrophy of the adrenal cortex of the rat produced by the administration of large amounts of cortin. *Science* 86, 245 (1937).

Ito, T. Experimental studies on the hypothalamic control of the pars intermedia activity of the frog, *Rana nigromaculata*. *Neuroendocrinology* 3, 25-33 (1968).

Iversen, L.L. Role of transmitter uptake mechanisms in synaptic neurotransmission. *Brit. J. Pharmac.* 41, 571-591 (1971).

Iversen, L.L. Catecholamine uptake processes. *Brit. Med. Bull.* 29, 130-135 (1973).

Iversen, L.L. and Salt, P.J. Inhibition of catecholamine uptake$_2$ by steroids in the isolated rat heart. *Brit. J. Pharmacol.* 40, 528-530 (1970).

Jackson, I.M.D. Extrahypothalamic and phylogenetic distribution of hypothalamic peptides, in *The Hypothalamus*. S. Reichlin, R.J. Baldessarini, and J.B. Martin, eds. Raven Press, New York (1978), pp. 217-231.

Jackson, I.M.D. and Reichlin, S. Brain throtropin-releasing hormone is independent of the hypothalamus. *Nature* 267, 853-854 (1977).

Jacobs, L.S., Mariz, I.K., and Daughaday, W.H. A mixed heterologus radioimmunoassay for human prolactin. *J. Clin. Endocrinol.* 34, 484-490 (1972).

Jacobs, L.S., Snyder, P.F., Wilber, J.F., Utiger, R.O., and Daughaday, W.H. Increased serum prolactin after administration of synthetic thyrotropin releasing hormone (TRH) in man. *J. Clin. Endocrinol. Metab.* 33, 996-998 (1971).

Janowsky, D.S. and Davis, J.M. Progesterone-estrogen effects on uptake and release of norepinephrine by synaptosomes. *Life Sci.* 9, 525-531 (1970).

Jenkin, G. and Heap, R.B. Formation of oestradiol-17β from oestrone sulphate by sheep foetal pituitary *in vitro*. *Nature* 259, 330-331 (1976).

Johnson, J.A. and Assam, S. Use of betamethasone in reduction of cerebral edema. *Milit. Med.* 131, 44-47 (1966).

Jones, M.T., Tiptaft, E.M., Brush, F.R., Fergusson, D.A.N., and Neame, R.L.B. Evidence for dual corticosteroid-receptor mechanisms in the feedback control of adrenocorticotrophin secretion. *J. Endocrin.* 60, 223-233 (1974).

Jonsson, G., Fuxe, K., and Hokfelt, T. On the catecholamine innervation of the hypothalamus with special reference to the median eminence. *Brain Res.* 40, 271-281 (1972).

Joseph, S.A. and Kingge, K.M. The endocrine hypothalamus: recent anatomical studies, in *The Hypothalamus*. S. Reichlin, R.J. Baldessarini, and J.B. Martin, eds. Raven Press, New York (1978), pp. 15-47.

Jouan, P., Samerez, S., and Thieulant, M.L. Testosterone "receptors" in purified nuclei of rat anterior hypophysis. *J. Steroid Biochem.* 4, 65-74 (1973).

Kahwanago, I., Heinrichs, W.L., and Herrmann, W.L. Isolation of oestradiol "receptors" from bovine hypothalamus and anterior pituitary gland. *Nature* 223, 313-314 (1969).

Kakiuchi, S. and Rall, T.W. The influence of chemical agents on the accumulation of adenosine 3',5'-phosphate in slices of rabbit cerebellum. *Mol. Pharmacol.* 4, 367-378 (1968a).

Kakiuchi, S., and Rall, T.W. Studies on adenosine 3',5' - phosphate in rabbit cerebral cortex. *Mol. Pharmacol.* 4, 379-388 (1968b).

Kakiuchi, S., Rall, T.W. and McIlwain, H. The effect of electrical stimulation upon the accumulation of adenosine 3',5'-phosphate in isolated cerebral tissue. *J. Neurochem.* 16, 485-491 (1969).

Kalra, P.S. and McCann, S.M. Effect of CNS implants of ovarian steroids gonadotropin release. *IVth Int. Congress Endocrinol. Excerpta Medica ICS* 256, 118 (Abstract) (1972).

Kalsner, S. Mechanism in hydrocortisone potentiation of responses to epinephrine and norepinephrine in rabbit aorta. *Circulat. Res.* 24, 383-395 (1969a).

Kalsner, S. Steroid potentiation of responses to sympathomimetric amines in aortic strips. *Brit. J. Pharmacol.* 36, 582-593 (1969b).

Kamberi, I.A. Suppression of ovulation and the proestrou surge of LH and FSH by atropine amelatonis. *Physiologist* 15, 187 (1972).

Kamberi, I.A. The role of brain monamines and pineal indoles in the secretion of gonadotrophins and gonodotrophin-releasing factors, in *Drug Effects of Neuroendocrine Regulation: Progress in Brain Research*, Vol. 39. E. Zimmerman, W.H. Gispen, B.H. Marks, and D. DeWied, eds. Elsevier, Amsterdam (1973), pp. 261-280.

Kamberi, I.A. and McCann, S.M. Effect of biogenic amines, FSH-releasing factor (FRF) and other substances on the release of FSH by pituitaries incubated *in vitro*. *Endocrinology* 85, 815-824 (1969).

Kamberi, I.A., Mical, R.S., and Porter, J.C. Effect of anterior pituitary perfusion and intraventricular injection of calecholamines and indoleamines on LH release. *Endocrinology* 87, 1-12 (1970).

Kamberi, I.A., Mical, R.S. and Porter, J.C. Effect of anterior pituitary perfusion and intraventricular injection of catecholamines of FSH release. *Endocrinology* 88, 1003-1011 (1971a).

Kamberi, I.A., Mical, R.S., and Porter, J.C. Effect of anterior pituitary perfusion and intraventricular injection of catecholamines on prolactin release. *Endocrinology* 88, 1012-1020 (1971b).

Kansal, P.C., Buse, J., Talbert, O.R., and Buse, M.G. The effect of L-DOPA on plasma growth hormone, insulin and thyroxine. *J. Clin. Endocrinol. Metab.* 34, 99-105 (1972).

Kastin, A.J. and Ross, G.T. Melanocyte-stimulating hormone activity in pituitaries of frogs with hypothalamic lesions. *Endocrinology* 77, 45-48 (1965).

Kastin, A.J. and Schally, A.V. MSH activity in pituitary glands of rats treated with tranquilizing drugs. *Endocrinology* 79, 1018-1020 (1966).

Kastin, A.J., Arimura, A., and Schally, A.V. Mass action-type direct feedback control of pituitary release. *Nature* 231, 29-30 (1971).

Kastin, A.J., Kullander, S., Borglin, N.E., Dahlberg, B., Dyster-Aag, K., Krakau, C.E.T., Inguar, D.H., Miller, M.C., Bowers, C.Y., and Schally, A.V. Estrapigmentary effects of melanocyte-stimulating hormone in amenorrhoeic women. *Lancet* 1, 1007-1010 (1968).

Kato, J. Estrogen receptors in the hypothalamus and hypophysis in relation to reproduction, in *Hormonal Steroids, Proceedings of the Third International Congress, Hamburg, Int. Congr. Ser. No. 219*. V.H.T. James and L. Martini, eds., Excerpta Medica, Amsterdam (1970), pp. 764-773.

Kato, J. Steroid hormone receptors in brain, hypothalamus, and hypophysis, in *Receptors and Mechanism of Action of Steroid Hormones, Part 2.* J.R. Pasqualini, ed. Marcel Dekker, Inc., New York (1977), pp. 603-671.

Kato, J. and Minaguchi, H. Cholinergic and adrenergic mechanisms in the female rat hypothalamus with special reference to reproductive functions, in *Neurosecretion and Neural Control of Internal Secretion.* K. Kurosumi, ed. Sunmar University Press, Maebashi, Japan (1964), pp. 269-281.

Kato, J. and Onouchi, T. 5-α-Dihydrotestosterone receptor in the rat hypophysis. *Endocrinol. Japan* 20, 641-644 (1973).

Kato, J. and Villee, C.A. Preferential uptake of estradiol by the anterior hypothalamus of the rat. *Endocrinology* 80, 567-575 (1967).

Kawakami, M. and Sawyer, C.H. Induction of behavioral and electroencephalographic changes in the rabbit by hormone administration or brain stimulation. *Endocrinology* 65, 631-643 (1959)

Kawakami, M., Koshino, T., and Hattori, Y. Changes in the EEC of the hypothalamus and limbic system after administration of ACTH, SU-4885 and ACH in rabbits with special reference to neurohumoral feedback regulation of pituitary-adrenal system. *Japan J. Physiol.* 16, 551-569 (1966).

Kawakami, M., Seto, K., Terasawa, E., and Yoshida, K. Mechanisms in the limbic system controlling reproductive functions of the ovary with special reference to the postive feedback of progestin to the hippocampus. *Prog. Brain Res.* 27, 69-102 (1967).

Kawakami, M., Seto, K., and Yoshida, K. Influence of corticosterone implantation in limbic structure upon biosynthesis of adrenocortical steroid. *Neuroendocrinology* 3, 340-354 (1968).

Kendall, J.W. Feedback control of adrenocorticotropic hormone secretion, in *Frontiers in Neuroendocrinology.* L. Martini and W.F. Ganong, eds. Oxford University Press, London (1971), pp. 177-207.

Kendall, J.W. and Allen, C.W. Studies on the glucocorticoid feedback. *Endocrinology* 82, 397-405 (1968).

Kendall, J.W., Tang, L., and Cook, D.M. Sites of feedback control in the pituitary adrenocortical system, in *Anatomical Neuroendocrinology Int. Conf. Neurobiology of CNS. Hormone Interactions,* W.E. Stumpf and L.D. Grant, eds. S. Karger, Basel (1975), pp. 276-283.

Kissel, J.H., Rosenfeld, M.G., Chase, L.R., and O'Malley, B.W. Response of chick oviduct adenyl cyclase to steroid hormones. *Endocrinology* 86, 1019-1023 (1970).

Kitay, J.I. and Altchule, M.D. *The Pineal Gland.* Harvard University Press, Cambridge (1954).

Kitay, J.I., Holub, D.A., and Jailer, J.W. Inhibition of pituitary ACTH release: an extra adrenal action of exogenous ACTH. *Endocrinology* 64, 475-482 (1959).

Klee, C.B. and Sokoloff, L. Mitochondrial differences in mature and immature brain; influence on rate of amino acid incorporation into protein and responses to thyroxine. *J. Neurochem.* 11, 709-716 (1964).

Kleinberg, D.L., Noel, G.L., and Frantz, A.G. Chlorpromazine stimulation and L-DOPA suppression of plasma prolactin in man. *J. Clin. Endocrinol.* 33, 873-876 (1971).

Knigge, K.M., Shock, D. and Sladek, J.R., Jr. Monoamines of median eminence, in *Brain-Endocrine Interaction II. The Ventricular System* 2nd Int. Symp., Shizuoka (K.M. Knigge, ed.). S. Karger, Basel (1975), pp. 282-294.

Kobayashi, H. and Matsui, T. Fine structure of the median eminence and its functional significance, in *Frontiers in Neuroendocrinology*. W.F. Ganong and L. Martini, eds. Oxford University Press, New York (1969), pp. 3-46.

Kobayashi, T., Takeyawa, S., Oshima, K., and Kawamura, H. Electrophysiological studies on the feedback mechanism of progesterone. *Endocrinol. Japan* 9, 302-320 (1962).

Koe, B.K. and Weissman, A. P-chlorophenylalanine: a specific depletor of brain serotonin. *J. Pharmacol. Exp. Ther.* 154, 499-516 (1966).

Kracier, J., Gosbee, J.L., and Benscome, S.A. Pars intermedia and pars distalis: two sites of ACTH production in the rat hypophysis. *Neuroendocrinology* 11, 156-176 (1973).

Krulich, L., Dhariwal, A.P.S., and McCann, S.M. Stimulatory and inhibitory effects of purified hypothalamic extracts on growth hormone release from rat pituitary in vitro. *Endocrinology* 83, 783-790 (1968).

Kuroshima, A., Ishida, Y., Bowers, C.P., and Schally, A.V. Stimulation of release of follicle-stimulating hormone by hypothalamic extracts *in vitro* and *in vivo*. *Endocrinology* 76, 614-619 (1965).

Lassman, M.N. and Mulrow, P.J. Deficiency of deoxycorticosterone binding protein in the hypothalamus of rats resistant to deoxycorticosterone-induced hypertension. *Endocrinology* 94, 1541-1546 (1974).

Latham, K.R., Ring, J.C., and Baxter, J.D. Solubilized nuclear "receptors" for thyroid hormones. Physical characteristics and binding properties, evidence for multiple forms. *J. Biol. Chem.* 251, 7388-7397 (1976).

Lawton, I.E. and Sawyer, C.H. Role of amygdale in regulating LH secretion in the adult female rat. *Amer. J. Physiol.* 218: 622-626 (1970).

Lebowitz, Harold E., Skyler, J.S., and Boyd, A.E. L-DOPA and growth hormone secretion in man. *Advanced Neurol.* 5, 461-469 (1974).

Lee, M.D. and Van Buskirk, E.F. Studies on vigor. XV. The effect of thyroidectomy on spontaneous activity in the rat, with a consideration of the relation of the basal metabolism to spontaneous activity. *Am. J. Physiol.* 84, 321-329 (1928).

Leonard, B.F. The effect of two synthetic ACTH analogues on the metabolism of biogenic amines in the rat brain. *Arch. Int. Pharmacodyn.* 207, 242-253 (1974).

Lewis, U.J., Singh, R.N.P., Sinhia, Y.N., and Van der Laan, W.P. Electrophoretic evidence for human prolactin. *J. Clin. Endocrinol.* 33, 153-156 (1971).

Li, C.H. Hormones of the adenohypophysis. *Proc. Am. Phil. Soc.* 116, 365-382 (1972).

Libertun, C., Timiras, P.S., and Kragt, C.L. Sexual differences in the hypothalamic cholinergic system before and after puberty: inductory effect of testosterone. *Neuroendocrinology* 12, 73-85 (1973).

Lichtigfeld, F.J. and Simpson, G.M. Hypocalcemis status. *New Engl. J. Med.* 276, 874-875 (1967).

Liddle, G.W., Nicholson, W.E., Island, D.P., Orth, D.N., Abe, K., and Lauder, S.C. Clinical and laboratory studies of ectopic humoral syndromes. *Rec. Prog. Horm. Res.* 25, 283-314 (1969).

Lieberburg, I. and McEwen, B. Estradiol-17β: a metabolite of testosterone recovered in cell nuclei from limbic areas of neonatal rat brains. *Brain Res.* 85, 165-170 (1975).

Leiberburg, I., Wallach, G., and McEwen, B.S. The effects of an inhibitor of aromatization (1, 4, 6-androstatriene-3, 17-dione) and an anti-estrogen (CI-628) on *in vivo* formed testosterone metabolites recovered from neonatal rat brain tissues and purified cell nuclei. Implications for sexual differentiation of the rat brain. *Brain Res.* 128, 176-181 (1977).

Lieberburg, I., Maclusky, N.J., Roy, E.J., and McEwen, B.S. Sex steroid receptors in the perinatal rat brain. *Am. Zool.* 18, 539-544 (1978).

Lipton, M.A., Prange, A.J., Dairman, W., and Udenfriend, S. Increased rate of norepinephrine biosynthesis in hypothyroid rats. *Federation Proc.* 27, 399 (1968).

Lisk, R.D. Estrogen-sensitive centers in the hypothalamus of the rat. *J. Exp. Zool.* 145, 197-208 (1960)

Lisk, R.D. Testosterone sensitive centers in the hypothalamus of the rat. *Acta Endocrinol.* 41, 195-204 (1962).

Littlejohn, B.M. and De Groot, J. Estrogen sensitive areas in the rat brain. *Federation Proc.* 22, 571 (Abstract) (1963).

Loewenstein, J., Mariz, E., Peake, G.T., and Daughaday, W.H. Prolactin bioassay by induction of ν-acetyllactosamine synthetase in mouse mammary gland explants. *J. Clin. Endocrinol.* 33, 217-224 (1971).

Loizou, L.A. The postnatal ontogeny of monoamine-containing neurons in the central nervous system of the albino rat. *Brain Res.* 40, 395-418 (1972).

Lorenzen, L.C. and Ganong, W.F. Effect of drugs related to α-ethyltryptamine on stress-induced ACTH secretion in the dog. *Endocrinology* 80, 889-892 (1967).

Lovenberg, W., Besselar, G.H., Bensinger, R.E., and Jackson, R.L. Physiologic and drug-induced regulation of serotonin synthesis, in *Serotonin and Behavior*. J. Barchus and E. Usdin, eds. Academic Press, New York (1973), pp. 49-54.

Lu, K.H. and J. Meites. Effects of L-Dopa on serum prolactin and PIF in intact and hypophysectomized, pituitary-grafted rats. *Endocrinology* 91, 868-872 (1973).

Lucy, I. Experimental models for biological membranes, in *Biological Membranes*. D. Chapman, ed., Academic Press, London (1968), pp. 223-288.

Luft, R., Efendic, S., Hokfelt, T., Johansson, O., and Arimura, A. Immunohistochemical evidence for the localizations of somatostatin-like immunoreactivity in a cell population of the pancreatic islets. *Med. Biol.* 52, 428-430 (1974).

Luine, V.N., Khylcherskaya, R.I., and McEwen, B.S. Effect of gonadal steroids on activities of monoamine oxidase and choline acetylase in rat brain. *Brain Res.* 86, 293-306 (1975).

Maayan, M.L., Shapiro, R., and Ingbar, S.H. Eprinephrine precursors: effect on the iodine and intermediary metabolism of isolated calf thyroid cells. *Endocrinology* 92, 912-916 (1973).

MacIndoe, J.H. and Turkington, R.W. Stimulation of human prolactin secretion by intravenous infusion of L-tryptophan. *J. Clin. Invest.* 52, 1972-1978 (1973).

MacLeod, K.M. and Baxter, J.D. Chromatin receptors for thyroid hormones: interactions of solubilized proteins with DNA. *J. Biol. Chem.* 251, 7380-7387 (1976).

MacLeod, R.M. and Lehmeyer, S.E. Regulation of the synthesis and release of prolactin, in *Lactogenic Hormones*. G.E.W. Wolstenholme and J. Knight, eds. Churchill Livingstone, Edinburgh (1972), pp. 55-82.

Maclusky, N.J., Turner, B.B., and McEwen, B.S. Corticosteroid binding in rat brain and pituitary cytosols: resolution of multiple binding components by polyacrylamide gel based isoelectric focusing. *Brain Res.* 130, 564-571 (1977).

Maickel, R.P., Westermann, E.O., and Brodie, B.B. Effects of reserpine and cold exposure on pituitary adrenocortical function in rats. *J. Pharmacol. Exp. Ther.* 134, 167-175 (1961).

Manganiello, V., and Vaughan, M. An effect of dexamethasone on adenosine 3',5'-monophosphate content and adenosine 3',5'-monophosphate phosphodiesterase activity of cultured hepatoma cells. *J. Clin. Invest.* 51, 2763-2767 (1972).

Marczynski, T.I., Yamaguchi, N., Ling, G.M., and Grodzinska, L. Sleep induced by the administration of melatonin (5-methoxy-N-acetyltryplamine) to the hypothalamus in unrestrained cats. *Experentia* 20, 435 (1964).

Martin, J.B. Inhibitory effect of somatostatin (SRIF) on the release of growth hormone (GH) induced in the rat by electrical stimulation. *Endocrinology* 94, 479-593 (1974).

Martin, J.B., Boshans, R., and Reichlin, S. Feedback regulation of TSH secretion in rats with hypothalamic lesions. *Endocrinology* 87, 1032-1040 (1970).

Martin, J.B. and Jackson, I.M.D. Neural regulation of pituitary TSH and GH secretion, in *Anatomical Neuroendocrinology*. Int. Conf. Neurobiology of CNS-Hormone Interactions. W.E. Stumpf and L.D. Grant, eds., S. Karger, Basel (1975) pp. 343-353.

Martini, L. Androgen reduction by neuroendocrine tissues: physiological significance, in *Subcellular Mechanisms in Reproductive Neuroendocrinology*. F. Natfolin, K.J. Ryan, and J. Davies, eds. Elsevier, Amsterdam (1976), pp. 327-345.

Massa, P., Justo, S., and Martini, L. Conversion of testosterone into 5 α-reduced metabolites in the anterior pituitary and in the brain of maturing rats, in *Sexual Endocrinology of the Perinatal Period*. M.G. Forest and J. Bertrand, eds. INSERM, Paris (1974), pp. 219-232.

Matsuda, K., Kendall, J.W., Puyck, C., and Greer, M.A. Neural control of ACTH secretion: effect of acute decerebration in the rat. *Endocrinology* 72, 845-852 (1963).

Matsuo, H., Baba, Y., Nair, R.M.G., Arimura, A., and Schally, A.V. Structure of the porcine LH-ana FSH-releasing hormone. I. The proposed amino acid sequence. *Biochem. Biophys. Res. Commun.* 43, 1334-1339 (1971).

Matsuyama, S. Effects of brain implantation of androgen, estrogen and androgen antagonist on the reproductive organs of male rat. *Nippon Seirigaku Zasshi* 32, 152-164 (1970).

Maxwell, R.E., Long, D.M., and French, L.A. The effects of glucosteroids on experimental cold-induced brain edema. *J. Neurosurg.* 34, 477-487 (1971).

McCann, S.M. A hypothalamic luteinizing hormone-releasing factor LH-RF) *Am. J. Physiol.* 202, 395-400 (1962).

McCann, S.M. and Ojeda, S.R. Synaptic transmitters involved in the release of hypothalamic releasing and inhibiting hormones, in *Reviews of Neuroscience*, Vol. 2. S. Ehrenpries and I.J. Kopin, eds. Raven Press, New York (1976), pp. 91-110.

McCann, S.M. and Ramirez, V.D. The neuroendocrine regulation of hypophyseal luteinizing hormone secretion. *Rec. Prog. Horm. Res.* 20, 131-181 (1964).

McCann, S.M., Dhariwal, A.P.S., and Porter, J.C. Regulation of the adenohypophysis. *Ann. Rev. Physiol.* 30, 589-640 (1968).

McCann, S.M., Taleisnik, S., and Friedman, H.M. LH-releasing activity in hypothalamic extracts. *Proc. Soc. Exp. Biol. Med.* 104, 432-434 (1960).

McEwen, B.S. Steroid receptors in neuroendocrine tissues: topography subcellular distribution, and functional implications, in *Subcellular Mechanisms in Reproductive Neuroendocrinology*. F. Naftolin, K.J. Ryan, and J. Davies, eds. Elsevier, Amsterdam (1976a), pp. 277-304.

McEwen, B.S. Endocrine effects on the brain and their relationship to behavior, in *Basic Neurochemistry*. G.J. Siegel, R.W. Albers, R. Katzman, and B.W. Agranoff, eds. Little, Brown and Co., Boston (1976b), pp. 737-764.

McEwen, B.S. Gonadal steroid receptors in neuroendocrine tissues, in *Receptors and Hormone Action*, Vol. II. B.W. O'Malley and L. Birnbaumer, eds. Academic Press, New York (1978), pp. 353-400.

McEwen, B.S. and Pfaff, D.W. Factors influencing sex hormone uptake by rat brain regions. I. Effects of neonatal treatment, hypophysectomy, and competing steroid on estradiol uptake. *Brain Res.* 21, 1-16 (1970).

McEwen, B.S., Dekoet, E.R., and Wallach, G. Interactions *in vivo* and *in vitro* of corticoids and progesterone with cell nuclei and soluble macromolecules from rat brain regions and pituitary. *Brain Res.* 105, 129-136 (1976).

McEwen, B.S., Magnus, C., and Wallach, G. Soluable corticosterone-binding macromolecules extracted from rat brain. *Endocrinology* 90, 217-226 (1972a).

McEwen, B.S., Pfaff, D.W., and Zigmond, R.E. Factors influencing sex hormone uptake by rat brain regions. II. Effects of neonatal treatment and hypophysectomy in testosterone uptake. *Brain Res.* 21, 17-28 (1970a).

McEwen, B.S., Pfaff, D.W., and Zigmond, R.E. Factors influencing sex hormone uptake by rat brain regions. III. Effects of competing steroids on testosterone uptake. *Brain Res.* 21, 29-38 (1970b).

McEwen, B.S., Weiss, J.M., and Schwartz, L.S. Uptake of corticosterone by rat brain and its concentration by certain limbic structures. *Brain Res.* 16, 227-241 (1969).

McEwen, B.S., Weiss, J.M., and Schwartz, L.S. Retention of corticosterone by cell nuclei from brain regions of adrenalectomized rats. *Brain Res.* 17, 471-482 (1970c).

McEwen, B.S., Zigmond, R.E., and Gerlach, J.L. Sites of steroid binding and action in the brain, in *Structure and Function of Nervous Tissue*, Vol. 5. G.H. Bourne, ed. Academic Press, New York (1972b), pp. 205-291.

McGuire, G.L. and Lisk, R.D. Estrogen receptors in the intact rat. *Proc. Nat. Acad. Sci.* (Wash.) 61, 497-503 (1968).

Meites, J. and Nicoll, C.S. Adenohypophysis: prolactin. *Ann. Rev. Physiol.* 28, 57-88 (1966).

Meites, J., Lu, K.H., Wuttke, W., Welsch, C.W., Nagasawa, H., and Quadri, S.K. Recent studies on functions and control of prolactin secretion in rats. *Recent Prog. Horm. Res.* 28, 471-526 (1972).

Merryman, W., Boiman, R., Barnes, L., and Rothchild, I. Progesterone "anesthesia" in human subjects. *J. Clin. Endocrinol. Metab.* 14, 1567-1569 (1954).

Mess, B., Tima, L., and Trentini, G.P. The role of pineal principles in ovulation, in *Drug Effects on Neuroendocrine Regulation (Prog. Brain Res. Vol. 39)*. B.H. Marks and D. De Wied, eds. Elsevier, Amsterdam (1973) pp. 251-259.

Millichap, J.G. Systemic electrolyte and neuroendocrine mechanisms, in *Basic Mechanisms of the Epilepsies*. H.H. Jasper, A.A. Ward and A. Pope, eds. Little, Brown and Company, Boston (1969), pp. 709-726.

Mittler, J.C. and Meites, J. In *vitro* stimulation of pituitary follicle-stimulating-hormone release by hypothalamic extract. *Proc. Soc. Exp. Biol. Med.* 117, 309-313 (1964).

Mittler, J.C. and Meites, J. Effects of hypothalamic extract and androgen on pituitary FSH release in *vitro*. *Endocrinology* 78, 500-504 (1966).

Moss, R.L. Role of hypophysiotropic nemohormones in mediating nemal and behavioral events. *Federation Proc.* 36, 1978-1983 (1977).

Moss, R.L., Dyball, R.E.J., and Cross, B.A. Excitation of antidromically identified neurosecretory cells of the paraventricular nucleus by oxytocin applied iontophoretically. *Exp. Neurol.* 34, 95-102 (1972).

Motta, M., Fraschini, F., Guilliani, G., and Martini, L. The central nervous system, estrogen and puberty. *Endocrinology* 83, 1101-1107 (1968a).

Motta, M.F., Fraschini, F., Pira, F., and Martini, L. Hypothalamic and estra-hypothalamic mechanisms controlling adrenocorticotrophin secretion. *Mem. Soc. Endocrin.* 17, 3-17 (1968b).

Mowles, T.F., Ashkanazy, B., Mix, E., Jr., and Sheppard, H. Hypothalamic and hypophyseal estradiol-binding complexes. *Endocrinology* 89, 484-491 (1971).

Mueller, G.P., Chen, H.J., and Meites, J. In *vivo* stimulation of prolactin release in the rat by synthetic TRH. *Proc. Soc. Exp. Biol. Med.* 144, 613-615 (1973).

Müller, E.E. Brain monoamines participation in the control of growth hormone secretion indifferent animal species. *Frontiers in Catecholamine Research*, E. Usdin and H. Solomon, eds., Pergamon Press, Oxford (1973), pp. 835-841.

Munson, P.L. Pharmacology of neuroendocrine blocking agents, in *Advances in Neuroendocrinology*. A.V. Nabaldov, ed. University of Illinois Press, Urbana (1963), pp. 427-444.

Nadler, R.D. Sexual differentiation following intrahypothalamic implantation of steroids, in *Influence of Hormones on the Nervous System* (Proc. Int. Soc. Psychoneuroendocrinology, Brooklyn). S Karger, Basel (1971) pp. 306-321.

Naess, O. and Attramadal, A. Uptake and binding of androgens by the anterior pituitary gland, hypothalamus, preoptic area, and brain cortex of rats. *Acta Endocrinol.* 76, 417-430 (1974).

Naftolin, F., Ryan, K.J., and Davies, I.J. Androgen aromatization by neuroendocrine tissues, in *Subcellular Mechanisms in Reproductive Neuroendocrinology*. F. Naftolin, K.J. Ryan, and J. Davies, eds. Elsevier, Amsterdam (1976), pp. 347-355.

Naftolin, F., Ryan, K.J., Davies, I.J., Reddy, V.V., Flores, F., Petro, Z., and Kuhn, M. The formation of estrogens by central neuro-endocrine tissues. *Rec. Prog. Horm. Res.* 31, 295-315 (1975).

Naidoo, S., Valcana, T., and Timiras, P.S. Thyroid hormone receptors in the developing rat brain. *Amer. Zool.* 18, 545-552 (1978).

Neckers, L. and Sze, P.Y. Regulation of 5-hydroxytryptamine metabolism in mouse brain by adrenal glucocorticoids. *Brain Res.* 93, 123-132 (1975).

Nir, I., Behroozi, K., Assael, M., Iwriani, I., and Sulman, F.G. Changes in the electrical activity of the brain following pinealectomy. *Neuroendocrinology* 4, 122-127 (1969).

Nixon, R.L., Janowsky, D.S., and Davis, J.M. Effects of progesterone, β-estradiol, and testosterone on the uptake and metabolism of ^{3}H-norepinephrine, ^{3}H-dopamine and ^{3}H-serotonin in rat brain synaptosomes, in *Research Communications in Chemical Pathology and Pharmacology*, D.V. Siva Sankar, ed., Westbury, N.Y. 1 (1974), pp. 233-236.

Noel, S.L., Sah, H.K., and Frantz, A.S. Induction of prolactin release by breast stimulation in humans, in *Fourth Int. Congr. Endocrinol.* Wash. D.C., Excerpta Med., Int. Congr. Series, Amsterdam, June 1972 (Abstract No. 256).

Notides, A.C. Binding affinity and specificity of the estrogen receptor of the rat uterus and anterior pituitary. *Endocrinology* 87, 987-992 (1970).

Nurberger, J.I. Clinical and metabolic effects of exposure to low environmental temperatures on the central nervous system and related visceral structures. *Res. Publ. Ass. Res. Nerv. Ment. Dis.* 32, 132-173 (1953).

Oliver, C., Eskay, R.L., Ben-Jonathan, N., and Porter, J.C. Distribution and concentration of TRH in the rat brain. *Endocrinology* 96, 540-546 (1974).

Oppenheimer, J.H., Schwartz, H.L., and Surks, M.I. Tissue differences in the concentration of triiodothyronine nuclear binding sites in the rat: liver, kidney, pituitary, brain, spleen and testis. *Endocrinology* 95, 897-903 (1974).

Oppenheimer, J.H., Koerner, D., Schwartz, H.L., and Surks, M.I. Specific nuclear triiodothyronine binding sites in rat liver and kidney. *J. Clin. Endocrinol. Metab.* 35, 330-333 (1972).

Palkovits, M., Arimura, A., Brownstein, M., and Saavedra. J.M. Lutenizing hormone releasing hormone (LH-RH) content of the hypothalamic nuclei in the rat. *Endocrinology* 95, 554-558 (1974).

Papaikonomou, E. Biocybernetics, biosystems analysis and the pituitary adrenal system. Biomedical Systems Engineer, Free University Medical Faculty, Amsterdam, The Netherlands, Nooy's Drukkerij, Purmerend 106 (1974).

Parker, M.L., Hammond, J.M., and Daughaday, W.H. The arginine provocative test: an aid in the diagnosis of hyposomatotropism. *J. Clin. Endocr. Metab.* 27, 1129-1136 (1967).

Parker, C.W., Huber, M.G., and Baumann, M.L. Alterations in cyclic AMP metabolism in human bronchia asthma. *J. Clin. Invest.* 52, 1342-1348 (1973).

Parlow, A.F. Bioassay of pituitary luteinizing hormone by depletion of ovarian ascorbic acid, in *Human Pituitary Gonadotrophins*. A. Albert, ed. Charles C. Thomas Co., Springfield, Ill. (1961), pp. 300-310.

Paul, S.M. and Axelrod, J. Catechol estrogens: presence in brain and endocrine tissues. *Science* 197, 657-659 (1977).

Pearse, A.G.E. and Takor, T. Embryology of the diffuse neuroendocrine system and its relationship to the common peptides. *Federation Proc.* 38, 2288-2294 (1979).

Pelletier, G., Leclerc, R., Dube, D., Labrie, F., Puviane, R., Arimura, A., and Schally, A.V. Localization of growth-hormone release inhibiting hormone (somatostatin) in the rat brain. *Am. J. Anat.* 142, 397-401 (1975).

Perkins, J.P. Regulation of responsiveness of cells to catecholamines: variable expression of the components of the second messenger system, in *Cyclic Nucleotides in Disease*. B. Weiss, ed. University Park Press, Baltimore (1975), pp. 351-376.

Pert, C.B. and Snyder, S.H. Opiate receptor: demonstration in nervous tissue. *Science* 179, 1011-1012 (1973).

Peter, R.E. Feedback effects of thyroxine on the hypothalamus and pituitary of goldfish, *Carassius auratus*. *J. Endocrinol.* 51, 31-39 (1971).

Pfaff, D.W. Uptake of ^{3}H-estradiol by the female rat brain. An autoradiographic study. *Endocrinology* 82, 1149-1155 (1968).

Pfaff, D.W. and Keiner, M. Atlas of estradiol-concentrating cells in the central nervous system of the female rat. *J. Comp. Neurol.* 151, 121-158 (1973).

Pfaff, D.W., Gerlach, J.L., McEwen, B.S., Ferin, M., Carmel, P., and Zimmerman, E.A. Autoradiographic localization of hormone-concentrating cells in the brain of the female rhesus monkey. *J. Comp. Neurol.* 170, 279-293 (1976).

Pfapinger, L. and McEwen, B.S. Ontogeny of estradiol-binding sites in rat brain. I. Appearance of presumptive adult receptors in cytosol and nuclei. *Endocrinology* 93, 1119-1128 (1973).

Pickering, B.T., Jones, C.W., and Burford, G.D. Biosynthesis and intraneuronal transport of neurosecretory products in the hypothalamo-neurohypophysial system. Ciba. Found. Study Group 39, G.E.W. Wolsteholme and J. Birch, eds. Churchill Livingstone, London (1971), pp. 58-74.

Pickering, B.T., Jones, C.W., and Burford, G.D. Biochemistry and general physiology biochemical aspects of the hypothalamo-neurohypophysial neurone, in *Neurosecretion - The Final Neuroendocrine Pathway*. F.G.W. Knowles and R. Vollrath, eds. Springer-Verlag, Berlin (1974), p. 72.

Pliska, V., Thorn, N.A., and Villhardt, H. *In vitro* uptake and breakdown of tritiated lysine-vasopressin by bovine neurohypophyseal and cortical tissue. *Acta Endocrinol.* (Copenhagen) 67, 12-22 (1971).

Porter, J.C., Mical, R.S., and Cramer, O.M. Effect of serotonin and other indoles on the release of LH, FSH, and prolactin: hormones and antagonists. *Gynec. Invest.* 2, 13-22 (1972).

Porter, J.C., Vanatta, J.C., and Dillon, H.T. Effects of plasma obtained from hypothalamico-hypophysial portal vessel blood on urinary electrolyte excretion by the rat. *Federation Proc.* 14, 116 (1955).

Prange, A.J., Meek, J.L., and Lipton, M.A. Catecholamines: diminished rates of synthesis in rat brain and heart after thyroxine pretreatment. *Life Sci.* 1, 901-907 (1970a).

Prange, A.J., Wilson, I.C., Breese, G.R., and Lipton, M.A. Behavioral effects of hypothalamic releasing hormones in animals and men, in *Progress in Brain Research, Vol. 42, Hormones, Homeostasis and the Brain*. W.H. Gispen, T.B. Van Wimersa Griedonus, B. Bohus, and D. DeWied, eds. Amsterdam, Elsevier (1975).

Prange, A.J., Wilson, I.C., Rabon, M., and Lipton, A. Enhancement of imipramine antidepressant activity by thyroid hormone. *Am. J. Psychiatry* 126, 457-469 (1969).

Prange, A.J., Wilson, I.C., Knox, A., McClane, T.K., and Lipton, M.A. Enchancement of imipramine by thyroid stimulating hormone: clinical and theoretical implications. *Am. J. Psychiatry* 127, 191-199 (1970b).

Ramirez, V.D., Abrams, R.M., and McCann, S.M. Effects of estrogen implants in the hypothalamo-hypophysial region on the secretion of LH in the brain. *Federation Proc.* 22, 506 (1963).

Ramirez, D.V. and McCann, S.M. Comparison of the regulation of luteinizing hormone (LH) secretion in immature and adult rats. *Endocrinology* 72, 452-464 (1963).

Ramirez, V.D. and McCann, S.M. Induction of prolactin secretion by implants of estrogen into the hypothalamo-hypophysial regions of female rats. *Endocrinology* 75, 206-214 (1964).

Reichlin, S. Growth hormone content of pituitaries from rats with hypothalamic lesions. *Endocrinology* 69, 225-230 (1961).

Reichlin, S., Martin, J.B., Mitnick, M., Boshans, R.L., Grimm, Y., Bollinger, J., Gordon, J., and Malacara, J. The hypothalamus in pituitary-thyroid regulation. *Rec. Prog. Horm. Res.* 28, 229-286 (1972).

Reichlin, S., Jackson, I.M.D., Seyler, L.E., and Grimm-Jorgenson, Y. Regulation of thyrotropin-releasing hormone (TRH) and luteinizing hormone-releasing hormone (LRH), in *Frontiers in Neurology and Neuroscience Research*. P. Seeman and G.M. Brown, eds. University of Toronto Press, Toronto (1974), pp. 48-59.

Reiter, R.J. and Fraschini, F. Endocrine aspects of the mammalian pineal gland: a review. *Neuroendocrinology* 5, 219-255 (1969).

Renson, J. Assays and properties of tryptophan-5-hydroxylase, in *Serotonin and Behavior*. J. Barchas and E. Usdin, eds. Academic Press, New York (1973), pp. 19-32.

Resko, J.A., Goy, R.W., and Phoenix, C.H. Uptake and distribution of exogenous testosterone-1, 2-^{3}H in neural and genital tissues of the castrate guinea pig. *Endocrinology*, 80, 490-498 (1967).

Rhees, R., Abel, J., and Haack, D. Uptake of tritiated steroids in the brain of the duck (*Anas. platyrhynchos*). *Gen. Comp. Endocrinol.* 18, 292-300 (1972).

Robel, P., Corpechol, C., and Baulieu, E.E. Testosterone and androstanolone in rat plasma and tissues. *FEBS Lett.* 33, 218-220 (1973).

Robison, G.A., Butcher, R.W., and Sutherland, E.W. Cyclic AMP. *Ann. Rev. Biochem.* 37, 149-174 (1968).

Robison, G.A., Butcher, R.W., and Sutherland, E.W. Cyclic AMP, in *The Catecholamines. General Considerations and Summary*. G. Robison, ed. Academic Press, New York (1971), pp. 224-228.

Royce, P.C. and Sayers, G. Purification of hypothalamic corticotropin releasing factor. *Proc. Soc. Exp. Biol. Med.* 103, 447-452 (1960).

Rumsfield, H.W., Jr. and Porter, J.C. Investigation of the release of ACTH. *Endocrinology* 64, 942-947 (1959).

Ryan, K.J., Naftolin, F., Reddy, V., Flores, F., and Petro, Z. Estrogen formation in the brain. *Am. J. Obstet. Gynec.* 114, 454-460 (1973).

Saavedra, J.M., Palkovits, M., Brownstein, M.J., and Axelrod, J. Serotonin distribution in the nuclei of the rat hypothalamus and preoptic region. *Brain Res.* 77, 157-165 (1974).

Sachar, E.J., Mushrush, G., Perlow, M., Weitzman, E.D., and Sassin, J. Growth hormone responses to L-DOPA in depressed patients. *Science* 178, 1304-1305 (1972).

Sachs, H., Fawcett, C.P., Takabate, Y., and Portanova, R. Biosynthesis and release of vasopressin and neurophysin. *Rec. Prog. Horm. Res.* 25, 447-491 (1969).

Sachs, H. Neurosecretion, in *Handbook of Neurochemistry*. A. Lajtha, ed. Plenum Press, New York (1970), pp. 373-428.

Sachs, H., Saito, S., and Sunde, D. Biochemical studies on the neurosecretory and neuroglial cells of the hypothalamo-neurohypophysial complex. *Mem. Soc. Endocrinol.* 19, 325-336 (1971).

Saffran, M., and Schally, A.V. The release of corticotrophin by anterior pituitary tissue *in vitro*. *Can. J. Biochem. Physiol.* 33, 408-415 (1955).

Sakamoto, A. Hypersensitivity induced in albino mice by melanocyte stimulating hormone. *Nature* (London) 211, 1370-1371 (1966).

Samuels, H.H., Tsai, J.S., and Casanova, J. Thyroid hormone action: *in vitro* demonstration of putative receptors in isolated nuclei and soluble nuclear extracts. *Science* 184, 1188-1191 (1974a).

Samuels, H.H., Tsai, J.S., Casanova, J., and Stanley, F. Thyroid hormone action: *in vitro* characterization of solubilized nuclear receptors from rat liver and cultured GH cells. *J. Clin. Invest.* 54, 853-865 (1974b).

Sassin, J.F., Frantz, A.G., Weitzman, E.D., and Kapen, S. Human prolactin 24-hour pattern with increased release during sleep. *Science* 177, 1205-1207 (1972).

Sattin, A. and Rall, F.W. The effect of adenosine and adenine nucleotides on the cyclic adenosine 3',5'-phosphate content of guinea pig cerebral cortex slices. *Mol. Pharmacol.* 6, 13-23 (1970).

Sawyer, C.H. Control of secretion of gonadotropins, in *Gonadotropins*. H.H. Cole, ed. Freeman, San Francisco (1947), pp. 113-159.

Sawyer, C.H. and Gorski, R.A. Steroid hormones and brain function; proceedings of a conference held May 24-27, 1970. Sponsored by the School of Medicine and Brain Res. Inst., University of California Press, Berkeley (1971) (CLSA Forum in Medical Sciences no. 15).

Sawyer, C.H., Markee, J.E., and Hollinshead, W.H. Inhibition of ovulation in the rabbit by the adrenergic-blocking agent dibenamine. *Endocrinology*, 41, 395-402 (1947).

Sawyer, W.H. Evolution of antidiuretic hormones and their functions. *Am. J. Med.* 42, 678-686 (1967).

Scapagnini, V. and Preziosi, P. Receptor involvement in the central noradrenergic inhibition of ACTH secretion in rat. *Neuropharmacology* 12, 57-62 (1973a).

Scapagnini, V. and Preziosi, P. Role of brain noradrenaline in the tonic A+ regulation of hypothalamic hypophysal adrenal axis. *Prog. Brain Res.* 39, 171-184 (1973b).

Schaaf, M. and Payne, C.A. Dystonic reactions to prochlorperazine in hypoparathyroidism. *New Engl. J. Med.* 275, 991-995 (1966).

Schally, A.V., Bowers, C.Y., and Redding, T.W. Presence of thyrotropic hormone-releasing factor (TRF) in porcine hypothalamus. *Proc. Soc. Exp. Biol. Med.* 121, 718-722 (1966a).

Schally, A.V., Bowers, C.Y., and Redding, T.W. Purification of thyrotropic hormone releasing factor from bovine hypothalamus. *Endocrinology* 78, 726-732 (1966b).

Schally, A.V., Lipscomb, H.S., and Guilleman, R. Isolation and amino acid sequence of a 2-corticotrophin-releasing factor from hog pituitary glands. *Endocrinology* 71, 164-173 (1962).

Schally, A.V., Bowers, C.Y., Redding, T.W. and Barrett, J.F. Isolation of thyrotropin releasing factor (TRF) from porcine hypothalamus. *Biochem. Biophys. Res. Commun.* 25, 165-169 (1966c).

Schneider, H.P.G. and McCann, S.M. Possible role of dopamine as transmitter to promote discharge of LH-releasing factor. *Endocrinology*, 85, 121-132 (1969).

Scott, A.P., Ratcliffe, J.G., Rees, L.H. and Landon, J. Pituitary peptide. *Nature* 244, 65-67 (1973).

Selye, H. Studies concerning the anesthetic action of steroid hormones. *J. Pharmacol. Exp. Ther.* 73, 127-141 (1941).

Shizume, K., Matsuda, L., Irie, M., Iino, S., Ishii, J., Nagataki, S.,

Matsuzaki, F., and Okinaka, S. Effect of electrical stimulation of the hypothalamus on thyroid function. *Endocrinology* 95, 968-977 (1974).

Silverman, A.J. Distribution of lutenizing hormone releasing hormone (LH-RH) in the guinea pig brain. *Endocrinology* 99, 30-46 (1976).

Silverman, A.J., Antunes, J.L., Ferin, M., and Zimmerman, E.A. The distribution of luteinizing hormone-releasing hormone (LHRH) in the hypothalamus of the rhesus monkey. Light microscopic studies using immunoperoxidase technique. *Endocrinology* 101, 134-142 (1977).

Slusher, M.A. Effects of cortisol implants in the brain stem and ventral hippocampus on diurnal corticosteroid level. *Exp. Brain Res.* 1, 184-194 (1966).

Smelik, P.G. The regulation of ACTH secretion. *Acta Physiol. Pharmacol. Neer.* 13, 370-371 (1965).

Smelik, P.G. and Sawyer, C.H. Effects of implantation of cortisol into the brain stem or pituitary gland on the adrenal response to stress in rabbit. *Acta Endocrinol.* 41, 561-570 (1962).

Smith, E.R. and Davidson, M. Differential response to hypothalamic testosterone in relation to male puberty. *Amer. J. Physiol.* 212, 1385-1390 (1967b).

Smith, E.R. and Davidson, M. Testicular maintenance and its inhibition in pituitary-transplanted rats. *Endocrinology* 80, 725-734 (1967a).

Smith, E.R. and Davidson, J.M. Role of estrogen in the cerebral control of puberty in female rats. *Endocrinology* 82, 100-108 (1968).

Stern, J. and Eisenfeld, A. Distribution and metabolism of ^{3}H-testosterone in castrated male rats: effects of cypertrone, progesterone, and unlabeled testosterone. *Endocrinology* 88, 1117-1125 (1971).

Stevens, W., Reed, D.J., and Grosser, B.I. Binding of natural and synthetic glucocorticoids in rat brain. *J. Steroid Biochem.* 6, 521-527 (1975).

Stevens, W., Reed, D.J., Erickson, S., and Grosser, B.I. The binding of corticosterone to brain proteins: dimnal variation. *Endocrinology* 93, 1152-1156 (1973).

Stith, R.D., Person, R.J., and Dana, R.C. Uptake and binding of ^{3}H-hydrocortisone by various pig brain regions. *Brain Res.* 117, 115-124 (1976).

Strott, C.A., Nugent, C.A., and Tyler, F.H. Cushing's syndrome by bronchial adenomas. *Am. J. Med.* 44, 97-104 (1968).

Stumpf, W.E. Estradiol concentrating neurons: topography in the hypothalamus by dry-mount autoradiography. *Science* 162, 1001-1003 (1968).

Stumpf, W.E. and Sar, M. Hormonal inputs of releasing factor cells, feedback sites, in *Drug Effects of Neuroendocrine Regulation Progress in Brain Research*, Vol. 39. E. Zimmerman, W.H. Gispen, B.H. Marks, and D. DeWied, eds. Elsevier, Amsterdam (1973), pp. 53-71.

Stumpf, W.E. and Sar, M. Hormone-architecture of the mouse brain with ^{3}H-estradiol, in *Anatomical Neuroendocrinology*. W.E. Stumpf and L.D. Grant, eds. S. Karger, Basel (1975a), pp. 82-103.

Stumpf, W.E. and Sar, M. Anatomical distribution of corticosterone concentrating neurons in rat brain, in *Anatomical Neuroendocrinology*. W.E. Stumpf and L.D. Grant, eds. S. Karger, Basel (1975b), pp. 82-103.

Stumpf, W.E. and Sar, M. Localization of thyroid hormone in the mature rat brain and pituitary, in *Anatomical Neuroendocrinology*. W.E. Stumpf and L.D. Grant, eds. S. Karger, Basel (1975c), pp. 318-327.

Sulman, F.S. *Hypothalamic Control of Lactation*. Springer-Verlag, New York (1970).

Surks, M., Koerner, D., Dillman, W., and Oppenheimer, J.H. Limited capacity binding sites for L-triiodothyronine in rat liver nuclei. *J. Biol. Chem.* 248, 7066-7072 (1973).

Sutherland, E.W., Oye, I., and Butcher, R.W. The action of epinephrine and the role of the adenyl cyclase system in hormone action. *Rec. Prog. Horm. Res.* 21, 623-646 (1965).

Sutherland, E.W., and Rall, T.W. The relation of adenosine-3',5'-phosphate and phosphoylase to the actions of catecholamines and other hormones. *Pharmacol. Rev.* 12, 265-299 (1960).

Tagliamonte, A., Tagliamonte, P., Gessa, G.L., and Brodie, B.B. Compulsive sexual activity induced by p-chlorophenylalanine in normal and pinealectomized male rats. *Science* 1433-1435 (1969).

Taleisnik, S., Orias, R., and de Olmos, J. Topographic distribution of the MSH-RF in rat hypothalamus. *Proc. Soc. Exper. Biol. Med.* 122, 325-328 (1966).

Tang, L.K.L. and Spies, H.G. Effect of synthetic LH-releasing factor (LRF) on LH secretion in monolayer cultures of the anterior pituitary cells of cynomolgus monkeys. *Endocrinology* 94, 1016-1021 (1974).

Tashjian, A.H., Jr., Barowsky, N.J., and Jensen, D.K. Thyrotropin releasing hormone: direct evidence for stimulation of prolactin production by pituitary cells in culture. *Biochem. Biophys. Res. Commun.* 43, 516-523 (1971).

Tata, J.R. and Windnell, C.C. Nucleic acid synthesis during the early action of thyroid hormones. *Biochem. J.* 92, 26 (1964).

Terasawa, E. and Timiras, P.S. Electrical activity during the estrous cycle of the rat: cyclic changes in limbic structures. *Endocrinology* 83, 207-216 (1968a).

Terasawa, E. and Timiras, P.S. Electrophysiological study of the limbic system in the rat at onset of puberty. *Am. J. Physiol.* 215, 1462-1467 (1968b).

Theodoridis, C.G., Brown, G.A., Chance, G.W., and Rayner, P.H.W. Growth-hormone response to oval glucose in children with simple obesity. *Lancet* 1, 1068-1069 (1969).

Thierry, A.M., Javory, F., Glowinski, J., and Rety, S.S. Effects of stress on the metabolism of norepinephrine, dopamine and serotonin in the central nervous system of the rat. I. Modification of norepinephrine turnover. *J. Pharmacol. Exp. Ther.* 163, 163-171 (1968).

Thoenen, H. Comparison between the effect of neuronal activity and nerve growth factor on the enzymes involved in the synthesis. *Pharmac. Rev.* 24, 255-267 (1972).

Thomas, J.A. and Singhal, R.L. Testosterone-stimulation of adenyl cyclase and cyclic 3', 5'-adenosine monophosphate-3H formation in rat seminal vesicles. *Biochem. Pharmacol.* 22, 507-511 (1973).

Thompson, R.G., Rodriquez, A., Kowarski, A., Migeon, C.J., and Blizzard, R.M. Integrated concentrations of growth hormone correlated with plasma testosterone and bone age in preadolescent and adolescent males. *J. Clin. Endocr. Metab.* 35, 334-337 (1972).

Timiras, P.S. Role of hormones in development of seizures, in *Basic Mechanisms of the Epilepsies*. H. Jasper, A.A. Ward, and A. Pope, eds. Little, Brown and Co., Boston (1969), pp. 727-736.

Timiras, P.S. and Luckock, A.S. Hormonal factors regulating brain developing, in *Psychoneuroendocrinology*. N. Hatotani, ed. S. Karger, Basel (1974) pp. 203-213.

Timiras, P.S., Woodbury, D.M., and Agarwal, S.L. Effect of thyroxine and triiodothyronine on brain function and electrolyte distribution in intact and adrenalectomized rats. *J. Pharmacol. Exp. Ther.* 115, 154-171 (1955).

Toivola, P. and Gale, C.C. Effect on temperature of biogenic amine infusion into hypothalamus of baboon. *Neuroendocrinology* 6, 210-219 (1970).

Toman, J.E.P., Swinyard, E.A., Merkin, N., and Goodman, L.S. Studies on the physiology and therapy of convulsive disorders. *J. Neuropathol. Exper. Neurol.* 7, 35-46 (1948).

Tomkins, G.M. and Maxwell, E.S. Some aspects of steroid hormone action. *Ann. Rev. Biochem.* 32, 677-708 (1963).

Touchstone, J.C., Kasparow, M., Hughes, P.A., and Horwitz, M.R. Corticosteroids in human brain. *Steroids* 7, 205-211 (1966).

Ungerstedt, U. Stereotaxic mapping of the monoamine pathways in the rat brain. *Acta Physiol. Scand.* 82, Suppl. 367, pp. 1-48 (1971).

Valcana, T. Developmental changes in ionic composition of the brain in hypo- and hyperthyroidism, in *Drugs and the Developing Brain*. A. Vernadakis and N. Weiner, eds. Plenum Press, N.Y. (1974) pp. 289-304.

Valcana, T. The role of triiodothyronine (T^3) receptors in brain development, in *Neural Growth and Development*. E. Meisami and M. Brazier eds. Raven Press, New York (1979).

Valcana, T. and Timiras, P.S. Nuclear triiodothyronine receptors in the developing rat brain. *Mol. Cell. Endocrinol.* 11, 31-41 (1978a).

Valcana, T. and Timiras, P.S. Changes in rat nuclear triiodothyronine receptors with age and thyroid activity, in *Hormones in Development*. L. Macho and V. Strbak, eds. VEDA, Publishing House of the Slovak Academy of Sciences (1978b), pp. 47-76.

Vale, W., Rivier, C., and Brown, M. Regulatory peptides of the hypothalamus. *Ann. Rev. Physiol.* 39, 473-527 (1977).

Valverde, R.C., V. Chieffo, and Reichlin, S. Prolactin-releasing factor in porcine and rat hypothalamic tissue. *Endocrinology* 91, 982-993 (1972).

Van Loon, G.R., Scapagnini, U., Cohen, R., and Ganong, W.F. Extraventricular administration of adrenergic drugs on the adrenal venous 17-hydroxycortico-steroid response to surgical stress in the dog. *Neuroendocrinology* 8, 257-272 (1971).

Verhoeven, G., Lamberigts, G., and DeMoor, P. Nucleus-associated steroid 5α-reductase activity and androgin responsiveness: a study in various organs and brain regions of rats. *Steroid Biochem.* 5, 93-100 (1974).

Verity, M.A., Su, C., and Bevan, J.A. Transmural and subcellular localization of monoamine oxidase and catechol-o-methyltransference in rabbit aorta. *Biochem. Pharmacol.* 21, 193-201 (1972).

Vernadakis, A. Neurotransmission: a proposed mechanism of steroid hormones in the regulation of brain function, in *Psychoneuroendocrinology*. N. Hatotani, ed. S. Karger, Basel (1974), pp. 251-258.

Vernadakis, A. and Culver, B. Uptake of ^{3}H-corticosterone in neuronal and glial cells in culture, in *Proceedings of International Society on Cellular and Molecular Basis of Neuroendocrine Processes*. Visegrad-Budapest, Publishing, House of the Hungarian Academy of Sciences (1976), pp. 1-13.

Vernadakis, A., Culver, B., and Nidess, R. Actions of steroid hormones on neural growth in culture, in *Proceedings of the 7th International Congress of the International Society of Psychoneuroendocrinology*, Strasbourg, France. *J. Pschoneuroendocrin.* 3, 47-64 (1978).

Vernadakis, A., Culver, B., and Nidess, R. Actions of steroid hormones on neural growth in culture, in *Proceedings of the 7th International Congress of the International Society of Psychoneuroendocrinology*,

Strasbourg, France. *J. Pschoneuroendocrinology*, Vol. 3 (1978), pp. 47-64.

Vernadakis, A., Nidess, R., Culver, B., and Arnold, E.B. Glial cells: modulators of neuronal environment. *Mechanisms of Aging and Development* 9, 553-566 (1979).

Verney, E.B. The antidiuretic hormone and the factors which determine its release. *Proc. Roy. Soc. (London), Ser. B.* 135, 25-106 (1947).

Vernadakis, A. and Nidess, R. Biochemical characteristics of C-6 glial cells. *Neurochem. Res.* 1, 385-402 (1976).

Verney, E.B. The antidiuretic hormone and the factors which determine its release. *Proc. Roy. Soc., Ser. B.* 135, 25-106 (1947).

Vernikos-Danellis, J. The pharmacological approach to the study of the mechanisms regulating ACTH secretion, in *Pharmacology of Hormonal Polypeptides and Proteins.* N. Back, L. Martini, and R. Paoletti, eds. Plenum Press, New York (1968), pp. 175-189.

Vernikos-Danellis, J. Effects of hormones on the central nervous system, in *Hormones and Behavior.* S. Levine, ed. Academic Press, New York (1972), pp. 11-62.

Vernikos-Danellis, J. Effects of hormones on the central nervous system, in *Hormones and Behavior.* S. Levine, ed. Academic Press, New York (1972), pp. 11-62.

Vernikos-Danellis, J. Effects of stress, adrenalectomy, hypophysectomy and hydrocortisone on the corticotropin-releasing activity of rat median eminence. *Endocrinology,* 76, 122-126 (1965).

Vernikos-Danellis, J. The pharmacological approach to the study of the mechanisms regulating ACTH secretion, in *Pharmacology of Hormonal Polypeptides and Proteins.* N. Back, L. Martini, and R. Paoletti, eds. Plenum Press, New York (1968), pp. 175-189.

Vernikos-Danellis, J. and Trigg, L.N. Feedback mechanisms regulating pituitary ACTH secretion in rats bearing transplantable pituitary tumors. *Endocrinology* 80, 345-350 (1967).

Versteeg, D.H.G. and Wurtman, R.J. Effect on $ACTH^{4-10}$ on the rats of synthesis of $\{^3H\}$ catecholamines in the brains of intact, hypophysectomized and adrenalectomized rats. *Brain Res.* 93, 552-557 (1975).

Versteeg, D.H.G. Effect of two ACTH-analogs on noradrenaline metabolism in rat brain. *Brain Res.* 49, 483-485 (1973).

Versteeg, D.H.G. and Wurtman, R.J. Effect of $ACTH^{4-10}$ on the rate of synthesis of (^3H) catecholamines in the brains of intact hypophysectomized and adrenalectomized rats. *Brain Res.* 93, 552-557 (1975).

Wagner, W., Erwin, W., and Critchlow, V. Androgen sterilization produced by intracerebral implants of testosterone in neonatal female rats. *Endocrinology* 79, 1135-1142 (1966).

Walter, R., Rudinger, J., Schwartz, I.L. Chemistry and structure-activity relations of the antidiuretic hormones. *Am. J. Med.* 42, 653-677 (1967).

Warberg, J., Eskay, R.L., Barnea, A., Reynolds, R.C., and Porter, J.C. Release of lutenizing hormone releasing hormone and thyrotropin releasing hormone from a synaptosome-enriched fraction of hypothalamic homogenates. *Endocrinology* 100, 814-825 (1977).

Watanabe, S. and McCann, S.M. Localization of FSH-releasing factor in the hypothalamus and neurohypophysis as determined by *in vitro* assay. *Endocrinology* 82, 664-673 (1968).

Weiner, N. Neurotransmitter systems in the central nervous system, in *Drugs and the Developing Brain.* A. Vernadakis and N. Weiner, eds. Plenum Publishing Company, New York (1974), pp. 105-131.

Weiss, B. and Craton, J. Gonadal hormones as regulators of pineal adenyl cyclase activity. *Endocrinology* 87, 527-533 (1970).

Weldon, V.V., Gupta, S.K., Haymond, M.W., Pagliara, A.S., Jacobs, L.S., and Daughaday, W.H. The use of L-dopa in the diagnosis of hyposomatotropism in children. *J. Clin. Endocr. Metab.* 36, 42-46 (1973).

Whalen, R.E. *Hormones and Behavior; An Enduring Problem in Psychology.* Van Nostrand Reinhold, New York (1967).

Whalen, R.E., Juttge, W.G., and Green, R. Effects of the anti-androgen cyproterone acetate on the uptake of the castrate male rat. *Endocrinology* 84, 217-222 (1969).

Whitby, L.G., Axelrod, J., and Weil-Matherbe, H. The fate of ^{3}H-norepinephrine in animals. *J. Pharmacol. Exp. Ther.* 132, 193-201 (1961).

Williams, P.B. and Hudgins, P.M. Actions of hydrocortisone desoxycorticosterone acetate and progesterone on ^{14}C-norepinephrine uptake and metabolism by rabbit corta. *Pharmacology* 9, 262-269 (1973).

Winokur, A. and Utiger, R.D. Thyrotropin-releasing hormone: regional distribution in rat brain. *Science* 185, 265-267 (1974).

Wirz-Justice, A., Hackmann, E., and Lichtstein, M. The effect of oestradiol dipropionate and progesterone on monoamine uptake in rat brain. *J. Neurochemistry* 22, 187-189 (1974).

Withrow, C.D. and Woodbury, D.M. Some aspects of the pharmacology of adrenal steroids and the central nervous system, in *Steroids and Brain Edema.* H.J. Reulen and K. Schurmann, eds. Springer-Verlag, Berlin (1972), pp. 41-55.

Woodbury, D.M. Effect of hormones on brain excitability and electrolytes. *Rec. Prog. Horm. Res.* 10, 65-107 (1954).

Woodbury, D.M. Relation between the adrenal cortex and the central nervous system. *Pharmacol. Rev.* 10, 275-357 (1958).

Woodbury, D.M. Biochemical effects of adrenocortical steroids on the central nervous system, in *Handbook of Neurochemistry*, Vol. 7, Chap. 13. A. Lajtha, ed. Plenum Publishing Company, New York (1972), pp. 255-287.

Woodbury, D.M. and Esplin, D.W. Neuropharmacology and neurochemistry of anticonvulsant drugs. *Res. Publ. Ass. Nerv. Ment. Dis.* 27, 1-56 (1959).

Woodbury, D.M., Timiras, P.S., and Vernadakis, A. Influence of adrenocortical steroids on brain function and metabolism, in *Hormones, Brain Function and Behavior*. H. Hoagland, ed. Academic Press, New York (1957), pp. 27-54.

Woodbury, L.A. and Davenport, V.D. Design and use of a new electroshock seizure apparatus, and analysis of factors altering seizure threshold and pattern. *Arch. Intern. Pharmacodynamie* 92, 97-107 (1952).

Woolley, D.E. and Timiras, P.S. The gonad-brain relationship: effects of female sex hormones on electroshock convulsions in the rat. *Endocrinology* 70, 196-209 (1962a).

Woolley, D.E. and Timiras, P.S. Gonad-brain relationship: effects of castration and testosterone on electroshock convulsions in male rats. *Endocrinology*. 71, 609-617 (1962b).

Wurtman, R.J., Axelrod, J., and Kelly, D.E. *The Pineal.* Academic Press, New York (1968).

Yates, F.E., Russell, S.M., and Maran, J.W. Brain-adenohypophysial communication in mammals. *Annu. Rev. Physiol.* 33, 393-444 (1971).

Zambrano, D. On the presence of neurons with granulated vesicles in the median eminence of rat and dog. *Neuroendocrinology* 2, 141-155 (1968).

Zigmond, R.E. and McEwen, B.S. Selective retention of estradiol by cell nuclei in specific brain regions of the ovariectomized rat. *J. Neurochem.* 17, 889-899 (1970).

Zimmerman, E.A. Localization of hypothalamic hormones by immunocytochemical techniques, in *Frontiers in Neuroendocrinology*, Vol. 4, L. Martini and W.F. Ganong, eds. Raven Press, New York (1976), pp. 25-62.

Zimmerman, E.A. Localization of hormone secreting pathways in the brain by immunohistochemistry and light microscopy: a review. *Federation Proc.* 36, 1964-1967 (1977).

Zimmerman, E. and Critchlow, V. Effects of intracerebral dexamethesone on pituitary-adrenal levels in female rats. *Amer. J. Physiol.* 217, 392-396 (1969).

Zimmerman, E.A., Hsu, K.C., Ferin, M., and Koslowski, G.P. Localization of gonadotropin releasing hormone (Gn-RH) in the hypothalamus of the mouse by immunoperoxidase technique. *Endocrinology* 95, 1-8 (1974).

Copyright©1982, Spectrum Publications, Inc.
Hormones in Development and Aging

Chapter 13
Biological Rhythms, Hormones, and Aging
Franz Halberg

INTRODUCTION

Many forms of life exhibit physiologic changes (recurring with a reproducible waveform) that can be validated statistically. This is how we define their biologic rhythms (Halberg, F. et al., 1977C). Rhythms can be viewed macroscopically—simply by the inspection of original data or averages in displays. Depending on the extent and kind of noise in the data, such viewing can lead to impressions, abstractions or intuitive inferences. This approach does not require any inferential statistical point-and-interval estimates of rhythm characteristics. At a risk of subjectivity and error discussed elsewhere (Halberg, F., 1969), the macroscopic approach is occasionally preferred by virtue of its apparent simplicity, yet sole reliance on it is not advocated herein. Christopher William Hufeland, private physician and friend of Goethe, Schiller and Herder, was one of its earliest author-protagonists. His book *Makrobiotik* (1797) earned extraordinary popularity in its time; in it he not only referred to what we now call the circadian period (about 24 hours) as the unit of our natural chronology, but also deemed it to be pertinent to "the art of prolonging life" (the subtitle of his book).

Any such macroscopic approach to rhythms has to be distinguished, however, from a (complementary) "microscopic" one which relies on the objective quantification of temporal characteristics in biologic data based on the fitting of appropriate mathematical models as one method of approach. One may not only estimate temporal parameters on this basis but may also obtain time-specified reference standards, "chronodesms" (i.e., time-qualified tolerance intervals or normal ranges), for interpretation of single sample values (Fig. 1).

During the past three decades a systematic temporal microscopy has revealed an impressive body of facts documenting the importance of circadian rhythms and has also uncovered a spectrum of intermodulating frequencies (Table 1) which interact with aging. Some of the findings—such as statistically significant alterations in circannual and circadian amplitude of human plasma aldosterone with age (Halberg, F. et al., 1980) are pertinent to many fields, in keeping with Hufeland's vision. It has now also become apparent that a complete study of aging must be concerned with the broad spectrum of rhythms not merely because it represents a confusing source of variability but rather because it may gauge sensitively (otherwise unattainable features in) the dynamics of senescence and senility.

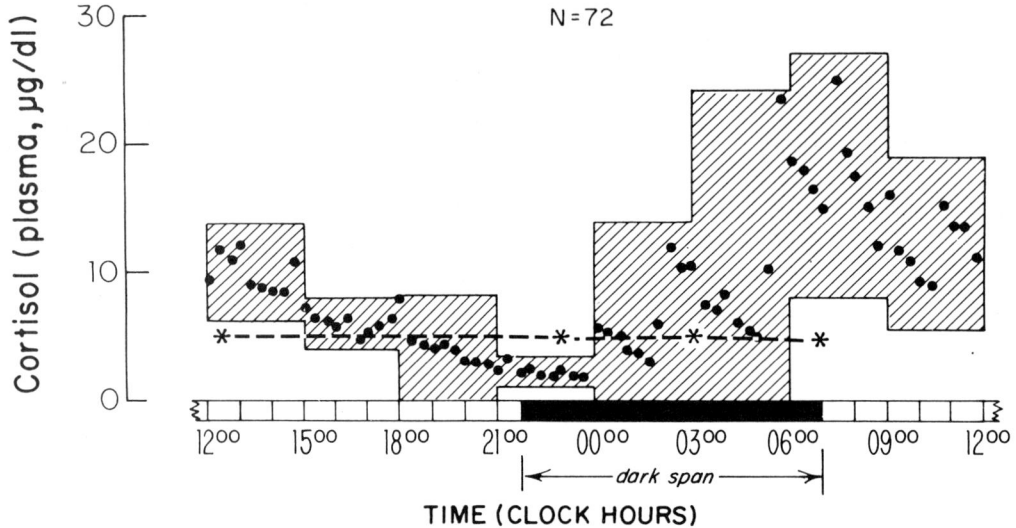

Fig. 1. Individualized circadian tolerance interval for plasma cortisol of a clinically healthy mature woman—determined separately for 3-h spans. This so-called merodesm (i.e., chronodesm with separate reference intervals computed for a number of timepoints throughout a rhythm) (Halberg, F. et al., 1980 and in press) indicates limits within which 90% of measurements would be expected to fall with 90% confidence. In relation to the hatched chronodesm, the value denoted by an asterisk can be normal at certain times and too high or too low at other times a few hours earlier or later. An unsatisfactory alternative to individualized or group chronodesms for plasma cortisol is the common practice of providing an 8 a.m., and 8 p.m. range (Scully et al., 1980). This practice ignores whether one deals with morningness or eveningness in day-active individuals (resting by night) and ignores even the fact that a night watchman's rhythms may be completely shifted in their timing along the 24-hour scale.

In this chapter we shall deal primarily, but not exclusively, with indices of the adrenal and reproductive hormones and body core temperature to illustrate approaches to some hormonal and vasculo-metabolic rhythms and aging. For such illustrative purposes, this short chapter relies mainly, if not exclusively, on data obtained or at least analyzed in the author's laboratory. After a consideration of background, it will be shown that chronobiologic reference standards are critical in the study of the body's many time-varying functions. By their use one may avoid controversy and secure new gauges of primary and secondary aging.

Table 1

PERIOD RANGES FOR TERMS DESCRIBING BIOLOGIC RHYTHMS[a]

Domain Region	Range
ultradian	$\tau < 20$ hr
circadian	20 hr $< \tau < 28$ hr
dian	23.8 hr $< \tau < 24.2$ hr
infradian	$\tau > 28$ hr
circadusdian	$\tau = 2 \pm 0.5$ d
circaseptan	$\tau = 7 \pm 3$ d
circadiseptan	$\tau = 14 \pm 3$ d
circavigintan	$\tau = 21 \pm 3$ d
circatrigintan	$\tau = 30 \pm 5$ d
circannual	$\tau = 1$ y ± 2 m

[a] τ = period; hr = hour; d = day, m = month; y = year. Terms coined by analogy to usage in physics. Just as frequencies higher than those audible or visible are called ultrasonic and ultraviolet, frequencies higher than one cycle per 20 h are designated as ultradian. By the same token, as frequencies lower than audible or visible are called ultrasonic and infrared, rhythms with a frequency lower than one cycle per 28 h are designated as infradian.

Chronobiology

Chronobiology (a section of biology) represents the objective description of biological time structure—the sum total of nonrandom, and thus predictable temporal aspects of organismic behavior including with the spectrum of biologic rhythms, developmental changes, growth and age trends. It is now recognized that biological time structure characterizes individuals and their subdivisions (organ systems, organs, tissues, cells and intracellular elements including electron-microscopic and molecular ultrastructure) as well as groups or populations of organisms. For a number of rhythms validated as a bioperiodic aspect of data, some basic characteristics have been measured, such as: the mesor (rhythm-adjusted mean), the amplitude (measure of the extent of predictable rhythmic changes), the acrophase (measure of the timing of predictable rhythmic changes), and, when multiple components are assessed, the waveform.

Figure 1 portrays the well-known circadian change of plasma cortisol. It also shows a much less known individualized and time-specified tolerance interval for the interpretation of single sample values. It can be seen that—in relation to this so-called chronodesm-the same value can be too low at one time, normal at other times and too high at a third time—for that individual. Such time-specified individualized usual ranges supersede 8 a.m. and 8 p.m. ranges that are currently used in some hospitals.

SENESCENCE AND SENILITY IN A CHRONOBIOLOGIC CONTEXT

Busse (1979) denotes the sum total of defects and disability resulting, with advancing age, from trauma and disease as senility or secondary aging; senescence or primary aging involves, in turn, detrimental time-related changes, relatively independent of trauma and acquired disease, apparently rooted in heredity. We document here that chronobiology is pertinent to both senescence and senility.

A practical point regarding senescence, which Busse further defines as inevitable, is that age-related changes in human beings may be not only delayed but reversed by the practice of autorhythmometry (Halberg, F. et al., 1972). Peak expiratory flow can be improved by "jogging of the lungs" over a span of 10 years (53-63 years of age) (Halberg, E. et al., 1981).

A specific strain of stroke-prone mesorhypertensive rat (Okamoto SHR-SP) may serve as a model of both senescence and senility (Halberg, J. et al., 1980); as this animal ages, it exhibits spontaneously a reduction in amplitude and an advance in acrophase of the circadian rhythm in its telemetered core temperature (Table 2) (Halberg, J. et al., 1981). The age-dependent changes are akin to those associated with a bilateral lesion of the suprachiasmatic nuclei in mature male Fischer rats (Table 2) (Halberg, F. et al., 1977b, in press; Walker and Timiras, in press).

Table 2

Age-Dependent Differences in Circadian Temperature Rhythm of Various Inbred Rat Strains[a]

Strain	Age or Treatment	P	Amplitude, °C (Confidence Limits)	Acrophase (Confidence Limits)
Sprague-Dawley	Young (female)	<0.001	1.13 (1.00, 1.27)	-11° (-3, -21)
	Old (female)	<0.001	0.70 (0.41, 4.20)	-10° (-0, -21)
Okamoto SHR-SP	Young (female)	<0.001	1.16 (0.98, 1.34)	-350° (-349, -8)
	Old (female)	<0.034	0.45 (0.08, 0.84)	-308° (-239, -3)
Fischer	A[b] (male)	<0.001	0.50 (0.43, 0.57)	-4° (-358, -9)
	B (male)	<0.009	0.66 (0.31, 1.14)	-6° (-345, -62)
	C (male)	<0.001	0.22 (0.16, 0.32)	-279° (-236, -307)

[a] Results from least-squares fit of 24-hour cosine curve; P-value from F-Test of zero-amplitude hypothesis; acrophase in angular degrees, with $360° \equiv 24$ hours, $0°$ midnight.

[b] A=Histologically validated sham-operated.
B=Histologically validated unilateral lesion of the suprachiasmatic nuclei.
C=Histologically validated bilateral lesion of the suprachiasmatic nuclei.

Female Sprague-Dawley rats 19 months of age exhibit amplitude-decrease without acrophase-change (as compared to same rats at about four months of age) and, in view of a much longer lifespan, may be physiologically younger than 14 months-old Okamoto rats. While a sex difference may contribute to this difference between Sprague-Dawley and Okamoto rats in effects of advancing age on the circadian temperature acrophase, the shorter lifespan of the Okamoto strain is emphasized as an important factor (see also Yunis et al., 1974). With respect to disability from diseases associated with secondary aging in human beings, correlations are found between the circannual amplitude of some hormonal rhythms and diseases of senility. In healthy subjects the rhythm-adjusted mean of blood pressure and the (epidemiologically-determined) risk of developing diseases associated with a high blood pressure both correlate negatively with the circannual amplitude of plasma aldosterone (Radke et al, 1979). The risk of developing breast cancer correlates negatively with the circannual amplitude of prolactin and positively with that of TSH, Fig. 2. The risks of developing prostatic cancer and of certain emotional states also reveal correlations with the circannual amplitude of some hormonal rhythms (Halberg, E. et al., 1979b; Mandel et al., 1979, Seal et al., in press). In prostatic cancer (a condition characterized by geographic differences in morbidity and mortality similar to those of breast cancer), the extent of circannual variation also changes as a function of risk and/or cancer. In blood sampled in the morning at different times of the year with serial independence, a prominent circannual rhythm in TSH of healthy subjects is lost in men with prostatic cancer (and perhaps even in men at high risk of prostatic cancer). For prolactin, a circannual rhythm becomes demonstrable in the case of men with prostatic cancer, while it is not demonstrable with serially independent sampling in healthy men of low or high prostatic cancer risk.

Thus, TSH and prolactin show opposite behavior along the one-year scale in cancers of both breast and prostate rather than responding in the same fashion, as is the case along the scale of minutes to hours—following the application of stimuli such as TRH (Hershman, 1974). It is yet more interesting that the circannual relations of plasma TSH and prolactin to the risk of breast cancer are opposite to those with respect to prostatic cancer (and, as noted, opposite to each other in each cancer).

The two correlations in breast cancer shown in Fig. 2 are but part of a large correlation matrix. The circumstance is noted that correlations emerged as statistically significant for the very hormones for which clinicians have long considered some relation to breast cancer, yet thus far could not rigorously establish such a relation as biologically significant, perhaps because of time-unspecified and in this sense compromised sampling.

Chronobiologically ascertained correlations are consistent with the view that changes in rhythm characteristics may be harbingers and possibly even determinants of diseases resulting in senility and may be indices of senescence. If these assumptions are true, the institution of measures to prevent or correct detrimental rhythm alterations may constitute a cost-effective approach to combating senility and senescence.

Features of Rhythms

Table 3 summarizes some features of rhythms. The term circadian (derived from Latin *circa dies;* about a day) refers to an average period of precisely 24 hours (also referred to as dian, if the period is longer than 23.8 hours but shorter than 24.2 hours) or any other duration between 20 and 28 hours. A rhythm of such a period is well evident in a number of functions, as part of a multifrequency web of cycles (Fig. 3), some of them *cellular* such as a sequence consisting of the circadian labelling of phospholipid (as a membrane phenomenon) followed by RNA labelling, DNA formation and mitosis in liver and other tissues, others *hormonal* (corticotropin-releasing factor, ACTH, 17-OH-corticosteroid) and yet others *neural* and *neuroendocrine*. The circadian amplitude- and pace-modulating role of the suprachiasmatic nucleus at these several levels of organization can be viewed in the context of a broader network involved in the modulation of several frequencies—ultradian to infradian, the latter including, with circatrigintan rhythms, not only the well-known menstrual cycle but rhythms with infradian periods found in women before menarche and after menopause and characterizing the spectrum of core temperature variability in men as well. All of these rhythms may be synchronized, modulated or influenced by the socioecologic day/night, weekly or yearly cycles, among others, acting, e.g., by alarm clocks, street noises, bird songs, and/or the effects of lighting, environmental temperature, food intake, electromagnetic phenomena, etc.

Sleep-wakefulness, body temperature, pulse and blood pressure, the composition of blood, the excretion rate of substances in urine, competence

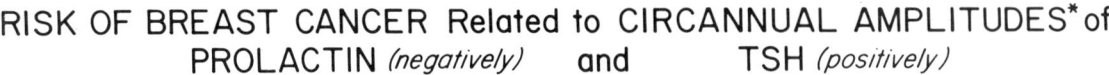
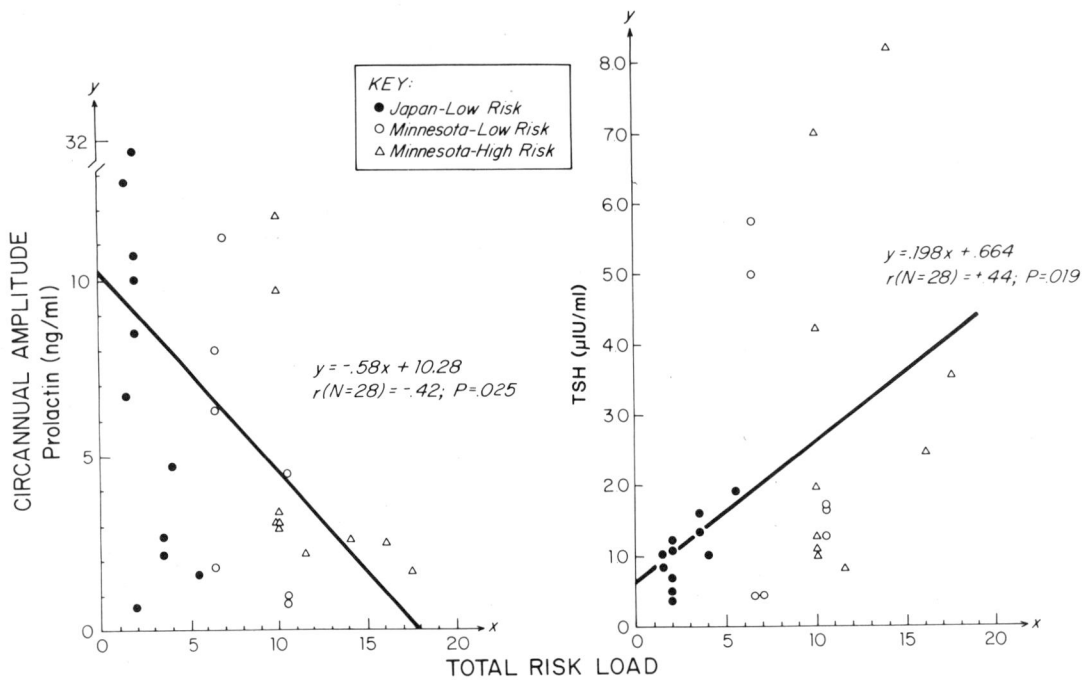

Fig. 2. Negative correlation between the total relative breast-cancer risk evaluated from epidemiologic criteria and the circannual prolactin amplitude. The data are in keeping with the finding of Tarquini et al. (1979) of a decrease in circannual amplitude or circannual rhythm loss in women with fibrocystic mastopathy and an associated elevation of breast cancer risk. Note further a positive correlation between epidemiologically assessed breast cancer risk and circannual amplitude of TSH. Clinical hypothyroidism has empirically been associated with breast cancer risk, a still controversial topic; this may, perhaps, be accounted for by the circannual-stage-dependence of the correlation.

While these conclusions rest on large samples, they describe only a small number of subjects. Moreover, all conditions required to apply to a linear regression between two variables are not satisfied. A test for lack of fit indicates that the model is not adequate for TSH: the error term is not normally distributed. In addition, the assumption of homogeneity of variance is not verified for prolactin as well as for TSH. In the case of prolactin, the circannual amplitude decreases with age and the breast cancer risk increases with age. This may contribute to, if not account for, the negative correlation illustrated in the figure. The correlations in the figure and other points in this paper are of ordering rather than documenting value. They are intended to emphasize that circannual rhythmicity deserves further study in relation to carcinogenesis and aging. If such correlations can be confirmed and if the circannual rhythms involved should prove to be determinants of carcinogenesis in the human breast, these same correlations will point to the possibility of a chemo-prevention of breast cancer.

Table 3

DURING MOST OF OUR LIFE MOST COMPONENTS IN A SPECTRUM OF PHYSIOLOGIC RHYTHMS:

  (A) ARE LOCKED INTO A SPECTRUM OF ENVIRONMENTAL CYCLES

     with synchronizer frequencies "acceptable" to ("resonant" with) the organism

  (B) YET MODULATED AND INFLUENCED

     from within and without (by other spectral components or noncyclic factors).

ENVIRONMENTAL FACTORS (SYNCHRONIZERS) MAY BE MANIPULATED SO THAT RHYTHMS:

  (C) ARE SHIFTED (in response to schedule shifts) with:

     (1) ASYMMETRIES (e.g., delay of rhythm faster than advance or vice versa)
     (2) POLARITIES (i.e., some rhythms advance while others delay in the same organism, responding to the same synchronizer shift)

  (D) OR PERSIST UNDER CERTAIN CONDITIONS OF ISOLATION with transients and eventually new stable internal time relations, in the absence of known external synchronizers, modulators and influencers—usually with "free-running" periods, different from known environmental ones.

THE ORGANISM MAY BE MANIPULATED BY VARIOUS MEANS INCLUDING SURGERY OR DISEASE, SO THAT RHYTHMS:

  (E) CHANGE CHARACTERISTICS (ALTER AMPLITUDE, MESOR OR ACROPHASE), DESYNCHRONIZE, TRANSFER VARIANCE, FREQUENCY-MULTIPLY AND/OR FREQUENCY-DIVIDE OR, MORE BROADLY, SPECTRALLY-COMPROMISE as does thermovariance in cancer

  (F) OR DISAPPEAR as does the circadian rhythm in blood eosinophils in certain patients with Addison's disease.

---

at work, susceptibility to drugs and even the probability of being born or dying, all depend to some extent on interactions (henceforth denoted as "spectral compromises" {Halberg, E. et al., 1979a,b}, or more broadly, "spectral solutions") among the several components in the spectra of external cycles and organismic rhythms, overlayered by rhythm-stage dependent responses to acyclic external factors.

FROM DESYNCHRONIZATION TO "SPECTRAL SOLUTION"

    A considerable number of studies indicates that human beings (and other animals as well as plants) continue to show certain about-daily, about-weekly, about-monthly, and/or about-yearly variations in the apparent absence of known environmental periodicity (Halberg, F., 1954; Halberg, F. et al., 1970, 1971). A woman or a man isolated in a cave for four or six months without time information continues to exhibit an about-24-hour rhythm in body temperature, pulse, adrenocortical and other

458 HALBERG

ENDOCRINE AND CELLULAR RHYTHM—WEBS

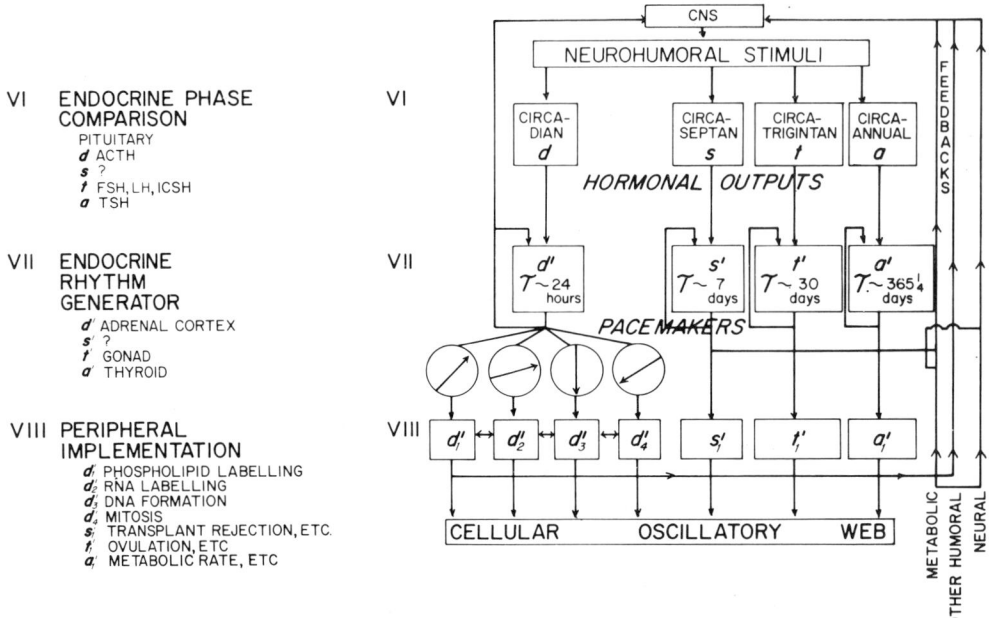

Fig. 3. Hypothetical mechanisms underlying a spectrum of biologic rhythms (Halberg, F. et al., 1979).

variables. Circadian rhythms also persist during flight in extraterrestrial space, for the few weeks of experience studied thus far, and may be desynchronized from both the solar day and 24.8-hour (lunar day) schedules. As another example of circadian desynchronization, the rhythm in intraperitoneal temperature, telemetered at intervals of approximately 10 minutes for 116 days, from blinded, mature inbred Minnesota Sprague-Dawley rats, kept singly housed with food freely available at 24°C environmental temperature, in light and darkness alternating at 12-h intervals, persisted with a 24.3-h period (rather than a 24-h period) and a 0.57°C amplitude. A similar amplitude (of 0.61°C) at a precise 24-h average period was found in five concomitantly evaluated intact (rather than blinded) rats kept under the same conditions. Circaseptan rhythms can desynchronize from precise seven-day schedules and circannual rhythms from a precisely one-year period. A circaseptan desynchronization has been reported in the case of a healthy man who for about 10 years showed a peak in urinary 17-ketosteroid excretion on Wednesdays or Thursdays, but following the taking of testosterone suppositories revealed a rhythm with a period slightly, yet statistically significantly different from precisely seven days, while his urine volume continued to exhibit a precise seven-day periodicity (Fig. 4) (Halberg, F. et al., 1965).

Thus, the entire spectrum of rhythms may be desynchronized from precise environmental schedules into which it may have been previously locked, once such schedules are: (1) rendered ineffective by disease or surgery, including removal of the eyes as after blinding of rodents previously locked into a precise 24-h schedule of alternating light and darkness; or (2) rendered unacceptable to the organism by virtue, e.g., of their cycle length; or (3) removed, by the institution of constancy as far as this is possible on earth.

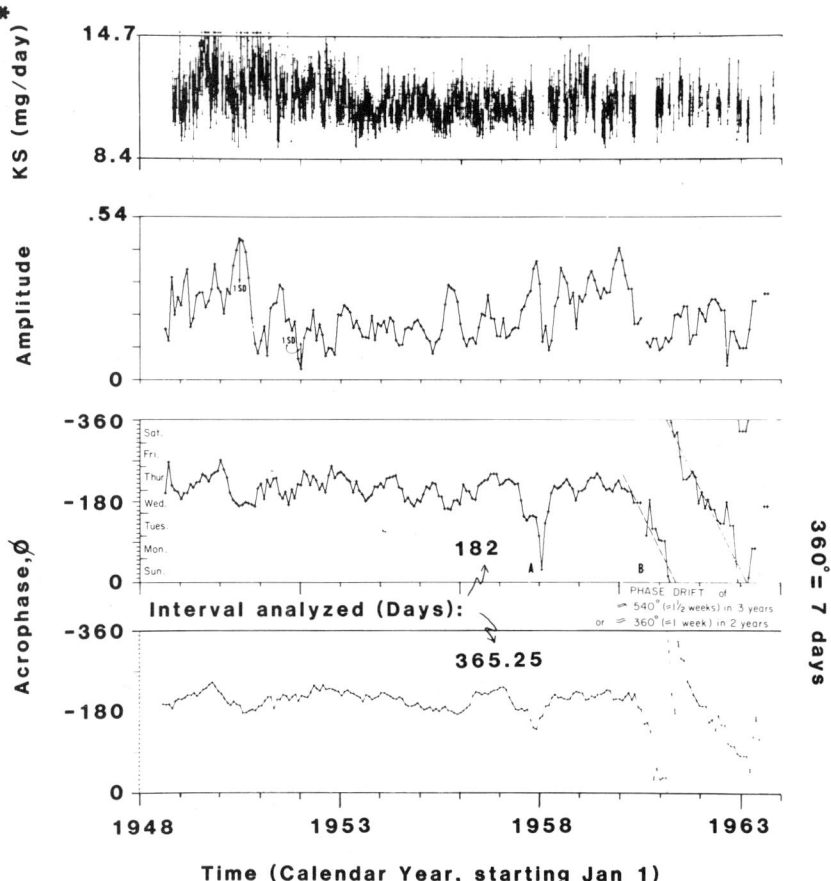

Fig. 4. 17-ketosteroid excretion, plotted as a function of time, evaluated in the top row and below by a temporal amplitude diagram and by two phase diagrams. Precise 7-day period fitted with increments of 28 days to consecutive overlapping intervals of 182 or 365 days. Note desynchronization of the about-weekly rhythm, revealed by phase drift. Testosterone in doses varying from approximately 200-9000 mg self-administered starting from January 1, 1956 and continued through 1963. Data during testosterone application and the week thereafter are omitted from analysis. Asterisk indicates outliers: dots above chronogram limit.

When rats in low constant illumination (Halberg, F. et al., 1971) or human beings in isolation—whether on the average (Wever, 1979) or singly (Colin et al., 1968)—tend to exhibit periods near 24.8 hours (the lunar-day length), the question might be raised as to whether a synchronizer change had occurred from environmental cycles associated with the solar or social day to those related to the lunar day. Such questions usually inquire into whether a single frequency is exogenous or endogenous in origin. By 1950 it was clear, however, that in contrast to human beings, blinded mice can exhibit free-running circadian periods near 23.5 hours (Halberg, F. 1954), a region without any known direct environmental counterpart. The term "desynchronization" then served to indicate some degree of endogenicity. This approach to the organism's time structure awaits further microscopic scrutiny (of the behavior of the period) of appropriate rhythms after chemical mutagenesis.

Subsequent studies documented not only for the environment (Stupfel et al., 1977) but also for the organism a spectrum of frequencies—in one and the same variable (urinary 17-ketosteroid excretion) in health. Moreover, in cases of breast cancer, periods reported for breast surface temperature tended toward 21 hours. Circadian desynchronization was again the initial consideration. When in a case studied for weeks rather than days, in addition to a 21-h period, periods of 42 and 84 hours also were noted in surface temperature of a cancerous breast (but not of the healthy one), the phenomenon of frequency division came to mind (Gautherie and Gros, 1977). On the other hand, one cycle per 84 hours also constitutes a frequency multiplied circaseptan rhythm. Thus, the question arose as to whether one dealt with circadian frequency division or circaseptan frequency multiplication, or perhaps with both as a so-called spectral compromise (Halberg, E. et al., 1979a,b). Whatever the answer to the foregoing question may be (apart from the particular case of breast temperature considered on the basis of analyses by others), the next logical step in dealing with spectra of both environmental and organismic cycles is to consider an additive scheme. This scheme includes components from within the organism—contributing a time structure which includes intermodulating frequencies—and age effects as well as cyclic and noncyclic external factors. Interaction between organismic and environmental factors would be amenable to "spectral solution" (or spectral accommodation or spectral adjustment). This concept is both mathematically formulatable (Halberg, F. et al., 1981) and experimentally verifiable and takes into account the current status of information in the field.

Acrophase Charts

Circadian, as well as circaseptan, circatrigintan and circannual, acrophase charts already are available to show the timing of rhythms in relation to pertinent time scales (e.g., in the case of circadian rhythms, a schedule consisting of about 16 h of activity alternating with about 8 h of rest each day). For example, along the 24-h scale, relatively high values of cortisol in human plasma or urine tend to occur early in the activity span (Halberg, F. et al., 1981), whereas in the same fluids high values of melatonin occur about the middle of the rest span (Fig. 5. and 6) (Wetterberg et al., 1979).

Schedule-Shifts

The temporal placement, along the 24-hour scale, of circadian rhythms investigated thus far can be shifted. Schedule shifts induce changes in both internal and external timing and in other relations; e.g., of human oral temperature and hormonal rhythms. Usually, but not invariably, circadian-rhythm-shifts occur more rapidly when the living schedule is delayed, as after a flight from Minnesota to Japan, than when the schedule is advanced, as after a flight from Japan to Minnesota. Figure 7 shows that internal timing, namely the relation among two rhythms themselves, may differ during schedule shifts; this is also documented for performance (for the tempo or count from one to 120) which may adjust much more slowly after an eastward flight than does either oral temperature or heart rate. Polarity of the circadian system is manifested, in turn, in that during an adjustment following a flight from Minnesota to Japan, the acrophases of 17-hydroxycorticosteroid, 17-ketosteroid and calcium excretion may delay (incidentally, at different rates—the phenomenon of intra-individual within-shift asymmetry), whereas the acrophases of sodium and potassium excretion advance (Halberg, F. 1969). Such a polarity may even be seen during a change in schedule without geographic displacement (Levine et al., 1974). Comparative schedule-shift studies, without and with geographic displacement into similar and different cultures and environments, assess the relative roles of different socioecologic synchronizers.

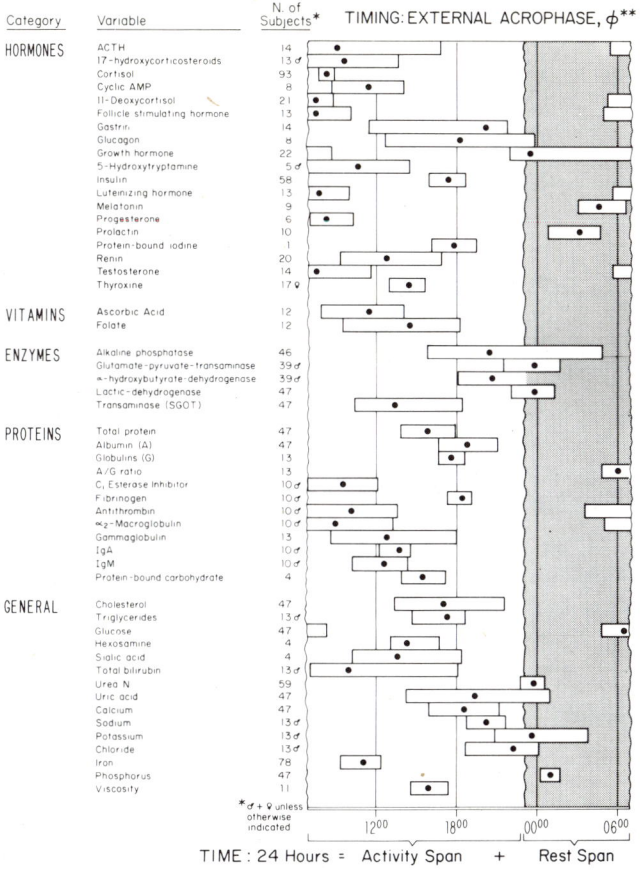

Fig. 5. Acrophasogram—display of sets of acrophases (•) with their 95% confidence intervals (□) characterizing human blood—qualified by the circumstance that in several cases results for the two sexes were pooled and that age is specified only as maturity. Changes with frequency other than circadian also are ignored. Nonetheless, differences in timing are apparent for the different variables being compared.

Some (but not all) schedule shifts can involve a deficit in performance, health and longevity (Aschoff et al., 1971; Halberg, F. et al., 1979; Halberg, J. et al., 1980; Halberg, F. and Nelson, 1978; Halberg, F. and Nelson, 1976; Luce, 1970). The direction as well as extent of change in performance and even longevity can be manipulated as a spectral solution—with due regard in particular for circadian-circaseptan interactions (Hayes et al., 1976; Hayes and Halberg, 1980). Is it coincidence that the famous nuclear accident at Three Mile Island reportedly would have been minor had it not been for mistakes starting about 0400, made by workers on a weekly rotation—days for one week, evenings for another week, then all night for a third week? Scrutiny of circadian-circaseptan interactions from this viewpoint seems to be mandatory, since it is known that the mammalian (including the human) spectral system can be manipulated by timing meals, other hygienic maneuvers or drugs.

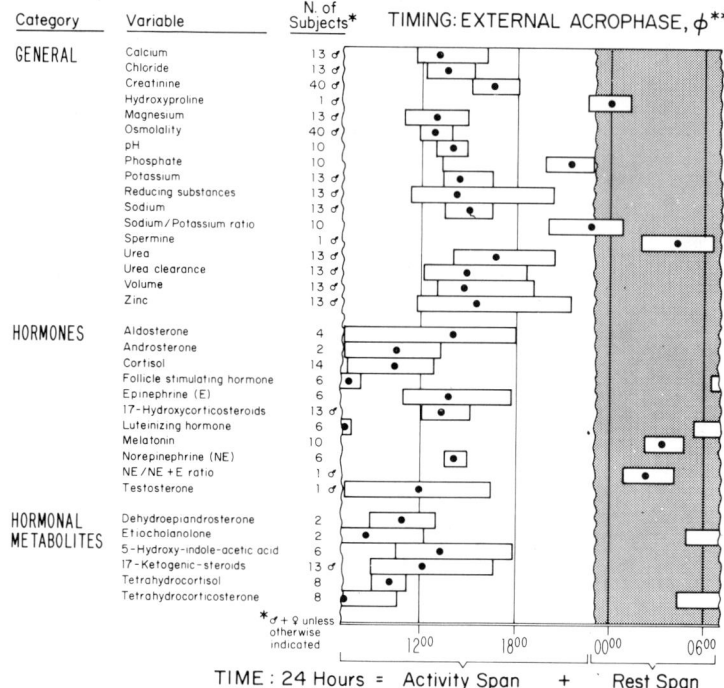

Fig. 6. Urinary human acrophasogram qualified as for Fig. 5.

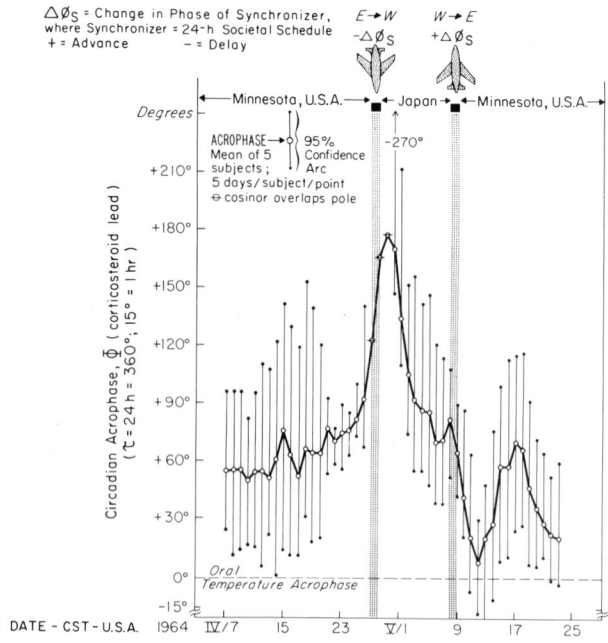

Fig. 7. Internal timing of two circadian indices—of the adrenal cortical cycle and overall metabolism, respectively, temporarily changed after shifts in schedule (transmeridian flight).

Basic Concept

The architectural section in the lower middle of Fig. 8 represents three major dimensions of organismic complexity: (1) the vertical concerns the different physiologic systems—neural systems, glandular secretions, and intracellular variations; (2) the horizontal represents, in three broad domains, the spectrum of frequencies with which the bioperiodicities occur in all three systems; (3) the arrow receding into the distance represents how this entire network of rhythmic variation evolves with the growth, development, maturation and aging of the individual. The criss-crossing lines represent the multiple influences and linkages that exist among the different physiologic systems and among rhythms of different frequencies.

The upper middle of the figure shows oversimplifications which result in misconceptions. A dynamic web of physiologic interaction among biochemical compounds in anatomical locations cannot be replaced by a light bulb, clock, and calendar, which represent simplistic views of environmental control and of unitary or isolated biological clocks. The simplifications all too often imply exclusive control by the environment or by heredity, when in fact both are pertinent. At the right of the figure are samples of quantitatively estimated characteristics of rhythms with a

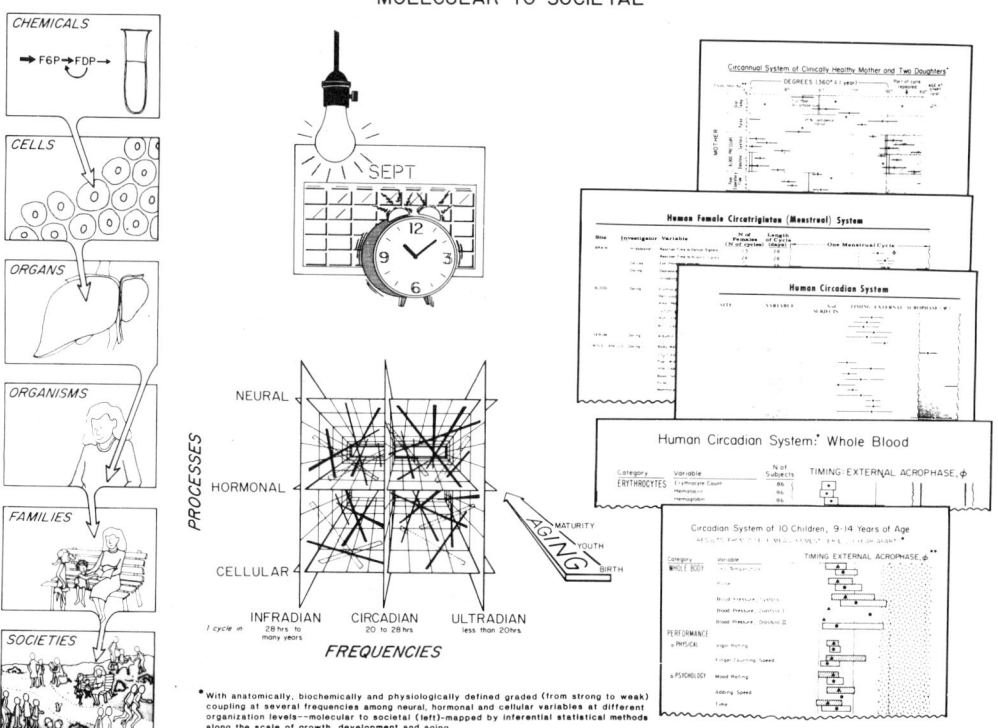

Fig. 8. Assessment of rhythms affecting chemicals, cells, organs, organisms, families and societies; at the neural, hormonal and cellular levels of integration; with infradian, circadian and ultradian frequencies as well as superimposed trends in the context of growth, development and aging, leads to charts of timing (see Fig. 5 and 6) at several frequencies and is the basis for a time-specified interpretation of single sample values (Fig. 1).

variety of frequencies. A series of many more detailed charts like those in Figs. 5 and 6 can now be drawn for human beings.

Controversy From Disregard For The Spectrum Of Rhythms

Disregard for rhythms over a wide range of frequencies can lead to controversy and confusion. For example, the literature contains contradictory reports about a possible effect of age on circulating prolactin (Jeffcoate, 1978). Vekemans and Robyn (1975) observed a progressive decline in women over the age range from about 20 to about 60, whereas Yamaji et al. (1976) found no statistically significant difference as a function of age over a similar span. Both studies involved sampling only during the daytime at clock-hours ordinarily associated with low prolactin concentrations in plasma even when allowances are made for rhythms with a wide range of periods (Halberg, F. et al., 1981).

The results in Table 4 were obtained in a study on 29 women in three age groups with blood sampling at 20-minute intervals through a 24-hour span in each of four menstrual stages (cycling subjects), distributed over four seasons (Halberg, F. et al., 1981; Haus et al., 1979; Nelson et al., 1980). These results suggest that age-related differences are likely to be smaller (and so perhaps less likely to be demonstrated reproducibly) during the day-time stage of the circadian rhythm in diurnally active subjects than at other stages. Assuming little if any age-dependent change in timing of this rhythm (an assumption supported by results of the recent chronoepidemiologic study (Halberg, F. et al., 1981), one sees that the low point of the rhythm (mesor minus amplitude) is on the average, 19.5 - 16.5 = 3.0 in group II (age 29-36) and 12.7 - 11.0 = 1.7 in group III (age 44-59)—for an age-associated difference of only 1.3 ng/ml. This can be contrasted with a difference of 12.3 ng/ml between these same groups when values at the high point

Table 4

AGE-RELATED DIFFERENCES IN MESOR AND AMPLITUDE OF CIRCADIAN PROLACTIN RHYTHM[a]
AVERAGED ACROSS FOUR CIRCATRIGINTAN STAGES
(FOR MENSTRUALLY CYCLING SUBJECTS) AND FOUR SEASONS

Age[b]			ANOVA[c]	
I	II	III	F	P
Mesor (ng/ml)				
22.9	19.5	12.7	10.37	0.001
Amplitude (ng/ml)				
12.4	16.5	11.0	4.34	0.027

[a] Subjects satisfying the criterion of having adequate data throughout a 24-hour span in each of four seasons and four menstrual stages: each subject contributed two values to the analysis: her average circadian mesor and her average circadian amplitude.

[b] I, II, and III indicate mean values for ages 15-21, 29-36, and 44-59, respectively.

[c] Results from two-way analysis of variance for effects of age on circadian mesor and amplitude.

of the circadian rhythm are compared (19.5 + 16.5 = 36.0 vs. 12.7 + 11.0 = 23.7).

The studies of Vekemans and Robyn (1975) and Yamaji et al. (1976) also apparently did not consider changes in circulating prolactin during the menstrual cycle or as a function of circannual rhythm stage, two other factors that may have contributed to the contradictory results in these previous reports.

Moreover, variability is likely to play a role. In our hands, the chronodesms for plasma prolactin (of diurnally active subjects) extended from 2.6 to 63 ng/ml between 0800 and 1000 (Yamaji's sampling time) and from 2.6 to 40.7 ng/ml in the span from 1000 to 1600 when Vekemans and Robyn (1975) sampled, Table 5.

AGE CHANGES IN CIRCADIAN HORMONAL RHYTHMS

Descovich et al. (1974) examined the circadian rhythm of urinary catecholamine excretion in men and women varying in age from 20 to 99 years. Both epinephrine and norepinephrine exhibit an age-related decrease in circadian amplitude and mesor (with most of the decrease occurring after 65 years) but no apparent change in acrophase, Fig. 9. Note that during the sampling span of Yamaji et al. (1976) (who did not find any age difference), namely between 0800 and 1000, 17.5% of values in a repeat study were found outside the chronodesm; by contrast, 10% outliers were found at the sampling time of Vekemans and Robyn (1975),

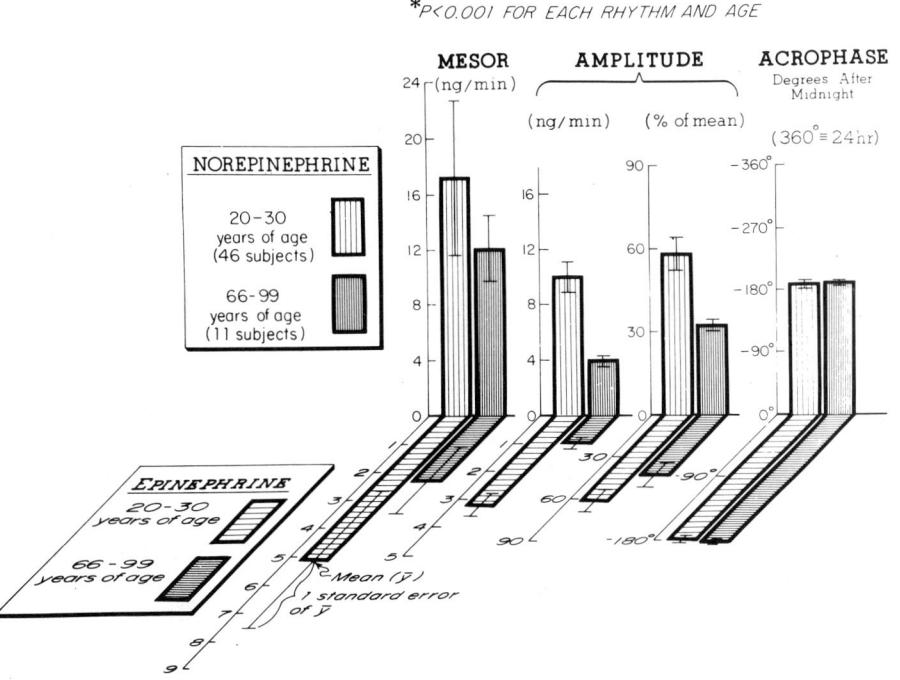

Fig. 9. Certain characteristics of neurohumoral rhythms change with age, in at least one (circadian) frequency region of the spectrum of catecholamines, which includes prominent circannual components (Descovich et al., 1974).

Table 5

Circadian Prolactin Concentration Chronodesm
(Group I) (ng/ml)

Time interval	Lower limit	Higher limit	Percent of data from Group II outside chronodesm
0000-0200	10.7	158.5	15 %
0200-0400	15.8	112.2	
0400-0600	10.7	134.9	0 %
0600-0800	10.7	102.3	
0800-1000	2.6	63.1	17.5%
1000-1200	2.6	40.7	
1200-1400	5.6	28.8	10 %
1400-1600	5.3	31.4	
1600-1800	5.0	33.9	10 %
1800-2000	4.3	89.1	
2000-2200	5.5	50.1	5 %
2200-2400	3.4	50.1	
			10.7% overall

Group I: 4 adolescent Japanese women sampled every 20 minutes for 24 hours in winter (1977);
Group II: 20 adolescent Japanese women sampled every 4 hours one year later in the same season (March 1978).

between 1000 and 1600. Variability not only within a series but also upon replication will be greater during certain times as compared to others. It seems further pertinent that in a recent issue of the New England Journal of Medicine, Zervas and Martin (1980) state that excessive prolactin secretion (by pituitary adenomas) can be readily detected by measuring basal (morning) plasma concentration. For these authors prolactin concentrations larger than 100 ng/ml are considered abnormal and indicative of a tumor, whereas normal values are considered to be less than 15 ng/ml. Time-specified tolerance intervals such as those reported in this table indicate that prolactin concentration in the morning is still decreasing from overnight high values. Prolactin assessment will be more discriminating, in the diagnosis of tumors and other disorders, if careful consideration is given to rhythms, i.e., if samples are obtained at appropriate times.

The problem on hand should also be taken into account. One must consider with the expected change (e.g., elevation of prolactin in patients with certain pituitary tumors) also the physiologic timing of the sleep-wakefulness or activity-rest pattern. By taking rhythmic variations into account, chronodesms will usually yield narrower intervals than conventional "normal ranges." The set of time-specified tolerance intervals shown in this table is expected to contain at least 90% of the distribution of data with 90% confidence. As expected, on the average about 90% of the data fall within the chronodesm. The small number of subjects involved in the construction of this chronodesm notwithstanding, one conclusion emerges:

concentrations as high as 150 ng/ml would still be considered normal if prolactin were sampled in the middle of the night, whereas an upper discriminant as low as 50 ng/ml is indicated if sampling occurs around noon.

Although the 24-hour urinary excretion of 17-ketogenic steroids (Jorgensen, 1957; Jensen and Blichert-Toft, 1970) as well as 17-ketosteroids (Hamburger, 1948) reportedly declines with age, no appreciable age-change in circadian characteristics of plasma cortisol has been observed (Serio et al., 1970; Krieger et al., 1971; Jensen and Blichert-Toft, 1970). The study of Serio et al. (1970) indicates a slightly delayed timing of the plasma cortisol rhythm in aged subjects.

The absence of a statistically significant effect of age, up to 54 years, on the circadian cortisol rhythm in blood plasma of women aged 15-59 years also was indicated in a concomitant study of over a dozen hormones (Halberg, F. et al., 1981). This investigation attempted to adjust for circadian, menstrual and circannual fluctuations by sampling at intervals of 20-100 minutes throughout 24-hour spans in each of four seasons and (in premenopausal women) four stages of the menstrual cycle. In addition to cortisol, prolactin, luteinizing hormone (LH), estrone, estradiol, 17-hydroxyprogesterone, aldosterone, dehydroepiandrosterone (DHEA-S), thyrotropic hormone, triiodothyronine, thyroxine and insulin were determined. Of these, hormones related to reproductive function (prolactin, LH, estrone, estradiol and 17-hydroxyprogesterone) exhibited a statistically significant age effect on their circadian mesor, with LH higher and the others lower in post-menopausal women, Fig. 10.

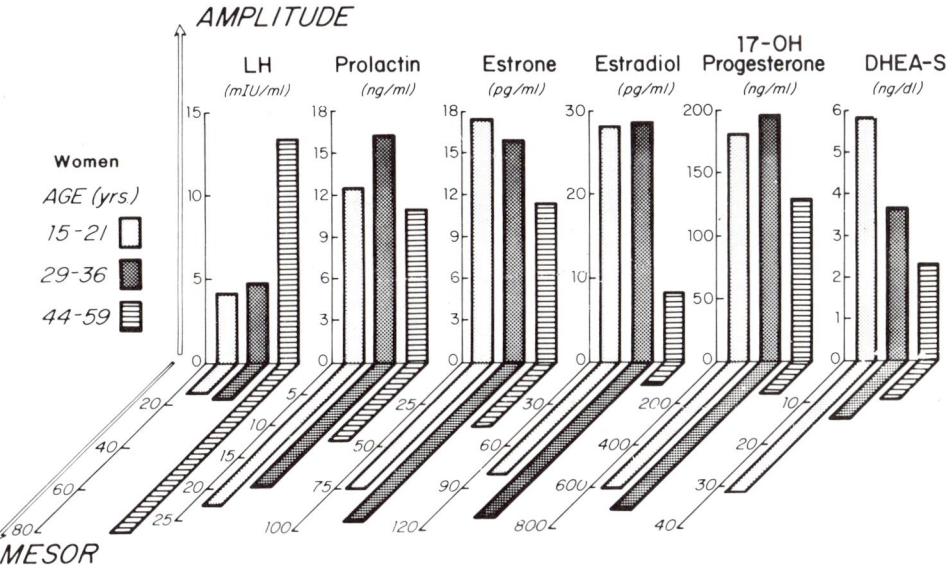

*Analyses of variance on average circadian parameter estimate for plasma hormones in women sampled in each of 4 menstrual stages and 4 seasons yielded P-value for null hypothesis (no age effect) <.001 for all mesors and for amplitudes of LH, estradiol, 17-OH progesterone and DHEA-S. P-values for amplitudes of prolactin and estrone were .027 and .03 respectively.

Fig. 10. Changes in hormonal mesors and amplitudes, notably after menopause.

In this same study, age also had a statistically significant effect on the circadian amplitude of LH, estradiol, 17-hydroxyprogesterone and DHEA-S; a probable effect of age on the circadian amplitudes of prolactin, estrone (Fig. 10), and aldosterone (Fig. 11), was also considered (Nelson et al., 1980; Haus et al., in press).

A more recent study of cortisol, aldosterone, prolactin, and growth hormone in blood of 2 groups of diurnally active, nocturnally resting men and women, sampled at 3-hour intervals during their waking span (0700-2200), indicated altered circadian rhythms of all four hormones in the older subjects (70-81 years of age) as compared with the young group (20-29 years of age). The results were qualified by the absence of (usually higher) values during sleep and limitation of focus upon a single (circadian) frequency. With this qualification, age-related reductions in the amplitudes of prolactin, aldosterone and cortisol and a changed acrophase of growth hormone were apparent. These age effects were more prominent in women than in men (Cugini et al., 1980a,b; Schramm et al., 1980).

An earlier investigation of growth hormone (Finkelstein et al., 1972) in men and women aged 8-62 years involved sampling at 20-minute intervals

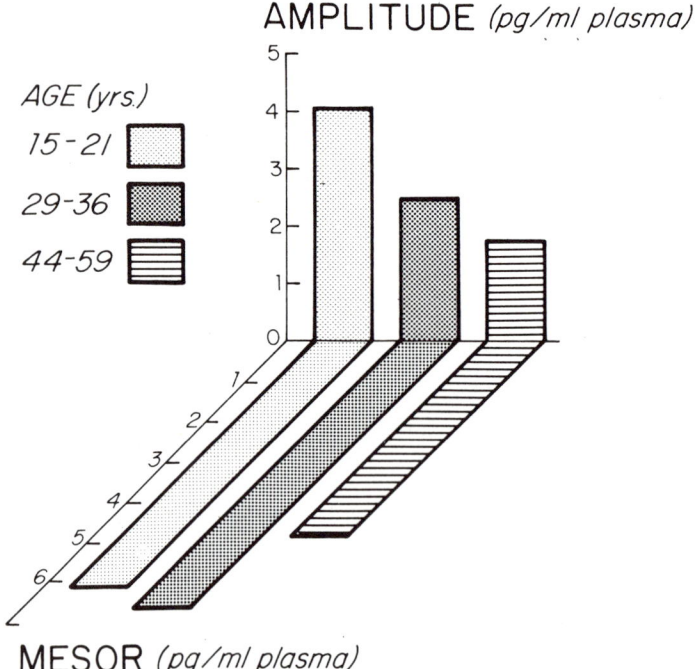

Fig. 11. Changes in both circadian and circannual amplitude (the latter shown only in Table 6) characterize aging in women.

throughout a 24-hour span. The average secretion per 24 hours (equivalent to circadian mesor) increased to a maximum in adolescent subjects and then declined to practically nothing in subjects 47-62 years old. These overall changes accompanied changes in ultradian variations, viewed macroscopically as the distribution of so-called secretory episodes throughout the 24-hour span.

AGE EFFECTS UPON CIRCANNUAL RHYTHMS

In a study already referred to (Halberg, F., et al., 1981) involving sampling of 13 plasma hormones throughout 24-hour spans in each of four seasons it was possible to examine any age effect on the amplitude of any circannual rhythms in these hormones. This was done by first fitting a 365-day cosine curve to the four circadian mesors (one in each season) obtained in each subject to obtain estimates of her circannual amplitude. It should be noted that the fit of the cosine curve to just four values was regarded as an imputation carried out to obtain an estimate for use only as an intermediate result. Differences among age-groups in the mean values of such amplitude imputations were then tested as summarized in Table 6. These tests suggested a statistically significant decrease with age in the circannual amplitude of prolactin, estrone, estradiol, 17-OH progesterone and aldosterone and an increase in the circannual amplitude of LH.

Table 6

CIRCANNUAL AMPLITUDE CHANGE OF PLASMA ALDOSTERONE, 17-OH PROGESTERONE, ESTRADIOL, ESTRONE, PROLACTIN, AND LH IN WOMEN AFTER MENOPAUSE

Hormone (units)	Age Group[a]		
	I	II	III
Aldosterone (ng/ml)	2.84	4.58	1.68
17-OH Progesterone (ng/ml)	491	615	73
Estradiol (pg/ml)	66.97	71.73	12.08
Estrone (pg/ml)	47.19	45.41	21.19
Prolactin (ng/ml)	7.11	8.82	2.38
LH (mIU/ml)	11.48	7.14	16.74

[a]I, II, and III indicate mean values for amplitude of subjects in age groups 15-21, 29-36, and 44-59 years. Comparison of circannual amplitudes of groups I and II consisting of 8 and 10 subjects, respectively, with those of group III (10 subjects) revealed a decrease in mean circannual amplitude as a function of age for five hormones and an increase for LH significant below the 5% level. These findings are qualified by possible effects of unbalanced representation of menstrual stages, limited numbers of circadian mesors provided for the fit of a one-year cosine curve and by any novelty effect biasing the data.

Such findings are the more important since they must be separated from the effects in healthy subjects of the risk of developing breast cancer, cancer of the prostate or diseases associated with a high blood pressure. The circannual amplitude appears to be a sensitive index of functional alteration, whether it be in relation to senescence or in relation to one of the diseases of senility. From the viewpoint of breast cancer risk it is pertinent that a circannual rhythm in serum prolactin is not demonstrable in fibrocystic disease in studies by Tarquini et al. (1979) in Italy. Studies involving strategically placed longitudinal sampling will be required to scrutinize the question as to whether age-related changes in hormonal rhythms are trivial corollaries or critical determinants of senescence and senility. Such further rhythmometry may be the most effective and, in the long range view, the least costly approach to an optimization of aging. If rhythm alteration underlies the age effects already recorded and here presented solely as illustrative, the manipulation of the timing as well as the kind of meals, exercise and other hygienic measures will be a first requirement. In addition, in after-the-fact considerations of age, hormones and rhythms, once diseases are established, chronobiology remains pertinent for a better interpretation of single samples (beyond screening for risk) in a more refined diagnosis and a more effective therapy (Halberg, F. et al., 1977a), Fig. 12.

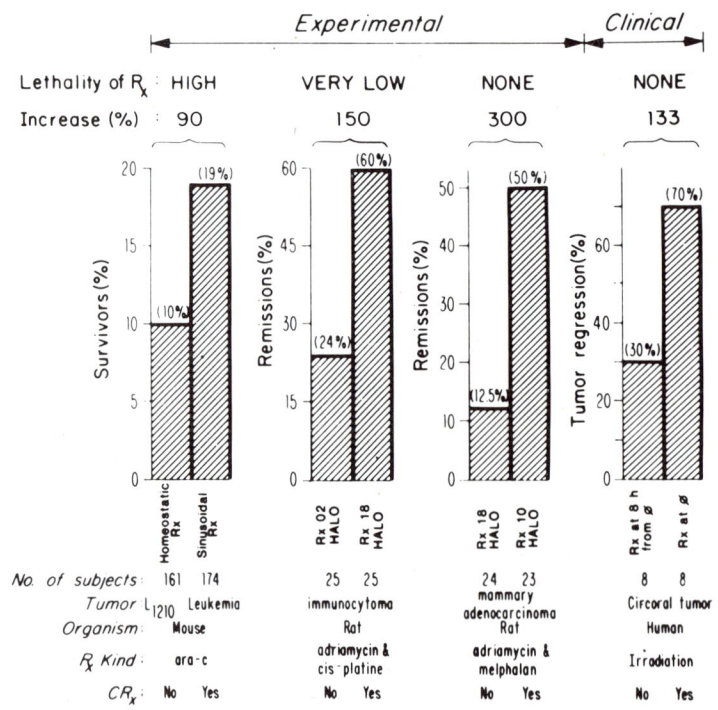

Fig. 12. Chronotherapy of elevated blood pressure (not shown) as well as of cancer gains particular importance with advancing age, timing being a cost-effective complement to dosing, and ignorance of either possibly leading to undue death or damage from treatment or forestalling full exploitation of a given desired treatment effect.

THEORETICAL RECONSTRUCTION OF PARTIAL SPECTRAL STRUCTURE OF HUMAN PLASMA PROLACTIN IN THREE AGE GROUPS[*]

Circadian mesor and amplitude both modulated by circatrigintan rhythms (except for post-menopausal group) and by circannual rhythms.

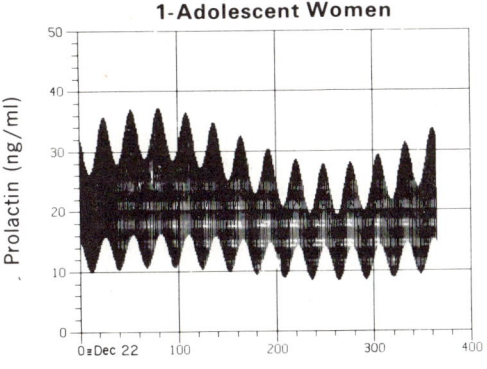

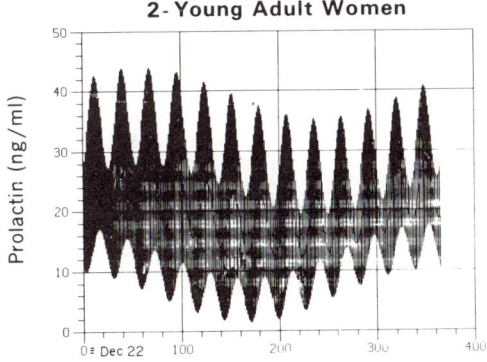

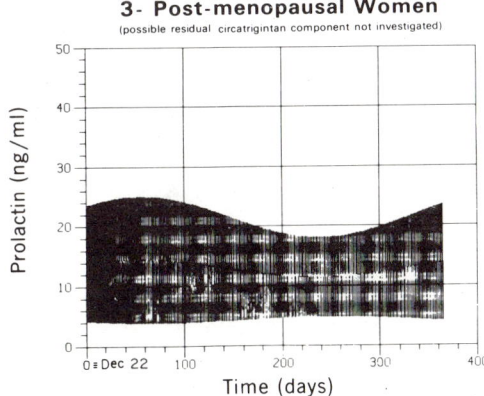

[*]Based on parameter estimates obtained from separate least-squares fittings of cosine curves with periods of 24 hours (circadian), 28 days (circatrigintan) and 365 days (circannual) to data on plasma prolactin obtained every 20' for 24 hours, ~4 times a year on a few women.

Fig. 13. Didactic example modeled on the basis of data collected in a chronoepidemiologic study carried out concomitantly in Japan and Minnesota suggests age-related differences in the spectrum of human plasma prolactin rhythms with different frequencies. The circatrigintan and circannual modulations of both the circadian mesor and the circadian amplitude are readily apparent in premenopausal women. Apart from changes in circadian amplitude and mesor, the lack of a circatrigintan group-modulation probably results from within-group frequency and phase desynchronization of infradian changes (with a period of about a month) known to persist in core temperature for over eight years after start of menopause.

Another aspect of the interrelationship in aging between hormones and chronobiology is represented by the possible involvement of hormonal factors in affective behavior, see Chapters 16-18; correlation of hormonal rhythm parameters to affective and mental optima, admittedly still uncertain, might open a new area of research of practical and theoretical value (Halberg, F.; and Wendt, 1980; Wendt et al., 1980a,b). This chapter may be concluded with the theoretical reconstruction of a partial spectral structure of human plasma prolactin in three age groups, Fig. 13. The actual number of components in a time structure already demonstrated for hormonal variability, as these may relate to affective optima or to the risk of major killer diseases leading to senility or just to senescence, is substantial. The longitudinal assessment from womb to tomb of these interwoven and intermodulating components in a time structure is overdue. It must be done in both experimental animal models and in at least a few human volunteers, not only in order to avoid waste and confusion from misinterpretation of time-unqualified single sample values, but also to quantify a new dimension of our statistical physiology, which remains pertinent from cradle to grave, for the optimization of health as well as the cure of disease.

ACKNOWLEDGMENTS

The collaboration of Erhard Haus and Walter L. Nelson of the Chronobiology Laboratories, University of Minnesota, Minneapolis, USA is gratefully acknowledged.

REFERENCES

Aschoff, J., Saint Paul, U.V., and Wever, R. Die Lebensdauer von Fliegen unter dem Einfluss von Zeit-Verschiebungen. *Naturwissenchaften* 58, 574 (1971).

Busse, E. (ed.). *Cerebral Manifestations of Episodic Cardiac Dysrhythmias* Excerpta Medica, Princeton (1979).

Colin, J., Timbal, J., Boutelier, C., Houdas, Y., and Siffre, M. Rhythm of the rectal temperature during a 6-month free-running experiment. *J. Appl. Physiol.* 25, 170-176 (1968).

Cugini, P., Scavo, D., Halberg, F., Schramm, A., Pusch, H.-J., and Franke, H. Age and sex difference in circadian amplitude of serum aldosterone, in *Proc. XXVIII Int. Cong. Physiol. Sci.*, Budapest, Hungary, 14, 366 (1980a).

Cugini, P., Scavo, D., Halberg, F., Schramm, A., Pusch, H.-J., and Franke, H. Circadian amplitude of prolactin, cortisol and aldosterone in human blood after menopause, in *Proc. Int. Symp. The Menopause*, Viareggio, Italy, May 26-28 (1980b), p. 52.

Descovich, G.C., Montalbetti, N., Kuhl, J.F.W., Rimondi, S., and Halberg, F. Age and catecholamine rhythms. *Chronobiologia* 1, 163-171 (1974).

Finkelstein, J., Roffwarg, H., Boyer, R., Kream, J., and Hellman, L. Age-related change in the twenty-four-hour spontaneous secretion of growth hormone. *J. Clin. Endocrinol.* 35, 665-670 (1972).

Gautherie, M. and Gros, C. Circadian rhythm alteration of skin temperature in breast cancer. *Chronobiologia* 4, 1-17 (1977).

Halberg, E., Halberg, F., Cornelissen, G., Simpson, H.W., and Taggett-Anderson, M.A. Toward a chronopsy: Part II. A thermopsy revealing asymmetrical circadian variation in surface temperature of human female breasts and literature review. *Chronobiologia* 6, 231-257 (1979a).

Halberg, E., Halberg, J., Halberg, Francine, Sothern, R.B., Levine, H., and Halberg, Franz. Familial and individualized longitudinal autorhythmometry for 5 to 12 years and human age effects. *J. of Gerontol.* 36, 31-33, 1981.

Halberg, E., Halberg, J., Halberg, F., Guillaume, F., Levine, H., Scheving, L.E., and Bartter, F.C. Chronoscopy estimates multiple disease risks as complement-substitute for arbitrarily scheduled (e.g., yearly) physical examination, in *Proc. 3rd Conf. Indian Soc. Chronobiol.* Varanasi, India, December 27-29 (1979b).

Halberg, F. Beobachtungen uber 24-Stunden-Periodik in standarsierter Versuchsanordnung vor und nach Epinephrektomie und bilateraler optischer Enukleation. *Ber. Ges. Physiol.* 354-355 (1954).

Halberg, F. Chronobiology. *Ann. Rev. Physiol.* 31, 675-725 (1969).

Halberg, F. and Nelson, W. Some aspects of chronobiology relating to the optimization of shift-work, in *Shift-Work and Health, NIOSH Research Symposium*, Cincinnati, Ohio (1976), pp. 13-47.

Halberg, F. and Nelson, W. Chronobiologic optimization of aging, in *Advances in Experimental Medicine and Biology*, Volume 108, H. Samis and S. Capobianco, eds. Plenum Press, New York (1978), pp. 5-56.

Halberg, F. and Wendt, H.W. Analysis of biological rhythms in psychology: Chairman's introductory remarks on the potential of chronoscopy, in *Proc. XXII Int. Cong. Psychol.* Leipzig, German Democratic Republic, July 6-12, 1980.

Halberg, F., Halberg, E., and Halberg, J. Collateral-interacting hierarchy of rhythm coordination at different organization levels, changing schedules and aging, in *Biological Rhythms and Their Central Mechanism*, M. Suda, O. Hayaishi and H. Nakagawa, eds. Elsevier North-Holland Biomedical Press, Amsterdam (1979), pp. 421-434.

Halberg, F., Engeli, M., Hamburger, C., and Hillman, D. Spectral resolution of low-frequency, small-amplitude rhythms in excreted ketosteroid; probable androgen-induced circaseptan desynchronization. *Acta Endocrinol.* (Suppl. 103) 50; 5-54 (1965).

Halberg, F., Haus, E., Nelson, W., and Sothern, R. Chronopharmacology, chronodietetics and eventually clinical chronotherapy. *Nova Acta Leopoldina* 46, 307-336 (1977a).

Halberg, F., Johnson, E.A., Nelson, W., Runge, W., and Sothern, R.B. Autorhythmometry–procedures for physiologic self-measurements and their analysis. *Physiol. Teacher* 1, 1-11 (1972).

Halberg, F., Reinberg, A., Haus, E., Ghata, J., and Siffre, M. Human biological rhythms during and after several months of isolation underground in natural caves. *Nat. Speleol. Soc. Bull.* 32, 89-115 (1970).

Halberg, F., Nelson, W., Runge, W., Schmitt, O.H., Pitts, G.C. Tremor, J., and Reynolds, O.E. Plans for orbital study of rat biorhythms. Results of interest beyond the biosatellite program. *Space Life Sci.* 2; 437-471 (1971).

Halberg, F., Powell, E.W., Lubanovic, W., Sothern, R.B., Brockway, B., Pasley, J.N., and Scheving, L.E. The chronopathology and experimental as well as clinical chronotherapy of emotional disorders. *Chronobiologia* (Suppl. 1) 4, (1977b).

Halberg, F., Carandente, F., Cornelissen, G., and Katinas, G.S.: Glossary of chronobiology. *Chronobiologia* (Suppl. 1) 4, 1977c, 189 pp.

Halberg, F., Haus, E., Halberg, E., Cornelissen, G., Nelson, W., Lakatua, D., Kawasaki, T., and Omae, T. Circannual amplitude change of plasma aldosterone, 17-OH progesterone, estradiol, estrone, prolactin and LH in women after menopause, in *Proc. Minn. Acad. Sci.*, Mankato, Minnesota, April 25-26, 1980, p. 5.

Halberg, F., Cornelissen, G., Sothern, R.B., Wallach, L.A., Halberg, E., Ahlgren, A., Kuzel, M., Radke, A., Barbosa, J., Goetz, F., Buckley, J., Mandel, J., Schuman, L., Haus, E., Lakatua, D.,

Sackett, L., Berg, H., Kawasaki, T., Ueno. M., Uezono, K., Matsuoka, M., Omae, T., Tarquini, B., Cagnoni, M., Garcia Sainz, M., Griffiths, K., Wilson, D., Wetterberg, L., Donati, L., Tatti, P., Vasta, M., Locatelli, I., Camagna, A., Lauro, R., Tritsch, G., Wendt, H. International studies of human host and tumor rhythms with multiple frequencies lead toward cost-effective sampling, in *Neoplasms—Comparative Pathology of Growth in Animals, Plants and Man*, H. Kaiser, ed. Williams and Wilkins Co., Baltimore (1981), pp. 553-593.

Halberg, J., Halberg, E., Hayes, D.K., Smith, R.D., Halberg, F., Delea, C.S., Danielson, R.S., and Bartter, F.C. Schedule shifts, life quality and quantity—modeled by murine blood pressure elevation and arthropod lifespan, *Int. J. Chronobiol.* 7; 17-64 (1980).

Halberg, J., Halberg, E., Regal, P., and Halberg, F. Changes with age characterize circadian rhythm in telemetered core temperature of stroke-prone rats. *J. Gerontol.* 36, 28-30, 1981.

Hamburger, C. Normal urinary excretion of neutral 17-ketosteroids with special reference to age and sex variation. *Acta Endocrinol.* 1; 19-37 (1948).

Haus, E., Halberg, F., Nelson, W., Lakatua, D.J., Kawasaki, T., Ueno, M., Uezono, K., and Omae, T. Age effects upon circadian amplitude in a concomitant study of 12 hormones in plasma of women. *Chronobiologia* 6; 266 (Abstract) (1979).

Haus, E., Halberg, F., Nelson, W., Lakatua, D.J., Kawasaki, T., Ueno, M., Uezono, K., and Omae, T. Age effects upon circadian rhythm parameters in a concomitant study of 12 hormones in plasma of clinically healthy women, in *Proc. XIVth Int. Conf. Int. Soc. Chronobiol.* (in press).

Hayes, D.K. and Halberg, F. Shifts in lighting regimen in the face fly suggest circadian-circaseptan interactions, in *Proc. Minn. Acad. Sci.*, Mankato, Minnesota, April 25-26, 1980, p. 4.

Hayes, D.K., Cawley, B.M., Halberg, F., Sullivan, W.N., and Schechter, M.S. Survival of the codling moth, the pink bollworm and the tobacco budworm after 90° phase-shifts at varied regular intervals throughout the lifespan, in *Shift-Work and Health, NIOSH Research Symposium*, Cincinnati, Ohio (1976), pp. 48-69.

Hershman, J.M. Clinical application of thyrotropin-releasing hormone. *N. Engl. J. Med.* 290; 886-890 (1974).

Hufeland, C.W. *Makrobiotik, The Art of Prolonging Life.* 2nd English Translation, printed for J. Bell, London (1797), p. 201.

Jeffcoate, S.L. Diagnosis of hyperprolactinemia. *Lancet* II, 1245-1247 (1978).

Jensen, H.K. and Blichert-Toft, M. Serum corticotrophin, plasma cortisol and urinary excretion of 17-ketogenic steroids in the elderly (age group 66-94 years). *Acta Endocrinol.* 66; 25-34 (1970).

Jorgensen, M. Determination of 17-ketogenic steroids in urine. *Acta Endocrinol.* 26; 424-442 (1957).

Krieger, D., Allen, W., Rizzo, F., and Krieger, H. Characterization of the normal temporal pattern of plasma corticosteroid levels. *J. Clin. Endocrinol.* 32; 266-284 (1971).

Levine, H., Halberg, F., Sothern, R.B., Bartter, F.C., Meyer, W.J., and Delea, C.S. Circadian phase shifting with and without geographic displacement, in *Biorhythms and Human Reproduction*, M. Ferin, F. Halberg, R.M. Richart, and R. Vande Wiele, eds. John Wiley and Sons, New York (1974), pp. 557-574.

Luce, G. *Biological Rhythms in Psychiatry and Medicine.* Public Health Service Publication No. 2088 (1970), p. 183.

Mandel, J., Halberg, F., Radke, A., Seal, U., and Schuman, L.

Circannual variation in serum TSH and prolactin of prostatic cancer patients, in *Proc. 3rd Conf. Indian Soc. Chronobiol.*, Varanasi, India, December 27-29 (1979).

Nelson, W., Bingham, C., Haus, E., Lakatua, D.J., Kawasaki, T., and Halberg, F. Rhythm-adjusted age effects in a concomitant study of twelve hormones in blood plasma of women. *J. Gerontol.* 35; 512-519 (1980).

Radke, A.Q., Halberg, F., Haus, E., Kawasaki, T., Mandel, J., Halberg, E., Schuman, L., Lakatua, D., Ueno, M., Uezono, K., Matsuoka, M., and Omae, T. Questionnaires complement tensopsy and hemopsy in chronoepidemiologic inquiry into risk of diseases associated with high blood pressure, in *Proc. 3rd Conf. Indian Soc. Chronobiol.*, Varanasi, India, December 27-29 (1979).

Schramm, A., Pusch, H.-J., Franke, H., Halberg, F., Cugini, P., and Scavo, D. Circadian exploration of serum cortisol (F) and pituitary growth hormone (GH) as function of age and sex, in *Proc. XXVIII Int. Cong. Physiol. Sci.*, Budapest, Hungary, 14; 688 (1980).

Scully, R.E., McNeely, B.V., and Galdabini, J.J. Normal reference laboratory values. *New Engl. J. Med.* 302; 37-48 (1980).

Seal, U., Mandel, J., Radke, A., Halberg, F., and Schuman, L. Circannual endocrine variations in prostatic cancer patients and matched hospital and neighborhood controls, in *Proc. XIVth Int. Conf. Int. Soc. Chronobiol.*, (in press).

Serio, M., Piolanti, P., Romano, S., DeMagistris, L., and Guisti, G. The circadian rhythm of plasma cortisol in subjects over 70 years of age. *J. Gerontol.* 25; 95-97 (1970).

Stupfel, M., Halberg, F., Mordelet-Dambrine, M., and Magnier, M. Perspectives in chronobiology of air pollution. *Chronobiologia* 4, 333-352 (1977).

Tarquini, B., Gheri, R., Romano, S., Costa, A., Cagnoni, M., Lee, J.K., and Halberg, F. Circadian mesor-hyperprolactinemia in fibrocystic mastopathy. *Am. J. Med.* 66; 229-237 (1979).

Vekemans, M. and Robyn, C. Influence of age on serum prolactin levels in women and men. *Brit. J. Med.* 4; 738-739 (1975).

Walker, R.F. and Timiras, P.S. Pacemaker insufficiency and the onset of aging, in *Cellular Pacemakers II*, D. Carpenter, ed. John Wiley and Sons, New York (in press).

Wendt, H., Halberg, F., Haus, E., Kawasaki, T., Lakatua, D., Halberg, E., Omae, T., Cornelissen, G., Ueno, M., Uezono, K., Matsuoka, M. Correlation of circadian subjective optimum with endocrine rhythm characteristics, in *Proc. XXVIII Int. Cong. Physiol. Sci.*, Budapest, Hungary, 14; 783 (1980a).

Wendt, H.W., Halberg, F., Halberg, E., Cornelissen, G., Haus, E., and Lakatua, D. Circadian subjective optimum and endocrine rhythms before and after menopause, in *Proc. Minn. Acad. Sci.*, Mankato, Minnesota, April 25-26, 1980b.

Wetterberg, L., Halberg, F., Tarquini, B., Cagnoni, M., Haus, E., Griffiths, K., Kawasaki, T., Wallach, L.A., Ueno, M., Uezono, K., Matsuoka, M., Kuzel, M., Halberg, E. and Omae, T. Circadian variation in urinary melatonin in clinically healthy women in Japan and U.S.A. *Experientia* 35; 416-419 (1979).

Wever, R.A. The Circadian System of Man. Results of Experiments Under Temporal Isolation. Springer-Verlag, New York (1979).

Yamaji, T., Shimamoto, K., Ishibashi, M., Kosaka, K., and Orimo, H. Effect of age and sex on circulating and pituitary prolactin levels in human beings. *Acta Endocrinol.* 83; 711-719 (1976).

Yunis, E.J., Fernandes, G., Nelson, W., and Halberg, F. Circadian temperature rhythms and aging in rodents, in *Chronobiology*. L.E. Scheving, F. Halberg, and J.E. Pauly, eds. Igaku Shoin Ltd., Tokyo (1974), pp. 358-363.

Zervas, N.T. and Martin, J.B. Current concepts in cancer: management of hormone-secreting pituitary adenomas. *New Engl. J. Med.* 302; 210-214 (1980).

Copyright©1982, Spectrum Publications, Inc.
Hormones in Development and Aging

Chapter 14
Hormones During Aging
Gregory M. Cole
Paul E. Segall
Paola S. Timiras

INTRODUCTION

The previous chapters have illustrated the essential role of hormones in regulating growth and development starting from fertilization and including fetal differentiation and maturation, survival at birth and adaptation to extrauterine life, somatic growth during childhood and adolescence, sexual maturation at puberty and, finally, attainment of physiological competence adequate for adaptation in adulthood. More difficult to evaluate is the role of hormones in senescence, partly because of the absence of clear-cut changes in endocrine function with advancing age and, partly because of the uncertainty of the impact that such changes, when present, may have on the aging process. Yet, endocrine dysfunctions have been implicated in the etiopathology of aging, and the administration of hormones has been advocated as a possible interventive measure to prolong the time of physiological competence and delay aging. In this chapter, we will consider changes with aging in endocrine structure, function and regulation (with the exception of the gonads, the aging of which is considered in Chapter 16). In the next chapter (Chapter 15) we will attempt to interpret these changes in the light of neuroendocrine theories of aging and of hormonal approaches to interventive gerontology.

Limitations and Expectations of Gero-endocrinology

The study of the changes which occur in endocrine function with old age presents considerable difficulties which are related to many variables partly dependent on the nature of the endocrine function, itself exquisitely responsive to a large variety of environmental (internal and external) stimuli and, partly on the accumulated stress and disease to which the individual is exposed throughout the life span. Thus, for the endocrines, as for other organs and tissues, competence at any given age depends on past and present biological and psychological events—the older the individual, the greater the influence of previous experience. Among these variables, some of the most important in terms of incidence or severity include: genetic status, sex and race differences, obesity, type of diet, extent of physical activity, hygienic habits (e.g., smoking, alcoholism), drugs (therapeutic or other), socioeconomic status, previous and concomittant nonendocrine diseases, exposure to stress, etc. Other factors that may contribute to the variability of endocrine responses depend on the methods of measurement, for example type of diagnostic test (e.g., secretory activity, hormone levels in plasma, free vs. conjugated hormones, hormone synthesis, catabolism and excretion, intracellular hormone transport and receptors) and the time of testing (e.g., time of day, nutritional state, sleep). Consequences of all these variables will be considered whenever possible in the course of this chapter.

During development, for example, endocrine ontogenesis takes place in the relatively constant environment of the maternal uterus, and therefore, is influenced primarily by factors related to the genetic makeup. In old individuals, endocrine function at any given time depends not only on its current competence and environment, but also on the sum total of all previous experiences accrued during the preceding lifetime. In the adult, hormones contribute to the maintenance of homeostasis, that is the constancy of the cellular environment ("le milieu interieur" of classical physiology) by acting on target tissues, cells and molecules which are functioning optimally; in the elderly, not only are endocrine organs themselves undergoing age-related changes, but they also act on tissues which have undergone aging changes (e.g., loss of cells, receptors, hormone-binding proteins; accumulation of age pigments, cross-linkages). In these senescent tissues, sensitivity to hormones (e.g., in terms of number and binding capacity of receptors, of membrane permeability, of energy-yielding molecules) and ability to respond to hormonal signals (e.g., in terms of secretory and metabolic activity, motility) may be significantly compromised.

One of the striking characteristics of the endocrinology of the elderly is the discrepancy between the presence of structural alterations and multiple signs and symptoms of hormonal dysfunction and the apparent persistence of endocrine competence well into old age. Thus, with the exception of the ovary, which exhausts its reproductive and endocrine functions at a specific time during the lifespan (see Chapter 16), most endocrine glands continue to synthesize and release hormones, albeit at a somewhat altered level, until death. For example, as discussed further on in this chapter, maintenance of normal blood glucose levels, a function dependent on multiple hormonal interactions, is well preserved in the healthy aged individual under steady-state conditions, in contrast to the increased incidence and severity of such disorders of glucose metabolism as diabetes in some genetically predisposed older people. The signs of endocrine insufficiency, clearly evidenced in older individuals, may not be related to inadequate hormone levels. For example, thyroid hormones which play a crucial role in body growth and brain maturation and influence basal metabolic rate and brain excitability in the adult, show no major decrease that can be unequivocally correlated with aging, but several investigators have related some of the features of senescence in humans and other animals (e.g., decreased metabolic rate, diminished motor activity, impaired muscular strength, dryness of skin, sparsity of hair) to an hypothyroid condition similar to that classically associated with lower than normal levels of circulating thyroid hormones. However, in old individuals various thyroid function tests do not appear to be significantly altered; rather, current studies seem to indicate that the sensitivity of target tissues to thyroid hormones is reduced perhaps by the presence of an inhibitory hormone or metabolic analogs competing for the receptor sites.

In view of the complexity of the hypothalamic-hypophyseal-target endocrine axes, aging may also affect endocrine function by acting on the individual components of a specific axis or, more likely by altering the interactions (e.g., feedback mechanisms) that relate one structure to the other. For example, it has been shown that, with progressive age, some of the hormones of the anterior pituitary (e.g., TSH) as well as those of other endocrines (e.g., insulin) may be present in altered forms, perhaps related to alterations in the activity of their metabolic enzymes (e.g., increase or decrease in different isoenzymes). Such altered anterior pituitary hormones might impart an altered signal to the target endocrine which would impair the peripheral hormonal response. Perhaps, more important to the overall endocrine function, are the changes in neuroendocrine relations which may occur with aging. Inasmuch as the neurosecretory function of the hypothalamic neurons is regulated by excitatory or inhibitory inputs relayed from other central nervous system structures (e.g., cerebral cortex, limbic system, spinal cord) via several neurotransmitters, alterations in synaptic structure and function in the hypothalamus and other brain centers may lead to impairment of both neurologic and mental competence and endocrine functions.

The importance of neuroendocrine regulation of homeostatic responses and of several biologic rhythms has been emphasized repeatedly in this book and related to specific events in the lifespan such as fertilization, birth, puberty, and adulthood (see Chapters 1, 2, 3 and 9). It is possible that aging, like growth and development, represents a programmed stage of the life span regulated by a "pacemaker" perhaps located in the brain. Alternatively, it is also possible that aging represents that period in the lifespan in which the activity of the pacemaker has ceased or is failing, thereby leading to disruption of adaptive neuroendocrine responses (Timiras, 1980; Walker and Timiras, 1981). Indeed, impairment of neuroendocrine control of adaptation is particularly evident when the old individual, exposed to stress conditions, is incapable of mustering the necessary neural and hormonal responses indispensable for survival. A more detailed discussion of neuroendocrine changes with aging is presented in the following chapter.

Another aspect of the aging of the endocrine system involves the well-known endocrine-endocrine relationships, the best example of which is represented by the interdependence between anterior pituitary and several peripheral endocrines. Alterations in one single endocrine with aging may generate a constellation of hormonal dysfunctions ultimately responsible for impaired adaptation and resistance to stress. The search for the primary cause of aging changes has not yet identified a key endocrine organ but has pointed to the anterior pituitary and/or the thyroid gland as likely candidates. The involvement of the anterior pituitary is obvious in view of the dominant role of this gland in regulating other endocrines. The thyroid gland has not only important actions during growth and development, but also regulates cellular metabolic activity throughout the life span; one of the universal signs of aging is a decline in basal metabolic rate consequent to decreased metabolic activity of each cell or to decreased number of cells. By implication, a decline in cellular metabolic activity would point to a lessening of thyroid control. Indeed, as we have already mentioned, and will discuss in more detail below, several signs of endocrine insufficiency in the aged individual may be related directly or indirectly to alterations in thyroid hormone effects. Despite the practical implications of ascribing a key role to a specific endocrine organ, further studies are needed to support the hypothesis that aging may be triggered by hormonal deficiency or excess.

Finally, in analyzing the changes in endocrine function which may occur with aging, impairment in metabolism and excretion of the hormones must also be considered as contributing and complicating factors. Such considerations not only involve organs (the liver and kidney) and tissues (adipose tissue) which are considerably altered with aging, but also more subtle changes in cell membranes and molecules. For example, it has been shown that such metabolically inactive materials as age pigments accumulate in the cells of the adrenal cortex where they may represent the results of altered intracellular metabolism. Clearly, changes related to old age in the metabolism of endocrine cells or in the metabolism, storage and excretion of hormones may significantly affect the level of the circulating hormones, and, for some hormones (e.g., corticoids, thyroid hormones, sex hormones), the proportion of the conjugated (biologically inactive) to the free (biologically active) forms.

Whereas there is a consensus on the key role of hormones throughout the lifespan, many important questions on the endocrinology of the aged remain to be answered. At which age do aging changes begin? What are the mechanisms responsible for these aging changes? In what sequence(s) do aging changes occur? Is there a primary, single endocrine affected or several endocrines affected simultaneously? What are the relationships between the endocrine and nervous systems with aging? None of these questions has been answered adequately thus far in humans or other animal species; until they are, it is impossible to present a coherent picture of the aging of endocrines. Rather, in this chapter, we have attempted, by the use of summary tables, to present a list of current data as complete as possible, letting the readers draw their own conclusions.

Maturational versus Aging Changes. The majority of age-related studies in the literature have dealt with changes generally considered to be "maturational," such as those occurring in the first year in the rat or between adolescence and 30 years in the human. While it is important to distinguish these early changes from those which occur in the last half, or quarter, of the lifespan (i.e., "true" aging), they are, nevertheless, of interest here for several reasons. First, on a theoretical level, "aging" may be viewed as a continuation of "developmental processes commonly associated with growth, terminal differentiation and cessation of growth" (Timiras, 1972) or as the consequence of a failure in old age of those neuroendocrine controls which insure optimal homeostasis during early life (Timiras, 1978; Walker and Timiras, 1981). Second, in agreement with this view, many age-related declines begin early in life, for example, the peak performance of athletes may be as early as 20 to 30 years of age and the age-related development of insulin resistance is most marked between 20 and 45 years and tends to level-off thereafter (DeFronzo, 1979). Third, early age-related changes are less likely to be complicated by the greatly increased incidence of pathology inevitably associated with old age. Therefore, "maturational" events will continue to be considered in this chapter particularly when they may play a causal role in other, late-onset, age-related alterations.

THE ADRENAL GLAND

The major function of the adrenal gland, both cortical and medullary portions, is to support homeostasis, that is, the constancy of the internal (cellular and molecular) environment, despite continuing changes in the external environment. Several endocrine, neural and neuroendocrine control systems are responsible for maintaining homeostasis and for providing the adult with an optimal capability for adaptation to a variety of stress conditions. As the individual passes from adulthood to senescence, adaptive capability to stress progressively declines. Accordingly, one of the most commonly accepted definitions considers old age the result of the "sum total of all changes that occur in a living organism with the passage of time and lead to functional impairment and decreasing ability to survive stress" and ultimately, results in death. Impairment of functional adjustments renders the aging organism susceptible to an increasing number

of challenges, each of which may be easily overcome by the adult but not
by the elderly. Thus, the accumulation and severity of environmental
stresses associated with a decreasing physiological competence lead to
increased death risk. The role that aging of the adrenal gland plays in
the age-related decrements of the adaptive capability remains to be
elucidated. We know that adequate function of the adrenal glands —
especially the adrenal cortex — is indispenable for adaptation to stress
and maintenance of life. Under normal, basal (i.e., nonstressful) conditions, none of the secretory products of the gland — corticosteroids
and catecholamines — seems to be significantly affected by aging. Nevertheless, the response of target tissues to adrenal hormones, as well as
the responsiveness of the gland to stress, may be impaired with aging.
It is this decreased capacity for adaptation to environmental changes and
stresses that constitutes one of the fundamental features of the aging
process.

Homeostasis and Stress. The capacity to adapt to an ever changing
environment is fundamental to all living organisms. With regard to human
and animal species, the ability to adjust to internal and external environmental changes has acquired broad significance and is implicated in almost
all specialties of biology and medicine. All the factors — e.g., bacterial,
viral, and fungal parasites, climatic and metereological crises, emotional
reactions — that tend to alter man's internal environment have been
designated "noxious" stimuli on the basis that their consequences may be
harmful, not only to the integrity of the tissue but also to individual
health and survival. Another term frequently used in this context is
"stress"; originally appled to mechanical phenomena, stress is described
by physicists as "a cohesive force or molecular resistance in a body
opposing the action of an applied external force." This definition provides
a convenient analogy which is useful for studying both the noxious agents
or "stressors" confronting the organisms and the physiologic responses
the body is capable of mobilizing against them.

The concept that physiologic adaptation represents a significant aspect
in the health and survival of living organisms is not new; it has been validated by numerous studies starting with the French physiologist Claude
Bernard, who coined the term "milieu interieur" and described the
importance of balanced states in maintaining its constancy. This constancy,
the so-called steady-state condition, has been designated by Walter B.
Cannon, one of the first American neuroendocrinologists, as "homeostasis,"
a term selected to indicate the sum of all physiologic adjustments necessary
to combat internal and external challenges. Cannon focused on the contribution of the sympathetic nervous system (particularly the adrenal
medulla) and its humoral effector substances (catecholamines) in the
adaptive responses. Investigators have now recognized the central role
of endocrine secretions, particularly those of the adrenal cortex and the
thyroid gland as well as of the hypothalamo pituitary complex in the
response of the organism to environmental influences. The importance of
neural and endocrine systems, and the influence of nervous imputs on these
systems, emphasized during development (see Chapters 5-7, 9, and 12)
continues to be pertinent to the adaptability and survival of the elderly.

The Adrenal Cortex

The adrenal cortex, like the thyroid, has long held a special interest
for gerontologists who have sought to ascribe many of the symptoms of
aging to the malfunction of a particular endocrine gland. Attention has
been focused on the adrenal cortex for several reasons. It was noted,
for example, that patients with hyperadrenocorticism (i.e., increased

secretion of corticosteroid hormones, particularly glucocorticoids) manifest a number of symptoms which bear a certain resemblance to aging, e.g., hyperglycemia, insulin resistance, hyperlipidemia, hypertension, thymic involution and lymphocytopenia, and extensive arteriosclerotic lesions. Thus, such adrenocortical diseases as the Cushing's syndrome could be viewed as cases of "premature aging" (Herman, 1976). Investigations of "genetically programmed senescence" in spawning salmonid fishes and eels have shown that the hyperadrenocorticism, followed by adrenocortical exhaustion may be the direct cause of the dramatic "senescent" degeneration that immediately precedes death in these animals (Wexler, 1976). The demonstration that prolonged and severe stress may shorten the lifespan with the apparent induction of at least some symptoms of "early aging" has reinforced the view that normal aging may be associated with alterations of the adrenal cortex and/or of the hypothalamic-pituitary axis that regulates its function (Selye and Tuchweber, 1976). Other studies have demonstrated that corticosteroids may influence gross brain activity (e.g., convulsive activity, see Chapters 7 and 12) as well as activity of discrete (e.g., hippocampus) brain areas (Landfield et al., 1978; Landfield, 1979). Thus, adrenocortical hormones could influence aging either directly by altering cellular homeostatic conditions and/or indirectly through their actions on the central nervous system. In either case, it appears important to a better understanding of the aging process to explain the nature of the structural and functional changes which may occur in the adrenal cortex with aging, their relations to the hypothalamic-pituitary axis and their eventual consequences in the responses to stress.

Structural and Functional Changes

The significance of changes in size in any endocrine gland, including the adrenal, under a variety of functional demands, may be difficult to interpret. This difficulty applies equally well to changes with aging. In the adrenals of humans and rats some investigators have reported an increase in weight, others a decrease and still others no change with aging (see Table 1). This discrepancy of results may depend on many factors, not only the period of the lifespan studied, but also the sex of the animal and the environmental conditions. Observations in humans and other animals attest to the presence of structural (histological) changes with aging which include: increased proliferation of connective tissue, decreased width of the zona glomerulosa and loss of steroid-containing lipid particularly in the zona fasciculata, increased incidence of hyperplastic nodules, and accumulation of lipofucsin pigments (see Table 1). The capsule enlarges to become a wide band of tissue and both cortical and medullary parenchyma undergo progressive interstitial fibrosis. It has been suggested that, in elderly humans, the nodularity is subsequent to focal ischemia due to local arteriopathy from systemic hypertension (Dobbie, 1969). In some species, tumorigenesis has been found to increase according to the sex of the animal and with aging: one study reported adrenocortical adenoma formation in 40% of WAG/Rij female rats but only in 6% of the males, the percent incidence rising sharply with age in both sexes, resulting in 60% of the females in the oldest group (over 37 months) showing adenoma formation (Burek, 1978).

No definitive alterations in basal levels (i.e., levels under steady-state, nonstressful conditions) of corticosteroids in plasma have been observed with aging in humans, laboratory or dairy animals (see Table 1 and reviews by Timiras, 1972; Andres and Tobin, 1977; Blichert-Toft, 1978). Evaluation of the reported data is complicated by the ready responsiveness of the adrenal cortex to a variety of stimuli (e.g., diet, temperature,

Table 1.

AGING-RELATED CHANGES IN THE STRUCTURE AND FUNCTION OF THE PITUITARY-ADRENOCORTICAL AXIS

PARAMETER STUDIED AND CHANGE REPORTED	REFERENCE (RAT)	REFERENCE (HUMAN)
Gross and Microscopic Structure of Adrenals		
weight (relative to body weight)		
unchanged	Tang and Phillips, 1978	Calloway et al., 1965
increased	Landfield et al., 1978	
absolute weight increased	Tang and Phillips, 1978	Calloway et al., 1965
cortical thickness unchanged		Khelimskii, 1964
width of zona glomerulosa	Dunihue, 1965	
nodular hyperplasia, increased	Lewis and Wexler, 1974	Dobbie, 1969
degenerative changes increased	Jayne, 1953	
lipofuscin accumulation, connective tissue proliferation, necrosis, lipid loss, capillary dilation	Bourne, 1967; Reichel, 1968 Szabo et al., 1970 Lewis and Wexler, 1974 Tang and Phillips, 1978 Landfield, 1979	Bourne, 1967
reduced membrane potential	Lymangrover et al., 1978	
Plasma ACTH Basal		
unchanged		Blichert-Toft, 1971
		Blichert-Toft, 1975
increased	Tang and Phillips, 1978	

Table 1. (Cont'd.)

PARAMETER STUDIED AND CHANGE REPORTED	REFERENCE (RAT)	REFERENCE (HUMAN)
Pituitary ACTH		
unchanged	Verzar, 1966	Jensen and Blichert-Toft, 1971
Circadian ACTH Periodicity		
unchanged		Jensen and Blichert-Toft, 1971
ACTH Response to Stress		
insulin induced unchanged		Friedman et al., 1969 Cartlidge et al., 1970
decreased		Hochstaedt et al., 1961
ether stress unchanged or decreased		
chronic restraint, decreased	Reigle, 1973	Tang and Phillips, 1978
Steroid-Induced Suppression of ACTH (Measured by corticosteroid response)		
decreased		Dilman et al., 1979
unchanged	Riegle and Hess, 1972	Gittler and Friedfield, 1962 Friedman et al., 1969
Adrenal Response to Exogenous ACTH		
adrenal androgen excretion (dehydropeiandrosterone) decreased		Antonini et al., 1968

Table 1. (Cont'd.)

PARAMETER STUDIED AND CHANGE REPORTED	REFERENCE (RAT)	REFERENCE (HUMAN)
Δ5-3(3-hydroxysteroid dehydrogenase) stimulation unchanged	Albrecht, 1979	
corticosterone, decreased unchanged	Hess and Riegle, 1970 Britton et al., 1975	
cortisol, unchanged		Friedman et al., 1969 West et al., 1961 Hochstaedt et al., 1961
17-OH-corticosteroids, unchanged		Gittler and Friedfield, 1962
Glucocorticoids Levels		
plasma corticoids, basal unchanged	Hess and Riegle, 1970 Britton et al., 1975 Landfield et al., 1978 Tang and Phillips, 1978	West et al., 1961 Gherondache et al., 1967 Blichert-Toft, 1975 Blichert-Toft, 1975 Friedman et al., 1969
increased (with aging)	Klug and Adelman, 1979	
increased (with maturation)	Britton et al., 1975 Tang and Phillips, 1978 Landfield et al., 1978 Barnett and Phillips, 1976	
Circadian Corticosteroid Periodicity		
peak, unchanged		Grad et al., 1971 Blichert-Toft, 1975
peak, delayed		Serio et al., 1970

Table 1. (Cont'd.)

PARAMETER STUDIED AND CHANGE REPORTED	REFERENCE (RAT)	REFERENCE (HUMAN)
higher midnight level		Friedman et al., 1969
broadened peak associated with higher basal levels	Klug and Adelman, 1979	
Volume of Cortisol Distribution Space		
unchanged		West et al., 1961
decreased		Samuels, 1956
Cortisol Half Life ($T_{\frac{1}{2}}$)		
increased	Hess and Riegle, 1972	West et al., 1961
		Serio et al., 1969
		Samuels, 1956
		Tyler et al., 1955
Cortisol Secretion Rate		
decreased		Romanoff et al., 1961
17-OH-Corticosteroid Secretion Rate		
decreased		Romanoff et al., 1961
		Gittler and Friedfield, 1962
		West et al., 1961
		Moncloa et al., 1963
		Grad et al., 1967
unchanged when measured per creatine excreted		Romanoff et al., 1961

Table 1. (Cont'd.)

PARAMETER STUDIED AND CHANGE REPORTED	REFERENCE (RAT)	REFERENCE (HUMAN)
Adrenal Androgens		
androsterone, etiocholanolone dehydroepiandrosterone (DHEA) urinary excretion, decreased		Gherondache et al., 1967
decreased plasma DHEA	Saksena and Lau, 1979	Migeon et al., 1957 Yamaji and Ibayashi, 1969
Aldosterone		
increased (with maturation)	Landfield et al., 1978	
decreased (with aging)	Landfield et al., 1978	Flood et al., 1967
renin, decreased		Weidmann et al., 1975 Horky et al., 1975 Noth et al., 1977
Glucocorticoid Receptors (Number)		
splenic lymphocyte, cerebral cortex, skeletal muscle, adipose tissue, isolated adipocytes, decreased	Roth, 1975 Roth, 1974	
cortical neurons, decreased	Roth, 1976	
liver, essentially unchanged	Petrovic and Markovic, 1975	
liver, decreased		Singer et al., 1973

emotional reactions) and the inherent difficulty of establishing standard basal levels. This difficulty is compounded by other variables which influence basal levels including circadian rhythms, sex differences and methods of steroid determination and blood sampling. Under basal conditions, none of the major secretory products of the adrenal cortex — glucocorticoids, mineralocorticoids and sex steroids — seem to be significantly affected by aging. Nevertheless, a decreased secretion of corticosteroid hormones with aging may be inferred from the progressive decline in the urinary excretion of 17-ketosteroid, metabolic products of testicular and adrenocortical hormones. In the male, this fall has been attributed primarily to a decline in testicular function (see Chapter 16) but adrenocortical involvement is entirely possible and, in fact, is indicated by the similar decline, although of a lesser magnitude, that occurs in the female. Changes in urinary 17-ketosteroids may also reflect alterations in hepatic metabolism and renal excretion of the hormones which may be impaired in the elderly. For example, the significantly prolonged cortisol halflife in older humans and rats (see Table 1) suggests a decreased metabolic clearance rate for the hormone or a decline in its ultization at the cellular level. In plasma, the levels of glucocorticoids and aldosterone do not change significantly from adulthood to 65 years of age and older, and changes, when detected, are slight. In rats, basal levels of plasma corticosterone increase with maturation; although, according to some, basal plasma levels continue to increase in old animals, the majority of reports indicate that levels remain unchanged beyond 12 months of age when senescent changes are quite obvious in other (gonadal) endocrine functions. Circadian rhythms are somewhat altered in these animals even though peak levels still occur at the same time of the day and are of similar height at all ages studied. These changes in corticosterone circadian rhythm are one example, among many, of the variations in cyclic activity that occur in aged animals. It has been suggested that disruption of one or several endocrine rhythms, perhaps consequent to a failure of a specific pacemaker, play a crucial role in the initiation and progression of aging changes (Walker and Timiras, 1981).

Beyond an early maturational increase and despite fluctuations from age to age, basal levels of glucocorticoids, and possibly of aldosterone, remain essentially unchanged with aging in other animal species such as cattle (Riegle and Nellor, 1967), goats (Riegle et al., 1968) and mice (Eleftheriou, 1974; Grad and Khalid, 1968). In mice, for example, basal corticosterone levels are similar at two and thirty months of age, but fluctuate between these ages, perhaps a reflection of the variables that may influence adrenocortical secretion (Eleftheriou, 1974; Grad and Khalid, 1968). Exposure of these animals to a mild stress increases hormone levels (Eleftheriou, 1974) and simultaneously decreases inter-age variations, suggesting that regardless of age the response to stress overrides other possible variables.

Despite the absence of clear-cut changes in plasma corticosteroid levels, some investigators believe that adrenocortical function in the elderly differs to some extent from that in the adult, perhaps in metabolism, target tissue responsiveness and efficiency of hypothalamo-pituitary regulation. For example, several studies have shown that the rate of secretion and metabolism of glucocorticoids declines with age, possibly in relation to a decrease in functional mass (see Table 1). Furthermore, even though aging seems to induce only a minor decline in cortisol production a significant drop seems to occur in some of its precursors (e.g., pregnenolone and progesterone) (Romanoff et al., 1961), but this reduction is no longer apparent after administration of ACTH (Romanoff et al., 1969). According to these and other studies, the adrenal cortex in the elderly has adequate although reduced, substrates and enzymes and is capable of responding as well as the young gland to an equivalent dose of ACTH or an environmental stimulus of comparable intensity; thus, the apparent reduction in glucocorticoid secretion could be interpreted as an indication of an homeostatic adaptation to a reduced peripheral demand for cortisol in old age (Blichert-Toft, 1975).

A decline in the responsiveness of target cells to glucocorticoids with aging is suggested by the reduction in the number of cortisol and corticosterone receptors in several tissues (see Roth, 1979a,b). Likewise, the ability of glucocorticoids to induce specific cellular metabolic reactions may be impaired with aging. For example, the ability of glucocorticoids to inhibit substrate entry into rat splenic leukocytes declines with age in proportion to the reduction in splenic glucocorticoid receptors (Roth, 1975). Similarly, reduction in glucocorticoid inhibition of glucose oxidation in rat epididymal fat pad adipocytes has been correlated with declining numbers of glucocorticoid receptors during aging (Roth and Livingston, 1976). In contrast, in the rodent liver where the number of specific high affinity glucocorticoid receptor numbers does not decline with aging, the induction of tyrosine-amino-transferase by corticosteroids remains unimpaired (Adelman and Freemen, 1972) while induction of the adaptive enzymes serine deaminase, ornithine aminotransferase, and glucose-6-phosphatase appears to be delayed but not ultimately diminished (Rahman and Peraino, 1973).

From the foregoing it emerges that changes in adrenocortical function with aging may be the consequence more of a slower rate of steroid metabolism and excretion by the liver and kidney, a decline in tissue utilization, or changes in steroid binding than of alterations in the secretory activity of the gland. Additionally, inasmuch as the function of the adrenal cortex reflects and is regulated by its interrelations with the anterior pituitary and the hypothalamus, alterations in old age may result from temporal or quantitative disruption of these interrelations. Under basal conditions, such alterations, although only marginally operative, would be sufficient to insure an adequate secretory function, but under stressful conditions, the increased demands imposed upon the hypothalamo-pituitary-adrenocortical axis would be associated with a reduced ability to support adaptation.

Regulation of Adrenocortical Responses to Stress. The physiological adjustments that become operative under stress depend on the integration of endocrine and nervous mechanisms. Stated simply, stress would alter the higher brain centers, particularly those involved in the perception and integration of environmental stimuli; from these centers, stimuli are relayed to the hypothalamus where they induce the appropriate release of the corticotropin releasing factor (CRF) which, in turn, directs the anterior pituitary to secrete ACTH; the latter then stimulates the adrenal cortex to increase its secretion of adrenocortical steroids, primarily glucocorticoids. This increase is essential for survival. Hypophysectomized or adrenalectomized animals treated with doses of glucocorticoids sufficient for maintenance of the animals under steady-state conditions die when exposed to noxious stimuli.

Increases in ACTH secretion to meet emergency situations are mediated through the hypothalamus, particularly the median eminence-portal system. It has been shown experimentally that if the median eminence is destroyed some basal glucocorticoid secretion continues and the adrenals do not atrophy, but the capacity to increase secretion in response to many different stresses is blocked. Afferent nervous pathways from many parts of the brain converge on the median eminence. Fibers from limbic nuclei such as the amygdala, mediate responses to emotional stresses and fear, anxiety and apprehension cause marked increase in ACTH secretion. Impulses ascending to the hypothalamus in the reticular formation trigger ACTH secretion in response to injury. Most of the stressful stimuli that cause an increase in ACTH secretion also activate the sympathetic medullary system. Thus, part of the function of the circulating glucocorticoids may be to maintain the vascular reactivity to the catecholamines. Glucocorticoids are also necessary for the catecholamines to mobilize free fatty acids which represent an important source of energy supply for adaptation to stress.

Comparative studies in several species have demonstrated that glucocorticoids also influence the synthesis of medullary catecholamines. Corticoids reach the medulla via a capillary net similar to the portal system in the pituitary, through which blood from the cortical veins is drained in the medullary sinuses. Thus, the blood that reaches the medulla is richer in glucocorticoids than systemic blood (Wurtman et al., 1972) and these hormones regulate the development and activity of such enzymes as phenylethanolamine-N-methyl-transferase (Wurtman and Axelrod, 1966) and tyrosine hydroxylase (Otten and Thoenen, 1976) necessary for the synthesis of epinephrine and norepinephrine. In this manner, cortical and medullary secretions mediate the role of the adrenal glands in response to stress.

Response of the Adrenal Cortex to ACTH and to Stress

In view of the progressive inadequacy of the elderly to respond to stress, considerable attention has been focused on whether aging may alter plasma and/or pituitary ACTH levels, ACTH circadian rhythmicity, ACTH response to stress and adrenocortical steroid responses to ACTH, and finally, whether feedbacks controlling hypothalamus, pituitary and adrenal integration change with advancing age. For the basal glucocorticoid levels, many of the questions raised remain unanswered or the answers show wide disparity from one study to another. Thus, pituitary content, plasma basal levels and circadian rhythmicity seem to remain unchanged despite some reports of increased ACTH levels with old age in rats (see Table 1). On the other hand, the smaller elevation of plasma corticosteroid levels in old as compared with young animals after administration of ACTH – rats (Hess and Riegle, 1970); goats (Riegle et al., 1968); cattle (Riegle and Nellor, 1967); rhesus monkeys (Bowman and Wolfe, 1969) – and exposure to ether stress – rats (Hess and Riegle, 1970) suggests a diminished responsiveness of the adrenal cortex to ACTH.

Similarly, the inadequate response to geriatric patients to the stress of surgery is interpreted as due to an impairment of ACTH stimulation of the adrenals (Blichert-Toft et al., 1967). However, when the pituitary-adrenocortical function was tested by the administration of ACTH or of metyrapone, no differences were observed between old and young individuals (see Blichert-Toft, 1975). Metyrapone is a drug capable of reducing corticoid production through synthesis inhibition; as a consequence, a compensatory increase in ACTH normally follows metyrapone administration and the secretion of cortisol from the adrenal cortex as well as its urinary excretion is increased. In patients with an insufficiency of the pituitary-adrenocortical axis, administration of metyrapone does not result in a compensatory response in the rate of ACTH and, reciprocally in an increase in corticoid production. Inasmuch as the response to metyrapone does not always correspond to the stimulation induced by surgical stress, a negative metyrapone test does not exclude the possibility that the ACTH response to stress (i.e., to surgery) may be adequate. It was originally thought that all types of stress induced a stereotypic activation of the adrenal cortical and medullary secretions (Selye, 1952). This may not always be the case; various types of noxious stimuli induce different patterns of response of the pituitary-adrenocortical axis (Dallman and Jones, 1973) and some types of stress are ineffective in activating the adrenal gland secretion (Katz et al., 1977). Therefore, within the relatively large spectrum of adrenal reactions to stress, aging may affect selected responses only.

An explanation for the declining adaptive capacity of the aged in the face of an apparently unchanged adrenocortical function has not yet been generated despite efforts from several investigators. Most of these efforts emphasize the possibility that endocrine requirements and metabolism as well as sensitivity of the target tissues may differ at different ages. Thus, it has been argued, and rightly so, that measurements of total plasma hormone are not a reliable guide to the concentration of that portion of the hormone that is biologically active, or that radioimmunoassay of a protein hormone does not necessarily represent the total amounts of the hormone secreted. With corticosteroids, it is possible that the binding affinity of these hormones for plasma protein changes with aging and that the ensuing alterations in the proportions of the free to the bound hormone influence hormonal efficacy. As we have reported above, the capacity of glucocorticoids to influence enzyme activity and cellular metabolism, as well as the number and affinity of glucocortoid receptors, are diminished with aging. Other sites where the aging process may be operative are the pituitary and/or the hypothalamus. For example, in the rat, hypophysectomy seems to delay physiologic and pathologic age-related changes but, at the same time, it shortens the lifespan (Everitt, 1980). The lifespan is also shortened by oversecretion of ACTH and glucocorticoids, as occurs in animals repeatedly stressed and showing an acceleration of aging (Everitt, 1976). Another mechanism by which the hypophyseal-adrenocortical axis could be influenced by aging involves an age-induced shift in the enzymes which cleave the large polypeptide molecule precursor for ACTH (Segall, 1979; Margules, 1979). This large precursor molecule, known as pro-opiocortin, yields not only ACTH but also β-lipotropin, the immediate precursor to β-endorphin. β-endorphin levels have been found to be elevated in individuals with both excess (Cushing's disease), or insufficiency (Addison's disease) as well as in obese humans, rats and *ob/ob* mice. With aging, alteration in the enzyme cleavage could result in different levels or ratios of the two hormones, and the ensuing imbalance, together with a genetic predisposition and repeated stress challenges, would lead to obesity, hypertension and Cushingoid symptoms found in some elderly individuals (Margules, 1979).

It is possible that the deviation from normal of ACTH and corticosteroid levels may be more important in the etiology of aging than the direction (excess or deficit) of the change. This concept would underlie the role or corticosteroids as "permissive" or "conditioning" hormones, i.e., whose presence in "optimal" levels is required for other factors to exert their actions. Among these factors, nervous stimuli from the hypothalamus and other central nervous system centers may initiate the cellular and systemic changes that occur with aging. Alterations in the structure, electrical activity and function of the hypothalamus have been reported (Frolkis, 1976; Dilman, 1976) and a summary of the major changes which occur in the central nervous system is presented in Chapter 15. On the other hand, little is known of the nature of the CRF and even less of changes it may undergo during development and aging. Whether or not aging changes in the hypothalamus will also involve the neurosecretory cells releasing CRF and in what manner, remains to be elucidated.

"Permissive" Actions of Corticosteroids. Under optimal conditions, a small dose of corticosteroid maintains the adrenalectomized animal in a state of well-being. Under adverse conditions, a relatively large dose is needed if the animal is to survive. This same large dose given under optimal conditions induces hypercorticism, a sign of excess corticosteroid. The fluctuations in the secretory activity of a normal subject are presumed to reflect the varying needs of the organism for corticosteroids.

Corticosteroids, like other hormones, probably have two major types of action: one action, pleiotropic in nature, regulates the machinery of the cell and is shared by most hormones. The other type of action, organizational in nature, is exerted on selected macromolecules, generating discrete responses specific for each hormone (Tomkins and Gelehrter, 1972). Organizational actions are operative primarily during ontogenesis, whereas pleiotropic actions persist throughout the lifespan. Some of these latter actions are "necessary" for a certain cellular activity to take place, but are not "sufficient" by themselves to trigger the activity. They have been designated by the term "permissive" (Ingle, 1954). For example, this permissive action may influence the rate of a reaction; in the absence of glucocorticoids, hepatic cells, although capable of glycogen formation, perform this activity at a slow rate; when cortisol is given to an adrenalectomized subject, the rate of glycogen deposition is proportional to the dose of cortisol. Permissive actions are also involved in the complex interrelations between corticosteroids and other hormones. For example, in vitro, in the absence of lipolytic enzymes, cortisol even in large concentrations has virtually no effect on the rate of lipolysis in adipose tissue. Likewise, a sympathomimetic amine has only a slight effect on the rate of lipolysis. However, if a necessary minimal amount of cortisol is added together with a range of sympathomimetic amine, the rate of lipolysis will be proportional to the concentration of the amine. When one compares older with younger rats, the rate of induction of certain enzymes by corticosteroids may be slowed or remain unchanged. Whether this permissive action of corticosteroids is affected by aging, to what extent, and by what mechanisms, is currently the subject of several investigations (see Adelman, 1979).

Aldosterone

While age-related changes in glucocorticoid metabolism have been extensively investigated, scant information is available in the literature on the mineralocorticoids, probably because serum electrolyte concentrations continue to be regulated within narrow limits in older human subjects and experimental animals. Nevertheless, it has been demonstrated that both secretory rate and plasma level of aldosterone are reduced by approximately 50% in elderly subjects despite a reduction in the metabolic clearance rate of the hormone (Flood et al., 1961, 1967). The cause of the reduction in aldosterone secretion apparently lies in a reduced plasma renin activity (PRA) (Noth et al., 1977). The primary factor regulating aldosterone secretion is Angiotensin II derived from the renin-angiotensin system. While the aldosterone response to ACTH is unaltered with aging, the response to sodium and blood volume depletion (mediated by juxtaglomerular renin secretion) is diminished as is the PRA (Weidmann et al., 1975). In support of these findings, Horky et al. (1975) found that both basal and orthostatically stimulated PRA was significantly reduced in normotensive older subjects. That plasma electrolyte balance is maintained despite the drop in aldosterone output and levels may reflect a reduction in ADH secretion and/or a decline in glomerular filtration rate. In the rat, basal aldosterone levels increase in middle age (in parallel with corticosterone) and decline thereafter (while corticosterone remains unchanged) (Landfield et al., 1978) consistent with the histological picture of a reduction in the width of the zone glomerulosa (Dunihue, 1965).

If the reduction in aldosterone output stems from a reduction in PRA, the age-related defect may lie in the sympathetic regulation of the juxtaglomerular apparatus. The presence of an elevated urinary dopamine/ norepinephrine excretion ratio in subjects with reduced PRA suggests a possible causal role for increased dopaminergic inhibition and decreased adrenergic stimulation of juxtaglomerular renin secretion with aging, (possibly secondary to a decline in serum dopamine β-hydroxylase activity) (Horky et al., 1975). However, urinary excretion does not directly reflect plasma catecholamine values and, in aging humans, plasma dopamine β-hydroxylase activity is not reduced but rather appears to be increased (Freedman et al., 1972) as are plasma norepinephrine levels (Ziegler et al., 1976). Since the higher values of PRA in younger subjects are reduced to the levels found in old age after the administration of a β blocker (Horky et al., 1975), it is possible that the age-related reduction in PRA is secondary to a reduction in juxtaglomerular β-adrenergic receptor sensitivity. This hypothesis is supported by the generalized decline in β-receptor activity demonstrated in aged rats (Roth 1979a,b). As discussed in the section on the thyroid, the decline in beta receptor activity may itself be explained by an age-related hypothyroidism and aldosterone secretion rates are reduced in hypothyroid patients (Luetscher et al., 1963).

The Adrenal Medulla

Degenerative changes of the chromaffin cells of the adrenal medulla increase with aging together with the appearance of fibrosis and occasional hyperplasia (Jayne, 1953). Pheochromocytoma, a tumor of the chromaffin cells, occasionally present in elderly humans, is an example of tumorigenesis concurrent with aging. The tendency towards neoplastic transformation with aging is also present in the rat in which one specific strain shows that 12% of the old males develop such tumors (Burek, 1978).

With respect to medullary function in humans, urinary excretion of catecholamines is increased with aging (Giorgino et al., 1969) as are plasma norepinephrine levels, an increase which becomes more marked with standing and isometric exercise (Ziegler et al., 1976). In rats, synthesizing enzymes for catecholamines — tyrosine hydroxylase and phenylethanolamine-N-metyltransferase — increase steadily as the animal ages; adrenal content of epinephrine and norepinephrine increases as well (Kvetnansky et al., 1978). The elevation in catecholamine levels and in their synthesizing enzymes may represent a compensatory reaction to the apparent increasing refractoriness of target tissues to catecholamines with aging (Schocken and Roth, 1977; Puri and Volicer, 1977; Parker et al., 1978; Greenberg and Weiss, 1978; Roth, 1979a,b; Abrass and Scarpace, 1979). For example, the ability of dopamine to stimulate adenylate cyclase in the rat brain (more specifically the corpus striatum) is progressively reduced with increasing age (Puri and Volicer, 1977); this reduction may be associated with a decrease in dopamine receptors between 7 and 25 months of age in the rat (Roth, 1979a,b) and, at corresponding ages, in mice (Finch, 1979) and rabbits (Makman, 1980; Makman et al., 1978). In view of the influence of corticosteroids on the development and activity of the metabolic enzymes for medullary catecholamines, possibly some of the changes in catecholamine metabolism observed with aging are related to alterations in adrenocortical function (see above). Thyroid hormones also seem to influence the activity of these enzymes (Vaccari et al., 1977) and catecholamine levels (Ito et al., 1977) in the developing and adult brain. The stimulatory action of these hormones on β-adrenergic receptors in the myocardium of young animals is also demonstrated in old animals (Abrass and Scarpace, 1979).

Dopaminergic receptors in the brain are also influenced by the thyroid status (Timiras and Vaccari, 1981; Vaccari and Timiras, in press). These observations suggest that thyroid hormones, like corticosteroids, may have permissive actions on target tissues such as the adrenal medulla where they would regulate catecholamine metabolism. This permissive action would be sufficient for the thyroid hormones to support an adequate response to isoproterenol by myocardial adrenergic receptors (Abrass and Scarpace, 1979). An alternate interpretation, is that an excess or deficit of thyroid hormones differentially influences receptors by other mechanisms which remain to be investigated. In view of current interest in hormone receptors and in the influence of hormones on other membrane receptors, current directions of research in this area include studies on receptor synthesis, structure and control as well as investigations of those molecular events resulting from receptor-hormone binding.

THE PANCREAS: INSULIN, GLUCAGON, AND DIABETES MELLITUS

Inasmuch as the endocrine function of the pancreas, mediated through the secretion of insulin, glucagon, and somatostatin, is concerned primarily with glucose metabolism, most studies have utilized various tests of glucose metabolism as indices of pancreatic changes with old age. With aging, tests of glucose metabolism show moderate, but progressive alterations. For example, more than 50% of randomly selected subjects over 60 years of age have significantly impaired glucose tolerance tests when compared with younger controls, that is, blood glucose levels take longer to return to normal after oral or parenteral glucose administration. Such findings have led to hundreds of publications and considerable controversy resulting from the apparent necessity of labeling more than half of the older population as diabetic (for reviews see Andres, 1971; Andres and Tobin, 1977; Davidson, 1979). This clinical dilemma may be neatly resolved by introducing age-corrected nomograms for the glucose-tolerance test (Andres and Tobin, 1977) or by substituting elevated fasting glucose, rather than decreased glucose tolerance, as the principal criterion for diagnosis (Davidson, 1979). The progressive increase in glucose intolerance with age cannot, however, be denied. It remains to account for the principal factors underlying the increased glucose intolerance with aging and whether they are physiologic or pathologic in nature. In this section, we will consider both aspects and present the available data in summary form in Table 2.

Changes in Pancreatic Structure and Function

The majority of studies have found that the increased insulin secretion ("insulin response") to oral or intravenous glucose administration is not significantly impaired with aging in humans or rats (see references Table 2). However, some investigators have argued that this type of test cannot reveal minor decrements in the insulin response to glucose because, by definition, glucose intolerance implies the prolonged maintenance of elevated blood glucose following a standard glucose load. Therefore, as long as glucose, the principal signal for insulin release, remains elevated, this elevation would lead to an apparently adequate insulin response. In support of their view of a deficit in pancreatic function with aging, Andres and Tobin (1977) compiled a list of 17 studies which indicate that the initial insulin response to glucose is depressed with aging; normal or elevated insulin follows only later in the test (the second hour) when the blood glucose continues to be elevated. Further, Andres and collaborators have attempted to directly test their postulate of an impaired (beta cell) insulin response by means of an "intravenous glucose clamp technique." With this technique, the blood glucose concentration can be maintained at any chosen level for a period of hours by monitoring its concentration at short intervals and infusing an appropriate amount of glucose to maintain the desired concentration. Utilizing this method, these investigators have demonstrated a small decline in insulin response with age (Andres and Tobin, 1977), but the magnitude of the deficit observed is not sufficient to account for the observed decrease in glucose tolerance with age (McGuire et al., 1979; DeFronzo, 1979). The absence of a major decrement in insulin secretion is in agreement with the lack of significant age-related deterioration and unchanged insulin content of the pancreas (Table 2).

In elderly humans, the observation of only minor alterations in circulating insulin levels and of lack of changes in insulin metabolic clearance rate or half-life probably reflects essentially adequate beta-cell output. In the rat, however, the tissue capacity for insulin degradation declines markedly and progressively with age (Runyan et al., 1979); this progressive decline in turnover may play a role in the hyperinsulinemia reported to occur with age in these animals, although the elevated glucose level associated with decreased glucose tolerance is probably the major factor involved. Insulin release per beta-cell is decreased with age in the rat in vitro, but the deficit is overcome by a compensatory increase in beta-cell numbers which insures a normal or elevated insulin response to glucose in tolerance tests. In conclusion, despite the observed reduction in pancreatic sensitivity to glucose and apparent delayed insulin response, the majority of investigations indicate that insulin release is adequate in both humans and rats (Davidson, 1979).

Insulin Resistance

If insulin secretion and circulating level are considered normal even in advanced age, then insulin resistance may be responsible for the observed alterations in glucose metabolism. However, despite the efforts of many investigators to attribute the depressed glucose metabolism of aging to insulin resistance, the literature on this subject is contradictory. As summarized in Table 2, four studies reported a reduced effectiveness of injected insulin in disposing of a glucose load with old age, whereas five similar studies showed no changes. All of these studies employed "nonphysiologically" high doses of insulin. To avoid exogenous administration of high doses, two more recent investigations have used a "hyperglycemic glucose clamp technique" to assess the tissue response to endogenously secreted insulin (well within physiological limits). The basis for this technique rests on the assumption that the ratio of glucose infused, to maintain a given elevated level of glucose, divided by the amount of endogenously secreted insulin, is a measure of the effectiveness of the secreted insulin. Utilizing this method, DeFronzo (1979) has observed increased insulin resistance with age, while Andres and coworkers have reported results suggestive of insulin resistance only in the second hour of glucose infusion (reviewed in Davidson, 1979). One explanation offered for this increased resistance is an elevated proinsulin secretion during the second hour; however, an increased secretion of hyperglycemic lipolytic hormones, for instance, glucagon or growth hormone,

Table 2.
AGING-RELATED CHANGES IN INSULIN AND GLUCAGON METABOLISM

PARAMETER STUDIED AND CHANGE REPORTED	REFERENCE (RAT)	REFERENCE (HUMAN)
Islet Histology		
increased size and beta cell numbers	Reaven et al., 1979 Hellman, 1959	
decreased beta/alpha cell ratio		Seifert, 1954
increased amyloid deposition		Schwartz et al., 1965
no significant degeneration		Andrew, 1944 Feldman, 1955
no significant difference in insulin content		Wrenchall et al., 1952 Jorpes and Rastgeldi, 1954
Glucose Tolerance		
decreased	Klimas, 1968 Gommers and de Gasparo, 1972 Gold et al., 1976 Brancho-Romero and Reaven, 1977 Yoshino et al., 1979	Andres, 1971 Davidson, 1979
unchanged in nonobese men		Kimmerling et al., 1977
Fasting Serum Insulin		
increased	Gommers and de Gasparo, 1972 Lewis and Wexler, 1974 Olefsky and Reaven, 1975	
unchanged		Dudl and Ensinck, 1972, 1977

Table 2. (Cont'd.)

PARAMETER STUDIED AND CHANGE REPORTED	REFERENCE (RAT)	REFERENCE (HUMAN)
increased (with maturation) in portal vein	Freeman et al., 1973	
decreased (with aging) in portal vein	Freeman et al., 1973	
Insulin Response to Glucose (in vitro)		
reduced per beta cell and decreased per islet despite increased beta cell numbers	Reaven et al., 1979	
Insulin Response to Glucose (in vivo)		
increased after oral glucose	Brancho-Romero and Reaven, 1977 Hoffman et al., 1972 Gold et al., 1976	Andres and Tobin, 1977 Davidson, 1979
unchanged after oral glucose	Yoshino et al., 1979	Davidson, 1979
decreased acute response, delayed peak		Andres and Tobin, 1977 Davidson, 1979
reduced (intravenous glucose clamp)		Andres and Tobin, 1977
Insulin Half-life ($T_{\frac{1}{2}}$) and Metabolic Clearance Rate (MCR)		
unchanged MCR		Andres and Tobin, 1977 McGuire et al., 1979 Barbagallo-Sangiorgi et al., 1979 DeFronzo, 1979
increased $T_{\frac{1}{2}}$		Orskov and Christensen, 1969

Table 2. (Cont'd.)

PARAMETER STUDIED AND CHANGE REPORTED	REFERENCE (RAT)	REFERENCE (HUMAN)
reduced tissue degradation	Runyan et al., 1979	
Plasma Pro-Insulin after Oral Glucose		
increased		Duckworth and Kitabchi, 1972, 1976
Basal (Fasting) Serum Glucagon		
unchanged		Dudl and Ensinck, 1972, 1977
Glucagon Levels after Glucose or Arginine		
unchanged arginine response		Dudl and Ensinck, 1972 Fedele et al., 1977
unchanged glucose response		Dudl and Ensinck, 1977 Nonaka and Tarui, 1977
increased (paradoxical) glucagon in response to glucose	Klug et al., 1979 Yoshino et al., 1979	
Insulin Effects on Glucose Metabolism (in vitro)		
decreased (with maturation) adipocyte response	Gries and Steinke, 1967 Holm et al., 1977 Jeanjean et al., 1977 Olefsky, 1977	Gries and Steinke, 1967
decreased (with maturation) diaphragm response	Gommers et al., 1977 Elsas et al., 1971	

Table 2. (Con'td.)

PARAMETER STUDIED AND CHANGE REPORTED	REFERENCE (RAT)	REFERENCE (HUMAN)
Insulin Binding		
decreased (with maturation) adipocyte	Olefsky and Reaven, 1975	
decreased (with maturation) hepatocyte	Freeman et al., 1973	
Glucagon Effects (in vitro)		
decreased (with maturation) adipocyte lipolysis	Holm et al., 1977	
abolition (with maturation) of adipocyte adenylate cyclase	Cooper et al., 1977	
unchanged hepatocyte adenylate cyclase stimulation (with maturation and aging)	Kalish et al., 1977	
Glucagon Binding		
unchanged (with maturation) hepatocyte binding	Lockwood and East, 1978	
decreased (with maturation) adipocyte binding	Livingston et al., 1974	

could also account for such a result (see below).

Insulin Resistance, Glucose Intolerance, and Obesity. In contrast to most insulin-dependent juvenile-onset diabetics who are typically underweight, late-onset diabetics are overweight. This fact and the prolonged hyperglycemia in older individuals, despite the apparently normal or only slightly elevated (second-hour) levels of insulin, have generated several attempts to ascribe a causal role to insulin resistance secondary to increased adiposity in the age-related decline in glucose tolerance.

The well-documented increase in the percentage of body fat with aging is characterized by an increase in the average adipocyte size (Bjorntorp, 1974) and elevated circulating free fatty acids (Davidson, 1979). Both of these factors have been related to insulin resistance and reduced glucose disposal. Strong correlative evidence for a relationship between increased adiposity, hyperinsulinemia, and insulin resistance has been provided by Bjorntorp and collaborators (1971), who suggested that aging-related changes in the glucose-insulin responses could be almost entirely accounted for by an increase in fat cell size. This work has received indirect support from a recent study by Kimmerling and collaborators (1977) who failed to observe any changes in aging-related glucose tolerance or insulin resistance in a group of 100 nonobese male subjects. The relationship between fat cell size and insulin resistance is further discussed below.

A role for elevated free fatty acids in abnormal glucose tolerance is suggested by the work of Randle and collaborators (1963) who have postulated a "glucose-fatty-acid cycle," where increased oxidation of free fatty acids leads to a reduction of glucose utilization. Although the clinical relevance of Randle's animal work has not yet been fully established (Ruderman, et al., 1969) available evidence suggests that elevated free fatty acids, secondary to an increased adipose tissue mass may play a significant role in the age-related decline in glucose tolerance (Davidson, 1979).

Inasmuch as obesity alone cannot explain all the effects of aging on glucose disposal, additional factors must be involved (Crockford et al., 1966; Seltzer, 1970; DeFronzo, 1979). Nevertheless, increased adiposity is without doubt an important factor in the decline of glucose tolerance. It is possible that the explanation for the increased adiposity evident in aging is to be found in other neuroendocrine disturbances, such as those suggested to account for a similar syndrome in the obese hyperglycemic mouse discussed below.

Alterations in Tissue Binding and Sensitivity to Insulin and Glucagon

Following maturation in humans and rats, adipocytes show a low *in vitro* response to insulin (see Table 2). In the rat, this decreased response may be related to reduced insulin binding because postreceptor mechanisms appear intact (Olefsky, 1977). There is a similar maturational loss of adipocyte glucagon-stimulated adenylate cyclase (Cooper et al., 1977) and lipolysis (Holm et al., 1977) which may be related to decreased glucagon binding (Livingston et al., 1974). The reduction in insulin and glucagon binding might be in part a consequence of receptor down-regulation due to increased circulating insulin and abnormal elevations of glucagon. However, this cannot be a complete explanation in the case of glucagon because glucagon-stimulated adenylate cyclase (Kalish et al., 1977) and glucagon binding (Lockwood and East, 1978) are unchanged over the same time period in the rat hepatocyte.

Various investigators have attempted to demonstrate that the decreased insulin responsiveness of rat adipocytes is a consequence of the increased fat cell size that occurs with maturation and aging (Davidson, 1979). In his recent review, Davidson discusses three different experimental approaches to clarify the potential association between size and responsiveness to insulin of fat cells: (1) Salans and Dougherty (1971) isolated fat cells of different sizes from the same animal and showed a reduced insulin response in the larger cells, arguing for a primary effect of cell size. (2) Other authors have shown that, regardless of isolated cell size, cells derived from older animals exhibit a reduced insulin response which argues for a primary age effect (Hansen et al., 1974; Holm et al., 1977). (3) In an attempt to resolve this controversy, caloric restriction was used to prevent or reduce the age-related increase in cell size (Davidson, 1979). These authors have demonstrated that the small fat cells in the caloric-restricted older animal retain their responsiveness to insulin and glucagon. These studies have been generally held to substantiate the view that the decrement in hormonal responsiveness with age is secondary to increased fat cell size and not due to aging. However, this interpretation neglects to take into account the well-established effects of early caloric restriction on the aging process itself. Numerous investigations have demonstrated that caloric restriction in early postnatal development (similar to that employed by Davidson and other authors) results in a retardation of the aging process, in general, and a greatly increased lifespan (reviewed by Ross, 1976). Hence, rather than ruling out an aging effect, these studies may merely have succeeded in extending the long list of age changes which may be retarded by caloric restriction. Therefore, the role of increased adipocyte cell size per se in the insulin resistance of aging remains uncertain.

Available estimates of glucose uptake following a standard glucose load indicate that adipose tissue is responsible for the disposal of less than 5% of the injected glucose in both humans (Bjorntorp et al., 1971) and rats (Bjorntorp et al., 1970). Since the primary sites of glucose disposal are the liver and muscles, it is in these tissues that the age-related defect in transport must develop. This conclusion is supported by the observation of maturation-associated low insulin binding in the rat hepatocyte (Freeman et al., 1973) and muscle response to insulin (Elsas et al., 1971; Gommers et al., 1977). Since the decline in glucose disposal and the development of insulin resistance occur primarily before 50 years of age in the human and 12 months in the rat, it is unlikely that an age-related loss of tissue could play a major role as some authors have suggested.

At present, the best explanation for insulin resistance in the maturing and aging rat is the down-regulation of insulin receptors caused by hyperinsulinemia secondary to increased glucose levels. The explanation for the elevated glucose may lie in the increased obesity and elevated free fatty acids or elevated glucagon release. These alterations may in turn be secondary to changes in hypothalamic-neuroendocrine function which occur as part of a developmental program as discussed in Chapter 15.

The "ob/ob" Hyperglycemic Mouse. The most extensively investigated animal model for human late-onset diabetes mellitus is the ob/ob C57B1/6J obese hyperglycemic mouse. These mice exhibit a number of symptoms, including hypertrophied pancreatic islets, hyperinsulinemia, insulin resistance, reduced physical activity, elevated plasma lipids, and obesity (see Herberg and Coleman, 1977). While the etiology of the syndrome is not well understood, a number of neuroendocrine defects have been demonstrated and suggested to play a major role in its genesis. These include a hypothalamic defect in the control of satiety (Baile et al., 1970), a paradoxical increase in glucagon levels in response to glucose (Laube et al., 1974), elevated central serotonin levels leading to increased ACTH release and elevated serum corticosterone (Garthwaite et al., 1979), and defective temperature regulation and reduced oxygen consumption secondary to a defect in the induction of Na/K ATPase by thyroid hormone (York et al., 1978) which may, in turn, be due to a reduced number of nuclear T_3 receptors (Guernsey and Morishige, 1979). Many of these observations are similar to those made in normal aging rodents, discussed elsewhere in this chapter, and invite the speculation that the ob/ob mouse might prove to be an interesting model of other aspects of endocrine aging. Indeed, an early report (Lane and Dickie, 1958) indicated a greatly reduced lifespan in one strain of these mice. But, more recently, the intriguing observation has been made that the complex of diabetic symptoms spontaneously abate after seven months of age, and the ob/ob mouse shows only a slightly reduced lifespan in comparison with its lean littermates (Westman, 1968). This remission of diabetic symptomatology suggests that the genetic defects in the ob/ob mouse may be primarily of a neuroendocrine character rather than due to an irreversible structural mutation. Spontaneous remission of diabetic or prediabetic symptoms is also a common finding in aging humans (Davidson, 1979).

Proinsulin

Duckworth and Kitabchi (1972, 1976) have observed an increase in circulating levels of proinsulin after oral glucose in older subjects. Because proinsulin has been reported to possess a reduced biological activity and is indistinguishable from insulin by conventional radioimmunoassays, the increase in proinsulin may contribute to the glucose intolerance of old age. However, the role of increased proinsulin in aging remains to be demonstrated (Davidson, 1979).

Glucagon

Recent investigations have demonstrated an apparently paradoxical rise in portal vein glucagon after glucose administration in maturing and aging rats (Klug et al., 1979; Yoshino et al., 1979). This abnormal glucagon secretion has also been observed in patients with late-onset diabetes and may be the most significant factor in the poor performance of aging rats in glucose tolerance tests. Available reports indicate no abnormalities of glucagon secretion in humans (Dudl and Ensinck, 1972, 1977; Fedele et al., 1977; Nonaka and Tarui, 1977). However, no detailed investigations of glucagon metabolism in aging humans have been reported. It is possible that a minor elevation of portal vein glucagon cannot be detected in the peripheral circulation without a careful investigation directly designed to reveal such a change.

Accelerated Aging and Diabetes Mellitus

While historically the majority of authors have been concerned with the increased incidence of diabetes with aging, in recent years a growing number of investigators have reversed this focus to ask whether there might not be an acceleration of aging in diabetes. Patients with diabetes display an increased incidence of several features commonly associated with aging: cataracts, microangiopathy, neuropathy, dystrophic skin changes, and accelerated atherosclerosis (Levine, 1979). Accelerated atherosclerosis is a major feature of the various genetic syndromes reported to resemble premature aging, and all of these syndromes include abnormal glucose tolerance (Goldstein, 1978). Further, in normal aging, in patients with progeria, and in diabetics, there is a reduction in the proliferative capacity of cultured fibroblasts (Goldstein et al., 1969; Vracko and Benditt, 1975; Goldstein, 1978). Vracko and Benditt have evoked this reduced proliferative capacity as a causal explanation of microangiopathy. This finding may also be related to a reduced response to insulin and to growth factors, such as nonsuppressible insulin-like activity (NSILA). Insulin resistance has been reported in cells from patients with Werner's syndrome (Kissebah et al., 1979) and in normal aging and progeria (Rosenbloom et al., 1976; Villee et al., 1979); a decreased response to NSILA has been reported in normal aging and progeria (Goldstein, 1978). In addition, in both juvenile-onset and maturity-onset diabetics preliminary results have indicated an accelerated rate of collagen aging (the aging of collagen having been represented as the fundamental aging process) which may be linked to certain clinical symptoms of diabetes associated with aging (Kohn, 1978, 1979). And finally, the putative autoimmune etiology of juvenile onset diabetes, observations of immune dysfunction in aging and in diabetes, and the reports of increased pancreatic anyloidosis in senile humans and animals have excited the interest of proponents of an immunogenesis of aging (Walford et al., 1978). Collectively, these studies point to the possibility of an intriguing relationship between diabetes and aging.

Microangiopathy. A widening of the capillary basement membrane thickness (CBMT) is one of the more important pathological changes in diabetic patients (Siperstein et al., 1968; Williamson and Kilo, 1977), and has been implicated in retinopathy, renal failure, and limb gangrene, depending on the primary tissue affected. Whether or not such basement membrane thickening also occurs in normal human aging is currently a matter of controversy. Siperstein and coworkers (1968) report no change in muscle CBMT with age in normal controls, while other authors report a progressive increase in muscle CBMT with age (Jordan and Perley, 1972; Kilo et al., 1972; Camerini-Davalos et al., 1979). An increase in CBMT with age has also been reported to occur in the rat kidney, (Yagihashi et al., 1978). The etiology of CBMT is also disputed (Osterby, 1976; Wolf and Berle, 1974).

Pathological versus Physiological Considerations

The relationship between pathological and aging processes is complicated by several issues. There is obviously a great deal of genetic variability in human populations with respect to both lifespan and predisposition to diabetes. Twin studies have demonstrated a high degree of concordance in late onset diabetics which would make the incidence of a putative recessive diabetic gene greater than 40%. However, genetic factors in diabetes are at present poorly understood, probably owing to the considerable heterogeneity in this difficult to define disorder (Cruetzfeldt, 1976). Juvenile onset diabetes has been recently linked to several specific major histocompatibility complex (HLA) phenotypes which may predispose selected individuals to viral infection or autoimmune reactions, but the basis of genetic factors in late-onset diabetes remains obscure. The apparent high heritability of late-onset diabetes may indicate a pathology complicated by the effects of "normal aging," or alternatively, "diabetes" may represent an acceleration of basic aging processes in a large, genetically predisposed percentage of the population.

Not all members of a population or even all populations can be demonstrated to develop coronary heart disease, yet this pathology is clearly

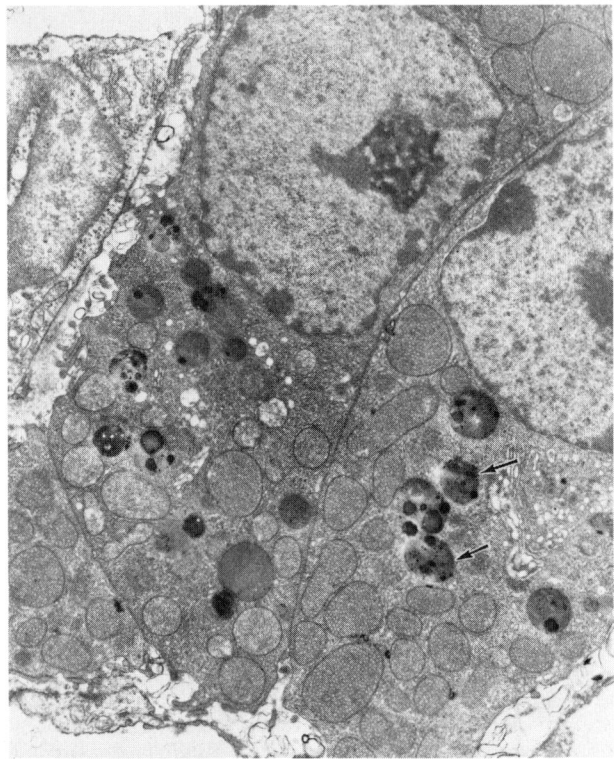

Fig. 1. Adrenal cortical cells from a 24-month-old male Long-Evans rat containing numerous lysosomes filled with granules (some indicated by arrows). Lysosomes are cytoplasmic organelles containing hydrolytic enzymes for intracellular digestion. The observed increase in waste removal capability. Stained with lead aspartate (x 18,900). Courtesy of Dr. J.R. Walton, Lawrence Livermore Laboratory, Livermore, California.

age-related, and remains the largest single cause of death in the elderly in the industrialized nations. Similarly, while diabetes (or abnormal glucose tolerance) may not be demonstrable in all aging individuals or populations, it remains unequivocally age-related, and ranks as the sixth largest killer in the USA (Dept. HEW, 1978). Whatever genetic and pathological components ultimately prove to underly selected specific manifestations of aging or the evident heterogeneity in diabetes and atherosclerosis, it is perhaps significant in this regard to note that the currently recommended treatment for late-onset diabetes mellitus is carefully restricted diet and regular exercise, and both are commonly considered to be the normal healthy individual's best defenses against atherosclerosis and senility, while the best assurance of a long lifespan remains the thoughtful choice of long-lived parents.

THE PARATHYROID GLAND

A widespread change that occurs with aging involves alterations in bone metabolism. This metabolism is dependent on many factors – e.g., nutrition, exercise, hormones – and among them, parathyroid hormone is one of the most important. Parathyroid hormone increases the rate of osteoclast-mediated resorption and decreases the rate of bone-matrix formation. During growth and development, skeletal mass increases, reaching its maximum in the third decade. In the fourth decade, the skeletal mass begins to decrease. This decrease occurs in both sexes, in all populations studied, and involves the entire skeleton. Racial differences are present; blacks generally have a larger bone mass than whites (Raisz, 1979), and black women show a reduced incidence of osteoporosis and higher serum calcium levels after menopause (Roof et al., 1976). Parathyroid hormone levels of these black women tend to remain low after the menopause, whereas they tend to increase in aging white women (Roof et al., 1976). Whether serum parathyroid hormone levels change significantly with aging and in which direction in normal, and, particularly, osteoporotic individuals remains unclear; however, in a small proportion of osteoporotic patients, parathyroid hormone levels appear to be increased and these higher hormonal levels probably are harmful and contribute to bone loss (Riggs et al., 1978).

As discussed in Chapter 16, declining estrogen titers subsequent to the menopause are implicated in the genesis of osteoporosis in the human female (Heaney et al., 1978a). Estrogen enhances intestinal absorption efficiency of calcium and improves renal calcium conservation and both of these actions would be lost or extremely reduced at menopause. Also, estrogen loss at menopause would be associated with a partial release from inhibition of skeletal resorption (Heaney et al., 1978b). It has been proposed that estrogen effectively antagonizes the stimulatory action of parathyroid hormone on bone resorption. The release from this inhibitory effect of estrogen on parathyroid hormone at menopause has been implicated in increased bone remodeling in post-menopausal women as well as in decreased peripheral calcium conservation (Heaney, 1979).

Studies of structural changes in the aging human parathyroid are few and studies from laboratory animals reveal only minor changes. Morphologic changes in the parathyroid glands of senile as compared to mature control dogs have been characterized by light and electron microscopy (Setoguti, 1977). Syncytial cells and so-called colloid follicles (which are thought to be a necrotic substance derived from degenerating parenchymal cells) are much more abundant in the older than younger animals. Oxyphil cells, and mitochondria-rich cells, with the mitochondria frequently showing bizarre patterns, are found in all senile dogs, but not in mature controls. Senile dogs also show interstitial tissue with increased fat cell content.

Thyrocalcitonin

In a study of 131 humans aged 20 to 79, basal levels of calcitonin, measured by radioimmunoassay, decreased progressively with aging, the decrease being more substantial in males than in females (Deftos et al., 1979). The increase in calcitonin secretion following calcium chloride infusion into the blood revealed a significant negative correlation between age and calcitonin levels. Since calcitonin inhibits bone resorption, this age-related decline in its secretion may also play a role in the development of bone pathology with aging. Thyrocalcitonin secretion and thyroid C-cell histology in the aging rat, showed a maturational increase in serum and thyroid thyrocalcitonin from ten weeks to six months in two different strains of rats (Peng et al., 1976). However, no statistically significant changes were reported between 6 and 18 months. Histologically, the older group of rats showed a higher area ratio of C-cells to follicular cells in the thyroid gland.

THE PINEAL GLAND

The human pineal gland undergoes calcification with aging (Kitay and Altschule, 1954). However, in a relatively large study (33 autopsies terminal to various pathological etiologies) no correlation could be drawn among gland weight, subject age, degree of calcification and activity of the major enzymes (5-hydroxytryptophan decarboxylase, monoamine oxidase and hydroxyindole-O-methyl transferase) involved in the metabolism of pineal neurotransmitter substances (Otani et al., 1968). These data are in agreement with a previous study which showed that throughout life, levels of monoamine oxidase, hydroxy-indole-O-methyl transferase and histamine-N-methyl transferase remained unchanged in the pineal (Wurtman et al., 1964). They suggest that the capacity of the human pineal to synthesize melatonin and to inactivate histamine and serotonin does not vary significantly throughout the lifespan. A markedly lower incidence of pineal calcification with aging was reported in a study of a Nigerian population, and it was proposed that this equatorial country of dark-skinned people has populations with lower pineal rates of activity due to light-related melatonin-inhibition and abundant epidermal pigmentation (Daramola and Olowu, 1972). A similar result has been reported from a study of a Japanese population (Chiba and Yamanda, 1948).

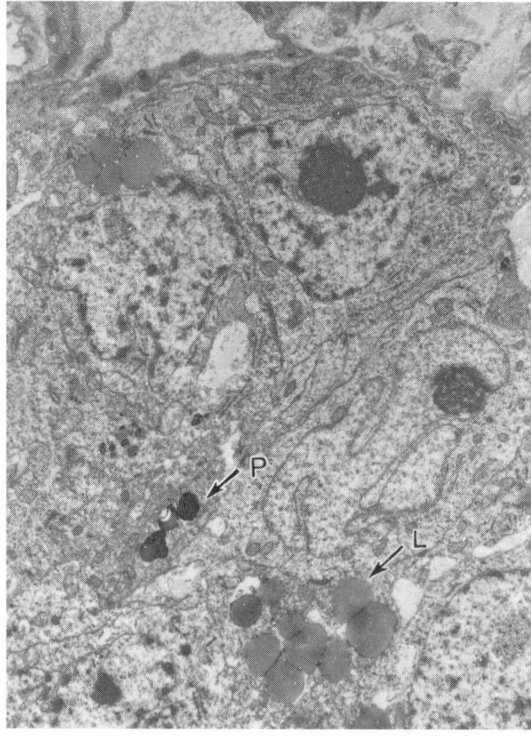

Fig. 2. Pineal gland cells from a 24-month-old male Long-Evans rat. Cells contain large lipid droplets (L) and aggregates of age pigment (lipofuscin?) (P). This pigment not normally seen in young animals usually accumulates in old pineal cells (x 18,900). Courtesy of Dr. J.R. Walton, Lawrence Livermore Laboratory, Livermore, California.

In the pineal of aged rats, an increment of connective tissue elements, an increase in the frequency of pleiomorphic dense bodies in pineal parenchymal cells, and no significant change in the relative distribution of vesicles in post-synaptic sympathetic nerve endings were observed (Bondareff, 1965). Pineal glands of male rats lose cells with aging, showing a 30% loss of pinealocytes at 360 days of age: underfeeding from 22 days of age was shown to prevent this loss (Walker et al., 1978).

THE PITUITARY OR HYPOPHYSIS

The pituitary supports the survival of the species through its regulation of reproduction and the survival of the individual through its regulation of homeostasis. These important functions are carried out by the secretion of several hormones from the gland and the integration of these secretory responses with neural inputs from the hypothalamus as well as feedbacks from target endocrine and effector tissues. Hypophysectomy in the young animal (rat) arrests or retards growth and development and delays several age-related changes involving almost all body systems. However, despite this apparent anti-aging action of hypophysectomy, the lifespan of the animal is shortened (Everitt, 1976). Thus, the pituitary may secrete both anti-aging and aging factors. The multiplicity of the pituitary functions and their interactions with other endocrines and with the central nervous system, make the exploration of the possible role of the pituitary in aging extremely complex. No resolution of the dual action of pituitary hormones in accelerating or delaying aging is available at present. We will attempt here only to delineate the many facets of the problems with respect to both anterior and posterior portions of the gland.

The Anterior Pituitary

Changes with aging in several hormones of the anterior pituitary have been discussed in a previous chapter (i.e., gonadotropins in Chapter 12) or are discussed in this chapter together with the respective target endocrine (i.e., ACTH, with the adrenal cortex and TSH, with the thyroid). Consequently, we will consider in this section the remaining two major hormones, growth hormone and prolactin.

Growth Hormone

Plasma levels of growth hormone (GH) seem to undergo little change with aging in humans and a number of other animal species under resting and stressful conditions (see Table 3). Findings in plasma parallel observations in the pituitary in which neither GH content nor somatotropes are altered with advancing age. However, the characteristic peak in GH secretion observed during sleep in young and adult subjects is reduced or absent in the elderly. In rats, the pulsatile nature of GH is preserved but the amplitude of the cycles is diminished. Also, in rats, some investigators have reported reduced GH plasma levels, reduced hypothalamic content of GH and reduced pituitary responsiveness to GRF (Pecile et al., 1965). Several possibilities, including increased somatostatin release, possibly reflected by decreased content (see Table 3), and decreased norepinephrine activity, have been advanced as possible explanations for the reduction in GH levels and cyclicity (Steger et al., 1979). Responses of GH to hypoglycemia, glucose load or arginine administration do not show any clear-cut differences between old and young subjects or animals. As in the case of insulin,

Table 3.

AGING-RELATED CHANGES IN THE SECRETION AND FUNCTION OF GROWTH HORMONE AND PROLACTIN

PARAMETER STUDIED AND CHANGE REPORTED	REFERENCE (RAT)	REFERENCE (HUMAN)
Pituitary		
somatotropes, unchanged		Calderon et al., 1978
		Pasteels et al., 1972
GH content, unchanged	Solomon and Greep, 1958	Gershberg, 1957
Plasma GH Levels		
unchanged	Florini et al., 1979	Cartlidge et al., 1970
		Dudl et al., 1973
		Lazarus and Young, 1966
decreased	Steger et al., 1979	Dilman, 1976
		Vidalon et al., 1973
		Danowski et al., 1969
GH Releasing Factor (GRF)		
reduced hypothalamic GRF	Pecile et al., 1965	
reduced response to GRF		
GH Response to Stimuli		
insulin-induced hypoglycemia unchanged		Kalk et al., 1973
		Sachar et al., 1971
arginine unchanged		Dudl et al., 1973
		Blichert-Toft, 1975
		Root and Oski, 1969
suppression by glucose load unchanged		Benjamin et al., 1970
		Root and Oski, 1969

Table 3. (Cont'd.)

PARAMETER STUDIED AND CHANGE REPORTED	REFERENCE (RAT)	REFERENCE (HUMAN)
reduced		Dilman, 1976
		Sandberg et al., 1973
		Buckler, 1969
protein "Bovril" ingestion reduced		
anesthesia reduced		Blichert-Toft, 1975
exercise reduced		Bazzare et al., 1976
L-Dopa reduced		Bazzare et al., 1976
Total 24-hour GH Secretion and Sleep-Related GH Release		
reduced	Steger et al., 1979	Carlson et al., 1972
		Finkelstein et al., 1972
		Thompson et al., 1972
		Bazzare et al., 1976
Somatomedin Plasma Levels		
constant to early senescence	Florini and Roberts, 1978	
decreased in senescence		
Tissue Response		
decreased growth response	Emerson, 1955	
decreased lipolysis	Jelinkova and Hruza, 1964	
unchanged nitrogen retention		
increased free fatty acids		Root and Oski, 1969
decreased Na^+ ("metabolic response")		Rudman et al., 1971

Table 3. (Cont'd.)

PARAMETER STUDIED AND CHANGE REPORTED	REFERENCE (RAT)	REFERENCE (HUMAN)
reduced hydroxyprolinuria		Root and Oski, 1969
reduced inhibition of glucose consumption by red cells in vitro		Root and Oski, 1969
reduced binding to thymocytes		Talwar et al., 1976
Metabolic Clearance Rate		
unchanged		Taylor et al., 1969
Hypothalamic Somatostatin		
decreased	Steger et al., 1979	
Prolactin		
Pituitary		
mammotropes involuted unchanged		Baker and Yu, 1977 Kovacs et al., 1977 Pasteels et al., 1972
increased incidence of prolactin-secreting microadenomas		Kovacs et al., 1979
prolactin content elevated	Kovacs et al., 1977	
unchanged	Bethea and Walker, 1979	Yamajii et al., 1976
Circulating Prolactin Levels		
females, increased		Ascheim, 1962 Ascheim and Pasteels, 1963 Bethea and Walker, 1979

Table 3. (Cont'd.)

PARAMETER STUDIED AND CHANGE REPORTED	REFERENCE (RAT)	REFERENCE (HUMAN)
males, increased	Clemens and Meites, 1971 Huang et al., 1976 Shaar et al., 1975	Vekemans and Robyn, 1975
unchanged	Meites et al., 1978 Simpkins et al., 1979	Yamaji et al., 1976 Frantz et al., 1972
females, decreased		Vekemans and Robyne, 1975 Vermeulen, 1978
Prolactin Inhibitory Factor		
reduced	Riegle et al., 1977	
Response to TRH		
delayed, prolonged elevation of serum prolactin		Yamaji et al., 1976

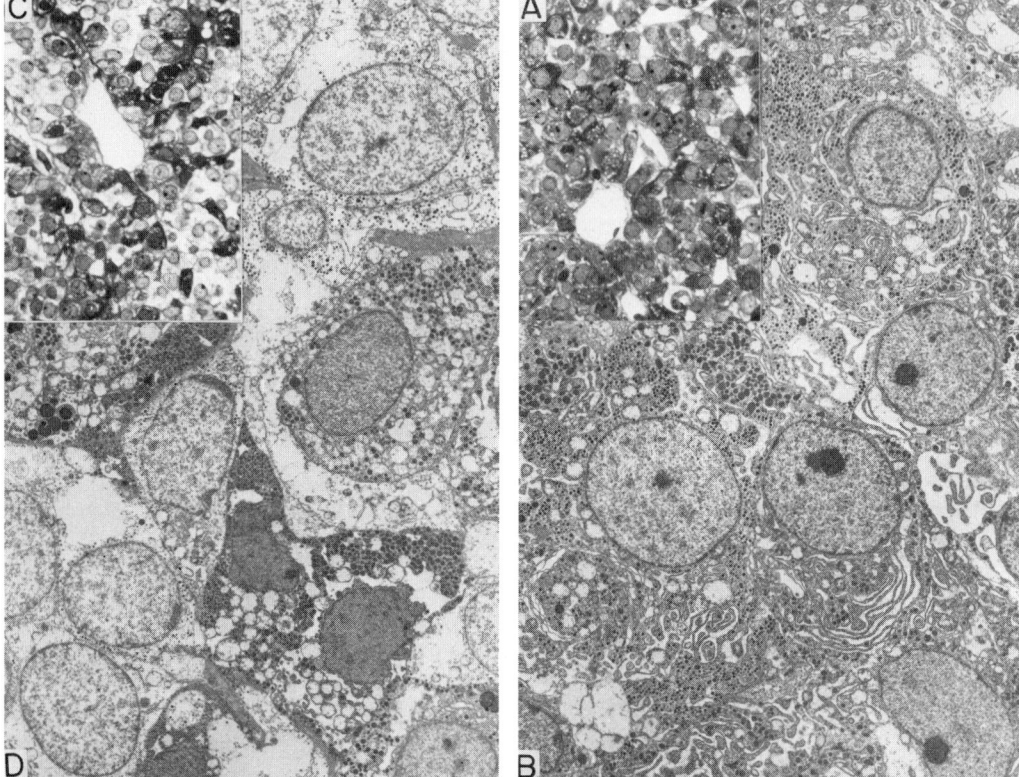

Fig. 3. Light and electron micrographs of cells of the anterior pituitary of a 3-month-old (A and B) and a 24-month-old male Long-Evans rats (C and D). The cells of the younger tissue (A and B) have a more homogeneous appearance, containing granules of various sizes. The cells of the older tissue (C and D) contain either more dense or abnormally sparse cytoplasm. The electron micrographs are stained with lead aspartate and the light micrographs are stained with azur II-methylene blue. (x 5,400 and x 617 respectively). Courtesy of Dr. J.R. Walton, Lawrence Livermore Laboratory, Livermore, California.

obesity may represent a complicating factor in GH regulation and lead to reduced GH responses to hypoglycemia.

Studies in rats have shown that somatomedin levels remain unchanged until 18 months of age but decrease significantly thereafter. Somatomedins are known to mediate the actions of GH; therefore, their reduction with old age may explain at least in part the decrease in the responses of several target tissues to the actions of GH (see Table 3).

Prolactin

Among the pituitary hormones that seem to show the most pronounced changes with aging is prolactin. In a recent study of pituitary tumor incidence in old age, 20 of 152 patients autopsied (13%) over 80 years of age had pituitary adenomas, the majority of which were prolactinomas (Kovacs et al., 1980). This high incidence of prolactin-secreting tumors has also been reported in old C57B1/6J mice (Clayton et al., 1979). In several strains of rats, specifically Long-Evans females, mammary tumors develop after middle age (Segall and Timiras, 1976), and may be related to hypersecretion of prolactin by the aging pituitary. The dopamine concentration of the median eminence and the arcuate nucleus decreases in male rats with aging (Simpkins et al., 1979) and a decrease in dopamine activity has been linked to increased prolactin secretion. The rise in prolactin secretion in aging female rats can be inhibited using dopamine agonists such as lergotrile mesylate (Clemens and Bennett, 1977; Clemens et al., 1978). It is possible that alteration of the neurosecretory activity of the aging median eminence predisposes the anterior pituitary to the formation of prolactin-rich adenomas.

Pituitary prolactin content is higher in old female rats when compared with young, and prolactin release is higher from pituitaries of old female rats when the glands are incubated in vitro. Blood levels are markedly increased in old female rats and slightly elevated in old male rats (Table 3).

In a study of human pituitary glands in subjects past 80, it was found by immunocytochemical techniques that prolactin-containing cells were numerous and well developed. Blood prolactin levels in human females tend to diminish with age, whereas in the human male they rise (Vekemans and Robyn, 1975). The decline in females has been attributed to the lack of estrogen after menopause as estrogen has been shown to be capable of stimulating prolactin release (Kovacs et al., 1977).

The Posterior Pituitary

Physiological and morphological studies suggest that with aging the ability of the large (magnocellular) neurons that comprise the hypothalamic portion of the hypothalamo-neurohypophyseal system to support the oxytocic and antidiuretic functions of the neurohypophysis is reduced. Most studies concerned with the effects of senescence on neurohypophyseal function have been conducted in rats (Turkington and Everitt, 1976); however, data from other species – including humans (Helderman et al., 1978), dogs (Scharrer, 1954), and mice (Machado-Salas et al., 1977; Davies and Fotheringham, 1980) – suggest that altered neurohypophysial function with age may be more universal.

Studies using histochemical and immunochemical detection of the neurosecretory substances show that the functional activity of the neurosecretory nuclei of the hypothalamus falls off with age (Frolkis et al., 1972); Machado-Salas et al., 1977; Rodeck et al., 1960; Turnington and Everitt, 1976) and the secretory discharge of the neurohypophysis is modified (Morrison and Staroscik, 1964; Frolkis et al., 1972; Turkington and Everitt, 1976; Friedman and Friedman, 1957). Although the majority of the data indicates an impairment of neurohypophysial function, further studies are required to identify the locus of this impairment which remains unknown.

Normal aged rats often suffer from mild diabetes insipidus which produces marked changes in salt and water metabolism (Friedman and Friedman, 1957; Friedman et al., 1960). This condition is not due to kidney dysfunction for the kidney of these rats is more sensitive to vasopressin than normal (Friedman and Friedman, 1957). A simple explanation for the redistribution of water between intracellular and extracellular sites, characteristic of old age, is a decline in neurohypophsial responsiveness.

Marked species differences in the control of water balance and in vasopressin metabolism make extrapolations to human physiology inadvisable (Lauson, 1974). Studies in humans have revealed an increased sensitivity of vasopressin secretion to inhibitory and excitatory stimuli (Helderman et al., 1978). Both young and old subjects respond to alcohol infusion with inhibition of vasopressin secretion. Unlike the younger subjects whose vasopressin levels returned to baseline after the infusion, the older subject, despite continued infusion of alcohol, showed a rebound of vasopressin levels to twice the basal level by the end of the study period. It is possible that osmoreceptor sensitivity to hyperosmolality is greater in older men. This heightened vasopressin response may compensate for the reduced renal ability to conserve salt and water in the aging human (Helderman et al., 1978).

In considering the possible implications of the aging of the neurohypophysis, much attention has been focused, in humans, on the direct antidiuretic action of arginine vasopressin (AVP) on the kidney as well as on its indirect participation in the control of blood volume. For example, AVP secretion is markedly increased in severe hemorrhage (e.g., as may occur during surgery), wherein baroreceptors in the aorta and carotid arteries send signals to the vasomotor medullary center of the brain and from there are relayed, in turn, to the hypothalamus to stimulate AVP release. In old individuals, baroreceptor sensitivity to changes in blood volume seems to be impaired (Robertson et al., 1979). This decreased sensitivity has been related to a reduction or loss of the aortic wall distensibility (probably as a consequence of atherosclerosis) and may present a serious risk factor during surgery when a reduction in blood volume could remain uncompensated because of the impaired water retention. Thus, decreased baroreceptor stimulation and consequent reduction in AVP release may represent predisposing factors to hypovolemic shock in the elderly. The baroreceptors generally provide a restraining influence to AVP secretion and, in young adults, this influence is decreased after reduction of blood volume and leads to increased AVP secretion, water retention and restoration of blood volume. With increasing age and decreasing baroregulatory input, the restraining influence of the baroreceptors on AVP may be dampened; this would explain the heightened responsiveness to hyperosmolar stimulation described in some older subjects and mentioned above (Helderman et al., 1978). Further studies may clarify the relations between the effects of aging on AVP regulation and the propensity of the elderly patient to develop abnormalities in salt and water balance and blood pressure.

Some attempts have been made to prevent or slow down the physiological age-associated decline in the activity of the hypothalamo-neurohypophyseal system; whether this decline is due to a decline in hormonal synthesis, or a decline in hormone release, cannot be said with certainty. Friedman and Friedman (1964) found that posterior pituitary extracts administered to old rats produced an improvement in health and a reduction in mortality. Further experiments pointed to oxytocin as the most probable factor in these extracts responsible for this phenomenon (Bodanszky and Engel, 1966). There is immunohistochemical and biochemical evidence which corroborates that the oxytocin-producing components of the hypothalamo-neurohypophyseal system are more susceptible to aging.

Little information is available on the relationship between the synthesis and release of posterior pituitary hormones and the monoaminergic innervation of the hypothalamic nuclei. The presence of numerous synapses on the perikarya and dendrites of Gomori-positive cells (Polenov and Senchik, 1966) would indicate that fluctuations in the hypothalamic noradrenaline level and release may influence the functioning of these magnocellular nuclei. Depletion of catecholamines (after administration of reserpine or chlorpromazine) is followed by a distinct increase in the secretory activity of these Gomori-positive cells (Aleshin, 1978). As evidence accumulates that aging is accompanied by alterations in monoaminergic neurotransmitter balance, it is possible that observed senescent alterations in the function of the neurohypophysis are dependent on synaptic events occurring in these and higher brain regions.

There is ample evidence, for example, that posterior pituitary hormones may be present in extrahypothalamic areas of the brain. Although these areas are anatomically separate from the neurohypophysis, they are worth considering here because of the presence of oxytocin consolidation and retrieval processes (Kovacs et al., 1979). Treatment of human patients with 8-lysine vasopressin improves memory and substantially alleviates the clinical symptoms of amnesia (Legros et al., 1978; Oliveros et al., 1978). Oxytocin has the opposite effect of vasopressin. It interferes with both memory consolidation and retrieval processes. In spite of the above well-documented actions of vasopressin and oxytocin, relatively little is known about the biochemical mechanisms by which these substances affect the central nervous system.

THE "ENDOCRINE" THYMUS AND AGING

Numerous attempts have been made to include the thymus among the endocrine glands. Its sensitivity to a variety of hormones, particularly corticosteroids and estrogens, its significant decline in size and function at puberty, and the isolation of one or several putative secretory products qualified the thymus for an endocrine role. However, the difficulty of identifying a definitive thymic hormone, the structural resemblance of the thymus to a lymphoid organ, and the clear demonstration of its function in immune responses led to the classification of the thymus within the lymphoid system. Currently the isolation and purification of at least one thymic product — the polypeptide thymosin — has reopened the question of whether or not the thymus should be considered an endocrine organ (Makinodan, 1978). Thymosin, originally detected in a partially purified thymus extract (Klein et al., 1966; Goldstein et al., 1966) has been isolated, sequenced (Goldstein et al., 1977), and utilized experimentally in clinical replacement studies in which the hormone has been administered to promote maturation of T-lymphocytes (thymus-derived) (Wara et al., 1975, p. 46; Kent, 1980).

Several investigators have described a relationship between autoimmunity, immunodeficiency and aging in man and several laboratory animals (Good and Yunis, 1974). As the thymus undergoes profound involution with aging it is possible that a decrease in thymosin production may be a source of increasing immunological dysfunction. Because of the current interest in the immunological theories of aging, the thymus may be found to play a role in aging far beyond its possible activity as an endocrine organ (Kay et al., 1979).

THE THYROID

The thyroid and its hormones have traditionally excited the interest and imagination of investigators seeking to define hormonal determinants of senescence. The work of early researchers was primarily sparked by the observation that individuals affected by hypothyroidism developed certain signs of "precocious senility" including a reduced metabolic rate, hyperlipidemia and atherosclerosis, dry skin, loss and greying of hair, slow reflexes and mental deterioration. Since these patients showed marked improvement following administration of thyroid extract, it was felt that similar symptoms in normally aging individuals might represent effects secondary to a thyroidal involution with age. Thyroid hormone activity controls the rate or state of development (see Chapters 4-7) and it is, therefore, logical to inquire whether it might not also control the rate or state of aging. In the 1920s, optimistic experimenters attempted rejuvenation or prolongation of life in animals and men by feeding hormone extracts. While chronic administration of thyroid hormones to rats seemed, if anything, to shorten the lifespan (Robertson, 1928), administration to aged fowl appeared to bring about a dramatic rejuvenation (Crewe, 1924-1925). Replacement therapy in elderly myxedemic patients sometimes produced "rejuvenating effects" such as a restoration of hair color and quality (Kerr et al., 1926), but despite numerous attempts, marked improvement was never observed with replacement therapy in the euthyroid elderly (Pittman, 1962). Hence, if, as early investigators believed, the thyroid axis is somehow centrally involved in aging processes, it is in some more subtle way than simply as a consequence of reduced circulating hormone levels. Unfortunately, as the following discussion will show, despite the current advances in our knowledge of both thyroid hormones and aging processes, we can still give no definite answer to the question of thyroid involvement in aging.

Thyroid Structure and Function

The majority of studies indicate that a histologic picture characteristic of a decreasingly active thyroid gland emerges with aging in both man and the rat (see Table 4). The appearance of distended follicles with pale colloid and flattened epithelium together with decreased mitosis and increased connective tissue fibrosis marks an apparently less active, involuting gland. However, in some aging individuals, the thyroid appears "normal" indicating that these degenerative changes and any effect they might have on secretory capacity are "a consequence and not a cause of aging" (Korenchevsky, 1961). Despite this relatively quiescent state, the thyroid is capable of greatly increased hormonal output in elderly patients with acute illness (Gregerman and Solomon, 1967) indicating that primary thyroidal failure is not the cause of reduced thyroid function in humans. This view is supported by the observation of an increased incidence of autonomous micronodules with aging and the characteristic modular hyperplasia or multinodular goiter (Plummer's disease) typical of hyperthyroidism in the elderly (Davis and Davis, 1974); this demonstrates the continued maintenance of an intrinsic capacity to secrete even excessive amounts of hormone when the endocrine gland is freed from normal regulation. Nevertheless, the histological evidence for decreased hormonal output with aging is confirmed by the majority of studies which indicate a decreased thyroidal radioactive iodine uptake and a reduction in the thyroxine secretion rate in both humans and rats (Table 4).

In humans thyroxine secretion rate decreases by 50% with advancing age; despite this decrease, the serum PBI and T_4 are normal or only slightly reduced due to an equivalent large decrease in the metabolic clearance rate demonstrated by the large decline in T_4 deiodination (see Table 4). The T_4 distribution space (per unit body weight) is essentially unchanged, thus, the decline in tissue degradation compensates for the decreased secretion and serum PBI is maintained. In the rat, the T_4 secretion rate is also significantly reduced but the metabolic clearance rate is increased (principally owing to an age-related increase in T_4 distribution space) and therefore, as one would anticipate, T_4 and T_3 levels are reduced (Gregerman, 1963; Frolkis and Valueva, 1978) (Table 4). The reported reductions in serum T_4 and T_3 in the rat are large (typically on the order of 50%) and apparently begin around the time of sexual maturation (Azizi, 1979). The possible significance of these

Table 4.

AGING-RELATED CHANGES IN THE STRUCTURE AND FUNCTION OF THE PITUITARY-THYROID AXIS

PARAMETER STUDIED AND CHANGE REPORTED	REFERENCE (RAT)	REFERENCE (HUMAN)
Histology		
flattened follicular epithelium, decreased mitosis, increased connective tissue, distended follicles	Jackson, 1916 Korenchevsky et al., 1950 Garner and Bernick, 1975	Cooper, 1925 Dogliotti and Nuzzi, 1935 Stoffer et al., 1961
increased micronodules		Dogliotti and Nuzzi, 1935 Irvine and Hodkinson, 1978
unchanged follicles in some individuals		Korenchevsky, 1961 Mustacchi and Lowenhaupt, 1950
unchanged thyrotrope		Ryan et al., 1979
hypertrophied thyrotrope	Garner and Bernick, 1975	
Thyroid Radioactive Iodine Uptake		
essentially unchanged	Johnson et al., 1966	Gaffney et al., 1962 Oddie et al., 1968
decreased	Frolkis et al., 1973 Verzar and Freydberg, 1956 Wilansky et al., 1957	Quimby et al., 1950 Perlmutter and Riggs, 1949
Protein Bound Iodine (PBI)		
unchanged	Wilanksy et al., 1957	Braverman et al., 1966 Ohara et al., 1974 Taylor et al., 1974 Wenzel et al., 1974

Table 4. (Cont'd.)

PARAMETER STUDIED AND CHANGE REPORTED	REFERENCE (RAT)	REFERENCE (HUMAN)
decreased		Azizi et al., 1975 Hesch et al., 1976
Serum T_4		
unchanged free T_4		Braverman et al., 1966 Hansen et al., 1975 Hesch et al., 1976 Ohara et al., 1974 Wenzel et al., 1974
reduced T_4	Chen and Walfish, 1978, 1979 Frolkis et al., 1973 Klug and Adelman, 1979 Sartin et al., 1977 Wostmann and Bruckner-Kardoss, 1979	Bermudez et al., 1975 Herman et al., 1974
elevated T_4	Latham and Tseng, 1979	Burrows et al., 1975
unchanged in germ-free animals	Wostmann and Bruckner-Kardoss, 1979	
elevated after parathyroidectomy	Peng and Garner, 1979	
Serum T_3		
unchanged		Braverman et al., 1966 Olsen et al., 1978
reduced T_3	Klug and Adelman, 1979 Sartin et al., 1977 Wostman and Bruckner-Kardoss, 1979	Bermudez et al., 1975 Hansen et al., 1975 Herman et al., 1974 Hesch et al., 1976

Table 4. (Cont'd.)

PARAMETER STUDIED AND CHANGE REPORTED	REFERENCE (RAT)	REFERENCE (HUMAN)
reduced only in illness	Chen and Walfish, 1978, 1979	Rubenstein et al., 1973 Burrows et al., 1975 Olsen et al., 1978
unchanged in germ-free animals	Wostmann and Bruckner-Kardoss, 1979	
T_4 Secretory Rate		
(reduced)	Grad and Hoffman, 1955 Johnson et al., 1966 Kumaresan and Turner, 1967 Narang and Turner, 1966 Wilansky et al., 1957	Gregerman and Solomon, 1967
can be greatly increased in acute phase of illness in elderly		Gregerman and Solomon, 1967
T_4 I^{131} Deiodination		
increased	Frolkis and Valueva, 1978 Gregerman, 1963	
greatly decreased (in vivo)		Stern-Nielsen and Friis, 1973 Gregerman et al., 1962 Inada et al., 1964 Oddie et al., 1966
can be decreased in illness		Gregerman and Solomon, 1967
T_4 I^{131} Distribution Space		
increased	Frolkis and Valueva, 1978 Gregerman, 1963	

Table 4. (Cont'd.)

PARAMETER STUDIED AND CHANGE REPORTED	REFERENCE (RAT)	REFERENCE (HUMAN)
depends only on weight, not on age		Oddie et al., 1966
Binding of Serum T_4 to Thyroid Binding Globulin		
decreased	Frolkis et al., 1973 Frolkis and Valueva, 1978	
essentially unchanged		Braverman et al., 1966
T_3 I131 Distribution Space (TDS) and Metabolic Clearance Rate (MCR)		
essentially unchanged TDS	Frolkis and Valueva, 1978	
unchanged MCR	Frolkis and Valueva, 1978	Wenzel and Horn, 1975
Serum TSH, Basal		
unchanged	Chen and Walfish, 1978 Klug and Adelman, 1977 D'Angelo, 1961	
slightly increased		Mayberry et al., 1971
slightly reduced		Lemarchand-Beraud and Vanotti, 1969 Cuttleod et al., 1974 Ohara et al., 1974 Olsen et al., 1978
TSH Periodicity		
abolished in older animals	Klug and Adelman, 1979	
unchanged		Blichert-Toft et al., 1975

Table 4. (Cont'd.)

PARAMETER STUDIED AND CHANGE REPORTED	REFERENCE (RAT)	REFERENCE (HUMAN)
TSH and Stress		
reduced secretion after stress (cold exposure) and reduced sensitivity in central TSH inhibition	Simpkins et al., 1978	
no change in increased TSH output in response to thyroidectomy or cold exposure	Huang et al., 1978	
TSH Response to TRH		
unchanged in females, decreased in males	Chen and Walfish, 1978, 1979	Azizi et al., 1975 Blichert-Toft et al., 1975
increased		Ohara et al., 1974
unchanged in women, decreased in men		Snyder and Utiger, 1972 Snyder and Utiger, 1972
decreased in women, unchanged in men		Wenzel et al., 1974
Thyroid Hormone Output in Response to Exogenous TSH		
unchanged		Baker et al., 1959 Einhorn, 1958
reduced	Sartin et al., 1977	Lederer and Bataille, 1969 Scazzigga et al., 1968
TSH Synthesis and Release		
decreased	Valueva and Verzhikovskaya, 1977	

Table 4. (Cont'd.)

PARAMETER STUDIED AND CHANGE REPORTED	REFERENCE (RAT)	REFERENCE (HUMAN)
Thyroid Hormone Output Response to Endogenous TSH		
reduced	Klug and Adelman, 1977	
T$_4$ to T$_3$ Peripheral Conversion		
reduced *in vitro*, (liver and kidney)	Ooka, 1979 Kohrle et al., 1979	
reduced *in vivo*		Wenzel and Horn, 1975
T$_4$ I^{131} Tissue Binding		
reduced	Frolkis and Valueva, 1978	
unchanged		Holm et al., 1975
Nuclear Receptors (for Thyroid Hormones)		
increased in liver; reduced in kidney, heart and cerebellum in males; unchanged or increased in liver, kidney, heart and cerebellum in females	Latham and Tseng, 1979	
reduced in liver	McCool et al., 1979	
unchanged in liver, cerebral hemispheres	Valcana and Timiras, 1979	

changes is considered below.

Most of the serum T_3 is derived from extrathyroidal monodeiodination of T_4, principally in the liver and kidney (Chopra et al., 1978) and, therefore, serum T_3, the more active metabolite of T_4, may vary independently of T_4 secretion rates and levels. T_3 levels in aging should reflect production rates because the metabolic clearance rate for T_3 is apparently unchanged in both humans and rats (Wenzel and Horn, 1975; Frolkis and Valueva, 1978).

Most studies have found about a 20% reduction in serum T_3 in older human subjects, but some investigators claim that this drop is simply an expression of increased nonspecific illness in the elderly population. When screening procedures have been used to select only healthy subjects, no age-related drop in T_3 was found (Burrows et al., 1975; Olsen et al., 1978). That stress and nonspecific illness can reduce peripheral T_4 to T_3 conversion, with a corresponding increase in reverse T_3 (a less biologically active isomer of T_3) formed, is well documented (Chopra et al., 1978). Therefore, reduced serum T_3 in the elderly may be secondary to increased nonthyroidal disease.

While T_3 drops markedly with age in the rat, it seems unlikely that nonspecific pathology alone could account for deficits of the observed magnitude (50%). Nevertheless, one must consider this possibility, particularly in view of a recent report of the absence of age changes in serum T_3 or T_4 in "germ-free" rats (Wostmann and Bruckner-Kardoss, 1979). An alternate hypothesis is that the decrease of thyroid hormone secretion rates with age in both humans and rats may be a more generalized phenomenon. This decline begins early in life, when it is most pronounced, and appears to parallel the fall in metabolic rate (Fisher and Dussault, 1972). Similar declines in "thyroid hormone secretion" (or activity) are indicated by the results of the relatively nonspecific "goiter prevention assay" (which merely measures the dose of hormone required to reestablish the euthyroid condition) in a wide variety of mammalian species (see Kumaresan and Turner, 1967). Thus, one must suppose that the early and progressive decline in T_4 secretion rate has some fundamental developmental significance which we will consider late in this discussion.

Response to TSH

In the rat, serum TSH increases with maturation (Panda and Turner, 1967; Valueva and Verzhikovskaya, 1977; Azizi, 1979) as the thyroid hormone levels fall (Sartin et al., 1977; Azizi, 1979; Valueva and Verzhikovskaya, 1977). A maturational decline is observed in the secretory response of the thyroid to exogenous (bovine) TSH and is associated with a reduction in the TSH-induced increase in cyclic AMP, suggesting that the age change might involve the membrane receptor adenylate cyclase complex. Other investigators have found essentially no change in immunoreactive TSH levels (Table 4), and in this case the fall in thyroid hormone levels could be ascribed to a primary failure of the thyroid. This phenomenon might also be ascribed to structural alterations in TSH as suggested by the precipitous decline with aging in the ratio of bioassayable to immunoreactive TSH and increasing proportions of a biologically ineffective "big TSH," which may block the effects of normal or exogenous TSH at the receptor level (Klug and Adelman, 1977). Such a mechanism would predict increased TSH output (if the negative feedback suppressive mechanism were intact) whereas a modest decline in total immunoreactive TSH occurs. Increased accumulation of big TSH subunits has been reported in the pituitaries of hypothyroid rats but the precise chemical identity and physiologic significance of "big TSH" is currently unknown (Ponsin et al., 1979). In hypothyroidism, the increased release of TSH would favor the formation of a large proportion of the uncleaved prohormone with resulting reduction in biologic activity. Without kinetic information about the secretion rate and metabolic clearance rate of this putative big TSH, it is impossible to tell whether it may serve as a cause or simply an effect of age-related hypothyroidism in this species.

On the basis of the foregoing, three possibilities may account for the reduction in T_4 secretion observed in the aged rat: (a) the accumulation of an inactive big TSH or some other extrinsic inhibitory factor; (b) intrinsic thyroid failure at the level of the receptor complex; or (c) a primary increase in hypothalamic-pituitary sensitivity to feedback suppression of TSH release. Although the identification of the exact cause is not possible at present, the fact that basal TSH is either elevated or unchanged and TSH responses to TRH and to cold stress are probably unchanged (Table 4), allow us to rule out a reduction in TSH secretion as a factor in the hypothyroidism of aging rats. The elevated or unchanged TSH levels are consistent with the histologic appearance of hypertrophied thyrotropes in aging rats (Garner and Bernick, 1975).

In humans, the majority of studies report slightly increased serum TSH levels associated with a slightly reduced serum T_3. Others report no alteration or a slight reduction (Table 4). Recent studies suggest that the majority of pituitary T_3 (effective in the negative feedback inhibition of TSH) may be derived from intrapituitary deiodination of T_4 (Larsen et al., 1979). If this mechanism proves to be physiologically operative, then systemic monodeiodination may decline independently of intrapituitary T_4 to T_3 conversion leaving T_3 hypothyroidism (with unaltered TSH and T_4) a distinct possibility. Indeed, a recent abstract indicates independent maturational age changes in rat pituitary and hepatic 5'-monodeiodination rates (Cheron et al., 1979).

The circadian periodicity of TSH release is abolished in aging rats (Klug and Adelman, 1979), while available clinical studies do not support a similar loss in the human (Blichert-Toft et al., 1975). In view of the relatively long serum half-life of thyroid hormones, it is not clear what physiological significance the pulsatile release of thyroid hormones could have. However, in view of the recent reports of "down-regulation" of TSH receptors (Field et al., 1979; Uchimura et al., 1980) it is conceivable that the phasic presentation of TSH might be important for thyroidal response. Nevertheless, until a physiologic role for the TSH rhythm can be demonstrated, these results are more useful as an index of changes with aging in hypothalamic function than in thyroid status.

Peripheral Deiodination

As previously noted, the maintenance of normal serum T_4 levels in older humans, despite a greatly reduced secretory rate, reflects a sharp reduction in peripheral deiodination. Most of the circulating T_4 is deiodinated to either T_3 or reverse T_3, primarily in the liver and kidney, although a small fraction is directly conjugated to glucuronides and sulfates for disposal through biliary-intestinal transport (Chopra et al., 1978). Therefore, the progressive decrease in T_4 degradation with age may primarily reflect reduced peripheral monodeiodination. The amount of peripheral deiodination is a function of the free hormone available to the tissues since 99.94% of T_4 circulates tightly bound to serum proteins, primarily thyroid-binding-globulin (TBG) and pre-albumin (TBPA). Although synthesis of both these proteins is partially dependent on sex steroids and, therefore, altered with age, the effects are in opposite directions, the net result being a preservation of free hormone availability (Braverman et al., 1966). Thus, the probable focus of the age-related decline in T_4 deiodination is the activity of the 5'-monodeiodinase of the tissues. Indeed, T_4 to T_3 conversion has been shown to be significantly reduced with age in rat liver and kidney (Ooka, 1979; Köhrle et al., 1979) and in the human in vivo (Wenzel and Horn, 1975), although the latter study in the human may not be completely conclusive (Ingbar, 1978).

Hepatic Metabolism of Thyroid Hormones. The hepatic 5'-monodeiodinase that catalizes the conversion of T_4 to T_3 is regulated by four major factors: thyroid status, nonspecific stress or systemic illness, glucose availability, and a reduced thiol-dependent co-factor (Chopra et al., 1978). The availability of reduced hepatic thiol groups is determined by the ratio of the reduced (GSH) to the oxidized (GSSG) form of gluthathione. This ratio decreases with maturation (Harisch and Schole, 1974) and aging (Hazelton and Lang, 1978). Therefore, the decline in 5'-monodeiodinase activity with aging may be attributed, in part, to a fall in hepatic GSH/GSSG ratio, although other factors such as a decline in plasma thyroid hormone levels may also be operative (Köhrle et al., 1979). An additional complexity is the observation that, under physiologic conditions, the limiting factor in the conversion of T_4 to T_3 may be represented by the uptake of T_4 in the perfused rat liver (Jennings et al., 1979). At present, information is not available concerning age-related alterations in T_4 uptake by the liver (or any other tissue) and the physiological significance of the newly described plasma thyroid hormone receptors in the uptake process remains to be established. It appears, however, that at least during maturation, thyroid hormone uptake processes are not altered as suggested by the persistence of hepatic nuclear T_3 receptor occupancy (as determined by in vivo tracer injection of T_3) at an age when peripheral deiodination declines sharply (Schwartz et al., 1979).

Peripheral Refractoriness

Recently it has been suggested that the resemblance between certain aspects of hypothyroidism and aging stems not from reduced circulating hormone levels but from a reduced tissue response. Denckla (1974) has published evidence for the existence of a previously unrecognized pituitary factor which purportedly inhibits the calorigenic response to thyroid hormones under carefully controlled in vivo assay conditions. Further work has demonstrated that long-term hypophysectomy and hormone replacement therapy can restore age-related declines in physiological function in the rat in a variety of systems (minimal oxygen consumption, xenograft rejection, colloidal carbon clearance, isoproterenol-mediated aortic strip relaxation, lymphocyte mitogen responsiveness and antibody forming capacity, and hepatic initiation of RNA synthesis (see Bolla and Denckla, 1979). In all of these systems, hormone replacement without hypophysectomy did not substantially stimulate the measured parameter toward a more elevated "juvenile" value, while in hypophysectomized animals replacement therapy produced restorations, supporting Denckla's hypothesis of a pituitary-mediated decline in the rat's response to thyroid hormones and its role in aging. However, the significance of this intriguing work awaits the isolation of the putative antithyroidal factor and confirmation of its effects.

Evidence from other laboratories concerning the potency of thyroid hormones in aging animals is controversial. For example, while some groups report a clear maturational decline in the induction of hepatic

α-glycerophosphate dehydrogenase (alpha-GPA) by T_3 (Hemon, 1967; Schapiro and Percin, 1966; Schwartz et al., 1979) others report no change with further aging (Bulos et al., 1972a), nor any change in the induction of hepatic succinoxidase (Bulos et al., 1972b). Although these discrepancies may reflect differences in the ages of the animals, it must be noted that the maturational decline in alpha-GPD response to T_3 detected by Schwartz et al. (1979) was gradually overcome by the continued injection of T_3, a result at variance with Denckla's work which would predict a decreased response to T_3 after continued administration to intact animals (Denckla, 1974). Furthermore, there is no evidence for a significant decline with aging in T_3 induction of increased cardiac β-adrenergic receptors (Abrass and Scarpace, 1979) nor any reports suggesting a gross reduction in levels of thyroid-dependent proteins such as growth hormone or serum albumin. Thus, the capacity of thyroid hormones to induce synthesis of specific proteins remains unchanged with age. This conclusion is supported by the lack of evidence for any clear-cut effect of age on tissue binding or nuclear T_3 receptor capacity and affinity in either rat or man (Table 4). Additional evidence for the lack of a relationship between a general alteration in nuclear T_3 hormone receptors and aging is offered by the absence of any obvious resemblance between aging and the syndromes of peripheral thyroid hormone resistance (secondary to nuclear receptor defects) which have been described in humans (Verhoeven and Wilson, 1979). (The specific defect in Na/K ATPase induction in the ob/ob mouse may represent an intriguing exception. See above section on Pancreas.)

Consequences of Reduced Serum Thyroid Hormones

Given that the aging rat shows a highly significant reduction in serum levels of both T_4 and T_3, it is reasonable to anticipate that at least some of the changes reported to occur with age in these animals are simply secondary to age-related hypothyroidism. For example, β-adrenergic receptor mediated function is lost in aging rats in heart, submaxillary glands, aorta, adipocytes, and erythrocytes, but not in the liver where epinephrine-stimulated adenylate cyclase has been shown to rise with age (Roth, 1979a). In hypothyroidism, the same tissue specific pattern of decline occurs with a decrease in the β-adrenergic receptor mediated responses in the majority of tissues (Fregly et al., 1979) but an increased response in hepatocytes (Preiksaitis and Kunos, 1979; Malbon et al., 1978). Similarly, thyroid hormones stimulate glucagon and insulin receptor binding in rat adipocytes (Madsen and Sonne, 1976) and binding is lost with aging (Livingston et al., 1974) while in the hepatocyte, glucagon-stimulated adenylate cyclase activity is unchanged with age (Kalish et al., 1977) and apparently also in hypothyroidism (Malbon et al., 1978). In addition, the rodent adipocyte shows an elevated alpha-adrenergic response in both aging (Lafontan, 1979) and hypothyroidism (Reckless et al., 1976). Thus, the general pattern of membrane receptor alterations seems identical in aging and in hypothyroidism. Furthermore, the cardiac beta-adrenergic receptors may be effectively induced by T_3 replacement in the aged rat (Abrass and Scarpace, 1979), while procedures which elevate serum thyroid hormone levels in aging Balb/c mice also result in a restoration of isoproterenol-stimulated DNA synthesis in the submaxillary gland (Piantelli et al., 1978). Collectively, these results suggest that reported age changes in rodent peripheral adrenergic receptor number and function may be secondary to age-related hypothyroidism.

Thyroid Hormones and Cholesterol Metabolism. Serum cholesterol levels rise with aging in both the human and the rat despite a fall in hepatic cholesterol synthesis indicating a net reduction in overall turnover (Story and Kritchevsky, 1978). A similar situation obtains in hypothyroidism; this effect has prompted several attempts to treat hyperlipedemia with thyroid hormones and their analogs (Burrow, 1978). Several studies suggest that a causal relationship exists between decreased thyroid activity and elevated serum cholersterol. Hruza (1971) found that in young rats thyroidectomy resulted in reduced cholesterol turnover and elevated serum levels while in old rats there was no significant postoperative change. Similar results have been observed in a wide variety of mammalian species (Fleischmann, 1951). The implication is that there is a reduction in thyroid hormone secretion and/or a decrease in the sensitivity of cholesterol metabolism to thyroid hormones (Hruza, 1971). The observation in aging rats of an inverse relation between thyroid secretory rate and cholesterol levels and of reduced cholesterol levels in response to the increased thyroid secretory rate induced by exercise (Story and Griffith, 1974), suggest that an increase in cholesterol levels may be dependent on reduction of thyroid hormones (Hruza, 1971). In humans, a hyperbolic fall in serum cholesterol occurs with increasing serum T_3 concentrations (Bantle et al., 1980) and a similar inverse correlation is seen between declining serum T_3 and the age-related increase in serum cholesterol levels (Ikejiri et al., 1978). Taken together, these results support the inference that age changes in cholesterol metabolism may be secondary to age changes in the thyroid axis.

The specific defect in cholesterol metabolism which develops in the hypothyroid state would result from a reduction in turnover of the serum low density lipoproteins elevated in both hypothyroidism (Walton et al., 1965) and aging (Slack et al., 1977), while metabolism of high density lipoprotein remains unchanged. The significance of altered lipoprotein metabolism in the development of atherosclerosis suggests that early attempts by investigators to relate declining thyroid function to factors predisposing to atherosclerotic lesions (Wren, 1968) may not be entirely without foundation.

Is Aging Associated with Alterations in Hormone Metabolism and Tissue Demand rather than Inadequate Secretion? Several of the major alterations in endocrine function described in this chapter seem to involve the peripheral tissues rather than the gland themselves. In general, the thyroid, adrenals and pancreas appear well able to maintain their functional capacity to manufacture and secrete "adequate" amounts of hormones in old age; that is, adequate in the sense that these endocrines are competent to maintain normal (basal) serum levels. The peripheral metabolism of the hormones, however, is altered; insulin resistance appears to develop with age and thyroxine and corticosteroid turnover rates decline by as much as 50 percent. Because the hormone levels are maintained constant, several authors have concluded that peripheral demand for the metabolic hormones declines with increasing age. The most common explanation offered for this decline in demand is a loss of metabolically active tissue mass with aging. The concept that the age-related reduction in endocrine function is the consequence of age-incurred cell loss cannot be accepted because: (1) The observed declines in endocrine function begin early in life, before any significant cell loss has been proposed. (2) The putative loss of lean body mass may simply not occur in healthy aging individuals. When time-related (secular) trends for increased height and body weight and selective survivorship are taken into account, fat-free body mass being used as the reference standard, intracellular or total-body-water measurements in humans do not reveal significant cell loss (Lesser and Markovsky, 1979). Similar conclusions were attained in longitudinal studies of aging rats (Lesser et al., 1980). (3) Reduced total body potassium and creatine excretion with aging are often taken as indices of cell loss and reduced total body mass without considering that the endocrine alterations described in this chapter (e.g., insulin resistance, reduced levels of T_3 production, declining secretion of aldosterone and vasopressin) may produce similar changes. Therefore, reduced potassium and creatine excretion may reflect a consequence and not a cause of altered endocrine function.

If a smaller lean body mass cannot account for the reduction in peripheral hormone metabolism, perhaps an explanation may be found elsewhere, for example in some metabolic alteration in the target tissues such as an overall decline with aging in protein synthesis. A selective impairment of the general anabolic effects of thyroid hormones may occur with cessation of growth and subsequent aging, without a concomitant loss of catabolic effects. For example, injections of high doses of thyroxine in young animals are tolerated quite well: the animals respond with increased appetite and more rapid growth. The same doses injected into adult animals result in muscle wasting and weight loss (Korenchevsky, 1961) — (a condition comparable with the symptoms of thyrotoxicosis in humans who also show an increased sensitivity to thyrotoxic wasting with aging). Similarly, the capacity of thyroid hormones to promote protein synthesis in rat liver appears to be impaired with maturation and aging (Bolla and Denckla, 1979). Thus, it seems reasonable to propose that the ability of thyroid and perhaps other hormones to promote an overall net protein synthesis (viz. growth) is selectively lost when animals reach their characteristic species-specific sizes.

Conversion of T_4 to T_3 is generally reduced in conditions where catabolism predominates, for example, in fasting, nonspecific illnesses, and dexamethasone therapy. Furthermore, experiments involving inhibition of T_4 to T_3 conversion have demonstrated that in contrast to reverse T_3 or T_4 alone, T_3 has potent catabolic effects. These findings have led to the suggestion that "there is a homeostatic 'wisdom' in the arrangement whereby conversion of T_4 to T_3 is inhibited when catabolism is already overactive" (Chopra et al., 1978). From this perspective, the age-related decline in thyroid hormone metabolism would reflect not so much a reduced demand for tissue utilization as an increased demand for "protection."

Similarly, the apparent reduction in glucocorticoid utilization may reflect a homeostatic adjustment required to protect the tissues from the potent catabolic effects of 17-hydroxycorticosteroids. Such an adjustment could explain the correlation between declining corticosteroid metabolism and creatinine excretion. In addition, an homeostatic reduction in glucocorticoid output would presumably entail reduced ACTH secretion resulting in the observed diminution in the production of adrenal androgens.

Thus, one might hypothesize that several of the endocrine changes which occur with aging may ultimately be associated with the decline in overall protein synthesis which begins with the cessation of growth and continues to fall throughout the duration of the lifespan. The parallel fall in metabolic rate, whole body protein synthesis, and thyroid hormone "utilization" lends support to this hypothesis. Thyroid hormones, necessary for growth and development when energy requirements are very high, may become detrimental when the only energy needed is for homeostasis. In addition to adverse catabolic effects, they may induce successive alterations in neuroendocrine pacemakers leading to a loss of temporal organization of endocrine and metabolic functions and consequent degenerative changes (see Walker and Timiras, 1981).

OVERVIEW

Although the data are not entirely consistent, we may summarize the major endocrine age changes as follows:

1. Adrenal corticoid secretion rates are reduced, but circulating levels remain unchanged because of a prolonged corticoid half-life which is suggestive of a reduced "peripheral demand".

2. Aldosterone secretion and plasma levels are substantially reduced, apparently secondary to defective renin secretion owing to a juxtaglomerular beta-adrenergic receptor-adenylate cyclase complex defect.

3. Adrenal androgen secretion is reduced leading to depressed levels of plasma dehydroepiandrosterone (DHEA).

4. Adrenal medullary secretion of catecholamines is apparently increased, perhaps secondary to increased tissue refractoriness to catecholamines.

5. The decreased glucose tolerance observed in aging populations is probably the result of increased insulin resistance secondary to elevated free fatty acid levels and obesity and/or inappropriate glucagon secretion.

6. Parathyroid hormone secretion has not been demonstrated to be consistently altered with age while available data do show a decline in calcitonin secretion rates.

7. There is no clear evidence for altered pineal function.

8. The sleep-related peak of growth hormone secretion is diminished or lost with aging. Changes in prolactin levels are sex and species dependent and probably secondary to the balance of changes in circulating estrogen and hypothalamic dopamine.

9. Regulation of vasopressin release by the posterior pituitary is altered, generally in the direction of impaired secretion.

10. Thymic involution results in reduced circulating levels of thymic hormones with consequent defective T lymphocyte maturation, possibly a major factor in immunosenescence. Unfortunately, the endocrine relations of the thymus remain poorly characterized.

11. Thyroid hormone secretion rates decline with maturation and aging, but circulating thyroxine levels are generally maintained and reductions in circulating triiodothyronine (T_3) in healthy aged humans remain problematic. In the rodent, T_3 levels decline markedly with maturation and aging, perhaps as a result of reduced peripheral conversion. One consequence may be that some of the reported age changes in rodent models may be secondary to an age-related hypothyroidism.

In conclusion, although we have offered several tentative speculations, it is not yet possible to draw a single coherent picture of the nature of age changes in the endocrine system. As several alternative views of the causes and consequences of reported alterations may be entertained at present, we have included summary tables of the available data which should serve to emphasize the need for further, more definitive studies of age-related endocrine changes.

REFERENCES

Abrass, I.B. and Scarpace, P. Catecholamine receptors in aging, in *Endocrine Aspects of Aging*, S.G. Korenman, ed. Summary of papers presented at a conference jointly sponsored by the National Institute on Aging and the Endocrine Society. Bethesda, Md., October 18-20, (1979).

Adelman, R.C. Loss of adaptive mechanisms during aging. *Federation Proc.* 38(6), 1968-1971 (1979).

Adelman, R.C. and Freeman, C. Age-dependent regulation of the glucokinase and tyrosine aminotransferase activities of rat liver in vivo by adrenal, pancreatic and pituitary hormones. *Endocrinology* 90, 1551-1560 (1972).

Albrecht, E.D. Adrenal responsiveness to adrenocorticotropin in aged male rats. Gerontological Society, 32nd Annual Meeting, Abstract, Washington, D.C., November 25-29 (1979), p. 37.

Aleshin, B.V. Adrenergic regulation of hypothalamic neurosecretory functions, in *Neurosecretion and Neuroendocrine Activity: Evolution, Structure and Function*, N. Bargmann, A. Aksche, A. Polenov, and B. Sharrer, eds. Springer-Verlag, New York (1978), pp. 117-121.

Andres, R. Aging and diabetes. *Med. Clin. North Am.* 55, 835-845 (1971).

Andres, R. and Tobin, J.D. Endocrine systems, in *Handbook of the Biology of Aging*, C.E. Finch and L. Hayflick, eds. Van Nostrand Reinhold Company, New York (1977), pp. 357-378.

Andrew, W. Senile changes in the pancreas of WISTAR Institute rats and of man. *Am. J. Anat.* 74, 97-127 (1944).

Antonioni, F.M., Porro, A., Serio, M., and Tinti, P. Gas chromatographic analysis of urinary 17-ketosteroids response to gonadotropin and ACTH in young and old persons. *Exp. Gerontol.* 3, 181-192 (1968).

Ascheim, P. Oestrus permanent et prolactine. *C.R. Acad. Sci.* 255, 3053-3055 (1962).

Ascheim, P. and Pasteels, J.L. Etude histophysiologique de la secretion de prolactine chez les rattes senile. *C.R. Acad. Sci.* 257, 1373-1375 (1963).

Azizi, F., Vagenakis, A.G., Portnoy, G.F., Rapoport, B., Ingbar, S.H., and Braverman, L.E. Pituitary-thyroid responsiveness to intramuscular thyrotropin-releasing hormone based on analyses of serum thyroxine, triiodothyronine, and TSH concentrations. *New Engl. J. Med.* 292, 273-277 (1979).

Baile, C.A., Herrera, M.G., and Mayer, J. Ventromedial hypothalamus and hyperphagia in hyperglycemic obese mice. *Am. J. Physiol.* 218, 857-863 (1970).

Baker, B.L. and Yu, Y.Y. An immunocytochemical study of human pituitary mammotropes from fetal life to old age. *Am. J. Anat.* 148, 217-249 (1977).

Baker, S.P., Gaffney, G.W., Shock, N.W., and Landowne, M. Physiological response of 5 middle-aged and elderly men to repeated administration of TSH. *J. Gerontol.* 14, 37-47 (1959).

Bantle, J.P., Dillman, W.H., Oppenheimer, J.H., Bingham, C., and Runger, G.C. Common clinical indices of thyroid hormone action; relationships to serum free 3,5,3'-triiodothyronine concentration and estimated nuclear occupancy. *J. Clin. Endocrinol. Metab.* 50, 286-293 (1980).

Barbagallo-Sangiorgi, G., Laudicina, E., Bompiano, G.D., and Durante, F. The pancreatic beta-cell response to intravenous administration of glucose in elderly subjects. *J. Am. Geriatr. Soc.* 18, 529-538 (1970).

Barnett, J.L. and Phillips, J.B. Age-dependent changes in body weight, hematocrit and corticosteroids following sequential castration in rats with a high incidence of spontaneous tumors. *Exp. Gerontol.* 11, 217-230 (1976).

Bazarre, T.L., Johanson, A.J., Huseman, C.A., Varman, M.M., and Blizzard, R.M. Human growth hormone changes with age, in *Growth Hormone and Related Peptides*, A. Pecile and H. Muller, eds. Proc. of the Third Intl. Symp. on Growth Hormone, Milan (1975), *Excerpta Medica*, Amsterdam (1976), pp. 261-270.

Benjamin, F., Casper, D.J., Sherman, L., and Kolodny, H.D. Growth hormone, age and the endometrium. *New Engl. J. Med.* 283, 375 (1970).

Bermudez, F., Surks, M.I., and Oppenheimer, J.H. High incidence of decreased serum T_3 in patients with non-thyroidal disease. *J. Clin. Endocrinol. Metab.* 41, 27-40 (1975).

Bethea, C.L. and Walker, R.F. Age-related changes in reproductive hormones and in Leydig cell responsivity in the male Fischer 344 rat. *J. Gerontol.* 34, 21-27 (1979).

Bjorntorp, P. Effects of age, sex and clinical condition on adipose tissue cellularity in man. *Metabolism* 23, 1091-1102 (1974).

Bjorntorp, P., Berchtold, P., and Tibblin, G. Insulin secretion in relation to adipose tissue in men. *Diabetes* 20, 65-70 (1971).

Bjorntorp, P., Krotkiewskis, M., Larsson, B., and Somlo-Szucs, Z. Effects of feeding states on lipid radioactivity in liver, muscle, and adipose tissue after injection of labeled glucose in the rat. *Acta Physiol. Scand.* 80, 29-38 (1970).

Blichert-Toft, M. Assessment of serum corticotrophin concentration and its nyctohemeral rhythm in the aging. *Clin. Gerontol.* 13, 215-220 (1971).

Blichert-Toft, M. Secretion of corticotrophin and somatotrophin by the senescent adenohypophysis in man. *Acta Endocrinol.* 78 (Suppl. 195), 1-157 (1975).

Blichert-Toft, M. The adrenal glands in old age, in *Geriatric Endocrinology* (Aging, Volume 5). R.B. Greenblatt, eds., Raven Press, New York (1978), pp. 81-102.

Blichert-Toft, M., Hippe, E., and Jensen, H.K. Adrenal cortical function as reflected by the plasma hydrocortisone and urinary 17-ketogenic steroids in relation to surgery in elderly patients. *Acta Chir. Scand.* 133, 591-599 (1967).

Blichert-Toft, M., Hummer, L., and Dige-Petersen, H. Human TSH level and response to TRH in the aged. *Geront. Clin.* 17, 191-203 (1975).

Bodanszky, M. and Engel, S.L. Oxytocin and the lifespan of male rats. *Nature* 210, 751 (1966).

Bolla, R. and Denckla, W.D. Effect of hypophysectomy on liver nuclear ribonucleic acid synthesis in aging rats. *Biochem. J.* 184, 669-674 (1979).

Bondareff, W. Electron microscope study of the pineal body in aged rats. *J. Gerontol.* 20, 321-327 (1965).

Bourne, G.H. Aging changes in the endocrines, in *Endocrines and Aging*, L. Gittman, ed. Charles C.Thomas, Springfield, Ill. (1967), pp. 66-75.

Bowman, R.E. and Wolfe, R.C. Plasma 17-hydroxycorticosteroid response to ACTH in *M. mulatta*; dose age, weight and sex. *Proc. Soc. Exp. Biol. Med.* 130, 61-64 (1969).

Brancho-Romero, E. and Reaven, G.M. Effect of age and weight on plasma glucose and insulin responses in the rat. *J. Am. Geriatr. Soc.* 25,

Braverman, L.E., Dawber, N.A., and Ingbar, S.H. Observations concerning the binding of thyroid hormones in sera of normal subjects of varying ages. *J. Clin. Invest.* 45, 1273-1279 (1966).

Britton, G.W., Rutenberg, S., Freeman, C., Britton, V.J., Karoly, K., Ceci, L., Klug, T.L., Lacko, A.G., and Adelman, R.C. Regulation of corticosterone levels and liver enzyme activity in aging rats, in *Explorations in Aging*, Adv. Exp. Med. V. 6, V.J. Cristfeld, J. Roberts, and R.C. Adelman, eds. Plenum Press, New York (1975) pp. 209-228.

Buckler, J.M. The effect of age, sex and exercise on the secretion of growth hormone. *Clin. Sci.* 37, 765-774 (1969).

Bulos, B., Shukla, S., and Sacktor, B. Effect of thyroid hormone on respiratory control of liver mitochondria from adult and senescent rats. *Arch. Biochem. Biophys.* 151, 387-390 (1972a).

Bulos, B., Shukla, S., and Sacktor, B. The rate of induction of the mitochondrial α-glycerolphosphate dehydrogenase by thyroid hormone in adult and senescent rats. *Mech. Ageing Dev.* 1, 227-231 (1972b).

Burek, J.D. *Pathology of Aging Rats.* CRC Press, West Palm Beach, Fl. (1978), p. 230.

Burrow, G.N. Thyroid hormone therapy in nonthyroid disorders, in *The Thyroid: A Fundamental and Clinical Text,* 4th ed., S.C. Werner and S.H. Ingbar, eds. Harper and Row, Hagerstown, Md. (1978) pp. 974-980.

Burrows, A.W., Shakespear, R.A., Hesch, R.D., Cooper, E., Aickin, C.M., and Burker, C.W. Thyroid hormones in the elderly sick: "T_4 euthyroidism." *Brit. Med. J.* IV, 437-439 (1975).

Calderon, L., Ryan, N., and Kovacs, K. Human pituitary growth hormone cells in old age. *Gerontology* 24, 441-447 (1978).

Calloway, N.O., Foley, C.F., and Lagerbloom, P. Uncertainties in geriatric data II. Organ size. *J. Am. Geriatr. Soc.* 13, 20-28 (1965).

Camerini-Davalos, R.A., Oppermann, W., Rebagliati, H., Glasser, M., and Bloodworth, J.M.B. Muscle capillary basement membrane width in genetic prediabetes. *J. Clin. Endocrinol. Metab.* 48, 251-259 (1979).

Carlson, H.E., Gillin, J.C., Gordon, P., and Snyder, F. Absence of sleep-related growth hormone peaks in aged normal subjects and acromegaly. *J. Clin. Endocrinol. Metab.* 34, 1102-1105 (1972).

Cartlidge, N.E.F., Black, M.M., Hall, M.R., and Hall, R. Pituitary function in the elderly. *Geront. Clin.* (Basel) 12, 65-70 (1970).

Chen, J.J. and Walfish, P.G. Effects of age and ovarian function on the pituitary-thyroid system in female rats. *J. Endocrinol.* 78, 225-232 (1978).

Chen, J.J. and Walfish, P.G. Effects of age and testicular function on the pituitary-thyroid system in male rats. *J. Endocrinol.* 82, 53-59 (1979).

Cheron, R.G., Kaplan, M.M., and Larsen, P.R. Contrasting age-related changes in rat pituitary and hepatic thyroxine 5'-monodeiodination rates. *Clin. Res.* 27(3), A572 (1979).

Chiba, M. and Yamanda, M. About the calcification of the pineal gland in the Japanese. *Folia Psychiatr. Neurol. Japan* 2, 301-303 (1948).

Chopra, I.J. Alterations in monodeiodination of iodothyronines in the fasting rat: effects of reduced nonprotein sulfhydryl groups and hypothyroidism. *Metabolism* 29, 161-167 (1980).

Chopra, I.J., Solomon, D.H., Chopra, O., Wu, S.Y., Fisher, D.A., and Nakamura, Y. Pathways of metabolism of thyroid hormones. *Rec. Progr. Horm. Res.* 34, 521-567 (1978).

Clayton, C.J., Schechter, J., Sladek, J.R., Jr., and Finch, C.E. Morphological characterization of pituitary tumors in aged female mice, in *Endocrine Aspects of Aging,* Summary of papers presented at a conference sponsored jointly by the National Institute on Aging and the Endocrine Society, Bethesda, Md., October 18-20, (1979).

Clemens, J.A. and Bennett, D.R. Do aging changes in the preoptic area contribute to loss of cyclic endocrine function? *J. Geront.* 32, 19-24 (1977).

Clemens, J.A. and Meites, J. Neuroendocrine status of old constant-estrus rats. *Neuroendocrinology* 7, 249-256 (1971).

Clemens, J.A., Fuller, R.W, and Owen, N.V. Some neuroendocrine aspects of aging, in *Experimental Medicine and Biology, Volume 113, Parkinson's Disease – II, Aging and Neuroendocrine Relationships,* C.E. Finch, D.E. Potter, and A.D. Kenny, eds. Plenum Press, New York (1978), pp. 77-100.

Cooper, B., Wernblatt, F., and Gregerman, R.I. Enhanced activity of the hormone-sensitive adenylate cyclase during dietary restriction in the rat: dependence on age and relation to cell size. *J. Clin. Invest.* 59, 467-474 (1977).

Cooper, E.R.A. *The Histology of the More Important Human Endocrine Organs at Various Ages.* Oxford University Press, London (1925), pp. 1-119.

Creutzfeldt, W. Etiological and promoting factors in the pathogenesis of diabetes mellitus, in *The Genetics of Diabetes Mellitus,* W. Creutzfeldt, J. Kobberling, and J.V. Neel, eds. Springer-Verlag, Berlin (1976), pp. 37-42.

Crewe, F.A.E. Rejuvenation of the aged fowl through thyroid medication. *Proc. Roy. Soc.* (Edinburgh) V. 45, 252-260 (1924-1925).

Crockford, P.M., Harbeck, R.J., and Williams, R.H. Influence of age one intravenous glucose tolerance and serum immunoreactive insulin. *Lancet* 1, 465-467 (1966).

Cuttelod, S., Lemarchand-Beraud, T., Magenat, P., Perret, C., Poli, S., and Venotti, A. Effect of age and role of the kidneys and liver on TSH turnover in man. *Metabolism* 23, 101-113 (1974).

Dallman, M.F. and Jones, M.T. Corticosteroid feedback control of ACTH secretion; effect of stress-induced corticosterone secretion on subsequent stress responses in the rat. *Endocrinology* 92(5), 1367-1375 (1973).

D'Angelo, S.A. TSH rebound phenomenon in the rat adenohypophysis. *Endocrinology* 69, 834-843 (1961).

Danowski, T.S., Tsai, T., Morgan, C., Sieracki, J., Alley, R., Robbins, T., Sabeh, G., and Sunder, J.H. Serum growth hormone and insulin in females without glucose intolerance. *Metabolism* 18, 811-820 (1969).

Daramola, G.F. and Olowu, A.O. Physiological and radiological implications of a low incidence of pineal calcification in Nigeria. *Neuroendocrinology* 9, 41-57 (1972).

Davidson, M.B. The effect of aging on carbohydrate metabolism; a review of the English literature and a practical approach to the diagnosis of diabetes mellitus in the elderly. *Metabolism* 28, 688-705 (1979).

Davies, I. and Fotheringham, A.P. The influence of age on the hypothalamo-neurohypophyseal system of the mouse: A quantitative ultrastructural analysis of the supraoptic nucleus. *Mech. Ageing Dev.* 12, 93-105 (1980).

Davis, P.J. and Davis, F.B. Hyperthyroidism in patients over the age of 60 years. *Medicine* 53(3), 161-181 (1974).

DeFronzo, R.A. Glucose intolerance and aging. Evidence for tissue insensitivity to insulin. *Diabetes* 28, 1095-1101 (1979).

Deftos, L.J., Weisman, M.H., Williams, G.H., Karpf, D.B., Frumar, A.M., Davidson, B.H., Parthemore, J.G., and Judd, H.L. Age- and sex-related changes of calcitonin secretion in humans, in *Endocrine Aspects of Aging*, S.G. Korenman, ed. Summary of papers presented at a conference jointly sponsored by the National Institute on Aging and the Endocrine Society. Bethesda, Md., October 18-20 (1979).

Denckla, W.D. Role of the pituitary and thyroid glands in the decline of minimal O_2 consumption with age. *J. Clin. Invest.* 53, 572-581 (1974).

Department of Health, Education and Welfare (HEW). Facts of life and death. *HEW* publication #PHS79-1222, National Center for Health Statistics, Hyattsville, MD, November (1978).

Dilman, V.M. The hypothalamic control of aging and age-associated pathology. The elevation mechanism of aging, in *Hypothalamus, Pituitary and Aging*, A.V. Everitt and J.A. Burgess, eds. Charles C. Thomas, Springfield, Ill. (1976), pp. 634-667.

Dilman, V.M., Ostroumova, M.N., and Tsyrlina, E.V. Hypothalamic mechanisms of ageing and of specific age pathology – II. On the sensitivity threshold of hypothalamo-pituitary complex to homeostatic stimuli in adaptive homeostasis. *Exp. Geront.* 14, 175-181 (1979).

Dobbie, J.W. Adrenocortical nodular hyperplasia: the aging adrenal. *J. Pathol.* 99, 1-18 (1969).

Dogliotti, G.C. and Nuzzi, N.G. Thyroid and senescence: structural transformations of the thyroid in old age and their functional interpretation. *Endocrinology* 19, 289-292 (1935).

Duckworth, W.C. and Kitabchi, A.E. Direct measurement of plasma proinsulin in normal and diabetic subjects. *Am. J. Med.* 53, 418-427 (1972).

Duckworth, W.C. and Kitabchi, A.E. The effect of age on plasma proinsulin-like material after oral glucose. *J. Lab. Clin. Med.* 88, 359-367 (1976).

Dudl, R.J. and Ensinck, J.W. The role of insulin, glucagon, and growth hormone in carbohydrate homeostasis during aging. *Diabetes* 21, 357 (1972).

Dudl, R.J. and Ensinck, J.W. Insulin and glucagon relationships during aging in man. *Metabolism* 26, 33-41 (1977).

Dudl, R.J., Ensinck, J.W., Palmer, H.E., and Williams, R.H. Effect of age on growth hormone secretion in man. *J. Clin. Endocrinol.* 37, 11-16 (1973).

Dunihue, F.W. Reduced juxtaglomerular cell granularity, pituitary neurosecretory material and width of the zona glomerulosa in aging rats. *Endocrinology* 77, 948-951 (1965).

Einhorn, J. Studies on the effect of thyrotropic hormone on thyroid function in man. *Acta Radiol. (Suppl.)* (Stockholm) 160, 1-107 (1958).

Eleftheriou, B.E. Changes with age in pituitary-adrenal responsiveness and reactivity to mild stress in mice. *Gerontologia* 20, 224-230 (1974).

Elsas, L.J., MacDonnell, R.C., and Rosenberg, L. Influence of age on insulin stimulation of amino acid uptake in rat diaphragm. *J. Biol. Chem.* 246, 6452-7459 (1971).

Emerson, J.D. Development of resistance to growth promoting action of anterior pituitary growth hormone. *Am. J. Physiol.* 181, 390-394 (1955).

Everitt, A.V. Hypophysectomy and aging in the rat, in *Hypothalamus, Pituitary and Aging*, A.V. Everitt and J.A. Burgess, eds. Charles C. Thomas, Springfield, Ill. (1976), pp. 68-85.

Everett, A.V. Comparison of the long-term anti-ageing actions of hypophysectomy at 70 and 400 days with those of food restriction begun at the same ages in the rat. VI International Congress of Endocrinology, Abstract. Melbourne, Australia (1980), p. 338.

Fedele, D., Valerio, A., Molinari, M., and Crepaldi, G. Glucose tolerance, insulin, and glucagon secretions in aging. *Diabetologia* 13, 392 (1977).

Feldman, M. The pancreas in the aged: an autopsy study. *Geriatrics* 10, 373-374 (1955).

Field, J.B., Titus, G., and Chou, M. Further characterization of thyroid stimulating hormone induced refractoriness in thyroid tissues. *Proceedings of the Endocrine Society*, Abstract 88, 61st Annual Meeting, Anaheim, Ca., June 13-15 (1979).

Finch, C.E. Neuroendocrine mechanisms and aging. *Fed. Proc.* 38, 178-183 (1979).

Finklestein, J.W., Roffwarg, H.P., Boyar, H.M., Kream, J., and Hellman, L. Age-related changes in the 24-hour spontaneous secretion of growth hormone. *J. Clin. Endocrinol.* 35, 665-670 (1972).

Fisher, D.A. and Dussault, J.H. Development of the mammalian thyroid in *Handbook of Physiology* (Endocrinology, Volume 3), J. Field, ed. Williams and Wilkins, Baltimore, Md. (1972), pp. 30-38.

Fleischmann, W. Mammals, in *Comparative Physiology of the Thyroid and Parathyroid Glands*, Charles C. Thomas, Springfield, Ill. (1951), pp. 43-44.

Flood, C., Gherondache, C., Pincus, G., Tait, J.F., Tait, S.A.S., and Willoughby, S. The metabolism and secretion of aldosterone in elderly subjects. *J. Clin. Invest.* 46, 960-966 (1967).

Flood, C., Layne, D.S., Rancharan, S., Rossipal, E., Tait, J., and Tait, S. An investigation of the urinary metabolites and secretion rates of aoldosterone and cortisol in man and a description of methods for their measurement. *Acta Endocrinol.* 36, 237-264 (1961).

Florini, J.R. and Roberts, S.B. Age-related changes in somatomedin levels. *The Gerontologist* 18(5), A172 (1978).

Florini, J.R., Harned, J.A., Weiss, J.P., and Richman, R.A. Effect of age on levels of growth hormone and somatomedins in rat serum. Gerontological Society, 32nd Annual Meeting, Abstract. Washington, D.C. November 25-29 (1979), p. 73.

Frantz, A.G., Kleinberg, D.L., and Noel, G.L. Studies on prolactin in man. *Rec. Prog. Horm. Res.* 28, 527-573 (1972).

Freedman, L.S., Ohuchi, T., Goldstein, M., Axelrod, F., Fish, I., and Dancis, J. Changes in human serum dopamine-β-hydroxylase activity with age. *Nature* 236, 310-311 (1972).

Freeman, C., Karoly, K., and Adelman, R.C. Impairments in availability of insulin to liver *in vivo* and in binding of insulin to purified hepatic plasma membrane during aging. *Biochem. Biophys. Res. Commun.* 54, 1573-1580 (1973).

Fregly, M.J., Field, F.P., Katovich, M.J., and Barney, C.C. Catecholamine-thyroid hormone interaction in cold-acclimated rats. *Federation Proc.* 38, 2162-2169 (1979).

Friedman, M., Green, M.F., and Sharland, D.E. Assessment of hypothalamic-pituitary-adrenal function in the geriatric age group. *J. Gerontol.* 24, 292-297 (1969).

Friedman, S.M. and Friedman, C.L. Salt and water balance in ageing rats. *Gerontology* 1, 107-121 (1957).

Friedman, S.M. and Friedman, C.L. Prolonged treatment with posterior piutitary powder in aged rats. *Exp. Gerontol.* 1, 37-48 (1964).

Friedman, S.M., Friedman, C.L., and Nakashima, M. Effect of pitressin on old-age changes of salt and water metabolism in the rat. *Am. J. Physiol.* 199, 35-38 (1960).

Frolkis, V.V. The hypothalamic mechanisms of aging, in *Hypothalamus, Pituitary and Aging*, A.V. Everitt and J.A. Burgess, eds. Charles

C. Thomas, Springfield, Ill. (1976), pp. 614-633.

Frolkis, V.V. and Valueva, G.V. Metabolism of thyroid hormones during aging. *Gerontology* 24, 81-94 (1978).

Frolkis, V.V., Verzihkovskaya, N.V., and Valueva, G.V. The thyroid and age. *Exp. Geront.* 8, 285-296 (1973).

Frolkis, V.V., Bazrukov, V.V., Duplenko, Y.K., and Genis, E.D. The hypothalamus and ageing. *Exp. Gerontol.* 7, 169-184 (1972).

Gaffney, G.W., Gregerman, R.I., and Shock, N.W. Serum protein bound iodine concentration in blood of euthyroid men aged 18 to 94 years. *J. Clin. Endocrinol. Metab.* 22, 784-794 (1962).

Garner, H.S. and Bernick, S. Effect of age upon the thyroid gland and pituitary thyrotrophs of the rat. *J. Gerontol.* 30(2), 137-148 (1975).

Garthwaite, T.L., Kalkhoff, R.K., Guansing, A.R., Hagen, T.C., and Manahan, L.A. Plasma free tryptophan, brain serotonin, and an endocrine profile of the genetically obese hyperglycemic mouse at 4-5 months of age. *Endocrinology* 105, 1178-1182 (1979).

Gershberg, H. Growth hormone content and metabolic actions of human pituitary glands. *Endocrinology* 61, 160-165 (1957).

Gherondache, C.N., Romanoff, L.P., and Pincus, G. Steroid hormones in aging men, in *Endocrines and Aging*, L. Gitman, ed. Charles C. Thomas, Springfield, Ill. (1967), pp. 76-101.

Giorgino, R., Scardapane, R., Nardelli, G.M., and Tafaro, E. Surrene e senescenza. Aspetti dell' attiva della midollare. *Folia Endocrinologica* 22, 215-224 (1969).

Gittler, R.D. and Friedfield, L. Adrenocortical responsiveness in the aged. *J. Am. Geriatr. Soc.* 10, 153-159 (1962).

Gold, G., Karoly, K., Freeman, C., and Adelman, R.C. A possible role for insulin in the altered capability for hepatic enzyme adaptation during aging. *Biochem. Biophys. Res. Commun.* 73, 1003-1010 (1976).

Goldstein, A.L., Slater, F.D., and White, A. Preparation and partial purification of a thymic lymphocytopoietic factor (thymosin). *Proc. Nat. Acad. Sci.* 56, 1010-1017 (1966).

Goldstein, A.L., Low, T.L.K., McAdoo, M., McClure, J., Thurman, G.D., Rossio, J., Lai, C.-Y., Chong, D., Wang, S.-S., Harvey, C., Ramel, A.H., Meienhofer, J., and Burns, J.J. Thymosin: Isolation and sequence analysis of a immunologically active thymic polypeptide. *Proc. Nat. Acad. Sci.* 74, 725-729 (1977).

Goldstein, S. Human genetic disorders that feature premature onset and accelerated progression of biological aging, in *The Genetics of Aging*, E.L. Schneider, ed. Plenum Press, New York (1978), pp. 171-224.

Goldstein, S., Littlefield, J.W., and Soeldner, J.S. Diabetes mellitus and aging: diminished plating efficiency of cultured human fibroblasta. *Proc. Nat. Acad. Sci.* 64, 155-160 (1969).

Gommers, A., and de Gasparo, M. Variation de l'insulinemie en fonction de l'age chez le rat male non traite. *Gerontologia* 18, 176-184 (1972).

Gommers, A., Dehez-Delhaye, M., and Jeanjean, M. The effect of age on the *in vitro* response to insulin in the rat. I. Glucose metabolism on the diaphragm. *Gerontology* 23, 134-141 (1977).

Good, R.A. and Yunis, E. Association of autoimmunity, immunodeficiency and aging in man, rabbits and mice. *Federation Proc.* 33, 2040-2049 (1974).

Grad, B. and Hoffman, M.H. Thyroxine secretion rates and plasma cholesterol levels of young and old rats. *Am. J. Physiol.* 182, 497-502 (1955).

Grad, B. and Khalid, R. Circulating corticosterone levels of young and old, male and female C57BL/6J mice. *J. Gerontol.* 23, 522-528 (1968).

Grad, B., Kral, V.A., Payne, R.C., and Berenson, J. Plasma and urinary corticoids in young and old persons. *J. Gerontol.* 22, 66-71 (1967).

Grad, B., Rosenberg, G.M., Liberman, H., Trachtenberg, J., and Kral, V.A. Diurnal variation in serum cortisol level of geriatric subjects. *J. Gerontol.* 26, 351-357 (1971).

Greenberg, L.H. and Weiss, B. β-adrenergic receptors in aged rat brain: reduced number and capacity of pineal gland to develop supersensitivity. *Science* 201, 61-63 (1978).

Gregerman, R.I. Estimation of thyroxine secretion rate in the rat by the radioactive thyroxine turnover technique: influence of age, sex and exposure to cold. *Endocrinology* 72, 382-392 (1963).

Gregerman, R.I., Gaffney, G.W., and Shock, N.W. Thyroxine turnover in euthyroid man with special reference to changes with age. *J. Clin. Invest.* 41, 2065-2074 (1962).

Gregerman, R.I. and Solomon, N. Acceleration of thyroxine and triiodothyronine turnover during bacterial pulmonary infections and fever: implications for the functional state of the thyroid during stress and in senescence. *J. Clin. Endocrinol. Metab.* 27, 93-105 (1967).

Gries, F.A. and Steinke, J. Comparative effects of insulin on adipose tissue segments and isolated fat cells of rat and man. *J. Clin. Invest.* 46, 1413-1421 (1967).

Guernsey, D.L. and Morishige, W.K. Sodium pump activity and nuclear T_3 receptors in tissues of genetically obese (ob/ob) mice. *Metabolism* 28, 629-632 (1979).

Hansen, F.M., Nielsen, J.H., and Gliemann, J. The influence of body weight and cell size on lipogenesis and lipolysis of isolated rat fat cells. *Eur. J. Clin. Invest.* 4, 411-418 (1974).

Hansen, J.M., Skovsted, L., and Siersback-Nielsen, K. Age-dependent changes in iodine metabolism and thyroid function. *Acta Endocrinol.* 79, 60-65 (1975).

Harisch, G. and Schole, J. Der Glutathionstatus der Rattenleber in abhängigkeit vom Lebensalter and und von akuter Belastung. *Z. Naturforsch.* 29c, 261-266 (1974).

Hazelton, G.A. and Lang, C.A. Glutathionine changes in aging C57BL/6J mice. *Federation Proc.* 37(3), 880-881 (1978).

Heaney, R.P. Age-related changes in calcium metabolism in perimenopausal women and thier relationship to the development of osteoporosis, in *Endocrine Aspects of Aging*, S.G. Korenman, ed. Summary of papers presented at a conference jointly sponsored by the National Institute on Aging and the Endocrine Soceity, Bethesda, Md., October 18-20 (1979).

Heaney, R.P., Recker, R.R., and Saville, P.D. Menopausal changes in calcium balance performance. *J. Lab. Clin. Med.* 92, 953-963 (1978a).

Heaney, R.P., Recker, R.R., and Saville, P.D. Menopausal changes in calcium balance performance. *J. Lab. Clin. Med.* 92, 964-970 (1978b).

Helderman, J.H., Vestal, R.E., Rowe, J.W., Tabin, J.D., Andres, R., and Robertson, G.L. The response of arginine vasopressin to intravenous ethanol and hypertonic saline in man: the impact of aging. *Gerontology* 33, 39-47 (1978).

Hellman, B. The total volume of the pancreatic islet tissue at different ages of the rat. *Acta Pathol. Microbiol. Scand.* 47, 35-50 (1959).

Hemon, P. L'activite α-glycerophosphate oxydase du foie de rat au cours du developpement. *Biochim. Biophys. Acta* 132, 175-178 (1967).

Herberg, L., and Coleman, D.L. Laboratory animals exhibiting obesity and diabetes syndromes. *Metabolism* 26, 59-99 (1977).

Herman, E. Senile hypophyseal syndromes, in *Hypothalamus, Pituitary and*

Aging, A.V. Everitt and J.A. Burgess, eds. Charles C. Thomas, Springfield, Ill. (1976), pp. 157-170.

Herman, J., Rusche, H.J., Kroll, H.J., Hilger, P., and Kruskemper, H.L. Free triiodothyronine (T_3) and thyroxine (T_4) levels in old age. *Hormone Metab. Res.* 6, 239-240 (1974).

Hesch, R.D., Gatz, J., Pope, J., Schmidt, E., and von zur Huhlen, A. Total and free 3',5'-triiododthyronine and thyroid hormone binding globulin concentration in elderly human persons. *Eur. J. Clin. Invest.* 6, 139-145 (1976).

Hess, G.D. and Riegle, G.D. Adrenocortical responsiveness to stress and ACTH in aging rats. *J. Gerontol.* 25, 354-358 (1970).

Hess, G.D. and Riegle, G.D. Effects of chronic ACTH stimulation on adrenocortical function in young and aged rats. *Am. J. Physiol.* 222, 1458-1461 (1972).

Hochstaedt, B.B., Schneebaum, M., and Shadel, M. Adrenocortical responsitivity in old age. *Gerontol. Clin.* 3, 239-246 (1961).

Hoffman, C.C., Carroll, K.F., and Goldrick, R.B. Studies on lipid and carbohydrate metabolism in the rat. Effects of diet on body composition, plasma glucose and insulin concentrations, insulin secretion *in vitro* and tolerance to intravenous glucose and intravenous insulin. *Aust. J. Exp. Biol. Med. Sci.* 50, 267-287 (1972).

Holm, A.C., Lemarchand-Beraud, T., Scazziga, B.R., and Cuttlerod, S. Human lymphocyte binding and deiodination of thyroid hormones in relation to thyroid function. *Acta Endocrinol.* 80, 642-656 (1975).

Holm, G., Jacobsson, B., Bjorntorp, P., and Smith, V. Effects of age and cell size on rat adipose tissue metabolism. *J. Lipid Res.* 16, 461-464 (1977).

Horky, K., Manek, J., Kopecka, J., and Gregorova, I. Influence of age on orthostatic changes in plasma renin activity and urinary catecholamines in man. *Physiol. Bohemoslov.* 24(6), 481-488 (1975).

Hruza, Z. Effect of endocrine factors on cholesterol turnover in young and old rats. *Exp. Gerontol.* 6, 199-204 (1971).

Huang, H.H., Marshall, S., and Meites, J. Capacity of old versus young female rats to secrete LH, FSH, and prolactin. *Biol. Reprod.* 14, 538-543 (1976).

Huang, H.H., Steger, R.W., and Meites, J. Capacity of old vs. young male rats to secrete thyroxine and thyrotropin. *Gerontologist* 18(15), 84 (1978).

Ikejiri, K., Yamada, T., and Orgura, H. Age-related glucose intolerance in hyperthyroid patients. *Diabetes* 27, 543-549 (1978).

Inada, M., Koshiyame, K., Torizuka, K.A., Kagi, H., and Miyake, T. Clinical studies of the metabolism of ^{131}I-thyroxine. *J. Clin. Endocrinol. Metab.* 24, 775-784 (1964).

Ingbar, S.H. The influence of aging on the human thyroid hormone economy in *Geriatric Endocrinology* (Aging, Volume 5), R.B. Greenblatt, ed. Raven Press, New York (1978), pp. 13-31.

Ingle, D.J. Permissive action of hormones. *J. Clin. Endocrinol. Metab.* 14, 1272-1274 (1954).

Irvine, R.E. and Hodkinson, H.M. Thyroid disease in old age, in *Textbook of Geriatric Medicine and Gerontology*, 2nd edition. J.C. Brocklehurst, ed. Churchill Livingston, Edin. (1978), pp. 451-494.

Ito, J.M., Valcana, T., and Timiras, P.S. Effect of hypo- and hyperthyroidism on regional monoamine metabolism in the adult brain. *Neuroendocrinology* 24, 55-64 (1977).

Jackson, L.M. Histological changes in rat thyroid with age. *Amer. J. Anat.* 19, 305-352 (1916).

Jayne, E.P. Cytology of the adrenal gland of the rat at different ages.

Anat. Rec. 115, 459-483 (1953).

Jeanjean, M., Dehez-Delhaye, M., and Gommers, A. The effect of age on the *in vitro* insulin response in the rat. II. Glucose metabolism in epididymal adipose tissue. *Gerontology* 23, 127-133 (1977).

Jelinkova, M. and Hruza, Z. Decreased effects of NE and GH on the release of free fatty acids in old rats. *Physiol. Bohemoslov.* 13, 327-332 (1964).

Jennings, A.S., Ferguson, D.C., and Utiger, R.D. Regulation of the conversion of thyroxine to triiodothyronine in the perfused rat liver. *J. Clin. Invest.* 64, 1614-1623 (1979).

Jensen, H.K. and Blichert-Toft, M. Serum corticotrophin, plasma cortisol and urinary excretion of 17-ketogenic steroids in the elderly (age group: 66-94). *Acta Endocrinol.* 74, 511-523 (1971).

Johnson, H.D., Ward, M.W., and Kibler, H.H. Heat and aging effects on thyroid function of male rats. *J. Appl. Physiol.* 21, 689-694 (1966).

Jordan, S.W. and Perley, M.J. Microangiopathy in diabetes mellitus and aging. *Arch. Path.* 93, 261-265 (1972).

Jorpes, E., and Rastgeldi, S. Insulin content of human pancreas. *Acta Physiol. Scandinav.* 29, 163 (1954).

Kalish, M.I., Katz, M.S., Pineyro, M.A., and Gregerman, R.I. Epinephrine and glucagon-sensitive adenylate cyclases of rat liver during aging: evidence for membrane instability associated with increased enzymatic activity. *Biochim. Biophys. Acta* 483, 452-466 (1977).

Kalk, W.J., Vinik, A.I., Pimstone, B.L., and Jackson, W.P.U. Growth hormone response to insulin hypoglycemia in the elderly. *J. Gerontol.* 28, 431-433 (1973).

Katz, J.L., Weiner, H., Gallagher, T.F., and Hellman, L. Stress, distress and ego defenses: psychoendocrine response to impending breast tumor biopsy, in *Stress and Coping,* A. Monat and R.S. Lazarus, eds. Columbia University Press, New York (1977), pp. 228-243.

Kay, M.M.B., Mendoza, J., Diven, J., Denton, T., Union, N., and Lajiness, M. Age-related changes in the immune system of mice of eight medium and long-lived strains and hybrids. *Mech. Ageing Dev.* 11(5&6), 295-362 (1979).

Kent, S. *Life Extension Revolution.* William H. Morrow, New York (1980), p. 465.

Kerr, W.J., Hosford, G.N., and Shepardson, M.C. Treatment of senile cataract with thyroid extract. *Endocrinology* 10, 126-144 (1926).

Khelimskii, A.M. Age changes in dimensions of adrenal cortex. *Federation Proc.* (Translation Suppl.) 23, T1250-T1252 (1964).

Kilo, V., Vogler, N.J., and Williamson, J.R. Muscle capillary basement membrane changes related to aging and diabetes mellitus. *Diabetes* 21, 881-905 (1972).

Kimmerling, G., Javorski, C., and Reaven, G.M. Aging and insulin resistance in a group of nonobese male volunteers. *J. Am. Geriatr. Soc.* 25, 349-353 (1977).

Kissebah, A.H., Tulloch, B., and Vydelingum, N. Insulin resistance in Werner's syndrome, a postreceptor defect. *American Diabetes Assoc.,* Abstract 177, 39th Annual Meeting, July (1979), *Diabetes* 28 (Suppl.) p. 389.

Kitay, J.I, and Altschule, M.D. *The Pineal Gland.* Harvard University Press, Cambridge, Mass. (1954), p. 280.

Klein, J.J., Goldstein, A.L., and White, A. Effects of the thymus lymphocytopoietic factor. *Ann. N.Y. Acad. Sci.* 135, 485-495 (1966).

Klimas, J.E. Oral glucose tolerance during the life-span of a colony of

rats. *J. Gerontol.* 23, 31-34 (1968).
Klug, T.L. and Adelman, R.C. Evidence for a large thyrotropin and its accumulation during aging in rats. *Biochem. Biophys. Res. Commun.* 77(4), 1431-1437 (1977).
Klug, T.L. and Adelman, R.C. Altered hypothalamic-pituitary regulation of TSH in male rats during aging. *Endocrinology* 104, 1136-1142 (1979).
Klug, T.L., Freeman, C., Karoly, K., and Adelman, R.C. Altered regulation of pancreatic glucagon in male rats during aging. *Biochem. Biophys. Res. Commun.* 89, 907-912 (1979).
Kohn, R.R. *Principles of Mammalian Aging*, 2nd ed. Prentice-Hall, Englewood Cliffs, N.J. (1978), p. 240.
Kohn, R.R. The question of accelerated collagen aging in diabetes mellitus, in *Endocrine Aspects of Aging*, S.G. Korenman, ed. Summary of papers presented at a conference jointly sponsored by the National Institute of Aging and the Endocrine Society, Bethesda, Md., October 18-20 (1979).
Köhrle, J., Ködding, R., Wong, C.C., and Hesch, R.D. Age-dependent changes of thyroxine-deiodination in rat liver. *Acta Endocrinol.* (Suppl.) 225, 20-00 (1979).
Korenchevsky, V. *Physiological and Pathological Aging*, G.H. Bourne, ed. S. Karger, Basel (1961), p. 311.
Korenchevsky, V., Paris, S.K., and Benjamin, B. Treatment of senescence in female rats with sex and thyroid hormones. *J. Gerontol.* 5, 120-157 (1950).
Kovacs, G.L., Bohus, B., Versteeg, D.H.G. Commentary. The effects of vasopressin on memory processes: the role of noradrenergic neurotransmission. *Neuroscience* 4, 1529-1537 (1979).
Kovacs, K., Ryan, N., Horvath, E., Penz, G., Ezrin, C. Prolactin cells of the human pituitary gland in old age. *J. Gerontol.* 32, 534-540 (1977).
Kovacs, K., Ryan, N., Horvath, E., Singer, W., and Ezrin, C. Pituitary adenomas in old age. *J. Gerontol.* 35(1), 16-22 (1980).
Kumaresan, P. and Turner, C.W. Effect of advancing age on thyroid hormone secretion rate of male and female rats. *Proc. Soc. Exp. Biol. Med.* 124, 752-754 (1967).
Kvetnansky, R., Jahnova, E., Torda, T., Strak, V., Balaz, V., and Macho, L. Changes of adrenal catecholamines and their synthesizing enzymes during ontogenesis and aging in rats. *Mech. Ageing Dev.* 7, 209-216 (1978).
Lafontan, M. Inhibition of epinephrine-induced lipolysis in isolated white adipocytes of aging rabbits by increased alpha-adrenergic responsiveness. *J. Lipid Res.* 20, 208-216 (1979).
Landfield, P.W. Manipulation of some brain aging correlates by long-term adrenal hormone administration and adrenalectomy in rats, in *Endocrine Aspects of Aging*, S.G. Korenman, ed. Summary of papers presented at a conference jointly sponsored by the National Institute on Aging and the Endocrine Society. Bethesda, Md., October 18-20 (1979).
Landfield, P.W., Waymire, J.C., and Lynch, G. Hippocampal aging and adrenocorticoids: quantitative correlations. *Science* 202, 1098-1102 (1978).
Lane, P.W. and Dickie, M.M. The effect of restricted food intake on the lifespan of genetically obese mice. *J. Nutr.* 64, 549-554 (1958).
Larsen, P.R., Dick, T.E., Markowitz, B.P., Kaplan, M.M., and Gard, T.G. Inhibition of intrapituitary thyroxine to 3,5,3'-triiodothyronine conversion prevents the acute suppression of thyrotropin release by thyroxine in hypothyroid rats. *J. Clin. Invest.* 64, 117-128 (1979).

Latham, K.R., and Tseng, Y.L. Nuclear thyroid hormone receptors and serum thyronine levels in aging rats. *Proc. Endocr. Soc.*, 61st Annual Meeting, Abstract. Anaheim, Ca., June 13-15 (1979), p. 108.

Laube, H., Gussgaenger, R.D., and Pfeiffer, E.F. Paradoxical glucagon release in obese hyperglycemic mice. *Horm. Metab. Res.* 6, 426-000 (1974).

Lauson, H.D. Metabolism of the neurohypophysial hormones, in *Handbook of Physiology*, Section 7: Endocrinology, Vol. IV. The Pituitary Gland and its Neuroendocrine Control, Part 1. E. Knobil and W.H. Sawyer, eds. William and Wilkins, Baltimore, Md. (1974), pp. 287-393.

Lazarus, L. and Young, J.D. Radioimmunoassay of human growth hormone in serum using ion-exchange resin. *J. Clin. Endocrinol. Metab.* 26, 213-218 (1966).

Lederer, J. and Bataille, J.P. Senescence et fonction thyroidienne. *Ann. Endocrinol.* (Paris) 30, 598-603 (1969).

Legros, J.L., Gilot, P., Seron, X., Claessens, J., Adams, A., Morglen, J.M., Audibert, A., Berchier, P. Influence of vasopressin on learning and memory. *Lancet* 1, 41-42 (1978).

Lemarchand-Beraud, T., and Vanotti, A. Relationships between blood thyrotropin level, protein bound iodine and free thyroxine in man under normal physiological conditions. *Acta Endocrinol. (Kbh.)* 60, 315-326 (1969).

Lesser, G.T. and Markovsky, J. Body water compartments with human aging using fat-free mass as the reference standard. *Am. J. Physiol.* 236, R215-R220 (1979).

Lesser, G.T., Deutsch, S., and Markovsky, J. Fat-free mass, total body water, and intracellular water in the aged rat. *Am. J. Physiol.* 238, R82-R90 (1980).

Levine, R. Relation of diabetes to aging and atherosclerosis, in *Physiology and Cell Biology of Aging (Aging, Vol. 8)*, A. Cherkin, ed. Raven Press, New York (1979), pp. 95-98.

Lewis, B.K. and Wexler, B.C. Serum insulin changes in male rats associated with age and reproductive activity. *J. Gerontol.* 29, 139-144 (1974).

Livingston, J.W., Cuatrecasas, P., and Lockwood, D.H. Studies of glucagon resistance in large rat adipocytes: ^{125}I-labelled glucagon binding and lipolytic capacity. *J. Lipid Res.* 151, 26-32 (1974).

Lockwood, D.H. and East, L.E. ^{125}I-glucagon binding by liver membrane from young and adult rats. *Diabetes* 27, 589-591 (1978).

Luetscher, J.A., Cohn, A.O., Camargo, C.A., Dowdy, A.J., and Callaghan, A.M. Aldosterone secretion and metabolism in hyperthyroidism and myxedema. *J. Clin. Endocrinol.* 23, 874-880 (1963).

Lymangrover, J., Saffran, M., and Matthews, E.K. Developmental changes in rat adrenocortical cell membrane potential. *Mech. Ageing Dev.* 8, 377-382 (1978).

Machado-Salas, J., Scheible, M.E., and Scheible, A.B. Morphological changes in the hypothalamus of the old mouse. *Exp. Neurol.* 57, 102-111 (1977).

Madsen, S.N. and Sonne, O. Increase of glucagon receptors in hyperthyroidism. *Nature* 262, 793-795 (1976).

Makinodan, T. The thymus in aging, in *Geriatric Endocrinology (Aging, Volume 5)*, R.B. Greenblatt, ed. Raven Press, New York (1978), pp. 217-230.

Makman, M. Changes in receptors during aging, in *Neural Regulatory Mechanisms During Aging*. Alan R. Liss, Inc., New York (1980) (In press.)

Makman, M.H., Ahn, H.S., Thal, L.J., Dvorkin, B., Horowitz, S.G.,

Sharpless, N.S., and Rosenfield, M. Biogenic amine in stimulated adenylate cyclase and spiroperidol-binding sites in rabbit brain: Evidence for selective loss of receptors with aging, in *Parkinson's Disease II, Advances in Experimental Medicine and Biology, Volume 113,* C.E. Finch, D.E. Potter, and A.D. Kenny, eds., Plenum Press, New York (1978), p. 211.

Malbon, C.C., Li, S.Y., and Fain, J.N. Hormonal activation of glycogen phosphorylase in hepatocytes from hypothyroid rats. *J. Biol. Chem.* 253, 8820-8825 (1978).

Margules, D.L. The obesity of middle age: a common variety of Cushing's syndrome due to a chronic increase in adrenocorticotrophin (ACTH) and beta-endorphine activity, in *Modulators, Mediators and Specifiers in Brain Function, Advances in Experimental Biology and Medicine, Volume 116,* Y.H. Ehrlich, J. Volavka, L.G. Davis, and E.G. Brumigraber eds. Plenum Press, New York (1979), pp. 270-290.

Mayberry, W.E., Gharil, H., Bilshel, J.M., and Sizmore, G.W. Radioimmunoassay for human thyrotrophin, clinical value in patients with normal and abnormal thyroid function. *Ann. Int. Med.* 74, 471-480 (1971).

McCool, M.D., Gruol, D., Tolmasoff, J.M., Cutler, R.G. Relative number of chromatin bound T_3 receptors in liver from both normal and hypophysectomized rats as a function of age. 9th Annual Meeting, American Aging Association, Abstract, Washington, D.C. September 20-22 (1979), p. 6.

McGuire, E.A., Tobin, J.D. Berman, M., and Andres, R. Kinetics of native insulin in diabetic, obese, and aged men. *Diabetes* 28, 110-120 (1979).

Meites, J., Huang, H.H., and Simpkins, J.W. Recent studies on neuroendocrine control of reproductive senescence in rats, in *The Aging Reproductive System,* E.L. Schneider, ed. Raven Press, New York (1978), pp. 213-235.

Migeon, C.J., Keller, A.R., Lawrence, B., and Shepard, T.H. Dehydroepiandrosterone and androsterone levels in human plasma. Effect of age and sex; day to day and diurnal variations. *J. Clin. Endocrinol. Metab.* 17, 1051-1062 (1957).

Moncloa, F., Gomez, R., and Pretell, E. Response to corticotrophin and correlation between secretion of creatinine and urinary steroids and between the clearance of creatinine and urinary steroids in aging. *Steroids* 1, 437-444 (1963).

Morrison, A.B. and Staroscik, R.N. The neurosecretory substance in the neurohypophysis of the rat during maturation and aging. *Gerontologia* 9, 65-70 (1964).

Mustacchi, P.O. and Lowenhaupt, E. Senile changes in the histologic structure of the thyroid gland. *Geriatrics* 5, 268-273 (1950).

Narang, G.D. and Turner, C.W. Effect of advancing age on thyroid hormone secretion rate of female rats. *Proc. Soc. Exp. Biol. Med.* 121, 203-205 (1966).

Nonaka, K. and Tarui, S. Aging and endocrine pancreas. *Folia Endocrinol. Japan* 53(12), 1321-1327 (1977).

Noth, R.H., Lassman, N., Tan, S.Y., Fernandez-Cruz, A., and Mulrow, P.J. Age and the renin-aldosterone system. *Arch. Intern. Med.* 137, 1414-1417 (1977).

Oddie, T.H., Meade, J.H., and Fisher, D.A. An analysis of published data on thyroxine turnover in human subjects. *J. Clin. Endocrinol. Metab.* 26, 425-436 (1966).

Oddie, T.H., Myhill, J., Pirnique, F.G., and Fisher, D.A. Effect of age and sex on the radioactive iodide uptake in euthyroid subjects.

J. Clin. Endocrinol. Metab. 28, 776-782 (1968).

Ohara, H., Kobayaki, T., Shiraishi, M., and Wada, T. Thyroid function of the aged as viewed from the pituitary thyroid system. Endocrinol. Japan 21, 377-386 (1974).

Olefsky, J.M. Insensitivity of large rat adipocytes to the antilipolytic effects of insulin. J. Lipid Res. 18, 459-464 (1977).

Olefsky, J.M. and Reaven, G.M. Effects of age and obesity on insulin binding to isolated adipocytes. Endocrinology 96, 1486-1498 (1975).

Oliveros, J.C., Jandali, M.K., Jinsit-Berthier, M., Remy, R., Benghezal, A., Audibert, A., and Moeglen, J.M. Vasopressin in amnesia. Lancet 1, 42 (1978).

Olsen, T., Laurberg, P., and Weeke, J. Low serum triiodothyronine and high serum reverse triiodothyronine in old age: an effect of disease not age. J. Clin. Endocrinol. Metab. 47, 1111-1115 (1978).

Ooka, H. Changes in extrathyroidal conversion of thyroxine (T_4) to $3,3',5'$-triiodothyronine (T_3) in vitro during development and ageing of the rat. Mech. Ageing Dev. 105, 151-156 (1979).

Orskov, H. and Christensen, N.J. Plasma disappearance rate of injected human insulin in juvenile diabetes, maturity onset diabetic and nondiabetic subjects. Diabetes 18, 653-659 (1969).

Osterby, R. Microangiopathy in diabetics of different etiologies, in The Genetics of Diabetes Mellitus. W. Creutzfeldt, J. Kobberling, and J.V. Neel, eds. Springer-Verlag, Berlin (1976), pp. 203-214.

Otani, T., Gyorkey, F., and Farrel, G. Enzymes of the human pineal. J. Clin. Endocrinol. Metab. 28, 349-354 (1968).

Otten, U. and Thoenen, H. Selective induction of tyrosine hydroxylase and dopamine beta-hydroxylase in sympathetic ganglia in organ culture: role of glucocorticoids as modulators. Mol. Pharmacol. 12, 353-361 (1976).

Panda, J.N. and Turner, C.W. Effect of advancing age thyrotropin content of the pituitary and blood of the rat. Proc. Soc. Exp. Biol. Med. 124, 711-714 (1967).

Parker, R.J., Berkowitz, B.A., Lee, C.H., and Denckla, W.D. Vascular relaxation, aging and thyroid hormones. Mech. Ageing Dev. 8, 397-405 (1978).

Pasteels, J.L., Gausset, P., Danguay, A., Ectors, F., Nicoll, C.S., and Varavudhi, P. Morphology of the lactotropes and somatotropes of man and rhesus monkeys. J. Clin. Endocrinol. Metab. 34, 959-967 (1972).

Pecile, A., Muller, E., Falconi, G., and Martini, C. Growth hormone-releasing activity of hypothalamic extracts at different ages. Endocrinology 77, 241-246 (1965).

Peng, T.C. and Garner, S.C. Parathyroidectomy reverses the fall in serum thyroxine associated with advancing age in the rat. Proc. Endocrine Soc., Abstract, 61st Annual Meeting, Anaheim, Ca., June 13-15 (1979), p. 108.

Peng, T.C, Cooper, C.W., and Garner, S.C. Thyroid and blood thyrocalcitonin concentrations and C-cell abundance in two strains of rats at different ages. Proc. Soc. Exp. Biol. Med. 153, 268-272 (1976).

Perlmutter, M. and Riggs, D.S. Thyroid collection of radioactive iodide and serum protein-bound-iodine concentration in senescence, in hypothyroidism, and hypopituitarism. J. Clin. Endocrinol. Metab. 9, 430-439 (1949).

Petrovic, J.S. and Markovic, R.Z. Changes in cortisol binding to soluble receptor proteins in rat liver and thymus during development and aging. Devel. Biol. 45, 176-182 (1975).

Piantelli, L., Basso, A., Muzzioli, M., and Fabris, M. Thymus-dependent reversibility of physiological and isoporterenol evoked age-related parameters in athymic (nude) and old normal mice. *Mech. Ageing Dev.* 7, 171-182 (1978).

Pittman, J.A. The thyroid and aging. *J. Amer. Geriatr. Soc.* 10, 10-25 (1962).

Polenov, A.V. and Senchik, J.S. Synapses on neurosecretory cells of the supraoptic nucleus in white mice. *Nature* 211(5056), 1423-1424 (1966).

Ponsin, G., Poncet, C., and Mornex, R. Accumulation of a large component related to thyrotropin subunits in the pituitary of thyroidectomized rats. *Biochem. Biophys. Res. Commun.* 89, 1135-1140 (1979).

Preiksaitis, H.G. and Kunos, G. Adrenoreceptor-mediated activation of liver glycogen phosphorylase: effects of thyroid state. *Life Sciences* 24, 35-42 (1979).

Puri, S.K. and Volicer, L. Effect of aging on cyclic AMP levels and adenylate cyclase and phosphodiesterase activities in the rat corpus striatum. *Mech. Ageing Dev.* 6, 53-58 (1977).

Quimby, E.H., Werner, S.C., and Schmidt, C. Influence of age, sex, and season upon radioactive iodide uptake by the human thyroid. *Proc. Soc. Exp. Biol. Med.* 75, 537-540 (1950).

Rahman, Y.E., and Peraino, C. Effects of age on enzyme adaption in male and female rats. *Exp. Gerontol.* 8(2), 93-100 (1973).

Raisz, L. Hormonal regulation of skeletal turnover, in *Endocrine Aspects of Aging*, S.G. Korenman, ed. Summary of papers presented at a conference jointly sponsored by the National Institute of Aging and the Endocrine Society. Bethesda, Md., October 18-20 (1979).

Randle, P.J., Garland, P.B., and Hales, C.N. The glucose-fatty acid cycle. Its role in insulin sensitivity and the metabolic disturbances of diabetes mellitus. *Lancet* 1, 785-789 (1963).

Reaven, E.P., Gold, G., and Reaven, G.M. Effect of age on glucose-stimulated insulin release by the beta-cell of the rat. *J. Clin. Invest.* 64, 591-599 (1979).

Reckless, J.P.D., Gilbert, D.H., and Galton, D.J. Alpha-adrenergic receptor activity, cyclic AMP, and lipolysis in adipose tissue of hypothyroid man and rat. *J. Endocrinol.* 68, 419-430 (1976).

Reichel, W. Lipofuscin pigment accumulation and distribution in five rat organs as a function of age. *J. Gerontol.* 23, 145-155 (1968).

Riegle, G.D. Chronic stress effects on adrenocortical responsiveness in young and aged rats. *Neuroendocrinology* 11, 1-10 (1973).

Riegle, G.D. and Hess, G.D. Chronic and acute dexamethasone suppression of stress activation of the adrenal cortex in young and aged rats. *Neuroendocrinology* 9, 175-187 (1972).

Riegle, G.D. and Nellor, J.E. Changes in adrenocortical function during aging in cattle. *J. Gerontol.* 22, 83-87 (1967).

Riegle, G.D., Przekop, F., and Nellor, J.E. Changes in adrenocortical responsiveness to ACTH infusion in aging goats. *J. Gerontol.* 23, 187-190 (1968).

Riegle, G.D., Meites, J., Miller, A.E., and Wood, S.M. Effect of aging on hypothalamic LH-releasing and prolactin inhibiting activites and pituitary responsiveness to LHRH in the male laboratory rat. *J. Gerontol.* 32, 13-18 (1977).

Riggs, B.L., Gallagher, J.C., DeLuca, H.F., Edis, A.J., Lambert, P.W., and Arnaud, C.D. A syndrome of osteoporosis, increased serum immunoreactive parathyroid hormone, and inappropriately low serum 1,25-dihydroxyvitamin D. *Mayo Clin. Proc.* 53(11), 701-706 (1978).

Robertson, G.L., Rowe, J., Helderman, H., and Andres, R. The effect of aging on the regulation of vasopressin secretion, in *Endocrine Effects of Aging*, S.G. Korenman, ed. Summary of papers presented at a conference sponsored jointly by the National Institute on Aging and the Endocrine Society. Bethesda, Md., October 18-20 (1979).

Robertson, T.B. The influence of thyroid alone and of thyroid administered together with nucleic acids upon the growth and longevity of the white mouse. *Aust. J. Exp. Biol. Med. Sci.* 5, 69-88 (1928).

Rodeck, H., Lederis, K., and Heller, H. The hypothalamo-neurohypophysial system in old rats. *J. Endocrinol.* 21, 225-228 (1960).

Romanoff, L.P., Baxter, M.N., Thomas, A.W., and Ferrechio, G.B. Effect of ACTH on the metabolism of pregnenolone-7α-^{3}H and cortisol 4-^{14}C in young and elderly men. *J. Clin. Endocrinol.* 29, 819-830 (1969).

Romanoff, L.F., Morris, C.W., Welch, P., Rodriguez, R.M., and Pincus, G. The metabolism of cortisol-4-^{14}C in young and elderly men. I. Secretion rate of cortisol and daily excretion of tetrahydrocortisol, allotetrahydrocortisol, tetrahydrocortisone and coratolone (20α and 20β) *J. Clin. Endocrinol.* 21, 1413-1425 (1961).

Roof, B.S., Piel, C.F., Hansen, J., and Fudenberg, H.H. Serum parathyroid hormone levels and serum calcium levels from birth to senescence. *Mech. Ageing Dev.* 5, 289-304 (1976).

Root, A.W. and Oski, F.A. Effects of human growth hormone in elderly males. *J. Gerontol.* 24, 97-104 (1969).

Rosenbloom, A.L., Goldstein, S., and Yip, C.C. Insulin binding to cultured human fibroblasts increase with normal and precocious aging. *Science* 193, 415-417 (1976).

Ross, M.H. Nutrition and longevity in experimental animals, in *Nutrition and Aging*, M. Wineck, ed. John Wiley & Sons, New York (1976) pp. 43-57.

Roth, G.S. Age-related changes in specific glucocorticoid binding by steroid responsive tissues of rats. *Endocrinology* 94, 82-90 (1974).

Roth, G.S. Age-related changes in glucocortioid binding by rat splenic leukocytes — possible cause of altered adaptive responsiveness. *Federation Proc.* 34, 83-85 (1975).

Roth, G.S. Reduced glucocorticoid binding site concentration in cortical neuronal perikarya from senescent rats. *Brain Res.* 107, 345-354 (1976).

Roth, G.S. Hormone action during aging: alterations and mechanisms. *Mech. Ageing Dev.* 9, 497-514 (1979a).

Roth, G.S. Hormone receptor changes during adulthood and senescence: Significance for aging research. *Federation Proc.* 38, 1910-1914 (1979b).

Roth, G.S. and Livingston, J.N. Reductions in glucocorticoid inhibition of glucose oxidation and presumptive glucocorticoid receptor content in rat adipocytes during aging. *Endocrinology* 99, 831-839 (1976).

Rubenstein, H.A., Butler, V.P., and Werner, S.C. Progressive decrease in serum T_3 with human aging: RIA following extraction of serum. *J. Clin. Endocrinol. Metab.* 37, 347-353 (1973).

Ruderman, N.B., Toews, C.J., and Shafner, K. Role of free fatty acids in glucose homeostasis. *Arch. Intern. Med.* 123, 299-313 (1969).

Rudman, D., Dhyatte, S., Patterson, J., Gerron, G., O'Beime, I., Barlow, J., Ahmann, P., Jordan, A., and Mosteller, R. Observations on the responsiveness of human subjects to human growth hormone. *J. Clin. Invest.* 50, 1941-1949 (1971).

Runyan, K., Duckworth, W.C., Kitabchi, A.E., and Huff, G. The effect of age on insulin-degrading activity in rat tissue. *Diabetes* 28, 324-325 (1979).

Ryan, N., Kovacs, K., and Ezrin, C. Thyrotrophs in old age. An immunocytologic study of human pituitary glands. *Endokrinologie* 73(2), 191-198 (1979).

Sachar, E.J., Finkelstein, J., and Hellman, L. Growth hormone responses in depressive illness. *Arch. Gen. Psych.* 25, 263-269 (1971).

Saksena, S.K. and Lau, I.F. Variation in serum androgens, estrogens, progestins, gonadotropins, and prolactin levels in male rats from prepubertal to advanced age. *Exp. Aging Res.* 5(3), 179-194 (1979).

Salans, L.B. and Dougherty, J.W. The effect of insulin upon glucose metabolism by adipose cells of different size. Influence of cell lipid and protein content, age and nutritional state. *J. Clin. Invest.* 50, 1399-1410 (1971).

Samuels, L.T. Effects of aging on steroid metabolism, in *Hormones and the Aging Process*, E.T. Engle and G. Pincus, eds. Academic Press, New York (1956), pp. 21-33.

Sandberg, H., Yoshime, N., Maeda, S., Symonds, D., and Zavodnick, J. Effects of an oral glucose load on serum immunoreactive insulin, free fatty acid, growth hormone and blood sugar levels in young and elderly subjects. *J. Am. Geriatr. Soc.* 21, 433-439 (1973).

Sartin, J.L., Pritchett, J.P., and Marple, D.N. TSH, theophylline, and cyclic AMP: *in vivo* thyroid activity in aging rats. *Mol. Cell. Endocrinol.* 9, 215-222 (1977).

Scazzigga, B., Lemarchang-Beraud, T., and Vanotti, A. Problemes de geriatrie, in *Problems de Geriatrie*, Sandoz, Paris (1968), p. 15.

Schapiro, S. and Percin, C.J. Thyroid hormone induction of α-glycerophosphate dehydrogenase in rats of different ages. *Endocrinology* 79, 1975-1978 (1966).

Scharrer, E. The maturation of the hypothalamo-hypophyseal neurosecretory system in the dog. *Anat. Rec.* 118, 437-438 (1954).

Schocken, D.D. and Roth, G.S. Reduced β-adrenergic receptor concentrations in aging man. *Nature* 267, 856-858 (1977).

Schwartz, H.L., Forciea, M.A., Mariash, C.N., and Oppenheimer, J.H. Age-related reduction in response of hepatic enzymes to 3,5,3'-triiodothyronine administration. *Endocrinology* 105, 41-46 (1979).

Schwartz, P., Kurucz, J., and Kurucz, A. Fluorescence microscopy demonstration of cerebrovascular and pancreatic insular amyloid in presinile and senile states. *J. Am. Geriatr. Soc.* 13, 199-205 (1965).

Segall, P.E. Interrelations of dietary and hormonal effects in aging. *Mech. Ageing Dev.* 9, 515-525 (1979).

Segall, P.E. and Timiras, P.S. Patho-physiologic findings after chronic tryptophan deficiency in rats: a model for delayed growth and aging. *Mech. Ageing Dev.* 5, 109-124 (1976).

Seifert, G. Zur Orthologie und Pathologie des qualitativen Inselzellbilder (nach Bensley-Terbruggen). *Virchows Arch.* 325, 379-385 (1954).

Seltzer, H.S. Diagnosis of diabetes, in *Diabetes Mellitus, Theory and Practice*, M. Ellenberg and H. Rifkin, eds. McGraw-Hill, New York (1970), pp. 436-507.

Selye, H. *The Story of the Adaptation Syndrome*. Acta, Inc. Montreal (1952), p. 225.

Selye, H. and Tuchweber, B. Stress in relation to aging and disease, in *Hypothalamus, Pituitary and Aging*. A.V. Everitt and J.H. Burgess, eds. Charles C. Thomas, Springfield, Ill. (1976), pp. 553-569.

Serio, M., Piolanti, P., Capelli, G., DeMagistris, L., Ricci, F., Anzalone, M., and Guisti, G. The miscible pool and turnover rate of cortisol in the aged, and variations in relation to time of day. *Exp. Gerontol.* 4, 95-101 (1969).

Serio, M., Piolanti, P., Romano, S., DeMagistris, L., and Giusti, G. The circadian rhythm of plasma cortisol in subjects over 70 years of age. *J. Gerontol.* 4, 95-97 (1970).

Setoguti, T. Electron microscopic studies of the parathyroid gland of senile dogs. *Am. J. Anat.* 148, 65-83 (1977).

Shaar, C.J., Euker, J.S., Riegle, G.D., and Meites, J. Effects of castration and gonadal steroids on serum luteinizing hormone and prolactin in old and young rats. *J. Endocrinol.* 66, 45-51 (1975).

Simpkins, J.W., Hodson, C.A., and Meites, J. Differential effects of stress on release of TSH in young and old male rats. *Proc. Soc. Exp. Biol. Med.* 157, 144-147 (1978).

Simpkins, J.W., Estes, K.S., Kalra, P.S., and Kalra, S.P. Age-related alterations in catecholamine and LHRH concentration in brain nuclei of the male rat, in *Endocrine Aspects of Aging*, S.G. Korenman, ed. Summary of papers presented at a conference sponsored jointly by the National Institute on Aging and the Endocrine Soceity. Bethesda, Md., October 18-20 (1979).

Singer, S., Ito, H., and Litwack, G. $^3(H)$ cortisol binding in young and old human liver cytosol proteins *in vivo*. *Int. J. Biochem.* 4, 569-573 (1973).

Siperstein, M.D., Unger, R.H., and Madison, L.L. Studies of muscle capillary basement membranes in normal subjects, diabetic and prediabetic patients. *J. Clin. Invest.* 47, 1973-1999 (1968).

Slack, J., Noble, N., Meade, J.W., and North, W.R.S. Lipid and lipoprotein concentrations in 1604 men and women in working population in northwest London. *Brit. Med. J.* 2(6083), 353-357 (1977).

Synder, P.J. and Utiger, R.D. Response to thyrotropin releasing hormone in normal men. *J. Clin. Endocrinol. Metab.* 34, 1096-1098 (1972).

Solomon, J. and Greep, R.O. Relationship between pituitary growth hormones content and age in rats. *Proc. Soc. Exp. Biol. Med.* 99, 725-727 (1958).

Steger, R.W., Sonntag, W.E., Huang, H.H., and Meites, J. Dimished pulsatile growth hormone release in old male rats, in *Endocrine Aspects of Aging*, S.G. Korenman, ed. Summary of papers presented at a conference sponsored jointly by the National Institute on Aging and the Endocrine Society. Bethesda, Md., October 18-20 (1979).

Stern-Nielsen, L.K. and Friis, T. Age dependence of thyroxine metabolism in euthyroid patients. *Ugeskr. Laeg.* 135, 640-644 (9173).

Stoffer, R.P., Hellwig, C.A., Welch, J.W., and McCusker, E.N. The thyroid gland after age 50. *Geriatrics* 16, 435-443 (1961).

Story, J.A. and Griffith, D.R. Effect of exercise on thyroid hormone secretion rate in aging rats. *Horm. Metab. Res.* 6, 403-406 (1974).

Story, J.A. and Kritchevsky, D. Age-related changes in cholesterol metabolism, in *Liver and Aging*, K. Kitani, ed., Elsevier/North Holland Biomedical Press, New York (1978), pp. 193-202.

Szabo, R., Dzsinich, D., Okros, I., and Stark, E. The ultrastructure of the aged rat zona fasciculata under various stressing procedures. *Exp. Gerontol.* 5, 335-337 (1970).

Talwar, G.P., Hanjan, S.N., and Kidway, Z. Growth hormone action on thymus and limphoid cells, in *Growth Hormone and Related Peptides*, A. Pecile and E.E. Muller, eds. Proc. of the Third Int'l Symposium on Growth Hormone, Milan (1975). *Exerpta Medica 181*, Amsterdam (1976), pp. 104-115.

Tang, F. and Phillips, J.G. Some age-related changes in pituitary-adrenal function in the male laboratory rat. *J. Gerontol.* 33, 377-382 (1978).

Taylor, A.L., Finster, J.L., and Mintz, D.H. Metabolic clearance and production rates of human growth hormone. *J. Clin. Invest.* 48, 2349-2358 (1969).

Taylor, B.B., Thompson, J.A., and Caird, F.I. Further studies of thyroid function tests in the elderly at home. *Age and Ageing* 3, 1220125 (1974).

Thompson, R.G., Rodriguez, A., Kowarski, A., and Blizzard, R.M. Growth hormone: Metabolic clearance rates, integrated concentrations, and production rates in normal adults and the effect of prednisone. *J. Clin. Invest.* 51, 3193-3199 (1972).

Timiras, P.S. *Developmental Physiology and Aging.* McMillan, New York (1972), p. 692.

Timiras, P.S. Biological perspectives on aging: in search of a masterplan. *Am. Sci.* 66, 605-613 (1978).

Timiras, P.S. Physiology of agin, in *Medical Physiology,* Volume II, V.B. Mountcastle, ed. The C.V. Mosby, St. Louis, Mo. (1980), pp. 1986-1999.

Timiras, P.S. and Vaccari, A. Thyroid dysfunction and striatal dopamine receptors. *Brit. J. Pharmacol.* 72, 125-126 (1981).

Tomkins, G.M. and Gelehrter, T.D. The present status of genetic regulation by hormones, in *Biochemical Actions of Hormones,* Volume II, G. Litwack, ed. Academic Press, New York (1972), pp. 1-20.

Turkington, M.R. and Everitt, A.V. The neurohypophysis and aging with special reference to the antidiuretic hormone, in *Hypothalamus, Pituitary and Aging,* A.V. Everitt and J.A. Burgess, eds. Charles C. Thomas, Springfield, Ill. (1976), pp. 123-136.

Tyler, F.H., Eik-Nes, K., Sandberg, A.A., Florentin, A.A., and Samuels, L.T. Adrenocortical capacity and the metabolism of cortisol in elderly patients. *J. Am. Geriatr. Soc.* 3, 79-84 (1955).

Uchimura, H., Nagataki, S., and Ingbar, S.H. Down regulation of TSH receptor by TSH in rats. *Program of the Sixth International Conference of Endocrinology,* Abstract 305. Melbourne, February (1980), p. 362.

Vaccari, A. and Timiras, P.S. Alterations in brain dopamine receptors in developing hypo- and hyperthyroid rats. *Neurochem. Intl.* In press.

Vaccari, A., Valcana, T., and Timiras, P.S. Effects of hypothyroidism on the enzymes for biogenic amines in the developing rat brain. *Pharmacol. Res. Commun.* 9, 763-780 (1977).

Valcana, T. The role of triiodothyronine (T_3) receptors in brain development, in *Neural Growth and Differentiation,* E. Meisami and M.A.B. Brazier, eds. Raven Press, New York (1979), pp. 39-58.

Valcana, T. and Timiras, P.S. Nuclear triiodothyronine receptors in brain and liver of developing and aging rats, in *Hormones in Development,* L. Macho and V. Strbak, eds. Bratislava, Czechoslovakia (1979), pp. 47-76.

Valueva, G.V., and Verzhikovskaya, N.V. Thyrotropic activity of hypophysis during aging. *Exp. Gerontol.* 12, 97-105 (1977).

Vekemans, M. and Robyn, C. Influence of age on serum prolactin levels in women and men. *Brit. Med. J.* 4, 738-739 (1975).

Verhoeven, G.F.M. and Wilson, J.D. The syndromes of primary hormone resistance. *Metabolism* 28: 253-288 (1979).

Vermeulen, A. Sex hormone levels in postmenopausal women: influence of number of years since the menopause, age and weight. *J. Endocrinol.* 77, 2P (1978).

Verzar, F. Anterior pituitary function in age, in *The Pituitary Gland,* Volume 2, G.W. Harris and B.T. Donavan, eds. Butterworth, London (1966), pp. 444-459.

Verzar, V. and Freydberg, F. Changes of thyroid activity in the rat in old age. *J. Gerontol.* 11. 53-57 (1956).

Vidalon, C., Khurana, R.C., Chae, S., Gigick, C.G., Stephan, T., Nolan, S., and Danowski, T.S. Age-related changes in growth hormone in non-diabetic women. *J. Am. Geriatr. Soc.* 21, 253-255 (1973).

Villee, D.B., Berger, R., and Wenniger, N. Insulin responsiveness in aging and progeric fibroblasts. *Proc. Endocrinol. Soc.*, 61st Annual Meeting, Abstract 145. Anaheim, Ca., June 13-15 (1979).

Vracko, R. and Benditt, E.P. Restricted replicative lifespan of diabetic fibroblasts *in vitro:* its relation to microangiopathy. *Federation Proc.* 34, 68-70 (1975).

Walford, R.L., Smith, G.S., Meredith, P.J., and Cheney, K.E. Immunogenetics of aging, in *The Genetics of Aging,* E.L. Schneider, ed. Plenus Press, New York (1978), pp. 383-401.

Walker, R.F. and Timiras, P.S. Pacemaker insufficiency and the onset of aging, in *Cellular Pacemakers,* D.O. Carpenter, ed. John Wiley & Sons, New York (1981).

Walker, R.F., McMahon, K.M., and Pivorum, E.B. Pineal gland structure and respiration as affected by age and hypercaloric diet. *Exp. Gerontol.* 13, 91-99 (1978).

Walton, K.W., Scott, P.J., Dykes, P.W., and Davies, J.W.L. Significance of alterations in serum lipids in thyroid dysfunction. 2. Alterations of metabolism and turnover of ^{131}I-low-density lipoproteins in hypothyroidism. *Clin. Sci.* 29, 217-238 (1965).

Wara, D.W., Goldstein, A.L., Doyle, N.E., Amman, A. Thymosin activity in patients with cellular immunodeficiency. *New Engl. J. Med.* 292, 70-74 (1975).

Weidmann, P., Myttenaere-Bursztein, S., Maxwell, M.H., and Lima, J. Effect of aging on plasma renin and aldosterone in normal man. *Kidney Int.* (Journal of the International Society of Nephrology) 8, 325-333 (1975).

Wenzel, K.W. and Horn, W.R. Triiodothyronine (T_3) and thyroxine (T_4) kinetics in aged men, 7th International Thyroid Conference, Abstract, Boston (1975). *Excerpta Medica,* International Congress Series, 361, 89-00 (1975).

Wenzel, K.W., Meinhold, H., Herpich, M., Adlkofer, F., and Schleusner, H. TRH-stimulation test mit alters-und geschlechsabhangigem TSG-ansteig bei normalpersonen. *Klin. Wochenschr.* 52, 721-727 (1974).

West, C.D., Brown, H., Simons, E.L., Carter, D.B., Kumagai, L.I., and Englert, E., Jr. Adrenocortical function and cortisol metabolism in old age. *J. Clin. Endocrinol. Metab.* 21, 1197-1207 (1961).

Westman, S. Development of the obese-hyperglycemic syndrome in mice. *Diabetologia* 4, 141-149 (1968).

Wexler, B.C. Comparative aspects of hyperadrenocorticism and aging, in *Hypothalamus, Pituitary and Aging,* A.V. Everitt and J.A. Burgess, eds. Charles C. Thomas, Springfield, Ill. (1976), pp. 333-361.

Wilansky, D.L., Newsham, G.S., and Hoffman, M.M. The influence of senescence on thyroid function: functional changes evaluated with I^{131}. *Endocrinology* 61, 327-336 (1957).

Williamson, J.R. and Kilo, C. Current status of capillary basement-membrane disease in diabetes mellitus. *Diabetes* 26, 65-73 (1977).

Wolf, S. and Berle, B.B. (eds.) *Dilemmas in Diabetes.* Plenum Press, New York (1974), pp. 87-95.

Wostmann, B.S. and Bruckner-Kardoss, E. Thyroid hormones in older germ free rats and mice. Abstract. *Federation Proc.* 38(3), 1030 (1979).

Wren, J.C. Thyroid function and coronary atherosclerosis. *J. Am*

Geriatr. Soc. 16, 696-704 (1968).
Wrenchall, G.A., Bogoch, A., and Ritchie, R.C. Extractable insulin of the pancreas. *Diabetes* 1, 87-107 (1952).
Wurtman, R.J. and Axelrod, J. Adrenaline synthesis: control of enzymatic synthesis of adrenaline in the adrenal medulla by adrenal cortical steroids. *J. Biol. Chem.* 241, 2301-2305 (1966).
Wurtman, R.J., Axelrod, J., and Barchas, J.D. Age and enzyme activity in the human pineal. *J. Clin. Endocrinol. Metab.* 24, 299-301 (1964).
Wurtman, R.J., Phorecky, L.A., and Baliga, B.S. Adrenocortical control of the biosynthesis of epinephrine and proteins in the adrenal medulla. *Pharmacol. Rev.* 24, 411-426 (1972).
Yagihashi, S., Goto, Y., Kakizaki, M., and Kaseda, N. Thickening of glomerular basement membrane in spontaneously diabetic rats. *Diabetologia* 15, 309-312 (1978).
Yamaji, T. and Ibayashi, H. Plasma dehydroepiandrosterone sulfate in normal and pathological conditions. *J. Clin. Endocrinol. Metab.* 29, 273-278 (1969).
Yamaji, T., Shimamoto, K., Ishibashi, M., Kosaka, K., and Orimo, H. Effect of age and sex on circulating and pituitary prolactin levels in human. *Acta Endocrinol.* 83, 711-719 (1976).
York, D.A., Bray, G.A., and Yukimura, Y. An enzymatic defect in the obese (ob/ob) mouse: loss of thyroid-induced sodium- and potassium-dependent adenosine triphosphatase. *Proc. Nat. Acad. Sci.* 75, 477-481 (1978).
Yoshino, G., Kazumi, T., Kobayashi, N., Terashi, K., Morita, S., and Baba, S. Insulin and glucagon relationships during aging in rats. *Endocrinol. Japan* 26, 325-329 (1979).
Ziegler, M.G., Lake, C.R., and Kopin, I.J. Plasma noradrenaline increases with age. *Nature* 261, 333-335 (1976).

Copyright©1982, Spectrum Publications, Inc.
Hormones in Development and Aging

Chapter 15
Neuroendocrine Theories of Aging: Homeostasis and Stress
Paola S. Timiras

INTRODUCTION

While the range of hormonal changes in the elderly varies considerably from slight to severe depending on the endocrine gland, the capacity of the organism to adapt to the environment irreversibly declines with advancing age and ultimately fails, resulting in disease and death (see Chapter 14). Hence, the failure of homeostasis—one of the major events concomitant with old age—may be ascribed only in part to quantitative changes (e.g., in hormone levels) in endocrine function. Rather, in view of the crucial role that nervous inputs play in the regulation of homeostatic responses as well as the complex interactions between nervous and endocrine systems repeatedly emphasized throughout this book, the decrement in adaptive capability which is characteristic of old age may be related to impairment of nervous function, in general, and its control of endocrine function, in particular. In parallel with other events of the lifespan that are regulated by neuroendocrine signals, such as growth, development and reproduction, aging may similarly be considered as dependent on neuroendocrine regulation, or alternatively, as consequent to failure of this regulation. Programmed controls (pacemakers) of growth, development and aging, probably located in the hypothalamus (a CNS center important for its endocrine and autonomic functions) or other integrative suprahypothalamic areas (such as the limbic system, important for its endocrine and behavioral functions) would, through the release of specific neurotransmitters, activate or inhibit one or several of the hypothalamo-pituitary-target endocrine axes. Such an action would trigger a variety of nervous, visceral and somatic, and endocrine actions which would mark the time for the passage from one life period to the next (Timiras, 1978). Although the evidence to support the validity of such a "program" is still tentative, similar programs regulating "timed" physiologic conditions are well demonstrated (see Chapter 13). Alternatively, a program for growth and development might exist until the individual has achieved optimal sexual function, but once the survival of the species has been guaranteed, the program would cease to operate. Therefore, after culmination of sexual maturity, the "disorganization" of physiological processes that follows the exhaustion of the program would be responsible for aging and death (Walker and Timiras, in press, b). Regardless of whether or not ongoing and future research will definitively substantiate one or the other of these hypotheses or formulate new ones, the neuroendocrine theories of aging provide a useful framework within which to design experiments to test the involvement of neuroendocrine controls in growth, development and aging. Current data indicate that it is possible to delay growth and maturation, to postpone the onset of reproductive and physiologic aging and to extend the lifespan in some laboratory animals by nutritional (e.g., restriction of total calories or a specific dietary constituent) and pharmacologic (e.g., administration of neurotropic drugs or hormones) intervention, initiated during development or in adulthood which act through neural and/or endocrine mechanisms (Timiras, 1972, 1978; Everitt, 1980; Everitt and Burgess, 1976; Finch, 1977; Segall, 1979; Walker and Timiras, in press,b).

In this chapter, some of the aspects of CNS aging that appear to be currently more relevant to neuroendocrine regulation and alterations in this regulation, with advancing age will be reviewed briefly. In recent years, a considerable body of literature has been accumulating on the morphological, chemical and physiological changes associated with old age in the CNS, and the interested reader is referred to the several books available on this topic. Affective and other CNS (psychiatric) disorders, their relation to hormones and aging, are presented in Chapters 18 and 19. Our purpose, here, is to summarize our present knowledge of the aging of neuroendocrine functions and to discuss the implications of such aging for the decrement in homeostatic competence of the aged individual.

AGING OF THE CENTRAL NERVOUS SYSTEM

Brain Size and Cellular Changes

Morphologic, biochemical, physiologic, and behavioral data on the developing CNS, presented in the preceding chapters (see Chapter 5-7) indicate that neurons, fixed in number at birth in humans and several other mammalian species, undergo postnatal changes (e.g., increase in cell body size, in dendritic branching and spines, in synaptic membranes and myelin) associated with the progressive acquisition of more efficient and complex endocrine and behavioral functions as the animal matures. The timetable and pattern of these changes is influenced by a variety of external (e.g., environmental) and internal (e.g., hormonal) factors, thereby attesting to the "plasticity" of the developing CNS. With aging, a moderate decrease in the weight of the brain has been described in several animal species. This finding which had first been demonstrated with autopsy material has now been confirmed in patients with the use of computerized tomography which measures, in addition to brain mass, also the subarachnoid compartment and the ventricular size (Roberts and Caird, 1976; deLeon et al., 1980; Yamaura et al., 1980). The progressive loss of brain weight with age has been and still is erroneously considered by some to be due to a catastrophic loss of neurons involving the entire brain and leading to atrophy, degeneration and dementia. While neuronal loss does occur with aging, it is limited to specific areas and cannot be related to cognitive decrements, for we know that variations in brain size among the general population do not account for individual variations in learning, memory and special talents. The clinical significance of cerebral atrophy as an index of the severity and progression of brain degenerative diseases (e.g., senile dementia, Alzheimer's disease) is also discounted inasmuch as severe cases of dementia are associated with little brain atrophy and, vice versa, significant decrements in size are not necessarily associated with a significant loss of function.

Senile Dementia and Alzheimer's Disease. Both diseases, characterized by profound impairment of behavior, memory and overall mental function, are briefly considered here in view of recent evidence that they may be induced by environmental factors acting, perhaps, on a predisposing genetic background. Thus, in the old as in the young, the extreme vulnerability of the CNS to environmental influences would modify the timetable of aging changes. Furthermore, in the old as in the young, manipulations of the environment may damage or may nurture CNS function and thereby accelerate or delay the pathological concomitants of CNS aging.

Dementia, a genetic designation for mental deterioration, represents a condition of high risk to a large percentageo of the aged population. It is often, uncorrectly equated with "senility" to indicate "the most dehumanizing of diseases leaving a functioning organism without a mind" (Liss, 1979). Naming this condition "senility" implies that it is a "natural" correlate of aging (from the Latin senium: old age). This is not the case, however, and the many older individuals who retain and, in fact, continuously augment their wisdom, clearly contradict the inevitability of mental deterioration with advancing age. Current research tends to view senile dementia as a disease process and to focus, therefore, on the detection of predisposing and etiologic factors responsible for its occurrence in a selected population; such studies should lead to a better understanding of the disease and permit the identification and implementation of preventive and corrective treatments.

Considered as a disease, senile dementia presents some analogy to other forms of dementia, including "juvenile" dementia, one form of which is known as Alzheimer's disease. This type of dementia which affects younger individuals (in their thirties or forties) is attracting considerable attention because of the large number of individuals (1 to 2 million in the United States) apparently affected, and because of increasing evidence for the correlation of its etiology to environmental and genetic factors. Alzheimer's disease may be diagnosed, in addition to its clinical picture of progressive mental deterioration, electroencephalographic findings and cerebral atrophy, by two characteristic pathologic findings: the "neurofibrillary tangles," on which extensive chemical and morphologic studies are being conducted, and the less well investigated, "neuritic or neurogenic plaques" (Terry, 1979, 1980). Neurofibrillary tangles, consisting of paired helical protein filaments wrapped around each other, are not exclusive to Alzheimer's disease but represent a non-specific type of pathological response of neurons to chemical toxins and biological agents. In Alzheimer's disease, neurofibrally tangles are not only quantitatively proportional to the severity of the mental deterioration, but also correlate with high aluminum content in several brain areas. This association between neurofibrillary tangles and high aluminum content of brain has prompted renewed efforts in the establishment and clarification of the disease etiology. Aluminum injections into the cerebral tissues of experimental animals result in the formation of neurofibrillary tangles, which, however, show some structural and chemical differences to those occurring naturally, and produce decreased learning ability (Arieff et al., 1979). Inasmuch as fluoride competes with aluminum for absorption in the gut, it has been hypothesized that a high level of fluoride in the community water would protect its residents not only from dental caries but also from the accumulation of aluminum in their brains and, eventually, from dementia (at least of the Alzheimer type).

Other factors implicated in the etiology of Alzheimer's disease either as causative or predisposing agents, include: circulatory factors such as arteriosclerotic involvement of cerebral vessels leading to vascular occlusion, brain ischemia or hemorrhage; infectious factors such as viruses; genetic factors such as the high incidence of Alzheimer's disease in patients with Down's syndrome. The possible role for viral and immunoreactive responses in the etiology of Alzheimer's disease is also reflected in the pathogenesis of the neurogenic plaque, the histological structure other than the tangle, characteristic of the disease. The plaque is believed to be formed from persisting cellular debris and to contain immunoglobulins and amyloid (the latter, extracellularly accumulating by-product of incompletely metabolized antibodies found frequently in animals suffering from wasting diseases). This structure has also been reproduced experimentally by injection of slow (scrapie) virus into mice; however, the plaques' etiology, relationship to the neurofibrillary tangle and role in dementia remain still very little understood.

Even though studies in humans and several other animal species indicate that, in the absence of brain pathology, the overall number of brain cells remains relatively stable well into old age, some loss does occur in specific and discrete brain regions (Brody and Vijayashankar, 1977a,b). Thus, in the brain of quite elderly (90-year-old) individuals a decrease of approximately 40% in cells has been reported in the frontal cortex and more moderate, but still significant, losses of neurons in areas of the tmeporal and occipital lobes, prefrontal gyrus, and brain stem, particularly the locus coeruleus (Brody, 1970, 1978). Since the locus coeruleus is comprised of numerous noradrenergic neurons, the decreased cell number may correlate with the decreased norepinephrine levels reported in this area in the aging mouse (Finch, 1973). In the rat, cell counts in selected areas of the cerebral cortex show that the postnatal reduction in neuronal number occurs at a young age and that, once sexual maturation has been attained, the number of neurons remains essentially constant in the adult and old animal (Brizzee et al., 1968, Connor et al., 1980a,b; Diamond, et al., 1977). In the hypothalamus, although neurons are lost from areas that regulate temperature and gonadotropic secretion, their number remains unaffected in neighboring hypothalamic areas (Hsu and Peng, 1978). On the other hand, glial cells, particularly astrocytes, seem to increase significantly in all the brain areas where they have been counted (Brizzee et al., 1968; Geinisman et al., 1978). The resulting gliosis has been interpreted as compensatory to the loss of neurons and/or the decline in neuronal function suggested by the progressive changes in the structure and metabolism of the neurons with aging. Included as signs of progressive alterations in neural cells (both neurons and glial cells) with advancing age are: accumulation of lipofuscin (i.e., so-called age pigments), loss of Nissl substance (i.e., nucleic acids), appearance of neurofibrillary tangles and neurogenic plaques, and loss of dendrites (see Timiras and Bignami, 1976). Of particular functional significance appears to be the loss of dendrites which results in a concomitant loss of synaptic connections and may be responsible, at least in part, for some of the neurologic deficits (e.g., alterations in spontaneous and evoked electrical activity, impairment of memory, slower reaction time) accompanying old age.

Age Pigments or Lipofuscin. These pigments are so called because of their yellow to deep brown color ("lipo" for the Greek "fat"; "fuscin" from the Latin "dusky") and their conspicuous presence in aging tissues. They are particularly apparent in postmitotic tissues such as cardiac muscle and nervous tissue. For example, in human myocardium, in which their presence has been carefully recorded during the lifespan, age pigments are scant or absent until early adolescence but thereafter accumulate, independent of any cardiac pathology, at a constant rate of approximately 0.6% of the intracellular myocardium volume per decade. Thus, the heart of a person living to be 90 years of age would be expected to have as much as 10% of its intracellular volume occupied by age pigments (Strehler et al., 1959). Age pigments also accumulate in nervous tissue where they have been demonstrated in increasing amounts with advancing age both in neurons and glial cells of a number of brain areas, e.g., cerebral cortex, cerebellum, brain stem (Brizzee et al., 1969; Hasan and Glees, 1972, 1973; Zeman, 1974; Rogers et al., 1980). (See Figs. 1 and 2).

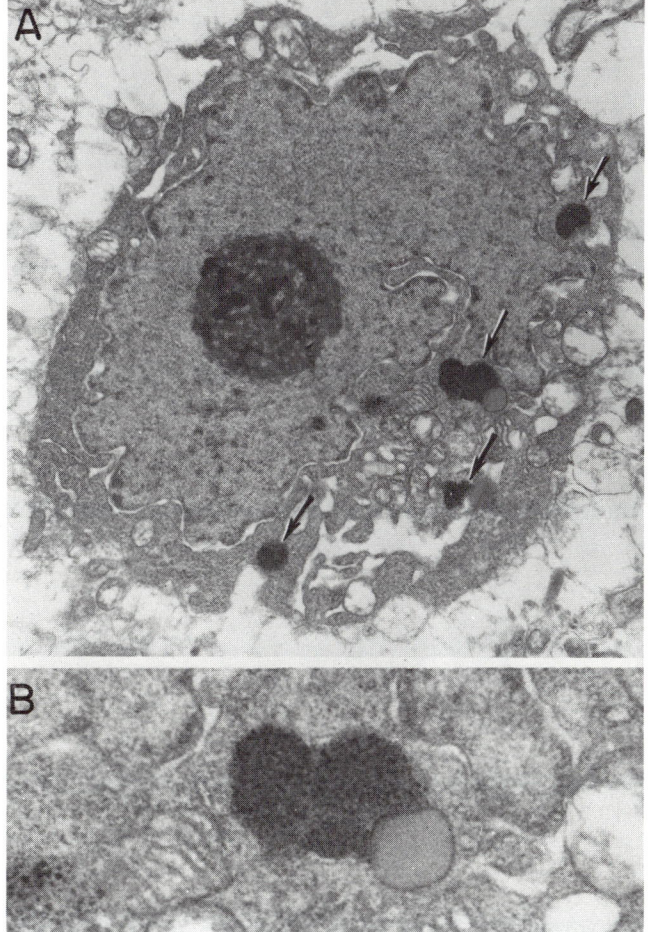

Fig. 1. (A) Electromicrograph of Purkinje cell from the cerebellum of a 24-month-old male Long-Evans rat. Arrows point to major accumulations of age pigments (lipofuscin?) (x 12,120).

(B) At higher magnification (x 19, 185) age pigment (lipofuscin?) appears granular with lipid component.

Courtesy of Dr. J.R. Walton, Lawrence Livermore Laboratory, University of California.

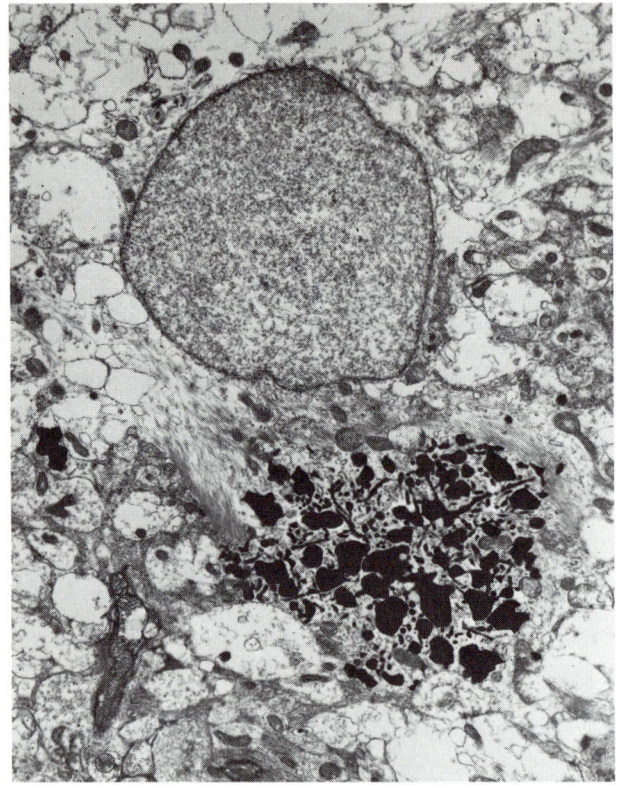

Fig. 2. Electronmicrograph of hypothalamic neuron from a 17-month-old female Long-Evans rat. Abundant microfilaments surround a cytoplasmic area containing pigment depositis, probably melanin (x 13,187). Melanin, a dark pigment, localized, among other tissues, in the brain (substantia nigra) and thought to be a residual product of cellular activity, appears to accumulate intracellularly with aging.
Courtesy of Dr. J.R. Walton, Lawrence Livermore Laboratory, University of California.

Viewed with the light microscope, the pigment appears as round or oblong, yellow or brown granules varying in size and either evenly diffused throughout the cytoplasm or aggregated in perinuclear or polar clusters. In general, the predominant pattern of pigment distribution seems to be related more to the age of the animal than to the type of cell. The deposition appears to progress from diffuse pigment granules, to perinuclear pigment cluster, to polar or axonal pigment aggregations and, finally, to bipolar pigment aggregations in the regions of the axon hillock and principal dendrites. When visualized with the electron microscope, the pigments exhibit a complex, polymorphic internal structure forming vacuoles, round bodies and laminated bodies which resemble myelin. Biochemical analyses of lipofuscin have suggested that it is derived from a lipoprotein complex similar to that found in membranes such as myelin. The pigments are characterized by a high affinity for lipid soluble dyes and a bright fluorescence under ultraviolet light. Because of their essential insolubility in both organic and inorganic solvents, these pigments are believed to be insoluble residues of proteins that become cross-linked; in vitro studies have shown that substances similar to lipofuscin can be produced by perioxidation of lipid and protein (Tappel et al., 1963). Of the many cellular membrane systems suggested as the source of the pigment, lysosomes appear to be the organelles most frequently implicated in the production of such intracellular deposits. Lipofuscin granules contain some of the lytic enzymes found in lysosomes and the accumulation or deficiency of these enzymes in aged tissues has been taken as supportive of the involvement of lysosomal alterations in lipofuscin formation (deDuve, 1966). However, the association of lytic enzymes with age pigments may represent only a transitory stage in the progressive transformation and accumulation of the pigments rather than a definitive cause of pigment production. Similar objections can be raised against other theories linking the formation of lipofuscin to malfunction of other cell organelle. Thus, the suggestion that the pigment originates from mitochondria is based primarily on the correlation of high lipofuscin content with high concentration of mitochondria in neurons of certain brain nuclei (e.g., the mesencephalic nucleus of the trigeminal nerve) (Hasan and Glees, 1972). The pigment formation would be preceded by an increased concentration and clumping of mitochondria resulting in disturbance of normal metabolic activity and the accumulation of insoluble fatty acids within the pigment (Glees and Gopinath, 1973). Whether these changes in mitochondria are the cause or the manifestation of a specific stage of development of lipofuscin cannot be ascertained at present.

Despite the considerable volume of the cell which the age pigments may occupy, the degree to which their accumulation impairs cellular function remains controversial. While there is an intuitive feeling that the accumulation of large amounts of lipofuscin may be harmful, if only because these pigments occupy such a large volume of the cell and thereby displace other cell structures more metabolically active, current evidence does not support this feeling because no generalized impairment of function has been consistently reported in various tissues despite progressive accumulation of pigments. With respect to nervous tissue, even though degeneration of nerve cells has been associated with pigment accumulation sufficiently high to displace the cell nucleus, it cannot be definitely stated that the degeneration is due to excessive accumulation of lipofuscin; it is equally possible that the cell has degenerated due to other causes and that the pigment masses were formed as a result of such degeneration. A rare example of nerve cell degeneration associated with deposition of large amounts of lipofuscin is represented by Batten's disease (juvenile amaurotic idiocy) which is characterized by short lifespan and neural and mental malfunction starting with blindness and failure of mental development and ending in loss of vegetative function (Zeman and Rider, 1975). On the other hand, lipofuscin is present in large amounts in certain brain stem nuclei in which neurons seem to retain a normal function well into old age (Brody, 1976; Rogers et al., 1980).

The presence of lipofuscin in aging organisms is phylogenetically a universal phenomenon, the pigment having been detected in all animal classes — from Protozoa to Chordata — where it has been sought. On the other hand, its distribution in animals is not universal; for example, in humans, although lipofuscin has been found to accumulate in the nervous system, cardiac, skeletal and smooth muscle, liver, spleen, epididymis, seminal vesicle, thymus, pancreas and adrenals, it is not found in other tissues and displays a differential pattern of deposition within a given tissue. In this respect, it is interesting to note that age pigments seem to demonstrate certain cellular specificites. In the adrenal gland of the senile rat, the zona reticularis has five times more pigment than any other zone of that structure; in the ovary, the pigment is localized exclusively in the perifollicular macrophages and not in the ovarian follicles themselves; and, in muscle, lipofuscin seems to be more abundant in muscles concerned with locomotion than those implicated with maintenance of posture (Kny, 1937; Reichel, 1968). Given such cellular and regional differences within a single structure, and the fact that some tissues do not accumulate pigment at all, the assumption that age pigments are a homogeneous group of substances whose accumulation in different cells of the body results from the same process in every case should be questioned.

It has been suggested that the formation of age pigments is caused, or at least, influenced, not only by intrinsic factors which are essentially genetic, but also by extrinsic factors such as drugs, hypoxia, vitamin E deficiency, and cirrhosis. For example, adverse environmental conditions (e.g., hypoxia) during fetal development or at birth may lead to an impairment of neurologic function associated with increased deposition of lipofuscin at an early age. In this case, a decreased ability of neurons to perform an adequate function in storing and transmitting information would be coincident with neurochemical alterations leading to metabolic damage and accelerated accumulation of age pigments (Liss, 1979). Similarly, disturbances in cerebral blood flow as occur in hydrocephalus are consistently accompanied by increased lipofuscin accumulation in cortical neurons, a quite striking observation in view of the young age of the affected individuals (Liss, 1979). In the elderly, atherosclerotic lesions of the cerebral and coronary vessels are responsible for ischemic episodes and progressively less efficient vascularization of nervous or cardiac tissues and the adverse effects of this decline in blood supply could induce or could aggrevate the age-related accumulation of pigments.

Even though the origin, nature and function of age pigments remain for the most part obscure, attempts have been made to reduce the levels of these substances in the brain (and other tissues) with the hope, still rather empirically, of alleviating (or halting, or possibly reversing) the effects of aging, in general, and of CNS aging in particular. Studies by Nandy and collaborators (1978a,b; 1979) have shown an improvement in learning and memory in aging mice following treatment with meclofenoxate (centrophenoxine) a central nervous system stimulant (Nandy and Lal, 1978). Pigment accumulation was markedly reduced in the CNS of the guinea pig following eight or more weeks of treatment with centrophenoxine. The alterations in enzyme activity accompanying this reduction in lipofuscin suggest that the effects of centrophenoxine may be due to a diversion of glucose metabolism through the pentose cycle and effects on lysosomes which are intimately related to the genesis of these pigments (Nandy, 1968a,b). Conversely, a decrement in learning and memory with increased lipofuscin in these neurons was found in mice on a Vitamin E-deficient diet. From such pharmacologic studies, it has been argued that a relation between brain lipofuscin content and neuronal function may exist. However, controphenoxine could affect aging of the brain by other mechanisms besides reduction in lipofuscin. Thus, studies in rats using electromicroscopic stereology, have shown that treatment with this drug prevents the age-related deterioration of synaptic structures (Giuli et al., 1980). Likewise, it is not known whether other psychoneuroleptic drugs (e.g., phenytoin, amphetamine derivatives, Valium) which are purported to improve the mental state of the elderly and some cognitive functions in old animals, also induce a reduction in lipofuscin. Furthermore, in conditions in which the physiologic competence of the animal persists until advanced age and the lifespan is extended (e.g., after severe restriction of total caloric intake or of specific dietary components such as tryptophan), it remains to be established whether the progressive accumulation of lipofuscin is concomitantly arrested or slowed. Further attempts to modify the age pigment content in brain and other tissues must undoubtedly await until additional investigations provide a clearer understanding of lipofuscin formation, structure and function.

Synaptic and Neurotransmitter Changes

Whereas reports in the literature vary regarding the loss of neurons and the compensatory gliosis found in old brains as indicated above, studies on dendritic branching and synaptic structure are consistent in showing impairment and decrease with aging. In rats (Feldman and Dowd, 1974), dogs (Mervis, 1978), and humans (Scheibel et al., 1975), the reduced dendritic branching with old age is associated with apparent loss of dendritic spines and reduction in apical shaft diameter and, in some cases (e.g., Alzheimer's disease), with dendritic tufting (Scheibel and Tomiyasu, 1978). A loss of synapses with advanced age has been reported in the hippocampus, dentate gyrus and lateral vestibular nucleus (see Bondareff, 1980). However, the ability to regenerate dendrites and synapses seems to persist well into old age, albeit at a reduced rate, not only in rats (Sotelo and Palay, 1971; Hinds and McNelly, 1977; Cotman, 1978; Uylings et al., 1978; Cotman and Scheff, 1979) but also in humans (Buell and Coleman, 1979). Indeed, throughout the lifespan, synapses would appear to undergo continuing processes of regression and regeneration and the "denudation" of neurons in old brains may result from a slowing down of regenerative processes rather than an absolute loss of current dendrites and synapses. Synapses in the human frontal cortex were described as unchanged until 74-90 years, an age period during which most activities, both physical and intellectual, become considerably restricted (Cragg, 1975; Huttenlocher, 1979). Whether the slower rate of synaptic regeneration with old age is intrinsic or related to progressively reduced sensory inputs due to aging of the sensory organs, remains to be established. The effects of an "enriched" environment – i.e., a more complex and possibly stimulating environment – on the number, length, diameter and spines of dendrites and the ultrastructure of the synapse are now being investigated to clarify the contribution of environmental factors to synaptic aging (Connor et al., 1980a,b).

If synaptic losses of either a specific or general nature occur during aging, as suggested by morphologic studies, then it may be expected that such losses are reflected in alterations – decrement or compensatory enhancement – of neurotransmitter systems. Data of age-related neurotransmitter changes during brain maturation show, indeed, a close correlation between structural and chemical patterns (see Chapter 7). However, such a correlation cannot be demonstrated as convincingly in the brain of old individuals due in part to technical difficulties, particularly in the sampling and preserving of human brains for chemical analysis. In fact, studies of neurotransmitter levels and metabolism associated with old age are few, even in animals.

A major difficulty in assessing aging of CNS neurotransmission is the necessity to follow changes in the large number of neurotransmitters and putative neurotransmitters demonstrated so far and continuously being augmented by the identification of new candidates (e.g., amino acids, gut-brain peptides, hypothalamic-releasing hormones, pituitary peptides and other peptides) (Snyder, 1980). Neurotransmission must be viewed not as a static function, but rather as a dynamic process responding with great sensitivity to a multitude of intrinsic and extrinsic stimuli; consequently, if, as already suggested, aging results from/in a decline of homeostatic competence, then another difficulty is the selection among the many responses, of the one most critical for continual maintenance of homeostasis. In addition, levels of a single neurotransmitter may be less critical to a specifid brain function than the relative amounts of different neurotransmitters with varying excitatory or inhibitory properties, or the rates of the enzymatic reactions that regulate their synthesis and degradation (turnover). Finally, basic to level, turnover and action of a neurotransmitter are the availability of the substrate necessary for its formation as well as the presence, properties and binding affinity of appropriate receptors. Therefore, a systematic study of the changes in neurotransmission with aging should take into consideration, besides overall CNS chemistry, many aspects – level, distribution, turnover, availability of substrate, relation to other neurotransmitters and receptors – and their evaluation with respect to function and behavior. At present, such a complete study does not exist for any of the neurotransmitter systems; little is known about changes in development and aging of the newly identified (peptides) neurotransmitters and only limited information is at hand on the more classical, cholinergic and monoaminergic transmitters. An example of the complex relationships of neurotransmitter defects, disease processes, and aging is well demonstrated by Parkinsonism, a condition characterized by low levels of dopamine in a discrete brain region and by overt motor abnormalities. This condition may be viewed as a disease entity of still unknown origin that can, occur at all ages. Nevertheless, because its incidence is highest in the aged, it has been suggested that there may be cumulative risk between the disease process and the effects of CNS aging.

Parkinson's Disease is a degenerative brain disorder which represents one of the more frequently occurring CNS diseases in middle and later life and affects about one percent of the population over 50 years of age. Its occurrence before 40 years of age is extremely rare. The etiology is varied ranging from viral to arteriosclerotic, however, the brain area involved is constant. From whatever cause, the dopaminergic cells of the substantia nigra degenerate reducing the activity of the nigrostriatal pathway and lowering the levels of dopamine in the brain of Parkinsonian patients; the consequent disruption in the transmission of nerve impulses causes the neurologic symptoms of the disease (e.g., difficulty in initiating voluntary movements, slowness of movements, muscular rigidity and tremors).

Many neuroregulators seem to work in parallel with or in opposition to one another. In Parkinson's disease there appears to be an imbalance between dopaminergic (inhibitory) and cholinergic (excitatory) systems. Anatomically, the major areas involved are the substantia nigra and the basal ganglia, especially the corpus striatum. Both areas are connected by an afferent (striopetal) pathway mediated by dopamine and an efferent (striofugal) pathway mediated by acetylcholine, gamma-aminobutyric acid and other as yet unidentified neurotransmitters. The earliest treatment of the disorder was the administration of anticholinergic-type drugs and some degree of control was achieved. Current knowledge of neurochemistry of Parkinson's disease led to the idea of replacement of dopamine as the prime therapy (often in combination with anticholinergic drugs). Dopamine itself cannot be given because it does not pass from the blood into the brain; instead, its immediate metabolic precursor, levodopa (L-Dopa) is given. This chemical can enter the brain where it is enzymatically converted to dopamine. This method of replenishing the brain dopamine concentrations of parkinsonian patients works dramatically when L-Dopa is first given. But for many patients, the efficacy of the drug declines after several years, suggesting the degenerative process progresses, with the continued loss of the enzyme necessary for the conversion of L-Dopa to dopamine. The limitations of the treatment and its failure in certain instances, underline the fact that L-Dopa administration corrects only a single neurotransmitter defect in the nigro-striatal system and further research should be directed toward reestablishing a proper balance among all neurotransmitters operating in the system.

Despite the technical and progmatic difficulties mentioned above, some studies are available and suggest that several aspects of central neurotransmission undergo pronounced changes in old age as they do during early maturation (see Chapters 7, 12). Changes with aging in the activity of monoaminergic and cholinergic systems have been demonstrated in discrete brain regions of various animal species, particularly rats and mice (see Timiras and Hudson, 1980). In the hypothalamus, data in humans show a decrement with aging in the activity of enzymes associated with the metabolism of catecholamines, γ-aminobutyric acid and acetylcholine (McGreer and McGreer, 1975). Changes in the cholinergic system with aging involve primarily the activity of the synthesizing enzyme, choline acetyltransferase, which is decreased in selected brain areas and the spinal cord, whereas the hydrolyzing enzyme, acetylcholinesterase, remains essentially unchanged (Valcana and Timiras, 1969; Finch, 1977; Unsworth et al., 1980). The observation that cognitively impaired or demented patients show definite reduction in choline acetyltransferase a activity as well as muscarinic receptor binding (Perry et al., 1977) without change in the number of receptors (Drachman and Leavitt, 1974) presents interesting therapeutic applications. Pharmacologic or dietary treatment of cognitive impairments – including those associated with old age – with agents which replace or stimulate cholinergic functions seems reasonable and current observations in rhesus monkeys (Bartus, 1979), aged mice (Bartus et al., 1980), and aged rats (Lippa et al., 1980) seem promising enough to be extended to humans, particularly in those individuals suffering from various types of dementia in which a deficiency in the cholinergic system (particularly in the hippocampus) has been invoked (Davies, 1978; Mohs et al., 1980; Drachman et al., 1980).

In the hypothalamus, norepinephrine content declines significantly with age in the rhesus monkey (Ordy, 1975) and the male rat (Miller et al., 1976; Simpkins et al., 1977). While in the rhesus monkey, dopamine and serotonin also decrease with advancing age (Ordy, 1975), in the male rat, dopamine is depressed as is norepinephrine, but serotonin metabolism seems to be enhanced (Simpkins et al., 1977). Indeed, a moderate increase in serotonin levels (Weil-Fugazza et al., 1980) and perhaps, more importantly, a definite increase in the serotonin to dopamine or norepinephrine ratio in several brain areas including the mesodiencephalon, have been reported in female rats with aging and related to aging of the reproductive function, in particular, and of inherent rhythmic processes regulating the lifespan sequence, in general (Timiras and Hudson, 1980; Walker and Timiras, in press, Walker et al., 1980a,b).

Autoxidation of Catecholamines. One of the theories of aging holds that accumulated cellular damage is the result of free radicals generated in the course of normal oxidative metabolism. These free radicals and other highly reactive substances may alter cell membranes and, consequently, lead to cell impairment and death. Changes in neurotransmission with aging may be ascribed to a loss of neurons or of dendrites, as discussed above, or to alterations in metabolism of the neurotransmitters themselves.

Autoxidation of the catecholamines does occur and the products thereof are themselves free radicals which may be toxic to the cell (Cohen, 1978). The brain possesses inducible enzymes which defend against oxidative damage and whose activity is higher in diencephalic neurons to which a "pacemaker" function has been ascribed (McKenna et al., 1976; Brannan et al., 1980). The fact that these enzymes are induced by their substrates suggests that pacemaker neurons experience a disproportionately high level of free radical production. Rates of free radical radical production may be of significance to aging inasmuch as a good correlation exists between the ratio of whole brain superoxide dismutase to specific metabolic rates and maximal life span in primates (Tolmasoff et al., 1980). Likewise, the age-related increase in melanin pigment may derive from a dopamine polymer formed by the nonenzymatic autoxidation of dopamine (Graham, 1979). Dopamine selectively inhibits the growth of mouse neuroblastoma cells in culture, an effect that can be enhanced by the addition of ascorbate to the medium but reversed by the addition of catalase (Prasad, 1979). Similarly, comparison of the growth inhibitory effects of dopamine in vitro among adrenergic and cholinergic lines of neuroblastoma cells and fibroblasts shows that these effects are more marked in the adrenergic lines, suggesting again that the toxic effects on growth are mediated by products of dopamine autoxidation (Cole, unpublished).

In view of the known involvement of neurotransmitters in the production of hypothalamic hypophysiotropic hormones, some of the alterations in the synthesis and secretion of pituitary hormones with old age have been ascribed to disturbances of neurotransmission in the hypothalamus and suprahypothalamic centers (see Chapters 14 and 16). Both the increase and decrease in some of the pituitary hormones with aging are associated with alterations in neurotransmission. For example, the increase in prolactin blood levels reported in older individuals has been related to decreased levels of the prolactin-inhibitory hormone (PIH) which, in turn, would depend on decreased dopamine levels in the hypothalamus and other brain areas (such as the corpus striatum). Indeed, dopamine concentration is decreased and/or dopamine turnover and transport are slowed down not only in the striatum and other brain areas of the aged mouse (Finch, 1973) and in the hypothalamus of the aged rat (Hoffman and Sladek, 1980), but also in the human caudate-putamen (Carlsson and Winblad, 1976). Dopamine decarboxylase, dopamine uptake, and dopamine-stimulated adenylate cyclase have all been reported to be lower in old than in young animals (Jonec and Finch, 1975; Makman et al., 1980; Govoni et al., 1980). Dopamine receptors also appear to decrease with aging, in number but not in affinity in the limbic cortex and striatum in rabbits (Makman et al., 1980) or in affinity but not in number in rats (Govoni et al., 1980), a discrepancy which may depend on species differences but also on other, as yet unclarified, factors.

The high LH levels characteristic of menopausal women and aged animals have also been related to desynchronization of hypothalamic signals. In old rats, failure of the LH surge, normally associated with mature reproductive cyclicity, would be consequent to disruption of the serotonin signal (Chapter 16). On the other hand, no firm correlation has been established between the decline which has been reported in GH, ACTH, and TSH pituitary and blood levels in some old individuals, alterations in GH inhibiting or releasing hormones, CRF or TRH levels and hypothalamic neurotransmitters. This lack of correlation is, however, dependent more on the paucity and uncompleteness of the available observations at all ages than on the negativity of the data. Clearly additional work is urgently needed to clarify intereactions between neurotransmitters neurohormones and hormones during development and aging.

Another group of substances with CNS transmitter function which have become the subject of intensive study are the endorphins and, within the endorphinergic system, age-related changes have been reported not only with respect to developmental ages but also old age. In aged rats, the number of opiate receptors would be decreased in various brain areas while, in some areas, the existing receptors would show enhanced opioid affinity (Jensen et al., 1980). This increased affinity parallels the increased sensitivity to some neurotropic drugs which appears to be associated with aging and is supported by several observations such as greater sensitivity to phenothiazines, dopamine receptor inhibitors (Finch, 1977), ethanol (Sun and Samorajski, 1975), amphetamine, an adrenergic agonist (Doty and Doty, 1966), and physostigmine, a cholinergic agonist (Doty and Johnston, 1966). Besides endorphins, other neuropeptides, such as fragments of the pituitary hormones ACTH or MSH, appear to affect cognitive functions in elderly humans and aged animals (Miller et al., 1980). While research in this area is still in its formative stages, the differential effects of pharmacologic challenges at different ages and with different neuropeptides (and their agonists or antagonists) offer both theoretical and practical means for a better understanding and more effective treatment of behavioral changes with senescence.

Sensitivity to Hormones and Other Physiologically Active Substances: Receptors, Membranes and Metabolism

CNS sensitivity, interpreted as the capacity to respond to stimuli, varies considerably with age. This statement, well supported by clinical and experimental data during the maturational period of the lifespan, attests to the "plasticity" and, hence, also the "vulnerability" of the developing CNS particularly during the so-called critical periods of accelerated growth and development (Meisami and Timiras, in press). Although CNS sensitivity and plasticity may persist throughout the life span, the degree of responsiveness is altered in old age. Evidence for this latter case is scarce and conflicting. It is becoming increasingly evident with the extensive use of neuropharmacological agents that the specific age of the subject msut be carefully weighed when prescribing a particular drug and establishing a treatment regimen because it is very difficult to differentiate the abnormal responsiveness of the elderly patient due to altered CNS sensitivity from the consequence of related alterations in the effector organ or tissue. The responses of several CNS areas to a variety of stimuli usually appear to be reduced in old age and even in a single area such as the hypothalamus, the large number of functions to be considered and the lack of uniformity of their changes with aging prevent the formulation of a coherent profile of aging changes (Frolkis and Bezrukov, 1979). The data of Dilman (1971, 1976) in humans indicate that the hypothalamus becomes less sensitive to glucose and glucorticoids with increasing age. Corresponding studies in rats, show that the hypothalamo-pituitary response to adrenocortical and gonadal steroid negative feedback is reduced in old age (Riegle, 1973; Shaar et al., 1975). The responses are not consistent and CNS sensitivity to barbiturates, catecholamines, acetylcholine and estrogens appear to increase or decrease with old age depending on a number of variables (e.g., parameter of action measured, brain area considered, agent tested, animal species used) (Frolkis and Bezrukov, 1979).

In old age, changes in cell and tissue responses to hormonal (and humoral) signals are determined by many factors and extend to all body systems including CNS. Essential to modulation of hormone action are the number, affinity, binding characteristics, location and translocation of receptors for hormones as well as for other substances (e.g., neurotransmitters) important for CNS function. Receptors for hormones and neurotransmitters in the senescent brain have been the subject of relatively few studies and the available data are somewhat contradictory, as with the dopamine receptors discussed above. As observed by some investigators with dopamine receptors, catecholamine receptors in rats (Puri and Volicer, 1977) and serotonin receptors in humans (Shih and Young, 1978) have been found to be reduced. It is proposed in these studies that selective, age-dependent decreases in number or alterations in the properties of postsynaptic neurotransmitter receptors occur in the absence of or independent of neuronal cell loss, possibly by mechanisms including desensitization, blockage, or decreased synthesis of receptors (Roth, 1979).

With respect to thyroid hormones (primarily, T3), binding to nuclei of cerebral hemispheres, as assessed by experiments *in vitro*, is similar in old (2 years) and young (3 months) rats (Valcana, 1979; Valcana and Timiras, 1979). However, some differences with aging are observed in *in vivo* experiments (Table 1), in which the nuclear binding of T3 in cerebral hemispheres appears higher in the old than young animals (Margarity et al., 1981). Increased nuclear T3 binding *in vivo* and *in vitro* is also observed in the cerebral hemispheres and liver of rats made hypothyroid either neonatally or in adulthood (Valcana and Timiras, 1978, 1979) as well as in neuroblastoma and glioma cells cultured in the absence of thyroid hormones (Draves and Timiras, 1980). The increase in nuclear binding in these examples was represented by an increase in the number of receptors rather than by changes in their affinity and was interpreted as a compensatory response to the low circulating thyroid hormones. In the old rat, T4 blood levels are significantly decreased, therefore, the increase in receptor number may also be viewed as compensatory to the age-related functional hypothyroidism. That hormones influence the composition of chromatin in terms of their own receptors is known not only for the thyroid hormones but also for other hormones such as estrogen and insulin. Inasmuch as thyroid binding may also occur in the other cellular components, whether changes in number or affinity occur with aging in these sites remains to be determined. It should be stressed at the conclusion of this brief review of hormone-receptor changes with aging that both an increase and a decrease in binding are possible in old age; alterations in binding properties may reflect either an absolute or relative loss of receptor sites and binding capacity or an attempt to compensate for altered hormonal levels and efficacy.

Another mechanism whereby the binding of molecules to their receptors may be influenced by old age is the changing properties of the cellular membranes themselves. A number of receptors and enzymes are membrane-bound and any variation in membrane structure and function may alter binding or enzyme activity with consequent failure or abnormality of the expected response. For example, changes in membrane fluidity due probably to qualitative changes in the lipid composition of microsomal membranes of cat liver and chick heart, were reported immediately after birth or hatching and were shown to continue until adulthood (Katchai et al., 1976; Kapitulnik et al., 1979). Fluidization of the microsomal membrane with development has been interpreted as contributing to maturational metabolic changes (e.g., increased uptake of sugars, amino acids, and urea by cardiac cells and increased enzyme activity in hepatic cells). Plasma membrane fluidity was also compared in lymphocytes from young and old mice and found to be decreased in older animals (Rivnay et al., 1979). The increased viscosity of the membrane in the

Table 1

Serum Thyroid Hormone Levels and Nuclear Binding in
Cerebral Hemispheres and Liver of Adult and Aged Female Rats

Age	Weight (grams)			Serum		T3 Nuclear Binding (fmoles/mg DNA)	
	Body	Liver	Thymus	T4 (µg/100ml)	T3 (ng/100ml)	Cer. Hem.	Liver
Adult (3 mos)	252[a] ± 9 (5)[b]	10.0 ±0.4 (5)	0.28 ±0.02 (5)	5.20 ±0.26 (5)	173 ±11 (5)	2.69 ±0.36 (4)	8.12 ±0.63 (4)
Aged (20 mos)	334 ± 9 (25)	13.5 ±0.4 (25)	0.12 ±0.01 (25)	2.25 ±0.13 (25)	178 ±10 (25)	3.36 ±0.58 (4)	9.99 ±0.72 (4)
P	<0.0001	<0.001	<0.001	<0.001		not significant	
and % change (adult vs aged)	↑32%	↑35%	↓57%	↓57%	↑3%	↑25%	↑23%

[a] Mean ± standard error
[b] Numbers in ( ) represent number of rats.

Adult and old Long-Evans rats were compared in terms of body and organ weight, serum T3 and T4 levels, and T3 nuclear binding in cerebral hemispheres (cer. hem.) and liver. Body and liver weights were taken as indicators of growth (in the rat, continuing throughout the lifespan) and thymus involution, of age-related changes. T3 and T4 serum levels were determined using a RIA kit from Antibodies, Inc. (Davis, CA). The nuclear binding was determined in vivo in animals injected intraperitoneally with 5ng ^{125}I-T3 (sp. ac. 1200 µCi/µg)/100g body weight and sacrificed after 4 hours-- the time of maximal ^{125}I-T3 uptake by brain tissue. Liver and brain nuclei were prepared according to the procedure of Eberhardt et al., 1978, and the activity found in nuclei after 0.5% Triton X-100 washing of the nuclear pellet was expressed per mg DNA and was considered as specifically bound hormone.

---

aged animals was associated with high serum cholesterol and altered lymphocyte behavior and immune response characteristic of these old animals. Current theories relate the fluidity of the membrane to the kinetics of neurotransmitter and hormone binding. For example, in the so-called two-step fluidity hypothesis, the ligand binds to the receptor on the membrane (step one) and then the ligand-receptor complex diffuses laterally (step two) (Cuatrecasas, 1974). In this two-step process, a "lag" period may explain some of the discrepancies between kinetics of hormone binding and activation of the biological response. Such a lag period would depend on the degree of fluidity, specific phospholipid and sterol composition of the membrane and temperature, all items that are sensitive to a variety of hormones; thus, the actions of hormones on the membrane — particularly of thyroid hormones, well known for their effects on myeline and other specialized membranes (see Chapters 7, 12) — may explain the synergism of one hormone with others and with catecholamines in glycogenolysis, colorigenesis, lipolysis and smooth muscle activity (Hock, 1974). With advancing age the progressive increase in membrane viscosity would prolong the lag period between signal and response and impair the optimal timetable of homeostatic adjustment.

NEUROENDOCRINE THEORIES OF AGING

As examined in the corresponding chapters on the aging of the endocrine system (see Chapters 14 and 16) and as briefly reviewed at the beginning of this chapter on aging of the CNS, most endocrine and nervous functions decline or are altered with advancing age but such decrements, taken individually, do not appear sufficient to endanger life. Thus, although the ability of the elderly to withstand stress is severely restricted, major changes have not been demonstrated in the physiologic competence of the adrenals, the endocrine glands most actively involved in adaptive responses (see Chapter 14). Rather, the view has been advanced that changes with aging may be due to progressive alterations in the articulation of one system to the other. Primary changes in neuroendocrine control would be expected to produce widespread secondary changes in the large number of functions they regulate and both primary and secondary changes would, in turn, lead to the impairment of optimal adjustments to the environment, the decline of physiological competence and the progressive increase in the incidence and severity of pathological processes characteristic of old age, and ultimately responsible for death.

The theories that have been presented to explain the aging process are numerous and will be encapsulated here for brevity under a few headings, among which the best known include: (a) cellular wear and tear (b) structural and functional changes in vascular and immunologic systems (in addition to changes in endocrine and nervous systems specifically considered in this chapter) (c) progressively increasing autointoxications (e.g., cross-linking of proteins and nucleic acids, formation of free radicals) (d) cessation of growth (e) loss of proliferative capacity within a finite lifespan, and (f) genetic hypotheses regarding the accumulation of somatic mutations and the loss of biologic information. The interested reader is referred to the large number of textbooks and reviews on aging which have been published in recent years and in which molecular, cellular and organismic theories are examined in detail.

Despite increasing research in the broad field of aging no definitive theory, including neuroendocrine, has gained general acceptance. Most of the current theories are equally tenable, and efforts to relate one to another give rise to certain unavoidable ambiguities at this stage of our knowledge. One of the major unresolved problems is to identify which, among multiple factors is the "cause" of aging and which are the "consequences." Possibly the various theories, each focusing on a different cause, are all correct.

Several current neural and endocrine theories emphasize the aspect of controlled aging or the failure of this control. Most of the theories are additive rather than exclusive. Despite considerable overlap, they can be grouped, for didactic purposes, into three major categories: (1) those theories which propose that aging is controlled by "clocks" or "pacemakers" which regulate growth, development and aging (2) those that regard aging as the consequence of the breakdown of neuroendocrine control and the consequent disorganization of structures and functions necessary for adaptation and (3) those that ascribe to hormone and neurotransmitters, themselves, aging effects.

Clocks or Pacemakers as Regulators of Aging

It has been proposed that aging may be the result of a genetically determined, precisely timed, process in which clocks or pacemakers provide signals for progression through a life program which terminates in senescent decline and death. Herein, aging would represent one last event in the timetable of passages from one period of the life span to another in the same manner as growth and development signal the age-related changes in the first part of life. Accordingly, growth (e.g., fetal, postnatal), development (e.g., organogenesis, birth, puberty) and aging (e.g., cessation of reproductive function, decline of physiologic competence) follow in an orderly sequence of changes determined from fertilization when the genetic code is laid down (see Chapter 1) and continuing until death. The precise timing in this program would be controlled by a single clock or by several independent or interdependent clocks regulating simple or complex body functions (see Chapter 13). The concept of pacemaker or "command" cells, possibly located in specific brain areas (e.g., hypothalamus, limbic system) provides a useful guideline for systematically investigating the timetable and, eventually, the nature of the neural signals responsible for hypothalamic neurosecretion, pituitary hormone secretions and control of automonic and sensorimotor functions. If, as is commonly accepted, the major function of the CNS is to assure transfer of information, specific impulses, translated through the release of stimulatory or inhibitory neurotransmitters would modulate the activity of the neurosecretory cells of the hypothalamus. Neurosecretions would then, induce secretion or inhibition of pituitary hormones and these, in turn, would either stimulate target cells to secrete their own hormones (e.g., ACTH, TSH and gonadotropin which, in a waterfall effect, then regulate adrenocortical, thyroid and hormone, prolactin, antidiuretic hormone, oxytocin). The involvement of the classical monaminergic and cholinergic neurotransmitters and the more recently considered peptidergic neurotransmitters (e.g., substance P, enkephalins, endorphins) in the regulation of many endocrine, visceral, and sensorimotor functions (e.g., sleep patterns, locomotor activity, reproductive functions, pain and pleasurable sensations, thermoregulation, regulation of blood pressure) are well known. With aging, qualitative and quantitative changes do occur in the release and metabolism of some of these neurotransmitters, and in the secretion of hypothalamic neurohormones as well as of pituitary hormones. These changes would trigger such decrements in physiologic performance as characterize the "passage" from adulthood to senescence and signal the terminal period of the lifespan.

Other clocks proposed to be operative in causing aging are those which act on endocrine glands to produce failure of key systems for survival, such as the immune and circulatory systems (Denckla, 1975), or those which are located in specific organs such as the pineal gland to induce alteration of biological rhythms (Quay, 1972) or the thymus to induce alterations in immune responses (Makinodan, 1980).

An alternative to the hypothesis that aging is a programmed period of the life span, is the view that aging may be the consequence of the cessation of a pacemaker-regulated program which is exclusively limited to the developmental period. In this context, the lifespan would be divided into two distinct but sequential periods; in the first, encompassing development until adolescence, physiological performance improves continuously according to a genetic program expressed through neural (e.g., mediated through the neurotransmitter, serotonin) and endocrine (e.g., mediated through thyroid hormones) signals which regulate ontogenic transformations and selective mechanisms promoting the survival of the individual until reproductive maturity has been attained. In the second period, including late adulthood and senescence, performance is progressively impaired due to failure of optimal integration of homeostatic control systems because they are no longer under any programmed regulation (Walker and Timiras, in press).

As amply documented in Chapter 13, cyclic processes are influenced by environmental (e.g., light, temperature) and endogenous factors (e.g., hormonal and neural stimuli). Among the latter, thyroid hormones and serotonin, substances involved in general body functions and with wide philogenetic distribution, have been suggested as potential regulators of development and expression of the clock responsible for the timing of developmental events. As already extensively reviewed (see Chapters 2, 5-7), thyroid hormones are essential for proper maturation and survival during early critical periods of development. After adulthood has been attained and continuation of the species has been insured, animals continue to respond to the actions of thyroid hormone; however, in the absence of a program to integrate those functions necessary for homeostasis, these responses become suboptimal and this leads or contributes to the "disorganization" of the "internal order." A primary locus of action of thyroid hormones is the brain (see Chapters 5-7, 12). In the extensively studied rat brain, thyroid hormones have been shown to promote the development and maturation of serotonergic transmitter systems and influence the activity of the system in adult animals (Walker and Timiras, 1980; in press). Serotonin levels follow cyclic patterns (e.g., serotonin circadian rhythms in the pineal and in the hypothalamic suprachiasmatic nucleus) and serotonin cyclicity modulates, in turn, other biological rhythms (e.g., corticosterone circadian cycle and LH surge in mammals; electrical activity of optic nerve in invertebrates) (See review by Walker and Timiras, in press, a). Since thyroid hormones influence the maturation of serotonergic rhythms in the pineal and hypothalamus (Walker and Timiras, in press) and levels of serotonin in the adult brain (Ito et al., 1977), failure of thyroid-serotonin interactions upon termination of the program for development would result in the disorganization of physiologic competence and thereby lead to aging and death. Development of senile-type syndromes after manipulation of serotonin metabolism in brain pacemaker areas and alterations with age in the pituitary-thyroid axis as well as in the peripheral actions of thyroid hormones, support this hypothesis.

The theories based on the view that aging may be due to the action of a pacemaker—whether such a pacemaker follows a specific program for aging or becomes "disorganized" after adulthood has been attained, and whether this pacemaker is situated in specific areas of the brain or in an endocrine gland—remain more speculative than factual. They offer, however, some useful models in which to further study aging processes in higher organisms and also provide some guidelines for possible interventions. If further evidence can substantiate that aging results from hormone-mediated disorganization at specific neural integrating centers, it may be possible in the future to enhance adult vitality by stabilization or reduction in the rate of change with age of these areas (see also next section).

Neural and Endocrine Alterations as Causes of Aging

A defect produced at any level of the neuroendocrine pathways may be expected to produce secondary aging changes. Several sites have been proposed where alterations which lead to aging may originate as well as several mechanisms by which they may act. These include:

(1) The Hypothalamic "Disregulation" Hypotheses which suggest that age-related changes in the hypothalamus may be responsible for physiologic and pathologic correlates of aging. According to one view, aging of the

hypothalamus does not proceed at the same pace in all of its nuclei; rather, as for most other body systems, the timetable of aging varies from one hypothalamic nucleus to the other (Frolkis and Bezrukov, 1979). This lack of uniformity would produce unequal changes in the numerous functions regulated by the hypothalamus; such functional disequilibrium would lead to a breakdown of the synchronized communication within the hypothalamus and between the hypothalamus and other neural and endocrine centers.

Another view proposes that, with aging, the sensitivity of the hypothalamus to the negative feedback of hormones is decreased or lost. Consequently, because of the loss or impairment of negative feedback, the secretory activity of endocrines is increased and the ensuing oversecretion may accelerate aging phenomena (Dilman, 1976). One clear-cut example of increased hormonal secretion with age, is represented by the high levels of pituitary gonadotropins in women after menopause (see Chapter 16). An age-related elevation of the hypothalamic threshold to negative feedback has also been reported for corticosteroids (Riegle, 1973) and gonadal steroids (Shaar et al., 1975) in the rat. The two hypotheses of hypothalamic disregulation may be further combined to suggest that there may be with aging a "shift" in the sensitivity of the hypothalamus to specific hormones with some nuclei becoming more sensitive and others less sensitive to the hormones. Such irregular patterns of change, when associated with concomitant changes in selected extrahypothalamic areas of the CNS and in endocrine function, would be sufficient to account for the etiopathogenesis of aging processes.

(2) The Neurotransmitter Hypotheses. As mentioned earlier in this chapter, changes in synaptic number, structure and function do occur in the aged brain despite the complexity of distinguishing differences which characterize animal species, separate brain regions and specific neurotransmitter systems. Thus, at least in experimental animals, the overall picture of the aged brain appears to be one of neurotransmitter imbalance. Preferential decrements may involve certain neurotransmitters while others remain unaffected with advancing age. In this case, excess or deficit of a specific neurotransmitter (e.g., dopamine) in a specific region (e.g., nigro-striatal pathways) can be directly implicated in a specific functional alteration (e.g., Parkinson's disease). Conversely, minor changes in several neurotransmitter levels in various brain regions may be sufficient to produce imbalances leading to marked functional alterations. Normal brain function depends on a delicate balance between inhibitory and excitatory impulses and any disequilibrium may have severe repercussions on homeostasis and survival. In the aging brain, disruption of this balance would lead to significant deviations which can be further aggravated by the normal reciprocity between neural and endocrine actions; thus, neurotransmitter imbalances would lead to endocrine imbalances and these, in turn, induce alterations in those aspects of brain function influenced by hormones.

Based on the availability of neurotransmitter agonists and antagonists, this hypothesis, by focusing on alterations in neurotransmission, suggests that aging may be amenable to pharmacological interventions (Ordy, 1979). Indeed, several reports show that the administration of certain centrally-acting drugs, including some which are metabolic precursors, alters various aspects of aging. The dopamine precursor, L-Dopa, administered in relatively large doses and from early development is capable of retarding growth, reducing overall tumor incidence and delaying the onset of aging (e.g., as manifested by maintenance of coat color, general appearance and vigor) in mice (Cotzias et al., 1974; 1977) as well as showing beneficial effects in the treatment of the early stages of pre-senile dementia (Alzheimer's disease) (Jellinger et al., 1980). Similarly, iproniazid, a monoamine oxidase inhibitor, and lergotrile or bromocryptine, alkaloids capable of mimicking the neuroendocrinological effects of dopamine, have been found capable of reinitiating estrous cyclicity in old rats, and, at least for the first two, of restoring some aspects of physiological competence as well as extending the life span (Quadri et al., 1973; Clemens and Fuller, 1977). Reduction in the levels of brain serotonin by parachlorophenylalanine (PCPA), an inhibitor of its synthesis, or restriction of the precursor amino acid, tryptophan, in the diet, also seems to retard growth and maturation as well as to delay the onset of aging and possibly prolong the lifespan in rats (Segall and Timiras, 1976; Segall et al., 1978). With respect to the cholinergic system, while normal individuals show little change in acetylcholine in old age, in dementias (including Alzheimer's disease), the acitvity of choline acetyltransferase, the synthesizine enzyme of acetylcholine, is decreased in the cerebral cortex (particularly the hippocampus). In addition, pharmacological, biochemical and electrophysiolotical studies in humans and other species have shown that cholinergic dysfunction may be related to the memory impairments of old age and have suggested, at least as an empirical approach, that replacement therapy with choline or lecithin, its natural dietary source, might be beneficial in restoring memory deficits. Although positive effects with dietary or pharmacological replacement therapy have been reported in some elderly subjects and aged animals, the nature of these effects is relatively transient and inconsistent among subjects and with respect to optimal doses (Bartus et al., 1980). An issue of critical importance here as in the previous cases of monoamine replacement or blockage, is whether choline (or monoamine or other precursor) manipulations induce alterations in CNS cholinergic (or other neurotransmitter) system capable of influencing behavior and function. Further interventive approaches based on the neurotransmitter hypothesis of aging will be described in more detail later in this chapter.

(3) The Pituitary Hypotheses include some contrasting points of view. According to some investigators, aging phenomena would be the consequence of a deficiency of pituitary hormones; according to others, pituitary hormones could accelerate the aging process and cause the pathology associated with old age. The first hypothesis, based on hypopituitarism as a cause of aging, is supported by the reports of senile changes occuring prematurely in a number of patients whose pituitary has been destroyed by disease or its hypofunctioning due to unknown (genetic?) factors (see Herman, 1976). In several cases, symptoms of hypopituitarism can be corrected by appropriate hormonal replacement therapy but whether or not the aging processes are slowed down as well, remain to be demonstrated. Furthermore, this hypothesis appears to be in direct contradiction with the reports by several investigators that drastic reduction in pituitary function, either by surgical removal of the gland (Everitt and Burgess, 1976), by pharmacological or dietary hypophysectomy; in both cases, the effects of drugs (e.g., by blocking or stimulating neurotransmission) or of dietary restrictions (e.g., by inducing general metabolic or specific neurotransmitter deficiencies) would act by affecting primarily the synthesis and release of pituitary hormones and, secondarily, the secretory activity of target endocrines and their hormones.

Besides quantitative changes, qualitative changes in pituitary secretion have been described with advancing age. As mentioned above, in aged rats the pituitary might secrete a factor (perhaps a new or transformed hormone?) which would act peripherally by reducing the responsiveness of tissues to the metabolic actions of thyroid hormones (Denckla, 1974). Although confirmation of the existence of such a factor is still pending, the dramatic effects of hypophysectomy, combined with appropriate replacement therapy, in reducing tumor incidence, restoring a number of functions to their juvenile level, and/or preventing age-related declines in old rats have been ascribed to the removal of this postulated inhibitory factor (see Chapter 14). Another mechanism by which qualitative changes may occur in pituitary secretion with aging may involve alterations in the synthesis and metabolism of hormones. If, as accepted for several hormones (e.g., ACTH, insulin), the active form is the result of the cleavage of a larger precursor molecule (e.g., pro-opiocortin, precursor of both ACTH and MSH, and pro-insulin), then changes in the activity of enzymes capable of breaking down the larger molecules to smaller active peptides would give rise to hormonal forms with reduced or altered activity (Segall, 1979). Finally, most pituitary hormones show a certain polymorphism manifested by physico-chemical differences (e.g., molecular weight, electrical charge) and reflected in different biological potency. Although such a polymorphism is particularly evident at young ages, it seems to increase in old rats (Klug and Adelman, 1979). Similarly, a decrease in carbohydrate content of glycoprotein hormones such as LH (Conn et al., 1980) and TSH (Choy et al., 1981) has been reported in rats with advancing age.

A corollary to the preceding view that ascribes aging to excess, deficit or transformation of pituitary hormones is the so-called hypothyroid hypothesis. This hypothesis suggests that many of the signs and symptoms of aging depend on decrements in thyroid hormone actions (see Chapter 14). These decrements may be due not only to a decrease in thyroid hormone levels, impaired conversion of T4 to T3, increased TSH polymorphism with advancing age, but also, as mentioned above, to the presence of a factor, possibly of pituitary origin, which reduces the effectiveness of thyroid hormones. An evolutionary precedent of interference in thyroid hormone action by a pituitary factor may be found in some anurans and urodels in which mammalian prolactin inhibits the developmental actions of thyroxine, perhaps by interfering with NaK-ATPase activity, a major target of thyroid action (Platt, et al. 1978). Growth and thyroid hormones interact on growth processes in mammals. It is possible that a factor similar to GH or prolactin, or a protein or peptide containing the same family of GH-related sequences, may modulate the developmental effects of thyroid hormones but exert an inhibitory action once maturity has been attained. Inasmuch as prolactin secretion is regulated by brain levels of dopamine (either directly or through control of the hypothalamic prolactin inhibiting hormone), the apparent decrease in brain dopaminergic activity with aging may be reflected in an increased secretion of prolactin (and eventually of GH and related peptides) and in alterations in their relationship to thyroid hormones. As noted in the previous chapter the ob/ob mouse, a currently widely accepted model of maturity-onset diabetes, shows a markedly altered regulation of GH and prolactin secretion and a selectively diminished NA+K+-ATPase response to thyroid hormones (see Chapter 14). The complexity of the consequences of the above described disorder points out that it may be unrealistic to ascribe aging to a single endocrine deficit, such as hypothyroidism. Aging, rather may result from multi-endocrine disturbances.

Progeria and Progeroid Syndromes. The role of genetic factors in determining the length of the life span, while undoubtedly important, remains to be clearly defined. Longevity is considered, at least in part, hereditary under stable environmental conditions (Martin, 1978). On the other hand, genetic variants may result in accelerations or decelerations of changes in the genetic phenotype. No single gene has been described that results in a true change in life duration; however, there are combinations of genes which may also direct, among other traits of the individual, some which may lead to abnormalities resulting in a shorter or longer life. Among the genetic syndromes in man with potential relevance to the pathobiology of aging, several seem to involve neuroendocrine components, thereby affording useful models for isolating neuroendocrinological variables which may be responsible for accelerated aging or for providing guidelines to study methods for preserving functional competence and, eventually, prolonging life. Progeria (premature old age) and progeroid (progeria-like) syndromes resemble but do not quite duplicate all the pathophysiology of aging, rather, each syndrome represents an acceleration of some characteristics associated with normal aging.

Progeria (premature senility/Hutchinson-Gilford syndrome) is a very rare type of dwarfism combined with premature senility. Growth is normal until about one year of age but progresses very slowly thereafter resulting in dwarfs with a characteristic senile somatic appearance (birdlike features, wrinkled and atrophic skin, baldness) associated with a number of pathological signs reminiscent of old age (e.g., increased lipofuscin deposition, osteoporosis, hypertension, atherosclerosis decreased number of cell doublings). Death occurs generally around 15 years of age and is commonly due to coronary occlusion. The etiology is obscure; among the several causes proposed, that of neuroendocrinological dysfunction is based on the stunted growth, the failure of gonadal maturation and the metabolic alterations observed in these subjects. Other endocrinopathies which may also be included in the category of progeroid conditions are those characterized by hypogonadism (Turner and Klinefelters syndromes) and diabetes mellitus (see Chapters 2 and 14). Indeed, a common denominator in most of these syndromes is the development of insulin resistance, a common feature of normal aging described in the preceding chapter (Chapter 14). Another syndrome that may be included in this group is a generalized lipodystrophy (Seip syndrome) characterized by accelerated development (e.g., earlier than usual dentition, acquisition of adult hair, genital enlargment) as well as accelerated aging (e.g., severe atherosclerosis, greater susceptibility to neoplasia and infection) (Seip, 1971). While this condition is usually congenital, it may also appear in a acquired form following diencephalic disturbances and resembles other similar disturbances with tumors of the pineal (e.g., Medenhall syndrome). As for progeria, the etiology of this syndrome is unknown; however, several peptides with diabetogenic, TRH-like and prolactin-like activity seem to be implicated (Berge et al., 1976; Trygstad and Foss, 1977). These observations provide tentative clinical support to the hypothesis that the pituitary may secrete factors other than the classical hormones and that these factors may act to accelerate development and produce selected aspects of the aged phenotype.

The Stress Theory of Aging

This concept applies the cellular hypotheses of "wear and tear" and autointoxication to the organismic level and amalgamates them with neuroendocrinologic theories of aging involving the hypothalamo-pituitary-adrenal axis. According to this theory, exposure to environmental stimuli, most often of a detrimental nature, but also, in some cases, with a positive influence, may decrease and eventually exhaust the ability of the organism to maintain homeostasis.

Central to this concept is the role of the adrenal and particularly the adrenal cortex, as the endocrine gland indispensable for adaptation and survival. Around this role of the adrenals, Selye (1950) formulated a concept called the "general adaptation syndrome" which postulates that individuals are born with a fixed quantity of a "adaptive energy" which is progressively reduced with each exposure to stress (the term stress being used by Selye to indicate both the stressing agent or stimulus and the response of the organism to it). This theory which has found numerous supporters as well as detractors is based on four fundamental observations: (1) (now classic) adrenocortical hormones, primarily mineralocorticoids such as aldosterone, are needed for life (2) the adrenalectomized (or hypophysectomized) animal can survive provided it is protected from environmental changes and receives appropriate replacement therapy (3) the capacity to survive various types of stress is severely impaired in the adrenalectomized (and hypophysectomized) animal, and (4) adrenocortical hormones, although sufficient to maintain life under nonstress conditions, are incapable of supporting adaptive reactions when stress is present—a finding that underlines the necessity for activation of the hypothalamo-pituitary-adrenal system above normal levels in order for adaptation to occur. Adaptation itself would follow a specific timetable characterized by three phases; an initial phase in which defense mechanisms are acutely challenged (alarm reaction); a period of enhanced adaptive capacity (stage of resistance); and loss of the capacity to adapt (stage of exhaustion). For example, when rats are exposed to moderate to severe cold (or to a variety of physical, chemical or psychological stimuli) they respond immediately with the alarm reaction, which comprises the sum of all damage and defense responses that take place upon first exposure to the environmental stress. Depending on the severity of the stress (e.g., cold) the animals will either adapt or die. Should they adapt, the animals will exhibit appropriate adjustments (e.g., increased thermogenesis) and give no indication that life expectancy is shortened. However, after a variable period of time (generally inversely proportional to the severity of the stress), the animals begin to lose weight and shortly die, again manifesting all signs of the alarm reaction. This general response pattern assumes that some form of energy, finite in amount, is necessary for the performance of adaptive work and can be exhausted with stress. Adaptive responses can occur as well in hypophysectomized or adrenalectomized animals with the difference that the stage of resistance is absent or very short. The simplest explanation proposed for this phenomenon is that in the absence of a normal hypothalamo-pituitary-adrenal axis, the organism uses most or all of its energy for normal adaptive responses and has little left for resistance against damaging agents.

This triphasic course of the adaptation syndrome, insofar as resistance to stress is concerned, seems to resemble the life course of the individual. In the first period of human life, adaptive capacity is not completely developed. Even though the newborn is generally capable of successful adaptation to the extrauterine environment, the period of birth itself represents the first important challenge and is associated with very high mortality. As the child grows its capacity to adapt also develops, although the child remains generally less resistant to variation in external temperature, muscular work, starvation, and so forth than the adult. When the organism reaches adulthood, its resistance reaches an optimum and thereafter declines until death. In old age, not only is resistance low (e.g., in terms of immunological competence), but when adaptive energy is used for resistance against one stress, resistance to other types of stress is diminished and the progressive loss and ultimate exhaustion of adaptive energy may explain the exponential increase in death risk. "Perhaps," as Selye writes, "the classic general adaption syndrome, as we see it in the laboratory or the clinic, is merely a 'telescoped' version of essentially the same adaptive phenomenon" (Selye, 1950).

Interesting as the stress theory may appear, it is difficult to reconcile it with the apparent adequate adrencortical function both under normal (basal) conditions and after ACTH administration in the elderly (see Chapter 14). As discussed in the previous chapter, even the decline with age in adrenocortical responses to injury appears to be relatively slight, if present at all. It is possible that the age-related decline in resistance to stress is dependent on qualitative rather than quantitative factors in the same manner as mentioned above for the pituitary hormones. Other factors that may play a role is such "derailment" of normal (adult) physiological patterns are age-related changes in the sensitivity of target organs to adrenal hormones, or in the efficiency with which the target organs can make the functional adjustments required for adaptation, or in the articulation of neural and endocrine inputs in the optimal control of these adjustments, or to a combination of all these factors. Systematic studies need to be pursued in order to critically evaluate the consequences of repeated stimulation of the hypothalamo-pituitary-adrenal system, in response to stress, on the complex integrative processes involved in homeostatic responses not only in the aged but also in the young and the adult.

Diseases of Adaptation. Repeated exposure to stress and the ensuing "disorganization" of adaptive mechanisms would lead to a number of diseases, designated by Selye as "diseases of adaptation." Many of the symptoms of these diseases bear some similarity to the conditions attendant an excess or deficiency of endocrine secretions and the direct consequences of stress on target tissues. Clinical and experimental evidence is available in support of this etiopathology. Clinically, it is generally recognized, although not definitely proven that cardiovascular accidents and pathology are somewhat related to the "lifestyle" and hence the amount of stress to which individuals are exposed. There is a general clinical impression that psychic and other emotional stress and anxiety are associated with sudden death, often resulting from coronary heart disease, and that compulsive, striving and deadline-conscious individuals have nearly three times the incidence of hypertension and coronary heart disease as less energetic and more passive individuals. The "anguished inner life of the patient" can be a prime factor in coronary thrombosis. Many studies are witnessing the profound effects of rage, grief and despair on blood and heart. Experimentally, animals exposed to stress conditions for a prolonged period of time demonstrate such pathologic alterations as arthritis, periarteritis nodosa, myocarditis, nephrosclerosis, hypertension, EEG and behavioral changes, increased incidence of autoimmune diseases, all diseases commonly associated with old age. Indeed, the induction of disease by a variety of procedures acting as stress is one method employed to induce "precocious aging," that is, decrement in one or several functions at an earlier than expected chronological age.

Disease as a Model for the Study of Aging. A useful approach to better understand aging, per se, and its accompanying pathology, as well as to efficiently test prophylactic and therapeutic measures, is the induction of various types of disease in young and adult animals to serve as models for those processes normally associated with advancing age. These models cover a wide range of complex intrinsic and extrinsic conditions wherein hormonal and neural factors interact with other determinants. For example, a progeria-like syndrome can be induced in rats by the administration of hydrotachcholesterol (a derivative of viosterol used in the treatment of rickets and other manifestations of faulty calcium and phosphate assimilation) which promotes the deposition of large amounts of calcium and phosphate in ectopic regions (e.g., in the skin) and induces a number of changes resembling some of the signs of old age. These age-associated alterations, kyphosis, hair loss, wrinkling of the skin, atrophy of thymus and lymphatic organs, involution of sex organs, anomalies of teeth, can be prevented if the animals are given small doses of calciphylactic challengers (e.g., certain metallic salts or chelates) together with anabolic steroids (e.g., testosterone and other androgens) (Selye et al., 1961). Other models with a strong hormonal or neuroendocrine component are considered in other chapters; the ob/ob mouse in Chapter 14; the induction of reproductive aging by the topical (hypothalamic) administration of monoamine antagonists, in Chapter 16; the relationship between hormones and neoplasia in Chapter 20.

Other animal models for the study of aging are concerned with mimicking the aging of the cardiovascular system whose competence is crucial to the survival of the individual. One of the most active areas in this respect is the experimental induction of atherosclerosis, the accelerated production of naturally, slowly-occurring atherosclerotic lesions. Without entering into discussion of the etiology and pathogenesis of the atherosclerotic lesion which remain controversial and well beyond the scope of this text, it may be stated that one of the most common and effective approaches of reproducing atherosclerosis, experimentally, in animals, is to implement dietary changes that elevate the plasma levels of lipids, lipoproteins and/or carbohydrates. However, the degree of dietary manipulation necessary to increase plasma lipids (or carbohydrates), the duration of the regimen, the interspecies variability in the effectiveness of this procedure and the structural and chemical differences between naturally-occurring and experimentally-induced lesions make these experiments difficult to interpret or correlate with the human disease. Various predisposing or aggravating factors have been added to the experimental protocol to accelerate the production of the lesion, to increase its severity, to accentuate a special sign of symptom, or to increase its similarity to the human lesion. Hormones have proved to be such a factor. Significant changes in the pathology of the advanced atherosclerotic lesion have been produced in rabbits by adding to the usual high cholesterol diet, mechanical traumas to the arterial wall and injections of cortisol (to mimic, perhaps, the increased secretion of this hormone under stress conditions) (Constantinides, 1965, Constantinides et al., 1980). The high-cholesterol diet is more effective in rats in producing the lesion when the animals are also made hypothyroid, reducing cholesterol metabolism and thereby increasing endogenous levels of cholesterol. Similarly, in rats, atherosclerotic lesions occur more rapidly and are more severe when associated with partial pancreatectomy wherein the resulting diabetes simultaneously increases the hyperlipemic and atherogenic effect of the diet itself. This effect is not reversible by insulin. In many cases, however, spontaneous regression of the lesions have been reported even while the administration of the high-lipid diet was continued. In those cases in which spontaneous regression did not occur, therapeutic regression could be induced either by withdrawal of the high-lipid diet, by administration of a diet rich in polyunsaturated fatty acids (and capable of lowering plasma cholesterol), or by the manipulation of the endocrine environment (e.g., administration of thyroxine, estrogens or insulin). Other interventive measures capable of modifying the aging process are presented in the following section.

NEUROENDOCRINE APPROACHES TO INTERVENTIVE GERONTOLOGY

As suggested by the experiments of Selye and other investigators, stress, by acting on the hypothalamo-pituitary-adrenal axis, may induce alterations in homeostasis which are conducive to disease and/or shortening of the lifespan. It may be argued that, reciprocally, interventions at any levels of control – i.e., the hypothalamus and related integrative nervous centers, the pituitary, the peripheral endocrine – strengthen the homeostatic responses to stress and thereby delay the onset of deteriorative aging changes and eventually prolong the duration of life. Alternatively, improvement of the ecological and socioeconomic environment, and correction of the lifesytle, by reducing the amount of stress to which the organism is exposed, may also beneficially influence the epiphenomena of aging. Similarly, the efficiency of the neurotransmitter system that acts as the putative pacemaker may be modified by agonists, precursors or antagonists capable of enhancing or inhibiting its actions, and, in this manner, influence the life program. When limited to the period of growth and development, this program may be prolonged or substituted for by the administration of specific hormonal or neurotropic replacement substances. Whichever neuroendocrine theory is adopted, current advances in our knowledge of brain chemistry, psychoneuropharmacology and endocrinology make it possible to envision, in the near future, interventions which may focus directly on the aging process rather than age-associated diseases. The shift from disease-oriented care to one centered on fortifying the physiological competence will enable us to discard a multitude of specific, costly, but purely symptomatic, approaches to treatment.

It is well-known that the average lifespan has significantly lengthened in the last 50 years and that the several millions of elderly – 65 years of age and older – continue to increase in number in the United States and other economically advanced countries. While we tend to emphasize the pathology of old age and focus on such causes of morbidity and mortality as cancer and cardiovascular diseases, it is well recognized that treatment and eventual cure of many diseases, commonly responsible for death in the elderly, would prolong life by only a few years (if vascular diseases were cured overnight, for example, the lifespan would be extended by about five years). While we hear repeatedly that we are facing an "epidemic of cancer," only the incidence of lung cancer has in fact significantly increased in the last 50 years in the United States. One form of cancer, cancer of the stomach, has markedly decreased. Similarly, the incidence of cardiovascular accidents has also begun to decline. Indeed, one could take the revolutionary view that "the United States is not suffering today of any epidemic of cancer but rather, it is experiencing an 'epidemic of life' in that an ever greater fraction of the population survives to the advanced ages at which cancer (or hypertension, or coronary heart disease) has always been prevalent" (Handler, 1979). To the extent that medicine will be able to control the leading causes of death, slowly evolving diseases now experienced only in their subclinical stages will reach proportions that warrant medical attention. It is possible that, contrary to the prophecies of Huxley's *Brave New World* (1932), man will never die of old age alone but always from disease. However, the terminal disease of the future will undoubtedly be different from that we face today, changing from damage wrought by the environment (e.g., infections, toxic substances) to the consequences of progressive debilitation associated with increasing old age.

Contemporary medicine has made remarkable strides in curing and preventing the effects of pathogens and/or in neutralizing environmental impacts, but little progress in optimizing and prolonging physiological competence. To sustain a balanced effort for new basic and applied knowledge in the area of aging, we must pursue the search for both the treatment of age-related diseases and the clarification of basic mechanisms of aging. Either approach has benefitted from neuroendocrinological studies. As already indicated in this chapter, encouraging results in delaying the onset of aging and/or in prolonging life have been obtained with the administration of neurotransmitter precursors (e.g., L-Dopa, choline or lecithin for the synthesis of dopamine and acetylcholine, respectively) or neurotransmitter inhibitors (e.g., PCPA for inhibition of serotonin synthesis) or of neurotransmitter agonists (e.g., lergotrile, bromocryptine for dopamine-like activity and iproniazid for inhibition of monoamine catabolism) or CNS-stimulatory drugs (e.g., centrophenoxine for behavioral effects and reduction of lipofuscin deposition). However, the complexity of aging changes points out that it may be unrealistic to expect to effectively fortify or replace neuroendocrine controls by intervening at a single level. Under favorable conditions, the progressive decline in function of some neurons with advancing age is compensated by a variety of mechanisms such as increased firing rate of neurons, upward or downward regulation of receptors for neurotransmitters and hormones, gliosis; therefore, changes with aging are gradual and do not result in overt functional and behavioral deficits. When this compensation is no longer adequate due to genetic predisposition, adverse environmental effects or disease, neurologic, psychiatric and hormonal signs and symptoms develop. In this case, best results could be obtained by a multiple approach, that is, by strengthening the normal compensatory mechanisms and/or replacing them when they fail (for example, by the administration of a "cocktail" so-to-speak of neurotransmitter precursors which would attempt to restore optimal neurotransmitter balance, so important to normal CNS function). The potential usefulness of such multifactorial intervention is supported by the consistent promising results obtained with deitary restriction, a type of intervention which induces profound alterations in neuroendocrine functions. The effects of total caloric reduction and of tryptophan restriction will serve, here, to illustrate some of the beneficial effects of this type of intervention.

Perhaps the best documented method for prolongation of the lifespan involves total caloric restriction during development. Underfed rats grow slower, retain youthful characteristics longer and demonstrate a lower incidence of neoplasma than animals fed a complete diet and libitum (McCay, 1952; Berg and Simms, 1960; Ross, 1976; Masoro et al., 1980). Little is known about the mechanism of life prolongation by underfeeding, but the changes in brain chemistry and in endocrine function (particularly hypothyroidism) associated with caloric restriction could account for delayed expression of the developmental program and retardation in its postmaturational decline (Shambaugh and Wilber, 1974). Other experiments involve the deficiency of the essential amino acid tryptophan capable of mimicing both protein deficiency and decreased serotonin levels. In these experiments, immature rats maintained for up to two years on low levels of tryptophan (ranging from 25% to 33% of the amount of tryptophan necessary for an optimal diet) and then returned to normal diets, showed delay in reproductive aging and in the onset of tumorogenesis, preservation of homeostatic (i.e., thermoregulatory) function, improved coat conditions and, possibly, increased longevity (from 3.5 to 3.75 years (Segall and Timiras, 1975, 1976). Accompanying observations with electron microscopy show that tissues normally involved in aging such as connective tissue (e.g., bone collagen), appear to be chronologically younger in the tryptophan-restricted animals than in controls (Figure 3). Even when the low tryptophan diet was introduced at three or 13 months of age, there were visible indications of aging retardation (Segall and Timiras, 1976).

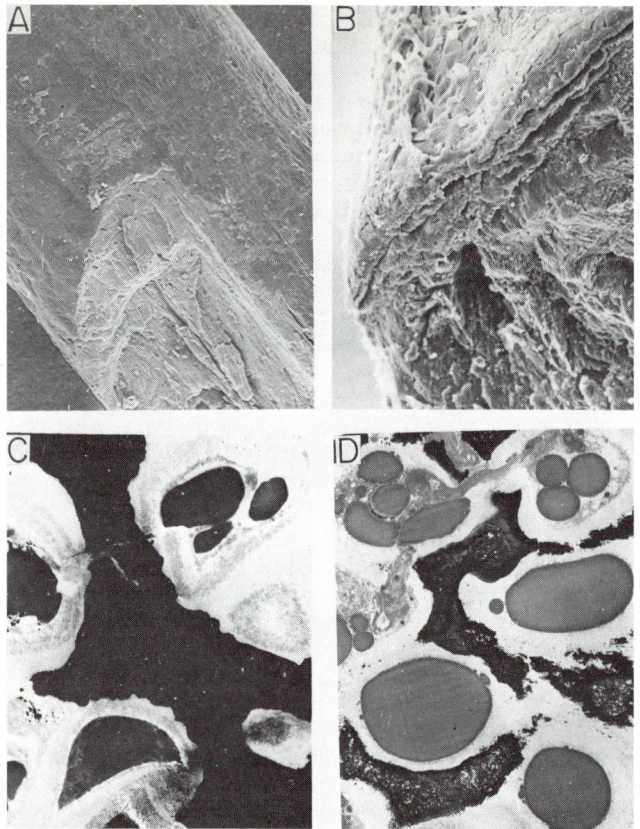

Fig. 3. Scanning electromicrograph of bone (A and B) and electronmicrograph of cartilage (C and D) in male Long-Evans rats at 9 months of age. Tissues on the life (A and C) were obtained from control rats fed from weaning until sacrifice a standard Purina rat chow (containing sufficient tryptophan to ensure adequate growth and maintenance); tissues on the right (B and D) were from rats fed over the same period, a diet deficient in tryptophan (containing approximately 33% of the tryptophan present in the control diet). In A, note flattened appearance of periosteocytes, perhaps related to extension of cells to compensate for cell loss (x 100). In B, note rounded appearance of periosteocytes reminiscent of the appearance of a much younger (2-3 months) bone (x 250). Comparison between C and D focuses on calcium deposition (dark areas). Cartilage calcification is much less advanced in the tryptophan-deficient (D) (x 2580) than in control (C) (x 2580) rats. The younger appearance of bone and cartilage of the tryptophan-deficient rats is associated with other youthful signs with respect to coat, reproductive activity, temperature regulation, etc. (See Segall et al., 1978; Segall and Timiras, 1975, 1976.) Courtesy of Dr. J.R. Walton, Lawrence Livermore Laboratory, California.

In these animals, dramatic changes were observed in the histology, ultrastructure and histochemistry of the anterior pituitary (Segall et al., 1978). Blood levels of TSH, thyroxine and triiodothyronine, weights of endocrine target organs such as the ovary, adrenal, and hormonally-dependent phenomena such as overall body size, rate of hair growth, and vaginal opening were all depressed or delayed in appearance (Segall et al., 1978). Upon refeeding a tryptophan-adequate diet, many, but not all, signs of endocrine function returned to normal (Segall et al., 1978; Ooka et al., 1978); in fact, the thyroid axis remained altered for several months following cessation of the experimental treatment (Ooka et al., 1978). Clearly, these approaches to intervention in the onset, course and outcome of the aging process represent only preliminary attempts and must be evaluated in the context of a larger program of "life-extension" sciences. In this program, several strategies besides those briefly reviewed here – e.g., nucleus manipulation and genetic reconstruction, tissue regeneration and transplantation of natural and artifical organs, suspended animation and resuscitation – are being developed and may become applicable in a not too distant future.

REFERENCES

Arieff, A.I., Cooper, J.D., Armstrong, D., and Lazarowitz, V.C. Dementia, renal failure and brain aluminum. *Ann. Int. Med.* 90, 741-747 (1979).

Bartus, R.T. Physostigmine and recent memory: effects in young and aged nonhuman primates. *Science* 206, 1087-1089 (1979).

Bartus, R.T., Dean, R.L., Goas, J.A. and Lippa, A.S. Age-related changes in passive avoidance retention and modulation with dietary choline. *Science* 209, 301-303 (1980).

Berg, B.N. and Simms, H.S. Nutrition and longevity in the rat. II. Longevity and onset of disease with different levels of food intake. *J. Nutr.* 71, 255-263 (1960).

Berge, T., Brun, A., Hansing, B., and Kjellman, B. Congenital generalized lipodystrophy. *Acta Path. Microbiol. Scand.*, Sect. A., 84, 47-54 (1976).

Bondareff, W. Synaptic organization as a function of aging in *Neural Regulatory Mechanisms During Aging*. R.C. Adelman, J. Roberts, G.T. Baker, III, S.I. Baskin, and V.J. Cristofalo, eds. Alan R. Liss, Inc., New York (1980) pp. 143-158.

Brannan, T.S., Maker, H.S., Weiss, C., and Cohen, G. Regional distribution of glutathion peroxidase in the adult rat brain. *J. Neurochem.* 35, 1013-1014 (1980).

Brizzee, K.R., Sherwood, N., and Timiras, P.S. A comparison of cell populations at various depth levels in cerebral cortex of young adult and aged Long-Evans rats. *J. Gerontol.* 23, 289-298 (1968).

Brizzee, K.R., Cancilla, P.A., Sherwood, N., and Timiras, P.S. The amount and distribution of pigments in neurons and glia of the cerebral cortex. Autofluorescent and ultrastructural studies. *J. Gerontol.* 24, 127-135 (1969).

Brody, H. An examination of cerebral cortex and brainstem aging, in *Neurobiology of Aging*. R.D. Terry and S. Gershon, eds. Raven Press, New York (1976), pp. 177-181.

Brody, H. Structural changes in the nervous system, in *The Regulatory Role of the Nervous System in Aging, Interdisciplinary Topics in Gerontology*, Vol. 7. H. Blumenthal, ed. S. Karger, Basel (1970), pp. 9-21.

Brody, H. Cell counts in cerebral cortex and brainstem, in *Alzheimer's Disease Senile Dementia and Related Diseases*. R. Katzman, R.D. Terry, and K.L. Bick, eds. Raven Press, New York (1978), pp. 349-355.

Brody, H. and Vijayashankar, N. Cell loss with aging, in *The Aging Brain and Senile Dementia*. K. Nandy and I. Sherwin, eds. Plenum Press, New York (1977a), pp. 15-21.

Brody, H. and Vijayashankar, N. Anatomical changes in the nervous system, in *Handbook of the Biology of Aging*. C.E. Finch and

L. Hayflick, eds. Van Nostrand Reinhold Co., New York (1977b), pp. 241-261.

Buell, S.Y. and Coleman, P.D. Dendritic growth in the aged human brain and failure of growth in senile dementia. *Science* 206, 854-856 (1979).

Carlsson, A. and Winblad, B. Influence of age and time interval between death and autopsy on dopamine and 3-methoxytyramine levels in human basal ganglia. *J. Neural. Transm.* 38, 271-276 (1976).

Choy, V.J., Klemme, W.R., and Timiras, P.S. Thyrotropin polymorphism during aging in the male rat. XII International Congress of Gerontology, Abstract, Hamburg, July, 1981.

Clemens, J.A. and Fuller, R.W. Chemical manipulation of some aspects of aging, in *Pharmacological Interventions in the Aging Process, Advances in Experimental Medicine and Biology*, Vol. 9, J. Roberts, R.C. Adelman and V.J. Cristofalo, (eds.). Plenum Press, New York (1977), pp. 187-206.

Cohen, G. The generation of hydroxyl radicals in biologic systems: toxicological aspects. *Photochem. Photobiol.* 28, 669-675 (1978).

Cole, G. Dopamine toxicity and neuroblastoma cells as a model for cellular aging of dopaminergic pacemaker neurons (unpublished).

Conn, P.M., Cooper, R., McNammara, C., Rogers, D.C., and Shoenhart, L. Qualitative change in gonadotropin during normal aging in the male rat. *Endocrinology* 106, 1949-1993 (1980).

Connor, J.R., Diamond, M.C., and Johnson, R.E. Occipital cortical morphology of the rat: alterations with age and environment. *Exp. Neurol.* 68, 158-170 (1980a).

Connor, J.R., Diamond, M.C. and Johnson, R.E. Aging and environmental influences on two types of dendritic spines in the rat occipital cortex. *Exp. Neurol.* 70, 371-379 (1980b).

Constantinides, P. *Experimental Atherosclerosis*. Elsevier, Amsterdam (1965).

Constantinides, P., Pratesi, F., Cavallero, C., and Di Perri, T. (eds.). *Immunity and Atherosclerosis Symposia*, Vol. 24, Academic Press, London (1980).

Cotman, C.W. (ed.) *Neuronal Plasticity*, Raven Press, New York (1978).

Cotman, C.W. and Scheff, S.W. Compensatory synapse growth in aged animals after neuronal death. *Mech. Ageing Dev.* 9, 103-117 (1979).

Cotzias, G.C., Miller, S.T., Nicholson, A.R., Maston, W.M., and Tong, L.C. Prolongation of the lifespan in mice adapted to large amounts of L-Dopa. *Proc. Nat. Acad. Sci.* 7, 2466-2469 (1974).

Cotzias, G.C., Miller, S.T., Tong, L.C., Papavasiliou, P.S. and Wang, Y.Y. Levodopa, fertility and longevity. *Science* 196, 549-551 (1977).

Cragg, B.C. The density of synapses and neurons in normal, mentally defective and ageing human brains. *Brain* 98, 81-90 (1975).

Cuatrecasas, P. Membrane receptors, in *Annual Review of Biochemistry*, Vo. 43, E.E. Snell, P.D. Boyer, A. Meister and C.C. Richardson (eds.) Annual Reviews, Inc., Palo Alto (1974), pp. 169-214.

Davies, P. Studies on the neurochemistry of central cholinergic system in Alzheimer's disease, in *Alzheimer's Disease — Senile Dementia and Related Disorders*. R. Katzman, R.D. Terry and K.L., Bick, eds. Raven Press, New York (1978), pp. 453-459.

de Duve, C. From cytases to lysosomes. *Fed. Proc.* 23, 1045-1049 (1966).

de Leon, M.J., Ferris, S.H., George, A.E., Reisberg, B., Kricheff, I.I., and Gershon, S. Computer tomography evaluations of brain-behavior relationships in senile dementia of the Alzheimer's type. *Neurobiol. Aging* 1, 69-79 (1980).

Denckla, W.D. Role of the pituitary and thyroid glands in the decline of minimal O_2 composition with age. *J. Clin. Invest.* 53, 572-581 (1974).

Denckla, W.D. A time to die. *Life Sci.* 16, 31-44 (1975).

Diamond, M.C., Johnson, R.E., and Gold, M.W. Changes in neuron and glia number in the young, adult and aging rat occipital cortex. *Behav. Biol.* 20, 409-418 (1977).

Dilman, V.M. Age-associated elevation of hypothalamic threshold to feedback control and its role in development, aging and disease. *Lancet.* ii, 1211-1291 (1971).

Dilman, V.M. The hypothalamic control of aging and age-associated pathology: The elevation mechanism of aging, in *Hypothalamus, Pituitary and Aging.* A.V. Everitt and J.A. Burgess, eds. C.C. Thomas Co., Springfield, Illinois (1976), pp. 634-667.

Doty, B.A. and Doty, L.A. Facilitative effects of amphetamine in avoidance conditioning in relation to age and problem difficulty. *Psychopharmacologia* 9, 234-241 (1966).

Doty, B.A. and Johnston, M.M. Effects of eserine administration, age and task difficulty on avoidance conditioning in rats. *Psychon. Sci.* 6, 101-102 (1966).

Drachman, D.A. and Leavitt, J. Human memory and the cholinergic system: a relationship to aging. *Arch. Neurol.* 30, 113-121 (1974).

Drachman, D.A., Noffsinger, D., Sahakian, B.J., Kurdziel, S., and Fleming, P. Aging, memory and the cholinergic system: a study of dichotic listening. *Neurobiol. Aging.* 1, 39-43 (1980).

Draves, D.J. and Timiras, P.S. Thyroid hormone effects in neural (tumor) cell culture: Differential effects on triiodothyronine nuclear receptors Na+K+ATPase activity and intracellular electrolyte levels, in *Tissue Culture in Neurobiology,* E. Giacobini, A. Vernadakis, and A. Shahar, eds. Raven Press, New York (1980), pp. 291-301.

Eberhardt, N.L., Valcana, T., and Timiras, P.S. Triiodothyronine nuclear receptors: an in vitro comparison of the binding of triiodothyronine to nuclei of adult rat liver, cerebral hemisphere and anterior pituitary. *Endocrinology* 102, 556-561 (1978).

Everitt, A.V. The neuroendocrine system and aging. *Gerontology* 26, 108-119 (1980).

Everitt, A.V. and Burgess, J.A. *Hypothalamus Pituitary and Aging,* C.C. Thomas Co., Springfield, Illinois (1976).

Feldman, M. and Dowd, C. Aging in rat visual cortex: light microscopic observations on layer V pyramidal apical dendrites. *Anat. Rec.* 178, 355-356 (1974).

Finch, C.E. Catecholamine metabolism in the brains of aging male mice. *Brain Res.* 52, 261-276 (1973).

Finch, C.E. Neuroendocrine and autonomic aspects of aging, in *Handbook of the Biology of Aging.* C.E. Finch and L. Hayflick, eds., Van Nostrand Reinhold, New York (1977), pp. 262-280.

Frolkis, V.V. and Bezrukov, V.V. *Aging of the Central Nervous System Interdisciplinary Topics in Gerontology,* Vol. 16, H.P. von Hahn, ed. S. Karger, Basel (1979).

Geinisman, Y., Bondareff, W., and Dodge, J.T. Hypertrophy of astroglial processes in the dentate gyrus of the senescent rat. *Am. J. Anat.* 153, 537-544 (1978).

Giuli, C., Bertoni-Freddari, C. and Pieri, C. Morphometric studies of synapses of the cerebellar glomerulus: the effect of centrophenoxin treatment in old rats. *Mech. Ageing Dev.* 14, 265-271 (1980).

Glees, P. and Gopinath, G. Age changes in the centrally and peripherally located sensory neurons in rat. *Z. Zellforsch* 141, 285-298 (1973).

Graham, D.G. On the origin and significance of neuromelanin. *Arch. Pathol. Lab. Med.* 103, 359-362 (1979).

Govoni, S., Memo, M., Saiani, L., Spano, P.F., and Trabucchi, M.

Impairment of brain neurotransmitter receptors in aged rats. *Mech. Ageing Dev.* 12, 39-46 (1980).

Handler, P. Some comments on risk assessment The National Research Council in 1979, *Nat. Acad. Sci.* 3-24 (1979).

Hasan, M. and Glees, P. Genesis and possible dissolution of neuronal lipofuscin, *Gerontologia* 18, 217-236 (1972).

Hasan, M. and Glees, P. Ultrastructural age changes in hippocampal neurons, and neuroglia. *Exp. Geront.* 8, 75-83 (1973).

Herman, E. Senile hypophyseal syndromes, in *Hypothalamus, Pituitary and Aging*, A.V. Everitt and J.A. Burgess, eds. C.C. Thomas Co., Sprungfield, Illinois (1976), pp. 157-170.

Hinds, J.W. and McNeilly, N.A. Aging of the rat olfactory bulb: growth and atrophy of constituent alyers and changes in size and number of mitral cells. *J. Comp. Neurol.* 171, 345-367 (1977).

Hoch, F.L. Metabolic effects of thyroid hormones, in *Handbook of Physiology*, Vol. III, Section 7. American Physiological Society, Washington, D.C., (1974), pp. 391-412.

Hoffman, G.E. and Sladek, J.R. Age-related changes in idopamines, LHRH and somatostatin in the rat hypothalamus. *Neurobiol. Aging* 1, 27-38 (1980).

Hsu, H.K. and Peng, M.T. Hypothalamic neuron number of old female rats. *Gerontology* 24, 434-440 (1978).

Huttenlocher, P.R. Synaptic density in human frontal cortex-developmental changes and effects of aging. *Brain Res.* 163, 195-205 (1979).

Huxley, A.L. *Brave New World*. Doubleday, New York (1932).

Ito, J.M., Valcana, T., and Timiras, P.S. Effect of hypo- and hyperthyroidism on regional monoamine metabolism in the adult brain. *Neuroendocrinology* 24, 55-64 (1977).

Jellinger, K., Flament, H., Riederer, P., Schmid, H., and Ambrozi, L. Levodopa in the treatment of (pre) senile dementia. *Mech. Ageing Dev.* 14, 253-264 (1980).

Jensen, R.A., Messing, R.B., Martinez, J.L., Vasquez, B.J., and McGaugh, J.L. Opiate modulation of learning and memory in the rat, in *Aging in the 80's: Psychological Issues*. L. Poon, ed., Section IV: Psychopharmacological Issues. American Psychological Association, Washington, D.C. (1980), pp. 191-200.

Jonec, V.F. and Finch, C.E. Senscence and dopamine uptake by subcellular fractions of the C57BL/6J male mouse. *Brain Res.* 91, 197-215 (1975).

Kanungo, M.S., Patnaik, S.K. and Kowl, O. Decrease in 17-β-oestradiol receptor in brain of ageing rats. *Nature* 253, 366-367, (1975).

Kapitulnik, J., Tshershedsky, M., and Barenholz, Y. Fluidity of rat liver microsomal membrane: increase at birth. *Science* 206, 843-844 (1979).

Katchai, H., Barenyolz, Y., Ross, T.F., and Wermer, D.E. Developmental changes in plasma membrane fluidity in check embryo heart. *Biochim. Biophys. Acta* 436, 101-112 (1976).

Klug, T.L., and Adelman, R.C. Altered hypothalamic-pituitary regulation of thyrotropin in male rats during aging. *Endocrinology* 104, 1136-1142 (1979).

Kny, W. Uber die verteilung des lipofuscins in der skeletmuskulator in ihrer begiehung zur funktion. *Virchows Arch.* (Pathol. Anat.) 299, 468-478 (1937).

Lippa, A.S., Pelham, R.W., Beer, B., Critchett, D.J., Dean, R.L., and Bartus, R.T. Brain cholinergic dysfunction and memory in aged rats. *Neurobiol. Aging* 1, 13-20 (1980).

Liss, L. Aging brain and dementia, in *Aging-Its Chemistry*. A.A. Dietz, (ed.). The American Society for Clinical Chemistry, Washington, D.C. (1979), pp. 183-206.

Makinodan, T. Nature of the decline in antigen-induced humoral immunity with age. *Mech. Ageing Dev.* 14, 165-172 (1980).

Makman, M.H., Gardner, E.L., Thal, L.J., Hirschhorn, I.D., Seeger, T.F., and Bhargava, G. Central monoamine receptor systems: influence of aging, lesion and drug treatment, in *Neural Regulatory Mechanisms During Aging.* C. Adelman, J. Roberts, G.T. Baker, III, S.I. Baskin, and V.J. Cristofalo, eds. Alan R. Liss, Inc. New York (1980), pp. 91-127.

Margarity, M., Veskoukis, M., Matoskis, N., Valcana, T., Miller, C. and Timiras, P.S. Changes with aging in serum thyroid hormone levels and thyroid hormone binding in brain tissue. 63rd Annual Meeting of the Endocrine Society, Cincinnati, Ohio, June, 1981.

Martin, G.M. Genetic syndromes in man with potential relevance to the pathobiology of aging, in *Genetic Effects on Aging.* D. Bergsma, D.E. Harrison, and N.W. Paul, eds. Alan R. Liss, Inc., New York (1978), pp. 5-39.

Masoro, E.J., Yu, B.P., Bertrand, H.A. and Lynd, F.T. Nutritional probe of the aging process. *Fed. Proc.* 39, 3178-3182 (1980).

McCay, C.M. Chemical aspects of aging and the effect of diet upon aging, in *Cowdry's Problems of Aging,* III edition. A.I. Lansing (ed.). Williams and Wilkins, Baltimore (1952), pp. 139-202.

McGreer, E.G. and McGreer, P.L. Age changes in human for some enzymes associated with metabolism of catecholamines, GABA and acetylcholine, in *Neurobiology of Aging,* J.M. Ordy and K.R. Brizzee, eds. Plenum Press, New York (1975), pp. 287-305.

McKenna, O., Arnold, G., and Holtzmann, E. Microperoxisome distribution in the central nervous system of the rat. *Brain Res.* 117, 181-194 (1976).

Meisami, E. and Timiras, P.S. Normal and abnormal biochemical brain development, in *Biochemical Development of the Fetus and Neonate,* C. Jones, ed. Elsevier, Amsterdam, (in press).

Mervis, R. Structural alterations in neurons of aged canine neocortex: a Golgi study. *Exp. Neurol.* 62, 417-432 (1978).

Miller, A.E., Shaar, C.J. and Riegle, G.D. Aging effects on hypothalamic dopamine and norepinephrine content in the male rat. *Expl. Aging Res.* 2, 475-480 (1976).

Miller, L.H., Groves, G.A., Bopp, M.J., and Kastin, A.J. A neuroheptapeptide influence on cognitive functioning in the elderly. *Peptides* 1, 55-57 (1980).

Mohs, R.C., Davis, K.L., Tinklenberg, J.R., and Hollister, L.E. Choline chloride effects on memory in the elderly. *Neurobiol. Aging* 1, 21-26 (1980).

Nandy, K. Further studies on the effects of centrophenoxine on the lipofuscin pigment in the neurons of senile guinea pigs. *J. Gerontol.* 23, 82-92 (1968a).

Nandy, K. Histologic and histochemical study of motor neurons with special reference to experimental degeneration, aging and drug actions, in *Contemporary Neurology Symposium,* Vol. II. *Motor Neuron Diseases: Research on Amyotrophic Lateral Sclerosis and Related Disorders.* F.H. Norris and L.T. Kurland, eds. Grune and Stratton, New York (1968b), pp. 319-334.

Nandy, K. The neuropathology of aging and dementia, in *Geriatric Dentistry,* C.J. Toga, K. Nancy, and H.H. Chauncey, eds. D.C. Heath and Co., Lexington, Massachusetts (1979), pp. 85-102.

Nandy, K. and Lal, H. Neuronal lipofuscin and learning deficits in aging mammals, in *Neuropsychopharmacology,* P. Deniker, C.

Radouco-Thomas, and A. Villeneuve, eds. Pergamon, New York (1978), pp. 1633-1645.

Ooka, H., Segall, P.E. and Timiras, P.S. Neural and endocrine development after chronic tryptophan deficiency in rats. II. Pituitary-thyroid axis. *Mech. Ageing Dev.* 7, 19-24 (1978).

Ordy, J.M. Neurobiology and aging in nonhuman primates, in *Neurobiology of Aging*, J.M. Ordy and K.R. Brizzee, eds. Plenum Press, New York (1975), pp. 575-597.

Ordy, J.M. Geriatric psychopharmacology: Drug modification of memory and emotionality in relation to aging in human and non-human primate brain, in *Brain Function in Old Age*. F. Hoffmeister and C. Muller, eds. Springer-Verlag, Berlin (1979), pp. 435-455.

Perry, E.K., Gibson, P.H., Blessed, G., Perry, R.H., and Tomlinson, B.E. Neurotransmitter enzyme abnormalities in senile dementia: choline acetyltransferase and glutamic acid decarboxylase in necropsy brain tissue. *J. Neurol. Sci.* 34, 247-265 (1977).

Platt, J.E., Christofer, M.A. and Sullivan, C.A. The role of prolactin in blocking thyroxine-induced differentiation of tail tissue in larval and neotenic Ambystoma tigrinum. *Gen. Comp. Endocrinol.* 35 402-408 (1978).

Prasad, K.N., Sinha, P.K., Ramanuja, M., and Sakamoto, A. Sodium ascorbate potentiates the growth inhibitory effect of certain agents on neuroblastoma cells in culture. *Proc. Nat. Acad. Sci.* 72, 829-832 (1979).

Puri, S.K. and Volicer, L. Effects of aging on cyclic AMP levels and adenylate cyclase and phosphodiesterase activities in the rat corpus striatum. *Mech. Ageing Dev.* 6, 53-58 (1977).

Quadri, S.K., Kledzik, G.S. and Meites, J. Reinitiation of estrous cycles in old constant estrous rats by centrally acting drugs. *Neuroendocrinology* 11, 248-255 (1973).

Quay, W.B. Pineal homeostatic regulation of shifts in the circadian activity rhythm during maturation and aging. *Trans. N.Y. Acad. Sci.* 34, 239-254 (1972).

Reichel, W. Lipofuscin pigment accumulation and distribution in five rat organs as a function of age. *J. Gerontol.* 23, 145-153 (1968).

Reigle, G.D. Chronic stress effects on adrenocortical responsiveness in young and aged rats. *Neuroendocrinology* 11, 1-10 (1973).

Rivnay, B., Globerson, A., and Shinitzky, M. Viscosity of lymphocyte plasma membrane in aging mice and its possible relation to serum cholesterol. *Mech. Ageing Dev.* 10, 71-79 (1979).

Roberts, M.A. and Caird, F.I. Computerized tomography and intellectual impairment in the elderly. *J. Neurol. Neurosur. Psych.* 39, 986-989 (1976).

Rogers, J., Silver, M.A., Shoemaker, W.J., and Bloom, F.E. Senescent changes in a neurobiological model system: cerebellar Purkinjie cell electrophysiology and correlative anatomy. *Neurobiol. Aging* 1, 3-11 (1980).

Ross, M.H. Nutrition and longevity in experimental animals, in *Nutrition and Aging*, M.D. Winick, ed. John Wiley and Sons, New York (1976), pp. 43-57.

Roth, G.S. Reduced glucocorticoid binding site concentration in cortical neuronal perikarya from senescent rats. *Brain Res.* 107, 345-354 (1976).

Roth, G.S. Hormone action during aging: alterations and mechanisms. *Mech. Ageing Dev.* 9, 497-514 (1979).

Scheibel, A.B. and Tomiyasu, U. Dendritic sprouting in Alzheimer's presenile dementia. *Exp. Neurol.* 60, 1-8 (1978).

Scheibel, M.E., Lindsay, R.D., Tomiyasu, U., and Scheibel, A.B.

Progressive dendritic changes in aging human cortex. *Exp. Neurol.* 47, 392-403 (1975).

Segall, P.E. Interrelations of dietary and hormonal effects in aging. *Mech. Ageing Dev.* 9, 515-525 (1979).

Segall, P.E. and Timiras, P.S. Age-related changes in thermoregulatory capacity of tryptophan-deficient rats. *Fed. Proc.* 34, 83-85 (1975).

Segall, P.E. and Timiras, P.S. Pathophysiologic findings after chronic tryptophan deficiency in rats: a model for delayed growth and aging. *Mech. Ageing. Dev.* 5, 109-124 (1976).

Segall, P.E., Ooka, H., Rose, K., and Timiras, P.S. Neural and endocrine development after chronic tryptophan deficiency in rats. I. Brain monoamine and pituitary responses. *Mech. Ageing Dev.* 7, 1-17 (1978).

Seip, M. Generalized lipodistrophy, in *Ergebnisse der Inneren Medizin und Kinderheilkunde*, Vol. 31. P. Frick, G.A. von Harnack, A.F. Muller, A. Prader, R. Schoener, and H.P. Wolff, eds. Springer-Verlag, New York (1971), pp. 59-95.

Selye, H. *Stress – The Physiology and Pathology of Exposure to Stress*, Acta Inc., Medical Publishers, Montreal (1950).

Selye, H., Gentile, G., and Prioreschi, P. Cutaneous molt induced by calciphylaxis in the rat. *Science* 134, 1876-1877 (1961).

Shaar, C.J., Euker, J.S., Riegle, G.D., and Meites, J. Effects of castration and gonadal steroids on serum LH and prolactin in old and young rats. *J. Endocrinol.* 66, 45-51 (1975).

Shambaugh, G.E. and Wilber, J.F. The effect of caloric deprivation upon thyroid function in the neonatal rat. *Endocrinology* 94, 1145-1149 (1974).

Shih, J.C. and Young, H. The alterations of serotinin binding sites in the aged human brain. *Life Sci.* 23, 1441-1448 (1978).

Simpkins, J.W., Mueller, G.P., Huang, H.H., and Meites, J. Evidence for depressed catecholamine and enhanced serotonin metabolism in aging male rats: possible relation to gonadotropin secretion. *Endocrinology* 100, 1672-1678 (1977).

Snyder, S.H. Brain peptides as neurotransmitters. *Science* 209, 976-983 (1980).

Sotelo, C. and Palay, S.L. Altered axons and axon terminals in the lateral vestibular nucleus of the rat. *Lab. Inv.* 25, 653-671 (1971).

Strehler, B.L., Mark, D.D., Mildvan, A.S., and Gee, M.W. Rate and magnitude of age pigment accumulation in the human myocardium. *J. Gerontol.* 14, 430-439 (1959).

Sun, A.Y. and Samorajski, T. The effects of age and alcohol on Na+K+ATPase activity of whole homogenate and synaptosomes prepared from mouse and human brain. *J. Neurochem.* 24, 161-164 (1975).

Tappel, A.L., Sawant, P.L., and Shibko, S. Lysosomes: distribution in animals, hydrolytic capacity and other properties in *Lysosomes*, A.V.S. de Reuck and M.P. Cameron, eds. Little, Brown & Co., Boston (1963), pp. 98-113.

Terry, R.D. Morphological changes in Alzheimer's disease-senile dementia: ultrastructural changes and quantitative studies, in *Congential and Acquired Cognitive Disorders*. R. Katzman, ed. Raven Press, New York (1979), pp. 99-105.

Terry, R.D. Some biological aspects of the aging brain. *Mech. Ageing Dev.* 14, 191-202 (1980).

Timiras, P.S. *Developmental Physiology and Aging*, Macmillan, New York (1972).

Timiras, P.S. Biological perspectives on aging: in search of a masterplan. *Am. Sci.* 66, 605-613 (1978).

Timiras, P.S. and Bignami, A. Pathophysiology of the aging brain, in *Special Review of Experimental Aging Research, Progress in Biology*. M.F. Elias, B.E. Eleftheriou, and P.K. Elias, eds. EAR Inc., Bar Harbor, Maine (1976), pp. 351-378.

Timiras, P.S. and Hudson, D.B. Changes in neurohumoral transmission during aging of the central nervous system, in *Neural Regulatory Mechanisms During Aging*. R.C. Adelman, J. Roberts, G.T. Baker, III, S.I. Baskin, and V.J. Cristofalo, eds. Alan R. Liss, Inc., New York (1980), pp. 25-51.

Tolmasoff, J.M., Ono, T., and Cutler, R.G. Superoxide dismutase: correlation with lifespan and specific metabolic rate in primate species. *Proc. Nat. Acad. Sci.* 77, 2777-2781 (1980).

Trygstad, O. and Foss, I. Congenital generalized lipodystrophy and experimental lipoatrophic diabetes in rabbits treated successfully with fenfluramine. *Acta Endocrinol.* 85, 436-448 (1977).

Unsworth, B.R., Felming, L.H., and Caron, P.C. Neurotransmitter enzymes in telecephalon, brain stem and cerebellum during the entire life span of the mouse. *Mech. Ageing Dev.* 13, 205-217 (1980).

Uylings, H.B.M., Kuypers, K., Diamond, M.C., and Veltman, W.A.M. Effects of differential environments on plasticity of dendrites of cortical pyramidal neurons in adult rats. *Exp. Neurol.* 62, 658-677 (1978).

Valcana, T. The role of triiodothyronine (T3) receptors in brain development, in *Neural Growth and Differentiation*, E. Meisami and M.A.B. Brazier (eds.) Raven Press, New York (1979), pp. 39-58.

Valcana, T. and Timiras, P.S. Choline acetyltransferase activity in various brain areas of aging rats. VIII International Congress of Gerontology, (abstract) 2,24 (1969).

Valcana, T. and Timiras, P.S. Nuclear triiodthyronine receptors in the developing rat brain. *Mol. Cell. Endocrinol.*, Vol. II, pp. 31-41 (1978).

Valcana, T. and Timiras, P.S. Changes in rat liver nuclear triiodothyronine receptors with age and thyroid activity, in *Hormones and Development*, L. Macho and J. Strbak, eds. Slovak Academy of Sciences, Bratislava, Czechoslovakia (1979), pp. 47-75.

Walker, R.F. and Timiras, P.S. Pacemaker insufficiency and the onset of aging, in: *Cellular Pacemakers II*, P. Carpenter, ed. Wiley Interscience, New York, (in press, b).

Walker, R.F., Cooper, R.L., and Timiras, P.S. Constant estrus: role of rostral hypothalamic monoamines in development of reproductive dysfunction in aging rats. *Endocrinology* 107, 249-255 (1980).

Walker, R.F. and Timiras, P.S. Serotonin in development of cyclic reproductive function, in *Serotonin: Current Aspects of Neurochemistry and Function, Advances in Experimental Biology and Medicine*, E. Haber, ed. Plenum Press, New York (in press a).

Weil-Fugazza, J., Godefroy, F. and Stupfel, M. Relation of sex and ageing to serotonin metabolism in rats. *Mech. Ageing Dev.* 13, 199-204 (1980).

Yamaura, H., Masatoshi, I., Kubota, K. and Matsuzawa, T. Brain atrophy during aging: a quantitative study with computer tomography. *J. Gerontol.* 35, 492-498 (1980).

Zeman, W. Studies on the neuronal ceroid-lipofuscinosis. *J. Neuropath. Exp. Neurol.* 33, 1-12 (1974).

Zeman, W. and Rider, J.A. (eds.) *Dissection of a Degenerative Disease: Proceedings of Four Round-Table Conferences on the Pathogenesis of Batten's Disease*, New York, American Elsevier (1975).

Copyright©1982, Spectrum Publications, Inc.
Hormones in Development and Aging

Chapter 16
Physiological Aspects of Menopause: Clinical and Experimental Studies
Eugene Eisenberg
Richard F. Walker

CLINICAL STUDIES

Introduction

The cessation of menses is a dramatic event that is but one feature of the physiological changes occurring as ovarian function ceases. At menopause, changes occur in a number of organs. In some, the fall in estrogen levels is the primary cause of the changes observed. In others, the change in endocrine environment interacts with the continuing aging process. The interactions of these processes are as yet incompletely understood. Genital tract atrophy and impaired ability of autonomic control of vascular tone are clearly related to fall in estrogen levels occurring at menopause. The skin and bone changes appear to be affected by estrogen lack interacting with other factors. Understanding of the physiological aspects of the menopause in humans is limited by the small number of clearly defined and specific observations on the biological changes in the menopause as well as our limited basic knowledge of the aging process. The changes which occur at menopause will be discussed here from a clinical point of view under the following categories: (a) the hormonal events responsible for the menopause (b) the hormonal milieu resulting after menopause (c) the effects of these hormonal changes on genital tissues (d) the effects of these hormonal changes on nongenital tissues, and (e) the interrelations of the changes in (c) and (d) with the natural aging process that begins at different times in different tissues.

Hormonal Changes

Cessation of menses is the most dramatic sign of the loss of normal cyclic ovarian function and it is relatively abrupt. The climateric process–that is, the process resulting ultimately in the cessation of ovarian function–is more gradual and begins a few years prior to the menopause. There is gradual reduction in ovarian responsiveness to the pituitary gonadotropins for about two to five years. This is seen as a gradual decrease in ovarian estrogen (estradiol-17β) and progesterone production in spite of increased (but still "normal") levels of gonadotropins (Sherman and Korenman, 1975). Thus, many of the changes described below probably begin imperceptibly and progress gradually as a response to decreasing levels of ovarian hormones over a variable number of years prior to menopause. The final cessation of cyclic ovarian function results in marked fall in plasma estrogen levels (Longcope, 1971). Release of the normal negative feedback response of the hypothalamo-pituitary system from the inhibitory action of estrogens results in the very high levels of both follicle-stimulating hormone (FSH) and lutenizing hormone (LH) characteristic of the postmenopausal state.

The postmenopausal ovary while secreting almost no estrogen continues to produce small but significant amounts of testosterone and androstenedione (Judd et al., 1974). This augments the circulating levels of androgenic hormones produced as the adrenals continue to secrete these hormones at about the same rate as they did premenopausally (Poortman et al., 1973). An additional androgen, dehydroepiandrosterone, is produced solely by the adrenal. Most of the small amount of estrogen found in plasma of postmenopausal women results from conversion of androstenedione by "nonglandular" tissues, especially subcutaneous adipose tissue, to estrone (Grodin et al., 1973). Several conditions may increase the rate of adipose tissue conversion of androstenedione to estrone (e.g., liver disease, obesity, and thyrotoxicosis).

A number of other alterations in endocrine function have been noted in postmenopausal women, but these are slight. Furthermore, similar changes may be noted in older men, as well, as so that these may be primarily the effects of aging (see Chapter 14). However, the magnitude and direction of some of these effects of aging may be influenced by the fall in estrogen levels. As an example, one may cite loss of increased morning growth hormone secretion in postmenopausal women (Frantz and Rabkin, 1965).

The fall in ovarian hormone secretion results, of course, in profound atrophy of those genital tissues that they stimulate most directly, namely the uterus, vaginal epithelium and vulvar skin. The vaginal atrophy and decrease in vaginal secretions may result in dryness, burning, and other forms of vaginal discomfort (atrophic vaginitis) and in increased frequency of urethritis. The degree of atrophy varies considerably, and is largely but not entirely dependent on the residual estrogen level. The symptoms described are also only partially dependent on the degree of atrophy since nutritional state, concurrent infections, frequency of intercourse and muscle tone may affect them.

Sex Hormone Binding Globulin and Receptors. During the menopause, although the reduction in progesterone is permanent, estrone levels of ovarian origin may remain sufficiently high to elicit responses of the uterine endometrium, breast, bone and other tissues (Siiteri, 1975). This postmenopausal production of estrone does not exhibit cyclicity or negative feedback regulation and may be increased in association with obesity, hepatic disease and hyperthyroidism. It has been also shown that the major portion of circulating, unconjugated steroids is specifically bound to high affinity serum proteins such as sex hormone binding globulin (SHBG). Though changes in SHBG could hypothetically affect the amount of physiologically active estrone which is present, no association between age and SHBG capacity is observed in normal women (Siiteri, 1979). However, obese postmenopausal women show an increase in the percent free estrogen which correlates with a reduction in serum SHBF capacity in these women. Since postmenopausal women with breast and endometrial cancer tend to have higher percent free estrogen than normal women, it is possible that the more frequent occurrence of these neoplastic diseases in obese menopausal women may be related to increased estrogen production and reduced SHBG capacity (Siiteri, 1979). The consequent prolonged stimulation of estrogen receptors in target cells (including the presence of estrogen receptors in about 70% of primary or metastatic breast cancers) would create a chronic proliferative state that may set the stage for neoplastic transformations. Chronic estrogen stimulation is further enhanced by the fact that progesterone, which normally reduces estrogen receptor number, is no longer produced by the ovary of menopausal women.

Alterations of Neural Controls

The other major portion of the symptom-complex called "the post-menopausal syndrome" are hot flushes or sweats that are the result of instability of autonomic control of vasomotor tone. The exact mechanism and location of these effects of estrogen withdrawal are unknown but some animal studies (discussed below) suggest an intermediary role for hypothalamic and pontine neurotransmitters such as norepinephrine, dopamine, serotonin, and prostaglandins (Donoso et al., 1967; Bapna et al., 1971; Brody and Kadowitz, 1974). This process is probably modulated by the cerebral cortex since psychological effects on the frequency and severity of these symptoms are very frequently reported by the women experiencing them. There may also be some effect of normal aging on this neuroendocrine process since the passage of time is usually accompanied by decreasing frequency and severity of the flushes. The overlay of nutritional status and other environmental factors on these processes has been studied but their importance remains conjectural (Vasquez et al., 1976). The same may be said for other central nervous system symptoms that may appear just before, during or after the menopause, such as depression, insomnia, and emotional instability (see Chapter 16).

Systemic Changes

Many of the changes in skin that occur with normal aging, (e.g., thinning, decreased mitosis, loss of elasticity) appear to be acutely accelerated by castration and presumably occur at a slower rate in natural menopause. The interrelations of aging and hormone-lack are complex; whereas estrogen therapy may retard some of these changes it does not reverse them as completely as it does the genital and neuroendocrine changes described above. Some of the improvement in skin texture seen with estrogen treatment may be secondary to increased hydration resulting from increased renal retention of salt and water and to increased hyaluronidase activity.

Osteoporosis is a disorder of bone associated with many endocrine disturbances but most commonly found in postmenopausal women. Osteoporotic bone has thinner trabeculae and cortex (too little bone = osteopenia) but the remaining bone has normal matrix and mineralization. The loss of bone substance in osteoporosis is thought to be due to changes of bone remodeling rates rather than to a primary disturbance in bone mineral metabolism as is the case in osteomalacia. The bone of all normal adults is constantly being remodeled. Small areas are resorbed by osteoclastic activity and the resorptive cavity is replaced by new bone formed by osteoblastic activity. These two processes are usually linked and their overall rates are normally very close. After the age of 20-25 there is slight predominance of bone resorption on endosteal surfaces and slight predominance of bone formation on periosteal surfaces. These slight differences on the two surfaces result in gradual thinning of the bone cortex and decreased endosteal trabecular bone as a concomitant of normal aging, thus, bone mass is constantly decreasing in all adults ("physiologic" osteoporosis). These processes are modulated by interaction of a number of endocrine influences including parathormone, vitamin D (or more specifically its metabolite 1,25-dihydroxycholecalciferol), estrogens, androgens, growth hormone and thyroxine. Bone turnover rates and bone density may also be affected by diet, race and physical activity. Therefore, estrogen deficiency may not be the primary or sole cause of postmenopausal osteoporosis. However, the striking association of osteoporosis with estrogen deficiency states (including congenital estrogen deficiency such as occurs in gonadal dysgenesis and hypopituitary patients) strongly implicates it as an important factor. Kinetic studies have distinguished two types of osteoporosis: one associated with decreased bone formation and one with increased bone breakdown. The first type is seen in postmenopausal osteoporosis and long-standing Cushing's disease, the second in thyrotoxicosis, hyperparathyroidism and acromegaly as well as in acute immobilization and acute administration of corticoids in high doses (Eisenberg and Gordan, 1961). The postmenopausal women may also have increased bone destruction since it has been shown that when estrogen levels decline, bone resorption rates increase slightly as evidence by increases in blood calcium levels and in urine calcium and hydroxyproline excretion (Nordin, 1977) and in number of bone resorption sites (Riggs et al., 1972). There is no accompanying compensatory increase in bone formation rate as occurs in thyrotoxicosis, hyperparathyroidism, etc. After a variable number of years these effects of estrogen deficiency combined with those of the normal aging process result in critical decreases in bone mass in some women. This decrease in bone mass occurs more rapidly in ovariectomized women than in women having a natural menopause. Minor trauma in women with this form of osteopenia may result in painful and debilitating fractures most commonly in the vertebral bodies, wrist and femur ("pathologic" osteoporosis). The ability of estrogen therapy to arrest or prevent postmenopausal osteoporosis remains a matter of some controversy. The estrogens do induce a positive balance of calcium (Gordan, 1964) apparently by reducing bone resorption rate (Eisenberg, 1966). Most workers agree that such treatment will reduce the bone resorption rate. Some studies suggest that there is a secondary fall in bone formation rate that may nullify the beneficial effects of treatment, but Heaney and Recker (1975) concluded that administration of estrogens early in menopause maintains normal bone turnover rates and normal calcium balance. Many studies demonstrate favorable effects of estrogen treatment in calcium balance, bone turnover rates, bond density and fracture rates but are all of too short duration to definitively answer these questions. Interpretation of many of the studies is also hampered by the heterogeneity of the patient groups and the small degree of change in the observations made. However, in the absence of other proven therapies, estrogen administration to women with proven osteoporosis or likely to develop it appears warranted with the fiats suggested below.

Estrogen Use in Postmenopausal Women. The benefits of estrogen administration (treatment of postmenopausal symptoms and possible prevention of osteoporosis) the hazards of estrogen use, the risks of benefits of various types of estrogen therapy and the accompanying socioeconomic issues, are the subject of current concern and controversy. At a recent (1979) conference sponsored by the National Institute on Aging to discuss several of these issues, some points of consensus were reached among experts and the most important of which are briefly summarized here.

Estrogens are more effective than placebo (i.e., inactive substance) in decreasing the frequency and/or severity of vasomotor symptoms (e.g., hot flashes and sweating), in overcoming the atrophy of the vaginal epithelium and associated symptoms (e.g., dryness, itching, burning, pain during intercourse) and in retarding bone loss and, by inferrence, preventing the ultimate development of osteoporosis and attendant fractures. There is no current evidence to justify the use of estrogens in the treatment of primary psychological problems associated with menopause; the psychological improvement in well-being reported by some women after estrogens may be secondary to alleviation of the physical symptoms.

Since some of the symptoms listed above (e.g., hot flashes) decline spontaneously with time, unnecessary prolongation of therapy for these symptoms should be avoided; others (e.g., bone loss) appear to require continuous treatment and stopping therapeutic doses of estrogen may result in loss of previously obtained favorable effects. In such cases, cyclic treatment with estrogens should be continued but with close surveillance for early detection of possible unwanted effects and immediate interruption of treatment. The possibility of reducing risk for endometrial cancer by appropriate insertion of progesterone in the treatment cycle requires additional research studies.

Considering untoward effects, there is no convincing evidence that estrogens in moderate doses increase the risk of thromboembolic phenomena, stroke or heart disease, in women who have undergone natural menopause. Although it was once hoped that estrogens would protect against heart disease in aging women, this effect has not yet been demonstrated. On the other hand, there is evidence to show that the incidence of endometrial cancer (and perhaps also of cystic hyperplasia of the endometrium, a premalignant condition) increases several fold beginning approximately two to four years after use of relatively moderate daily doses of estrogens, the risk of the endometrial cancer increasing with the duration of use and declining after discontinuation. Estrogen use is most strongly associated with lesions of the lowest grade and earliest stage. The temporal relationship of the number of estrogen prescriptions and the incidence of carcinoma of the endometrium is of interest, both rising steadily until 1976 and showing a parallel decline after the warnings of the possible complications of estrogen use. Although the incidence of the carcinoma rose, mortality from the disease did not, a discrepancy that may be explained in large part by early detection and high cure rate.

Whereas the association of estrogens to mammary cancer is well demonstrated in experimental animals (see Chapter 19), a careful review of several well-documented case-control studies in humans has not revealed such a relationship, e.g., incidence rates of breast cancer have not changed in parallel with those of estrogen use, as have those for endometrial carcinoma. On the other hand, an increase in breast cancer over that of the general population was recorded after 15 years of estrogen use. Because of the high incidence and relatively poor prognosis for breast cancer, any possible association with estrogen use remains a concern, and further studies are needed to determine whether a causal relationship exists between the hormone and breast cancer (see Chapter 19).

Women who have undergone menopause many years in advance of the normally expected age represent a special group of particular concern with respect to the use of estrogens. Although the approximation of the normal physiological state through hormone replacement therapy would be best, there are no carefully controlled studies comparing the risks and benefits in these circumstances. It is clear that much additional information is needed to better define the relationship of estrogen use to the aging process in women.

Specifically, we need systematic knowledge of the natural course of the menopause in the absence of hormonal therapy, of alternatives to estrogen use, of the optimal means of providing estrogens if they are to be used, and of all aspects of their beneficial and adverse effects. Special attention should be directed toward studies that can determine the proper use of estrogens in young women who have undergone oophorectomy ten or more years before the natural time of menopause. Currently, no general formulation regarding therapy can be given. Rather each individual patient must base her decision on the relative values that she assigns to relief of symptoms, to expectations for optimizing health and well-being, and to various risks sustained in the process. Socially and culturally based attitudes toward menopause may influence these values and require further definition.

EXPERIMENTAL STUDIES

Introduction

In addition to the above-mentioned limitations, moral and ethical restrictions of invasive analytical procedures in human subjects further limit the study of the mechanisms of ovarian aging in women. Opportunely, a wealth of experimental literature on reproductive physiology is available from laboratory animals, the rat in particular. Exploitation of this resource will undoubtedly aid in our understanding of human reproductive aging, as it does in development, as long as properties of the animal models which relate to human or mammalian aging in general, are carefully selected. In the rat, the estrous cycle sufficiently resembles the human menstrual cycle to be useful for investigating mechanisms of female reproductive function. Accordingly, valid comparisons of certain aspects of normal ovarian function, such as follicular development have been made between rats and humans (see Chapter 9). The following review will discuss functional aspects of reproductive senescence in rats and will make comparisons with the human condition when applicable.

Patterns of Reproductive Failure in Aging Rats

Constant Estrus

At 10-15 months of age (the lifespan of the rat is approximately three years), estrous cycles lengthen due to the persistance of two or more days of vaginal cornification (estrus) per cycle. In spite of the diminished numbers of total oocytes, the number of eggs ovulated and corpora lutea formed are normal during this period of presenile irregularity (Aschheim, 1976). With continued irregularity, increased anovulation and lengthening of the cycles, the animals finally become acyclic and enter a condition called "constant estrus" in which the vaginal epithelium remains cornified. The high percentage of cornified epithelial cells in the smears of rats at this stage of senescence appears to result from uninterrupted ovarian estrogen secretion with little or no progesterone (Beach et al., 1979; Huang et al., 1978).

Ovaries from rats in constant estrus are small in comparison to those of younger animals, contain mixed populations of growing, atretic and occasionally cystic follicles, but they lack corpora lutea. These morphological changes are accompanied by hormonal alterations: in agreement with similar observations in humans (see above), conversion of progesterone into estrogen is increased as is accumulation of testosterone and androstenedione (Weisz and Lloyd, 1965). Serum FSH levels are always elevated. Basal LH levels, depending on the methodology used for measurement (bioassay or immunoassay), appear to be either depressed (Aschheim, 1976), unchanged (Shaar et al., 1975), or elevated (Meites et al., 1976; Reigle and Miller, 1978; Walker et al., 1980). Elevation of gonadotropins in the constant estrous rat resembles that seen in postmenopausal women, except that the increases are comparatively much greater in humans than in rats.

Diestrus and Anestrus

Another syndrome involving aging of the reproductive function in rats generally occurs later in life (>20 months) than constant estrus, and may follow in sequence or occur spontaneously after a period of estrous cycle irregularity (Aschheim, 1976). Distinct vaginal histology and ovarian function readily identify this syndrome often also called "repetitive pseudopregnancy." In this condition, vaginal smears are characteristically of the diestrous type and the vaginal epithelium shows mucification, apparently in response to the preponderant influence of progesterone in the virtual absence of estrogen. When estrogen is injected into these animals, vaginal cornification does not occur, but, instead induces a gravid-type (pseudopregnancy) mucification suggestive of a progesterone/estrogen synergism. The ovaries of these diestrous rats are not atrophied, as in constant estrus, primarily because many generations of corpora lutea are retained for prolonged periods. Serum gonadotropins are similar to those of young diestrous rats. Every 14-20 days or more, the vaginal smear spontaneously becomes cornified and is accompanied by ovulation which may be qualitatively and quantitatively normal when compared with younger cycling rats.

Occasionally, very old rats (> 28 months) exhibit an anestrous condition which is neither constant estrus nor diestrus. In this senile condition, the atrophic state of the ovaries and uterus is striking, the vagina is unstimulated, without mucification or keratinization and gonadotropin levels are profoundly depressed.

The reproductive disturbances in senile diestrous or anestrous rats differ markedly from those of menopausal women: for example, diestrous rats show normal ovulation, though at abnormal intervals, while postmenopausal women are anovulatory. Also, the ovary of diestrous rats contains persistent populations of functional, progesterone-secreting corpora lutea, whereas corpora lutea are rarely found in postmenopausal ovaries and progesterone output declines dramatically at a relatively early age after menopause in humans (Everitt, 1976). Because of these differences, in contrast to the similarities between the constant estrous syndrome in rats and the menopause in women discussed below, constant estrous appears to be the condition of choice for the experimental study of aging in the female reproductive function.

Human Menopause in Comparison With The Rat Constant Estrus Syndrome

The human vagina contains three major cell types; parabasal, intermediate, and superficial, which reflect degrees of maturation under the influence of estrogen (Schiff and Wilson, 1978). In young adult females, high levels of estrogen at midcycle apparently account for a preponderance of cornified superficial cells. The cells tend to cluster and become vesicular and basophilic after ovulation, while the parabasal cells predominate during the proliferative portion of the cycle. Exfoliated cells from the vaginal epithelium of older humans indicate that a shift in cell types occurs with advancing age. Diminshing estrogen levels in aging women correlate with decreasing numbers of superficial cells. However, a certain percentage of postmenopausal women show a persistence of superficial cells in vaginal smears, indicating prolonged exposure to estrogen (McLennan and McLennan, 1971). Therefore, initial similarities between menopause and constant estrus lie in the fact that persistent, albeit reduced estrogen secretion, as indicated by vaginal cytology, continues well beyond failure of normal cycles both in humans and rats (Talbert, 1977).

In aging (premenopausal) women, ovaries show a reduced number of oocytes and a decline in growing follicles; but graffian follicles may persist into middle age, and disappear only during the fifth decade (Block, 1952). Ovulation is not guaranteed with the development of graffian follicles, and occasionally, a pathological condition characterized by the development of follicular cysts occurs (Riley, 1964). Corpora lutea are less stable than earlier in life, and luteal failure appears to be a common occurrence during the period immediately preceeding menopause and may result in shortening of cycles and irregularity (Collett et al., 1954). Ovarian estrogen production progressively diminishes with advancing age, though its "peripheral conversion" from androgen may continue well beyond menopause (Siiteri and McDonald, 1973). Production of androgen by the ovarian stroma continues in postmenopausal women (Judd et al., 1974).

A comparison of these characteristics which are common to the menopause with those observed in rats in constant estrus are presented in Table 1. These similarities, which are basically endocrine, support the hypothesis that there are common properties in selected types of senile reproductive dysfunction which transcend species boundaries. If the animal model is valid, it would provide an invaluable resource to explore the more complex aspects of neuroendocrine changes which accompany advancing age. As discussed below, there is growing evidence that the reproductive dysfunction associated with senescence occurs in response to altered neural regulation of the pituitary-ovarian axis. Changes in monoaminergic activation and temporal regulation of reproductive cycles may be "at the heart" of aging phenomena. Through the use of the rat constant estrous model, these neural changes may be investigated, characterized and manipulated. Such studies may provide the background for future clinical treatment of menopause through therapeutic and/or prophylactic intervention.

Neural and Endocrine Changes Associated With The Development Of Constant Estrus

Ovaries

The female reproductive system in the rat has been examined at all levels of control, from ovary to brain, to identify the primary lesion responsible for failure of function with old age. In humans, a theory linking age-related decrease of oocyte populations with declining fertility was developed after oocyte depletion was reported to occur at about the time of reproductive failure (Jones, 1970). However, the concept of oocyte depletion as a primary mechanism for reproductive failure in humans may be invalid for several reasons (Steger, 1976). Similarly, in rats the reproductive potential of the senescent ovary is retained until advanced old age, and in all mammalian species studied so far, age-related changes in ovarian function seem to derive primarily from alterations in gonadotropin stimulation. Thus, extraovarian changes,

Table 1.

Common properties of menopause and constant estrus which support the validity of using the female rat as an experimental tool to investigate general mechanisms of reproductive senescence.

Observation	Reference
1. Increased irregularity and anovulatory cycles preceed failure of periodic cyclicity.	Timiras and Meisami, 1972; Meites et al., 1976.
2. Progesterone stimulates ovulation and promotes cycling during the period of presenile irregularity, and during early stages of reproductive failure.	Caligaris et al., 1971; Clemens et al., 1969; Odell and Swerdloff, 1968.
3. Both menopause and constant estrus occur at times which are relatively similar in comparison with the lifespan of both species.	Aschheim, 1976; McKinlay et al., 1972.
4. Oocyte populations are diminished with advancing age.	Jones, 1970; Talbert, 1977.
5. Graffian follicles tend not to ovulate, accumulate in the ovary and sometimes become cystic.	Talbert, 1977; Jones, 1970; Riley, 1964.
6. Corpora lutea are absent after cycling ceases.	Aschheim, 1976; Krohn, 1959.
7. Vaginal cytology suggests the continued influence of estrogen beyond the period of normal cycling.	McLennan and McLennan, 1971; Reigle and Miller, 1977.
8. Androgen continues to be produced by the ovarian stroma after estrogen production declines.	Judd et al., 1974; Mattingly and Huang, 1969.
9. Basal levels of serum gonadotropin are elevated when compared with young, reproductively competent controls.	Talbert, 1977; Reigle and Miller, 1978; Walker et al., 1978.
10. Occasional ovulation does not usually occur after termination of regular cycles.	Aschheim, 1976; Talbert, 1977.

at the level of the pituitary and hypothalamus would underlie the loss of cycles with advancing age.

However, it was recently suggested that extra-ovarian changes leading to failure of the reproductive neuroendocrine axis may result from a cumulative impact of estrogen upon the brain, leading ultimately to loss of the steroid-induced gonadotropin surge (Finch, 1979). This suggestion was based upon various experiments showing that during development the dose and time of exposure to estrogen, whether of ovarian or exogenous sources, is inversely correlated with the length of the reproductive function (Harlan and Gorski, 1978; Brawer et al., 1978). Therefore, the cumulative impact of steroid exposure resulting from successive reproductive cycles could represent a "counting mechanism" to determine the duration of cyclicity in the lifespan. Hence, the extraovarian changes leading ultimately to reproductive senescence in rodents, may result, in part, from ovarian influence (Finch, 1979).

Pituitary and Hypothalamic Hormones

Keeping in mind the eventual involvement of altered gonadotropin control in aging, the constant estrous syndrome could be viewed as resulting, in part, from changes in quantity and pattern of LH release from the pituitary. Indeed, rats in constant estrus ovulate (Meites et al., 1978) and cycle (Everett and Zeilmaker, 1979) after LH injection or are capable of increased LH secretion in response to injection of synthetic gonadotropin releasing hormone (Watkins et al., 1975). Additionally, serum FSH and LH levels are elevated in these aged rats, an observation of interest particularly when correlated with the increased gonadotropins of menopausal women (Talbert, 1977). Not only are gonadotropin levels elevated above basal values, but the cyclicity of their secretory pattern is lost, as shown by the disappearance of the LH surge. This loss of cyclicity occurs gradually and is manifested at first by a delay and attenuation of the LH surge in proestrus (van der Schoot, 1976).

The impairment of cyclicity of pituitary gonadotropin secretion in rats in constant estrus apparently results from changes in hypothalamic neurosecretion since specific (pulsatile) modes of LH-releasing hormone administration restore LH secretion to normal pattern values. This effect of LH-releasing hormone pulses suggests that the pituitary in the aged female rat is capable of sustaining greater gonadotropin release than that which normally occurs (Reigle and Miller, 1978). On the other hand, hypothalamic concentrations of LH-releasing hormones are not markedly reduced in aged rats and their levels appear to be sufficient to provide adequate control of anterior piutitary function and, thereby, support normal ovarian function (Reigle and Miller, 1978). The fact that they do not and that ovarian function is abolished, suggests that the functional deficit cannot be ascribed to alterations in pituitary gonadotropins and hypothalamic releasing hormones, but rather to alterations in higher brain centers. Thus, the condition of constant estrus in the rat and by extrapolation, menopause in women would develop from alterations in neural inputs controlling neurosecretion.

In addition to impairment of neural controls, both positive and negative feedback effectiveness of ovarian steroids on LH secretion is reduced with advancing age. Although basal gonadotropin levels are higher in constant estrous rats than in young rats, removal of steroid negative feedback by ovariectomy does not stimulate as great an additional rise of LH as it does after ovariectomy in young cycling rats in vaginal estrus (Huang et al., 1976a; Shaar et al., 1975). Also replacement of estrogen and progesterone in ovariectomized constant estrous rats is less effective in stimulating an LH surge than similar treatment in young rats (Meites et al., 1978). Recently, Miller and Reigle (1979) have shown that progesterone release is diminished in aging rats in constant estrus; they suggest that both progesterone deficiency in the ovaries and reduced capacity of the hypothalamo-pituitary system to respond to gonadal hormone feedback contribute to the development of the constant estrus syndrome of aged rats.

Hypothalamic Neurotransmitters

Norepinephrine and Dopamine. The reduced responsiveness of the hypothalamus to feedback by ovarian steroids has been attributed, in part, to a deficiency in catecholamine levels and metabolism in this brain area. Reciprocally, changes in brain neurotransmitter metabolism may initiate age-dependent disturbances in ovarian function (Meites et al., 1978). On the assumption that catecholamine deficiencies may be responsible for loss of reproductive function in aging rats, several investigators have administered catecholamine precursors, such as tyrosine or L-Dopa to constant estrus rats in an attempt to restore or prolong ovarian function (Quadri et al., 1973; Linnoila and Cooper, 1976). It is well-known that catecholamines, particularly norepinephrine and, in some cases, dopamine, control the phasic release of LH (Sawyer, 1975). Such catecholamine involvement in normal cyclicity was recently demonstrated when systemic treatment of old acyclic rats with lergotril mesolate, a direct dopamine receptor stimulant, reinstated regular vaginal cycles (Clemens and Bennett, 1977). Also, injection of L-Dopa into aged rats in constant estrus resulted in elevated levels of serum LH (Reigle and Miller, 1978) and reinstated cycling, presumably by affecting catecholamine metabolism in the brain, since concomitant treatment with drugs which reliably block the conversion of L-Dopa peripherally, but not centrally, failed to prevent the effect (Linnoila and Cooper, 1976). Current experiments have shown that when L-Dopa is applied locally to the medial preoptic area, but not to the dorsomedial septum or cortex of old rats in constant estrus, vaginal cycles and ovulation are reinstated (Cooper et al., 1979), thereby providing more direct support for the view that a localized neurotransmitter deficiency in a discrete brain area may be responsible for the reproductive failure associated with advancing age. Furthermore, actual measurements of catecholamine levels in the brain of aged rats, reveal significant depression in both norepinephrine and dopamine (Walker et al., 1980; Reigle and Miller, 1978; Simpkins et al., 1977b). Similar catecholamine deficits in menopausal women are suggested by the ability of L-Dopa to stimulate postmenopausal bleeding in some clinical cases (Wajsborth, 1972; Kruse-Larsen and Garde, 1971). Also decreased catecholamine levels and increased monoamine oxidase activity in several brain areas have been reported in old humans (Robinson et al., 1972) and may be in some way related to human reproductive senescence.

Serotonin. While catecholamines apparently stimulate gonadotropin release, the role of serotonin in control of anterior pituitary function is less clear. It has been suggested that alterations in serotonergic systems, either levels or turnover, may be involved in loss of phasic LH secretion in aged rats (Finch, 1978). Progesterone administration which reduces brain serotonin (Izquierdo et al., 1975), also stimulates ovulation and cycling in rats in constant estrus (Huang et al., 1976b). Also administration of serotonin inhibits LH secretion under certain conditions (Schneider and McCann, 1970). Thus, serotonin would play an inhibitory role in the release of LH from the pituitary. However, other studies contradict this hypothesis on the basis that blockade of serotonin synthesis with p-chlorphenylalanine (PCPA), prevents phasic LH release (Hery et al., 1976). Similarly, lesions which destroy the suprachiasmatic nucleus, a hypothalamic nucleus rich in serotonin (Saavedra et al., 1974), result in constant estrus (Brown-Grant and Raisman, 1977). Furthermore, when PCPA is applied directly to the medial preoptic/suprachiasmatic area in young cycling rats, subsequent cycles are blocked and the animals enter a state of persistant vaginal cornification (Walker et al., 1980). Additional information which suggests a positive effect for serontonin in promoting continuation of reproductive function comes from work with neonatally-androgenized female rats. When newborn female rats are injected with testosterone proprionate, forebrain serotonin levels become depressed (Ladosky and Gaziri, 1970) and estral cyclicity is abolished with the development of constant estrus (Gorski, 1968). These data suggest that serotonin is required for normal reproductive cycles to continue in the female, and contradict an inhibitory role for the amine on gonadotropin secretion. Furthermore, depressed levels of serotonin in the brain of androgenized rats are associated with loss of circadian rhythms in the pineal gland and hypothalamic serotonin levels (Walker and Timiras, 1979a,b). Apparently the persistence of these rhythms is correlated with estral cyclicity since the onset of constant estrus, either as the result of drug treatment or as the result of old age, is accompanied by loss of serotonin rhythms. These observations suggest that the daily rhythms in brain serotonin levels may be more important in control of cyclic reproductive function than the absolute concentrations of this indoleamine.

ROLE OF SEROTONIN IN CYCLIC GONADOTROPIN SECRETION

In the foregoing section, we suggested that serotonin plays an important role in maintaining reproductive rhythms. To test this hypothesis further, rats were treated with agents which deplete (PCPA) or enhance (5-hydroxytryptophan; 5-HTP, the precursor of serotonin) brain serotonin (Walker and Timiras, 1979a,b). Serotonin levels were then measured in the hypothalamus and correlated with gonadotropin levels. As is to be expected from the known actions of the substances used, serotonin levels in the hypothalamus and pineal decreased or increased. However, irrespective of the direction of the changes, serotonin circadian rhythms were depressed by both blocking and stimulatory agents of serotonin; simultaneously the surge of gonadotropin (LH), normally occuring in proestrus and required for ovulation was blocked (Fig. 1). Hence, loss of cyclic gonadotropin secretion is best correlated with loss of serotonin rhythms rather than alterations in absolute serotonin concentrations.

To test the significance of serotonin rhythms in maintaining normal phasic secretion of LH, cycling animals were subjected to altered environmental lighting, a more physiologic approach than drug administration for studying serotonin rhythms. It is well-known that serotonin rhythms, especially in the pineal gland, are extremely sensitive to changing light conditions, and even small photoperiodic alterations can induce immediate loss of cyclicity (Snyder et al., 1965). In experiments in which the night-day period was reversed for one day preceeding proestrus, the anticipated LH surge was delayed and markedly attenuated (Fig. 2). Interestingly, aging rats show a similar delay in onset and blunting of the LH surge prior to the age-related spontaneous loss of ovulation (van der Schoot, 1976), supporting the view that in old rats, reproductive

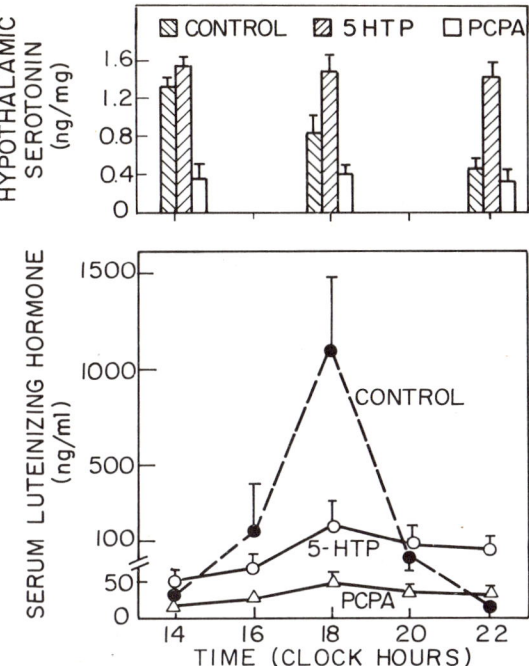

Fig. 1. Effect of p-cholorphenylalanine (PCPA; 20 mg/kg) or 5-hydroxytryptophan (5-HTP; 100 mg/kg) injections/(subcutaneous) on each day of diestrus on hypothalamic serotonin levels and the preovulatory LH surge. Serotonin levels are significantly different in control rats ($p<.05$) between 1400 and 2200 h, while no circadian changes are evident in PCPA and 5-HTP treated animals. Serum LH levels were significantly reduced ($p<.01$) by both drug treatments.

failure may result from modified serotonin metabolism. Reciprocally, in the young, serotonin rhythmicity in the hypothalamus and pineal gland may facilitate cyclic gonadotropin secretion. When the serotonin rhythms are damped by altered metabolism due to drugs, altered environment or age, then the LH surge is depressed and reproductive cycles fail.

Observations from the direct application of antagonists of neurotransmitter synthesis, or neurotoxins causing cell destruction, to discrete brain areas provides additional support for the hypothesis that changes in brain amines are responsible for reproductive senility. For example, depletion of norepinephrine in the hypothalamus by microinjection of 6-hydroxydopamine, a drug which destroys noradrenergic terminals, results in constant estrus (Benedetti et al., 1976). The same neurotoxin applied topically near the suprachiasmatic nucleus blocks the phasic release of LH from the pituitary (Simpkins et al., 1977a). From these studies it emerges that catecholamine and peptide-secreting neurons interact in this region (Simpkins and Kalra, 1979). Application through indwelling cannuale of serotonin synthesis inhibitors such as PCPA in the suprachiasmatic nucleus induces a condition of prolonged estrus which closely mimics the constant estrous syndrome of the aged animals (Walker et al., 1980). Blockade of catecholamines under the same experimental conditions alters some aspects of the reproductive cycle and

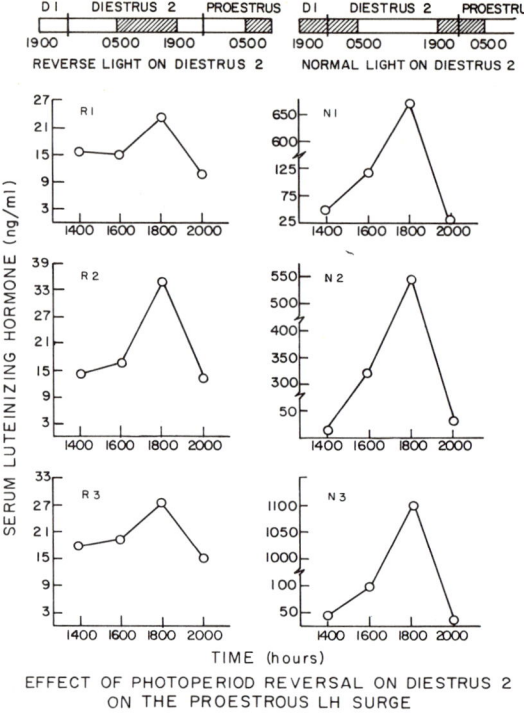

Fig. 2. Blockade of the preovulatory LH surge in rats after reversal of the photoperiod during diestrus. Rats were exposed to ten additional hours of light on diestrus 1, then alternating periods of dark: light were resumed. This caused the normally daylight hours of diestrus and proestrus to be spent in darkness. Treatment caused significant depression ($p<.001$) of serum LH in proestrus in photoperiod reversed rats (R_1, R_2, R_3). When compared with controls (N_1, N_2, N_3) which were kept in a standard photic environment.

produces prolonged vaginal cornification in some cases (Fig. 3), but the hormonal and neurochemical changes do not resemble those in old rats in constant estrus (Walker et al., 1980). From these observations, it may be postulated that aging of the female reproductive system follows failure of serotonergic input in the rostral hypothalamus to provide proper timing or activation cues for phasic gonadotropin secretion.

Overview

Morphological and functional similarities between the human menopause and the constant estrus condition of aged female rats suggest that our knowledge of reproductive aging in humans may be enhanced by experimental analysis of the rodent model. The validity of this model comes from common characteristics of menopause and constant estrus, including changes in follicular growth and development, steroid hormone synthesis, pituitary and hypothalamic hormones, and brain monoamines. The complete neuroendocrine axis from ovary to hypothalamus, appears to be functionally intact until old age, when certain aspects of its regulation by higher brain centers, possibly involving altered neurotransmission, become impaired.

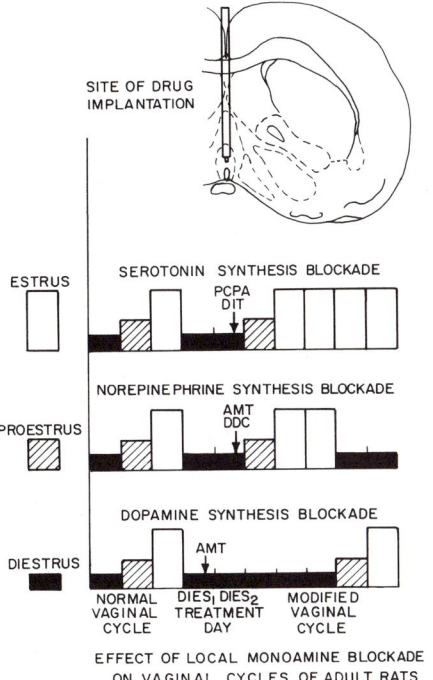

Fig. 3. Vaginal smear patterns in previously cycling rats after 10 µg pellets of antiaminergic neuroleptics were placed into the rostral hypothalamus on the afternoon of diestrus 2. Serotonin synthesis blocking drugs (PCPA; diiodotyrosine, DIT) caused the vagina to remain cornified for as much as two weeks after treatment. α-methyl-p-tyrosine (AMT) or diethyldithiocarbamate (DDC) which block catecholamine synthesis routinely caused vaginal estrus to persist for two days, though occasionally for as much as ten days. When given on diestrus 1, AMT prolonged diestrus.

Among the neurotransmitters, catecholamines have been implicated as modulators of tonic gonadotropin secretion and serotonin in concert with catecholamines as regulator of their cyclic release. Consequences of altered serotonin metabolism on continuance of cyclic reproductive function were tested by modifying the photic environment, which changes the periodicity of brain serotonin circadian rhythms, and by using drugs to elevate and depress levels of this amine. Levels of hypothalamic neurotransmitters in aged rats in constant estrus were compared with the values from animals in which constant estrus had been induced by the localized administration of monoamine inhibitors. While catecholamines may be primarily responsible for control of hypothalamic peptide and pituitary gonadotropin secretion, the activity of the catecholamine-peptide unit seems to be modulated by serotonergic inputs. Furthermore, circadian rhythmicity of brain serotonin seem to be more important than its absolute concentration for maintenance of cyclic reproductive function. Constant estrus, endocrinologically similar to that of aged rats, may be induced by depression of serotonin synthesis in the area of the suprachiasmatic nucleus which controls biological rhythms. Therefore, age-related disturbances in this area leading to altered or lost circadian serotonin rhythms may in part account for failure of cyclic reproductive function with advancing age.

MALE CLIMACTERIC

In contrast with the cessation of ovarian function at menopause, senescent changes in the reproductive system of the male occur gradually and at a later age than in the female, a difference that may be traced back to early developmental events; in women, gametogenesis is an embryonic event (Chapter 2) that terminates at menopause, whereas, in men, spermatogenesis begins at puberty (Chapter 9) and continues throughout life.

In parallel with the continuation of spermatogenic activity, normal secretion of testicular androgens may persist until advanced age (Harman et al., 1979), though reports to the contrary are published (Vermeulen, 1979). The discrepancies may result from health differences in the test participants since Harman's et al. (1979) survey population consists of healthy, nonobese middle-class, community-dwelling men, whereas Vermeulen's 1979 study uses hospital patients of various ages. In the highly selected healthy population, sex steroids did not change with advancing age, while a fall in serum testosterone levels started around the sixth decade of life and was preceeded by a fall in free testosterone concentrations beginning as early as age 50 in the general hospital population.

Although some clinicians, on the basis of a few cases of men suffering from abnormal hypofunction of the testes associated with loss of potency and libido, certain cardiovascular changes (e.g., flushes and tachycardia) chills and sudden perspiration, numbness and vertigo, as well as emotional instability have suggested that such males undergo a climacteric phase; these symptoms usually occur, if at all, at a later age (early sixties) and to a lesser degree than in women. Nevertheless, in the majority of men, spermatogenesis and androgen and gonadotropin secretion are only moderately slowed or altered and the competence of the reproductive function, including sex behavioral patterns, is retained until advanced age (generally in and after the sixties). Such quantitative assessments, however, of functional competence are very much tempered by qualitative considerations and the decline in hormonal, physiologic and reproductive capacity notwithstanding, the major determinants of sexual adequacy in later years—in the absence of pathology—are psychodynamic in nature (Chapter 16). In animals, more quantitative assessments have been made of physiologic and endocrine changes associated with aging of male reproduction. Plasma testosterone levels decrease with advancing age in the rat (Riegle and Miller, 1978), possibly as the result of loss of Leydig cells (Bethea and Walker, 1977) or of decrements in steroidogenic capacity (Leatham and Albrecht, 1974). As in aging human males, rats fail to show markedly depressed plasma levels of pituitary gonadotropins until they reach extreme old age (Bethea and Walker, 1977; Schaar et al., 1975). Similarly, it was recently demonstrated that age-related deficiencies in pituitary regulatory hormones are modest and that the amounts of hypothalamic peptides which are present in aging rats are sufficient to maintain pituitary secretion of a normal level. Hence, in male rats, as in females, the mild decrements in function with advancing age would result from deficient neural regulation of hypothalamic neurosecretory cells. This view is supported by the reduced catecholamine levels which occur in aged male rats when compared with young animals (Riegle and Miller, 1978; Simpkins et al., 1977b). Despite the qualitative similarity of the changes in neuroendocrine regulation of reproductive function between male and female rats, the greater organizational complexity of the femal reproductive system, requiring finer periodic adjustments, probably accounts, at least in part, for the early and more dramatic loss of normal function in this sex.

ACKNOWLEDGMENTS

The original experiments discussed in this chapter were supported by NIH Grant No. AG 00043 to Dr. P.S. Timiras and NIH Postdoctoral Fellowship No. 5-F32-AG05068 to Dr. R. Walker.

REFERENCES

Aschheim, P. Aging in the hypothalamic-hypophyseal ovarian axis in the rat, in *Hypothalamus, Pituitary and Aging,* A.V. Everitt and J.A. Burgess, eds. Charles C. Thomas, Springfield, Ill. (1976), pp. 376-418.

Bapna, J., Neff, N.H., and Costa, E. A method for studying norepinephrine and serotonin metabolism in small regions of the rat brain: effect of ovariectomy on amine metabolism in anterior and posterior hypothalamus. *Endocrinology* 89, 1345-1349 (1971).

Beach, J.E., Tyrey, L., Schomberg, D.W., and Everitt, J.W. Serum concentration of sex hormones in spontaneously persistent estrous rats at different times of day. (unpublished, in preparation), 1979.

Benedetti, W.L., Sala, M.A., and Otegui, J.T. Persistent estrus in rats after anterolateral hypothalamus micro injections of 6-hydroxydopamine. *Neuroendocrinology* 21, 297-303 (1976).

Bethea, C. and Walker, R.F. Age related changes in reproductive hormones and in Leydig cell responsivity in the Male Fischer 344 rat. *J. Gerontol.* 34, 21-27 (1977).

Block, E. Quantitative morphological investigations of the follicular system in women. Variations at different ages. *Acta Anat.* 14, 108-123 (1952).

Brawer, J.R., Naftolin, F., Martin, J., and Sonnenschein, C. Effects of single injection of estradiol valerate on the hypothalamic arcuate nucleus and on reproductive function in the female rat. *Endocrinology* 103, 501-512 (1978).

Brody, M.J. and Kadowitz, P.J. Prostaglandins as modulators of the autonomic nervous system. *Federation Proc.* 33, 48-60 (1974).

Brown-Grant, K. and Raisman, G. Abnormalities in reproductive function associated with destruction of the suprachiasmatic nuclei in female rats. *Proc. Roy. Soc. Lond. (Biol.)* 198, 279-296 (1977).

Caligaris, L., Astrada, J.J., and Taleisnik, S. Biphasic effect of progesterone on release of gonadotropins in the rat. *Endocrinology* 89, 331-341 (1971).

Clemens, J.A. and Bennett, D.R. Do aging changes in the preoptic area contribute to loss of cyclic endocrine function? *J. Gerontol.* 32, 19-24 (1977).

Clemens, J.A., Amenomori, Y., Jenkins, T., and Meites, J. Effects of hypothalamic stimulation, hormones and drugs on ovarian function in old female rats. *Proc. Soc. Exp. Biol. Med.* 132, 561-563 (1969).

Collett, M.E., Wertenberger, G.E., and Fiske, V.M. The effect of age upon the pattern of the menstrual cycle. *Fert. Ster.* 5, 437-448 (1954).

Cooper, R.L., Brandt, S.J., Linnoila, M., and Walker, R.F. Induced ovulation in aged female rats by L-DOPA into the medial preoptic area. *Neuroendocrinology* 28, 234-240 (1979).

Donoso, A.O., Stefan, F.J.E., Biscardi, A.M., and Cukier, J. Effects of castration on hypothalamic catecholamines. *Am. J. Physiol.* 212, 737-739 (1967).

Eisenberg, E. Effects of androgens, estrogens and corticoids on strontium kinetics in man. *J. Clin. Endocrinol. Metab.* 26, 566-572, 1966.

Eisenberg, E. and Gordan, G.S. Skeletal dynamics in man measured by nonradioactive strontium. *J. Clin. Invest.* 40, 1809-1825, 1961.

Everett, J.W. and Zeilmaker, G.H. Experimental restoration of ovarian cycles in spontaneously persistent estrous rats. *Anat. Rec.* 193, 533-000 (1979).

Everitt, A.V. The female climacteric, in *Hypothalamus, Pituitary and Aging*, A.V. Everitt and J.A. Burgess, eds. Charles C. Thomas, Springfield, Ill. (1976), pp. 419-430.

Finch, C.E. Reproductive senescence in rodents: factors in the decline of fertility and loss of regular estrous cycles, in *The Aging Reproductive System* (Aging Vol. 4), E.L. Schneider, ed. Roben Press, New York (1978), pp. 193-212.

Finch, C.E. Extraovarian mechanisms in reproductive aging: evaluation of possible pituitary and hypothalamic roles. Proc. of the Conf. on Endocrine Aspects of Aging. (Endocrine Society and NIA) Bethesda, Md. October 18-20 (1979).

Frantz, A.G. and Rabkin, M.T. Effects of estrogen and sex difference on human growth hormone. *J. Clin. Endocrinol. Metab.* 25, 1470-1480 (1965).

Gordan, G.S. Evaluation and use of anabolic steroids. *G.P.* 10, 86-102 (1964).

Gorski, R.A. Influence of age on the response to perinatal administration of a low dose of androgen. *Endocrinology* 82, 1001-1004 (1968).

Grodin, J.M., Siiteri, P.K., and MacDonald, P.C. Source of estrogen production in postmenopausal women. *J. Clin. Endocrinol. Metab.* 36, 207-214 (1973).

Harlan, R.E. and Gorski, R.A. Effects of postpubertal ovarian steroids on reproductive function and sexual differentiation of lightly androgenized rats. *Endocrinology* 102, 1716-1724 (1978).

Harman, S.M., Martin, C.E. and Tsitouras, P.D. Gonadotrophins, Sex Steroids and Sexual Activity in Healthy Aging Men. (Endocrine Society and NIA) Bethesda, Md. October 18-20 (1979).

Heaney, R.P. and Recker, R.R. Estrogen effects on bone remodeling at menopause. *Clin. Res.* 23, 535-000 (1975).

Hery, M., LaPlante, E., and Kordon, C. Participation of serotonin in the phasic release of LH. I. Evidence from pharmacological experiments. *Endocrinology* 90, 496-503 (1976).

Huang, H.H., Marshall, S., and Meites, J. Capacity of old versus young female rats to secrete LH, FSH and prolactin. *Biol. Reprod.* 14, 538-543 (1976a).

Huang, H.H., Marshall, S., and Meites, J. Induction of estrous cycles in old noncyclic rats by progesterone, ACTH, ether stress or L-DOPA. *Neuroendocrinology* 20, 21-34 (1976b).

Huang, H.H., Steger, R.W., Bruni, J.F., and Meites, J. Patterns of sex steroid and gonadotropin secretion in aging female rats. *Endocrinology* 103, 1855-1859 (1978).

Izquierdo, R., Chemerinski, R., and Acevedo, C. Effects of cortisol and progestins on trypotphan, serotinin and 5-hydroxyindole acetic acid in brain stem, in *Abstracts of Sixth International Congress of Pharmacology*, Helsinki, 1975. Pergamon Press, New York (1975), p. 166.

Jones, E.C. The aging ovary and its influence in reproductive capacity. *J. Reprod. Fert.* Suppl. 12, 17-30 (1970).

Judd, H.L., Judd, G.E., Jucas, W.E., and Yen, S.S.C. Endocrine function of the postmenopausal ovary: concentrations of androgens and estrogens in ovarian and peripheral vein blood. *J. Clin. Endocrinol. Metab.* 39, 1020-1024 (1974).

Krohn, P.S. Ageing process in the female reproductive tract. *Lect. Scient. Basis Med.* 7, 285-313 (1959).

Kruse-Larsen, C. and Garde, K. Postmenopausal bleeding: another side effects of levodopa. *Lancet* I, 707-709 (1971).

Ladosky, W. and Gaziri, L.C.J. Brain serotonin and sexual differentiation of the nervous system. *Neuroendocrinology* 6, 168-174 (1970).

Leatham, J.H. and Albrecht, E.O. Effect of age on testes $\Delta 5$-3β-hydroxy-steroid dehydrogenase in the rat. *Proc. Soc. Exp. Biol. Med.* 145 1212-1214 (1974).

Linnoila, M. and Cooper, R.L. Reinstatement of vaginal cycles in aged female rats. *J. Pharmacol. Exp. Therap.* 199, 477-482 (1976).

Longcope, C. Metabolic clearance rate and blood production rates of estrogens in postmenopausal women. *Am. J. Obstet. and Gynecol.* 111, 778-781 (1971).

Mattingly, R.F. and Huang, W.Y. Steroidogenesis of the menopausal and postmenopausal ovary. *Am. J. Obstet. Gynecol.* 103, 679-690 (1969).

McKinlay, S., Jeffreys, M., and Thompson, B. An investigation of the aged at menopause. *J. Biosoc. Sci.* 4, 161-173 (1972).

McLennan, M.T. and McLennan, C.E. Estrogenic status of menstruating and menopausal women assessed by cervicovaginal smears. *Obstet. Gynecol.* 37, 325-331 (1971).

Meites, J., Juang, H.H., and Reigle, G.D. Relation of the hypothalamo-pituitary-donadal system to decline of reproductive function in aging female rats, in *Hypothalamus and Endocrine Function*, A. Labrie, J. Meites, and B. Pelletier, eds. Plenum Press, New York (1976), pp. 3-19.

Meites, J., Huang, H.H., and Simpkins, J.W. Recent studies on neuroendocrine control of reproductive senescence in rats, in *The Aging Reproductive System* (Aging, Vol. 4), E.L. Schneider, ed. Raven Press, New York (1978), pp. 213-235.

Miller, A.E. and Riegle, G.D. Endocrine factors associated with the initiation of constant estrous in aging female rats. *Federation Proc.* 38, 1248 (1979).

Nordin, B.E. Bone resorption and oestrogen status at the menopause. *Probl. Actuels Endocrinol. Nutr.* 20, 147-154 (1977).

Odell, W.D. and Swerdloff, R.S. Progesterone-induced luteinizing and follicle-stimulatory hormone surge in postmenopausal women and stimulated ovulatory peak. *Proc. Nat. Acad. Sci. USA* 61, 529-536 (1968).

Poortman, J., Thijseen, J.H.H., and Schwarz, F. Androgen production and conversion to estrogens in normal postmenopausal women and in selected breast cancer patients. *J. Clin. Endocrinol. Metab.* 37, 101-109 (1973).

Quadri, S.K., Kledzik, G.S., and Meites, J. Reinitiation of estrous cycles in old constant estrous rats by central acting drugs. *Neuroendocrinology* 11, 248-255 (1973).

Reigle, G.D. and Miller, A.E. Aging effects of the hypothalamic hypophyseal gonadal control system in the rat, in *The Aging Reproductive System* (Aging, Vol. 4), E.L. Schneider, ed. Raven Press, New York (1978), pp. 159-192.

Riggs, B.L., Jowsey, J., Goldsmith, R.S., Kelly, P.J., Hoffman, D.L., and Arnaud, C.D. Short- and long-term effects of estrogen and synthetic anabolic hormone in postmenopausal osteoporosis. *J. Clin. Invest.* 51, 1659-1663 (1972).

Riley, G.M. Endocrinology of the climacteric. *Clin. Obstet. Gynecol.* 7, 432-450 (1964).

Robinson, D.S., Nies, A., Davis, J., Bunney, W.E., Davis, J.M., Colburn, R.W., Bourne, H.R., Shaw, D.M., and Coppen, A.J. Aging monoamines and monoamine oxidase levels. *Lancet* I, 290-291 (1972).

Saavedra, J.M., Palkovits, M., Brownstein, M.J., and Axelrod, J. Serotonin distribution in the nuclei of the rat hypothalamus and preoptic region. *Brain Res.* 77, 157-167 (1974).

Sawyer, C.H. First Geoffrey Harris Memorial Lecture. Some recent developments in brain-pituitary-ovarian physiology. *Neuroendocrinology* 17, 97-124 (1975).

Schiff, I. and Wilson, E. Clinical aspects of aging of the female reproductive system, in *The Aging Reproductive System* (Aging, Vol. 4), E.L. Schneider, ed. Raven Press, New York (1978), pp. 9-28.

Schneider, H.P.G. and McCann, S.M. Mono and indolamines in control of LH-secretion. *Endocrinology* 86, 1127-1133 (1970).

Shaar, C.J., Euker, J.S., Riegle, G.D., and Meites, J. Effects of castration and gonadal steroids on serum luteinizing hormone and prolactin in old and young rats. *J. Endocrinol.* 66, 45-51 (1975).

Sherman, B.M. and Korenman, S.G. Hormonal characteristics of the human menstrual cycle throughout reproductive life. *J. Clin. Invest.* 55, 699-706 (1975).

Siiteri, P.K. Postmenopausal estrogen production, estrogens in the postmenopause. *Front. Horm. Res.* 3, 40-44 (1975).

Siiteri, P.K. Interaction of androgens, estrogens and the sex hormone binding globulin. Proc. of the Conf. on Endocrine Aspects of Aging. (Endocrine Society and NIA). Bethesda, Md. October. 18-20 (1979).

Siiteri, P.K. and McDonald, P.C. Role of extraglandular estrogen in human endocrinology, in *Handbook of Physiology, Section 7: Endocrinology,* vol. II. Female Reproductive System, Part 1, R.O. Greep and E.B. Astwood, eds. Willaims and Wilkins, Baltimore (1973), pp. 615-629.

Simpkins, J.W., and Kalra, S.P. Central site(s) of norepinephrine (NE) and LHRH interaction. *Federation Proc.* 38, 1107 (1979).

Simpkins, J.W., Advis, J.P., Hodson, C.A., and Meites, J. Blockade of steroid-induced luteinizing hormone release by selective depletion of anterior hypothalamic norepinephrine activity. *Endocrinology* 104, 506-509 (1977).

Simpkins, J.W., Mueller, G.P., Huang, H.H., and Meites, J. Evidence for depressed catecholamine and enhanced serotonin metabolism in aging male rats: possible relation to gonadotropin secretion. *Endocrinology* 100, 1672-1678 (1977b).

Snyder, S.H., Zureig, M., Axelrod, J., and Fischer, J.E. Control of the circadian rhythm in serotonin content of the rat pineal gland. *Proc. Nat. Acad. Sci.* 53, 301-305 (1965).

Steger, R.W. Aging and the hypothalamo-hypophyseal-gonadal axis, in *Aging and Reproductive Physiology,* E.S.E. Hafez, ed. Ann Arbor Science Publishers, Inc., Ann Arbor (1976), pp. 51-65.

Talbert, G.B. Aging of the reproductive system, in *Handbook of the Biology of Aging.* C.E. Finch and L. Hayflick, eds. Van Nostrand Reinhold Company, New York (1977), pp. 318-356.

Timiras, P.S. and Meisami, E. Changes in gonadal function, in *Developmental, Physiology and Aging.* P.S. Timiras, ed. The MacMillan Company, New York (1972), pp. 527-541.

van der Schoot, P. Changing pro-oestrous surges of luteinizing hormone on aging 5-day cyclic rats. *J. Endocrinol.* 69, 287-288 (1976).

Vasquez, E., Casares, H.J., and Sereno, J.A. Influence of the nutritional status upon the response of menopausal women to estrogen therapy, in *Consensus of Menopause Research*, P.A. Van Keep, R.B. Greenblatt, and M. Albeaux-Fernet, eds. Univ. Park Press, Baltimore (1976), pp. 109-115.

Vermeulen, A. Reproductive physiology in the aging male.

Wajsborth, J. Post-menopausal bleeding after L-DOPA. *New Engl. J. Med.* 286, 784 (1972).

Walker, R.F. and Timiras, P.S. Serotonin circadian rhythm: a daily signal associated with luteinizing hormone release in the female rat. (Submitted for review, 1979a).

Walker, R.F. and Timiras, P.S. Sexual maturation in rats treated neonatally with antiaminergic drugs. *Federation Proc.* 38(3) 4657 (1979a).

Walker, R.F., Cooper, R.L., and Timiras, P.S. Constant estrus: role of rostral hypothalamic monoamines in development of reproductive dysfunction in aging rats. *Endocrinology* 107, 249-255 (1980).

Walker, R.F., Segall, P., and Timiras, P.S. Neuroendocrinology of aging, in *The Aging Nervous System*, G.J. Maletta and F.J. Priozzolo, eds. Praeger Publishers, New York (1980) pp. 89-109.

Watkins, B.E., Meites, J. and Reigle, G.D. Age-related changes in pituitary responsiveness to LHRH in the female rat. *Endocrinology* 97, 543-548 (1975).

Weisz, J. and Lloyd, C.W. Estrogen and androgen production *in vitro* from 7-^{3}H-progesterone by normal and polycystic rat ovaries. *Endocrinology* 77, 735-744 (1965).

Copyright©1982, Spectrum Publications, Inc.
Hormones in Development and Aging

Chapter 17
Psychological Aspects of Menopause
Dorothy Strauss

INTRODUCTION

The two major endocrinologic events in the life of the female are menarche and menopause. They define the boundaries of her reproductive career—her years of productivity in the strictly species-survival sense. Each of these events is identified by a series of observable physiologic changes and by attitudinal and behavioral changes as well. All cultures, as far as we know, have endowed these landmark events with varying degrees of significance. Historically, the gamut of roles available to women, and the social and economic consequences of those roles, have been determined more by ovarian function than by any other single factor.

This chapter is limited to the psychological aspects of menopause. By way of definition, the terms *menopause* and *climateric* will be used interchangeably. As has been discussed elsewhere in this book (Chapter 15) *menopause* refers to the physiologic cessation of menses. *Climacteric* (or climacterium) is the term applied to the span of years during which there is gradual decline in gonadal function. In most women, tapering off and final cessation of menstruation is a gradual process, resembling, in reverse, the hormonal changes of adolescence and the establishment of the menstrual cycle. This chapter will not include consideration of "premature" menopause, that which occurs spontaneously before the age of 40 and affecting about 8% of American women, nor menopause following surgery. Too little reliable data are available for both of these conditions to permit drawing even tentative conclusions about what may be unique, psychologically, to each or to both, or how they may be compared with menopause whose onset and course are within the normally anticipated range of time. There are persuasive reasons for increasing our understanding of the emotional and social factors that accompany and often complicate this physiologic process. There has been a dramatic increase in life expectancy during the past few decades, manifested less in actual extension of the life span than in the fact that living well into the later years may be anticipated as a matter of course by the average person. This trend, which is expected to continue, has had an impact upon the way menopause is being experienced. Where formerly it signified an endpoint, it has now become a transitional phase in more than the biological sense. The majority of women in the United States and in most of the developed nations of the world now expect to live fully one-third of their lives as postmenopausal. Hence, all that was once implied by the expression "change of life" has taken on new dimensions, such that the psychological aspects of menopause require a broader range of adjustments, not only by women but by the men who share their lives.

Although in this chapter the major concern will be with the climacteric process in the female, consideration of the male climacteric is included toward the end of the chapter. Increased reports of stress symptoms in men, related to life adjustment difficulties during the latter mid-life years in particular, suggest that, at the psychological if not the hormonal level, comparisons are appropriate. A brief review of traditional clinical theories of psychological development is included for the purpose of comparison with more recent efforts to view the entire life span as a developmental journey.

MENOPAUSE AS DESCRIBED IN SCIENTIFIC LITERATURE

The most striking aspects of much early scientific writing on cessation of menstruation are (1) brevity and (2) the implication that the most useful attitude is acceptance of the fact that biology cannot be reversed. These cautionary comments in older medical literature appeared with great consistency. Let us go back, for example, to the classic three volume work by Ploss et al., *Woman*, which first appeared in German in 1885 and ran through multiple editions until its appearance in English translation in 1935. The entire subject of menopause is disposed of in 14 pages of text, plus seven full-page illustrations of "matrons" showing "characteristic" signs of physical deterioration, ranging from excessive relaxation of facial muscles and hirsuitism to obese torsos with abnormally pendulous breasts and a redundancy of layered transverse swellings. The message is obvious: menopause results in extreme ugliness. Fully two-thirds of the textual content concerns itself with anthropological material comparing European with non-European women. The summarizing impression is conveyed by a simple statement: "The course of the climacteric is determined above all by the growing old of the organs, functionally and anatomically."

The perception of menopause as nature's way of sentencing the female to inevitable defeminization and decline was not limited to this comprehensive treatise on women, but is classically stated there. Incidentally, the chapters which follow *Menopause* in this oeuvre are *Woman as Grandmother*, *Woman in Old Age*, and *Woman in Death*. The psychological implications of this sequence of climacterium→senium→decease would be difficult to ignore.

As one examines the writing produced toward the middle of this century, modifications in point of view and interpretation of the psychological effect of menopausal phenomena are minimal. This fact holds true even when the writer is female. The two-volume study by Deutsch, which appeared in 1945, *Psychology of Women*, confines discussion of the climacteric to a chapter to which no sequential number is assigned, but merely the designation *Epilogue*--a message in itself. In brief, Deutsch stresses that, while the "endocrine system is deranged in its functioning," all individual problems arise from the basic personality structure. Her view is merely a restatement of the Freudian principle which holds that, following the oedipal stage, nothing new is to be expected. The climacteric is experienced as "a narcissistic mortification." Bursts of activity are regarded as part of a defense mechanism against having reached "partial death." Some women are seen as "quasihypomanic," seizing upon a need to act out all previously forbidden urges or to escape into fantasies which repeat psychologic pubescence. Other women are described as responding to the reality of impending old age by overemphasizing behaviors traditionally ascribed to aging, such as seeking isolation or an ascetic, self-sacrificing style of living.

The depression associated with menopause is compared with the depressed moods of adolescent girls. The analogies to puberty, according to Deutsch, constitute the most useful way of understanding the psychology of the menopausal woman. The climacterium is defined as "a third edition" of the infantile stage. (Freud referred to puberty as the second edition because the oedipal complex was defined as a remobilizing of early relationships to parents.) These analogies and similarities are utilized as the effective tool for understanding attitudes and behaviors considered typical for menopausal women. Intensified interest in sexuality, for example, is explained as the "too late" counterpart of the "too early" defense against "sexual assault" in puberty. While the young girl in the early 1900s resisted her sexuality, the older woman is seen as struggling to negate her loss (her reproductive capacity) by unsuccessful efforts to recapitulate that resistance, and finally throwing over any attempt to do so.

Good adjustment, according to Deutsch (1945), depends upon satisfactory sublimations, especially intellectual sublimations. Typical organic symptoms of the climacteric acquire secondary importance when "motherliness" is diverted from actual reproduction to the various other activities of life. A most interesting variation on this theme appears in Erikson's (1950) generativity model, elaborated further along in this chapter. Duetsch offers resignation without compensation as the only satisfactory psychological solution, an outlook that does little to whet an appetite for living.

In a general way, women today are moving into an expanded psychological and social world; however, the majority of those old enough to be entering or just emerging from their menopausal years are still subject to much of the traditional, stereotypic thinking. The slow pace of progress is born of early conditioning and reinforced by the cultural impacts upon the training of physicians and other professionals engaged in the helping services.

Textbooks published during the 1970s have presented some important modifications in regard to their treatment of menopause. Notice was taken of the fact that women now live well past their reproductive years. Previously, while there usually was some recognition of the role of hormonal imbalance in the overall picture of psychological disturbances attributed to menopause, the primary emphasis was on the symbolic meaning of the physiologic process. More recent treatment of the subject tends to stress that the relationship between the physical and the psychological may be coincidental in some respects, since the menopause occurs at a time of transition along several dimension of living. The psychological problems of menopause are seen as extending beyond the significance of discontinued ovarian function. The stresses attributed to "midlife crisis" have assumed greater importance.

The circularity of the relationship between physical symptoms and psychological stress is now generally recognized. Intensification of one tends to aggravate the severity of the other. The degree to which the physical process, itself, may be the triggering mechanism for certain psychological states remains an unanswered question. Little is known about the direct effect of gonadal hormones upon human behavior. Most of our observations are derived from animal studies and are primarily concerned with sexual behavior. Psychoendocrinologic studies of young married couples reported by Persky et al. (1977, 1978) have provided supportive data for close relationship between endocrine and sexual behavior not hitherto documented. An on-going project by the same researchers with couples where the wife is postmenopausal is anticipated to yield significant information for correlation with reported sexual behavior.

Some medical writing has focused upon menopause as a deficiency disease, causing not only discomfort but specific degenerative changes. This view appears, in some ways, to lead us away from the developmental perspective back toward the perception of the climacteric as a pathological process. The discussion of laboratory findings that portray the considerable variability among menopausal women is beyond the scope of this chapter. Of concern here is the damaging effect, psychologically, of a position which equates a normal developmental process with progressive disease. Psychosomatic complaints occurring at this time of life are readily ascribed to menopause. Such a "diagnosis" leads to greater stress and, frequently, a demand for active pharmacologic intervention which may not relieve the psychological symptoms, or do so only transiently, even though physical discomfort may be decreased.

THE INFLUENCE OF THEORIES OF PERSONALITY

The classical psychological view of crises during adult life is that they are manifestations of unresolved infantile or oedipal problems. In this context, only the years of infancy and childhood are psychologically significant. Continued development during adult life is perceived as the working through of earlier conflicts. While this concept allows for a "freeing up" of the personality, there is no provision for the possibility of growth through phase-appropriate tasks other than resolution of the previously unresolved. The emphasis upon infancy and childhood, a major tenet of classical Fruedian theory, has dominated psychological thinking in a large sector of the professional community over a long period of time. Variations and differences existed but not with sufficient acceptance to generate major trends until fairly recently. The most interesting of these divergent systems, from the point of view of periods of growth and development extending into adulthood, is contained in the writings of Jung particularly in the discussion of the stages of life (1960). The Jungian system singled out the middle years of life as the time when redirection of energy and values becomes essential in order to maximize opportunities for growth. While this Jungian concept was seen as taking the form of removal of the repressions of youth (thus not fully departing from the Freudian position), it was unique for its time insofar as it spelled out a phase-specific developmental task for the middle years. It is this characterization of midlife as a critical stage, where growth or regression are the alternatives, which transcends the narrower classical view. Yet, there was grudging acceptance of the Jungian concept until recently. One might speculate that an idea needs its time in order to take root and that the population shift of recent decades has generated increased interest in the problems of midlife and beyond. It is of interest to note that Jung has claimed considerable clinical success with middle-aged patients.

The Movement Toward a Lifespan Developmental Psychology—Significance For Menopause

The first major shift in the classical position comes from Erikson (1950) who extended the developmental process over eight stages with stage-specific tasks to be accomplished. This schema also emphasized the growth potential of crises. His extension of fundamental problem-solving over the total life cycle may be regarded as a time-expansion of what is basic in Freud rather than outright disagreement. The development of creative living is seen as requiring the accomplishment of the early tasks of separation and individuation in order to move on to the later tasks of achieving generativity and ego integrity. I can find no better way of applying this concept to the difficulties associated with the climacteric, female and male, than to turn directly to Erikson. He interprets the Freudian "Lieben und Arbeiten" to mean the capacity of the adult to move through life engaged in "general work productiveness which would not preoccupy the individual to the extent that he (*she*) loses his (*her*) right or capacity to be a genital and loving being."

The final stages of *Generativity vs. Stagnation* and *Ego Integrity vs. Despair* flow from one to the other. Either the individual continues personal enrichment by continued absorption in responsibility toward others in some appropriate manner or turns inward as obsessive efforts to maintain false and inappropriate lines of "pseudointimacy" fail. Generativity is seen not only as a constructive defense against the harsher realities of the climacteric years but as appropriate preparation for the final phase of life. Erikson defines ego integrity as "a post-narcissistic love of the human ego—not the self...and acceptance of one's one and only life cycle." These characteristics are the hallmarks of successful aging, for which the groundwork is laid during the menopausal years.

Very recent investigations by Gould (1972) and Levinson and associates (1974) focus upon the existence of postadolescent developmental steps and make specific what Erikson posed as generalizations. For example, inner dissatisfaction is seen as reaching serious proportions between the ages of 29 and 34 in most instances. From 35 to the mid 40's is the time when preoccupation with "have I done it right?" gives way to "do I still have time to change?" This, for some women, is also the time when the earliest indications of menopause may surface or when thoughts of menopause may begin to intrude.

Both Gould and Levinson suggest that passage from the mid-40s through the 50s is usually marked by an increased mellowing, greater self-acceptance, and emphasis upon the desirability of cultivating activities which generate satisfaction and contentment with life. While it may be premature, at this point, to regard these studies as a clear signal for restructuring our understanding of crises experienced by menopausal individuals, there can be no doubt about two implications; namely, that (1) a social time clock is superimposed upon the biological clock and (2) cultural factors largely determine the extent to which physiologic indices of impending age are over-experienced.

THE MENOPAUSAL WOMAN AND MIDLIFE ISSUES

As has already been stated, not all of the stresses experienced by women during the climacteric may be ascribed to the menopausal process itself. A scanning of current gynecological literature supports the assumptions that many menopausal women do not seek medical relief; and, of those who do, it is unlikely that more than 50% report complaints directly traceable entirely or partially to the hormonal shift.

There are psychosomatic complaints at this time which seem to be typical for the anxious woman. These may take the form of exaggeration of vague fears of cancer, morbid concern over presumed unavoidable obesity, worry about defeminization and loss of interest in sex, and alarm that "insanity" will strike. Even educated and sophisticated women have been known to cling to vestiges of an ancient mythology of menopause. Education and supportive treatment may be all that is required. In particular, feeling depressed in response to annoying vasomotor symptomatology or boredom with old and empty routines must be distinguished from psychotic depression. When the latter does occur at this time, it is coincidental with the menopause, not the result of the hormonal changes, and assessment of the prepsychotic personality is essential.

A relatively small percentage of otherwise healthy women seek professional help for menopause-associated problems. Whether or not therapy or counseling is sought, there are issues that are commonly confronted at this time of life, and some of these have a particular impact upon women.

Watching one's children reach for autonomy, thus requiring less parenting, can distress the woman whose life has been centered in home and children. However, this "empty nest" syndrome occurs less frequently as more women turn toward out-of-house interests such as jobs, careers, and resumed education.

The generalization may be made that women, particularly those whose lives have been narrowly circumscribed by confining circumstances, have poor tolerance for the menopause as a sign of aging. This observation is not limited to the housewife. It applies equally to the woman who has subordinated her personal existence to a life of service to a parent or other relative and to those chronically trapped in psychologically eroding job situations. The data are still lacking to support conclusions about the effect of menopause on women who have gained access to the competitive world of high professional and executive status.

Not all of the psychological aspects of the climacteric are negative. Some freedoms compensate for apparent losses. Modern women whose domestic responsibilities decrease tend to cultivate opportunities for intellectual and emotional growth and those whose earlier sexual pleasure was curtailed by fear of pregnancy and/or conditioned inhibitions frequently become liberated from both deterrents. It may be argued that these changed behaviors represent vigorous efforts at denial; but, although this explanation may hold true in some cases, it cannot be regarded as the significantly appropriate explanation for any particular percentage of women. While it is not possible to offer broad generalizations applicable to all women, it appears, nonetheless, that social and cultural factors have modified attitudes so that it is acceptable for women to regard their postreproductive years as a time for enrichment; sexual, social and intellectual.

CLIMACTERIC-ASSOCIATED SYMPTOMS IN THE MALE

To define the male climacteric narrowly, that is, solely in terms of an age-related hormonal shift, is no more satisfactory than limiting the discussion of the female menopause to symptoms directly attributable to hormonal imbalance or insisting that all indications of distress are physiologically produced. Men do not experience a comparably dramatic endocrinologic event. A gradual decline in testosterone level is normal with age but rarely becomes significant before the age of 70.

Some endocrinologists maintain that a *true* male climacteric can be established only by laboratory findings which demonstrate a rise in pituitary gonadotropins coincident with the drop in testosterone (Stearns et al., 1974; see also, Chapter 15). This event is presumed to occur in about 35% of middle-aged (45-60 years of age) and older men. A change in sexual function commonly reported by men in the midlife group is less a result of reduced production of testicular androgen than of psychological factors. Most researchers in this area agree that the lower testosterone level normal for this age group is not the major cause of impairment of potency or other sexual dysfunctions in the mid-life male. Emotional, social, and economic stresses are the significant contributory elements. In this connection, it should be noted that there is research evidence to suggest that tension, mental fatigue, anxiety, and depression can lower testosterone blood levels in the male. It may well be that the level considered normal for this age group is even further reduced by psychologically stressful situations, thus complicating the hormonal picture.

Culturally, the male has been taught to equate sexual potency with self-esteem. Hence, the grudging acceptance of physical "slowing down" in nonsexual activities is frequently rejected as having direct application to sexual functioning because of the threatening implications for self-worth. The anxiety a man may feel from any alteration in his earlier sexual pattern may be intensified by his need to please a menopausal wife whose erotic appetite has increased. Management of the problem relating to changing sexual patterns with aging is beyond the scope of this chapter. I suggest, however, that a broader view of the male climacteric, as extending beyond the biologic to the psychologic, is essential for treatment. It is important to recognize that many men have been conditioned to evaluate sexual activity in terms of coital frequency and endurance; thus, they may find it difficult to cultivate appreciation for relaxed sensuous pleasuring.

Work-related stresses are not unusual among men at this phase of life. There is an urgency to arrive at the highest possible rung of success before it is too late, coupled with awareness that the downward path is in the foreseeable future. Indeed the term "climacteric" is derived from the Greek work (klimax) meaning ladder. Its original usage as synonymous with the female menopause reflects the assumption that general decline was the expected sequel after the top rung of development had been achieved. Frantic restitutive efforts, sexual and nonsexual, are commonly reported in this age group. Men may have a lesser tendency to deny age than women for whom physical attractiveness and reproductive capacity have been coupled, historically, with desirability. On the other hand, men continue to measure themselves by achievement, a demand which becomes increasingly burdensome with age.

MENOPAUSE: A TIME FOR GROWTH

Consideration of the psychological aspects of menopause requires looking beyond the biologic process to the circumstances of life which exert influence upon the emotional and social well-being of the individual. The climacteric phase coincides with the stage of development referred to as mid-life. Thus, it is frequently complicated by issues which may aggravate hormonally induced symptoms or set in motion psychological symptoms unrelated to the hormonal imbalance but ascribed to it.

This is a critical time of life for both men and women, but the observable phenomenon of the menopause limits attention largely to women. Nonetheless, there are commonalities as well as differences between the sexes. Physiological aging is culturally unacceptable, and age-related decrements tend to be viewed as narcissistic insults. Signals which generate distress come from two directions: reminders of what once was and shadows of what lies ahead. The emergent sexuality of maturing children and the growing infirmities of aged parents both threaten the self-image of youth and sexuality, and the hope of having ultimate control over one's destiny.

These may occasion a reawakening of old, unresolved psychological problems. Conceptualization of human development as continuing throughout the life span, however, adds the dimension of stage-specific tasks to be accomplished. Successful negotiation of the menopausal years depends upon personal reassessment and the setting of realistic goals.

Recent literature contains many references to what may be the major task of life at this time but none more eloquent than the following (Jaques, 1965, p. 506) which states succinctly the reason for the psychological discomfort often experienced by those on the threshold of the "prime of one's life" years. There is, indeed, a level of ambivalence because the opportunities for fulfillment are seen as curtailed by a factor outside of one's control; namely: the number of "good" years reasonably anticipated.

"I believe that it is this fact of entry upon the psychological scene of the reality and inevitability of one's own eventual personal death, that is the central and crucial feature of the midlife phase."

This awareness can project the individual into frenzied efforts at denial consisting of a final thrust at success in any of several directions or in all of them. Or there can be a rounding out of the self. The climacteric phase is a time for personal growth, for self-discovery, for the cultivation of interpersonal relationships, and for continuing to make the most of available opportunities which expand rather than constrict the psychological horizon. Biological reproductivity may cease, and cease to be desired, but the capacity for psychological productivity continues.

REFERENCES

Deutsch, H. *Psychology of Women*, vol. II, Grune and Stratton, New York (1945), pp. 457-459.
Erikson, E. Eight stages of man, in *Childhood and Society*, W.W. Norton Company, New York (1950), pp. 229-232.
Gould, R. The phases of adult life: a study in developmental psychology. *Am. J. Psych.* 129, 521-531 (1972).
Jaques, E. Death and the mid-life crisis. *Int. J. Psychoanal.* 46, 502-506 (1965).
Jung, C.G. The stages of life, in *Collected Works*, Vol. 8, Princeton University Press, Princeton (1960).
Levinson, D.J., Darrow, C.M., Klein, E.B., Levinson, M.H., and McKee B. The psychosocial development of men in early adulthood and the mid-life transition, in *Life History Research in Psychopathology*, Vol. 3, A. Thomas and M. Roff eds. University of Minnesota Press, Minneapolis (1974).
Persky, H., Lief, H.I., Strauss, D., Miller, W.R. and O'Brien, C.P. Reproductive hormone levels and sexual behavior of young couples during the menstrual cycle, in *Progress in Sexology*, R. Gemme and C.C. Wheeler, eds. Plenum Press, New York (1977).
Persky, H., Lief, H.I., Strauss, D., Miller, W.R., and O'Brien, C.P. Plasma testosterone level and sexual behavior of couples, *Arch. Sex Behav.* 7, 157-172 (1978).
Ploss, H.H., Bartels, M., and Bartels, P. *Woman* vol. 3, E.J. Dingwall, ed. Heinemann, London (1935).
Stearns, E.L., MacDonnell, J.A., Kaufman, B.J. Padua, R. Lucman, T.S., Winter, J.S.D., and Faiman, C. Declining testicular function with Age — hormonal and clinical correlates, *Am. J. Med.* 57, 761-766 (1974).

FURTHER READING

Hull, R. and Reubsaat, H.J. *The Male Climacteric.* Hawthorne Books New York (1975).

Lidz, T. The middle years, Ch. 16, in *The Person.* Basic Books, New York (1968).

Masters, W.H. and Johnson, V.E. *Human Sexual Response,* Little, Brown and Company, Boston (1966).

Masters, W.H. and Johnson, V.E. *Human Sexual Inadequacy,* Little, Brown and Company, Boston (1970).

Money, J. and Ehrhardt, A. *Man & Woman — Boy & Girl,* The Johns Hopkins University Press, Baltimore (1972).

Novak, E.R., Jones, G.S., and Jones, H.W. Management of the menopause, Ch. 32, in *Novak's Textbook of Gynecology* 8th ed. Williams and Wilkins, Baltimore (1970).

Simons, R.C. and Pardes, H. (eds.). *Understanding Human Behavior in Health and Illness,* Williams and Wilkins, Baltimore (1977).

Copyright©1982, Spectrum Publications, Inc.
Hormones in Development and Aging

Chapter 18
Neuroendocrine Strategies in Psychiatric Research
Robert Freedman
Douglas B. Carter

INTRODUCTION

One of the most promising strategies in biological psychiatry is the use of neuroendocrine measurements to asses changes in brain physiology in human psychiatric patients. Despite a vast increase in biological knowledge over the past several decades, little of this knowledge has been applied to psychiatric problems. The direct application of basic science discoveries to other fields of medicien is well-known. Transplantation immunology, cancer chemotherapy, and antiarrhythmic drugs are but a few examples of the many contributions of basic science to other areas of medical practice. Although similar strides in basic neuroscience have occurred, such discoveries as the anatomic and physiologic characterization of central monoamine pathways (Bloom et al., 1972) or the existence of opiod substances (Hughes, 1975) and opiate receptors (Snyder, 1978) cannot be as readily used to answer questions in clinical psychiatry. Several reasons for this difficulty are obvious. First, the human brain, unlike other human organs, cannot be studied directly using techniques such as biopsy or invasive recording. Second, the human brain is unique; its behavioral function and psychiatric pathology cannot be studied in animal models. Third, human life is complex. Behavioral changes, such as depression and psychosis, do not solely reflect changes in brain function. Depressive reactions, for example, can be seemingly endogenously caused, as in manic depressive illness, or seemingly exogenously determined, as in the pathological grief response to the loss of a loved one. In practice, the symptomatology of both types of depressive reaction may be indistinguishable. In many cases, the clinical history points to both a loss and family history of endogenous depression, suggesting a multiplicity of causes. Thus, careful description of pathological states does not clearly identify brain malfunction, as a careful description of dropsy pointed to cardiac failure.

The use of neuroendocrine parameters is a research strategy which may overcome some of these difficulties. Basically, neuroendocrine parameters are thought to be dependent on physiological processes in the brain and, in particular, to reflect deviations from normal which also cause psychopathology. Several lines of evidence support such a hypothesis. First, putative brain neurotransmitters such as dopamine, norepinephrine, and serotonin control the activity of several neuroendocrine systems (Frohman, 1975). In some cases, the control is at the level of the hypothalamus and, in others, at the level of the pituitary. Providing that the physiology of these neurotransmitters in the hypothalamus is similar to their physiology elsewhere in the brain, neuroendocrine parameters would provide a direct indication of the activity of these important biogenic amine neurotransmitters. Second, there is increasing evidence that neuropeptides such as the endorphins and several of the hypothalamic releasing factors, which are secreted from hypothalamic neurons, may also function as central nervous system neurotransmitters (Guillemin, 1978). Neurosecretory cells in the hypothalamus may be thought of as specialized neurons or, alternatively, synaptic neurotransmission may be a specialized type of neurosecretion. This latter alternative has grown increasingly attractive as more and more "nonclassical" types of neurotransmission have been described (Dismukes, 1979). Norepinephrine, for example, is released from many nonsynaptic release sites within the brain, activates an adenylate cyclase in the neurons it reaches, and modulates the synaptic action of other, more classical pathways. These properties nonspecific distribution, a metabolic mechanism of action, and a modulatory function – resemble hormonal actions. Furthermore, many of the hormones released by the hypothalamus, such as thyrotropin releasing hormone (TRH) are widely distributed through the brain, suggesting that they may be used as classical or putative neurotransmitters. Findings such as these make the study of the neurosecretory cells of the hypothalamus more relevant to the study of brain function. Study of these hypothalamic neuroendocrine systems may, in fact, shed light on systems in other parts of the brain which are just becoming understood.

It is not the purpose of this chapter to review these findings in the basic understanding of neuroendocrine and brain function already reviewed in detail in Chapters 7 and 12. Rather, the chapter focuses upon the ways in which this knowledge is being applied to psychiatric research. Accordingly, this review is organized not by psychiatric disease or by neuroendocrine system, but rather by strategy of application. Six different uses of neuroendocrine tools can be delineated: (1) neuroendocrine parameters as indicators of pathophysiological mechanisms (2) neuroendocrine parameters as diagnostic aids (3) neuroendocrine substances as therapeutic agents (4) neuroendocrine parameters as indicators of brain response to psychotropic drugs (5) neuroendocrine mechanisms of action to explain side effects of psychotropic drugs, and (6) neuroendocrine indicators of problems in development. The following sections will give examples of studies which utilize each of thse neuroendocrine strategies.

PATHOPHYSIOLOGICAL MECHANISMS

Attempts to use neuroendocrine parameters to study the pathophysiology of psychiatric disease include studies of thyroid and adrenal cortex regulation in depression, studies of prolactin and endorphins in schizophrenia, and studies of the regulation of steroid sex hormones in sexual deviance.

Adrenal and Thyroid Regulation in Depression

Progress in this field has been recently summarized by Carroll et al. (1976a) and by Sachar (1975). The observation of increased cortisol secretion in depressive illness is among the oldest of neuroendocrine findings, first reported by Gibbons in 1964. The increased secretion is due to increased release of pituitary adenocorticotropin hormone (ACTH). ACTH is not secreted continuously from the pituitary, but rather in brief spurts throughout the day, with more secretion in the morning than in the evening. Although the circadian rhythm is maintained in depressed patients, there are more secretory episodes in the late evening and early morning than in normals. This change in pattern does not seem to relate to changes in the sleep cycle in individual patients, although changes in circadian rhythm of sleep and wakefulness during depression are well-known. Nor is there clear evidence relating the increase in ACTH secretion in ACTH secretion to stress.

Evidence for the mechanism of this increase in ACTH secretion comes from challenge tests with the synthetic glucocorticoid dexamethasone. Normal subjects suppress ACTH and glucocorticoid secretion for at least 24 hours following a midnight dose of 2 mg of dexamethasone. A depressed subject is suppressed the following morning, but during the remainder of the day is not suppressed. Thus, early escape from suppression is characteristic of depressed patients. These findings suggest a lack of inhibition of ACTH release, both during portions of the circadian rhythm and in response to dexamethasone. The lack of inhibition of ACTH release is in turn thought to reflect a hypothalamic defect leading to abnormal release of corticotropin releasing hormone (CRH), which is in turn controled by various limbic areas in the brain (Carroll et al., 1976a). Patterns of excess cortisol secretion similar to those seen in depressed patients are seen in patients who have Cushing's disease as a result of diencephalic lesions (Carroll and Mendels, 1976). Thus, excess cortisol secretion in depression points to an abnormal limbic-hypothalamic neuronal circuit, involving the hippocampus. Study of this circuit in animals has been limited and, as yet, is inconclusive in terms of precisely which neurotransmitters are involved. A major catecholaminergic projection, the noradrenergic input, which comes from the nucleus locus coeruleus, has been implicated. ACTH may exert its effect on hippocampal pyramidal neurons through norepinephrine (Segal, 1976). Thus, a depletion of norepinephrine might result in failure of ACTH to activate the proper limbic circuits, which control CRH secretion.

Another hypothalamic-pituitary endocrine interaction which seems to be altered in depressed patients involves the regulation of thyroid hormones (Hollister et al., 1976). Although depressed patients are generally euthyroid, unipolar patients secrete lower amounts of thyroid stimulating hormone (TSH), in response to the hypothalamic hormone thyrotropin releasing hormone (TRH). Whereas excess cortisol secretion is seen in both unipolar and bipolar depressed patients, a blunted TSH response is observed only in unipolar depressed patients. It generally correlates with the severity of the depression and the degree of somatization (Vogel et al., 1977). The most impressive finding is a significant negative correlation between the blunting of the response and the level of cerebrospinal fluid 5-hydroxyindolacetic acid, a principal metabolite of serotonin (Gold et al., 1977). Thus, the blunted TSH response seems to correlate with decreased metabolism of serotonin. Again, a derangement of neuroendocrine function points to a derangement function points to a derangement in central neurotransmission.

A similar type of finding has been made by Garver et al., (1975) in regards to growth hormone. Growth hormone is secreted from the pituitary in response to insulin-induced hypoglycemia. This secretion can be blocked by the alpha noradrenergic antagonist phentolamine in normal subjects, which suggests a noradrenergic mediation of this effect, probably within the hypothalamus. Sachar (1975) has found that growth hormone release can also be stimulated by the catecholamine precursor L-dopa and not blocked by the dopamine antagonist pimozide. This evidence also points towards noradrenergic mediation of growth hormone release. Garver and his associates found that the response to insulin-induced hypoglycemia was reduced in depressed patients and that this reduction correlated with a diminished level of central norepinephrine turnover. Twenty-four hour levels of 3-methoxy-4-hydroxy-phenylglycol (MHPG) were used to indicate central norepinephrine metabolism, since this metabolite is produced only in the central nervous system (Maas, in press).

In summary, alterations in the regulation of three hormonal systems in depressed patients have been reviewed. These alterations seem to point to derangements in monoamine physiology of the central nervous system. It is known that metabolites of these neurotransmitters, serotonin and norepinephrine, are low in certain depressed patients and that antidepressant drugs of various types increase the function of monoamine pathways as if in compensation for deficits in these transmitters. The neuroendocrine evidence reviewed here supplies important physiological evidence for a monoamine hypothesis of depression.

Prolactin and Endorphins in Schizophrenia

Biological research in schizophrenia has proven to be, to date, a long and fruitless search for a disordered molecule to explain disordered thinking. Two prevalent theories are the dopamine theory, which suggests that there is a pathologically increased activity of dopaminergic neuronal transmission, and the opioid theory, which suggests a dysfunction of unspecified type in the metabolism of opioid peptides. Both hypotheses have been investigated using neuroendocrine strategies.

There are three dopamine-containing neuronal systems in the brain. The substantia nigra, the largest source of dopamine-containing neurons, projects to the corpus striatum. This system is responsible for Parkinson's disease and other movement disorders, including those induced by the acute and chronic use of antidopaminergic neuroleptic drugs. The adjacent mesolimbic system is a dopaminergic projection from the midbrain to various limbic structures, including the amygdala and several septal nuclei, and to portions of frontal and temporal cortex. Because tumors and other lesions in these areas can produce symptoms which resemble schizophrenia, it has been suggested that this system is responsible for schizophrenia. Unfortunately, there is no direct proof for this hypothesis. Although dopamine metabolites can be measured in blood, urine, and spinal fluid, these metabolites are not abnormal in psychotic patients. Alterations in postsynaptic sensitivity to dopamine have been proposed, but there is no way of proving this hypothesis.

The third dopaminergic system has a neuroendocrine output and, therefore, is more amenable to direct investigation. Dopamine neurons in the arcuate nucleus of the hypothalamus release dopamine into the portal circulation, from which it reaches prolactin-secreting cells in the pituitary. This "tubero-infundibular system," therefore, has a measurable product, the hormone prolactin. Unfortunately, basal prolactin levels are quite low in normals, so that an increase in dopaminergic inhibition of prolactin release may not be readily measureable. In fact, attempts have been made to correlate prolactin level with schizophrenic symptomatology, but have not proven successful (Meltzer et al., 1974). It is also possible, of course, that this dopaminergic system in particular is not related to the dopaminergic system which may malfunction in psychosis, presumably the mesolimbic system.

Interest among many investigators has turned to the opioid systems in the brain (Watson et al., 1979). The hypothalamus makes a class of compound endorphins with opiate-like activity and cleaved from a parent compound which also includes the molecule ACTH (see Chapter 12). Further cleavage of the endorphin molecule yields a smaller peptide called enkephalin,

which also has opiate properties. Enkephalin-containing neurons are
widely distributed throughout the nervous system. Endorphins can cause
a rise in prolactin secretion, suggesting an anti-dopaminergic activity.
In other areas of the brain, however, enkephalins depend on dopamine
for their activity. In some areas, both catecholamines and enkephalins
seem to be found in the same neuron. These and other interactions of
enkephalins and endorphins with dopaminergic and other neuronal systems
are not yet fully characterized.

Enkephalins are rapidly degraded and cannot be isolated from the blood
or urine. Endorphins, however, are longer-lived and can be isolated.
Two strategies have been used to test the role of opioid peptides in schizophrenia. The first strategy which was tested assumed that an abnormal
or excess amount of endorphin was produced by schizophrenic patients.
One group has reported endorphins in the spinal fluid of schizophrenics,
but not of normals (Terenius et al., 1976). This preliminary report has
not been confirmed after investigation by other workers (Lewis et al.,
1979). The second strategy attempted to use specific antagonists of
opiates, such as naloxone and naltrexone, to block the hypothesized increased opioid peptide activity. The reports of the success of the opiate
antagonists as antipsychotic drugs is mixed. Although several groups
have reported effects, particularly in blocking auditory hallucinations,
other groups have not been able to confirm this finding (Gunne et al.,
1977); Kurland et al., 1977; Watson et al., 1979).

Although neither prolactin measurements nor the opioid peptides have
provided a solution to the problem of schizophrenia, each represents an
interesting but different use of neuroendocrine techniques. The prolactin
research used the production of a hormone as a measure of the activity of
a neurotransmitter. The endorphin research used the hormone as a prototype of an unmeasureable neurotransmitter, enkephalin.

DIAGNOSTIC TESTS

Only one neuroendocrine measure has so far been proposed as a diagnostic test for a particular mental illness. The early escape of cortisol
secretion from the dexamethasone suppression test, described in the
previous section, has been proposed as a diagnostic test for endogenous
depression (Carroll et al., 1976b; Brown et al., 1979). The test is positive
in 40% of patients with endogenous depression, and negative in diseases
which the clinician would wish to differentiate — reactive depression and
schizophrenia. Cushing's disease itself is not common enough to be a
significant problem in interpretation, although diencephalic pathology
should be considered in a patient whose cortisol levels do not return to
normal following recovery from the depression. The test itself, involving
a small dose of dexamethasone and several blood samples, can be done
quickly and with minimum risk to the patient. Most important, it would
seem to be clinically useful: although patients with reactive depressions
and patients with endogenous depressions both respond well to tricyclic
antidepressants, schizophrenic patients constitute a population for which
another class of drugs, the neuroleptics, is preferable. Also, the differential diagnosis between endogenous depression and psychosis is difficult,
and this test may prove to be a useful adjunct to clinical judgement.

An important issue in the treatment of depressions is the selection of
the type of the antidepressant drug. Some tricyclic antidepressants, such
as desipramine, are relatively specific for noradrenergic neurotransmission;
others wuch as amitriptyline are more specific for serotonergic neurotransmission (Maas, 1975). Although noradrenergic neural pathways play a
large role in the regulation of cortisol secretion, as described above,

serotonergic and other influences on this neuronal parameter have not been excluded. It is possible, however, that the relatively low positive rate of 40% for this test may reflect derangement in only noradrenergic neurotransmission. Patients with other defects, such as serotonergic deficiency, might be among the false negative group. Certainly, if the neurochemical as well as the diagnostic specificity of the test could be determined, this test would be more useful in directing antidepressant therapy.

TREATMENT

The use of endocrine and neuroendocrine hormones for the treatment of psychiatric disease is another neuroendocrine strategy. Although some of these treatments are highly empirical, others are grounded upon careful observation and theoretical conception. None of these courses of treatment, the use of thyroid hormones, endorphins, gonadal steroids, or calcitonin, is yet an accepted or proven means of treatment.

Treatment of Depression with Hormones

The use of tri-iodo-thyronine (T_3) as an adjunct to therapy with tricyclic drugs. Initial reports suggested that T_3 could potentiate the action of tricyclic drugs in euthyroid patients (Prange et al., 1977), but attempts to replicate these findings have been inconclusive. TRH and TSH have also been used to activate the thyroid axis. Again, initial reports have been promising in both unselected depressed patients and in patients resistant to antidepressant drugs alone (Prange et al., 1976). TRH itself has also been reported to have a euphoric effect in normal women (Wilson et al., 1973). Subsequent studies, however, have been unable to confirm a significant effect of TSH or TRH in depression (Hollister et al., 1974).

A limited number of studies with endorphin have been conducted in depressed patients. Kline et al. (1977) reported a lifting of depression in several psychotically depressed patients. The general applicability of this finding is yet to be determined.

Gonadal hormones have also been shown to reverse depression in patients who are resistant to therapy with conventional antidepressants (Vogel et al., in press). Estrogen was used in females and mesterolone in males. The rationale for this therapy was evidence of high estradiol production in depressed men and high testosterone production in depressed women.

Endorphins in the Treatment of Schizophrenia

Both endorphins and their putative antagonists nalaxone and naltrexone, have been examined as possible treatments for schizophrenia. Initially, because of preliminary evidence suggesting increased levels of endorphins in the spinal fluid of schizophrenic patients, opiate antagonists were chosen for study as therapeutic agents. Despite initial reports of success, neither naloxone or the longer-acting drug, naltexone, have been of consistent benefit to schizophrenics (Janowsky et al., 1977; Watson et al., 1979). More recent work has involved the endorphins themselves as antipsychotic agents. A recent report has shown in six patients, using double-blind evaluation, that (Des-Tyr[1])-γ-endorphin as antipsychotic properties (Verhoeven et al., 1979). The six patients chosen had been previously resistant to neuroleptics. Although the report is encouraging and a naturally occurring antipsychotic drug would have great utility, more evidence for the efficacy of this treatment is still required.

Sex Hormones in the Treatment of Sexual Deviance

Investigation of the effects of gonadal hormones in sexual deviance is in a very early stage of development. Little is known about the behavioral effects of these hormones, particularly in human beings. Furthermore, their use and study in homosexuals or in sexually deviant criminals bring up many ethical and social problems.

Attempts to characterize a population with abnormal levels of sexual hormones which, in turn results in an abnormal sexual behavior, have not been successful. Although some studies have identified a subpopulation of homosexual patients, those who have never had heterosexual experience as having lower urinary testosterone excretion, this characterization has not been uniformly found. Similarly, aggressive males and sexual criminals in particular, have been identified as having higher levels of urinary testosterone excretion in some studies, but the finding has not been uniform (Rose, 1975).

Well controlled attempts to treat sexual criminals with anti-testosterone agents have been few. Money and his colleagues (1976) treated 10 XY males and 13 XYY males, most of whom were impulsive sexual offenders, with an antiandrogen compound, medroxyprogesterone acetate. The threshold for erotic imagery and sexual function, measured as achievement of erection, was diminished in all cases. In five cases, sex-offending paraphilia also went into remission. There was no change in aggressive behavior itself.

One of the problems cited in this type of investigation is the multidetermined nature of human behavior. Homosexuality or sexual aggressiveness could represent many different factors, including neuroendocrine, environmental, and psychosocial factors. A similar argument can, of course, be made for depression, which can represent a reaction to the loss of a loved one, an inadequate character structure, or a derangement of brain biochemistry. Separating physiological from psychological causes is always difficult. The beginning of separation of types of disorders have been made for depression. Clinicians are generally able to distinguish between reactive depression, which reflects loss, various kinds of chronic character pathology, and endogenous depression, which is more likely to involve changes in neurochemical activity. Such a distinction has obvious prognostic and therapeutic implications. Sexual deviance may eventually be subjected to the same sort of differential diagnosis and treatment.

MONITORING DRUG EFFECTS

One of the many problems confronting modern psychiatrists is that after diagnosis has been made and treatment has been initiated, there are very few ways to monitor the biological effects of treatment. If the patient has the expected beneficial effects from a drug, it is difficult to know when to withdraw medication without risking relapse. Conversely, it is difficult to know if nonresponse means a failure of the biological treatment or whether it means that psychological factors are overwhelming. Examples of treatment problems in schizophrenia and depression are described below.

Monitoring of Antipsychotic Drug Treatment with Prolactin

The release of dopaminergic inhibition of the pituitary and the resulting secretion of prolactin has been proposed as a means of monitoring the action of antipsychotic drugs. Patients show an elevation of prolactin which parallels in time the onset of behavioral action of the drugs (Meltzer et al., 1974). Like the antipsychotic effect, there appears to be no tolerance to drug effect. Extrapyramidal motor system effects, like dystonia or pseudo-Parkinson's disease, on the other hand, show tolerance to the drug over the first several weeks of treatment. Although neuroleptic drugs reach steady state levels within one week, the antipsychotic effect continues for over six weeks. Prolactin levels parallel the course of the pharmacokinetic distribution of the drug, not the antipsychotic effect. Although that finding may indicate that the antipsychotic effect, particularly in the later phases of treatment, does not solely depend upon dopamine inhibition, neuroendocrine data are not sufficient to establish that hypothesis. Antipsychotic effect also, moreover, does not correlate with the amount of prolactin rise (Gruen, et al., 1978). Finally, the prolactin effect is maximal at doses which are only about half the therapeutic dose for psychosis (Meltzer et al., 1974). Thus, the prolactin system would not seem to provide a good parameter for monitoring the therapeutic effects of antipsychotic drugs.

A second strategy which used the prolactin system for monitoring the effects of antipsychotic drugs involved a side effect, rather than the therapeutic effect. Tardive dyskinesia is a movement disorder which occurs in from 5%-40% of patients who have been treated for several years or more with antipsychotic drugs. Patients generally show abnormal buccal-lingual movement, such as pouting or tongue protrusion. Because antipsychotic drugs both cause this syndrome and suppress its symptomatic appearance, it has been suggested that antipsychotic drug may be inducing supersensitivity to dopamine in basal ganglia neurons which receive dopaminergic synapses. The drug, by blocking dopaminergic neurotransmission, is hypothesized to cause a chemical denervation, leading the postsynaptic caudate neuron to increase its sensitivity to the neurotransmitter. Analogous changes in postsynaptic sensitivity to neurotransmitters have been observed in muscle tissue in which the nervous input has been lesioned. Evidence for this hypothesis comes predominantly from animal experimentation, in which supersensitivity can be demonstrated after three weeks of chronic drug administration. This time course is far shorter than the several years of drug use generally required in humans before even early signs of tardive dyskinesia are visible.

The prolactin system was used to study the supersensitivity hypothesis in patients with tardive dyskinesia. Tamminga et al., (1977) administered L-dopa and apomorphine to chronic schizophrenic patients, both with and without tardive dyskinesia, and to normal volunteers. L-dopa stimulates dopaminergic neurotransmission, probably because it is a precursor of dopamine. Apomorphine stimultes dopamine receptors as an agonist. The supersensitivity hypothesis would predict that the suppression of prolactin secretion would be greatest in patients with tardive dyskinesia. However, Tamminga et al. found that the response to L-dopa was significantly less than that of normal controls for both chronic patient groups, those with and those without tardive dyskinesia. Response to apomorphine was the same between groups.

The implication of these findings is not clear. Tamminga et al. point out that they were measuring magnitude of response, not threshold, so that a true measure of supersensitivity has not been performed. Other possible explanations include indirect effects of chronic neuroleptics on either dopamine metabolism or hormone feedback loops. The data would seem to suggest, however, that in this system, at least, there is not postsynaptic supersensitivity after chronic exposure to antipsychotic drugs.

TRH Stimulation in the Therapy of Depression

One of the problems in the pharmacologic treatment of depression is the determination of the length of treatment. Various clinical rules of thumb have been established, generally based on the estimate that untreated depressions usually last about six months. Since the onset of action of the tricyclic antidepressants is several weeks long and the drugs need to be tapered slowly when discontinued because of their anticholinergic properties, it is not practical to stop drug administration to test if the drug is still needed. A related problem is the determination of the need for tricyclic antidepressant medication after electroshock therapy. Double-blind studies show that about half of patients require medication to prevent relapse after shock therapy, but clinical grounds for determining the need in an individual case are lacking (Davis, 1976). Kirkegaard and Smith (1978) have tested a neuroendocrine solution, however. As discussed above, the TSH response to TRH is blunted in depression, but this parameter is normalized when the depression is over. Kirkegaard and Smith found that patients in whom the response had not normalized after shock therapy did significantly better with continuing tricyclic antidepressant treatment than without such treatment. Patients in whom the response had normalized, however, did not need drugs to prevent relapse. During drug treatment, a second TRH test was performed. Again, only those in whom the TSH test was abnormal required continued drug therapy. Thus, a neuroendocrine measurement could direct drug treatment. The patient sample in this study was thirty patients. Replication in a larger group would be important and would make this finding very valuable clinically. Antidepressant drugs have a low toxic to therapeutic ratio, particularly for patients with pre-existing cardiac problems. This neuroendocrine strategy would limit their use to patients requiring them and to the shortest period of treatment necessary.

SIDE EFFECTS OF PSYCHOTROPIC MEDICATION

Although neuroendocrine systems may not be intimately involved in the pathophysiology of most mental illness or even in the mechanisms of its treatment, all parts of the neuroendocrine system are chronically exposed to the medication. Since biogenic amines, including the catecholamines and serotonin, are profoundly affected by psychotherapeutic medication, it might be expected that there would be marked effects of these drugs on endocrine function.

Actually, in clinical practice, hormonal dysfunctions are few. In the treatment of schizophrenia, a low percentage of women will experience lactation due to the blockade of the dopamine inhibition of the pituitary by the neuroleptic drug. Other side effects of the neuroleptic drugs would appear to involve a locus of action in the hypothalamus; patients on these drugs cannot regulate food intake or body temperature as well as normal people. However, there are no other clinically relevant derangements in endocrine function. Similarly, the antidepressants do not cause endocrine dysfunction, despite a marked effect on neurotransmitters. Methylphenidate, which stimulates growth hormone release and is used in the treatment of childhood hyperactivity, causes only minor changes in growth rates in treated children (Satterfield et al., 1979).

There are probably several reasons why the endocrine effects of psychotropic medication are few. First, the drugs are given chronically, so that the hypothalamic neurons themselves can develop tolerance to their effect before neuroendocrine changes become clinically apparent. Second, since the drugs primarily affect neurotransmitter function, their effects are limited to the hypothalamus and connecting brain circuits. Most hormonal systems have feedback regulation at other sites as well, however. The regulation of cortisol secretion, for example, includes feedback in the limbic system, the hypothalamus, and the pituitary. Drug alteration of the response of the limbic and hypothalamic sites still leaves the pituitary capable of regulating cortisol levels. Prolactin would be expected then to be an exception, since its secretion is directly controlled by the neurotransmitter dopamine at the pituitary level.

Endocrine side effects are more common with abused drugs than with psychotherapeutic compounds. The best studied example is the inhibition of testosterone secretion in male alcoholics. This inhibition has been shown to be caused by alcohol's interference with a dehydrogenase reaction in the testis. Accordingly, levels of gonadotropics normes which normally control testosterone secretion are quite high. The effects of high levels of this pituitary hormone and personality traits of chronic alcoholics are unknown (Persky et al., 1977).

DISORDERS OF DEVELOPMENT

Disorders in development have long been the domain of both the psychiatrist and the endocrinologists. Recent neuroendocrine research has begun to clarify syndromes in which there is an interaction between behavior and endocrine function. Four stages in development will be examined here: development in utero, growth in early childhood, sexual development with puberty, and aging during the involutional period. (Additional information on these topics is presented in Chapters 2, 4, 8-11, and 14-16, respectively.)

Development in Utero

In both rodents and primates, it has been demonstrated that in utero or neonatal exposure to gonadal hormones can cause a profound effect on sexually determined behaviors. The hormonal exposure need only have occurred during a critical period of development, as short a period as several days in rats. Thereafter, the affected animal permanently displays some or all of the sexually determined behaviors of the opposite sex. It has been hypothesized that certain programmed behaviors, such as sexual aggressiveness or nest building are programmed into the animal's brain. At a critical and quite early time in development, sexual hormones permanently select which behaviors are to be expressed. Ehrhardt (1975) has examined the possible existence of this phenomenon in human beings. Two clinical syndromes exist in which genetic females are exposed to large amounts of masculinizing hormones in utero. One syndrome is the adrenogenital syndrome, in which a defect in the adrenal cortex of the fetus leads to production of an androgen steroid instead of cortisol. The newborn shows clitoral enlargement and may show labial fusion, but internal sexual organs are normal. Cortisol replacement therapy suppresses production of the abnormal androgenic steroid, and minor corrective surgery is performed soon after birth. These females then are endocrinologically normal, except for an excess exposure to androgens in utero. A second syndrome is iatrogenic. Some women are treated with progestinic drugs during pregnancy to prevent miscarriage. Female children of such pregnancies may show masculinizing changes at birth, suggesting that the drug acted an an androgen hormone in utero. Again, minor corrective surgery after birth is performed, but no other treatment is required.

Several surveys have been performed to determine if the prenatal exposure to androgens has any effect on sex-sterotyped behaviors. Subjects examined to date were in childhood or early puberty, so that all sexual and maternal behavior could not be examined. The study design used ratings by other members of the family and compared affected with nonaffected siblings. Females with in utero androgen exposure, in comparison with their siblings, showed more intense physical activity with an emphasis upon rough, outdoor play with males. Interest in dolls, infants, and maternal roles was markedly less than that of their sisters. Clothing choices were more functional, and interest in jewelry or cosmetics was low.

The adrenogential syndrome can also occur in males. These males then would be exposed to higher in utero levels of male hormone. If treated with cortisol early, there seem to be few effects on behavior, except for interest and excellence in athletics. There is one uncontroled study of males whose mothers received estrogens and progesterone during pregnancy to prevent miscarriage because of maternal diabetes. Twenty 6-year-old and twenty 16-year-old children were examined (Yalom et al., 1973). Both groups were less assertive and athletic than their peers.

Growth in Childhood

Retardation of growth and development in the face of environmental deprivation was initially described by Spitz in 1946 as "hospitalism." Current research has continued to attempt to define the endocrinological basis of this phenomenon. The behavioral syndrome, commonly referred to as "psychosocial dwarfism," is complex. The most striking feature is a retardation in growth in stature which is reversed when the child is removed from his home. Generally, there is no starvation, although there may be specific vitamin and mineral deficiencies. Sleep patterns are poor, with difficulty falling asleep and frequent night wakenings. There is frequent enuresis, encopnesis, and diminished response to pain. Puberty is delayed. The children are characterized as "distant" and may have frequent temper tantrums. These findings reverse when the child is removed from his home. The families of these children have not been well characterized. The abuse may range from physical trauma to more passive deprivations.

The most unusual finding in this syndrome is a deficit in growth hormone secretion (Brown, 1976). In various type of malnutrition syndromes, growth hormone is elevated. Growth hormone levels in these children are depressed, however. Insulin-induced hypoglycemia or arginine which normally evoke growth hormone secretion, are ineffective in these children, until after several week of hospitalization, when reversal of the syndrome has occurred. Growth hormone is also secreted in slow wave sleep in normals, but not in these children. Secretion of growth hormone has been accomplished in five cases of psychosocial dwarfism by treatment with epinrphrine and propranalol, and in one case, by propranalol alone (Parra, 1973; Imura et al., 1971). This finding suggests that the mechanism of suppression of growth hormone secretion in these children may involve a beta adrenergic pathway.

Anorexia Nervosa

Anorexia nervosa is another psychiatric disease which involves a clearly demonstrable retardation in physical development. The cardinal features of the syndrome are severe weight loss due to decreased food consumption and amenorhea in a pubescent girl. Numerous authors have pointed to the combination of an eating disorder and amenorhea as indicative of a hypothalamic site of pathology. Indeed, there are two case reports of hypothalamic tumors found at autopsy in patients with anorexia nervosa (Halmi, 1978). The entire hypothalamic-pituitary-endocrine axis is not affected, however. Thyroid and adrenal function is largely normal. Minor deviations from normal are adequately accounted for by the emaciation of the patient. At autopsy, pituitary cells which secrete ACTH, TSH, growth hormone or prolactin all appear normal (Halmi, 1976). The deficiency seem to be confined to luteinizing hormone (LH) and follicle-stimulating hormone (FSH) (Katz, 1975). These hormones are not secreted in patients with anorexia nervosa and, consequently, there is amenorhea and atrophy of the ovaries. Clomiphene, a drug which stimulated LH and FSH secretion in normal people by blocking feedback inhibition by estrogen at the pituitary, does not stimulate LH and FSH secretion in anorexia nervosa. The hypothalamic hormone gonadotrophic releasing hormone (GnRH) does cause normal LH and FSH secretion, suggesting that the primary deficit is hypothalamic or higher in the nervous system. However, more FSH than LH is released, which is typical of a prepubescent girl. Thus, there may be some "immaturity" at the pituitary level as well.

What are the psychological concomitant of this hypothalamic disorder? The disease is thought to represent a young girl's attempt to assert control over the size of her body, and the development of her sexuality. The attempt to control body size is seen in the form of relentless self-imposed dieting, self-imposed vomiting, and obsessive hyperactivity. Although the patients are not classically psychotic, they have a definitely altered sense of body image, often seeing themselves as obese despite their morbid emaciation. The family constellation generally includes a domineering mother who has made food an issue of control and an emotionally distant father. Not infrequently, the patient is initially obese, and begins the syndrome during dieting.

Family practitioners see another form of this syndrome as a more benign occurrence. The weight loss is generally much less severe and the presence of amenorrhea is more variable. The patients respond to a diet imposed by the physician.

It has been proposed that this disease may be a psychogenic, neuroendocrine disorder, like the psychosocial dwarfism disorder. There is also some preliminary evidence that catecholaminergic neurons may be involved. Halmi et al., (1978) have reported that 3-methoxy-4-hydroxypheneleglycol (MHPG), the metabolite of norepinephrine which is produced in the central nervous system is abnormally low in these patients. This evidence does not establish, however, that noradrenergic neurons form the link between the emotional conflict and the neuroendocrine disturbance.

Involutional Melancholia, Premenstrual Tension, and Postpartum Depression

The oldest syndrome described in which an endocrine change and a psychological disturbance are linked is involutional melancholia. A propensity to severe depression was thought to accompany the involution of the uterus during the menopause. Replacement therapy with estrogens to compensate for atrophy of the ovaries has often been recommended by clinicians, but has been shown to have no particular psychological benefit (Ripley et al., 1940).

A recent review of the literature by Winokur and Cadoret (1975) suggests that there is little reason to continue to regard involutional melancholia as a syndrome separate from other severe depressions. The incidence of depression in females rises during early adulthood, but by age 35-44, it is stable. Figures for suicide are similar. Hence, there seems to be no increase in illness, which would indicate an effect of the menopause. Winokur and Cadoret also point out that depressions occurring solely within the menopause have fewer delusional or psychotic features than recurrent depressions with episodes both during and after the menopause. These data imply that the menopause does not increase the severity of depression. However, the data could also be interpreted to suggest that the menopausal depression is different from recurrent depressions and, hence, a unique syndrome. Some supporting evidence for the latter proposition comes from the work of Majer (1941), who found that patients with menopausal depression had significantly fewer relatives with depressive disorder than other depressed patients. This finding suggests that the menopause itself may have a depressive effect on women who do not have a genetic predisposition to depression. However, a later study by Stenstedt (1959) does not replicate this finding. A third line of investigation pursued by Winokur was to look at patients with recurrent depression to see if the number of incidents of depression increase during menopause. There was no such increase, however. Winokur and Cadoret conclude that there is little evidence to favor the concept of involutional melancholia.

Interest has recently also been directed to premenstrual tension as a neuroendocrine disorder. Smith (1975) has pointed out that this disorder may consist of a multitude of syndromes, all of which are symptomatically exacerbated during the menstrual cycle. There is, however, some evidence that fluctuations in the level of one or more estrogenic steroids may correlate with some of the symptoms. Smith suggests that estriol, which blocks the effects of its metabolic precursor estradiol, may not be produced in high enough quantities in patients with premenstrual tension. He points out that there are many reports of euphoria in patients with premenstrual tension during the last two trimesters of pregnancy, when the placenta produces estriol. The mechanism may be an antagonism of noradrenaline by estrogens, which in turn is impeded by estriol. However, much work in both patients and animal models needs to be done to substantiate such an hypothesis. Smith cautions that there have been many etiological hypotheses and resulting treatments, all of which have been successful in uncontrolled, but not controlled studies. Such findings suggest that there is a large psychogenic compound, but do not rule out an interaction with neuroendocrine systems.

Postpartum depression and postpartum psychoses also may be the result of a multitude of etiologic factors. The profound psychological impact of the termination of pregnancy and the commencement of parental responsibility is accompanied by changes in a number of endocrine systems, including the adrenal, the thyroid, and sexual hormones. Hamilton (1962), for example, has pointed out that the regulation of thyroid hormone is abnormal for up to four months following partuition. He suggests that late-appearing postpartum depression may be a form of the depression seen in hypothyroidism (see Chapter 18), and has successfully treated some of these patients with thyroid hormones. The recurrences of symptoms in other patients during menstruation suggests that changes in levels of sexual hormones may be of importance in other patients.

Although alterations in levels of sexual hormones have a profound effect on behavior in a number of developmental stages, from in utero through the menopause, the precise mechanism of interaction has not yet been characterized. The multiplicity of interactions suggest very profound interactions between psychological factors, the stage of development, and the hormones themselves. Although much work remains to be done in animal models to elucidate the effect of various sexual steroids on various brain functions, including but not limited to the biogenic amines, animal studies may not prove to be sufficient to fully characterize the mechanism of these disorders. Unlike other animal species, human behaviors are more independent of sexual hormones. In particular, sexual excitement and activity are not as closely linked to the oestral cycle. It may therefore be harder to detect endocrine effects on behavior in human beings. This problem may already be reflected in the data presented here: the studies on in utero influence of hormones, anorexia nervosa, and involutional melancholia do not provide the type of definitive data which would clarify the issue of endocrine involvement in these syndromes.

CHALLENGES AND NEW APPROACHES

This chapter has reviewed current progress in clinical neuroendocrine research under the rubric of possible research strategies for solving psychiatric problems. Six possible strategies were examined: (1) neuroendocrine parameters as indicators of pathophysiological mechanisms in mental disease (2) neuroendocrine parameters as diagnostic indicators (3) neuroendocrine hormones as therapeutic agents (4) neuroendocrine parameters to monitor the therapeutic action of psychotropic drugs (5) neuroendocrine side effects of psychotropic drugs, and (6) the action of neuroendocrine systems during development, as it relates to several pathological conditions.

Few of the examples reviewed above represent entirely solved problems. In most cases, the links between behavioral and endocrine changes are tenuous and circumstantial. The eventual elucidation of the linkage will require increased knowledge from basic nenroscience of the physiology of the hypothalamus, particularly its interrelationship with the rest of the brain. Such knowledge will have to include a developmental perspective. perspective. For example, changes in the metabolis, of norepinephrine were implicated in the change of steroid and growth hormone secretion in adult depression, the alteration of growth hormone secretion in psychosocial dwarfism, and the sexual hormone irregularities of anorexia nervosa. The endocrine, somatic, and psychological features of these three diseases are all quite different. Is an lateration in noradrenergic neurotransmission central to all diseases? If so, why are they so dissimilar? One obvious approach to answering these questions is to hypothesize that these diseases are similar and that much of their distinctive character comes from their occurrence at different critical stages of development: early childhood, pubescence, and adulthood. Such a hypothesis points to additional investigations in both animal models and in psychiatric patients.

Neuroendocrine studies of psychiatric problems raise important questions about the relationship of psychiatric disease to biological process. On one level, the problem is a problem of nosology. The term depression, for example, can mean everything from a feeling of diffidence and boredom to a complete withdrawal with suicidal intentions. The work on adrenal steroids described above suggests that these two syndromes represent different diseases, with only the more severe syndrome likely to have changes in the dexamethasone suppression test. The recent improvement in psychiatric nomenclature, which now clearly specify the diagnostic criteria for major affective disorder as opposed

to minor depressive illness, will improve both clinical practice and
research. Even with more precise diagnostic criteria, however, problems
remain. Only about half of patients with major affective disorders
have abnormal dexamethasone suppression tests. Similarly, not all
patients with Addison's disease become depressed (see Chapter 18).
Thus, while there is a good correlation between biologic state and
psychiatric disorder, the concordance is not complete.

While a lack of complete concordance makes the data look less than
perfect, the possibilities for future research and for clinical utility
are very exciting. Patients with different biological parameters but
identical psychiatric diagnoses may mean several things. Their existence
may imply that the nosological categories are too broad; closer examination
of biologically distinct groups may reveal different clinical syndromes
with different prognoses and response to treatment. Alternatively, the
existence of clinically similar patients with differing biological parameters
may imply that diseases like depression are a final common pathway for
diverse biological disturbances or even for both psychological and
biological dysfunction. This possibility is also raised by the material
in Chapter 18 on affective disorders accompanying different endocrine
disease. The possibility that clinicians may not be able to distinguish
various biological and psychological causes of mental illness from
clinically apparent symptomatology along makes neuroendocrine research
of particular importance.

It has become increasingly likely, for example, that major depressions
may be caused by either a defect in serotonin metabolism or a defect
in norepinephrine metabolism (Maas, 1975). Clinically, the depressive
syndromes caused by the two defects appear to be indistinguishable.
However, tricyclic drugs like chlorimipramine would be far better in the
serotonin-mediated depression and drugs like desipramine would be
better in the norepinpehrine-mediated depression. Although urine
measurements of central norepinephrine metabolites can be made, urine
measurements of serotonin metabolites are not possible, due to
contamination from peripheral sources. A neuroendocrine approach, as
described above, to determine which neurotransmitter system had
malfunctioned, would be extremely valuable to the clinicians. Similar
problems may well occur in the treatment of other mental disorders.

Although these advances would require more research and a greater
cost to the patient and society to support the increased technology,
society has come to realize the importance of accurate diagnosis and
rapid treatment to return the patient quickly to society as a functioning
individual. The avoidance of long term or even permanent disability
by better treatment would surely justify its cost.

REFERENCES

Bloom, F.E., Hoffer, B.J., and Siggins, G.R. Norepinephrine mediated
 synapses: a model system for neuropharmacology. *Biol. Psychiat.*
 4, 157-171 (1972).
Brown, G.W. Endocrine aspects of psychosocial dwarfism, in *Hormones,
 Behavior, and Psychopathology*, E.J. Sachar, ed. Raven Press,
 New York (1976), pp. 253-262.
Brown, W.A., Johnston, R., and Mayfield, D. The 24-hour dexamethasone
 suppression test in a clinical setting: relationship to diagnosis,
 symptoms. *Am. J. Psychiat.* 136, 512-514 (1979).
Carroll, B.J. and Mendels, J. Neuroendocrine regulation in affective
 disorders, in *Hormones, Behavior, and Psychopathology*, E.J. Sachar,
 ed. Raven Press, New York (1976), pp. 193-224.

Carroll, B.J., Curtis, G.C., and Mendels, J. Neuroendocrine regulation in depression. I. Limbic system-adrenocortical dysfunction. *Arch. Gen. Psychiat.* 33, 1039-1044 (1976a).

Carroll, B.J., Curtis, G.C., and Mendels, J. Neuroendocrine regulation in depression. II. Discrimination of depressed from nondepressed patients. *Arch. Gen. Psychiat.* 33, 1051-1058 (1976b).

Davis, J.M. Overview: maintenance therapy in psychiatry. II. Affective disorders. *Am. J. Psychiat.* 133, 1-13 (1976).

Dismukes, K. New concepts of molecular communication among neurons. *The Behavioral and Brain Sciences*, in press.

Ehrhardt, A.A. Prenatal hormone exposure and psychosexual differentiation, in *Topics in Psychoendocrinology*, E.J. Sachar, ed. Grune and Stratton, New York (1975), pp. 67-82.

Frohman, L.A. Neurotransmitters as regulators of endocrine function. *Hosp. Prac.* 12, 54-67 (1975).

Garver, D.L., Pandey, G.N., and Dekiremenjian, H. Growth hormone and catecholamines in affective disease. *Am. J. Psychiat.* 32, 1149-1154 (1975).

Gibbons, J.L. Cortisol secretion rate in depressive illness. *Arch. Gen. Psychiat.* 10, 572-575 (1964).

Gold, P.W., Goodwin, F.K., Wehr, T., and Rebar, R. Pituitary thyrotropin response to thyrotropin-releasing hormone in affective illness: relationship to spinal fluid amine metabolites. *Am. J. Psychiat.* 134, 1028-1031 (1977).

Gruen, P.H., Sachar, E.J., Altman, N., Langer, G., Tabrizi, M.A., and Halpern, F.S. Relation of plasma prolactin to clinical response in schizophrenic patients. *Arch. Gen. Psychiat.* 35, 1222-1227 (1978).

Guillemin, R. Biochemical and physiological correlates of hypothalamic peptides. The new endocrinology of the neuron. *Res. Publ. Assoc. Res. Nerv. Ment. Dis.* 56, 155-194 (1978).

Gunne, L.M., Lindstrom, L., and Terenius, L. Naloxone-induced reservsal of schizophrenic hallucinations. *J. Neurol. Transm.* 40, 13-19 (1977).

Halmi, K.A. Selective pituitary deficiency in anorexis nervosa, in *Hormones, Behavior, and Psychopathology*, E.J. Sachar, ed. Raven Press, New York (1976), pp. 285-290.

Halmi, K.A. Anorexia nervosa: recent investigations. *Ann. Rev. Med.* 29, 137-148 (1978).

Hamilton, J.A. *Post Partum Problems.* C.V. Mosby Co., St. Louis (1962).

Hollister, L.E., Davis, K.L., and Berger, P.A. Pituitary response to thyrotropin-releasing hormone in depression. *Arch. Gen. Psychiat.* 33, 1393-1396 (1976).

Hollister, L.E., Berger, P., Ogle, F.L., Arnold, R.C., and Johnson, A. Protirelin (TRH) in depression. *Arch. Gen. Psychiat.* 31, 468-474 (1974).

Hughes, J. Isolation of an endogenous compound from the brain with pharmacological properties similar to morphine. *Brain Res.* 88, 295-308 (1975).

Imura, H., Yoshimi, T., and Ikekubo, K. Growth hormone secretion in a patient with deprivation dwarfism. *Endocrinol. Japan* 18, 301-304 (1971).

Janowsky, D.S., Segal, D.S., Bloom, F., Abrams, A., and Guillemin, R. Lack of effect of naloxone on schizophrenic symptoms. *Am. J. Psychiat.* 134, 926-927 (1977).

Katz, J.L. Psychoendocrine considerations in anorexia nervosa, in *Topics in Psychoendocrinology*, E.J. Sachar, ed. Grune and Stratton

New York (1975), pp. 121-134.

Kirkegaard, C. and Smith, E. Continuation therapy in endogenous depression controlled by changes in the TRH stimulation test. *Psychol. Med.* 8, 501-503 (1978).

Kline, N.S., Li, C.H., Lehmann, H.E., Lajtha, A., Laski, E., and Coope, T. β-Endorphin-induced changes in schizophrenic and depressed patients. *Arch. Gen. Psychiat.* 34, 1111-1115 (1977).

Kurland, A.A., McCabe, O.L., Hanlon, T.E., and Sullivan, D. The treatment of perceptual disturbances in schizophrenia with naloxone hydrochloride. *Am. J. Psychiat.* 134, 1408-1410 (1977).

Lewis, R.V., Gerber, O.D., Stein, S., Stephen, R.L., Grosser, B.I., Verlick, S.F., and Udenfriend, S. On βH-Leu^5-Endorphin and schizophrenia. *Arch. Gen. Psychiat.* 36, 237-239 (1979).

Maas, J.W. Biogenic maines and depression. *Arch. Gen. Psychiat.* 32, 1357-1364 (1975).

Maas, J.W., Greene, N.M., and Hattox, S.F. Neurotransmitter metabolite production by human brain in *Catecholamines: Basic and Clinical Frontiers*, E. Usdin, ed. Pergamon Press, New York (in press).

Majer, O. Beitrag zur erbbiologie involutiver, klimakterischer und reactiver depressionen. *Zeit. Neurol. Psychiat.* 172, 737-753 (1941).

Meltzer, H.Y., Sachar, E.J., and Frantz, A.G. Serum prolactin levels in unmedicated schizophrenic patients. *Arch. Gen. Psychiat.* 34, 564-571 (1974).

Money, J., Wiedeking, C., Walker, P.A., and Gain, D. Combined antiadrogenic and counseling program for treatment of 46 XY and 47 XYY sex offenders, in *Hormones, Behavior, and Psychopathology*, E.J. Sachar, ed. Raven Press, New York (1976), pp. 105-120.

Parra, A. Discussion, in *Endocrine Aspects of Malnutrition, Marasamus, Kwashiokor, and Psychosocial Deprivation*, L.I. Gardner and P. Amacher, eds. Kroc Foundation, Santa Ynez, Calif. (1973), p. 155.

Persky, H., O'Brien, C., Fine, E., Howard, W.J., Khan, M.A., and Beck, R.W. The effect of alcohol and smoking on testosterone function and aggression in chronic alcoholics. *Am. J. Psychiat.* 134, 621-625 (1977).

Prange, A.J., Lipton, M.A., Nemeroff, C.B., and Wilson, I.C. The role of hormones in depression. *Life Sciences* 20, 1305-1318 (1977).

Ripley, H.S., Shorr, E., and Papanicolaou, G.N. Effect of treatment of menopause depression with estrogenic hormone. *Am. J. Psychiat.* 96, 905-911 (1940).

Rose, R.M. Testosterone, aggression, and homosexuality, in *Topics in Psychoendocrinology*, E.J. Sachar, ed. Grune and Stratton, New York (1975), pp. 83-104.

Sachar, E.J. Neuroendocrine abnormalities in depressive illness, in *Topics in Psychoendocrinology*, E.J. Sachar, ed. Grune and Stratton, New York (1975), pp. 135-156.

Satterfield, J.H., Cantwell, D.P., Schell, A., and Blaschke, T. Growth of hyperactive children treated with methylphenidate. *Arch. Gen. Psychiat.* 36, 212-219 (1979).

Segal, M. Interactions of ACTH and norepinephrine on the activity of rat hippocampus cells. *Neurochem.* 15, 329-333 (1976).

Smith, S.L. Mood and the menstrual cycle, in *Topics in Psychoendocrinology*, E.J. Sachar, ed. Grune and Stratton, New York (1975), pp. 19-58.

Synder, S.H. The opiate receptor and morphine-like peptides in the brain. *Am. J. Psychiat.* 135, 645-652 (1978).

Spitz, R. Hospitalism: an inquiry into the genesis of psychiatric conditions in early childhood, in *Psychoanalytic Study of the Child*,

Vol. 2. R.S. Eissler, ed. International Universities Press, New York (1946), pp. 113-117.

Stenstedt, A. Involutional melancholia. *Acta Psychiat. Neurol. Scand.* (Suppl. 127), 34, 1-71 (1959).

Tamminga, C.A., Smith, R.C., Pandey, G., Frohman, L.A., and Davis, J.M. A neuroendocrine study of supersenitivity in tardive dyskinesia. *Arch. Gen. Psychiat.* 34, 1199-1203 (1977).

Terenius, L., Wahlstrom, A., and Lindstrom, L. Increased CSF levels of endorphins in chronic psychosis. *Neurosci. Lett.* 3, 157-162 (1976).

Verhoeven, W.M.A., van Pragg, H.M., van Ree, J.M., and deWied, D. Improvement of schizophrenic patients treated with [Des-Tyr^1]-γ-endorphin (DTγE). *Arch. Gen. Psychiat.* 36, 294-302 (1979).

Vogel, H.P., Benkert, O., Illig, B., Muller-Oerlinghauser, and Popenberg, A. Psychoendocrinologic and therapeutic effects of TRH in depression. *Acta Psychiat. Scand.* 56, 223-232 (1977).

Vogel, W., Klaiber, E.L., and Broverman, D. Roles of the gonadal steroid hormones in psychiatric depression in men and women. *Prog. Neuro-Psychopharm.* (in press).

Watson, S.J., Akil, H., Berger, P.A., and Barchas, J.D. Some observations on the opiate peptides and schizophrenia. *Arch. Gen. Psychiat.* 36, 35-46 (1979).

Wilson, I.C., Prange, A.J., Lara, P.P., Alltop, L.B., Stikelcather, R., and Lipton, M.A. THR (Lopremore): psychobiological responses of normal women. *Arch. Gen. Psychiat.* 29, 15-21 (1973).

Winokur, G. and Cadoret, R. The irrelevance of the menopause to depressive disease, in *Topics in Psychoendocrinology*, E.J. Sachar, ed. Grune and Stratton, New York (1975), pp. 59-66.

Yalom, J.D., Green, R., and Fisk, N. Prenatal exposure to female hormones. *Arch. Gen. Psychiat.* 28, 554-561 (1973).

Copyright©1982, Spectrum Publications, Inc.
Hormones in Development and Aging

Chapter 19
Affective Disorders in Endocrine Disease
Douglas B. Carter

INTRODUCTION

When Hippocrates "published" his views on the etiology of depression over two thousand years ago, he named the disease for what he believed to be its cause — black bile or melancholia — and became the first psychoendocrinologist. The idea that a "humor" circulating through the body caused sad feelings, irritability, loss of appetite for food and sex, early morning awakening, and sometimes death via suicide has kept physicians, physiologists, biochemists, and pharmacologists on its trail ever since.

Each organism, no matter how simple or complex, must be able to send messages from one part to tell another part what to do. Our knowledge of the nervous system with its multiple neurotransmitters and pathways for rapid and relatively slower messages needs to be synthesized with our knowledge of the slower hormone messengers secreted from the endocrine glands in order to fully understand the causes of the affect disorders.

ENDOCRINPOATHIES OF THE THYROID

Hypersecretion: Thyrotoxicosis

When Parry and then Graves described thyrotoxicosis, they were on the trail. Here, a specific "humor" was found to be secreted in excessive quantities producing profound physical and emotional disturbances. The disease raised another fascinating question in that it was often preceded by a major emotional loss or stress. Could feelings of sadness and helplessness cause the thyroid to produce excessive hormone? Could this be the brain's attempt to make itself "feel better" by sending a message to the hypothalamus via noradrenergic or serotonergic neurons to the hypothalamus telling it to produce more thyrotropin-releasing hormone (TRF) which would circulate back to other brain receptors producing some improvement in mood and decreasing the vegatative symptoms? The TRF would also enter the hypothalamic-pituitary portal circulation, and on reaching the anterior pituitary would deliver its biochemical message, causing the release of thyrotropin (TSH) which would in turn enter the general circulation and relay the message to the thyroid gland for production of more thyroxine. This complex cortex-limbic-pituitary-thyroid system clearly has an equally

complex feedback system that ordinarily maintains the circulating levels
of thyroid hormone within a very narrow range. Does the feedback
system and general homeostatic system become deranged in thyrotoxicosis?
Hyperthyroid patients have been found to show an exaggerated increased
response to a stressful stimulus as compared to normal "euthyroid"
subjects (Robbins and Vinson, 1960). Could some cases of Grave's
disease be psychogenic? Are there predisposing genetic and/or psycho-
developmental factors? Several psychiatrists and endocrinologists have
investigated this disease to try to answer some of these questions. The
careful work of Lidz (1949) found that significant emotional loss had
preceded the onset of the disease. Kleinschmidt et al., (1956) found
emotional problems in 81 of 84 thyrotoxic patients. Whybrow et al.,
(1969) studied 14 female hyperthyroid patients; five were found to have
a family history of the disease, while only two had suffered a significant
emotional loss just prior to the disease onset. Multiple etiologic factors
are undoubtedly involved, including genetic predisposition, psycho-
developmental factors, and other as yet unknown factors.

Hyposecretion: Myxedema

Myxedema is a messenger-deficiency disease that also has an affect
disorder component. With a deficiency of the thyroid hormone, most
metabolic processes slow down, and this manifests itself in the brain in
multiple ways. The electroencephalogram shows a decreased amplitude
(voltage) and slowing of the frequency. Mild to severe defects are
apparent in recent memory and general intellectual functioning
(Reitan, 1953). The mood is down; most studies report apathy or
depression (Akelaitis, 1936). Whybrow and Hurwitz (1975) more recently
studied seven cases and found that all but one had a significant degree
of depression. These were not patients with long histories of previous
depressions, and they were referred medically for thyroid evaluation
not because of psychiatric symptoms. It is thus clear that a functional
deficiency of circulating thyroid hormone in most cases produces symptoms
of depression. Replacement therapy "cures" the metabolic disorder,
and the symptoms of depression clear. From this example, it is easy
to see why clinical and basic science researchers want to discover the
messengers, elucidate their actions, and after developing laboratory
methods to precisely measure their levels, determine their normal values.
We can then look for ways to treat those conditions mediated by
derangements of the messengers. In a subsequent section of the chapter,
we will attempt to answer the question, is there a relative deficiency of
thyroid hormone in other types of clinical depression?

ENDOCRINOPATHIES OF THE ADRENAL CORTEX

Hypersecretion: Cushing's Disease

Cushing's paper (1932) showed that what had appeared to be a
disease of the adrenal gland was caused by an adenoma of the anterior
pituitary. Previously, he had postulated that some disorders of the
endocrine glands might have "derangement of the nervous system" as
the primary etiology (Cushing, 1913). Today we know that Cushing's
syndrome is not a single disease, and that in some cases, the basic
defect is "central" in the limbic-hypothalamic-pituitary axis, while in
others, an adrenal or pituitary tumor is primary (Kreiger and Glick,
1972). The increased incidence of mental symptoms has not always been
recognized or reported. These symptoms were often viewed simply as

the emotional reaction to the marked change in the patient's body. Later, endocrinologists believed that the mental symptoms were part of the syndrome (Starr, 1952), or were caused by the increased circulating levels of adrenocortical hormones. The psychiatric manifestations are diverse, ranging from mild confusion to auditory hallucinations, but depression is the most common finding, and suicide is not rare (Trethowan and Cobb, 1952). The question remains as to whether some of the "central" type cases are caused or precipitated by severe emotional distress. A recent study (Gifford and Gundersen, 1970) utilizing the hospital records of 10 patients with proven Cushing's disease, found depression or the pathological defenses against depression in all cases. They had excluded many cases from their study because of inadequate psychiatric past history in the medical record, and this certainly biased their sample. They suggest that some cases of Cushing's disease may be a psychogenic or psychosamatic disorder with depression as the etiologic factor. Since these antecedent depressions are not found in all cases, we must continue to consider that the increased level of 17-hydroxycorticosteroids in some way causes the depression. Adrenocorticotropin and synthetic steroids have been available and used as pharmacologic agents to treat various conditions for many years, and many reports have been published describing their mental or emotional side effects. With high doses, the Cushing's-like physical changes do occur, but depression is quite rare, and elation or euphoria is a more common effect on the patient's affect state (Quarton et al., 1955). Are we dealing with an endocrine disease that shares a common central regulatory system with depressive disorders? Wolff et al. in 1964 reported the study of a patient with limbic system damage who demonstrated periodic hypothalamic discharge manifest by pituitary adrenocortical hypersecretion lasting three to five days, accompanied by depressive symptoms. A recent study by Kreiger and Glick (1972) provides further evidence for Nugent's 1960 suggestion that Cushing's disease is a defect in central regulation. The fact that some of the Cushing's disease cases described by Kreiger were in remission following pituitary irradiation or adrenalectomy and yet still showed the same lack of normal growth hormone response to the Insulin Tolerance Test, the Pitressin Test, and the Piromen Test, indicates a central defect that is not simply the effect of high circulating levels of the 17-hydroxy-corticosteroids. Unfortunately, they did not report the patients affect state either before or during remission.

It is clear that some cases of Cushing's disease seem to be precipitated by depressive affect disorder, that many cases have concomitant depressive illness, and that more and more evidence points to a central brain dysfunction that produces clinical depression and adrenocortical hypersecretion mediated through the hypothalamic-pituitary-adrenocortical axis.

Hyposecretion: Addison's Disease

There have been many articles since 1855 when Thomas Addison published his monograph on "the consitutional and local effects of disease of the suprarenal capsule." Most of the reports, including those of Kraepelin (1913), Engel and Margolin (1942), Cleghorn (1951), and Ebaugh and Drake (1957), in the excellent review by Michael and Gibbons (1963), and a recent review by Smith et al. (1972) depict the wide variability of the psychiatric symptoms that accompany the disease. They range from "childish" to frank psychosis during Addisonian crisis. Commonly mentioned are mild delirium and apathy or depressed mood.

The symptoms are much less frequent and severe than in Cushing's disease, and they usually clear with replacement therapy. The effects of adrenocortical insufficiency on glucose and sodium metabolism offer some explanation of the symptoms. The etiology of the majority of the Addison's disease cases in the literature was the tubercle bacillus invading the adrenal gland. Other cases do not yield evidence for a "central" hypothalamic-pituitary etiology, but this has not been carefully studied.

PARATHYROID DISEASE

What can we learn about depression and mania from reviewing the occurrence of these alterations of mood in patients with abnormal secretion of parathyroid hormone? Normal levels of this hormone are clearly required in the homeostatic mechanisms that maintain calcium ion concentration in the plasma, CSF and extracellular fluid. There is no trophic or releasing hormone from the hypothalamus, pituitary or elsewhere, and it is generally accepted that calcium concentration in the extracellular fluid of the parathyroid "glands" controls the precursor uptake, synthesis and release of the hormone into the general circulation.

Because of this simple, direct peripheral control mechanism, we have an in vivo system that can be studied for hormone effects on mood or affect. There is no complex cortical mechanism modulating limbic centers that act via monoamine neurones to control a hypothalamic release of the control hormone that circulates to the anterior pituitary causing the release of a trophic or inhibiting hormone that travels via the general circulation as with the adrenal or thyroid. Thus, in parathyroid disease, we can study the direct and secondary effects of the parathyroid hormone levels without being concerned with the brain's mechanisms that regulate the hormone that, in themselves, could be causing the mood change, rather than the hormone itself.

What happens in humans when there is a deficiency of parathyroid hormone?

Hypoparathyroidism

A classical clinical picture include paresthesias, muscle spasms, mild anxiety, emotional lability, cognitive impairment and *depression*. Again, we have the problem that most clinical studies have focused primarily on the physical and physiological aspects, and do not include or report careful documentation or objective testing of the psychiatric symptoms. In 1962, Denko and Kaelbling reviewed the literature. They excluded reports that did not contain psychiatric information, and further, excluded those cases that had a prior history of psychiatric illness. They found that 50% of the patients had a major degree of cognitive impairment, 5% were psychotic, and 30% had other psychiatric symptoms. Depression was not reported in enough cases to justify separate tabulation. Thus, there is no evidence that hypoparathyroidism and its resulting hypocalcemia produces a significant alteration in mood, even though it causes major neuropsychiatric symptoms in the majority of patients.

Hyperparathyroidism

With this disease, there is an affect disturbance in a significant number of the patients, while 30% show anxiety and irritability (Karpati and Frame, 1964). Prior to the superb work of Peterson in 1968, no one had attempted to study the correlations between the psychiatric manifestations in serum calcium levels. Peterson demonstrated a strong correlation between the calcium levels and the severity and type of psychiatric disturbance. Minor elevations were associated with depression, while moderate to large alterations were found in patients with organic psychoses. These findings have helped explicate the clinical observation that in the early phase of hyperparathyroidism, patients are often depressed, while later, as the disease progresses, they show severe psychiatric symptoms. The causal relationship between the degree of hypercalcemia and severity of the neuropsychiatric disturbances seems clear, and is supported by similar findings in other patients who do not have hyperparathyroidism but have hypercalcemia from other causes. The neuropsychiatric symptoms can be alleviated by correcting the calcium levels, even though the PTH levels remain high.

ENDOCRINOPATHIES OF THE PANCREATIC ISLETS

Diabetes Mellitus

Diabetes mellitus is not generally associated with severe endogenous depression, although stress, including depressive psychological reactions to disappointments and losses, has been implicated in the onset of the illness. Most evidence is anecdotal. One study by Slawson et al. (1963) showed a history of antecedent stress in 20 of 25 newly diagnosed diabetics.

Depressive reactions, again not usually of great severity, have also been observed following the diagnosis of diabetes (Treuting, 1962). Juvenile diabetics seem to be most susceptible, probably because of the stress of having a chronic disease during adolescence.

Insulinoma

The mental symptoms accompanying insulinoma are thought to be caused by hypoglycemia resulting from the increased insulin excretions (Marks and Rose, 1965). Acute reactions are due to reactions of the sympathetic nervous system and are predominantly anxiety reactions. Chronic hypoglycemia leads to somnolence, reduced motor activity, lassitude and negativism, including insidious personality changes. Dementia, occasionally with psychotic paranoid features, has also been reported.

OVERVIEW

Hippocrates' intuition of a neurohumoral factor in depression has been documented by the clinical observation of depression accompanying several endocrinopathies reviewed here, and by more recent biochemical demonstrations of changes in endocrine function in endogenous depression (see also Chapter 17). Although importance to clinical practice of recognizing depression as a feature of several of the endocrinopathies cannot be underestimated, it is disappointing that there have not been more attempts to use this clinical knowledge to further the understanding of the psychobiology of affective illness. For example, little work exists which attempts to differentiate the clinical features of the affective changes occurring in the various endocrine syndromes. An hypothesis of specific mental changes for each endocrine change, rather than a final common pathway of depression, could be tested using modern psychological techniques. Such information might point to specific brain functions for each hormone.

REFERENCES

Addison, T. *On the Constitutional and Local Effects of Disease of the Suprarenal Capsule.* Samuel Highly, London (1955).

Akelaitis, A.J.E. Psychiatric aspects of myxedema. *J. Nerv. Ment. Dis.* 83, 22-36 (1936).

Cleghorn, R.A. Adrenal cortical insufficiency. Psychological and neurological observation. *Canad. Med. Assn. J.* 65, 449-454 (1951).

Cushing, H. Psychic disturbances associated with disorders of the ductless glands. *Am. J. Insan.* 69, 965-989 (1913).

Cushing, H. The basophil adenomas of the pituitary body and their clinical manifestations (pituitary basophilism). *Bull. Hopkins Hosp.* 50, 137-195 (1932).

Denko, J.D. and Kaelbling, R. The psychiatric aspects of hypoparathyroidism. *Acta Psychiat. Scand.* 38, 1-70 (1962).

Ebaugh, F.G. and Drake, F.R. Neuropsychiatric-like symptomatology of Addison's disease: a review. *Am. J. Med. Sci.* 234, 106-113, (1957).

Engel, G.L. and Margolin, S.G. Neuropsychiatric disturbances in internal disease: metabolic factors and electroencephalographic correlations. *A.M.A. Arch. Int. Med.* 70, 236-259 (1942).

Gifford, S. and Gundersen, J.C. Cushing's disease as a psychosomatic disorder: a selective review of the clinical and experimental literature and a report of ten cases. *Perspect. Biol. Med.* 13, 169-221 (1970).

Karpati, G. and Frame, B. Neuropsychiatric disorders in primary hyperparathyroidism. *Arch. Neurol.* 10, 387-397 (1964).

Kleinschmidt, H.J., Waxenberg, S.E., and Cuker, R. Psychophysiology and psychiatric management of thyrotoxicosis: a two-year followup study. *J. Mount Sinai Hosp. N.Y.* 23, 131-153 (1956).

Kraeplin, E. *Psychiatrie,* 8th edition. Barth. Leipzig (1913).

Krieger, D.T. and Glick, S.M. Growth hormone and cortisol responsiveness in Cushing's syndrome. *Am. J. Med.* 52, 25-40 (1972).

Lidz, T. Emotional factors in the etiology of hyperthyroidism. *Psychosom. Med.* 11, 2-8 (1949).

Marks, V. and Rose, F.C. *Hypoglycemia.* Blackwell Scientific Publications, Oxford (1965).

Michael, R.P. and Gibbons, J.L. Interrelationships between the endocrine system and neuropsychiatry, in *International Review of Neurobiology,* Vol. 5, C.C. Pfeiffer and J.R. Smythies, eds. Academic Press, New York (1963), pp. 243-302.

Peterson, P. Psychiatric disorders in primary hyperparathyroidism. *J. Clin. Endocr.* 28, 1491-1495 (1968).

Quarton, G.C., Clark, L.D., Cobb, S., and Bauer, W. Mental disturbances associated with ACTH and cortisone; a review of explanatory hypotheses. *Medicine* 34, 13-50 (1955).

Reitan, R.M. Intellectual functions in myxedema. *Arch. Neurol. Psychiat.* 69, 436-449 (1953).

Robbins, L.R. and Vinson, D.B. Objective psychological assessment of the thyrotoxic patient and the response to treatment, preliminary report. *J. Clin. Endocr.* 20, 120-129 (1960).

Slawson, P.F., Flynn, W.R., and Kollar, E.J. Psychological factors associated with the onset of diabetes mellitus. *J.A.M.A.* 185, 166-170 (1963).

Smith, C.K., Barish, J., Correa, J., and Williams, R.H. Psychiatric disturbance in endocrinologic disease. *Psychosom. Med.* 34, 69-84

(1972).

Starr, A.M. Personality changes in Cushing's syndrome. *J. Clin. Endocr.* 12, 502-505 (1952).

Trethowan, W.H. and Cobb, S. Neuropsychiatric aspects of Cushing's syndrome. *A.M.A. Arch. Neurol. Psychiat.* 67, 283-309 (1952).

Treuting, T.F. The role of emotional factors in the etiology and course of diabetes mellitus: a review of the recent literature. *Am. J. Med. Sci.* 244, 93-100 (1962).

Whybrow, P.C. and Hurwitz, T. Psychological disturbances associated with endocrine disease and hormone therapy, in *Hormones, Behavior, and Psychopathology,* E.J. Sachar, ed. Raven Press, New York (1975), pp. 125-143.

Whybrow, P.C., Prauge, A.J., and Treadway, C.R. Mental changes accompanying thyroid gland dysfunction. *Arch. Gen. Psychiat.* 20, 48-63 (1969).

Wolff, S.M., Adler, R.C., Buskirk, E.R., and Thompson, R.H. A syndrome of periodic hypothalamic discharge. *Am. J. Med.* 36, 956-967 (1964).

Copyright©1982, Spectrum Publications, Inc.
Hormones in Development and Aging

Chapter 20
Hormones and Cancer
Evelyn M. Rivera
Howard A. Bern

INTRODUCTION

The impact of the news media and an increasingly informed public have confronted today's medical practitioners with perplexing problems, not the least of which concerns decisions in prescribing oral contraceptives to young women (Rinehart and Piotrow, 1979) and hormone replacement therapy for postmenopausal women (Lauritzen and van Keep, 1978). The clinician must be responsible for providing the patient with the most appropriate beneficial therapy and at the same time, assure the minimum of undesirable consequences. Yet there is considerable debate among practitioners on such questions as: Which of the 24 combinations of birth control pills available in the U.S. should be prescribed (Speroff, 1976)? Do the beneficial effects of hormone therapy in combating osteoporosis outweigh the risk of endometrial cancer in postmenopausal women (Greenblatt, 1977; Silverberg and Major, 1978)?

It is widely recognized that under certain conditions and in certain tissues hormones increase the risk of developing cancer. The International Agency for Research on Cancer devotes two of its monographs to hormonal agents: the sex steroids and their mimetics (IARC, 1974a, 1979) and chemicals which cause excessive thyrotropin secretion resulting in thyroid neoplasia (IARC, 1974b). Despite this awareness and concern about hormones and cancer, thyrotropin would hardly be considered a carcinogen per se, and no *direct* carcinogenic role can be assigned with certainty to natural or synthetic sex hormones, not even diethylstilbestrol (DES), although certain sex hormones are classified as carcinogens or potential carcinogens for humans in accordance with present IARC criteria (DES, estradiol, estrone, testosterone).

The purpose of this chapter is to discuss several ways in which hormones may be involved in carcinogenesis. In many instances, hormones must be present for neoplasia to occur; in other instances, their principal effect may be to promote the growth of transformed neoplastic cells (i.e., increase incidence, stimulate tumor growth, lower the age of onset, increase the likelihood of metastasis); in yet other instances, hormones may decrease the incidence or occurrence of tumors or, as is well known clinically, induce tumor regression (e.g., Huggins, 1967, 1978). It should also be noted that the "hormonal influence" involved in increasing tumor risk is not necessarily an increase in hormonal level. Neoplasia may in fact be favored by hormone levels that are normal or lower than normal, whereas neoplastic growth may be inhibited by chronic high hormone levels. Expressed most economically, an important factor contributing to neoplasia is "hormonal imbalance" (Gardner, 1953).

GENERAL CONSIDERATIONS

Before discussing in a more specific way the possible involvement of hormones in the etiology of cancer, it is useful to review a number of basic concepts and problems in endocrinology and oncology.

"Target Organs" of Hormones

Despite the dogma of specific target organs for hormones, there is probably no organ or tissue in the body that is not directly or indirectly affected by components of the endocrine system. As discussed in other chapters of this book, much has yet to be learned about the interactions among the endocrines and the nervous system and also among the endocrines and endocrine-responsive tissues. For example, the liver had not previously been considered a target tissue for either estrogen or prolactin, but recent findings of liver receptors for both hormones (Posner et al., 1975; Eisenfeld et al., 1976) raise the question of the functional significance of estrogen and prolactin on an organ that, until recently, had been considered principally a site of hormone catabolism. Noble et al. (1975, 1977) have also shown that estrogen administration results in tumors not only in sex-related organs but also in tissues not usually considered estrogen target organs, and "the pill" appears to be involved in the production of liver adenomas in women. For the present, it may be enough to state that normal physiological and biochemical processes in homeostasis depend upon a proper balance among the endocrines, so that any "imbalance," whether natural or experimentally-produced, could increase the risk of functional and developmental derangements, including cancer.

Mechanisms of Hormone Action

In other chapters in this book the mechanisms of actions of several hormones have been discussed in some detail as they relate to their roles in fertilization, sex differentiation, neural growth and differentiation, aging, and homeostatic mechanisms. A brief account of hormonal mechanisms is also considered in this chapter since they constitute the formulation of the cellular basis of the role of hormones in cancer. As discussed previously in this book, it is generally accepted that many hormones elicit their effects by interacting with specific receptors in the cell membrane (many peptide hormones, catecholamines) or in the nucleus (steroid hormones, thyroid hormones), and as a consequence of the hormone-receptor interaction, a series of events are set in motion which are known as the "hormonal response." The concept of a unitary receptor for each hormone is no longer tenable, since considerable heterogeneity and polymorphism of receptors for the same hormone have been demonstrated (Baulieu, 1975; Agarwal and Rossier, 1977). In addition, other mechanisms for hormonal effects on cells may exist (e.g., Szego, 1974 on lysosome mediation of hormone action).

The primary initial sites of hormone action are believed to be (1) the cell membrane or (2) the genome. The first model is proposed for peptide hormones and catecholamines, for which cell entry may not be essential, although internalization of peptides is now well established (Posner et al., 1979). These hormones interact with specific receptors at the cell surface with the resultant generation of one or more second messengers, notably cyclic AMP, which are in turn believed responsible for changes in cell function (Cuatrecasas, 1974; Butcher et al., 1972; Sutherland, 1972). The second model applies principally to steroid hormones, which enter the cell (presumably by diffusion) and combine with specific receptors in the cytoplasm to form a hormone-receptor complex. Upon interaction with their specific hormones, the receptors undergo a temperature-dependent conformational change and are transported as "transformed" hormone-receptor complexes to the nucleus, where they effect specific gene activation (Jensen and DeSombre, 1973; O'Malley et al., 1976; Baulieu, 1975; Liao, 1975).

The pattern of events following hormone-receptor interaction, at the membrane and in the nucleus, is currently the subject of intensive inquiry and speculation (Palmiter et al., 1977; O'Malley et al., 1977; Sterling, 1979; Catt et al., 1979; Tata, 1979). Whereas the second messenger concept of hormone action (i.e., where cyclic AMP and other nucleotides are secondary mediators of hormonal effects) appears to be valid for most of the hormonal peptides and amines, we know little of how these ubiquitous molecules regulate cell growth and function. Much current research places emphasis on the regulatory role of cyclic AMP-dependent protein kinases (Lincoln and Corbin, 1978; Jungmann and Russell, 1977). Perhaps the best-studied hormone-cyclic AMP model is that of epinephrine and glucagon action in the liver (Sutherland and Robison, 1966). These hormones, by activating the synthesis of cyclic AMP, set in motion an intricate cascade of allosteric enzyme alterations resulting in glycogenolysis.

Are the hormones under study actually the effective principles? There is now much evidence to indicate that most, if not all, of the growth-promoting effects of growth hormone are mediated by smaller serum peptides collectively known as somatomedins (Van Wyk et al., 1974; Chochinov and Daughaday, 1976). The growing number of low molecular weight growth factors being found in serum (Shields, 1977; Rechler and Nissley, 1977) poses the question of whether they might not play a role in the mediation of hormones possessing growth activity.

With regard to steroid hormones, estradiol and its biological mimetics remain unchanged in their bound form in the uterus (Jensen and DeSombre, 1973), indicating that estrogens are the proximate agents for the effects attributed to them, at least in this tissue. By contrast, testosterone is converted to 5α-dihydrotestosterone (DHT) in several androgen-responsive tissues (Liao, 1975). DHT is more potent than testosterone in several assays and competes with testosterone for specific androgen receptors, indicating that this androgen metabolite is the active principle. There is increasing evidence that steroid hormones undergo further metabolic transformation in tissues other than their sites of secretion. Epoxides appear to be intermediates in steroid metabolism in a variety of tissues (Kadis, 1978). These and other studies indicate that the metabolism of steroid hormones, like that of chemical carcinogens, may play a role in their biological activity.

Genetic Background

References to "genetic background" in the etiology or predisposition to endocrine diseases and cancer imply that the disease is genetically determined, although the mechanisms involved are still obscure. There are at least 50 phenotypically distinct and simply inherited endocrine disorders (including multiple endocrine adenomatosis and medullary thyroid carcinoma-pheochromocytoma) in humans, indicating at least 50 genes which affect endocrine function in a major way (Goldstein and Motulsky,

1974). However, the genetic element in most common disorders is rarely based on a single gene or chromosomal abnormality. Rather, multiple gene interaction *and* their interaction with environmental factors are most likely the "genetic background" for the development of cancer and other diseases.

"Inherited hormonal influences" have often been discussed in connection with the development of breast cancer in inbred strains of mice (Bern and Nandi, 1961; Nandi and McGrath, 1973). In certain mouse strains, e.g., C3H, there is a high incidence of spontaneous mammary tumors in both virgin and parous females, whereas in strain A mice, only parous females have a high incidence of mammary tumors. Attempts to reveal differences in the hormonal status of these two strains have been generally inconclusive except in mammary tissue *responsiveness* to growth hormones. A positive correlation appears to exist between the high incidence of mammary tumors in virgin mice and in the capacity of mammary tissue to respond to the lactogenic effects of growth hormones in vivo (Nandi and Bern, 1960; Bern and Nandi, 1961) and in vitro (Rivera, 1966; Rivera et al., 1967). It is possible that strain differences reside in the biochemical and cytostructural components of the responding cells rather than in circulating hormone levels.

Age

Except for certain types of cancer, such as neuroblastoma, leukemias, and other cancers of mesenchymal origin, most cancers occur late in life. However, the *susceptibility* to cancer varies in accord with the carcinogenic agent, age of exposure, sex, tissue, and species. Fetal and neonatal exposure to hormonal steroids produces effects in later life that are now well documented (see below). The two major effects of perinatal administration of sex steroids and DES are (1) permanent structural, physiological, and behavioral abnormalities and (2) altered risk of cancer.

Age is an important factor in the susceptibility to mammary tumor induction by mouse mammary tumor virus (MMTV), where resistance increases with advancing age (Nandi and McGrath, 1973). Nevertheless, mammary tumors do not develop until nearly one year after infection by MMTV, illustrating once again that a tumor-inducing event and the actual manifestation of the tumor may be widely separated in time.

There are also *changing susceptibilities* to chemical carcinogens in different tissues during fetal and postnatal development (Rice, 1973). Some carcinogens are effective both in transplacental carcinogenesis and in adult animals; urethane is an example, inducing pulmonary tumors with high frequency in mice. On the other hand, a polycyclic aromatic hydrocarbon, dimethylbenz(α)anthracene (DMBA) is most effective in inducing mammary tumors when administered to 50-60 days-old rats (Huggins et al., 1961; Dao, 1965; Nagasawa and Yanai, 1974) than in younger or older animals. Transplanted rat mammary hyperplasias also undergo neoplastic transformation more readily in older than in younger host rats (Rivera et al., 1981). Recently, Haslam (1979) has reported that the frequency of progressively growing tumors is not age-related, whereas the frequency of spontaneously regressing tumors (histologically indistinguishable) is strongly age-related as described above.

Latency

As noted above, long latencies are characteristic of most cancers. This implies that host factors prevent in some way the proliferation of neoplastic cells and/or that aging is accompanied by changes that favor the emergence or expression of neoplasia. A provocative hypothesis has been proposed by Nandi (1978a,b), which states that the long latency for overt tumor expression is the result of inhibition by the normal cell components of the tissue; reduction in the population of normal cells, such as that which occurs in age-associated tissue atrophy, favors the emergence of dormant tumor cells. In view of the close relationship between age and latency in neoplasia, a systematic investigation of other age-related factors is warranted. These include the immune system, changing capacities of tissues to metabolize or retain carcinogens, and age-associated nutritional and hormonal conditions that may promote or inhibit carcinogenesis.

Mechanisms of Action of Carcinogens

The diversity in the chemical structure of carcinogens poses the intriguing question of whether there might be a common mechanism triggering neoplastic transformation or whether multiple mechanisms exist, one for each major class of agent. How does such a vast array of compounds, natural and synthetic, alter a normal cell so that it becomes a cancer cell?

Three major theories prevail: (1) The *virus theory* states that carcinogens activate viral oncogenes (Todaro and Huebner, 1972; Todaro, 1978) or induce proto-virus (Temin, 1974, 1976), which are the ultimate carcinogens. In the first case, cancer genes are said to be present in normal cells but in inactive form; in the second case, cancer genes are absent from normal cells but arise as a result of "misevolution" (Temin, 1974). The proto-virus theory is thus a special case of the mutation hypothesis (see below). Some evidence does exist that irradiation may activate latent viruses in mice (Gross, 1961), and considerable evidence has also been obtained for the presence of oncornaviruses in chemically-induced tumors (Heidelberger, 1975), but the relationship of viruses and chemical carcinogens in tumor induction is far from clear, and such viruses may even be nontumorigenic "passengers" in the cells. An etiologic role of viruses in neoplasia may prove to be restricted to special types of tumors in a limited number on nonhuman species.

(2) The *somatic mutation theory* states that carcinogens cause a structural change in the cellular genome resulting in a mutation (or a sequence of mutations) which, in some way, give rise to cancer. This is by far the most widely-accepted working hypothesis, since it fits into currently understood ideas in molecular genetics and biology (Heidelberger, 1967, 1975; Farber, 1973; Miller, E. C., 1978; Weinstein, 1978) and is consistent with the present concern with environmental carcinogens. The mutation theory is supported by the observations that most chemical carcinogens, after metabolism to electrophilic reactants, bind covalently to DNA. As a result of this binding, heritable and irreversible changes are believed to occur in the genome. However, since carcinogens also bind to RNA, membranes, and cell proteins, it is still uncertain whether the primary alteration is in the genome or in some other macromolecule regulating gene structure or function. Furthermore, carcinogens are metabolized to more than one reactive agent, and there are multiple electrophilic sites in each of the cellular macromolecules. It is clear that a basic problem is to sort out the "key reactions" from this complexity of background "noise."

Assays indicating that many or most chemical carcinogens are mutagens (Ames et al., 1973a,b; 1975) lend support to the mutation hypothesis, although this assay system is not without its critics (Ashby and Styles, 1978; see also below).

(3) The *differentiation or epigenetic theory* states essentially that carcinogenesis is basically a problem of the control of gene expression and results from persistent abnormalities in the regulation of normal genes (Foulds, 1969; Pierce, 1974; Coggin and Anderson, 1974; Dustin, 1972; Potter, 1969). Holtfreter (1948) was among those who indicated many years ago that malignancy is a legitimate problem of the embryologist. Recognizing the range of chemicals capable of inducing neoplasia, he postulated that malignancy, like parthenogenesis, might be another case of a trigger mechanism where the pattern of the reaction is determined by the specific properties of the cell rather than of the stimulus. Today, even with our remarkable advances in molecular and developmental biology, the mechanisms of normal cell differentiation are still poorly understood. Despite the complexity of the problem, the study of neoplasia within the context of developmental principles does offer an alternative model to the mutation hypothesis. The construction of testable models to analyze gene activation (Britten and Davidson, 1969), progressive gene repression (Caplan and Ordahl, 1978), and alterations in regulatory signals and molecules are clearly needed. Alterations in the dynamic interplay between different cells should not be overlooked (Gilbert, 1974).

The Development of Cancer: a Single Step or a Progressive Process?

Researchers generally adhere to one of two theories of cellular transformation regardless of their different views of the initial inciting event. One school believes that once a normal cell becomes a cancer cell, it has all the essential characteristics of the definitive neoplastic state. The (overt) tumor cells may in time acquire additional properties, but the critical features of neoplasia are said to be present from the inception of the transformed neoplastic cell. The second view, best conceived by Foulds (1949, 1969, 1975), is that neoplastic transformation is one of discontinuous progression through *qualitatively* different stages. According to this concept, progression occurs independently both among tumors of the same general type and in the charactertistics of a particular tumor. Whereas the idea of tumor progression is widely accepted, its application has been generally limited to the overt properties of *recognizable* neoplasms. Thus, progression is said to underlie the alteration from "benign" or "precancerous" to "malignant" and from "dependent" to "autonomous" tumors. This emphasis on the pathological features overlooks subtle changes that may be occurring progressively long before the cells are recognized as altered from normal. Some recently proposed mechanisms of tumor progression include the sequential selection of variant subpopulations of cells with increasing survival advantage (Cairns, 1975; Nowell, 1976) and heritable changes in cytoplasmic membranes (Mekler, 1973; Pitot, 1974; Pitot et al., 1974).

Initiation and Promotion

A two-stage concept of initiation and promotion was proposed many years ago to account for observations in experimental skin carcinogenesis, namely, that cells transformed by chemical carcinogens could remain dormant for long periods of time but form visible tumors following application of noncarcinogenic agents (Berenblum, 1974). The basic idea put forth is that carcinogenesis is composed of an *initiating* process, responsible for the transformation of normal cells into "latent" tumor cells, and a *promoting* process, where these latent transformed cells replicate and become overt tumors. The "silent" and "overt" aspects of tumorigenesis have been emphasized, but researchers are often imprecise with respect to whether neoplastic transformation is conceived of as a single step or a progressive process.

With the acceptance of this two-stage concept, it is now customary to classify chemical carcinogens as being either "true" carcinogens, i.e., initiators of the neoplastic change, or tumor promoters, which are themselves not carcinogenic but which act to promote the growth of transformed neoplastic cells. The changes produced by initiators are considered to be irreversible, whereas those produced by promoters may be reversible. In this context, hormones may be considered promoters, especially as they apply to hormone-dependent tumors. However, the distinction between initiators and promotors is not always clear, since in high doses, some chemicals appear to act as both classes of agents. Problems in interpretations also arise in cases where the initiating agents are unknown.

The major weakness in the initiation-promotion concept is that there is no specific means available to recognize or detect initiated cells apart from their ability to form tumors (or to acquire the characteristics of transformed cells). That is, the *same criterion* used to recognize promotion is that applied in the recognition of initiation. Another difficulty with the concept is that it assumes that promotion affects *only* those cells which had been initiated, i.e., that tumors arise only in sites which had undergone neoplastic transformation. A carcinogenic stimulus could, in fact, establish a variety of cellular changes, some permanent and others reversible, and with differing potentialities to form tumors. In this case, the quality and nature of subsequent treatment could determine the ultimate fate in the different classes of altered cells.

Teratogenesis, Carcinogenesis, and Mutagenesis

R.W. Miller (1977) has recently discussed the relationship between teratogens and carcinogens, noting that whereas clinical observations show a close link between specific forms of human cancer and congenital malformations (e.g., the high risk of leukemias in Down's, Bloom's and Fanconi's syndromes), there is no such link when the causes are environmental rather than genetic or idiopathic. Miller also provides a list of 31 chemicals, two parasites, and two forms of radiation as agents for which there is some evidence (at times scanty) of human carcinogenicity but which have not been shown to be teratogenic in man. Nevertheless, it should be emphasized that while 23 of the 35 listed agents have not or cannot be tested in the human, approximately one-third of them were found to be mutagenic in the Ames test. More of the carcinogens may ultimately prove to be teratogenic as well.

There are only five agents known to be both carcinogenic and teratogenic in the human: alcohol, diphenylhydantoin, ionizing radiation, DES, and androgens. It is interesting that with the exception of DES, these agents produce their carcinogenic and teratogenic effects in different sites in the body; for example, androgens masculinize female embryos but produce liver carcinomas (Miller, 1977). It must be emphasized, however, that carcinogenicity cannot be defined in absolute terms. It is as much a function of the test system as it is a function of the agents being tested.

The Ames mutagenesis assay (Ames et al., 1973a,b; 1975) utilizes mutants of *Salmonella* in a back-mutation system with mammalian liver preparations to provide metabolic activation of the compounds tested. Ames and coworkers (McCann et al., 1975) claim a 90% correspondence between carcinogenicity and mutagenicity in over 300 widely diverse compounds. Results in other laboratories (e.g., Coombs et al., 1976), although supporting the mutagenicity of most carcinogens, reveal an overall lack of a linear relationship between carcinogenic potency and mutagenic potency for the compounds tested. One problem is that mutagenesis is assayed in one system (bacteria + rat liver microsomes) while carcinogenicity is assayed in another (e.g., mouse skin). Thus, species and organ variability may contribute to the lack of quantitative correspondence between mutagenicity and carcinogenicity. This problem may be resolved by the development of culture models (e.g., Barrett et al., 1978; Mishra et al., 1978; Kakunaga, 1978) enabling simultaneous study of mutagenesis and carcinogenesis (transformation) in the same cell system. Furthermore, a variety of in vitro screening tests for carcinogens is emerging (cf. Fisher and Weinstein, 1978; Devoret, 1979).

Ashby and Styles (1978) also emphasize that it is the absorption, distribution, metabolic activation, and half-life of derived active compounds rather than the in vitro mutagenic potency which determines a compound's carcinogenic potency. These authors further indicate quantitative problems posed by lack of standards and activity variations in the microsomal enzyme preparations used in different laboratories. Although the Ames test has proved to be a sensitive and important qualitative assay, studies on the relationships among carcinogenicity, teratogenicity, and mutagenicity are only just beginning.

ROLE OF HORMONES IN CANCER

The topic of hormonal carcinogenesis has been reviewed several times in recent years (e.g., Furth, 1975; Clifton and Sridharan, 1975; Hilf et al., 1976; Noble, 1977; Welsch and Nagasawa, 1977; Welsch and Meites, 1978). The present surge of interest in this topic is not surprising in view of today's widespread use of hormones as oral contraceptives and therapeutic agents for a broad spectrum of clinical entities. These include cancer of a variety of organs, anemias, hypofunctional endocrine disorders, arthritis and related diseases, allergies, delayed puberty, postmenopausal symptoms, and muscular atrophy. Hormones are also used as life-long therapy by individuals who have selected sexual reassignment (transsexuals) and by athletes attempting to improve muscular development and performance.

In fact, there is no satisfactory evidence for a direct cause-effect relationship between hormones and cancer. There is only evidence that derangements in endocrine homeostasis, naturally-occurring or experimentally-produced, *can result* in cancer. Our own view is that hormones play an indirect rather than a direct role in neoplasia and act primarily, by mechanisms still poorly defined, as promoting or permissive agents. Nevertheless, it is not our intent to minimize the role of hormones as high-risk factors in neoplasia but rather to consider, in a necessarily selective fashion, several possible ways that hormones may mediate the neoplastic response.

Hormones Versus Chemical Carcinogenesis in Tumorigenesis

The tumorigenic effects of hormones and chemical carcinogens differ in a number of ways (Clifton and Stridharan, 1975; Furth, 1975; Nandi, 1975). In general, tumors can be produced by chemical carcinogens after a single administration, whereas prolonged and continued exposure is required of hormones. The distinction does not always prevail, since chemical carcinogen dosage, method, and route of administration, etc., all affect latency and tumor incidence. However, this overall difference does raise the possibility that the two classes of agents produce tumors by different mechanisms.

Chemical carcinogens usually give rise at the start to "autonomous" tumors (the term generally refers to transplantability and hormone-independence), whereas endocrine-related tumors are usually initially hormone-dependent but invariably become autonomous (Furth, 1975). An exception may be the mammary tumors induced by DMBA (Huggins et al., 1961), which are considered to be initially hormone-dependent. Recent observations, however, indicate that the DMBA-induced mammary tumor system is more heterogeneous than had been supposed (Rivera et al., 1980; Rivera and Vijayaraghavan, in press).

The strongest argument against hormones being direct inducers of neoplasia is their lack of carcinogenic or transforming activity in vitro, whereas chemical carcinogens are effective both in vivo as well as in vitro (Heidelberger, 1975). Nevertheless, some authorities in the field refer to some hormone-associated cancers as being caused by "carcinogenic hormones" (cf. Lingemann, 1978).

Hormonal Derangements Resulting In Neoplasia

Endocrinological procedures shown to be effective in producing tumors include endocrine ablation, prolonged administration of hormones, ectopic pituitary transplantation, and blocking of hormone synthesis. Specific examples are given in the reviews of Clifton and Sridharan (1975) and Furth (Furth et al., 1973; Furth, 1975) and will not be repeated here. It is sufficient to reiterate only that all these procedures result in a derangement of hormonal homeostasis, with physiologic and metabolic consequences on virtually all cells in the body. Thus, the resultant "substrate" on which tumors arise and develop is the sum total of the effects of a profoundly altered and complex environment. It is of interest that most of the hormones implicated in carcinogenesis are those whose circulating levels are regulated by other hormones. Although disturbances in feedback regulation are certainly involved, we still do not know the fundamental nature of the "hormonal imbalance" which predisposes tissues to cancer.

Hormonal Modulation of Tissue Substrate for Carcinogenesis

One way in which hormones may play a supportive role in carcinogenesis is to maintain a sufficiency of tissue upon which other factors operate to initiate neoplasia (Foulds, 1965). The inference is that a critical mass of tissue is required for the neoplastic change. This may explain why, after castration and hypophysectomy, mammary glands are refractory to spontaneous carcinogenesis (Lemon, 1977) and to carcinogenesis by polycyclic aromatic hydrocarbons and mammary tumor virus (Dao, 1969, 1971; Nandi and McGrath, 1973). However, the limiting factor is probably not mammary atrophy per se but the specific loss or decline of particular cells in the population. These may be hormone-dependent stem cells or sets of cells at high risk for neoplastic transformation. Apart from protecting against cell loss, hormones are probably required during critical stages of neoplastic transformation, at least until the transformed cells are provided with a selective growth advantage over adjacent normal cells. Thus, a proper hormonal environment may be essential not only for the "fixation" of the critical neoplastic lesion (Nandi, 1978b; Farber et al., 1979) but also for the stabilization of the altered cells.

From studies in chemical carcinogenesis (Berenblum, 1974), it is evident that transformed cells can remain dormant or unexpressed for long periods of time. Under these circumstances, they may retain many properties of their tissue of origin, including the need for hormonal support and other factors of the microenvironment. The hormonal influence would thus ensure their persistence in the population and allow them to accumulate despite restraining protective mechanisms of the host.

In addition to their supportive role in transformation, hormones are capable of altering the sensitivity of tissues to carcinogenesis. For example, the chemical induction of liver carcinomas is more effective in male rats than in female rats and less effective in castrated males than in intact males (Reuber, 1975). Women who have early first pregnancies are at lower risk for breast cancer than those who have late pregnancies or never become pregnant (MacMahon et al., 1973). In laboratory rats, stimulation of the mammary glands by pregnancy or by exogenous hormones before or shortly after exposure to chemical carcinogens protects against the development of mammary tumors (Dao et al., 1960; Huggins et al., 1962). Conversely, mammary glands exposed first to the carcinogen and subsequently stimulated by pregnancy or hormones develop mammary tumors sooner than the glands of carcinogen-treated virgin rats (Dao and Sutherland, 1959; Dao, 1969). Thus, depending upon the sequence of mammary stimulation and exposure to chemical carcinogen, the hormonal milieu of pregnancy can either inhibit or enhance the carcinogenic potential of the tissue.

These opposing effects of hormones on mammary tumorigenesis support the argument that hormones modulate rather than induce de novo cell properties. Although the influence of hormones in enhancing the development of mammary tumors may be considered one of promotion in accord with Berenblum's concept (1974), their inhibitory effects are difficult to explain. Indeed, theories of promotion are still highly speculative (Weinstein et al., 1979), but our understanding of inhibitory mechanisms is even more fragmentary. In a discussion relating to renewing epithelial cell populations, Cairns (1975) proposes 3 possible protective mechanisms against the accumulation of potentially dangerous mutants: (1) restriction of stem cells to limited territories in the epithelium, which would prevent their expansion and competition with normal cells, (2) segregation during DNA replication, so that daughter cells destined for elimination from the population are those which carry the mutation, and (3) restricting the number of stem cells.

The third possibility may be of some relevance in explaining the protective effects of an early first pregnancy. Assuming that mutations occur in stem cells and that the chance of a mutation is proportional to the number of stem cell generations, one way of protecting against mutations is to restrict the number of stem cells. Since the risk of breast cancer increases as the interval between puberty and the first pregnancy increases (MacMahon et al., 1973), an *early* first pregnancy may protect against mutations (i.e., the spontaneous appearance of mammary tumors) by establishing a final, unchanging population of stem cells, which is unaffected by successive pregnancies (Cairns, 1975).

The tumor-promoting activity of hormones appears to be closely linked to the ability of hormones to stimulate proliferation. Indeed, hyperplasia is one of the earliest recognized alterations in endocrine-associated cancers, especially after chronic hormonal stimulation (Furth, 1967; Dao, 1967; Foulds, 1975). This proliferative aspect of hormone action is particularly noteworthy since 80% of malignant tumors in humans occur in tissues in which there is continual proliferation (Oehlert, 1973). Although the relationship between hyperplasia and malignancy has not been precisely defined, there is now considerable evidence to support the hypothesis that hyperplasias are an early stage in neoplastic development of a wide variety of tissues (see symposium on Early Lesions and the Development of Epithelial Cancer, 1976). However, it must also be noted that hyperplasias are remarkably heterogeneous in their morphological, biochemical, and tumorigenic properties. Their diffuse and multiple character contrasts with the focal origin of tumors. Their occurrence is not specific to the neoplastic response, since they occur in response to injury and other circumstances not involving neoplasia. Hence, although cancers appear to arise on a background of hyperplasia, they are probably distinct processes. In spite of these reservations, the positive correlation between cell proliferation (and hyperplasia) and tumor incidence is well documented (Oehlert, 1973) and deserves further study.

A significant characteristic of hormone-induced hyperplasias is their reversibility following the cessation of hormonal stimulation (Furth, 1967). This lability of hormonal effect would support the proposition that hormones are promoters rather than initiators of irreversible changes characteristic of neoplasia. How then does one account for the formation of adenomas and carcinomas after long, sustained hormonal stimulation of tissues such as the mammary gland, thyroid, testis, and ovary (Bonser and Robson, 1940; Noble et al., 1940; Guthrie, 1959; Haran-Ghera et al., 1960)? In the absence of identifiable carcinogens, do hormones under these circumstances play a "causative" role in the etiology of cancer? Whatever the explanation, it must take into account the more fundamental and still unresolved question of whether *any treatment* resulting in the formation of tumors involves the emergence of pre-existing transformed cells or the formation de novo of such cells during the course of treatment. Since neither possibility has been unequivocally refuted for endocrine-associated neoplasia, speculations about the origin of tumors following chronic hormone administration are necessarily limited.

In accord with other speculations (Miller and Miller, 1974; Nowell, 1976), it is possible that hormone-stimulated proliferation increases the probability of errors in DNA replication, giving rise to genetically-altered variants, which may propagate and eventually express the tumor phenotype. Alternatively, and as discussed earlier, hormonal stimulation may enhance the emergence of latent or unexpressed variants, transformed previously by agents such as virus and chemical carcinogens. Evidence for this possibility is provided by the studies of DeOme et al. (1978), who showed by dissociation and transplantation techniques that virus-transformed cells are present but inapparent in mouse mammary glands long before the spontaneous appearance of lesions. Still another possibility is that cell proliferation increases the sensitivity to neoplastic transformation by chemical and physical carcinogens present in the environment but not immediately identifiable. Such a process could be operative in "spontaneous transformation" in vitro, where cells propagated for many generations in culture give rise to tumors on inoculation into host animals (Sanford, 1965).

It is well known that proliferating cells are particularly sensitive to transformation by chemical and viral carcinogens (Bertram and Heidelberger, 1974; Butel and Estes, 1975; Nagasawa and Yanai, 1974; Lin et al., 1976. A round of cell division is considered essential for the stabilization of transformed cells (Butel and Estes, 1975; Farber et al., 1979). Cell proliferation is also a constant feature of the response to known promoters of carcinogen-induced tumors (Diamond et al., 1978; Weinstein et al., 1979). The efficiency of chemical transformation of synchronized mouse fibroblasts is dependent upon the cell cycle, the sensitive phase reported to be somewhere between four hours before S phase and the G1-S boundary (Bertram and Heidelberger, 1974).

Recent studies have drawn attention to the old embryological problem concerning the mutual exclusivity of proliferation and differentiation.

Promotors of carcinogen-induced tumors, such as phorbol esters, which stimulate cell proliferation in vivo and in vitro, apparently inhibit cell differentiation in vitro (Weinstein and Wigler, 1977; Marx, 1978a). By inhibiting differentiation, tumor promoters are believed to prevent cell loss from the population, thereby keeping cells in an "immature" state, where they continue to divide and in some manner give rise to malignant variants. Although highly provocative, this proposition must take into account certain inconsistencies: There is no evidence that phorbol esters or other promoters inhibit differentiation in vivo; many tumors are highly differentiated (Dustin, 1972), especially those associated with hormonal etiology; agents which are not tumor promoters are capable of stimulating cell growth in vitro.

Another possible role of cell proliferation in neoplasia is based on the assumption that normal cells, co-existing with transformed neoplastic cells, inhibit in some way the proliferation of the latter cell type (Nandi, 1978a,b). Since normal epithelial cells undergo a finite number of cell divisions (i.e., they have a limited life span unlike their transformed counterparts) (Daniel, 1973), the hormonal stimulation of division hastens their senescence and removal from the population. This resultant diminution of normal cells would remove their inhibitory influences and thus favor the emergence of the transformed neoplastic cells.

Transplacental and Perinatal Effects of Hormones and the Risk of Cancer

The syndrome of vaginal adenocarcinoma in the daughters of women who ingested the synthetic estrogen diethylstilbestrol (DES) during the first trimester of their pregnancy would seem to provide the clearest instance of carcinogenesis by a hormone-mimetic substance (Herbst, 1978). This statement is based on the confidence, for which there is in fact only minimal evidence, that similar exposure to natural steroidal estrogen during the critical period of genital tract development would give rise to a similar pathology. This confidence is derived in part from the analysis of an analogous tumorigenic pattern using mouse models in which antenatal or neonatal animals are exposed to estrogen (17β-estradiol) as well as to DES (Bern, 1979; Forsberg, 1979; McLachlan, 1979) and on the likelihood that the initial effect of the hormonal stimulus is teratogenic rather than carcinogenic (Ulfelder, 1976).

In both the human female and the female mouse, early exposure to estrogen (and possibly other sex hormones) may lead to the retention of epithelial cell populations in regions from which they should normally disappear. The adenocarcinoma-adenosis situation in women and the adenosis and possible adenocarcinoma in the mouse can be ascribed to retention of gland-forming Mullerian epithelium in the upper vagina (especially in the fornix) and the ectocervix from which it is normally absent in the adult. This plausible explanation leaves the ultimate carcinogenic stimulus undefined. Is it the hormonal milieu of the menarche which serves as a promoting agent? Does the "natural history" of misplaced (ectopic) Mullerian epithelium include occasional "inevitable" carcinogenesis? Is there an increased susceptibility of the ectopic epithelium to low-level carcinogenic stimuli: viral, chemical, radiation?

There is one other possibility, of course, and this would be the action of DES as a chemical carcinogen, after metabolic activation. There is evidence for epoxide formation in the metabolism of DES (Metzler, 1979), and it is possible that these epoxide(s) are directly acting carcinogens, but this has not been demonstrated. Underlying this possibility is the still-ongoing controversy about the direct carcinogenic status of DES — an unresolved problem with much environmental and emotional impact.

Sex Hormone Administration and the Risk of Cancer In Adult Humans

In addition to the consequences of intrauterine exposure to DES discussed above, there are several other instances where the contraceptive or therapeutic utilization of sex hormones may be related to cancer risk in the human species. A World Health Organization Scientific Group (WHO, 1978) reviewed the data on contraception and the risk of cancer exhaustively but largely avoided conclusions. Organs considered for possible influence on cancer risk by steroidal contraceptive agents include the breast (cf. Montague et al., 1977), uterine cervix and corpus, ovary, pituitary, and liver.

The recent IARC Monographs on Sex Hormones and Cancer (1979) place emphasis on the following: (1) *Endometrial cancer*. Adult women receiving estrogen treatment (including DES and the use of estrogens for treating postmenpoausal disorders) show an increased incidence of endometrial cancer (Hoover et al., 1976a; Greenwald et al., 1977). Even this categorical, seemingly well documented statement, strongly supported by a recent, extensive study (Antunes and Stolley, 1977), has been challenged on sampling grounds (Feinstein and Horwitz, 1978). At this writing, seven of ten epidemiological studies favor a link between estrogens and endometrial cancer in women receiving treatment for menopausal symptoms (for discussion of the problems involved in finalizing conclusions, see Marx, 1978b). Factors which favor increased endogenous estrogen levels for a prolonged period may also result in a high risk of endometrial cancer. Sequential oral contraceptives (estrogen followed by progestin)—no longer in use—may also increase its incidence (cf. Silverberg et al., 1977). (2) *Cervical cancer*. Long-term use of oral contraceptives may increase the risk of cervical cancer in situ (Stern et al., 1977), but the effect on invasive cervical cancer remains uncertain. (3) *Breast cancer*. Adult women receiving estrogen may show an increased risk of breast cancer, but the incidence is not conclusive (cf. Hoover et al., 1976b). The risk of benign breast disease is decreased by oral contraceptives (i.e., the contraceptives appear to have a protective effect) (cf. Vessey et al., 1975). On the other hand, the risk of breast cancer may be increased by the interaction of contraceptive effects with other factors favoring higher breast-cancer risk (cf. Paffenbarger et al., 1977). Interpretation of the epidemiologically complex data in this area is obviously difficult. (4) *Liver tumors*. Hepatocellular tumors may be favored by prolonged androgen therapy, but the data are not conclusive (cf. Antunes and Stolley, 1977). On the other hand, there is an increased risk of the development of such tumors in women taking oral contraceptives for long periods of time. The older the women and the larger the amounts of steroids to which they are exposed, the greater the risk (cf. Vessey et al., 1977; Mahboubi and Shubik, 1977; Christopherson and Mays, 1977; Nissen et al., 1978). (5) *Other tumors*. Available data suggest but are not conclusive that oral contraceptives may increase the risk of malignant melanoma (Beral et al., 1977) and pituitary adenoma (e.g., Sherman et al., 1978) and may decrease the risk of ovarian cancer (Newhouse et al., 1977).

From the extensively qualified statements discussed above, it is difficult to come to definitive conclusions about the pro- or anti- carcinogenic influence of sex hormones, and different conclusions are reached by other epidemiologists (cf. Thomas, 1978). The sex hormones, both endogenous and exogenous in source, do not operate in an otherwise cancer-neutral environment. The external environment (including the diet—cf. Miller, A.B., 1978) and the internal environment (including subtle hormonal imbalances) may operate in synergism with or in antagonism to the sex hormone(s) in question to complicate the interpretation of the answer to what would otherwise seem to be a simple question.

Ectopic Hormone Production by Tumors

One of the possible consequences of neoplastic transformation is the production of small or large amounts of peptide hormones which are not ordinarily produced by the tissue from which the tumor originates. Other nonhormonal proteins may also be produced by tumors and the general explanation for their occurrence has been based on the notion of derepression of components of the genome, which would normally be inactive in the tissue concerned. However, recent studies indicate the presence of small amounts of ectopic peptides (e.g., human chorionic gonadotropin-like substance—Yoshimoto et al., 1977) in normal tissue, as well as the absence of gene derepression as indicated by nucleic acid hydridization techniques. Apparently, the abnormal tissues produce large amounts of the mRNA coding for the ectopic protein. "Dedifferentiation" of the rapidly growing tumor cell population may allow the formation of peptides characteristic of the embryonic cells from which the normal tissue took its origin (cf. Shields, 1978). In addition to peptide hormones, tumors may produce characteristic embryonic macromolecules such as carcinoembryonic antigen and α-fetoprotein. The "aging" of tumors as proliferation occurs in these situations is accompanied by the re-expression of embryonic abilities.

The production of hormones in unusual amounts and/or kinds may have consequences on distant organs affected by these hormones and give rise to minor or major metabolic and morphological changes directly attributable to the presence of the tumor ("paraneoplastic syndromes"—see Hall, 1974). Ectopic hormones have also been considered as possible "tumor markers" (Liddle, 1978).

Proof of the ectopic secretion of hormones by non-endocrine tumors involves one or more of the following criteria (Rees, 1975), which need, however, to be judged critically: (1) elevated hormone levels associated with presence of the tumor (but the tumor could be stimulating excess activity by the normal endocrine source) (2) fall in hormone level after removal of tumor (same problem as in (1)) (3) elevated hormone level after removal of normal source (4) arteriovenous gradient of hormone across the tumor (5) demonstration of hormone in tumor (but the level may be very low if there is rapid turnover), and (6) demonstration of hormone secretion in vitro.

The types of ectopic hormones which have been reported from human tumors include the ACTH-lipotropin-MSH complex (with the recent recognition of endorphins as an entity, one wonders whether some tumors may not produce their own analgesic), parathormone, the vasopressin-oxytocin-neurophysin complex (one wonders whether arginine vasotocin, the lower vertebrate and mammalian fetal "antidiuretic" octapeptide, is sometimes secreted), human chorionic gonadotropin, human placental lactogen, growth hormone (?), immunoreactive insulin and insulin-like substances, calcitonin, human chorionic thyrotropin, prolactin (?), erythropoietin, etc. It is possible that some tumors may produce hypothalamic releasing hormones, and certain individual tumors are ectopic sources for the production of a variety of hormonal and hormone-associated peptides.

More consideration needs to be given to the possibility that tumors may produce ectopic growth factors which in some cases could stimulate local tumor growth. There is some evidence that breast tumors and even normal mammary cells may produce small amounts of steroids which again could act on the tumor cells themselves (e.g., Raju et al., 1978; Abul-Hajj et al., 1979; MacIndoe, 1979; Miller et al., 1979). However, steroid-producing tumors have only been encountered to date in organs which are normally steroidogenic (Hertz, 1976).

Enhancement of Tumor Virus Production By Hormones

One other way in which hormones could affect the carcinogenic process would be by the stimulation of the production oncogenic viruses. Thus, the hormone itself would not be directly involved but exert its influence by increasing the amount of an active carcinogenic stimulus (or permitting its full expression) (cf. McGrath and Jones, 1978).

Evidence for the role of hormones in tumor virus production is especially strong in the case of the well-studied murine mammary tumor virus (MTV). The ability of natural and synthetic corticoids (including dexamethasone) to increase MTV expression in normal, preneoplastic and neoplastic mammary cells seems well established, although different investigators do not agree on all details (cf. Varmus et al., 1979). Prolactin, acting with a glucocorticoid, increases MTV expression by normal cells (Yang et al., 1977; Michalides et al., 1978). Tumor cell populations in culture show increased MTV production when other hormones are added to the media; a glucocorticoid appears to be sine qua non, but insulin and estradiol supplementation resulted in more virus production according to Cardiff et al. (1976) and insulin supplementation alone is effective in serum-free medium (Nagle and Fine, 1978). Thyroid hormone may have a stimulatory effect (with insulin and cortisol—Cardiff et al., 1976) or an inhibitory effect (Nagle and Fine, 1978).

It is evident from the discussion above that a variety of hormones can influence oncogenic virus production in this particular system. Existence of this information makes it mandatory to consider other instances of tumorigenesis in terms of hormonal involvement. Hormones appear not only to support cell populations wherein virus can multiply and mature, but may also directly stimulate viral synthesis.

Hormonal Therapy of Cancer

In dealing with the subject of the role of hormones in cancer, it is important not to ignore those situations where endocrine therapy may alleviate and conceivably even "cure" cancers. Huggins (1978) points out that two varieties of therapy may be of utility in treating human and animal cancers: hormone deprivation and hormone interference. In the treatment of human prostatic cancer, in some cases castration will result in regression and this response is also seen following injection of estrogens (including DES). Huggins feels that in seven kinds of cancer, modifications of endocrine status can result in "shrinkage of the cancer" and improvement of the host: breast (rat and human), prostate (human), endometrium (human), kidney (hamster), seminal vesicle (human), lymphoma (mouse), and leukemia (human). In the last two instances, the regression is clearly only temporary, and remission often occurs also in others of the tumor types listed.

Some tumors are "hormone-dependent," that is they fail to grow in the absence of specific hormonal stimuli; other tumors arising in tissues dependent on hormones for their normal development are nevertheless "hormone-independent." In the case of metastasizing human breast cancers, some are hormone-responsive, and others are not. Some breast cancers respond to ovariectomy, adrenalectomy and/or hypophysectomy; later, however, there may be remission and the cancer grows despite the absence of stimulatory hormones: the hormone-dependent tumor cell population has been replaced by a hormone-independent population. Human prostatic cancer may also transform from an androgen-dependent state to a hormone-independent state.

Experimental mammal models for breast cancer include tumors arising spontaneously (as in the dog and in the mouse, in the latter case virus-induced) or as a result of external stimuli (as in the rat where aromatic chemical carcinogens, ionizing radiation and prolonged estrogen treatment are all effective). Mammary adenocarcinomas induced by dimethylbenz(α)-anthracene or methylcholanthrene in the rat are generally hormone-dependent. Virus-induced mammary adenocarcinomas in the mouse are generally hormone-independent. On the other hand, in the rat and the mouse, respectively, hormone-independent cancer (not regressing after ovariectomy) and hormone-dependent cancer (regressing in the absence of the stimulus of pregnancy or in the absence of a pituitary transplant secreting prolactin) also can develop. In the mouse, the initially pregnancy-dependent tumor apparently "progresses" into a pregnancy-independent tumor resembling those arising spontaneously.

Hormone Receptors and Responsiveness to Hormone Therapy

This section will be restricted to discussion of hormone receptors in breast cancer, which has been the focus of extensive study (cf. King, 1979). It was Jensen et al. (1967) who first proposed that estrogen receptors (ER) in breast cancers might indicate hormone-responsiveness and that such cancers would regress after estrogen therapy. The underlying assumption is that responsive tumors contain specific hormone receptors while nonresponsive tumors do not. To some extent, the results obtained have supported this prediction. Patients whose breast cancers are low or negative for estrogen receptors (ER-) generally respond poorly or not at all to estrogen therapy; those positive for estrogen receptors (ER+) respond in about 55%-60% of the cases, regardless of the form of therapy, i.e., endocrine ablation, pharmacological doses of ovarian steroids, anti-estrogen therapy (McGuire et al., 1975, 1976). In contrast, only one-third of the overall population of breast cancers studied respond to endocrine therapy, irrespective of ER content (McGuire et al., 1976; Leung, 1978).

Patients with primary breast cancers which are ER- suffer recurrence earlier than those with ER+ cancers, regardless of tumor size or location, axillary node status, or age (McGuire et al., 1978). Premenopausal women are more likely to have ER- breast cancers than postmenopausal women, presumably because receptor sites are masked by endogenous estrogens (McGuire et al., 1976). Whether low ER content is necessarily indicative of autonomy is uncertain and may be clarified by the use of techniques (Katzenellenbogen et al., 1973; Daehnfeldt, 1974) which permit the measurement of ER occupied by endogenous estrogen.

Despite their usefulness, estrogen receptor assays are not unambiguously predictive of hormone responsiveness, since nearly one-half of ER+ breast cancers still fail to respond to hormone therapy. Perhaps the greatest value of ER assays is the high probability of predicting nonresponsive breast cancers. It is the lack of complete correspondence between ER+ cancers and responsiveness to estrogen therapy that remains a puzzling problem.

One limitation of receptor assays is their failure to reveal defects in receptor structure and function. Whereas negative ER measurements are usually predictive of tumor insensitivity to estrogen therapy, positive measurements of ER do not guarantee that ER is functional. For example, mammary tumor cells in culture can lose their responsiveness to androgens, but this loss of response is not accompanied by loss of androgen receptor (Yates and King, 1978). Defects in translocation of ER from cytosol to nuclear binding sites or alterations in nuclear ER have also been postulated (Shyamala, 1972). An unusual aspect of receptor function was revealed by the recent discovery in a hormone-independent human breast cancer line, of nuclear ER which is biologically active in the *absence* of estrogen (Zava et al., 1977). A similar observation was reported for nuclear androgen receptors in an autonomous line of a mouse mammary tumor (Bruchovsky and Rennie, 1978). Although this line is deficient in cytosol and nuclear androgen-binding sites, it contains significant amounts of *displaceable* androgen binding sites in the nucleus. It remains to be determined whether these anomalies are peculiar to autonomous cell lines of mammary tumors or whether they exist as well in breast tumors in vivo.

The wide range in ER content of experimental and human breast cancers (Leung and Sasaki, 1975; McGuire et al., 1976; DeSombre et al., 1978) may indicate that critical levels of ER are essential for responsiveness. Although the studies of DeSombre et al. (1978) suggest that critical levels may differ among breast cancers of premenopausal and postmenopausal women, some patients were still unresponsive regardless of their ER levels. Tumor heterogeneity may also account for the wide range of ER values and differential sensitivities to hormone therapy (King et al., 1976; McGuire et al., 1974). Within individual tumors there can be enormous variation in the relative composition of epithelium, fibrous stroma, and cells of the immune system. The epithelium alone may be composed of functionally distinct cell populations (Markland et al., 1978), which may account for partial and temporary remissions after hormone therapy.

The multiplicity of hormones involved in mammary tumor growth warrants the study of other hormones and their receptors, since they are potentially capable of regulating ER. Prolactin is of particular interest because of its important role in the induction and growth promotion of mammary tumors in experimental mammals (Meites et al., 1972; Welsch and Rivera, 1972; Furth, 1975; Kim and Furth, 1976). Receptors for prolactin have been demonstrated in mammary tumors (Nagasawa et al., 1979) but there is no consistent correlation between receptor content and hormone-dependence for growth. For example, Costlow et al. (1974) found no significant differences between normal and tumor tissues of the rat mammary gland with respect to prolactin binding sites, receptor affinity, or specificity of binding. In a subsequent study of estrogen-progesterone induced mammary tumors in GR mice, a decline in prolactin receptors was correlated with the transition from hormone-dependence to autonomy (Costlow et al., 1977b). However, no such correlation was found in the DMBA-induced rat mammary tumor, where hormone-independent and hormone-dependent forms were indistinguishable on the basis of prolactin receptors (Costlow et al., 1977a). These results are compatible with observations that mammary tumors undergo regression despite high levels of serum prolactin (Sinha et al., 1973) and high levels of prolactin receptors in the tumors (Smith et al., 1977). Thus, there is no compelling evidence to date that prolactin receptors are predictive of hormone-responsiveness.

There are now some reports that hormones can influence the receptors of other hormones, which may explain the ability of pharmacological doses of steroids to inhibit tumor growth. Leung (1978) proposes that progesterone may be a negative regulator of ER, since at high doses it reduces ER content and the prolactin-induced increase of ER in mammary tumors (Sasaki and Leung, 1975). High doses of estrogen do not appear to inhibit prolactin secretion (Blake et al., 1972) but may inhibit tumor growth by interfering with the peripheral actions of prolactin (Meites, 1972), of which one important action may be to stimulate the synthesis of ER (Leung, 1978).

As noted above, the failure of a considerable proportion of ER+ breast cancers to respond to estrogen therapy appears inconsistent with what is generally understood about the significance of hormone receptors. However, the binding of hormone to receptor is only one process in the chain of events associated with the action of the hormone. A more reliable test for predicting estrogen-responsiveness should include one or more of the biochemical markers of estrogen action. Progesterone receptors (PgR) are a potential marker since their synthesis is dependent upon estrogen (Freifeld et al., 1974). Hence, the presence of PgR in ER+ breast cancers might prove more reliable in predicting responsiveness than ER measurements alone (cf. McGuire, 1979). Preliminary reports are promising since they indicate that breast cancers which are ER+ and PgR+ respond to hormone therapy at double the rate of ER+ and PgR- patients (McGuire et al., 1978). Although important, these studies require further investigation since some ER+ and PgR- breast cancers do respond to therapy. Other receptors, such as those for glucocorticoids (Wittliff et al., 1978), also deserve careful scrutiny as part of the total hormonal milieu regulating tumor growth. In this connection, approximately 80% of human breast cancers which are positive for glucocorticoid receptors are also ER+ (Fazekas and MacFarlane, 1977).

Finally, some recent observations should be mentioned, since they open up a new perspective concerning the growth promoting actions of estrogen. In a study of several estrogen-responsive rodent tumor cell lines, Sirbasku (1978) found that although estrogen is required for optimal tumor formation from these lines in host animals, estrogen has no mitogenic effect on the lines in vitro. However, extracts of rodent liver, kidney, and uterus did stimulate growth in culture, especially if the rodents had been pretreated with estrogen. Taken together, these results indicate that the effects of estrogen on tumor growth in vivo are probably mediated by specific growth factors from other organs.

ACKNOWLEDGMENTS

The authors wish to acknowledge the support of the National Cancer Institute for their research relevant to areas covered in this chapter (Grant number CA-17862 to E.M.R. and CA-05388 to H.A.B.). Part of this chapter was written during tenure of a National Research Service Award to E.M.R. from the National Cancer Institute.

REFERENCES

Abul-Hajj, Y.J., Iverson, R., and Kiang, D.T. Aromatization of androgens in human breast cancer. *Steroids* 33, 205-222 (1979).
Agarwal, M.K. and Rossier, B.C. Multiplicity of steroid hormone receptors. *FEBS Lett.* 82, 165-168 (1977).
Ames, B.N., Lee, F.D., and Durston, W.E. An improved bacterial system for the detection and classification of mutagens and carcinogens. *Proc. Nat. Acad. Sci.* 70, 782-786 (1973a).
Ames, B.N., McCann, J., and Yamasaki, E. Methods for detecting carcinogens and mutagens with the Salmonella/mammalian-microsome mutagenicity test. *Mutation Res.* 31, 347-364 (1975).
Ames, B.N., Durston, W.E., Yamasaki, E., and Lee, F.D. Carcinogens are mutagens: a simple test system combining liver homogenates for action and bacteria for detection. *Proc. Nat. Acad. Sci.* 70, 2281-

2285 (1973b).
Antunes, C.M.F. and Stolley, P.D. Cancer induction by exogenous hormones (possible androgen-induced cancer). *Cancer* 39, 1896-1898 (1977).
Ashby, J. and Styles, J.A. Does carcinogenic potency correlate well with mutagenic potency in the Ames assay? *Nature* 271, 452-455 (1978).
Barrett, J.C., Bias, N.E., and Ts'o, P.O.P. A mammalian cellular system for the concomitant study of neoplastic transformation and somatic mutation. *Mutation Res.* 50, 121-136 (1978).
Baulieu, E.-E. Steroid receptors and hormone receptivity: new approaches in pharmacology and therapeutics. *J.A.M.A.* 234, 404-409 (1975).
Beral, V., Ramcharan, S., and Faris, R. Malignant melanoma and oral contraceptive use among women in California. *Brit. J. Cancer* 36, 804-809 (1977).
Berenblum, I. *Carcinogenesis as a Biological Problem. Frontiers of Biology*, vol. 34, North-Holland, Amsterdam and Oxford (1974).
Bern, H.A. The neonatal mouse—tumorigenesis after short-term exposure to hormones and its possible relevance to human syndromes, in *Proceedings, Symposium on Endocrine-Induced Neoplasia*, Omaha, Eppley Institute for Cancer Research, (1979), pp. 31-37.
Bern, H.A. and Nandi, S. Recent studies of the hormonal influence in mouse mammary tumorigenesis. *Prog. Exp. Tumor Res.* 2, 90-144 (1961).
Bertram, J.S. and Heidelberger, C. Cell cycle dependency of oncogenic transformation induced by N-methyl-N'-nitro-N-nitrosoguanadine in culture. *Cancer Res.* 34, 526-537 (1974).
Blake, C.A., Norman, R.L., and Sawyer, C.H. Effects of estrogen and/or progesterone on serum and pituitary gonadotropin levels in ovariectomized rats. *Proc. Soc. Exp. Biol. Med.* 141, 1100-1103 (1972).
Bonser, G.M. and Robson, J.M. The effects of prolonged oestrogen administration upon male mice of various strains: development of testicular tumours in the Strong A strain. *J. Pathol. Bact.* 51, 9-22 (1940).
Britten, R.J. and Davidson, E.H. Gene regulation for higher cells: a theory. *Science* 165, 349-357 (1969).
Bruchovsky, N. and Rennie, P.S. Classification of dependent and autonomous variants of Shionogi mammary adenocarcinoma based on heterogenous patterns of androgen binding. *Cell* 13, 273-280 (1978).
Butcher, R.W., Robison, G.A., and Sutherland, E.W. Cyclic AMP and hormone action, in *Biochemical Actions of Hormones*, vol. 2, G. Litwack, ed. Academic Press, New York (1972), pp. 21-54.
Butel, J.S. and Estes, M.K. Properties of cells transformed by DNA tumor viruses. *In Vitro* 11, 142-150 (1975).
Cairns, J. Mutation selection and the natural history of cancer. *Nature* 255, 197-200 (1975).
Caplan, A.I. and Ordahl, C.P. Irreversible gene repression model for control of development. *Science* 201, 120-130 (1978).
Cardiff, R.D., Young, L.J.T., and Ashley, R.L. Hormone synergism in the *in vitro* production of the mouse mammary tumor virus. *J. Toxicol. Environ. Health* Suppl. 1, 117-129 (1976).
Catt, K.J., Harwood, J.P., Aguilera, G., and Dufau, M.L. Hormonal regulation of peptide receptors and target cell responses. *Nature* 280, 109-116 (1979).
Chochinov, R.H. and Daughaday, W.H. Current concepts of somatomedin and other biologically-related growth factors. *Diabetes* 25, 994-1004 (1976).

Christopherson, W.M. and Mays, E.T. Liver tumors and contraceptive steroids: experience with the first one hundred registry patients. *J. Nat. Cancer Inst.* 58, 167-171 (1977).

Clifton, K.H. and Sridharan, B.N. Endocrine factors and tumor growth, in *Cancer. A Comprehensive Treatise*, vol. 3, F.F. Becker, ed. Plenum Press, New York (1975), pp. 249-285.

Coggin, J.H. and Anderson, N.G. Cancer, differentiation, and embryonic antigens: some central problems. *Adv. Cancer Res.* 19, 106-165 (1974).

Coombs, M.M., Dixon, C., and Kossonerghis, A.-M. Evaluation of the mutagenicity of compounds of known carcinogenicity, belonging to the benz(α)anthracene, chrysene, and cyclopenta(α)phenanthrene series, using Ames's test. *Cancer Res.* 36, 4525-4529 (1976).

Costlow, M.E., Buschow, R.A., and McGuire, W.L. Prolactin receptors in an estrogen-receptor deficient mammary carcinoma. *Science* 184, 85-86 (1974).

Costlow, M.W., Buschow, R.A., and McGuire, W.L. Prolactin receptors in 7,12-dimethylbenz(α)anthracene-induced mammary tumors following endocrine ablation. *Cancer Res.* 36, 3941-3943 (1977a).

Costlow, M.E., Sluyser, M., and Gallagher, P.E. Prolactin receptors in mammary tumors of GR mice. *Endocrine Res. Commun.* 4, 285-295 (1977b).

Cuatrecasas, P. Membrane receptors. *Ann. Rev. Biochem.* 43, 169-214 (1974).

Daehnfeldt, J.L. Endogenously blocked high affinity estradiol receptors in the immature and mature rat uterus. *Proc. Soc. Exp. Biol. Med.* 146, 159-162 (1974).

Daniel, C.W. Finite growth span of mouse mammary gland serially propagated *in vivo*. *Experientia* 29, 1422-1424 (1973).

Dao, T.L. Mammary cancer induction by 7,12-dimethylbenz(α)anthracene: relation to age. *Science* 165, 810-811 (1965).

Dao, T.L. Endocrine environment and neoplasia, in *Endogenous Factors Influencing Host-Tumor Balance*, R.W. Wissler, T.L. Dao, and S. Wood, eds. Univ. of Chicago Press, Chicago (1967), pp. 75-97.

Dao, T.L. Studies on mechanism of carcinogenesis in the mammary gland. *Prog. Exp. Tumor Res.* 11, 235-261 (1969).

Dao, T.L. Inhibition of tumor induction in chemical carcinogenesis in the mammary gland. *Prog. Exp. Tumor Res.* 14, 59-88 (1971).

Dao, T.L. and Sutherland, J. Mammary carcinogenesis by 3-methylcholanthrene. I. Hormonal aspects in tumor induction and growth. *J. Nat. Cancer Inst.* 23, 567-585 (1959).

Dao, T.L., Bock, F.G., and Greiner, M.J. Mammary carcinogenesis by 3-methylcholanthrene. II. Inhibitory effect of pregnancy and lactation. *J. Nat. Cancer Inst.* 25, 991-1003 (1960).

DeOme, K.B., Miyamoto, M.Y., Osborn, R.C., Guzman, R.C., and Lum, K. Detection of inapparent nodule-transformed cells in the mammary gland tissues of virgin female BALB/cfC3H mice. *Cancer Res.* 38, 2103-2111 (1978).

DeSombre, E.R., Greene, G.L., and Jensen, E.V. Estrophilin and endocrine responsiveness of breast cancer, in *Hormones, Receptors, and Breast Cancer. Prog. Cancer Res. and Therapy*, vol. 10, W.L. McGuire, ed. Raven Press, New York (1978), pp. 1-14.

Devoret, R. Bacterial tests for potential carcinogens. *Sci. Amer.* 241, 40-49 (1979).

Diamond, L., O'Brien, T.G., and Rovera, G. Tumor promoters: effects on proliferation and differentiation of cells in culture. *Life Sci.* 23, 1979-1988 (1978).

Dustin, P., Jr. Cell differentiation and carcinogenesis: a critical review. *Cell Tissue Kinet.* 5, 519-533 (1972).

Eisenfeld, A.J., Aten, R., Weinberger, M., Haselbacher, G., Halpern, K., and Krakoff, L. Estrogen receptor in the mammalian liver. *Science* 191, 862-865 (1976).

Farber, E. Carcinogenesis—cellular evolution as a unifying thread. *Cancer Res.* 33, 2537-2550 (1973).

Farber, E., Cameron, R.G., Laishes, B., Lin, J.-C. Medline, A., Ogawa, K., and Solt, D.B. Physiological and molecular markers during carcinogenesis, in *Carcinogens: Identification and Mechanism of Action*, A.C. Griffin and C. R. Shaw, eds. Raven Press, New York (1979), pp. 319-335.

Fazekas, A.G. and MacFarlane, J.K. Macromolecular binding of glucocorticoids in human mammary carcinoma. *Cancer Res.* 37, 640-645 (1977).

Feinstein, A.R. and Horwitz, R.I. A critique of the statistical evidence associating estrogens with endometrial cancer. *Cancer Res.* 38, 4001-4005 (1978).

Fisher, P.B. and Weinstein, I.B. In vitro screening tests for potential carcinogens, in *Carcinogens in Industry and the Environment*, J.M. Sontag, ed. Marcel Dekker, Inc. New York (1981), pp. 113-166.

Forsberg, J.-G. Studies on the developmental mechanism of estrogen-induced irreversible changes in the mouse cervicovaginal epithelium, in *Perinatal Carcinogenesis, Nat. Cancer Inst. Mongr.* 51, 41-56 (1979).

Foulds, L. Mammary tumours in hybrid mice: growth and progression of spontaneous tumours. *Brit. J. Cancer* 3, 345-375 (1949).

Foulds, L. Multiple etiologic factors in neoplastic development. *Cancer Res.* 25, 1339-1347 (1965).

Foulds, L. *Neoplastic Development*, vol. 1, Academic Press, New York (1969).

Foulds, L. *Neoplastic Development*, vol. 2, Academic Press, New York (1975).

Freifeld, M.L., Feil, P.D., and Bardin, C.W. The in vivo regulation of progesterone "receptor" in guinea pig uterus: dependence on estrogen and progesterone. *Steroids* 23, 93-103 (1974).

Furth, J. The role of mammosomatotropin in tumorigenesis of the mammary gland, in *Endogenous Factors Influencing Host-Tumor Balance*, R.W. Wissler, T.L. Dao, and S. Wood, eds. University of Chicago Press, Chicago (1967), pp. 49-62.

Furth, J. Hormones as etiological agents in neoplasia, in *Cancer. A Comprehensive Treatise*, vol. 1, F.F. Becker, ed. Plenum Press, New York (1975), pp. 75-120.

Furth, J., Ueda, G., and Clifton, K.H. The pathophysiology of pituitaries and their tumors: methodological advances. *Methods in Cancer Res.* 10, 202-277 (1973).

Gardner, W.U. Hormonal aspects of experimental tumorigenesis. *Adv. Cancer Res.* 1, 173-232 (1953).

Gilbert, D.A. The temporal response of the dynamic cell to disturbances and its possible relationship to differentiation and cancer. *South African J. Sci.* 70, 234-244 (1974).

Goldstein, J.L. and Motulsky, A.G. Genetics and endocrinology, in *Textbook of Endocrinology*, 5th ed. R.H. Williams, ed. W.B. Saunders Co., Philadelphia (1974), pp. 1004-1029.

Greenblatt, R.B. Estrogens and endometrial cancer—gross exaggeration or fact? *Geriatrics* 32, 60-72 (1977).

Greenwald, P., Caputo, T.A., and Wolfgang, P.E. Endometrial cancer after menopausal use of estrogens. *Obstet. Gynecol.* 50, 239-243 (1977).

Gross, L. *Oncogenic Viruses,* Pergamon Press, New York and Oxford (1961).

Guthrie, M.M. Tumorigenesis in ovaries of mice transplanted to the liver. *Nature* 184, 916-917 (1959).

Hall, T.C. Ectopic synthesis and paraneoplastic syndromes. *Cancer Res.* 34, 2088-2091 (1974).

Haran-Ghera, N., Pullar, P., and Furth, J. Induction of thyrotropin-dependent tumors by thyrotropes. *Endocrinol.* 66, 694-701 (1960).

Haslam, S.Z. Age as a modifying factor of 7,12-dimethylbenz(α)anthracene-induced mammary carcinogenesis in the Lewis rat. *Int. J. Cancer* 23, 374-379 (1979).

Heidelberger, C. Some reflections and speculations about chemical carcinogenesis. *Canad. Cancer Conf.* 7, 326-350 (1967).

Heidelberger, C. Chemical carcinogenesis. *Ann. Rev. Biochem.* 44, 79-121 (1975).

Herbst, A.L., ed., *Intrauterine Exposure to Diethylstilbestrol in the Human,* American College of Obstetricians and Gynecologists, Chicago (1978).

Hertz, R. Steroid-induced, steroid-producing, and steroid-responsive tumors, in *Steroid Hormone Action and Cancer,* K.M. Menor and J.R. Reel, eds. Plenum Press, New York (1976), pp. 1-14.

Hilf, R., Harmon, J., Matusik, R.J., and Ringler, M.B. Hormonal control of mammary cancer, in *Control Mechanisms in Cancer,* W.E. Criss, T. Ono and J.R. Sabine, eds. Raven Press, New York (1976), pp. 1-23.

Holtfreter, J. Concepts on the mechanism of embryonic induction and its relation to parthenogenesis and malignancy. *Symposia of the Society for Experimental Biology,* no. 11, Growth (1948), pp. 2-49.

Hoover, R., Fraumeni, J.F., Jr., Everson, R., and Myers, M.H. Cancer of the uterine corpus after hormonal treatment for breast cancer. *Lancet* i, 885-887 (1976a).

Hoover, R., Gray, L.A., Cole, P., and MacMahon, B.L. Menopausal estrogens and breast cancer. *New Eng. J. Med.* 295, 401-405 (1976b).

Huggins, C. Endocrine-induced regression of cancers. *Science* 156, 1050-1054 (1967).

Huggins, C. Two principles in endocrine therapy of cancers: hormone deprival and hormone interference, in *Endocrine Control in Neoplasia,* R.K. Sharma and W.E. Criss, eds. Raven Press, New York (1978) pp. 1-9.

Huggins, C., Grand, L.C., and Brillantes, F.P. Mammary cancer induced by a single feeding of polynuclear hydrocarbon and its suppression. *Nature* 189, 204-207 (1961).

Huggins, C., Moon, R.C. and Morii, S. Extinction of experimental mammary cancer. I. Estradiol-17β and progesterone. *Proc. Nat. Acad. Sci.* 48, 379-386 (1962).

IARC Monographs on the Evaluation of Carcinogenic Risk of Chemicals to Man. International Agency for Research on Cancer, Lyon, vol. 6, Sex Hormones (1974a).

IARC Monographs on the Evaluation of Carcinogenic Risk of Chemicals to Man. International Agency for Research on Cancer, Lyon, vol. 7. Some Anti-thyroid and Related Substances, Nitrofurans and Industrial Chemicals (1974b).

IARC Monographs on the Evaluation of Carcinogenic Risk of Chemicals to Man. International Agency for Research on Cancer, Lyon, vol. 21, Sex Hormones, 2nd ed. (1979).

Jensen, E.V. and DeSombre, E.R. Estrogen-receptor interactions. *Science* 182, 126-134 (1973).

Jensen, E.V., DeSombre, E.R., and Jungblut, P.W. Estrogen receptors in hormone-responsive tissues and tumors, in *Endogenous Factors Influencing Host-Tumor Balance*, R.W. Wissler, T.L. Dao, and S.W. Wood, eds. University of Chicago Press, Chicago (1967), pp. 15-30.

Jungmann, R.A. and Russell, D.H. Cyclic AMP, cyclic AMP-dependent protein kinase, and regulation of gene expression. *Life Sci.* 20, 1781-1797 (1977).

Kadis, B. Steroid epoxides in biologic systems: a review. *J. Steroid Biochem.* 9, 75-81 (1978).

Kakunaga, T. Neoplastic transformation of human diploid fibroblast cells by chemical carcinogens. *Proc. Nat. Acad. Sci.* 75, 1334-1338 (1978).

Katzenellenbogen, J.A., Johnson, H.J., Jr., and Carlson, K.E. Studies on the uterine cytoplasmic estrogen binding protein. Thermal stability and ligand dissociation rate. An assay of empty and filled sites by exchange. *Biochem.* 21, 4092-4099 (1973).

Kim, U. and Furth, J. The role of prolactin in carcinogenesis. *Vitamins and Hormones* 34, 107-136 (1976).

King, R.J.B. (ed.), *Steroid Receptor Assays in Human Breast Tumours*. Alpha Omega, Cardiff, Wales (1979).

King, R.J.B., Cambray, G.J., and Robinson, J.H. The role of receptors in the steroidal regulation of tumour cell proliferation. *J. Steroid Biochem.* 7, 869-873 (1976).

Lauritzen, C. and van Keep, P. *Estrogen Therapy. The Benefits and Risks*, S. Karger, Basel (1978).

Lemon, H.M. Experimental basis for multiple primary carcinogenesis by sex hormones. A review. *Cancer* 40, 1825-1832 (1977).

Leung, B.S. Hormonal dependency of experimental breast cancer, in *Hormones, Receptors, and Breast Cancer, Prog. Cancer Res. and Therapy*, W.L. McGuire, ed. Raven Press, New York (1978), pp. 219-261.

Leung, B.S. and Sasaki, G.H. On the mechanism of prolactin and estrogen action in 7,12-dimethylbenz(α)anthracene-induced mammary carcinoma in the rat. II. *In vivo* tumor responses and estrogen receptor. *Endocrinology* 97, 564-572 (1975).

Liao, S. Cellular receptors and mechanisms of action of steroid hormones. *Intern. Rev. Cytol.* 41, 87-172 (1975).

Liddle, G.W. Ectopic hormones as tumor markers: an overview, in *Biological Markers of Neoplasia: Basic and Applied Aspects*, R.W. Ruddon, ed. Elsevier, New York (1978), pp. 257-264.

Lin, F.K., Banerjee, M.R., and Crump, L.R. Cell cycle-related hormone carcinogen interaction during chemical carcinogen induction of nodule-like mammary lesions in organ culture. *Cancer Res.* 36, 1607-1614 (1976).

Lincoln, T.M. and Corbin, J.D. On the role of the cyclic AMP and cyclic GMP-dependent protein kinases in cell function. *J. Cyclic Nucleot. Res.* 4, 3-14 (1978).

Lingemann, C.H., ed., *Carcinogenic Hormones*, Springer-Verlag, New York (1978).

MacIndoe, J.H. Estradiol formation from testosterone by continuously cultured human breast cancer cells. *J. Clin. Endocrinol.* 49, 272-277 (1979).

MacMahon, B., Cole, P., and Brown, J. Etiology of human breast cancers: a review. *J. Nat. Cancer Inst.* 50, 21-42 (1973).

Mahboubi, E. and Shubik, P. Epidemiological relationship between steroid hormones and liver lesions. *J. Toxicol. Environ. Health* 3, 207-230 (1977).

Markland, F.S., Chopp, R.T., Cosgrave, M.D., and Howard, E.G.

Steroid hormone receptor characterization of several histologic variants of a rat prostatic adenocarcinoma. *J. Supramol. Struct.* 9, 509-524 (1978).

Marx, J.L. Tumor promoters: carcinogenesis gets more complicated. *Science* 201, 515-518 (1978a).

Marx, J.L. Estrogens: hormones' link to cancer. *Science* 202, 1270-1272 (1978b).

McCann, J., Choi, E., Yamasaki, E., and Ames, B.N. Detection of carcinogens as mutagens in the Salmonella/microsome test: assay of 300 chemicals. *Proc. Nat. Acad. Sci.* 72, 5135-5139 (1975).

McGrath, C.M. and Jones, R.F. Hormonal induction of mammary tumor viruses and its implications for carcinogenesis. *Cancer Res.* 38, 4112-4125 (1978).

McGuire, W.L. Steroid hormone receptors and disease: breast cancer. *Proc. Soc. Exp. Biol. Med.* 162, 22-25 (1979).

McGuire, W.L., Carbone, P.P., and Wollmer, E.P. (eds.), *Estrogen Receptors in Human Breast Cancer*, Raven Press, New York (1975).

McGuire, W.L., Chamness, G.C., Costlow, M.E., and Shepherd, R.E. Hormone dependence in breast cancer. *Metabolism* 23, 75-100 (1974).

McGuire, W.L., Horwitz, K.B., Chamness, G.C., and Zava, D.T. A physiological role for estrogen and progesterone in breast cancer. *J. Steroid Biochem.* 7, 875-882 (1976).

McGuire, W.L., Horwitz, K.B., Zava, D.T., Garola, R.E., and Chamness, G.C. Hormones in breast cancer: update 1978. *Metabolism* 27, 487-501 (1978).

McLachlan, J.A. Transplacental effects of diethylstilbestrol in mice. *Nat. Cancer Inst. Monogr.* 51, 67-72 (1979).

Meites, J. Relation of prolactin and estrogen to mammary tumorigenesis in the rat. *J. Nat. Cancer Inst.* 48, 1217-1224 (1972).

Meites, J., Lu, K.H., Wuttke, W., Welsch, C.W., Nagasawa, H., and Quadri, S.K. Recent studies on functions and control of prolactin secretion in rats. *Rec. Prog. Hormone Res.* 28, 471-516 (1972).

Mekler, L.B. Chemical carcinogenesis. A new approach to the molecular and cellular mechanisms. *Oncology* 28, 63-82 (1973).

Metzler, M. Diethylstilbestrol: evidence for metabolic activation in man, rat, and hamster. *Nat. Cancer Inst. Monogr.* 51, 73-76 (1979).

Michalides, R., Van Deemter, L., Nusse, R., Ropcke, G., and Boot, L. Involvement of mouse mammary tumor virus in spontaneous and hormone-induced mammary tumors in low-mammary-tumor mouse strains. *J. Virol.* 27, 551-559 (1978).

Miller, A.B. An overview of hormone-associated cancers. *Cancer Res.* 38, 3985-3990 (1978).

Miller, E.C. Some current perspectives on chemical carcinogenesis in humans and experimental animals. *Cancer Res.* 38, 1479-1496 (1978).

Miller, E.C. and Miller, J.A. Biochemical mechanisms of chemical carcinogenesis, in *The Molecular Biology of Cancer*, H. Busch, ed. Academic Press, New York (1974), pp. 377-402.

Miller, R.W. Relationship between human teratogens and carcinogens. *J. Nat. Cancer Inst.* 58, 471-474 (1977).

Miller, W.R., Stewart, R., and Hawkins, R.A. Hormonal status and steroid metabolism in two transplantable rat mammary tumours. *Brit. J. Cancer* 39, 200-204 (1979).

Mishra, N.K., Wilson, C.M., Pant, K.J., and Thomas, F.O. Simultaneous determination of cellular mutagenesis and transformation by chemical carcinogens in Fischer rat embryo cells. *J. Toxicol. Environ. Health* 4, 79-81 (1978).

Montague, A.C.W., Stonesifer, G.L., Jr., and Lewison, E.F. *Breast*

Cancer, Alan R. Liss, New York (1977).

Nagasawa, H. and Yanai, R. Frequency of mammary cell division in relation to age: its significance in the induction of mammary tumors by carcinogen in rats. *J. Nat. Cancer Inst.* 52, 609-610 (1974).

Nagasawa, H., Sakai, S., and Banerjee, M.R. Prolactin receptors. *Life Sci.* 23, 193-208 (1979).

Nagle, S.C. and Fine, D.L. Demonstration of components of serum-free culture medium effecting maximum *in vitro* expression of mouse mammary tumor virus. *In Vitro* 14, 218-226 (1978).

Nandi, S. Comparison of the tumorigenic effects of chemical carcinogens and hormones. *J. Animal Sci.* 40, 1263-1266 (1975).

Nandi, S. Role of hormones in mammary neoplasia. *Cancer Res.* 38, 4046-4049 (1978a).

Nandi, S. Hormonal carcinogenesis: a novel hypothesis for the role of hormones. *J. Environ. Pathol. Toxicol.* 2, 13-20 (1978b).

Nandi, S. and Bern, H.A. Relation between mammary gland responses to lactogenic hormone combinations and tumor susceptibility in various strains of mice. *J. Nat. Cancer Inst.* 24, 907-931 (1960).

Nandi, S. and McGrath, C.M. Mammary neoplasia in mice. *Adv. Cancer Res.* 17, 353-414 (1973).

Newhouse, M.L., Pearson, R.M., Fullerton, J.M., Bossen, E.A.M., and Shannon, H.S. A case control study of carcinoma of the ovary. *Brit. J. Prevent. Soc. Med.* 31, 148-153 (1977).

Nissen, E.D., Nissen, S.E., and Kent, D.R. Liver neoplasia and oral contraceptives, in *Risks, Benefits, and Controversies in Fertility Control*, J.J. Sciarra, G.I. Zatuchni, and J.J. Speidel, eds. Harper and Row, Baltimore (1978), pp. 176-184.

Noble, R.L. Hormonal control of growth and progression in tumors of Nb rats and a theory of action. *Cancer Res.* 37, 82-94 (1977).

Noble, R.L., Hochachka, B.C., and King, D. Spontaneous and estrogen-produced tumors in Nb rats and their behavior after transplantation. *Cancer Res.* 35, 766-780 (1975).

Noble, R.L., McEuen, C.S., and Collip, J.B. Mammary tumours produced in rats by the action of oestrone tablets. *Canad. Med. Assoc. J.* 42, 413-417 (1940).

Nowell, P.C. The clonal evolution of tumor cell populations. *Science* 194, 23-28 (1976).

Oehlert, W. Cellular proliferation in carcinogenesis. *Cell Tissue Kinet.* 6, 325-335 (1973).

O'Malley, B.W., Schwartz, R.J., and Schrader, W.T. A review of regulation of gene expression by steroid hormone receptors. *J. Steroid Biochem.* 7, 1151-1159 (1976).

O'Malley, B.W., Towle, H.C., and Schwartz, R.J. Regulation of gene expression in eucaryotes. *Ann. Rev. Genetics* 11, 239-275 (1977).

Paffenbarger, R.L., Jr., Fasal, E., Simmons, M.E., and Kampert, J.B. Cancer risk as related to use of oral contraceptives during fertile years. *Cancer* (suppl.) 39, 1887-1891 (1977).

Palmiter, R.D., Mulvihill, E.R., McKnight, G.S., and Senear, A.W. Regulation of gene expression in the chick oviduct by steroid hormones. *Cold Spring Harbor Symp.* 42 (part 2), 639-647 (1977).

Pierce, G.B. Neoplasms, differentiations, and mutations. *Am. J. Pathol.* 77, 103-118 (1974).

Pitot, H.C. Neoplasia: a somatic mutation or a heritable change in cytoplasmic membranes? *J. Nat. Cancer Inst.* 53, 905-911 (1974).

Pitot, H.C., Shires, T.K., Moyer, G., and Garrett, C.T. Phenotypic variability as a manifestation of translational control, in *The Molecular Biology of Cancer*, H. Busch, ed. Academic Press, New York (1974),

pp. 523-534.
Posner, B.I. Polypeptide hormone receptors: characteristics and applications. *Canad. J. Physiol. Pharmacol.* 53, 689-703 (1975).
Posner, B.I., Josefberg, Z., and Bergeron, J.J.M. Polypeptide hormone receptors—recent studies on localisation, regulation, and internalisation in *Clinical Neuroendocrinology: A Pathophysiological Approach*, G. Tolis, F. Labrie, J.B. Martin, and F. Naftolin, eds. Raven Press, New York (1979).
Posner, B.I., Kelley, P.A., and Friesen, H.G. Prolactin receptors in rat liver: possible induction by prolactin. *Science* 188, 57-59 (1975).
Potter, V.R. Recent trends in cancer biochemistry: the importance of studies on fetal tissue. *Canad. Cancer Conf.* 8, 9-30 (1969).
Raju, U., Sklarew, R.J., Post, J., and Levitz, M. Steroid metabolism in human breast cancer cell lines. *Steroids* 32, 669-680 (1978).
Rechler, M.M. and Nissley, S.P. Somatomedins and related growth factors. *Nature* 270, 665-666 (1977).
Rees, L.H. The biosynthesis of hormones by nonendocrine tumors—a review. *J. Endocrinol.* 67, 143-175 (1975).
Reuber, M.D. Evolution of hyperplasia, hyperplastic nodules, and carcinomas of the liver induced in rats by N-2-fluorenyldiacetamide, in *Recent Topics in Chemical Carcinogenesis. Monograph in Cancer Res. no. 17*, University Park Press, Baltimore (1975), pp. 301-342.
Rice, J.M. An overview of transplacental chemical carcinogenesis. *Teratol.* 8, 113-135 (1973).
Rinehart, W. and Piotrow, P.T. Oral contraceptives—an update on usage, safety and side effects. Populations Reports (Oral Contraceptives) Ser. A., no., 5, Part 1, A-133 - A-186, The Johns Hopkins University (Population Information Program), Baltimore (1979).
Rivera, E.M. Strain differences in mouse mammary tissue sensitivity to prolactin and somatotropin in organ culture. *Nature* 209, 1151-1152 (1966).
Rivera, E.M., Hill, S.D., and Taylor, M. Differential growth patterns of carcinogen-induced rat mammary tumors in mammary fat pads. *Proc. Amer. Assoc. Cancer Res.* 21, 43 (1980).
Rivera, E.M., Hill, S.D., and Taylor, M. Organ culture passage enhances the oncogenicity of carcinogen-induced hyperplastic mammary nodules. *In Vitro* 17, 159-166 (1981).
Rivera, E.M., and Vijayaraghavan, S. Rat mammary tumors from carcinogen-induced nodules and their responsiveness to ovariectomy. *Europ. J. Cancer*, in press.
Rivera, E.M., Forsyth, I.A., and Folley, S.J. Lactogenic activity of mammalian growth hormones *in vitro*. *Proc. Soc. Exp. Biol. Med.* 124, 859-865 (1967).
Sanford, K.K. Malignant transformation of cells *in vitro*. *Intern. Rev. Cytol.* 18, 249-311 (1965).
Sasaki, G.H. and Leung, B.S. On the mechanism of hormone action in 7,12-dimethylbenz(α)anthracene-induced mammary tumor. *Cancer* 35, 645-651 (1975).
Sherman, B.M., Harris, C.E., Schlechte, J., Duello, T., Halmi, N.S., Van Gilder, J., Chapter, F.K., and Granner, D.K. Pathogenesis of prolactin-secreting pituitary adenomas. *Lancet ii,* 1019-1021 (1978).
Shields, R.J. Growth hormones and serum factors. *Nature* 267, 308-310 (1977).
Shields, R.J. Ectopic hormone production by tumors. *Nature* 272, 494 (1978).
Shyamala, G. Estradiol receptors in mouse mammary tumors: absence of the transfer of bound estradiol from the cytoplasm to the nucleus.

Biochem. Biophys. Res. Commun. 46, 1623-1630 (1972).

Silverberg, S.G. and Major, F.J. (eds.), *Estrogens and Cancer,* John Wiley, New York (1978).

Silverberg, S.G., Makowski, E.L., and Roche, W.D. Endometrial carcinoma in women under 40 years of age. *Cancer* 39, 592-598 (1977).

Sinha, D., Cooper, D., and Dao, T.L. The nature of estrogen and prolactin effect on mammary tumorigenesis. *Cancer Res.* 33, 411-414 (1973).

Sirbasku, D.A. Estrogen induction of growth factors specific for hormone-responsive mammary, pituitary, and kidney tumor cells. *Proc. Nat. Acad. Sci.* 75, 3786-3790 (1978).

Smith, R.D., Hilf, R., and Senior, A.E. Prolactin binding to 7,12-dimethylbenz(α)anthracene-induced mammary tumors and liver in diabetic rats. *Cancer Res.* 37, 4070-4074 (1977).

Speroff, L. Which birth control pill should be prescribed? *Fertil. Steril.* 27, 997-1008 (1976).

Sterling, K. Thyroid hormone action on the cell level. *New Engl. J. Med.* 300, 117-123 (1979).

Stern, E., Forsythe, A.B., Voukeles, L., and Coffelt, C.F. Steroid contraceptive use and cervical dysplasia: increased risk of progression. *Science* 196, 1460-1462 (1977).

Sutherland, E.W. Studies on the mechanism of hormone action. *Science* 177, 401-408 (1972).

Sutherland, E.W. and Robison, G.A. The role of cyclic 3',5'-AMP in response to catechol amines and other hormones. *Pharmacol. Rev.* 18, 145-161 (1966).

Symposium: *Early Lesions and the Development of Epithelial Cancer. Cancer Res.* 36, no. 7, part 2 (1976).

Szego, C.M. The lysosome as a mediator of hormone action. *Rec. Prog. Horm. Res.* 30, 171-233 (1974).

Tata, J.R. Control by oestrogen of reversible gene expression: The vitellogenin model. *J. Steroid Biochem.* 11 (1 Part B), 361-371 (1979).

Temin, H.M. On the origin of the genes for neoplasia. *Cancer Res.* 34, 2835-2841 (1974).

Temin, H.M. The DNA hypothesis. *Science* 192, 1075-1080 (1976).

Thomas, D.B. Role of exogenous female hormones in altering the risk of benign and malignant neoplasms in humans. *Cancer Res.* 38, 3991-4000 (1978).

Todaro, G.J. RNA-tumor-virus genes and transforming genes: patterns of transmission. *Brit. J. Cancer* 37, 139-158 (1978).

Todaro, G.J. and Heubner, R.J. The viral oncogene hypothesis: new evidence. *Proc. Nat. Acad. Sci.* 69, 1009-1015 (1972).

Ulfelder, H. DES–transplacental teratogen and possibly also carcinogen. *Teratology* 13, 101-104 (1976).

Van Wyk, J.J., Underwood, L.E., Hintz, R.L., Clemmons, D.R., Voina, S.J., and Weaver, R.P. The somatomedins: a family of insulin-like hormones under growth hormone control. *Rec. Prog. Horm. Res.* 30, 259-318 (1974).

Varmus, H.E., Ringold, G., and Yamamoto, K.R. Regulation of mouse mammary tumor virus gene expression by glucocorticoid hormones, in *Glucocorticoid Hormone Action,* Baxter, J.D. and G.G. Rousseau, eds. Springer-Verlag, Berlin (1979), pp. 253-278.

Vessey, M.P., Doll, R., and Jones, K. Oral contraceptives and breast cancer. *Lancet* i, 941-943 (1975).

Vessey, M.P., Kay, C.R., Baldwin, J.A., Clarke, J.A., and MacLeod, I.B. Oral contraceptives and benign liver tumors. *Brit. Med. J. i*

1064-1065 (1977).
Weinstein, I.B. Current concepts on mechanisms of chemical carcinogenesis. *Bull. N.Y. Acad. Sci.* 54, 366-383 (1978).
Weinstein, I.B. and Wigler, M. Cell culture studies provide new information on tumour promoters. *Nature* 270, 659-660 (1977).
Weinstein, I.B., Yamasaki, H., Wigler, M., Lee, L.-S., Fisher, P.B., Jeffrey, A., and Grunberger, D. Molecular and cellular events associated with the action of initiating carcinogens and tumor promoters, in *Carcinogens: Identification and Mechanism of Action*, A.C. Griffin and C.R. Shaw, eds. Raven Press, New York (1979), pp. 399-418.
Welsch, C.W., and Meites, J. Prolactin and mammary cancerigenesis, in *Endocrine Control in Neoplasia*, R.K. Sharma and W.E. Criss, eds. Raven Press, New York (1978), pp. 71-92.
Welsch, C.W., and Nagasawa, H. Prolactin and murine mammary tumorigenesis: a review. *Cancer Res.* 37, 951-963 (1977).
Welsch, C.W. and Rivera, E.M. Differential effects of estrogen and prolactin on DNA synthesis in organ cultures of DMBA-induced rat mammary carcinoma. *Proc. Soc. Exp. Biol. Med.* 139, 623-626 (1972).
WHO Scientific Group. Steroid contraception and the risk of neoplasia. *Technical Report Series* 619, WHO, Geneva (1978).
Wittliff, J.L., Lewko, W.M., Park, D.C., Kute, T.E., Baker, D.T., and Kane, L.N. Steroid binding proteins of mammary tissues and their clinical significance in breast cancer, in *Hormones, Receptors, and Breast Cancer, Progress in Cancer Res. and Therapy*, vol. 10, W.L. McGuire, ed. Raven Press, New York (1978), pp. 325-359.
Yang, J., Enami, J., and Nandi, S. Regulation of mammary tumor virus production by prolactin in BALB/cfC3H mouse normal mammary epithelial cells *in vitro*. *Cancer Res.* 37, 3644-3647 (1977).
Yates, J. and King, R.J.B. Multiple sensitivities of mammary tumor cells in culture. *Cancer Res.* 38, 4135-4137 (1978).
Yoshimoto, Y., Wolfsen, A.R., and O'Dell, W.D. Human chorionic gonadotropin-like substance in nonendocrine tissues of normal subjects. *Science* 197, 575-577 (1977).
Zava, D.T., Chamness, G.C., Horwitz, K.B., and McGuire, W.L. Human breast cancer: biologically active estrogen receptor in the absence of estrogen? *Science* 196, 663-664 (1977).

Index

Acetylcholine, 152
Acromegaly, 50
Acrophase charts, 460, 461-463
Acrosomal reaction, 19
ACTH (adrenocorticotropic hormone), 30-31, 39, 53, 405
  hypothalamic control of the release of, 373, 376
  secretion, 380, 381, 382-385 386, 392-393, 489-491, 620-621
Actinomycin-D, 288
Addison's disease, 639-640
Adenohypophyseal hormones, 379
  See also specific hormones
Adenohypophysis, 372
Adenosine monophosphate (AMP), 48
Adenosine triphosphate (ATP), 412
Adipose cell number, changes in, 132
Adolescence, 349-367
  distinction between puberty and, 332
Adrenal cortex, 319
  in depressive illness, 620-621
  endocrinopathies of, 638-639
  in old age, 481-482
  response to ACTH and stress, 490-491
  structural and functional changes, 482, 483-487, 488-490
Adrenal cortical cells from male rat, 504
Adrenal cortical hormones, 187
  morphological effects on nervous system, 187-191
  See also specific hormones
Adrenal gland, 28
  in old age, 480-481

Adrenal hyperplasia, 340
Adrenal medulla, 493
Adrenal steroids, 418-419
Adrenalectomy, and hypothalamus, 188-189
Adrenocortical hormones, 208, 482
  in CNS, 154-156
  and pregnancy, 66-67
  See also specific hormones
Adrenocortical steroids
  in brain excitability, 415-416
  and stress, 489-490
Adrenocorticoids, 52
Adrenocorticotropic hormone, see ACTH
Age changes in circadian hormonal rhythms, 465, 467, 468
Age-dependent differences in circadian temperature rhythm, 454
Age effects upon circannual rhythms, 469
Age pigments, 554-559
  in nervous tissues, 555, 556

Age-related differences in circadian rhythm, 464
Age-related hypothyroidism in rats, 565
Aging, 551-552
  neuroendocrine therories of, 566-575
Alanine, 232
Alcohol, 360-361, 628
Aldosterone, 26, 28, 67, 404-405
  age-related changes in, 492-493
Alloxan diabetes, 195-196
Alpha fetoprotein (AFP), 39, 294, 317
Alterations in skull and face from

newborn to adulthood, 135
Aluminum, as factor in mental deterioration, 553
Alzheimer's disease, 553-554
Amine precursor uptake and decarboxylation (APUD), 379-380
Amino acids, in brain maturation, 231-233
Amino acid incorporation into protein, 235
Amitriptyline, 367, 623
Amnesia, in aging, 515
Ampicillin, 109-110
Androgen localization in brain, 160-162
Androgens, 16, 26, 39, 136, 151, 387
  in CNS tissue, 398-400
  role in sex differentiation process, 297
Androst-4-ene-3,6,17-trione (ADT), 293
Androstenedione, 39, 292-293, 294, 588
Anorexia nervosa, 365, 629-630
Anterior pituitary cells of two male rats, 512
Antidiuretic hormone (ADH), 57, 372
Antiestrogens, 288
Antisocial behavior, in adolescence, 358-361
Anxiety neurosis, 362
Arginine, 48, 387
Arginine vasopressin, 57, 379
Arginine vasotocin, 379
Aromatization hypothesis, 295, 297
Asceticism, 351
Aspartic acid, 231, 233
Atresia, 4, 37-38
Atropine, 394
Autoradiographs illustrating sex hormone uptake, 167, 168
Autorhythmometry, 454
Autosomes, 36

Barr bodies, 36
Batten's disease, 557
Behavior modification, 367
Bentolamine, 394
β-endorphin, 31
β-lipotropin, 31
Bethamethasone, 188
Biologic pacemakers, 479
  as regulators of aging, 566-568
Biologic rhythms, 451
Biological identity of I-labeled compounds in brain tissue of guinea pig, 154
Biosynthesis of the neurohypophyseal hormones, 373
Blood-testis barrier, 10
Body composition, 130-134
Body weight, 128, 130
Bone diseases, and growth retardation, 144
Brain activity, effects of hormones on, 415-421
Brain degenerative diseases, 552-554
Brain maturation
  behavioral effects of hormones on, 252-255
  critical periods, of 207
  general metabolic processes, 231-243
  hormonal influence on biochemical aspects of, 219-221
  hormonal influence on physiological aspects of, 243-252
  ionic composition of, 241-243
  neurotransmitter metabolism, 221-231
Brain maturation, effect of estradiol on, 246
Brain metabolic functions, effects of hormones on, 207-255, 404-415
Breast cancer, 454, 470
  hormone receptors in, 660-662
  and hormones, 653, 654, 657
  and inherited hormonal influences, 648
Bromocriptine, 570
Bulbourethral (Cowper's) gland, 16

Calcitonin, 52-53, 58
Calcium, 52-53, 136, 372
  skeletal, 133, 590
Cancer, 647-651
  hormonal therapy for, 659-660
Capacitation, process of, 14
Capillary basement membrane thickness (CBMT), 503
Carcinogenesis, and hormones, 645-662
Carcinogens, 645, 648
  mechanisms of action, 649-650
Castration, 589
  effect of, on protein metabolism in CNS, 236
  and hypothalamus, 189
Catecholamine-estrogen target neuron interrelationships in the rat hypothalamus, 172
Catecholamines, 33, 35, 190, 393,

autoxidation of, 561-562
  synthesis, turnover, and enzyme levels, 409-412
  uptake of, 407-409
Cell body of arcuate neuron from female rat, 186
Cell division, 2
Cell proliferation, and neoplasia, 655-656
Cell structure, 1-2
Central nervous system (CNS), 151
  changes in, due to aging, 552-565
  role in endocrine function, 391
  seizure activity in, 243-247
  uptake, distribution, and metabolism of hormones in, 396-404
Cerebellar cortex
  effect of adrenal cortical hormones on, 187-188
  effect of thyroid hormones on, 183, 184, 185
  synaptic changes with hypothyroidism, 211-212
Cerebral cortex
  effect of adrenal cortical hormones on, 187-188
  effect of thyroid hormones on, 182-183
Cervical cancer, and oral contraceptives, 657
Chlordiazepoxide, 367
Chlorpromazine, 367, 396
Cholesterol, 41, 87, 293
Choline, 231, 570
Choline acetyltransferase, 570
Chromaffin cells, 33, 34
Chromatin, 2
  binding, in sex differentiation, 284, 285, 289, 290, 291-292
Chromatin binding of estradiol, 290, 291
Chromosomes, 2-3
  Y and X, 36, 43
Chronobiology, 453
  senescence and senility in, 453-457
Chronobiology in diagnosis and therapy, 470
Chronodesm, 451
Chronodesms for interpretation of single sample values, 452
Circadian change of plasma cortisol, 452
Circadian desynchronization, 458-460
Circadian hormonal rhythms, age changes in, 465-469

Circadian period, 451, 455
Circadian prolactin concentration chronodesm, 466
Circadian rhythms, 455-456, 457
  controversy over, 464-465
Circannual rhythms, age effects upon, 469-471
Circaseptan desynchronization, 459
Citrate, 16
Climacteric, see Menopause
Cloaca, 43
Clomiphene, 7
Cloning, 21
Cognitive development, in adolescence, 350-351
"Common peptides," 380
Contribution of organs and tissues to body weight at various ages, 130
Controls in the piuitary-adrenocortical system, 381
Conversion hysteria, 362-363
Corpus luteum, 6, 13
Corticosteroids, 187, 482
  in neural cell formation, 210
  permissive actions of, 492
  two major types of action, 492
  See also specific steroids
Corticosteroids on DNA concentration of cerebellar explants, effects of, 215
Corticosterone, 39, 65, 151, 154
Corticosterone in brain after intravenous injection, 158
Corticotropin, 30
Corticotropin-releasing factor (CRF), 373, 376, 489
Corticotropin releasing hormone (CRH), 30, 621
  development and role of, 31
Cortisol, 26, 154, 188, 405
  excess secretion in depression, 621
Cranio-facial growth, 135
Cretinism, 181, 207
Cumulus oophorus, 4
Cushing's syndrome, 142, 482, 621, 623, 638-639
Cyclic adenosine-3'-5' monophosphate (cAMP), 15
Cyclic AMP, interactions with hormones, 412-415
Cyclic 3',5'-AMP, 412-415
Cycloheximide, 234
Cystathionine, 232
Cytochalasin B, 217

Defeminization, 275-276, 292-295
Dehydrocorticosterone, 405

Dehydroepiandrosterone, 100, 588
Delinquency, juvenile, 358-359
Demasculinization, 275-276
Dental maturation, 134
Deoxyribonucleic acid (DNA) 2, 20, 182, 188, 237, 238, 239
Depression, pathophysiological mechanisms in, 620-622
   and endocrinopathies, 641
   reactive vs. endogenous, 625
   TRH test in, 627
   two syndromes in, 633
Depressive neurosis, 361-362
Desipramine, 623
Desoxycorticosterone (DOC), 404, 415
Development in utero, disorders in, 628-629
Dexamethasone, 106, 155, 190, 621
   suppression test, 623
Diabetes mellitus, 195-196, 572
   and accelerated aging, 502-503
   and depression, 641
   and "ob/ob" hyperglycemic mouse, 501-502
   pathological vs. physiological considerations in late onset of, 503, 504
   as threat to fetus and newborn, 49-50, 99
Diazepam, 367
Dibutyryl cyclic AMP (DBcAMP), 215
Diethylstilbestrol (DES), 645, 656
Dihydrotestosterone (DHT), 44, 293, 294
Diiodothyronine (DIT), 208
Diiodotyrosine, 152
Diploid complement of chromosomes, 2
Disease, as model for study of aging, 574-575
Diseases of adaptation, 574
Distance growth chart, 137, 138
Dizygotic (fraternal) twins, 7
DNA, see Deoxyribonucleic acid (DNA)
Dopamine, 54, 98, 158, 366, 392, 395, 493, 562
   in Parkinson's disease, 560-561
   role in controlling release of gonadotrophins, 393-394
   theory of schizophrenia, 622
Down's syndrome, 9
Dry-mount autoradiographic procedure, 152, 154, 156

Ectopic hormones, and tumors, 658
Effect of cortisol on electroshock threshold, 245
Ejaculatory ductus, 16
Electroshock seizure threshold (EST), 415
Emotional development, in adolescence, 351-352
Endocrine function, 26-27
   age-related reduction in, 527-528
Endocrine glands, in senescence, 478-480
   See also specific glands
Endocrine interrelations during labor, model of possible, 111
Endocrine pathologies, in puberty, 346-347
Endocrinopathies, 637-641
Endometrial cancer, and estrogen, 591, 657
Endometrium, 15
Endorphins, 379, 563, 623
Enkephalin, 622-623
Environmental factors, in fetal growth, 68-69
Environmental manipulation, in adolescence, 367
Enzymes
   in brain maturation, 227-231
   in general metabolic processes, 239-241
Epididymis, 16
Epinephrine, 33, 34, 35, 67
Estetrol, 89
Estradiol, 38, 87, 236, 279, 386
   binding of, 283-284, 290, 291, 292
   as growth stimulant, 137
   and male brain pattern, 192
   neural binding of, 284-292
Estradiol dipropionate, effect of, on seizure thresholds, 418, 419
Estriol, 87, 89
   in pregnancy, 94-95
   and premenstrual tension, 631
Estrogen, 5, 13, 38, 39, 45, 61, 157, 343, 386-387, 590, 656
   as active agent in sexual differentiation, 292-294
   in aging, 505
   catechol, 312
   in CNS tissue, 397-398
   as growth stimulant, 137
   and parturition, 109
   placental secretion of, 87-88
   in pregnancy, 66, 94-95, 97, 102
   spurious, 312

Estrogen implants in female rat brain
  brain-effects on reproductive
    function, 388-390
Estrogen localization in brain,
  163-165
Estrogen therapy, 341-342
  in postmenopausal osteoporosis,
    589-590
  in postmenopausal women, 590-592
Estrone, 87, 588
Evoked cortical activity, in brain
  maturation, 248-251
Excessive height, as problem at
  puberty, 339-340
  in the female, 341-342
  in the male, 340-341
Excitability, brain, 415-418
External genitalia, 42-43

Fallopian tubes, 42
Feedback interrelations among
  maternal, placental and fetal
  hormones, 108
Female sex organs, 14
Feminization, 45, 275-276
Fertility, 40, 345
Fertility and antifertility agents,
  21-22
Fertilization, 1, 17-18
  errors of, 18
  process of, 19-21
  role of the ovum in, 7
  role of sperm in, 12
Fertilization cone of sea urchin, 20
Fetal adrenal cortex, 28-33, 99-101
Fetal adrenal medulla, 28, 33
  embryogenesis of, 33
  functional development of, 33-35
Fetal anencephaly, 102, 103
Fetal apituitarism, 102
Fetal hormones, 25-27
  role of, 67-69
Fetal hypoglycemia, 47-48, 50
Fetal sex steroids, role of, 45
Fetectomy, 98, 99, 103
Fetoplacental unit, in steroid
  synthesis, 88-89
Fetus, changes in weight from 25
  weeks to birth, 129
Fluoride, 553
Follicle, 4, 5, 37
Follicle-stimulating hormone (FSH),
  4, 6, 10, 38, 41, 53, 377, 386
Follicle-stimulating hormone
  releasing factor (FRF), 377
Follicular antrum, 4
Freemartin, 7
Fructose, 16

γ-aminobutyric acid (GABA), 231-232
Gametes, 1
Gametogenesis, 3, 305
Gastrin, 46, 65
Gastrointestinal hormones, 65
General adaptation syndrome,
  573-574
Genes, 2
Genetic code, 2
  as regulator of aging, 567
Genetic factors
  in aging, 571-572
  in etiology of endocrine diseases
    and cancer, 647-648
  in fetal growth, 68
Genetic short stature, growth
  curve for male with, 145
Genetic short stature and
  constitutional delay in growth,
  144-145
Germ cells, 1, 2
Gero-endocrinology, 477-480
Gigantism, 135, 340
Glial cell, 217, 415
Glucagon, 46, 48, 65
  and age-related alterations in
    tissue binding, 500-501
  secretion in aging, 502
Glucocorticoids, 47, 188, 190-191
  behavioral effects of, 254-255
  and brain excitability, 245-247
  and cerebral edema, 405
  in CNS tissue, 397
  decline in utilization of, with
    aging, 528
  in evoked cortical activity, 251
  excess as factor in growth
    retardation, 142-143
  in nucleic acid metabolism, 238-239
Glucose, 47, 67, 387
  increased intolerance with age,
    494, 500
  receptors, 48
  role of, 47-48
Glumitocin, 379
Glutamine, 232
Glutamic acid, 231, 233
Glycerol phosphate dehydrogenase
  (GPDH), 217
Glycine, 232
Goiter, 386
Gonadal dysgenesis, 144, 331, 338-339, 347
Gonadal hormones, 420
  behavioral effects of, 255
  in sexual and somatic differentiation
    43-46

INDEX

See also Sex hormones
Gonadal sex hormones, role of, 45
  in puberty onset, 315-316
Gonadostat, 319
Gonadostat theory, of puberty
  onset, 316-317
Gonadotrophin profiles in immature
  female rat, 310
Gonadotropin releasing hormone
  (hCGnRH), 87, 377, 630
Gonadotropin secretion, 386-387
  role of biogenic amines in,
    393-395
Gonadotropins, 38, 54, 66, 319
  secretion, 386-387
Gonads, 1, 35-37
Graafian follicle, 4
Granulosa cells, 8, 305
Grave's disease, 638
Greulich and Pyle atlas standards,
  133
Growth, 125, 127
  excessive, 339-342
  hormonal effect on, 135-147
  and short stature at puberty,
    335-339
Growth curve during fetal life,
  126
Growth curves of three children
  with growth hormone deficiency,
  141
Growth hormone (GH), 31, 53
  55-56, 65, 187
  deficiency of, in growth retarda-
    tion, 140-142, 629
  in depression, 621-622
  morphological effects of nervous
    system of, 194
  secretion, 387, 391, 396
  as stimulant of growth, 136
Growth hormone-inhibiting factor
  (GIF), 378
  See also Somatostatin
Growth hormone-releasing factor
  (GRF), 378
Growth hormone-releasing hormone
  (GRH), 55
Growth retardation
  assessment of, 137-138
  causes of, 146-147
  hormonally induced, 139-143
  laboratory evaluation for, 141
  nonhormonal causes of, 144-146
Growth retardation, causes of,
  146
Guanethedine, 393
Gynecomastia, 343-344

^{3}H-norepinephrine (^{3}H-NE), 221-224
Haloperidol, 367
Haploid complement, of chromosomes,
  2
Height prediction methods, 134
Heroin, 360
High altitude, as affecting
  pregnancy, 109
Histidine, 232
Homeostasis, 551
Homosexuality, 355, 366, 625
  fears of, 343, 344
Hormonal effect on growth, 135-147
Hormonal imbalance, 646
Hormonal-induced differentiation of
  neural systems, 286-287
Hormonal inputs and feedback sites,
  380-391
Hormonal neurosecretion mechanisms,
  371-396
  in aging, 569
Hormonal response, 646
Hormonal secretion, regulation of,
  391-396
Hormonal uptake, distribution and
  metabolism in CNS tissue, 396-404
Hormonally induced growth retardation, 139-143
Hormone action, mechanisms of,
  646-647
Hormone receptors, 27, 281-282
  and breast cancer, 660-662
  interaction of, 646-647
Hormones, 207-208
  behavioral effects of, 253-255, 422
  and brain metabolic functions,
    404-415
  and carcinogenesis, 645-646
  effects on developing and aging
    nervous systems, 181
  influence on neural cell growth and
    differentiation, 209-219
  role in cancer, 652-662
  in treatment of psychiatric disease,
    624
  uptake by CNS, 151
  See also specific hormones
Hormones, and brain metabolic
  functions, 207-255
  cyclic AMP interactions, 412-415
  neurotransmitter metabolism,
    407-412
  protein metabolism, 406-407
  water and electrolytes, 404-406
Human chorionic gonadotropin (hCG),
  7, 30, 41, 107
  placental secretion of, 82-83, 95-96

Human chorionic somatomammotropin (hCS), 86
  in pregnancy, 96, 97
Human chorionic thyrotropin (hCT), 86-87
Human chorionic thyrotropin releasing hormone (hCTRH), 87
Human embryo, reimplanatation of, 21
Human ovum and sperm, 13
Human sexual behavior, problems of reporting on, 353
Hydrocortisone, 188, 190
  and cell spreading, 217
Hyperadrenalism, 366
Hypercalcemia, 641
Hyperglycemia, 49, 387
Hyperparathyroidism, 640-641
Hyperplasia, 654-655
Hyperthyroidism, 63, 97
Hyperventilation, in anxiety neurosis, 362
Hypocalcemia, 52, 421
Hypoglycemia, 641
Hypogonadism, 339, 347, 572
Hypoparathyroidism, 421, 640
Hypophysectomy, 38, 507
Hypophysis, see Pituitary gland
Hypopituitarism, 570
Hypothalamic "disregulation" hypotheses, in aging, 568
Hypothalamic hormone localization, 170-171
Hypothalamic-hypopituitarism, classification of, 143
Hypothalamic peptides, 402-404
Hypothalamic-pituitary stimulating and inhibiting hormones, 371-378
Hypothalamic-portal-pituitary complex, development of, 54-55
  and masterplan for development and aging, 57-58
Hypothalamus, 54, 377
  effect of adrenal cortical hormones on, 188-190, 191
  effects of thyroid hormones on, 185
  role in release of gonadotropins, 386
  and secretion of ACTH, 392
Hypothesis for control of anterior lobe ACTH, 381
Hypothyroidism, 207-208, 346, 515
  and aging, 526, 571
  and myelin formation, 220
  neonatal, 64
  undernutrition and, 186-187

Hysterical neurosis, 362-363
Imipramine, 367
Immunocytochemical illustration of distribution of LH-RH in brain, 168
Incest, 353
Incorporation of amino acids into protein, effects of cycloheximide on, 236
Incremental height chart, 138, 139
Incremental weight gain for males from birth to 18 years, 129
Infertility, 17, 339
Inhibin, 42
Inhibitory theories, of puberty onset, 305
  neural input, 313-315
  steroid feedback, 315-316
Injection of labeled hormone in male rat, 155
Insulin, 13, 46, 47, 67
  age-related changes in, 495
  as factor in growth retardation, 143
  mechanisms of action, 48-49
  morphological effects on nervous system, 195-196
  role in fetal growth, 50
  secretion, 47-48
Insulin and glucagon metabolism, aging-related changes in, 496-499
Insulin resistance, 495, 500
Insulinoma, 641
Intellectualization, 351
Insterstitial cell stimulating hormone (ICSH), 10
  See also Luteinizing hormone (LH)
Interstitial cells of Leydig, 10, 39, 40
Interventive gerontology, 576-579
Intrauterine growth retardation, 144
Involutional melancholia, 630
Iodide, 59, 64, 152, 208
Iodine, 64
Iodotyrosines, 152
Iproniazid, 570
Irradiation, 134
Islets of Langerhans, 46
Isotocin, 379

Klinefelter's syndrome, 43, 45, 340, 343, 572

L-Dopa (levodopa), 387, 395, 396, 560-561, 569
Laron dwarfism, 143

Lean body mass/height ratio, 131
Lecithin, 570
Lergotrile, 570
Leucine, 48, 233, 234
Limb growth, 127
Lipofuscin, 554-559
Lipoproteins, 379
Lithium carbonate, 367
Liver, and endocrine system, 27
Liver tumors, and sex hormones, 657
Long-loop corticoid feedback sites, 382-385
Lordosis, 275, 289, 293
Lordosis behavior in adult gonadectomized rats, 289
Lordosis behavior in adult ovariectomized hamsters, 295, 296
Luteinizing hormone (LH), 5, 6, 38, 53, 377, 386
  surge, 6, 318, 319-320, 562
  See also Interstitial cell stimulating hormone (ICSH)
Luteinizing hormone-releasing factor (LRF), 377
Luteinizing hormone releasing hormone (LHRH), 402-403
Lysine, 232
Lysine vasopressin, 379

McCune-Albright syndrome, 346
Magnesium, 52, 412
Male climacteric, 602
  symptoms in, 614-615
Male sex organs, 16
Malnutrition, 144, 186
Mammalian ova, aging of, 8-9
Mammary cancer, and estrogen, 591
Manic-depressive psychosis, 365
Marfan's syndrome, 340
Masculinization, 275-276, 295-297
Masturbation, 354
Maternal endocrine dysfunction, 99
Maternal hormones, 66-67
Maternal malnutrition, in pregnancy, 109
Mature ovum, 7-8
Maturity scoring system for epiphysis of radius, 134
Maximow double coverslip assembly, diagram of, 214
Mean height gain from peripubertal to pubertal period, 127
Meclofenoxate (centrophenoxine), 558
Medroxyprogesterone acetate, 625
Meiosis, 2
  in reproduction, 3, 36

Melanin, 556
Melanocyte-stimulating hormone (MSH) 56-57, 421
Melanocyte stimulating hormone-releasing and -inhibiting factors, 378
Melatonin, effects on sleep mechanisms, 421
Menstrual cycle, 45
Menarche, 350, 609
  lack of, 44, 338
Menopause, 4, 587, 609
  and alterations of neural controls, 589
  atrophy of genital tissues after, 588
  and depression, 630-631
  hormonal changes leading to, 587-588
  hormonal milieu after, 588
  in scientific literature, 610-612
  and sex hormone binding globulin and receptors, 588
  systemic changes, 589
Menopause, psychological aspects of, 609
  and lifespan developmental process, 612-613
  and midlife issues, 613-614
  in scientific literature, 610-612
  and theories of personality, 612
  a time for growth and self-discovery, 615-616
Mental retardation, 43, 45, 346
MER-25 (antiestrogen), 293
Mesotocon, 379
Methionine, 231
Methylphenidate, 627
Metyrapone, 490
Midparental height chart, 138, 140
Migration rate of cells in cultured cerebellar explants, effects of hormones on, 216
Mineralocorticoids, 30
Missing link hypothesis, of puberty onset, 313
Mitosis, 2-3
Modeling, 367
Monoamine-estrogen target relationship in the central nervous system, 173
Monoaminergic-corticosteroid relationship in the central nervous system, 174
Monoiodothyronine (MIT), 208
Monoiodotyrosine, 152
Monozygotic (identical) twins, 7

Mosaics, 45
Mullerian ducts, 42
Mullerian regression factor, 42
Multiple personality syndrome, 363
Muscle cell population, changes in, 131
Myelination, 219-221
Myometrium, 15
Myxedema, 638

Nembutal, 8
Neomycin, 110
Neonatal hypothyroidism, effects of, on cerebral cortex, 242
Neonatal thyroid hyperactivity, 63
Neoplasia, 645
  and hormonal derangements, 653
Neural culture systems, 213
Neuraminidase, 83
Neuroblastoma N_2A culture, 219
"Neuroendocrine pacemakers," 371, 479, 551
Neuroendocrine parameters, in psychiatric research, 619-620
  challenges and new approaches, 632-633
  as diagnostic aid, 623-624
  in disorders of development, 628-632
  as indicators of pathophysiological mechanisms, 620-623
  in monitoring psychotropic drug effects, 625-627
  in treatment of psychiatric disease, 624-625
  and side effects of psychotropic drugs, 627-628
Neuroendocrine system
  in aging, 479-480
  homeostasis and stress, 481
  and maturational vs. aging changes, 480
  peripheral division of, 379-380
Neurofibromatosis, 346
Neurogenesis, in man, 209
Neurohumors, regulation of hormonal secretion by, 391-396
Neurohypophysis, 377, 513, 514, 515
Neurophysins, 57, 372
Neurosis, in adolescence, 361-363
Neutotransmitter hypotheses, in aging, 569
Neurotransmitters, 407
  aging of, 559-561
  metabolism, 221-231
Nitrogen, 136

Nocturnal emissions, 354
Nocturnal sex dreams, 354
Norepinephrine, 33, 54, 67, 158, 392, 561, 621

Obesity, 387
  in children, 132
  and effects of aging on glucose disposal, 500
  in postmenopausal women, 588-589
Oligopeptides, secreted by placenta, 82
Oocytes, 37
  aging of, 8-9
  maturation division of, 6
  primary, 3
Oogenesis, 3-6
Oogonia, 3, 37
Opioid peptides, in schizophrenia, 623
Opiod theory, of schizophrenia, 622
Organ culture, diagram of, 214
Ornithine, 232
Ossification, 133
  maturity scoring system, 134
Osteoporosis, 505
  in postmenopausal women, 589-590
Ouabain, 221
Ovariectomy, 99
Ovaries, reproductive role of, 13
Ovary, 37-39
Ovary and testis, 4, 5, 11
Oviduct, 13, 14
Ovulation, 6
Ovulatory delay, 8-9
Ovum, role of, 7
Oxytocin, 15, 57, 66, 372, 379, 515

Pancreas, 494
  changes in structure and function in aging, 495
  embryogenesis of, 46-47
  functional development of, 47-49
Pancreatic islets, endocrinopathies of, 641
Parachlorphenylalanine (PCPA), 570
Parathyroid, 51
  in aging humans, 504-505
  embryogenesis of, 51
  functional development of, 51
Parathyroid disease, 640-641
Parathyroid hormone, 52, 505
Parkinson's disease, 396, 560-561, 622

Parthenogenesis, 18
Peptides, within brain, 158, 402-404
  and immunocytochemical techniques, 157-159
Percent of mature height attained from one to eighteen years, 126
Peripheral nervous system
  effect of adrenal cortical hormones on, 190-191
  effect of thyroid hormones on, 185-186
Permissiveness, in child rearing, 356
Petting, 354
Phencyclidine (PCP) intoxication, 357
Phenoxybenzamine, 393, 394
Phentolamine, 621
Phenylethanolamine-N-methyltransferase (PNMT), 33, 34, 35
Phobias, 363
Phosphorus, 52, 136
Physical development, in adolescence, 350
Pineal gland
  and calcification with aging, 506-507
  in sexual maturation and gonadal function, 394
Pineal gland cells from male rat, 506
Pituitary-adrenocortical axis, aging-related changes in structure and function of, 483-487
Pituitary gland, 53, 66
  age changes in posterior lobe, 513-515
  anterior lobe, 53-57
  posterior lobe, 57-58
  in pregnancy, 66, 96-97
  role in aging, 507
Pituitary hormones, 379, 571
  See also specific hormones
Pituitary hypotheses, in aging, 570-571
Pituitary-thyroid axis, aging-related changes in the structure and function of, 517-522
Placenta, as endocrine gland, 81-82
  in adverse environments, 108-109
  developmental stages in, 89-95
  fetal endocrine influences on, 103-104
  interaction and integration with fetomaternal endocrine continuum, 95-106
  intraplacental interrelations, 106-107
  maternal endocrine influences on, 98-99
Placental hormones, 82-88
  effects on fetal endocrine development, 99-102
  effects on the mother, 95-98
  as indices, 104-106
  and parturition, 110-112
  secretory rhythms of, 94-95
Placental hormones, developmental pattern of, 90
Placental hormones, normal values of at various stages of gestation, 91-93
Placental peptide hormones, physiochemical and biological properties of, 84-85
Placentotropin, 95, 97, 98
Polypeptides, secreted by placenta, 82
Positive and negative correlations between LH-RH and enzymes, 169
Postmenopausal syndrome, 588-589
Postpartum depression, 631
Potassium, 16
Precocious puberty, 345, 346
Pregnenolone, 87, 89
Premarital sexual intercourse, 355-356
Premature adrenarche, 345
Premature thelarche, 345
Premenstrual tension, 631
Primary follicles, 3
Primary hypothyroidism, in growth retardation, 139
Progeria, 572
Progesterone, 6, 13, 45, 87, 157, 386-387
  and parturition, 109
  in pregnancy, 66, 94, 97
  in spontaneous electrical activity, 420
Progesterone localization in brain, 166
Proinsulin, 502
Prolactin (PRL), 31, 41, 50, 53, 56, 454-455
  changes in aging, 508-511, 512-513
  in monitoring effect of antipsychotic drugs, 625-626
  in pregnancy, 97
  and schizophrenia, 622
  secretion, 391, 395-396
Prolactin-inhibitory hormone (PIH), 562

Prolactin release-inhibiting factor (PIF), 378, 391
Prolactin-releasing factor (PRF), 391
Proline, 234
Propylthiouracil (PTU), 211
Prostaglandins, 15
Prostate gland, 16, 43
Prostatic cancer, 454, 470
Protein synthesis, 233-237
  decline in, with aging, 528
Proteins/glycoproteins, secreted by placenta, 82
Psychiatric disorders, classification of, 357-358
Psychiatric syndromes, in adolescence, 358-367
Psychologic factors, in growth retardation, 145, 147
Psychopathology, of adolescence, 357-358
  anorexia nervosa, 365
  antisocial behavior, 358-361
  brief considerations for treatment, 366-367
  neurosis, 361-363
  psychosis, 363-365
  sexual dysfunction, 366
  situational disturbances, 358
Psychosis, 363-365
Psychosocial dwarfism, 629
Psychotherapy, 366
Puberty, 305, 331-332, 334
  distinction between adolescence and, 332
  endocrine pathologies in, 346-347
  in humans, 319-320
  problems with sexual development in, 342-345
  psychological and educational management of, 347
  in rats, 306-319
  sequence of events in, 333
  short stature at, 335-339
Puberty onset, 305
  clock mechanisms in, 318
  environmental cues in, 318-319
  gonadostat theory of, 316-317
  inhibitory theories of, 313-316
  mechanism of, 312
  missing link hypothesis of, 313
  rat as model for, 305-312
Purkinje cells, 210, 211
Purkinje cells in normal and hypothyroid rats, 184

Racial differences in bodily proportions, 127

Radiation, effect on oocytes, 4
Rat, as model for puberty onset, 305-306
  correlations with human, 319-320
  reproductive development in, 306, 307, 308-312, 313-319
Rats, gonadotropic research in, 38
Regional distribution of radioactivity following injection of corticosterone in rat brain, 157
Relaxin, 13
Release mechanisms of cholinergic, adrenergic and endocrine neurons, similarities and differences between, 373, 374-375
Releasing hormones (RH), 6
Renin, 30
Reproduction, 1
  artificial, 21
  immunology and, 17
Reproductive capacity, in mammals, 193
Reproductive senescence in rats, experimental studies of, 592-594, 595, 596-601
  human menopause in comparison with, 593-594
  overview of, 600-601
Reproductive system
  female, 13-15
  male, 16-17
Reproductive system changes in rats, 307-308
Retention of 1,2-^{3}H-corticosterone, 218
Ribonucleic acid (RNA), 2, 20, 237
RNS, see Ribonucleic acid
RU-2858 (synthetic estrogen), 293, 294, 297
Runaways, juvenile, 359

Schedule shifts, 460-461, 462, 463
Schizophrenia, 364-365
  pathophysiological mechanisms in, 622-623, 627
School refusal, 359-360
Secondary sex characters in male and female, changes in, 334
Secondary sex organs, differentiation and development of, 42
Secretin, 65
Seip syndrome, 572
Seizure threshold curves during 2 estrous cycles, 417
Semen, 17
Seminal vesicle, 16
Senescence and senility, in chronobiologic context, 453-457

Senile dementia, 553
Serine, 233
Serotonin, 158, 231, 233, 561
Sertoli cells, 10, 40, 305
Sex cells, 1
Sex chromosomes, 36, 43
  anomalies of, 45-46
Sex determination, 36
Sex differences, neural correlates of, 276
  biochemical, 278-284
  morphological, 276-278
Sex hormone binding globulin (SHBG), 588
Sex hormones, 26, 255
  and brain activity, 416-418
  and cancer, 657
  in the CNS, 156-159
  effects on behavior, 631-632
  enchancement of tumor virus production by, 659
  morphological effects on nervous system, 191-194
  in protein synthesis, 235-237
  in treatment of sexual deviance, 625
  See also specific hormones
Sex-linked inheritance, 46
Sex steroids, 645
  in CNS tissues, 397-400
  excess secretion of, as factor in growth retardation, 143
  as stimulus of growth, 135-137
  See also specific hormones
Sexual development, in adolescence 352-356
  problems with, 342-346
Sexual differentiation, 273-274
  aberrant, 45-46
  conceptual aspects of, 274-276
  genetic and hormonal factors in, 43
  in humans, 297
  nature of differentiating substance in, 292-297
  neural binding of estradiol and, 284-292
  neural correlates of, 276-284
  somatic correlates of, 44-45
Sexual dysfunction, in adolescence, 366
Sexual maturation, 320
Short stature, as problem at puberty, 335
  in the female, 338
  in the male, 336-337
Sitting height and subischial leg length, changes in, 128

Situational disturbances, in adolescence, 358
Skeletal age, determination of, 133, 134
Skeletal age of three male children, 133
Smooth endoplasmis reticular (SER) whorls in arcuate nucleus, 189, 190-191
Social development, in adolescence, 356-357
Sodium, 16
Somatic cells, 2
Somatic sexual differentiation, 44-45
Somatomammotropin, 66
Somatomedins, 647
  as factor in growth retardation, 143
  as mediators of metabolic effects of growth, 194-195
  as putative growth factors, 136, 137
Somatostatin, 46, 50-51, 378, 403-404
  See also Growth hormone-inhibiting factor
Spectral solutions, 457, 460
Spectral structure of human plasma prolactin in three age groups, 471
Spectrum of intermodulating frequencies which interact with aging, 453
Spermatids, 10
Spermatogenesis, 3, 9-12
Spermatogonia, 9, 10, 40
Spermatozoon, mature, 12
Spermiogenesis, 12
Spontaneous electrical activity, in brain maturation, 251-252
Stature, as problem at puberty, 335
  excessive height, 339-342
  short, 335-339
Sterility, 45
Steroid hormone biosynthesis, functional interrelations among mother, placenta and fetus with regard to, 88
Steroid hormones, 151, 208, 278, 647
  and cell functions, 414
  influence on glial cell function, 217
  secreted by placenta, 82, 87-88
  See also specific hormones
Steroid receptors, 44, 278
  importance of, 44-45

Steroidogenesis, 43-44
Stigma, of ovary, 6
Stimulus-secretion coupling, 372
Stress theory, of aging, 572-574
Studies of fetal endocrinology, difficulties and limitations of, 28
Studies of hormonal actions using neural cultures, 212-219
Subcortical nuclei, effect of thyroid hormones on, 185
Substance abuse, 360-361
Suicide, among adolescents, 361-362
Synaptic aging, 559
Syngamy, 14

Tardive dyskinesia, 626
Taurine, 232
Teratogenesis, carcinogenesis, and mutagenesis, 651-652
Testes, 16
  descent of, 40
  embryogenesis of, 39-40
  functional development of, 40-43
Testicular feminization, 44
Testosterone, 10, 16, 39, 292, 387, 588
  as growth stimulant, 136-137
  in human gestation, 40
  production of by females, 38-39
  in virilization, 43, 44
Tetany of the newborn, 51-52
Theca externa, 5
Theca interna, 4-5
Thiodothyronine, as growth stimulant, 136
Thioridazine (Mellaril), 366, 367
Threonine, 231
Thymidine kinase, 238
Thymosin, 515
Thymus, as agent in aging, 515
Thyrocalcitonin secretion, in aging, 505
  See also Calcitonin
Thyroid deficiency, 181
Thyroid gland, 58, 139, 406
  endocrinopathies of, 637-638
  functional development of, 58-61
  and peripheral deiodination, 524-525
  and peripheral refractoriness, 525-526
  in pregnancy, 67
  role in aging, 515-516, 517-522, 523-528
Thyroid hormone receptors, 62-63
Thyroid hormones, 58-64, 151, 207-208, 386, 420
  behavioral effects of, 252-254
  and brain excitability, 247
  and cholesterol metabolism in aging, 527
  in the CNS, 152-154, 400-402
  in depression, 621
  and evoked cortical activity, 248, 249, 250, 251
  hepatic metabolism changes with aging, 525
  morphological effects on nervous system, 181-187, 188
  in neural cell formation, 209-210, 217, 219
  in nucleic acid metabolism, 237-238
  permissive actions of, 494
  in protein synthesis, 233-235
  and replacement therapy, 516
  in spontaneous brain electrical activity, 420-421
  as stimulants of growth, 136
  See also specific hormones
Thyroid insufficiency, 64
Thyroid stimulating hormone (TSH) 53, 58, 59, 139, 208, 455
  in aging, 523-524
  secretion of, 386
Thyroidectomy, 237, 239, 421
  and hypothalamus, 189
Thyrotoxicosis, 637-638
Thyrotropin, 645
Thyrotropin releasing factor (TSH-RF), 208
Thyrotropin-releasing hormone (TRH), 59, 60, 159, 185, 376, 391, 403
  used as test in therapy of depression, 627
Thyroxine, 58, 59, 60, 65, 139, 152, 208
  in aging, 516, 523
  and EST, 416
  as growth stimulant, 136
  in myelin formation, 220
  in protein synthesis, 233
Toxic psychoses, 364
Toxic substances, and pregnancy, 109-110
Triiodothyronine, 58, 152
  in brain areas of rats, 156
  reverse, 61
Tryptophan, 231, 233, 570
  in aging, 577-579
Tryptophan-restricted animals, indications of aging retardation in, 578

Tumorigenesis, differing effects of chemical carcinogens and hormones in, 652-653
Turner's syndrome, 36, 43, 45, 572
Twinning, 7
Two-messenger system of hormone interaction, 413
Tyrosine, 231, 234

Underfeeding, and life prolongation, 577
Uptake of ^{3}H-NE into cerebellum taken from chick embryos, 224
Uptake of ^{3}H-NE into cerebellum hemispheres taken from chick embryos, 223
Urethra, 16
Urethral gland, 16
Urogenital structures, 43
Uterine motility, substances affecting, 15
Uterus, 13, 14-15

Vagina, 13, 15
Vandalism, 359
Vas deferens, 16
Vasopressin, 372, 515
Vasotocin, 57
Virility, 40
Vitamin D, 53

Wolffian ducts, 42
Women's rights movement, 357

X-ray autoradiograph of guniea pig brain, 153

Zinc, 16
Zona pellucida, 4, 20, 38
Zona reaction, 19